The Minerals, Metals & Materials Series

Adamantia Lazou · Katrin Daehn ·
Camille Fleuriault · Mertol Gökelma ·
Elsa Olivetti · Christina Meskers
Editors

REWAS 2022: Developing Tomorrow's Technical Cycles (Volume I)

TMS

Springer

Editors
Adamantia Lazou
Norwegian University of Science and Technology
Trondheim, Norway

Camille Fleuriault
Eramet Norway
Sauda, Norway

Elsa Olivetti
Massachusetts Institute of Technology
Cambridge, MA, USA

Katrin Daehn
Massachusetts Institute of Technology
Cambridge, MA, USA

Mertol Gökelma
İzmir Institute of Technology
İzmir, Turkey

Christina Meskers
Norwegian University of Science and Technology
Trondheim, Norway

ISSN 2367-1181 ISSN 2367-1696 (electronic)
The Minerals, Metals & Materials Series
ISBN 978-3-030-92562-8 ISBN 978-3-030-92563-5 (eBook)
https://doi.org/10.1007/978-3-030-92563-5

This Springer imprint is published by the registered company Springer Nature Switzerland AG
The registered company address is: Gewerbestrasse 11, 6330 Cham, Switzerland

Preface

Commitments to achieve net zero emissions are growing in number and strength. The UK, France, Denmark, and New Zealand have written net zero emission mandates by 2050 into law (as well as Norway by 2030 and Sweden by 2045). Japan, South Korea, Canada, and the European Union have adopted decarbonization strategies to achieve net zero economies by 2050, and China has committed to this goal by 2060. U.S. President Biden recently announced a commitment to decrease emissions by 50 percent below 2005 levels by 2030. Further, a number of leading private corporations and organizations have promised net zero operations by 2050. Indeed, 2020 was the year of the "net zero by 2050" commitment. The COP26 annual summit reflects the sustained efforts in 2021 on how to meet these goals and tackle the climate change.

The participants of REWAS are essential to the execution of these goals. Materials, the bulk commodities (cement, steel, copper) as well as minor elements, are crucial to the infrastructure to generate and transmit clean energy. A recent report by the International Energy Agency[1] cautions that the responsible supply of these materials may be a rate-limiting step in economy-wide decarbonization. Further, emissions targets cannot be met unless the considerable greenhouse gas emissions associated with materials production, one of the hardest to abate sectors, are reduced in the midst of this historic increase in demand. The profound impacts of the COVID-19 pandemic have also exposed weak points within the global supply chain. The need for more diversified, resilient, and flexible production is woefully clear.

In this context, the theme of REWAS 2022 is *Developing Tomorrow's Technical Cycles*. The materials industry can only realize the goals above by establishing sustainable product loops. End-of-life materials must be re-used or recycled, decarbonized processes must be developed and scaled, by-product streams must be properly treated and utilized, and the increasing array of data from digitalized industry must be leveraged to set new standards of practice. The four symposia focus on these topics. In addition, the honorary symposium for Professor Diran Apelian highlights vital work that pioneered the field of sustainable metallurgy.

---

[1] IEA (2021), *The Role of Critical Minerals in Clean Energy Transitions*, IEA, Paris https://www.iea.org/reports/the-role-of-critical-minerals-in-clean-energy-transitions.

REWAS is a unique venue in which insights from novel bench-scale experiments are presented alongside the results of new industrial practices implemented on the million-ton scale. This mix of experts from academia and industry is highly adept at making rigorous connections between vastly different scales, from atoms and molecules to reactors and factories. We believe this community is thus equipped with the tools and ingenuity to lead the fundamental changes required across the industry, and we are honored to highlight this work. We must acknowledge that heartbreaking events related to the pandemic have occurred since the last time this international group was in the same room together. We look forward to reconvening with renewed purpose and with appreciation for the simple pleasure of exchanging scientific ideas and friendly conversation.

Diran Apelian Honorary Symposium

We are honored to showcase Professor Diran Apelian's outstanding contributions, from metal processing to education, mentorship, and service to the TMS community. This symposium traverses a wide range of technical topics, reflecting his key contributions to topics such as additive manufacturing and powder metallurgy, next-generation aluminum alloys and lithium-ion battery recycling. The works presented in this symposium all reflect a greater purpose beyond innovation or novelty simply for the sake of it. Apelian compelled the materials community to rigorously incorporate sustainability principles, leading by example and offering tangible solutions to close metal cycles. This symposium provides insight into his philosophy on how to conduct research with significant industrial impact, develop the next generation of leaders, as well as bridge silos across disciplines, and build communities amongst teachers and students, industry experts, and academics.

Decarbonizing the Materials Industry

Motivated by the commitments to achieve net zero emissions in the coming decades, this symposium focuses on developing materials processes with carbon-neutral power sources and feedstocks, optimizing energy usage and integrating negative emissions technologies. Various sectors within industry are investigated, including iron and steelmaking, silicon production, and ferroalloy smelting. The state-of-the-art for incorporating greater shares of hydrogen, renewable electricity, bio feedstocks, and even captured CO_2 to supplant fossil fuels are presented. Experts must grapple with daunting questions of scale and economics: how to retro-fit existing equipment and establish infrastructure and supply chains for alternative feedstocks, all while competing with existing processes. The contributions to this symposium reveal that these new processes may come with benefits beyond decarbonization, yielding better products with lower cost. By covering a range of processes and contexts, we hope synergies and opportunities to tackle common problems throughout industry may be revealed.

Automation and Digitalization for Advanced Manufacturing

The fourth industrial revolution is underway with the use of automation, complex numerical simulation and digitization becoming commonplace. This symposium seeks to leverage the increasing diversity and availability of data to better manage waste, reduce environmental impact, and optimize material production. The applications for these new modes of sensing, modeling, and computing are nearly limitless, allowing traditional limitations to be completely reimagined. This symposium provides many creative ideas and encouraging results. For example, the application of automation and machine learning to recycling may fundamentally change economic, safety, and quality concerns. Digital twins of single reactors, or whole supply chains, may enhance impact assessments and reveal optimal processing routes. The pipeline from laboratory to widescale production may be shortened and eased with greater predictive power from initial surveys and experiments. As the materials community pursues decarbonization, the development and widespread utilization of these tools will be invaluable.

Sustainable Production and Development Perspectives

Shifting toward a sustainable global system is one of the main goals of REWAS. In this symposium, we explore ways to achieve and maintain the balance between the three pillars of sustainability: economics, the environment, and social considerations. The contributions to this symposium delve into the complex trade-offs and unexpected outcomes that may arise from adopting more circular practices—for example, copper recycling may not displace primary mining 1:1 as we would assume. These papers show the need to carefully assess for environment and human health impacts as we adopt new materials and systems in the pursuit of sustainability, whether lithium-ion batteries, high-entropy alloys, or photovoltaic systems. This research must be conducted collaboratively between industry and academia, and this symposium details the goals and methodologies of comprehensive, multi-year initiatives, such as TransFIRe (UK) and REMADE (US). In total, this symposium brings together examples of the types of questions we must ask and efforts we must undertake to achieve the UN's Sustainable Development Goals.

Recovering the Unrecoverable

Due to increasing global production, waste is being generated at unprecedented rates. This symposium tackles material streams that are under-utilized, improperly treated, or otherwise accumulating in massive volumes, causing a slew of resource, environmental, and health issues. These material streams from end-of-life products or industrial by-products are highly challenging due to their variability in composition or form, as well as complexity, containing multiple elements and phases. Nevertheless, there is value embedded within these streams. This symposium includes innovative results related to some of the most challenging waste sources, such as the energy capture and storage components, electronic waste and complex scrap, metallurgical by-products, residues, and slags. In particular, the contributions to this symposium offer new insights into the recycling of lithium-ion batteries, photovoltaics, and electronics, as well as steel mill dusts and non-ferrous slags. Several papers consider interactions with ecosystems, proposing solutions to remediate contaminated soils or incorporate waste as soil amendment. Key considerations that this symposium attempts to address are scalability and high productivity, as well as flexibility to accommodate various feedstocks and recover multiple elements. Overall, by "recovering the unrecoverable" we hope to mitigate the environmental, economic and health issues related to waste while generating additional revenue.

Joint Sessions

The proceedings of several symposia organized in conjunction with REWAS 2022 will be published in separate volumes.

The Recycling and Sustainability in Cast Shop Technology is a joint session in partnership with the Light Metals Division and will be published in *Light Metals 2022*. The usage of light metals in the aerospace and automotive industries has been rising due to their outstanding properties. The importance of recycling and sustainable metal casting has risen in the last decades dramatically due to the increasing demand on primary resources and environmental concerns. Innovative and more efficient recycling technologies are needed for complex scraps to increase the recycled metal portion in the total metal production. This joint session within Cast Shop Technology explores economic and environmental impacts of recycling, recyclability of complex scraps, pre-treatment of scraps, new recycling processes, and life cycle assessment.

Aluminum Reduction Technology is a joint session with REWAS: Decarbonizing the Metals Industry and the Light Metals Division and will be published in *Light Metals 2022*. This session covers decarbonization efforts across the primary and secondary industries via development of alternative, renewable energies, and the optimization of fuel consumption for energy generation. Described concepts refer to recent technologies or policies used for the purpose of reducing CO_2 emissions. In particular, hydrogen reduction, inert anode smelting, and alternative sources of energy for production plants are of interest. Beyond the technologies described, emphasis should be made on the metrics used for the quantification of carbon reduction.

The Energy Technologies and CO_2 Management symposium is a partnership between the Energy Committee and REWAS 2022. The proceedings will be published in REWAS 2022: Energy Technologies and CO_2 Management (Vol. II). The topics covered include hydrogen technologies, thermal management, biomass, renewable energies, and combustion technologies.

We are honored to present the following papers that address urgent topics with a rigorous, scientific approach. We are so thankful to the authors and presenters for generously sharing their work and participating in REWAS 2022. We want to thank the outstanding TMS team: Kelcy Marini, Patricia Warren, and Matt Baker, as well as the editors, organizers, reviewers, session chairs, and committee chairs who brought this meeting to life and contributed their time and expertise:

Gerardo Alvear Flores
Alexandra E. Anderson
Gisele Azimi
Bart J. Blanpain
Camille Fleuriault
Kerstin M. Forsberg
Ioanna Giannopoulou
Mertol Gökelma
Nolan Goth
John A. Howarter
Chukwunwike Iloeje
Hojong Kim
Christina Meskers
Brajendra Mishra
Neale R. Neelameggham
Elsa Olivetti
Takanari Ouchi
Hong Peng
Adam C. Powell
Jesse F. White
Mingming Zhang

Adamantia Lazou
Katrin Daehn
Lead editors

Contents

About the Editors

Adamantia Lazou is a Researcher in the Department of Materials Science and Engineering at the Norwegian University of Science and Technology. Her research area is process metallurgy, and her current activities relate mainly to high-temperature processes. Her research interests focus on recycling and valorizing by-products and waste streams. Her vision is on coupling sustainable principles in the metallurgical approaches for resources efficiency and closing the materials loop. She holds an M.Sc. degree in Metallurgical Processes from the National Technical University of Athens, Greece, and a Ph.D. in Materials Science and Engineering from the Norwegian University of Science and Technology, Norway.

Katie Daehn grew up in Columbus, Ohio and earned her BS in Materials Science and Engineering at The Ohio State University in 2015. She then went on to a Ph.D. in Engineering at the University of Cambridge (completed in 2019) as a Cambridge Trust scholar, working with Prof. Julian Allwood on end-of-life steel recycling. She is currently working as a postdoctoral researcher at MIT with Professors Antoine Allanore and Elsa Olivetti, continuing in the realm of industrial ecology, but also gaining laboratory experience in high temperature electrochemistry. Achievements she is particularly proud of include receiving the Goldwater scholarship, authoring the Top Policy paper in Environmental Science and Technology (2017), and working with a team to extract copper and iron from natural concentrate using only electricity. She is a third-generation metallurgist and wants to spend her career improving the management of Earth's resources.

Camille Fleuriault is senior project manager at Eramet Norway AS in Sauda, Norway. Her work focuses on identifying and enabling zero carbon strategies for the production of manganese alloys. She previously worked on developing innovative and environmentally friendly recycling processes for the secondary metals industry. She holds a B.S. in Geological Engineering and an M.Eng. in Mineral Engineering from the National School of Geological Engineering in Nancy, France, and a M.Sc. in Metallurgical Engineering from Colorado School of Mines, USA. She is vice-chair of the TMS Pyrometallurgy Committee and a former *JOM* advisor for the same committee.

Mertol Gökelma is an academic researcher in the Department of Materials Science and Engineering at İzmir Institute of Technology (IZTECH). He finished his B.Sc. degree at Dokuz Eylul University in Metallurgical Engineering and Materials Science, Turkey, and his M.Sc. and Ph.D. in Metallurgical Engineering at RWTH Aachen University, Germany. During his doctoral research, he worked as a research assistant for four years at the Institute of Process Technology and Metal Recycling (IME), RWTH. Dr. Gökelma worked as a post-doctoral researcher at Norwegian University of Science and Technology for 2.5 years. His research interests focus on process metallurgy and recycling of nonferrous metals. He has been active in different R&D projects focused on recycling of magnesium black dross, recycling and refining of precious metals, powder synthesis of titanium alloys and silicon, metallothermic reduction of oxides, aluminum recycling, and melt treatment. He has been lecturing on aluminum technologies, mineral processing, extractive metallurgy, and metal recycling since 2017. Dr. Gökelma is a member of Aluminum Committee and Recycling and Environmental Technologies Committee at TMS. He has served TMS as a co-organizer and co-editor of REWAS 2019 and REWAS 2022 as well as a program organizer for Light Metals 2019. Dr. Gökelma is currently the head of the metallurgy laboratory at IZTECH.

Elsa Olivetti is the Esther and Harold E. Edgerton Career Development Professor in the Department of Materials Science and Engineering (DMSE) and co-director of the MIT Climate and Sustainability Consortium at the Massachusetts Institute of Technology. Her research focuses on reducing the significant burden of materials production and consumption through increased use of recycled and waste materials; informing the early-stage design of new materials for effective scale up; and understanding the implications of policy, new technology development, and manufacturing processes on materials supply chains. Dr. Olivetti received her B.S. degree in Engineering Science from the University of Virginia in 2000 and her Ph.D. in Materials Science Engineering from MIT in 2007.

Christina Meskers is currently senior adviser at the Industrial Ecology Programme, department of Energy Process Engineering at Norwegian University of Science and Technology (NTNU), Trondheim, Norway. She is also Chair of the Extraction and Processing Division and Member of the Board of Directors of The Minerals, Metals & Materials Society (TMS), USA.

Dr. Meskers graduated from Delft University of Technology with an M.Sc. (2004) in Resource Engineering and a Ph.D. (2008) in Materials Science and Engineering, including stays at McGill University, the Norwegian University of Science and Technology, and the University of Melbourne. At UMICORE she was Senior Manager Open Innovation and Senior Manager Market Intelligence & Business Research. Through innovation and research, her aim is to ensure the product and materials value chain becomes more sustainable. The United Nations' International Resource Panel report, *Metal Recycling-Opportunities, Limits, Infrastructure* (2013) is a key publication that she co-authored.

Dr. Meskers is a 2020 Brimacombe Medalist, co-recipient of the 2014 Ondernemers voor Ondernemers Award, 2013 Belgian Business Award for the Environment recipient, and received in 2008 the TMS Young Leaders Professional Development Award. She has a strong passion for innovation, strategy, and partnerships coupled with a focus on connecting people and ideas across disciplines, industries, organizations, and value chains. Dr. Meskers served on the industrial advisory boards of international master programs and innovation networks. She has accumulated over 15 years of experience in the (raw) materials sector.

Part I
REWAS 2022: Coupling Metallurgy and Sustainability: An EPD Symposium in Honor of Diran Apelian

Aluminum Roadmap to a Sustainable Future

John Weritz and Marcella Dudek

Abstract For more than a century, aluminum has played a major role in providing high-strength, low-weight mass reduction solutions to all forms of efficient transportation. As companies, governments, and consumers increasingly prioritize responsible environmental stewardship, the aluminum industry is collaborating to define the research and development goals necessary to ensure a sustainable future. At the helm of that collaboration is The Aluminum Association, its member companies, industry colleagues, and automotive customer base. Achieving measurable and impactful goals over the next decade will enable manufacturers to not only optimize aluminum applications in future passenger vehicles, but also maximize the value of the aluminum content in those vehicles as they are removed from service. The Aluminum Association's new *Automotive Aluminum Roadmap* outlines strategic priorities while recognizing how the changing landscape of mobility is increasingly connected and electric. Within that landscape, this manuscript and related presentation detail the reasons why aluminum is the fastest growing automotive material, in what applications aluminum content continues increasing its share, and the impact consumer demand and government regulations have in directing automobile producers away from conventional internal combustion engines to increasingly electrified fleets.

Keywords Aluminum · Automotive · Lightweighting · Materials · Metals · Sustainability

Introduction

The global transportation sector is evolving to meet the needs of the planet. An urgent climate crisis, combined with consumer demands for new, high-performing, and stylish vehicle options, is driving technological innovation at a rapid pace. This

J. Weritz (✉) · M. Dudek
Standards & Technology, The Aluminum Association, 1400 Crystal Drive Suite 430, Arlington, VA 22202, USA
e-mail: jweritz@aluminum.org

A. Lazou et al. (eds.), *REWAS 2022: Developing Tomorrow's Technical Cycles (Volume I)*, The Minerals, Metals & Materials Series,
https://doi.org/10.1007/978-3-030-92563-5_1

is underscored by a twofold commitment to clean energy priorities from policymakers in Washington, D.C., and leading global automakers. In Washington, the current administration's stated goal to position America as a world leader in electric vehicle manufacturing is clear. Nationwide and globally, automotive manufacturers are working with suppliers to continue to pursue ambitious decarbonization goals.

Federal regulations and corporate initiatives must align with consumer preference. For years, Americans have favored trucks and sport utility vehicles (SUVs) over sedans, with two SUVs sold for every one sedan in 2019. While leading automakers are setting aggressive net zero goals, they are simultaneously removing sedans from fleets in response to consumer demand. The shift to greater numbers of large vehicles on American roadways reinforces the need for mass reduction in those bigger, heavier vehicles to ensure vehicle occupants, pedestrians remain safe, and automakers have the ability to offer competitive vehicle performance characteristics. The equation is simple: lighter vehicles require less energy (fuel or battery range) to move and the strategic use of low-weight materials is imperative as the industry works toward a sustainable future mobility scenario.

Aluminum, already the fastest growing automotive material, is poised to continue steady growth over the next decade as design engineers continue to leverage the metal's inherent benefits in their pursuit to improve the EV driving experience. While projected growth is robust, many opportunities remain to be realized for aluminum. The Aluminum Association, its aluminum producer member companies, and automotive stakeholders recently collaborated to create *A Roadmap for Automotive Aluminum,* to recognize how the changing landscape of transportation as connectivity, hybridization, and the race to electrification all drive strategic priorities.

Ushering in Electric Vehicles

According to new data from Pew Research, nearly 1.8 million EVs were registered in the USA last year, soaring from fewer than 300,000 in 2016. It is projected future EVs will rely on an aluminum architecture to enable safety, value, and performance. A forecast released in 2020 by the Aluminum Association shows that aluminum content in vehicles could grow 12% by 2026—continuing a strong upward trend. This will be driven by greater use of the metal in vehicle closures, body-in-white, and chassis applications with notable increased use of aluminum in electrified powertrains and EV platforms. Recent vehicle launches, like the all-electric Ford F-150 lightning and plug-in electric Jeep Wrangler, demonstrate this growth trend.

While estimates of the pace at which EVs will displace ICE vehicles vary, the relative impact on aluminum shows a net increase in demand for key application areas, such as battery and motor housings as well as body structural components to improve crash management systems.

The aluminum industry is poised to take advantage of the wealth of opportunities arising from the expected explosion in EVs on the market and lasting changes in mobility. To reap the benefits, producers and manufacturers are exploring new materials and component designs that lower costs, increase performance, and enhance sustainability. Through industry collaboration, the Aluminum Association developed explicit goals and opportunities to kick-start the develoment of new innovations in support of an electrified future of mobility.

Challenges and Opportunities

The evolution of the automotive manufacturing sector comes with a unique set of challenges and opportunities. To pave the way, there is a need for advances in material research and development and the reengineering of fabrication and joining processes.

The Association's Roadmap uncovered several challenges and the importance of addressing these in the near term, for progress in recycling and carbon reduction. Pathways for optimizing the use of aluminum in automotive include.

- Design and product innovations;
- Next generation material and process research and development to lower costs and enhance performance;
- Targeted efforts to create alloys and components that meet OEM specifications for emerging EVs and other propulsion systems;
- Incorporating sustainability and recyclability in all aspects of aluminum production, including a focus on closed loop recycling and lower energy and carbon footprints;
- Greater understanding of how to optimize materials and production through state-of-the-art digital technologies and data; and
- Uniform testing and validation across the industry.

These goals as outlined in the Roadmap are ambitious and far-reaching, impacting all segments of the aluminum industry.

Going Full Circle

Reuse of aluminum has been growing for decades, with recent gains due to stronger demand for automotive aluminum in vehicle applications. Aluminum recovery from scrap in April 2021 amounted to 263,000 tons, up 8% from the year prior. In a 2020 paper published in Resources, Conservation and Recycling Journal (RCR), experts from the University of Michigan, Ford Motor Company, and Light Metal Consultants LLC used a dynamic flow analysis to estimate aluminum ABS scrap generated from today's highest selling vehicles in America. They found that if production continues at current volumes, "aluminum ABS scrap from these vehicles will increase to approximately 125,000 tons per year in 2035 and 246,000 tons per year in 2050. The majority of this scrap will be available for US processing with [approximately] 10 percent of deregistered vehicles exported or achieving vintage status." This new wave of auto body sheet (ABS) scrap available for recycling as the first mass-produced aluminum-intensive vehicles begin to reach end of life creates a plethora of recycling opportunities in the overall industry.

To better prepare for this wave, more efficient ways of separating, recovering, and reusing aluminum scrap from automotive components are required.

The aluminum industry is a leading partner in addressing decarbonization as more sustainable vehicles enter the market. Notably, the aluminum used in automotive applications already contains a great amount of recycled content. Aluminum castings are made largely from recycled material. In addition, aluminum in end-of-life vehicles has been well recycled, and mostly reused, again and again, to make new vehicle parts. The design of recycle-friendly aluminum alloys will help increase already high automotive aluminum recycling rates.

The aluminum industry's stated goals for future recycling and sustainability imperatives range from improved infrastructure for the recovery of aluminum scrap to greater capability for scrap use and widespread adoption of aluminum content as a sustainable solution for automotive design. Today, the industry starts with a strong foundation with various end-of-life recycling processes already in practice through leading automakers like Ford, Jaguar Land Rover, and Nissan. Near-term goals in the next one to five years include building and deploying the infrastructure needed to collect, sort, and reuse scrap. Key factors to achieving those goals will be the ability to return scrap to the mill, application to both pre- and post-consumer recycled materials, and the ability to identify, segregate, and use different types of alloys effectively. In the future, aluminum industry stakeholders aim to benchmark the current recycling technologies and assess the composition of final remelted scrap, including the full spectrum of dismantling, shredding, and sorting.

Collaboration for Sustainable Mobility

Ultimately, the aluminum industry identifies two key focus areas in its support of EVs and more sustainable mobility solutions: first, a deeper understanding of how to optimize materials and production for electrified platforms through state-of-the-art digital technologies and data and second, a unified commitment from the aluminum industry in standardization, testing, and validation across the board with close collaboration with automotive customers. Greater collaboration with automakers is foundational to achieving these goals.

While all materials will come together in the vehicles of tomorrow, one narrative persists: aluminum remains an essential part of the mix. No other current material offers the design flexibility and sustainability that aluminum does, and the industry is continually working to improve its impact. The *Roadmap for Automotive Aluminum* takes this already booming material to new heights, with clear objectives to collaborate with important stakeholders to create a new wave of sustainable mobility options for American consumers.

Electrochemical Pathways Towards Sustainable Energy

Donald R. Sadoway

Abstract A sustainable energy future is axiomatically an electric future whose realization depends in part upon electrochemical innovations. Two examples are stationary energy storage and carbon-free steelmaking. Grid-scale electricity storage not only treats the intermittency of renewable electric power generation (wind and solar) but also confers resilience to today's grid. For example, the liquid metal battery provides colossal power capability on demand and long service lifetime at requisite low cost. In 2019, worldwide steel production, 1.869 billion tonnes, generated 9% of total anthropogenic CO_2 emissions. As an example of novel approaches in this sector, molten oxide electrolysis represents an environmentally sound alternative to today's carbon-intensive thermochemical process which produces an average 1.83 tonnes CO_2 per tonne of steel. In the narratives of both of these emerging technologies, there are lessons more broadly applicable to innovation: pose the right question, engage young minds (not experts), establish a creative culture, and invent inventors.

Keywords Energy storage · Deep decarbonization · Steelmaking · Liquid metal battery

Stationary Storage

Batteries for this application space must meet very different requirements from those of batteries used in handheld portable devices. For massive stationary applications, the batteries must exhibit a long service lifetime (say, 20 years), they must be safe (free of threat of fire), operationally versatile (long-duration slow power output along with short-duration power burst delivery), all at super-low cost (competitive with electricity generated at the lowest present price point). So, what is the path forward for grid-level storage that can meet all these requirements? Two points: First, confine the chemistry to earth-abundant elements. To make something dirt cheap, make it out

D. R. Sadoway (✉)
Department of Materials Science and Engineering, Massachusetts Institute of Technology, Cambridge, MA 02139-4307, USA
e-mail: dsadoway@mit.edu

A. Lazou et al. (eds.), *REWAS 2022: Developing Tomorrow's Technical Cycles (Volume I)*, The Minerals, Metals & Materials Series,
https://doi.org/10.1007/978-3-030-92563-5_2

of dirt, preferably dirt that is locally sourced. This guarantees a secure supply chain. Secondly, make it easy to manufacture. Think about design at the early discovery stage. This path is what I call ***cost-informed discovery***. In inventing the liquid metal battery, I took inspiration from outside the field. No battery experts were consulted. Instead, I looked to electrometallurgy: the Hall-Héroult cell. This traffics in huge electric currents (500,000 A × 4 V) without interruption and yet converts dirt to metal for less than $1.00/kg. I reasoned that if I could re-train the Hall-Héroult cell rather than to consume vast quantities of electricity but instead to store vast quantities of electricity and on demand to discharge same, then I would have something big and cheap that meets the needs of grid-level storage. In this way I conceived the liquid metal battery. The original chemistry was Mg|$MgCl_2$|Sb. We moved on to Li|LiCl|Pb-Sb and Li|LiCl|Bi. Today, the chemistry is based upon Ca|$CaCl_2$| Sb. This is on the path to commercialization with installations exceeding 200 MWh in the plans. Boasting round-trip energy efficiency of ~80% and resistance to capacity fade (>4000 cycles, 4 years of continuous operation, retention of 99% nameplate capacity), no threat of fire or thermal runaway, the battery is projected to reach well below $50/kWh capital cost.

Green Steel

Now let us turn steel which is receiving considerable attention from regulators intent on achieving deep decarbonization of the manufacturing sector (In 2019 steel generated 9% of world CO_2 emissions!). When it comes to carbon-free ironmaking, there are two routes. One involves electrometallurgy; the other involves hydrogen reduction. While the latter has its ardent proponents who, in turn, have attracted investment from the private and public sectors (especially in Europe), it is the firm belief of this author that electricity is much more efficiently used in metal extraction when feedstock is electrolyzed directly to produce liquid metal. The production of hydrogen by water electrolysis followed by hydrogen reduction of iron oxide to produce solid iron for subsequent processing in an electric arc furnace is a tortuous route from dirt to metal. And downstream operations want liquid metal as the input, not solid metal. Molten oxide electrolysis (MOE) involves dissolution of iron oxide feedstock in a multicomponent melt comprising a plurality of metal oxides such as silica, calcia, alumina, magnesia, and the like. The electrolysis cell looks somewhat like a Hall-Héroult (HH) cell except that the operating temperature is 1600 °C, which sets it above the melting point of pure iron. As importantly, to distinguish it from HH, the MOE cell is fitted with an inert anode that supports the evolution of oxygen gas in contrast to the evolution of carbon dioxide on the carbon anode of the HH cell. At this writing, Boston Metal, the company that was founded to commercialize MOE steelmaking, is on the path to build a 200,000 A cell.

Thoughts About the Future

Neils Bohr is reputed to have said that prediction can be difficult, especially when it is about the future. But here are some miscellaneous ramblings of mine.

In a world with abundant, cheap electric power:

1. Almost all industrial chemistry will become industrial electrochemistry. In other words, anything that can be done electrochemically will be done electrochemically unless it is uneconomically feasible to do so.
2. Almost all thermal chemistry will convert from combustion to electrical heating.
3. It is a good time for electrochemistry.
4. It is a good time for TMS.
5. Best wishes to Diran Apelian on the occasion of this EPD symposium.

Informatics Driven Materials Innovation for a Regenerative Economy: Harnessing NLP for Safer Chemistry in Manufacturing of Solar Cells

Deepesh Giri, Arpan Mukherjee, and Krishna Rajan

Abstract The objective of this paper is to demonstrate the use of natural language processing tools to map the propensity of usage of specific solvents used for the fabrication of different perovskite photovoltaic (PV) systems. This article focuses on implementing contextual natural language processing (NLP) tool to identify different chemicals that appear in the synthesis of perovskite solar cells. This information is linked to the known level of hazard of those solvents. This work serves to demonstrate how by harnessing the tools of informatics, in this case NLP, offers a powerful framework to guide environmentally conscious selection of chemicals used in the manufacturing of perovskite solar cells.

Keywords Perovskite · Solvent · NLP

Introduction

Published data on current solar manufacturing lifecycle technologies are primarily retrospective and "forensic" in nature; they provide information on the impact of manufacturing and processing strategies *after the fact*. Important as it is, this approach can, at best, help with remediation, but not in preventing these problems. We need systemic changes in the design and development of renewable energy technologies to accelerate the development and discovery of safer materials without compromising performance. This paper provides an example of how by harnessing informatics, at the onset of technology design, data-driven strategies can be developed that reduce the footprint of materials and chemicals to build the infrastructure for a safe and sustainable environment in industrial communities. Such an approach offers the potential to develop and deploy technologies that are capable of delivering better processes, products, and outcomes.

D. Giri · A. Mukherjee · K. Rajan (✉)
Department of Materials Design and Innovation, University at Buffalo, Buffalo, NY 14260-1660, USA
e-mail: krajan3@buffalo.edu

A. Lazou et al. (eds.), *REWAS 2022: Developing Tomorrow's Technical Cycles (Volume I)*, The Minerals, Metals & Materials Series,
https://doi.org/10.1007/978-3-030-92563-5_3

Harnessing solar energy is a critical component in developing a sustainable and regenerative economy. In that context, over the last decade, perovskite-based solar cells have shown extraordinary promise traction as PV materials due to their higher power conversion efficiency, easy processing techniques, and low manufacturing cost [1–4]. Perovskites can be broadly categorized as organic and inorganic based on their constituent elements. Perovskites adopt the ABX_3 structure, and in the case of organic perovskites, typically the "A" cation is organic while "B" is a metal and X is a halogen (Cl, Br or I), e.g., $CH_3NH_3PbI_3$ [5].

While one aims in making perovskites with better performance metrics, it is equally important to account for the safety and environmental hazard impact in exploring processing and manufacturing pathways of next generation solar cells [6–11]. In this work, we identify different solvents associated with the common organic–inorganic hybrid perovskites and link them with occupational hazards. This paper serves as a guide for adopting safer solvents in synthesizing perovskites. Information on solvent use is complex, and details vary from material to material [12, 13]. This information is not organized via some theory or models, rather it is embedded in the scientific literature, which forms a vast and continuously growing library of unstructured information. Even with sophisticated tools, the key challenge remains in gathering useful information before employing any data-driven model to minimize human intervention [14]. NLP bridges this gap by converting the unstructured textual data to encoded variables. NLP provides a sophisticated tool that queries targeted information from this enormous corpus such that it forms an input to any ML paradigm such as classification or regression [15].

NLP can be both context agnostic [16, 17] and contextual [18–20]. Contextual NLP can be used for tasks such as sequence classification [21], question answering [22, 23], language modeling [24], and translation [25, 26]. Contextual NLP has already found its implementation in the study of chemical protein interaction [27]. NLP-based research in materials science domain mostly uses context-agnostic NLP to manually annotate whole articles and build classification tools to extract relevant keywords [28–32]. In this study, we utilize the context-aware algorithm BERT to extract information on different aspects of perovskites synthesis. Our approach of implementing contextual NLP allows the luxury of getting the most relevant paragraphs from thousands of articles without having to manually annotate the whole article.

Methods

Our text-mining pipeline begins with journal article collection. We downloaded around 1800 peer-reviewed articles primarily focused on perovskites. The NLP algorithms act upon these to extract relevant texts, which are further processed to obtain meaningful results. Our approach to data collection and analysis follows the pipeline shown in Fig. 1.

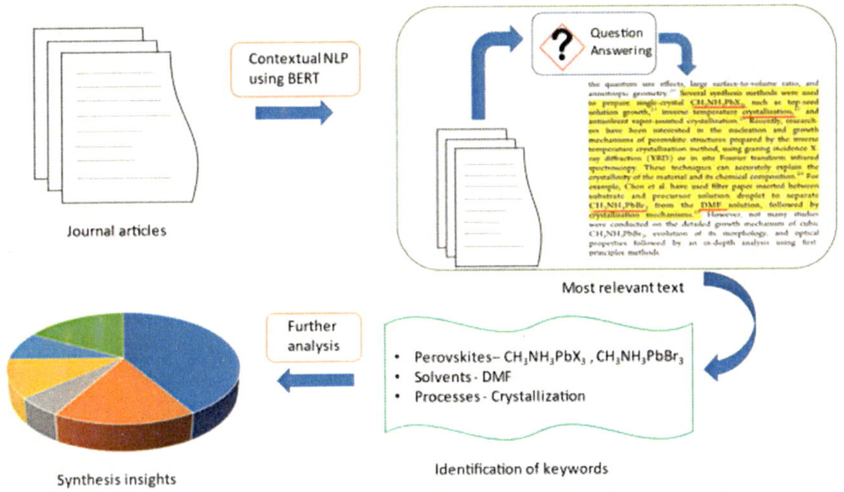

Fig. 1 Schematic representation of the pipeline for data extraction and analysis to understand perovskite synthesis. The pipeline starts with journal articles collection in pdf format which is processed using NLP to extract answers to the given question. These answers and the paragraph containing them are used to extract the perovskites, solvents, and the process keywords involved in the synthesis pipeline. The extracted information forms a synthetic dataset that can be used to create different data-driven models catering to different scientific questions

After downloading the articles, we employed BERT for question-answering task. BERT is a very large and complicated neural network architecture composed of extremely simple mathematical operations that can be utilized towards various NLP tasks. It has an encoder-decoder architecture that is modified to form a question-answering tool. Our database of 1835 articles acts as the input corpus based on which closed domain question answering (CDQA) system provides relevant answers. CDQA is a part of information retrieval in the question answering (QA) system within the field of NLP that is capable of correctly answering questions from a given database of information. To perfectly mimic the human way of reading a paragraph, one might need to ask specific questions, such as—“Which perovskite is being synthesized?” or, “What solvents are being used?”, etc. However, as pointed out by Kononova et al. [33], no material science-specific BERT model has been developed as yet. Hence, BERT is capable of answering only generic questions. This is the reason why we deliberately chose a very broad query that reproduces what we already know. This way we get to evaluate our test question and use this as a platform to fine tune our approach for future tasks. For this task, the question asked was—“How are perovskites synthesized?”. On average, we extracted three answers and the corresponding paragraphs where those answers were based. From these paragraphs, we extracted the keywords that are vital to perovskite synthesis.

Keyword extraction forms an important step in many NLP tasks. It is a subtask of an NLP pipeline that classifies keywords extracted from a given paragraph. In this case, we are classifying the extracted keywords as perovskites, solvents, and synthesis keywords. We took advantage of an existing named entity recognition (NER) tool, ChemDataExtractor [34], to extract the names of the chemicals. ChemDataExtractor is an NER toolkit that can identify the chemicals automatically from the given text. However, we could not benefit from the existing tools [35–37] to identify the solvents, precursors, and synthesis keywords. It appears that the NER tools which are trained on one set of data cannot be easily implemented for other tasks. Hence, we manually annotated the keywords in each paragraph and used them for further analysis. Although this is tedious and time consuming, we hope this effort will be beneficial for other research activities on perovskite synthesis.

Results

Our results indicate that methylammonium (MA)-based perovskites are the most popular organic perovskites while Cesium-based ones are their inorganic counterparts. MA along with lead halides make up the majority of organic perovskites while formamidinium (FA)-based perovskites come next. “Organometal halide” perovskites also gain a significant mentioning, thereby referring to other organic compounds that are being used in the synthesis of perovskites. Synthesis of organic perovskites requires organic solvents. Our results show that dimethylformamide (DMF) is the most popular organic solvent, followed by dimethylsulfoxide (DMSO). Both DMF and DMSO are commonly used for the dissolution of lead and MA salts [38, 39], and hence, it is no surprise that these two appear at the top of the list. The appearance of MA and MA-based compounds on top 3 also makes sense because hybrid perovskites shot into prominence after the work by Kojima et al. [3] in 2009 and most of the articles we collected are published after 2015. Figure 2 is a list of all the organic compounds reported in the literature with frequency 5 or more. Similarly, we also identified the most frequently appearing synthesis keywords, and it revealed that halide-based and solution-based syntheses are the top choices for making perovskites. This makes sense because the X in ABX_3 is generally a halogen anion, and the synthesis begins with the preparation of a precursor solution.

Perovskite synthesis involves different solvents and processing activities [40, 41]. We mapped the association between the high-frequency solvents and synthesis keywords with the perovskites. This gives an idea of perovskite-process-chemical relationship. Following Fig. 2 that shows the marginal frequency distribution of the identified organic compounds, we have presented a method for visualizing the joint distribution of the different extracted keywords. Figure 3 is a Sankey diagram representing the frequently observed organic perovskites with the solvents and process keywords [42]. Sankey diagram is a visualization tool that connects source and target entities with bands whose widths are directly proportional to the strength of the connection. The diagram shows the joint probability distributions of the perovskites and solvents and the conditional distribution of the process keywords based on their mutual occurrences in each paragraph. The figure shows that while oleic acid is the

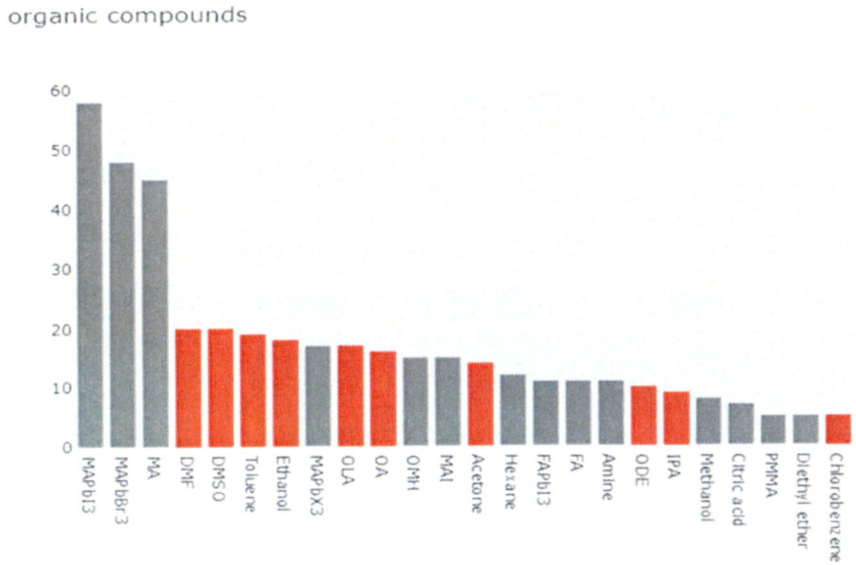

Fig. 2 List of high-frequency organic compounds appearing in the paragraphs that discuss synthesis of perovskites. Crimson red indicates the organic solvents while light gray color indicates the other organic compounds. Methylammonium (MA)-based perovskites have been discussed the most in the literature while dimethylformamide (DMF) and dimethylsulfoxide (DMSO) are the two most used solvents (Color figure online)

most used solvent in the synthesis of organic perovskites, solution-based synthesis is the most preferred method.

Next, we take these commonly used solvents and associate them with their hazardous nature. Although a separate study can be done on the toxicity of solvents, we have resorted to assessing their occupational, safety, and environmental hazards on this work. It is imperative to eliminate or reduce the usage of hazardous solvents in the industrial manufacturing of perovskites. Figure 4 presents a similar distribution as seen in Fig. 3a. Additionally, we present a hazard chart in the inset of Fig. 4 to clarify the associations between perovskites and the hazardous solvents. The data required for plotting the hazard chart are adopted from Ref. [11]. As seen in Fig. 4, the solvent which appears most frequently in the literature, DMF, is a hazardous solvent and requires substitution. Although the other two highly used solvents—DMSO and Toluene—are not as hazardous as DMF, they are also substitution advised. Only ethanol, from the top 4 list of solvents, has been identified as a recommended one. The usage of these solvents is going to render a corresponding hazardous ranking to the perovskite being manufactured. Both DMF and DMSO have maximum associations with methylammonium iodide (MAI) perovskites, while ethanol and acetonitrile have more associations with $MAPbI_3$ and MAI perovskites. This gives us a way to be able to categorize the manufacturing hazard associated with the perovskites of interest. At the same time, Fig. 4 also suggests if a recommended solvent can be used

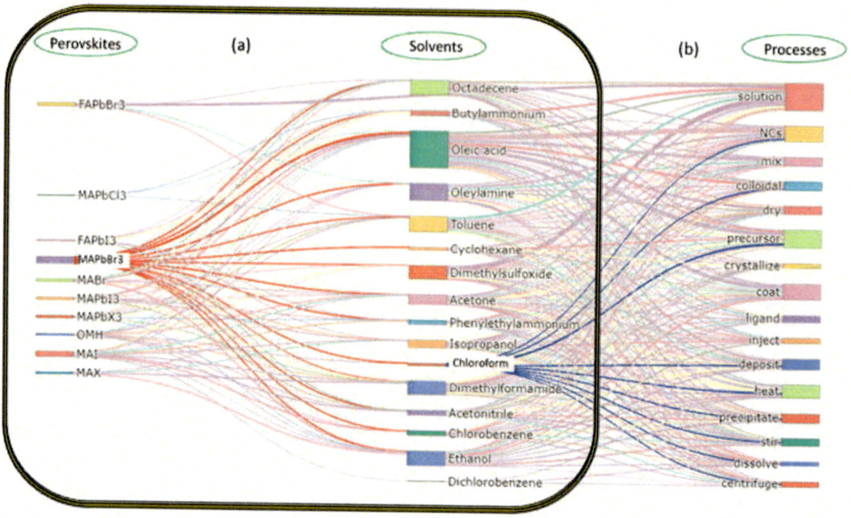

Fig. 3 Sankey plot depicting the choice of solvents for the organic–inorganic hybrid perovskites. Synthesis of each perovskite may involve different solvents, and each solvent might be associated with different synthesis keywords. The joint association between the perovskites and the solvents is shown in (**a**) and the conditional association with the process keywords in (**b**). We have highlighted in red—the solvents that are used with MAPbBr3 and in blue—the keywords that appear along with the solvent chloroform. The thickness of the lines indicates the association frequency between the connected entities while the box thickness corresponds to the total frequency of the entity (Color figure online)

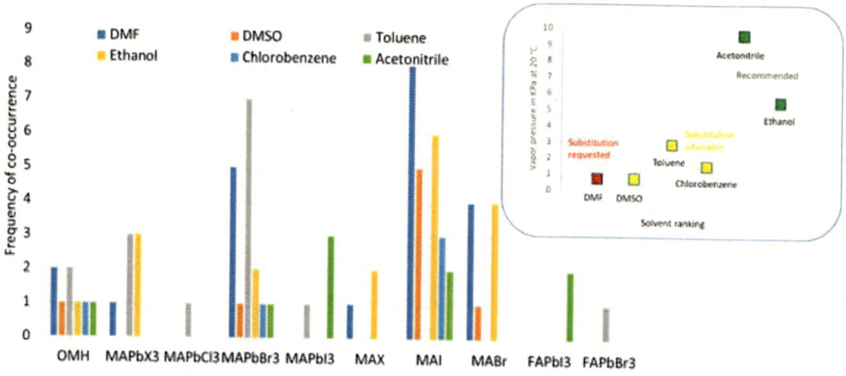

Fig. 4 Frequently occurring perovskites and the solvents associated with them. The inset at the top right corner is a ranking of the hazardous nature of the solvents and has been adopted from Ref. [11]. It helps to identify if a hazardous solvent is being used in the industrial manufacturing of perovskites. This information can further be used to rank the manufacturing hazard of the perovskites. From the graph, it is easily observable that methylammonium iodide perovskites use DMF the most, and hence, their industrial manufacturing is potentially more hazardous in nature than the others. However, as ethanol is also used in making the MAI perovskites, it provides an option to switch to the recommended alternative (Color figure online)

instead of the other solvents in the synthesis of the perovskites. For instance, MABr uses both DMF and ethanol; however, one is substitution requested while the other is recommended. Thus, this study helps identify recommended solvents to replace the substitution requested ones.

Conclusion

Developing strategies and solutions that improve chemicals management as well as accelerate sustainable material development for renewable energy technologies is critical for a regenerative/circular economy. Sustainable development of clean and safe technologies can best be achieved through simultaneous and integrated consideration of technical, environmental, and social factors in all aspects of the design, development, and adoption of renewable energy technologies. In this study, we have described how informatics tools based on text mining can help identify pathways to accelerate solutions for safe and clean technologies by anticipating the chemical footprint of manufacturing operations in the renewable energy sector.

Acknowledgements The authors acknowledge the support from NSF Award #1640867-DIBBs: EI: Data Laboratory for Materials Engineering and the Collaboratory for a Regenerative Economy (CoRE center) in the Department of Materials Design and Innovation, University at Buffalo.

References

1. Zhou Y, Yin X, Luo Q, Zhao X, Zhou D, Han J, Hao F, Tai M, Li J, Liu P, Jiang K, Lin H (2018) Efficiently improving the stability of inverted perovskite solar cells by implying polyethylenimine-modified carbon nanotubes as electrodes. Appl Mater Interfaces 10:37
2. Roy P, Sinha NK, Tiwari S, Khare A (2020) A review on perovskite solar cells: Evolution of architecture, fabrication techniques, commercialization issues and status. Solar Energy 198:665–688
3. Kojima A, Teshima K, Shirai Y, Miyasaka T (2009) Organometal halide perovskites as visible-light sensitizers for photovoltaic cells. J Am Chem Soc 131:17
4. Luo Q, Chen H, Lin Y, Du H, Hou Q, Hao F, Wang N, Guo Z (2017) Huang discrete iron(III) oxide nanoislands for efficient and photostable perovskite solar cells. Adv Func Mater 27:34
5. Weber D (1978) $CH_3NH_3PbX_3$, a Pb(II)-system with cubic perovskite structure. Zeitschrift für Naturforschung B 33:12
6. Hoefler SF, Trimmel G, Rath T (2017) Progress on lead-free metal halide perovskites for photovoltaic applications: a review. Montash Chem 148:795–826
7. Zuo C, Ding L (2017) Lead-free perovskite materials $(NH_4)_3Sb_2IxBr$ 9–x. Angew Chem Int Ed 56:23
8. Yang B, Chen J, Hong F, Mao X, Zheng K, Yang S, Li Y, Pullerits T, Deng W, Han K (2017) Lead-free, air-stable all-inorganic cesium bismuth halide perovskite nanocrystals. Angew Chem Int Ed 56:41
9. Green MA, Emery K, Hishikawa Y, Warta W, Dunlop ED (2016) Solar cell efficiency tables (version 48). Prog Photovolta Res Appl 2016(24):7

10. Ogawa K, Suzuki H, Zhong C, Sakamoto R, Tomita O, Saeki A, Kageyama H, Abe R (2021) Layered perovskite oxyiodide with narrow band gap and long lifetime carriers for water splitting photocatalysis. J Am Chem Soc 143:22
11. Wu C, Wang K, Li J, Liang Z, Li J, Li W, Zhao L, Chi B, Wang S (2021) Volatile solution: the way toward scalable fabrication of perovskite solar cells? Matter 4:3
12. Mitzi DB (2001) Thin-film deposition of organic-inorganic hybrid materials. Chem Mater 13(10):3283–3298
13. Park G, Oh IH, Park JMS, Jung J, You CY, Kim JS, Kim Y, Jung JH, Hur N, Kim Y, Kim JY, Hong CS, Kim KY (2018) Solvent-dependent self-assembly of two dimensional layered perovskite $(C_6H_5CH_2CH_2NH_3)_2MCl_4$ (M = Cu, Mn) thin films in ambient humidity. Sci Rep 8:4661
14. Kononova O, Huo H, He T, Rong Z, Botari T, Sun W, Tshitoyan V, Ceder G (2019) Text-mined dataset of inorganic materials synthesis recipes. Sci Data 2019(6):203
15. Olivetti EA, Cole JM, Kim E, Kononova O, Ceder G, Han TYJ, Hiszpanski AM (2020) Data-driven materials research enabled by natural language processing and information extraction. Appl Phys Rev 2020:7
16. Mikolov T, Chen K, Corrado G, Dean J (2013) Efficient estimation of word representations in vector space. arXiv preprint arXiv:1301.3781
17. Pennington J, Socher R, Manning CD (2014)Glove: global vectors for word representation. Proceedings of the 2014 conference on empirical methods in natural language processing (EMNLP):1532–1543
18. Peters M, Neumann M, Iyyer M, Gardner M, Clark C, Lee K, Zettlemoyer L (2018) Deep contextualized word representations. CoRR abs/1802.05365
19. Devlin J, Chang M, Lee K, Toutanova K (2018) BERT: pre-training of deep bidirectional transformers for language understanding. CoRR abs/1810.04805
20. Hegde C, Patil S (2020) Unsupervised paraphrase generation using pre-trained language models. CoRR abs/2006.05477
21. Chung J, Gulcehre C, Cho K, Bengio Y (2018) Empirical evaluation of gated recurrent neural networks on sequence modeling. CoRR abs/1412.3555
22. Sun H, Dhingra B, Zaheer M, Mazaitis K, Salakhutdinov R, Cohen W (2018) Open domain question answering using early fusion of knowledge bases and text. arXiv preprint arXiv:1809.00782arXiv:1809.00782
23. Sukhbaatar S, Szlam A, Wenston J, Fergus R (2015) Weakly supervised memory networks. CoRR abs/1503.08895
24. Cho K, Merrienboer B, Gulcehre C, Bougares F, Schwenk H, Bengio Y (2018) Learning phrase representations using {RNN} encoder-decoder for statistical machine translation. CoRR abs/1406.1078
25. Sutskever I, Vinyals O, Le Q (2014) Sequence to sequence learning with neural networks. CoRR abs/1409.3215
26. Bahdanau D, Cho K, Bengio Y (2015) Neural machine translation by jointly learning to align and translate. ICLR arXiv preprint arXiv:1409.0473arXiv:1409.0473
27. Sun C, Yang Z, Luo L, Wang L, Zhang Y, Lin H, Wang J (2019) A deep learning approach with deep contextualized word representations for chemical-protein interaction extraction from biomedical literature. IEEE Access 2019:7
28. Gao X, Tan R, Li G (2020) Research on text mining of material science based on natural language processing. In: IOP conference series: materials science and engineering, vol 768. No. 7. IOP Publishing
29. Kononova O, Huo H, He T, Rong Z, Botari T, Sun W, Tshitoyan V (2019) Ceder G (2019) Text-mined dataset of inorganic materials synthesis recipes. Sci Data 6(1):1–11
30. Krallinger M, Rabal O, Lourenco A, Oyarzabal J, Valencia A (2017) Information retrieval and text mining technologies for chemistry. Chem Rev 117(12):7673–7761
31. Huang S, Cole JM (2020) A database of battery materials auto-generated using ChemDataExtractor. Sci Data 7(1):1–13

32. Olivetti EA, Cole JM, Kim E, Kononova O, Ceder G, Yong-Jin Han T, Hiszpanski AM (2020) Data-driven materials research enabled by natural language processing and information extraction. Appl Phys Rev 7(4):041317
33. Kononova O, He T, Huo H, Trewartha A, Olivetti EA, Ceder G (2021) Opportunities and challenges of text mining in aterials research. iScience 24:3
34. Swain MC, Cole JM (2016) ChemDataExtractor: a toolkit for automated extraction of chemical information from the scientific literature. J Chem Inf Model 2016(56):10
35. Jessop DM, Adams SE (2011) OSCAR4: a flexible architecture for chemical text-mining. J Cheminform 2011(3):41
36. Hawizy L, Jessop DM, Adams N, Murray-Rust P (2011) ChemicalTagger: a tool for semantic text-mining in chemistry. J Cheminform 2011(3):17
37. Eltyeb S, Salim N (2014) Chemical named entities recognition: a review on approaches and applications. J Cheminform 2014(6):17
38. Doolin AJ, Charles RG, De Castro CSP, Rodriguez RG, Pean EV, Patidar R, Dunlop T, Charbonneau C, Watson T, Davies ML (2021) Sustainable solvent selection for the manufacture of methylammonium lead triiodide (MAPbI3) perovskite solar cells. Green Chem 2021(23):6
39. Wang J, Giacomo FD, Bruls J, Gorter H, Katsouras I, Groen P, Janssen RAJ, Andriessen R, Galagan Y (2017) Highly efficient perovskite solar cells using non-toxic industry compatible solvent system. Solar RRL 2017(1):11
40. Chen LC, Tien C, Jhou Y, Lin W (2020) Co-solvent controllable engineering of MA0.5FA0.5Pb0.8Sn0.2I3 lead-tin mixed perovskites for inverted perovskite solar cells with improved stability. Energies 13:10
41. Hameed MSNS, Aziz F (2021) Solvent engineering of lead-free bismuth-based perovskite material for potential application of solar cell. Mater Today: Proc 2021:46
42. Schmidt M (2008) The sankey diagram in energy and material flow management. J Ind Ecol 2008(12):2

The Discharge Crucible Method: Update on Experimental Design, Measurements, and Orifice Wetting

Hani Henein

Abstract The physicochemical properties, viscosity, density, and surface tension, are critical properties of liquid metals and alloys. These properties are needed for thermodynamics, solidification modelling, and materials properties databases. The discharge crucible method (DC) developed in 2003 has been used to measure and report these properties for a wide range of liquid metals and alloys, including Sb, Sn, Zn, Al, Al–Cu, Sb–Sn, Sn–Ag, and AZ91D. The results are compared with published data and models that are proposed to predict these property values. This method is based on a mathematical formulation that predicts the velocity of a stream draining from an orifice. The viscous losses are calculated using a discharge coefficient equation, and the gas–liquid surface tension is determined using the Young–Laplace overpressure induced in the jet. The model and experiments will be described along with the effect of nozzle shapes on the distribution of forces in the DC method, including the effect of wetting of the orifice. The aim is to define the optimal nozzle design for a good distribution of forces throughout a draining experiment.

Keywords Physiochemical properties · High temperature fluids · Surface tension · Viscosity

Introduction

With the advent of increased need for process modeling and simulation, the demand is increasing for accurate data for viscosity, gas–liquid surface tension, and density of metals and alloys. The values of these properties as a function of temperature, alloy composition, and gas atmosphere including the effect of impurities for recycled materials are needed. Numerous methods have been developed to measure one of these properties [1–14]. These methods are limited in providing the values of one or two of these properties from a single experiment. There is a need to have one

H. Henein (✉)
Department of Chemical and Materials Engineering, University of Alberta, Edmonton, AB, Canada
e-mail: hani.henein@ualberta.ca

A. Lazou et al. (eds.), *REWAS 2022: Developing Tomorrow's Technical Cycles (Volume I)*, The Minerals, Metals & Materials Series,
https://doi.org/10.1007/978-3-030-92563-5_4

method that can measure all three of them using one test. This makes the process of measurement much more efficient as these tests are very time consuming, and great care must be taken to ensure the accuracy of the results. Thus, the level of effort to generate all these values is very significant and costly.

Two methods have been developed that can measure all three properties using one test. The first is electromagnetic levitation (EML). It is a containerless technique where a droplet of about 6 mm is levitated and all three properties may be measured. The advantage of this method is that other fluid properties such as heat capacity may also be obtained. However, the distinct advantage of the EML is that it is containerless and that property values for undercooled liquids may be determined [15, 16]. The disadvantage of the EML is that it must be placed in microgravity environment in order to make these measurements [17, 18]. This comes at an obvious high cost and time overhead. There is presently an EML unit located on the International Space Station.

An alternative method to obtaining the three properties of a fluid, namely viscosity, gas–liquid surface tension, and density is the discharge crucible method (DC) [19–26]. In this paper, the principle of the DC method will be described. This will be followed by a description of the apparatus, some of the alloy results generated, and the effect of experimental design of the apparatus.

Mathematical Model

A detailed description of the model developed for the DC method is described elsewhere [19–26]. A brief overview will be presented here. The model is based on using the Bernouli equation to describe the flow of a fluid exiting the bottom of a crucible through an orifice. The forces that must be accounted for are gravity, inertial force of the exiting fluid, viscous forces of the fluid going through the nozzle, and the induced pressure differential due to interfacial phenomena according to the Young–Laplace equation. With the assumption that there is no wetting of the fluid exiting the orifice, and that quasi-steady state is achieved of the descent of the fluid from the crucible, the exit velocity of the fluid may be described by

$$u_2 = C_d\sqrt{2g\left(h - \frac{\sigma}{\rho g r_o}\right)}. \tag{1}$$

where u_2 is the velocity of the fluid exiting the orifice, C_d accounts for frictional losses to the orifice, g is the gravitational constant, h is the height of the fluid in the crucible, σ is the gas-fluid surface tension, ρ is the fluid density, and r_o is the radius of the orifice. Note that as the crucible empties u_2, C_d, and h are functions of time. Thus, as the crucible empties, Eq. (1) may be written in the following form:

$$M_{\text{exp}}(t) = \pi r_o^2 \rho C_d(t) \sqrt{2g\left(h(t) - \frac{\sigma}{\rho g r_o}\right)}. \quad (2)$$

where $M_{\text{exp}}(t)$ is the mass flowrate of the fluid exiting the crucible. Equation (2) may be written in dimentionless form as shown in Eq. (3):

$$F_r + B_o^{-1} = 1. \quad (3)$$

where F_r and B_o are the Froude and Bond numbers, respectively, and are given by

$$F_r = \frac{\left(\frac{M_{\text{exp}}}{\pi r_o^2 \rho C_d}\right)^2}{2gh}. \quad (4)$$

and

$$B_o = \frac{\rho g r h}{\sigma}. \quad (5)$$

The dimensionless form of Eq. (2) is useful in designing a particular apparatus for a specific fluid (e.g. nozzle size and crucible wall thickness). Note that as the crucible empties the balance between the viscous and surface tension forces changes. Apparatus design assists in maximing the accuracy of the data by ensuring that both forces play an important role during the conduct of the experiment.

Description of DC and Sample Results

A schematic and a photo of the current apparatus for conduting the DC experiment are shown in Figs. 1 and 2, respectively. The apparatus is composed of an induction furnace that heats a susceptor and crucible that contains the melt of interest. A thermocouple is inserted in the melt, and a radiation pyrometer is used to monitor the melt temperature. The crucible has an orifice machined in its bottom. To ensure that no flow of liquid occurs during heating and melting, a stopper rod plugs the bottom of the crucible and is removed once the melt reaches the desired temperature. A load cell is placed as close as possible near the bottom of the crucible to capture the stream and provide a mass flowrate as a function of time. More recently, a laser has been installed to measure the melt height as a function of time. Previously, the crucile geometry needed to be calibrated, and the height change with time was determined by back calculating the height knowing the mass flow measured by the load cell and the initial mass of metal placed in the crucible. An oxygen sensor provides a measurement of the oxygen content in the apparatus before starting to melt the charge in the crucible. During the melting stage, an oxygen getter, placed

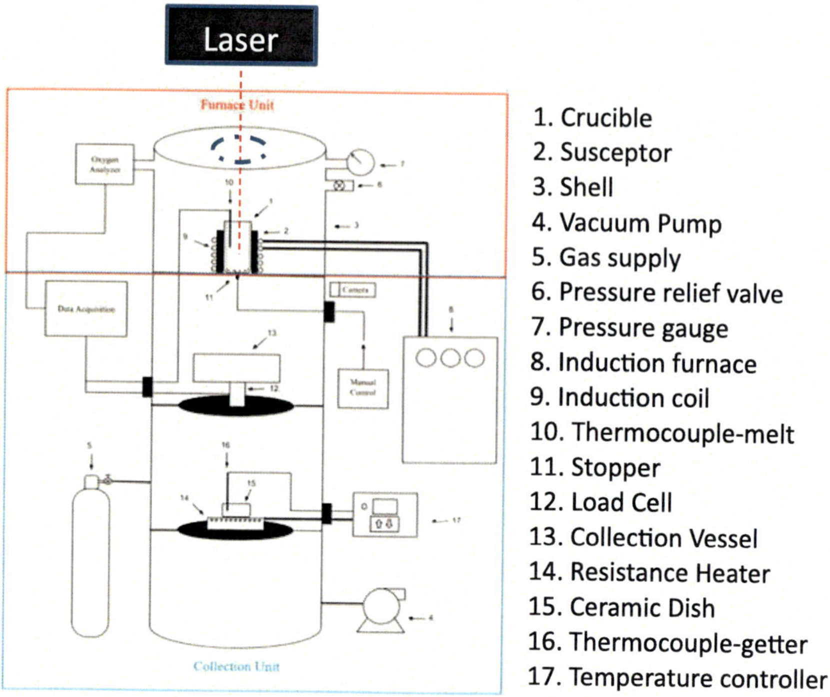

Fig. 1 Schematic of DC apparatus

at the bottom of the apparatus, ensures that the oxygen content remains constant throughout the experiment.

Prior to carrying out an experiment with a fluid with unknown properties, a set of calibration runs are carried out with a fluid of known properties. This provides the necessary data in order to obtain a C_d versus *Re* (Reynolds number) plot that characterizes the orifice. Typically, water or other fluid with known properties is used for these calibrations. Once a crucible and its orifice have been calibrated, an experiment can be conducted. The data obtained from a DC experiment is the $M_{\exp}(t)$ and corresponding $h(t)$. A nonlinear regression is then used to obtain the best fit values of σ, μ, and δ at the experimental temperature. For experiments carried out using aluminum, a maximum drop in temperature of the metal of 5 K was experienced during the draining of the crucible. This temperature drop does not introduce any significant error in the resulting property values calculated.

The DC method has been used to measure the thermophysical properties of Al, Sn, Sb, and Zn and various alloys, such as AZ91D, Al–Cu, Al–Sn–Ag, Al–Zn, Al–Li–Zn, Sb–Sn, Sb–Sn–Zn, Pb–Sb, Al–Mg, Al–Mg–Zn, Ga–Sn, and Ga–Sn–Zn [19–30]. Only a few sample results will be discussed here.

The surface tension of Al was found to be within 1% of averaged values determined by Mills for oxygen saturated elements [31]. With careful control of oxygen, Gancarz

Fig. 2 Photo of current apparatus for DC experiment

et al. [30] reported a viscosity of Al as a function of temperature that was in excellent agreement with literature values obtained using other techniques. Similar good results were reported for alloy systems such as Sb, Sn, Zn, and Sb–Sn [25], Sn–Ag [26], Al–Zn and Al–Li [27], Ga–Sn [28], Pb–Sb [29], and Al–Mg and Al–Mg–Zn [30]. In these results, the property values compared very favourably with other measurements reported in the literature using other techniques as well as with theoretical predictions of these properties. There was good agreement with the Butler model for surface tension [32] and reasonable agreement with the Hirai [33] and Kaptay [34] models for viscosity. The ideal soution model was found to be appropriate for alloy density.

Experimental Design

Most of the previous experiments reported were carried out using graphite crucibles. More recent work carried out using alumina crucibles for Al and Al-24 wt% Cu [23, 35] found that there was significant wetting of the orifice by the liquid metal. This was determined both visually as well as by determining the capillary number as a function of the height of the metal in the crucible. Even under these circumstances, the surface tension and the viscosity were determined within reasonable accuracy. The model converged on unrealistic values of density. It was found that the model was most sensitive to small changes in density but not those for surface tension or viscosity. Current research is aimed at extending the model for the DC method to include the prediction of dynamic contact angle [36].

More recent work [36] is focusing on further development of the DC model to incorporate the design of the orifice. Characteristics of the orifice such as bevel design for entrance of the fluid into the orifice and the length of the orifice can play a significant role in the contribution of the F_r and B_o numbers during the time of drainage of the crucible. This design flexibility can assist in enhancing the capabilities of the apparatus to yield increased accuracy in one of the property values of interest. Current research is also looking at extending the capabilities of the apparatus to carry our measurements on steels.

Summary

A new method for measuring the gas–liquid surface tension, viscosity, and density of liquid metals and alloys has been developed and evaluated on a range of alloys and using different gas atmospheres. The method was found to be reliable but sensitive to the wettability of the nozzle and its geometry. Nevertheless, with systems that do not wet the orifice of the crucible and under quasi-steady state for the draining of the crucible, the results are reliable and as accurate as any other method of measurement of these properties. Its distinct advantage is that presently gas–liquid surface tension, viscosity, and density are obtained simultaneously.

References

1. Yule AJ, Dunkley JJ (1994) Atomization of melts: for powder production and spray deposition. Oxford University Press, USA
2. Wessing JJ, Brillo J (2017) Density, molar volume, and surface tension of liquid Al-Ti. Metall Mater Trans A 48(2):868–882
3. Schick M, Brillo J, Egry I, Hallstedt B (2012) Viscosity of Al–Cu liquid alloys: measurement and thermodynamic description. J Mater Sci 47(23):8145–8152. https://doi.org/10.1007/s10853-012-6710-x
4. Crawley AF, Kiff DR (1972) The density and viscosity of liquid antimony. Metall Mater Trans B 3(1):157–159
5. Sato Y, Nishizuka T, Hara K, Yamamura T, Waseda Y (2000) Density measurement of molten silicon by a pycnometric method. Int J Thermophys 21(6):1463–1471
6. Cheng J, Gröbner J, Hort N, Kainer KU, Schmid-Fetzer R (2014) Measurement and calculation of the viscosity of metals—a review of the current status and developing trends. Meas Sci Technol 25(6):062001
7. Brooks RF, Dinsdale AT, Quested PN (2005) The measurement of viscosity of alloys—a review of methods, data and models. Meas Sci Technol 16(2):354
8. Iida T, Guthrie RI (1988) The physical properties of liquid metals. Clarendon Press, UK
9. Iida T, Guthrie RI (2015) The thermophysical properties of metallic liquids: fundamentals. Oxford University Press, USA

10. Matuyama Y (1927) On the surface tension of molten metals and alloys. Sci Rep Tohoku Univ 16:555–562
11. Staicopolus DN (1962) The computation of surface tension and of contact angle by the sessile-drop method. J Colloid Sci 17(5):439–447. https://doi.org/10.1016/0095-8522(62)90055-7
12. Hogness TR (1921) The surface tensions and densities of liquid mercury, cadmium, zinc, lead, tin and bismuth. J Am Chem Soc 43(7):1621–1628
13. Yoshikawa T (2013) Surface tensions of Fe–(30–40 mol%) Si–C alloys at 1523–1723 K. Mater Trans 54(10):1968–1974
14. Keene BJ (1993) Review of data for the surface tension of pure metals. Int Mater Rev 38(4):157–192
15. Hyers RW (2005) Fluid flow effects in levitated droplets. Meas Sci Technol 16(2):394
16. Paradis P-F et al (2014) Materials properties measurements and particle beam interactions studies using electrostatic levitation. Mater Sci Eng R Rep 76:1–53. https://doi.org/10.1016/j.mser.2013.12.001
17. Diefenbach A, Schneider S, Volkmann T (2020) Experiment preparation and performance for the electromagnetic levitator (EML) onboard the international space station. In: Pletser V (ed) Preparation of space experiments. IntechOpen
18. Fecht H-J, Wunderlich R, Ricci E, Egry I, Seetharaman S, Battezzati L (2010) The ThermoLab Project: thermophysical property measurements in space for industrial high temperature alloys. J Jpn Soc Microgravity Appl 27(4):190
19. Roach SJ (2002) Determination of the physical properties of melts. MSc Thesis, University of Alberta, Edmonton, Canada
20. Roach SJ, Henein H (2012) Physical properties of AZ91D measured using the draining crucible method: effect of SF 6. Int J Thermophys 33(3):484–494
21. Roach SJ, Henein H (2003) A dynamic approach to determining the surface tension of a fluid. Can Metall Q 42(2):175–186
22. Roach SJ, Henein H (2005) A new method to dynamically measure the surface tension, viscosity, and density of melts. Metall Mater Trans B 36(5):667–676
23. Flood RPS (2020) Thermophysical properties measurement of liquid Al and Al-Cu by the discharge crucible method. MSc Thesis, University of Alberta, Edmonton, Aleberta
24. Gancarz T, Gąsior W, Henein H (2014) The discharge crucible method for making measurements of the physical properties of melts: an overview. Int J Thermophys 35(9–10):1725–1748
25. Gancarz T, Gąsior W, Henein H (2013) Physicochemical properties of Sb, Sn, Zn, and Sb–Sn system. Int J Thermophys 34(2):250–266
26. Gancarz T, Moser Z, Gąsior W, Pstruś J, Henein H (2011) A comparison of surface tension, viscosity, and density of Sn and Sn–Ag alloys using different measurement techniques. Int J Thermophys 32(6):1210–1233
27. Trybula ME, Gancarz T, Gąsior W (2016) Density, surface tension and viscosity of liquid binary Al-Zn and ternary Al-Li-Zn alloys. Fluid Phase Equilibria 421:39–48. https://doi.org/10.1016/j.fluid.2016.03.013
28. Dobosz A, Gancarz T (2018) Density, surface tension and viscosity of Ga-Sn eutectic based alloys with Zn additions. J Mol Liq 264:600–606
29. Gancarz T, Gasior W (2018) Density, surface tension, and viscosity of liquid Pb–Sb alloys. J Chem Eng Data 63(5):1471–1479
30. Gancarz T, Jourdan J, Gasior W, Henein H (2018) Physicochemical properties of Al, Al-Mg and Al-Mg-Zn alloys. J Mol Liq 249:470–476. https://doi.org/10.1016/j.molliq.2017.11.061
31. Mills KC, Su YC (2006) Review of surface tension data for metallic elements and alloys: part 1—pure metals. Int Mater Rev 51(6):329–351
32. Butler JAV (1932) The thermodynamics of the surfaces of solutions. Proc R Soc Lond Ser A Contain Pap Math Phys Char 135(827):348–375
33. Hirai M (1993) Estimation of viscosities of liquid alloys. ISIJ Int 33(2):251–258
34. Kaptay G (2005) A unified equation for the viscosity of pure liquid metals. Zeitschrift für Met 96(1):24–31

35. Flood RPS, Henein H (2021) On the role of orifice wetting for Al and Al22.5wt%Cu with Al_2O_3 in the discharge crucible method. J Mol Liquids (in press)
36. Champdoizeau Q, Henein H (2021) Unpublished research, University of Alberta

Part II
REWAS 2022: Plenary

Actions of the Copper Industry Toward a Future Carbon-Neutral Society

Takashi Nakamura

Abstract The copper industry is striving to achieve sustainable developments while providing copper as an essential raw material for today's electrical/electronic society. Copper resources are limited, and copper production requires a large amount of energy, especially for ore processing. Meanwhile, the major end-product manufacturers are required to limit their carbon footprint of copper material, especially by using recycled copper. Whether recycled copper alloys have undergone a life cycle assessment (LCA) at the time of manufacture is doubtful. Furthermore, no metal material can be recycled without a primary metal. If the required amount of copper alloy can be covered by recycled copper, an LCA evaluation of the recycling process will be sufficient, but this situation seems unreasonable given the recent growth in copper consumption. Therefore, in this presentation, I propose an integrated LCA evaluation that includes copper ore extraction, smelting, and recycling processes.

Keywords Copper · Recycling · Carbon neutral

Introduction

The term "sustainable development goals" (SDGs) has become popularized over the past few years. The reports and websites of corporate social responsibility list the major companies that are meeting their SDGs to some extent. The SDGs incorporate 17 major goals and 169 concrete goals (called *targets*) that should be achieved through collaborative effort among all developed and developing countries [1]. These goals include our daily activities. Originally intended as the next-generation code of conduct in the United Nations (UN), this is linked to the movement of environmental, social, and governance investment [2] and greatly influences the activities of companies. Resource recycling, including waste treatment, is strongly stated in Goal 12 "Responsibility to use, responsibility to make".

T. Nakamura (✉)
Emeritus Tohoku University, 2-2-1 Katahira, Aoba, Sendai 980-8577, Japan
e-mail: ntakashi@sda.att.ne.jp

A. Lazou et al. (eds.), *REWAS 2022: Developing Tomorrow's Technical Cycles (Volume I)*, The Minerals, Metals & Materials Series,
https://doi.org/10.1007/978-3-030-92563-5_5

The wider handling of waste treatment is embodied in Goal 11 "Creating a city where you can continue to live".

Resource recycling has long been regarded as a concrete action of humankind. Especially, environmental protection requires not only the emission control of harmful substances and mitigation of resource waste, but also the reduction of carbon dioxide and chlorofluorocarbon emissions, which strongly influence global warming and eventually lead to obvious climate change.

The greatest role of the non-ferrous industry is the stable supply of non-ferrous metals, especially copper. Copper is indispensable in a carbon-neutral society because it is an excellent conductor of electricity and heat, is highly workable, and has sufficient mechanical strength. The recent production growth of copper is very large, exceeding 24,000,000 tons in 2019 [3]. This trend is expected to continue with the increasing global renewable energy supply, as shown in Fig. 1 [4].

All decarbonized activities are termed as "carbon neutral". We consider decarbonization in the non-ferrous smelting industry, which supplies copper, zinc, and lead. Unlike the iron industry, the non-ferrous industry does not directly generate a large amount of CO_2 in smelting. Although a certain amount of energy is consumed in pulverizing the ore during the mining-to-beneficiation process, the supply of natural copper resources does not strongly affect global CO_2 emissions. Under these conditions, setting a lower target for CO_2 emissions is a difficult task in the copper industry.

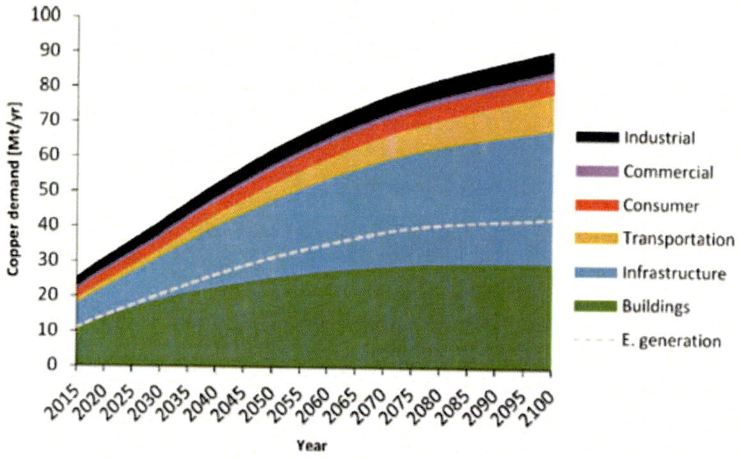

Fig. 1 Estimated global copper demand up to 2100 (from [4])

Contributions for Environmental Issues from Non-ferrous Industries

Although the copper industry does not seriously contribute to CO_2 emissions, it contributes to other environmental problems considered in SDGs, namely the management of the supplied heavy metals (mainly copper, zinc, and lead).

In copper smelting, iron sulfide is partially oxidized to produce a slag phase and a waste gas stream bearing sulfur dioxide. The melt, referred to as a matte, is then sulfur-oxidized to copper metal. The waste gas from further oxidation contains additional sulfur dioxide together with the oxide forms of less noble metals, which are ejected as fumes. Finally, the impure copper is purified in an electro-winning process and separated from more noble metals contained in the mineral, including precious metals such as silver and gold. The minor elements such as As, Sb, Bi, Se, and Te are emitted. These accompanying metals are sometimes beneficial but increase the manufacturing cost. Especially hazardous elements such as As and Hg always incur treatment and fixation costs. The copper grade in ores has been reducing, and the impurity grade has increased since ancient times. Therefore, treatment and fixation of these impurities have become more urgent, necessitating cooperation among copper smelting, zinc smelter, and lead smelting. Each smelting system treats the minor elements by several processes. The smelter processes can be combined as schematized in Fig. 2 [5].

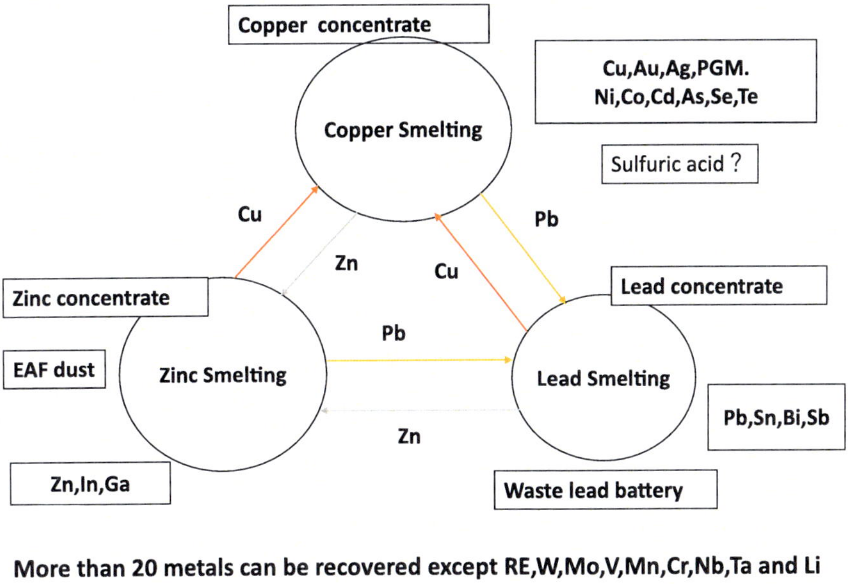

Fig. 2 Base and minor metals recovered from primary and secondary resources in the non-ferrous industry

Besides minor elements, copper concentrate contains a certain amount of lead and zinc. Copper is also included in other zinc and lead concentrates. Accordingly, the by-products of copper smelters contain lead and zinc, whereas those of zinc and lead smelters contain copper and lead. Therefore, a combination of non-ferrous smelters can sufficiently recover the minor elements, which is an important goal in the non-ferrous industry.

The non-ferrous industry must play a role in controlling heavy metals as a contribution to the environment and at the same time contribute to the construction of a carbon-free society.

Non-ferrous Metals in European Union (EU) Taxonomy [6]

The EU is committed to meeting the aims of the "European Green Deal" [7] and to transforming the EU into a modern, resource-efficient, and competitive economy. According to the European Green Deal 2, the EU will face and pursue various economic reforms, including decarbonization of the energy system, transition to a circular economy, and reversal of the alarming declines in ecosystems and biodiversity. Addressing the environmental challenges in the EU will help to achieve the broader international environmental objectives, such as those set out in the Paris Agreement [8], but will also require tremendous investments and innovations across sectors. The EU considers that the cost of decarbonization will be difficult to cover with public funds alone and recommends the promotion of private investment. For this purpose, it proposes a decarbonization taxonomy that can be evaluated for making new investments by each industry. Examples have been given in various fields, but the taxonomy is also being promoted in the fields of mining and metal resources. Table 1 summarizes the copper-related movements toward an eco-friendly society [9].

Modern extraction and processing of minerals and metals has greatly benefited society while reducing pressures on the environment, thereby mitigating greenhouse gas emissions, pollution, and waste production and improving the efficiency of natural resource use. The mineral industry will continue to enable these improvements by all downstream sectors. In this context, we believe that the Sustainable Finance Action Plan should adopt an integrated approach providing consistency, stability, and predictability along the whole value chain. The non-ferrous industry requires a large amount of capital for mine development, and the direction of corporate activities from a taxonomy perspective is necessary for favorable investments. One result of this approach is the promotion of recycling.

Table 1 Ecosystem contributions of the copper industry based on the EU taxonomy [9]

Contribution to low carbon technologies			
Transport	Wind	Solar	Enhancing recycling
By 2025: An additional 350 kt of copper will be required to meet energy infrastructure, charging, and storage needs An additional 950 kt of copper will be consumed by the automobile sector to meet EV requirements By 2030 An additional 1 Mt of copper will be required to meet energy infrastructure, charging, and storage needs An additional 3 Mt of copper will be consumed by the automobile sector to meet EV requirements	The insulated copper cables running down from the generator carry huge currents. A wind turbine contains an estimated 5 tons of copper. In a typical wind turbine, copper is used in the power cables, control cables, instrument cables, cooling and heating systems, the generator, transformer, and grounding system. Additionally, an average generator weighs 8.5 tons and is composed of 35% copper and 65% steel	Copper leads carry up to 8 times more heat than other materials Copper is ideal for heat exchangers and especially for solar thermal systems, which are more sustainable than traditional ones. Solar water heaters made of copper can save up to 34% of energy	Today, 50% of the EU's copper demand comes from recycled material because the life span of the materials can be decades and the demand is outpacing the return of the material

Consideration of LCA in Copper Industries

Circular Economy

The UN has been discussing means of increasing resource efficiency [10]. For this purpose, the efficient promotion of conventional reuse and recycling is demanded. The circular economy offers an economics-based solution to this problem. The essence of circular economy is the long-term use of the product (reuse, repair, and referral brush) and efficient recycling of a product that is ultimately difficult to use. In other words, when promoting a product, manufacturers must attempt to reduce the environmental load during the manufacture, assembly, and use of the materials from which the product is made, at the time of final disposal, and at all stages of the product life cycle. The UN has provided an action guideline for this purpose and a system for concrete action and its evaluation.

The development of the circular economy will lead to a transformation of the traditional mass production–mass consumption and the mass disposal industrial structure. However, as the mass production–mass disposal system is economically rational, a sustainable circular economy requires a change in human behavioral guidelines.

The minimum condition of carbon-neutral resource recycling is true lowering of the energy consumption of recycling. LCA has long been evaluated for this purpose. Although copper has been extensively researched, confirming the appropriate material flow and setting the overall boundary conditions in an LCA is more important for global assessment than acquiring detailed data.

Establishment of a Global Copper LCA

Copper has not yet been officially evaluated by LCA on a global scale. One purpose of this performance is proposing a global public LCA assessment of copper.

When establishing an LCA for copper, we can learn from the steel industry which has already established LCA on a global scale. Figure 3 shows the boundary of LCA evaluation examined by The World Steel Association [11]. In Panels (a) and (b), the flow of iron-product manufacturing by the conventional blast-furnace method is presented without and with recycling at the boundary, respectively.

In iron production, the blast-furnace and electric-furnace methods are responsible for 2/3 and 1/3 of the total generated CO_2, respectively. If circulation is omitted, the emissions of the blast and electric-furnace methods are not divided in a 2:1 ratio, but iron can be recycled innumerable times and recycled iron is first manufactured by the blast-furnace method. Numerical values are given [11].

I will omit the details, but one may reasonably suppose that the LCA for copper should be organized to include both the temporary raw material manufacturing process and the recycled manufacturing process.

Compiling similar world-standard LCA data for copper is required for understanding the true nature of copper recycling. Which organization will compile the data should be discussed, but organizations such as the World Copper Association or the International Copper Study Group can be nominated. Meanwhile, as in the case of total demand, copper must be discussed while also considering zinc, lead, and other by-products, which will likely complicate the boundary. I hope for sufficient discussion on this point.

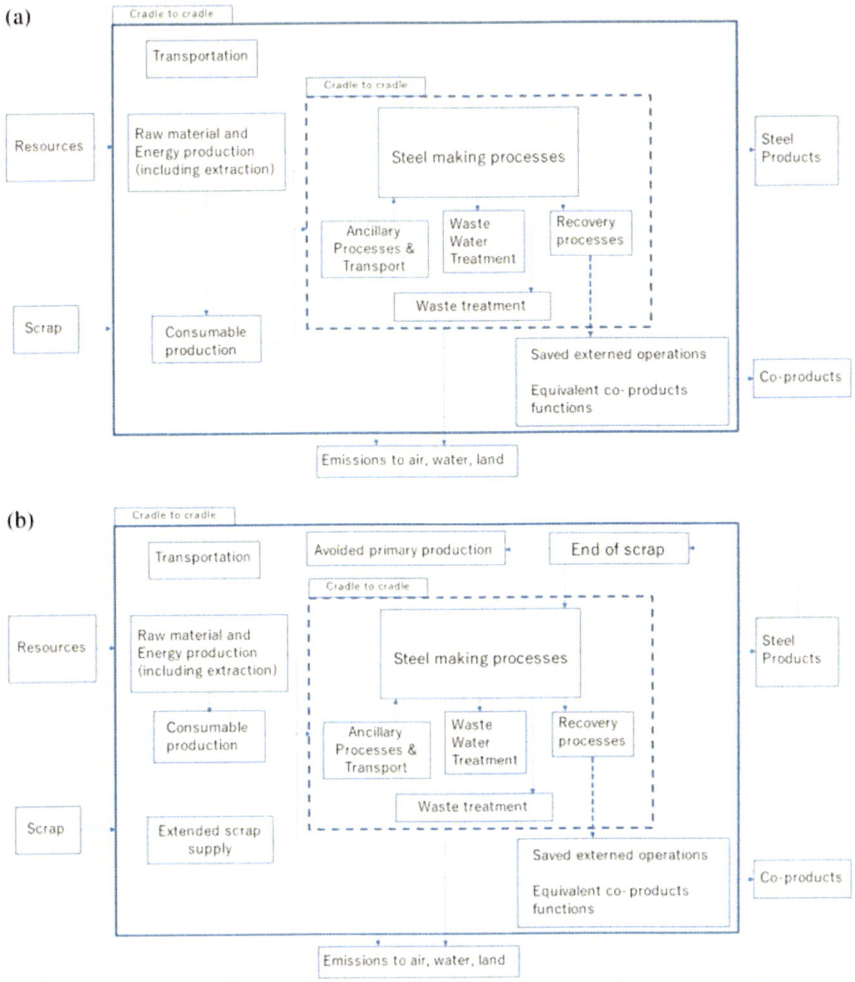

Fig. 3 **a** Boundary of LCA consideration of the iron and steel industry without recycling (World Steel Association). **b** Boundary of LCA consideration of iron and steel industry with recycling (World Steel Association)

Summary

Non-ferrous metals, especially copper, cannot be replaced in the future decarbonization of society. The goal is to introduce stable SDGs to the non-ferrous metal industry. To achieve decarbonization in a timely manner, all sectors involved in the collection and recycling of secondary raw materials and in the collection and smelting of primary resources must fully understand the overall flow and take actions. A comprehensive LCA assessment of the supply of non-ferrous metals, especially copper, should be conducted to avoid wrong actions. The establishment of a central organization for LCA evaluation can be considered as urgent.

References

1. https://www.un.org/development/desa/disabilities/envision2030.html
2. https://www.unepfi.org/annual-overview-2019-2020/
3. https://www.icsg.org/index.php/statistics/selected-data
4. Schippera BW, Lina H-C, Melonia MA, Wansleeben K, Heijungs R, van der Voet E (2018) Estimating global copper demand until 2100 with regression and stock dynamics. Resour Conserv Recycl 132:28–36
5. Nakamura T (2002) The role of the metallurgical industry in the recycling-based society. In: International conference on the sustainable processing of materials. Cares, Australia, pp 21–27
6. https://ec.europa.eu/info/business-economy-euro/banking-and-finance/sustainable-finance/eu-taxonomy-sustainable-activities_en
7. https://ec.europa.eu/info/strategy/priorities-2019-2024/european-green-deal_en
8. https://unfccc.int/process-and-meetings/the-paris-agreement/the-paris-agreement
9. http://www.wfw.com/articles/the-eu-taxonomy-for-sustainable-investmentmining-metals-minerals/
10. https://ec.europa.eu/environment/resource_efficiency/
11. World Steel Association (2012) Life cycle assessment methodology report

Part III
REWAS 2022: Recovering the Unrecoverable

BlueMetals Technology—Experience from Commissioning E-Scrap Recycling Plants

Timm Lux, Markus Reuter, Rolf Degel, Frank Kaussen, and Nikolaus Borowski

Abstract SMS group is innovating and developing future solutions in the non-ferrous metals sector to enable the circular economy (CE). The recycling of valuable metals is one of the key enablers for the CE. SMS group offers numerous recycling solutions to recovery metals and other valuable substances from metal scraps, electronic wastes, batteries, catalysts, etc., which are based on a sequence of pyro- and/or hydrometallurgical process steps. This paper shares experiences on recently commissioned solutions, for example, Aurus in Russia.

Keywords Anode furnace · Electronic waste · BlueMetals technology · BlueSmelter · WEEE recycling · TBRC · Tank house · PGM · Precious metals recovery · Dry slag granulation

Introduction and Market

Over the past years, the demand of electrical and electronic equipment has increased dramatically with technological progress. Innovations on the technology of electronic devices led to a shorter usage and lifespan and thus boosted the generation of waste electrical and electronic equipment (WEEE). WEEE is one of the fastest growing waste streams, growing with an average of 2–4% per year worldwide and is expected to reach approximately 75 million tons per year by 2035. Currently, about 55 Mt of these wastes are estimated to arise worldwide, and the trend is rising [1]. WEEE is a valuable urban resource, which contains significant amounts of precious metals and base metals with a high potential for metallurgical recycling [2]. Collection rates are often not sufficiently high; however, if collected for recycling, the BlueMetals plant allows the recycling of the complete range of PCBs and thereby provides a solution for low grade PCBs.

SMS provides various solutions for an efficient recycling of WEEE to recover the contained metals, such as gold, silver, platinum, palladium, copper, nickel, and

T. Lux (✉) · M. Reuter · R. Degel · F. Kaussen · N. Borowski
SMS Group GmbH, Eduard-Schloemannstrasse 4, Düsseldorf, Germany
e-mail: Timm.lux@sms-group.com

A. Lazou et al. (eds.), *REWAS 2022: Developing Tomorrow's Technical Cycles (Volume I)*, The Minerals, Metals & Materials Series,
https://doi.org/10.1007/978-3-030-92563-5_6

cobalt. Up to 98% of these metals can be completely recovered and converted to pure products that can be reused in the same consumer and other products these originated from. However, the wide range of WEEE qualities requires different types of recycling technologies. SMS group provides custom-fit solutions as are part of the BlueMetals family. The central component of SMS group's WEEE recycling technology is either the BlueSmelter and/or the top blown rotary converter (TBRC). Compared, the BlueSmelter can also handle lower grade WEEE with higher organic contents. The concept of the BlueSmelter was successfully demonstrated in an industrial size demonstration plant and resulted in a new generation of bath smelting furnace, applying a high-speed SMS injection system (SIS) which is substituting the burner and (submerged)lance. Currently, the BlueSmelter is designed in different sizes starting with an annual capacity of 3,000 tons up to a maximum annual capacity of 120,000 tons of e-scrap concentrate.

BlueMetals Technology

Processes for WEEE treatment in the copper industry are usually subdivided in pyrometallurgical pre-treatment followed by hydrometallurgical refining to recover base metals such as copper and nickel, as well precious metals such as gold, silver, platinum, and palladium.

Conventional pyrometallurgical technologies are commonly limited to a WEEE concentrate feed of approximately 30% of their total processing capacities. The main reasons for this are, on the one hand, the high energy input from the organic fractions of WEEE concentrates (e.g. plastics), which tends to overheat conventional smelter, and, on the other hand, the challenging gas cleaning process with regard to cleaning the off-gas from chlorine and bromine compounds which were brought into the process by flame retardants. In addition, established processes are mainly focused on the production of base metals (e.g. copper and nickel), and precious metals are often only a valuable by-product. Long lead times for the recovery of gold, silver, and platinum group metals are the inevitable result. Part of the process-encompassing BlueMetals solution of SMS to extend the limits of the maximum use of WEEE concentrate to 100% and to accelerate the recovery of precious metals is the use of a highly specialized melting furnace called BlueSmelter, which is based on bath smelting technology and is widely used in the copper, lead, and zinc industries [3]. The market and sources for e-waste are rather complex, and each country/region requires different recycling solutions.

In order to establish an effective recycling process, SMS offers various technology options and process combinations. Starting from a standalone technology to produce a copper bullion as an intermediate product, which can also be integrated in an existing copper production plant, over a compact plant up to a full BlueMetals plant. Thereby the BlueMetals technology always offers a variety of benefits for recycling ambitions, reflected by a flexible plant design with the focus on precious metals recovery.

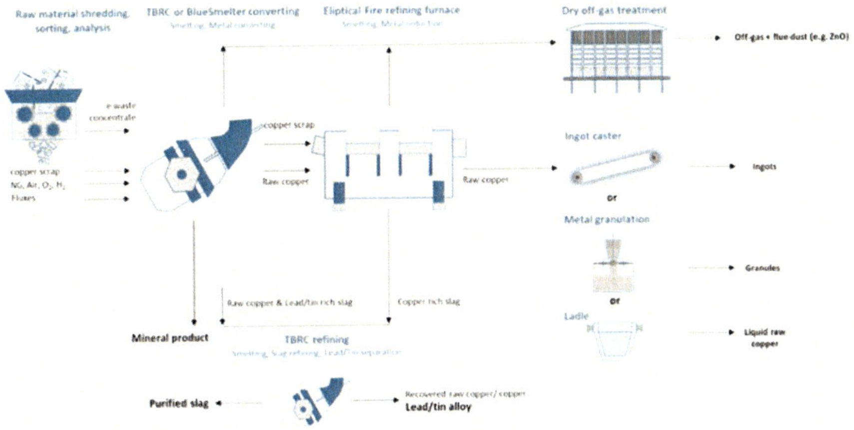

Fig. 1 Depiction of BlueMetals flowsheet for metal alloy production

In case the clients have WEEE material as well as copper scraps available, the flow sheet depicted by Fig. 1 is applicable [3]. In this first step, the plant produces the metals containing alloy in form of ingots, granules, or liquid metal as well as a Pb/Sn alloy.

Several customers are looking for a modular system, seeking for additional options to significantly boost the production. The BlueSmelter can be an ideal smelting unit and allows tripling the overall throughput of a WEEE recycling plant.

The plants can process up to 120,000 tpa of WEEE concentrate (see Table 1).

In case the plant shall also produce LME grade quality metals, the plant can be expanded with an anode casting wheel, electro refinery, and a precious metals refining plant. This principle was supplied to Aurus in Russia and realized with the shown technology and was successfully commissions (see Fig. 2).

For a further processing of the copper alloy, a tailor-made top blown rotary converter (TBRC) as a converting and refining unit is applied. The process is designed for converting "black copper" to "raw copper" and for separating heavy metals, e.g. lead and tin for a further processing. Developed in the 1930s, TBRCs smelt and convert various primary and secondary raw materials to extract Cu, Ni, Pb, Zn, Sn, and precious metals [6].

The anode furnace unit refines the "raw copper" into "anode copper" with a copper content >98.5%. The furnace is equipped with the patented PROX burner and an innovative charging system [7]. Overall, this reduces significantly the specific energy consumption compared to conventional fire refining furnaces. The adjustable burner flame length provides the process heat and is designed for heating up the vessel's refractory during start-up.

The anode copper is being shaped in anodes in the so-called anode casting wheel [8]. The process runs fully automatic and produces an anode with a weight deviation of only ±1%. This is achieved by the three-point weighing unit in the casting system.

Table 1 Exemplary types of BlueMetals plant configurations

Plant type	BlueMetals	BlueMetals compact	BlueMetals integrated
Application	Large e-scrap recycling plants and copper producers	Smaller and midsize e-scrap recycling plants mainly for PCBs	Existing copper and nickel producers with precious metal refinery options
Input	Low, medium, and high-quality WEEE concentrates	PCBs	Low, medium, and high-quality WEEE concentrates
Technology units	BlueSmelter/ TBRC-TBRC-anode furnace-casting wheel-slag holding furnace-slag granulation	BlueSmelter/ TBRC-Ladle stand/anode furnace-casting wheel or metal granulation	BlueSmelter/TBRC-TBRC-anode furnace-casting wheel or metal granulation
Semi-products	Raw copper	Raw copper	Black copper/raw copper
Further optional refining	Tank house-PMR	Tank house-PMR	Existing electrolysis
Final products	Cu, Ni, Co, Au, Ag, and PGM	Cu, Ni, Co, Au, Ag, and PGM	Depending on existing units
Capacity WEEE	30,000–120,000 tpa	1,000–30,000 tpa	3,000–120,000 tpa

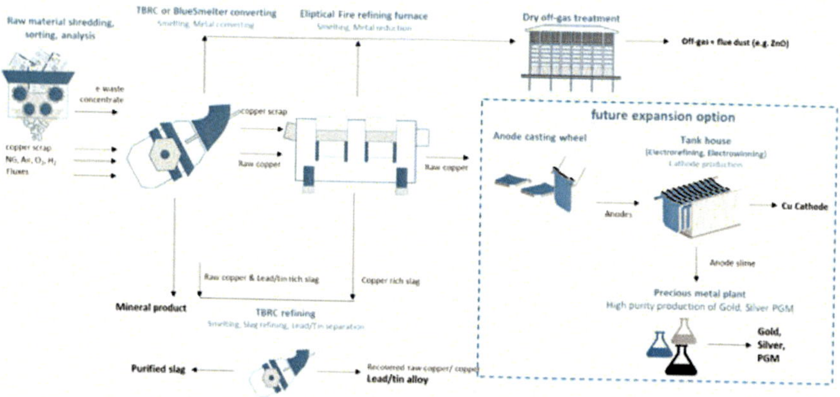

Fig. 2 Depiction of BlueMetals flow sheet for LME grade metals production

The dynamically controlled mold cooling reduces thermal loads, and a reliable wash coating process is providing an exceptional anode surface quality [8].

SMS group developed an innovative dry slag granulation system for certain types of non-ferrous slags. The aim is to produce a high-quality slag-product and to recover substantial quantities of energy. The system is proven for copper slags. We are also investigating to use the dry slag granulation technology for slags being produced in the ferro-alloy industry. The basic system—the so-called **Dry**Slag system—includes the granulation unit. An air jet, blowing diagonal into the slag stream, solidifies the liquid slag stream. The solidified granules settle in a collecting chamber.

The BlueSmelter

The BlueSmelter (see Fig. 3) is an innovative bath smelting technology, which is designed to process a wide range of copper and precious metals containing secondary raw materials, with the focus on handling low scrap qualities and high organic loads but with significant monetary value. The improved bath smelter furnace enables nearly maintenance-free blowing of media into the bath via the patented SIS [4] that provides an improved bath agitation and increased surface turbulence for rapid gas/melt reactions in the vessel [5]. SMS group successfully conducted pilot-scale tests to demonstrate its technical feasibility. The pyrometallurgical pre-treatment and smelting in the BlueSmelter ensures an enrichment of valuable metals in a metallic phase and a pre-separation of organic compounds as well as metallic impurities. Furthermore, it can handle high amounts of organics in the feed material and deal with the simultaneously increased input of halogens as part of the organic fractions. The features of the BlueSmelter include:

Fig. 3 Illustration of BlueSmelter

- "Modified" bath smelter/BOF technology with SIS media injection
- Flexible smelting (can run from 40 to 100% capacity with various input mixtures with high organic fractions)
- Flexible sealed charging systems for various input materials by injection and conveyer
- CO_2 reduced production process (with H_2-ready option)
- Low maintenance SIS burner/injector maintaining stable temperature during reduction
- Batch, semi-continuous operation (e.g. tap-to-tap 2–4 h).

The calculations and simulations confirmed the extraordinary results of the tests. Compared to conventional TSL or BOF-based bath smelting vessels, all reactions are significantly accelerated (see Table 2).

Table 2 Comparison of conventional TSL or BOF-based bath smelter technology to BlueSmelter

	Bath smelting/BOF type furnace	BlueSmelter
Feed mix	80–100% input	40–100% input
Organics	<20%	Up to 50%
Charging	Unsealed	Sealed
Hydrogen	Not possible (only enriched)	100% H_2 as fuel possible
Process	Batch or semi-continuous, high turbulence, and not adjustable	Batch or semi-continuous, high turbulence, and adjustable turbulence
Cooling/lining	For WEEE cooling required	No or simple cooling, simple lining concept
Vessel complexity	Complex in height + vibration compensation	Compact, robust, simple, and short vessel
Emissions	Higher fluctuating off-gas volumes with high NO_x level, larger dedusting system	Lower constant off-gas volumes with lower NO_x level
Maintenance	Lance consumption, lance maintenance, and splashing issue	Low maintenance SIS, simple technology
Efficiency, productivity	Medium—good metal yield (based on TSL know-how)	Higher—good metal yield, higher productivity per volume
OPEX	Higher due to increased media consumption and increased off-gas volume	Lower due to reduced media consumption, increased production efficiency
CAPEX	Complex vessel design, large foundation, and high buildings necessary + larger balance of plant items: high in CAPEX	Compact simple vessel design, smaller building height and reduced foundation + smaller balance of plant items: low in CAPEX

Recent References and Experiences

Over the past years, SMS supplied several units, which are in operation in numerous non-ferrous metals plants for the recycling of copper scrap and WEEE. Especially the TBRC unit is based on a strong track record. One of the largest TBRCs as well the first TBRC for the processing of WEEE was supplied by SMS. The first BlueMetals compact plant was installed in Malaysia for the client SPM and was commissioned in 2017. The bath smelter processes approximately 3000 tpa of PCB and produces a black copper. The smelter at SMS utilizes a top lance for the injection of oxygen-enriched air. The plant is operating well. SMS took this "smelter principle" and developed the next generation of the bath smelter, the so-called BlueSmelter. To investigate the smelting behavior, the slag chemistry, and the metallurgical reduction process, various material mixes of WEEE concentrates were pyrometallurgically processed at multiple lab- and pilot-scale tests. Additional industrial-scale demonstration tests of the BlueSmelter were conducted at MEFOS in Sweden in 2019 (Fig. 4).

The tested BlueSmelter included all optional features including a bottom purging with H_2 and N_2. Some conclusion of the tests is given below:

- MgCr-refractory works well (even without vessel cooling).
- Injection of fines tested and proven.
- Reducing smelting can be easily done; oxidation potential can be adjusted as desired.
- Cu content in slag of ~0.7 wt.% and even below can be achieved.
- Hydrogen purging technically works but is not required.
- Slag system $CaO–Al_2O_3–SiO_2$ works with a low viscosity + low Cu content.
- Separate reduction step is not necessary.
- SIS burner can also be operated without natural gas in the shroud gas.

Fig. 4 Photo of the BlueSmelter at Swerim in Sweden

Fig. 5 Anode furnace at Aurus during operation

- Low NO_x level achieved.

SMS supplied the world first Greenfield plant for WEEE to Aurus near Moscow, and the plant processes about 6,000 tons of printed circuit boards as well as 6,000 tons of copper scrap per year. It produces LME grade metals such as copper, nickel, gold, silver, and platinum. SMS group supplied all components, i.e. the system for mechanical preparation of the raw materials, TBRC converter (top blown rotary converter), refining furnace (Fig. 5), casting wheel, electrolysis, precious metals recovery plant, gas cleaning, and automation systems. SMS group provided the engineering, supervise erection and commissioning and trained the customer's personnel. Mettop supplied all components needed for electro-refining [9]. The plant was fully in hot-commissioned.

The plant demonstrated that it achieves a recovery rate of >98% for the most relevant metals such as Cu, Ni, Au, Ag, Pt, and Pd as well as Pb and Sn. The client already placed the order for the expansion of the plant. Basically, SMS will supply an additional tilting refining furnace, which will boost the overall production of this plant to 10,000 tpa of WEEE + 15,000 tpa of copper scrap.

Conclusion

SMS group offers various processes for e-waste recycling focusing on all grades and qualities of WEEE concentrates, the so-called BlueMetals processes. The company established a new standard in e-waste recycling, in which WEEE concentrate in general and including all qualities is considered a valuable resource rather than a problematic waste stream. The central component of SMS group's WEEE concentrate recycling concepts is either the BlueSmelter and/or the top blown rotary converter (TBRC). Compared, the BlueSmelter can also handle lower grade WEEE concentrates with higher organic shares. The concept of the BlueSmelter was successfully demonstrated at industrial scale and resulted in a new generation of bath smelting technology, applying a high-speed SIS-/injector. In these industrial-scale trials, various WEEE concentrate mixes of different qualities were treated. SMS group successfully commissioning a PCB recycling plant with a TBRC as primary smelter near Moscow, Russia. This plant is designed to process about 4,000 tons of printed circuit boards every year to LME/LBMA grade metals, such as copper, nickel, silver, gold, and platinum.

References

1. Forti V, Baldé CP, Kuehr R, Bel G (2020) The global e-waste monitor 2020, quantities, flows, and the circular economy potential, United Nations University (UNU)/United Nations Institute for Training and Research (UNITAR)—co-hosted SCYCLE Programme, International Telecommunication Union (ITU) & International Solid Waste Association (ISWA), Bonn/Geneva/Rotterdam
2. Goodship V, Stevels A (2012) Waste electrical and electronic equipment (WEEE) handbook. Woodhead Publishing Limited, Cambridge
3. Degel R, Lux T, Borowski N (2019) BlueTechnologies: our solutions in challenging times for the non-ferrous metals industry. Metall 74. Jahrgang
4. Odenthal H-J, Bader J, Nörthemann R, Reifferscheid M, Klioutchnikov I, Olivier H (2014) The optimized SIS injector for EAF application. In: AISTech 2014—the iron & steel technology conference and exposition, Indianapolis, USA
5. Degel R, Lux T, Joubert JL, Borowski N (2020) BlueTechnologies: our solutions for the non-ferrous metals industry. METAL ASIA
6. Degel R, Joubert JL, Filzwieser I, Filzwieser A, Hanel M (2017) Innovative solutions in copper production lines. In: Proceedings of the EMC, Leipzig, Germany
7. Degel R, Lux T, Hecker E, Filzwieser A, Filzwieser H-JKI (2019) Innovative solutions in non-ferrous productions production. In: Proceedings of the copper 2019, Toronto/Canada
8. Schatz T (2019) Anode casting wheel of SMS group. In: 23rd international conference on non-ferrous minerals & metals, July 12–13, 2019, Hotel Lalit Great Eastern, Kolkata
9. Hanel M, Filzwieser A, Lux T, Joubert JL, Degel R (2019) Innovative solutions in the non-ferrous metals production lines. In: Proceedings of the EMC, Düsseldorf, Germany

Recycling of Tungsten by Molten Salt Process

Tetsuo Oishi

Abstract Tungsten (W) is one of the most important metals in various industries, particularly in the machining industry. At least 60% of W is consumed in cemented carbide or super-hard alloys in Japan and the USA. The major recycling method for cemented carbides is the hydrometallurgical process combined with the roasting step, which oxidizes W in scrap into tungsten oxide. However, this conventional process requires repetition of roasting and dissolution in some cases, which makes the process costly and inefficient. In contrast, the molten salt process has specific advantages in terms of processing rate and simplicity. In this study, the recycling processes for W using molten salt are reviewed. Subsequently, our new recycling process using molten hydroxide is introduced, and recent data on this process discussed.

Keywords Rare earth elements · Recycling · Molten salt electrolysis · Alloy diaphragm

Introduction

Tungsten (W) is widely used in industries owing to its high hardness, melting point, and density. According to the literature, more than 60% of W is consumed in cemented carbide or super-hard alloys in Japan and the USA [1, 2]. Super-hard alloys are mainly used in machining industries and are composed of tungsten carbide (WC) and binder metals such as cobalt (Co) and nickel (Ni), as shown in Fig. 1. However, mineral sources of W are relatively scarce and unevenly distributed in the Earth's crust. Thus, the establishment of efficient W-recycling processes is one of the most important objectives for effectively utilizing the limited W resources.

Recycling processes for super-hard alloys are generally categorized into direct and indirect methods (Fig. 2) [3]. In the direct method, zinc (Zn) treatment is typically employed to make super-hard alloy scraps easy to crush in which cracks are

T. Oishi (✉)
National Institute of Advanced Industrial Science and Technology (AIST), 16-1 Onogawa, Tsukuba 305-8569, Ibaraki, Japan
e-mail: tetsuo.oishi@aist.go.jp

A. Lazou et al. (eds.), *REWAS 2022: Developing Tomorrow's Technical Cycles (Volume I)*, The Minerals, Metals & Materials Series,
https://doi.org/10.1007/978-3-030-92563-5_7

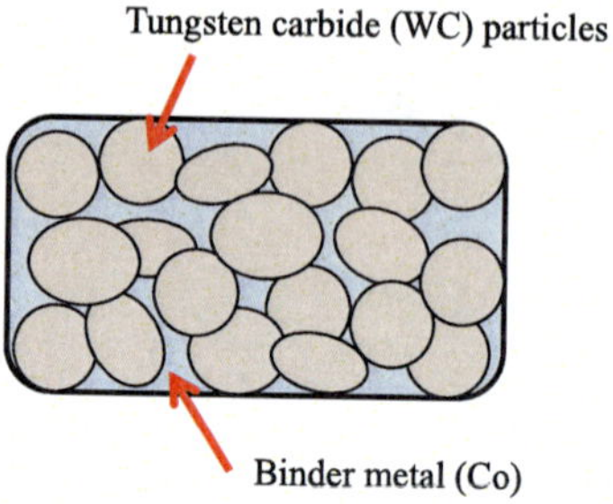

Fig. 1 Schematic drawing of cemented carbide (super hard alloy)

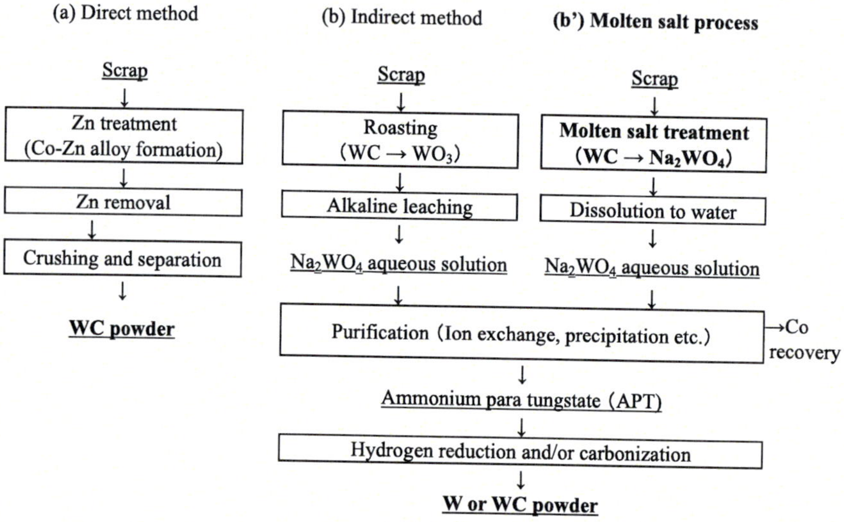

Fig. 2 Flow of typical direct and indirect methods

formed during the Zn–Co alloy formation step. After the evaporation of Zn from the Zn–Co alloy, the super-hard alloys are easily crushed, and WC powder is recovered via conventional physical separation steps. Although this method is simple and inexpensive, the use of recovered WC powder is limited because it essentially has no purification effect. The indirect method is also known as the chemical method. The super-hard alloy scrap is typically roasted in air to oxidize WC to WO_3. Thereafter, the roasted scraps are dipped in NaOH solutions, and W is dissolved. After purification of the NaOH solutions, W is recovered as ammonium paratungstate (APT), which is an important intermediate product from W ore during W metallurgy. WC and/or W powder are commonly produced from APT by roasting, hydrogen reduction, and carbonization. The indirect method has a high purification effect and is applicable to various super-hard alloy scraps; however, this method requires repetition of roasting and dissolution in some cases, which makes the process costly and inefficient.

Several other processes have been proposed and investigated from the past to date. Because numerous review papers have already been published on the W-recycling process [4–9], this paper focuses on the molten salt process proposed and/or investigated for W-recycling. The molten salt process has specific advantages in terms of processing rate and simplicity. First, some molten salt processes available in the literature were reviewed. Subsequently, our new recycling process using molten hydroxide was introduced, and the recent progress in this process discussed.

Molten Salt Processes from the Past to Date

Molten salt processes are mainly categorized as indirect methods. The conventional indirect method has an operational problem in that it is difficult to oxidize the entire scrap simultaneously when the method is applied to relatively large objects. This complicates the process by requiring repeated operations, such as roasting, alkali leaching, or peeling of the surface layer. The molten salt process is a solution to this problem (Fig. 2b). In other words, the main subject of the molten salt processes is the efficient oxidation of WC and W to water-soluble W species such as WO_3 and Na_2WO_4.

Some of the molten salt processes available in the literature are summarized in Table 1. More than half of them are patent documents, and because of the nature of patents, they tend to be broadly described. For instance, when sodium chloride (NaCl) is considered, it is often referred to as "alkali metal chloride." In commercial applications, however, it is difficult to consider options other than sodium salt from a cost perspective. Thus, all of them are described as sodium salts rather than "alkali metal salt" in this paper. As already mentioned, it is important to efficiently oxidize WC and W in the molten salt process, and the description is differentiated by the oxidant used.

Table 1 Molten salt processes for W-recycling

Oxidizing agent	Dilutant	Operation Temperature	Characteristics	References
$NaNO_3$ or $NaNO_2$ (with support by O_2 in some cases)	Na_2CO_3	800 °C	Rapid reaction Difficult to control Generation of NO_x gas	[4, 6]
	NaOH (+ NaCl)	400 ~ 700 °C		[10, 13, 14]
	MCl (M = Li, Na, K)	400 ~ 700 °C		[15]
	No	700 ~ 900 °C		[11, 16, 17]
O_2	Na_2CO_3	800 ~ 1000 °C	Low reactivity	[18, 19]
Na_2SO_4	NaOH	800 ~ 1100 °C	Relatively low reactivity Generation of SO_x gas	[20]
	NaOH + MO (M = Co etc.)	900 ~ 1100 °C		[21]

Sodium nitrate ($NaNO_3$) and sodium nitrite ($NaNO_2$) are the most popular oxidizing agents in this field. In the report by Shedd, which summarizes the situation of tungsten recycling in the USA, only the process using $NaNO_3$ and $NaNO_2$ as oxidizing agents and sodium carbonate (Na_2CO_3) as a diluent is introduced [6]. According to a review by Bhosale et al. [10], $NaNO_3$ or a mixture of $NaNO_3$ and Na_2CO_3 is widely used in the field of W metallurgy, at least by 1990. The main reaction in the presence of oxygen is explained by the following equation [11]:

$$WC + 3NaNO_3 + 1/4O_2 \rightarrow Na_2WO_4 + 3NO + CO_2 + 1/2Na_2O \quad (1)$$

Fs patent (US-1652646, 1927) as a processing method to obtain Na_2WO_4 from ores [12]. Tungsten in ores exists in a hexavalent state such as $CaWO_4$ and (Fe, Mn)WO_4, which is different from the treatment of WC and W to be recycled. However, the use of $NaNO_3$ as an oxidizing agent for various tungsten ores has been observed in several patents related to tungsten smelting in the 1910s–1920s and is assumed to have been widely recognized in this field. Although the molten salt process using $NaNO_3$ has a disadvantage in that NO_x gas is generated and exhaust gas treatment is required, it is superior in that the reaction proceeds quickly because of the strong oxidizing nature of $NaNO_3$. However, according to Lassner, this is an extremely violent reaction and "difficult to control [4]." The same has been highlighted in several patent documents, and methods using NaOH or alkali metal chlorides as diluents have been proposed to address this problem [10, 13–15]. Among these methods, a system using NaOH as a diluent was introduced in detail by Bhosale et al. [10]. According to them, the reaction can be well controlled, and NO_x generation can be suppressed under appropriate conditions. It has also been reported that Sandvik Asia Ltd. has already started operations on a semi-commercial scale. The combination of molten NaOH and $NaNO_3$ is superior to other molten salt processes in that it can be operated at a relatively low temperature of approximately 450 °C, and it also suppresses the generation of NO_x gases. However, if the following process flow is similar to that shown in Fig. 2b, the diluent will be dissolved in water at a later stage. Thus, the chemical cost increased as the amount of diluent increased, although the control of the reaction rate became easier.

As an alternative approach, Ikegaya and co-workers succeeded in controlling the reaction by accurately controlling the feeding rate of $NaNO_3$ [11, 16]. Notably, they succeeded in controlling the reaction, which had been considered "difficult to control" for a long time, without using a diluent such as NaOH. In addition to the molten salt process, they invented an efficient hydrometallurgical process to remove impurities and developed an exhaust gas treatment facility for NO_x gas. This process has been commercialized after further investigation and improvement [17].

Although $NaNO_3$ is the major oxidizing reagent in this field, as mentioned, other oxidizing agents such as O_2 gas and sulfide ions have also been investigated. Scott proposed a process for oxidizing and extracting the tungsten component to obtain Na_2WO_4 by contacting W-containing scraps with molten Na_2CO_3 at 800–1000 °C in air [18]. Recently, Yasuda et al. proposed the oxidation and dissolution of W from super-hard alloy scraps in molten Na_2CO_3 under controlled O_2 and CO_2 partial pressures. Although this study is in the fundamental stage, they found that CO_3^{2-} ions can also be an oxidizing agent for metallic W in addition to O_2 gas under specific conditions [19]. Lohse proposed a method of blowing air into a mixture of molten NaOH and sodium sulfate (Na_2SO_4) [20]. The problem here is that it is difficult to obtain a sufficient reaction rate with oxygen and sulfate. Therefore, it is necessary to maintain the reaction chamber at a temperature of 900 °C or higher, resulting in high energy consumption and serious damage to the equipment. Compared to oxygen (or air), Na_2SO_4 seems to oxidize faster, but SO_x is generated when the sulfate is reduced; therefore, an exhaust gas treatment facility is required, as in the case of $NaNO_3$ [7]. Itakura et al. reported the formation of soluble sulfides such as sodium sulfide (Na_2S) as a further problem when Na_2SO_4 was used [21]. This may lead to the generation of hydrogen sulfide (H_2S) gas during the subsequent hydrometallurgical process, and elemental sulfur may precipitate, which decreases the purity of the recovered tungsten compound. However, the formation of water-soluble sulfides can be avoided by the addition of metal oxides to molten salts. In this case, Na_2S reacts with other metal oxides, and water-insoluble metal sulfide is obtained according to the following reaction, which can avoid one of the problems of the Na_2SO_4 process [21].

$$Na_2S + MO \rightarrow Na_2O + MS \tag{2}$$

where M is typically iron, cobalt, or nickel. Among these metals, cobalt (cobalt oxide or cobalt tungstate) is suitable because the use of Co compounds is effective not only in reducing the abovementioned water-soluble sulfides but also in avoiding complications of the subsequent purification process; that is, the insoluble sulfide can be separated as a precipitate when the processed molten salt is dissolved in water, and the main component of the precipitates obtained here is cobalt when super-hard alloy scrap is processed. Therefore, the addition of Co compounds during the molten salt process has no negative impact on the subsequent steps.

These processes using oxygen and molten Na_2CO_3 or Na_2SO_4 are expected to be developed in the future, because they can significantly reduce the cost of chemicals compared to the process using $NaNO_3$, and the reaction can be easily controlled.

New Recycling Processes Using Molten Sodium Hydroxides

The aforementioned molten salt processes use chemical reactions to oxidize W or WC. Although these processes are simple and cost effective, the reaction rate is difficult to control, the process using molten Na_2NO_3 tends to proceed excessively fast, whereas the other processes experience difficulty in obtaining a sufficient reaction rate. Therefore, we proposed an electrochemical W-recycling process using molten NaOH from super-hard alloy scraps [22–24]. In this process, super-hard alloy scraps were used as the anode, and W was dissolved as tungstate ions in molten NaOH, as shown schematically in Fig. 3. Here, the anodic, cathodic, and total reactions are

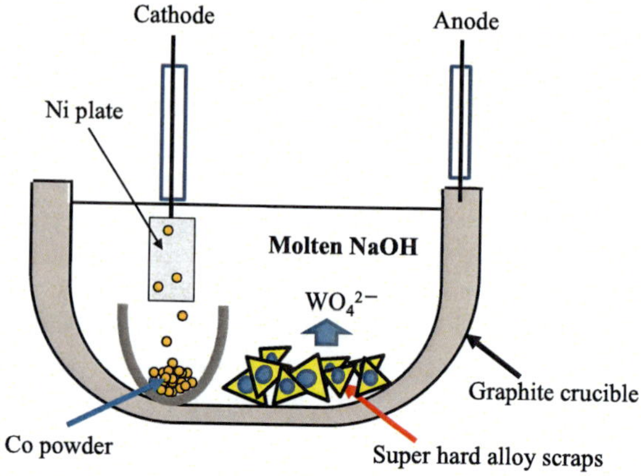

Fig. 3 Schematic drawing of anodic dissolution process of super hard alloy scraps in molten NaOH

described as follows:

$$WC + 7O^{2-} \rightarrow WO_4^{2-} + CO_3^{2-} + 10e^- \quad (3)$$

$$2OH^- + 2e^- \rightarrow H_2 + 2O^{2-} \quad (4)$$

$$WC + 10NaOH \rightarrow Na_2WO_4 + Na_2CO_3 + 5H_2 + 3Na_2O \quad (5)$$

Notably, the cathode reaction may result in Na precipitation, depending on the conditions. Cobalt, a major impurity, is also oxidized and dissolved, but most of the dissolved cobalt ions are reduced on the cathode and precipitated as Co powder. As reaction (5) proceeds, the tungsten content in the molten NaOH increases, and a Na_2WO_4 containing aqueous alkaline solution is obtained when the solidified NaOH melt is dissolved in water. Because this aqueous solution is essentially the same as that obtained using the conventional indirect method, W can be recovered by the usual process, as shown in Fig. 2b. Although this process has the disadvantage of consuming electric power, we believe it is advantageous for small-scale operations because the reaction is easy to control, and no harmful gas is generated. It has already been demonstrated that the oxidative dissolution of WC proceeds with a current efficiency of over 90% [22]. It was also confirmed that this process is greatly affected by the water content in the molten NaOH. For instance, when the water content is sufficiently low, the cathode reaction switches to Na electrodeposition. In addition, the solubility of Na_2WO_4 in molten NaOH significantly increases with water content [23]. We are currently investigating the precipitation of Na_2WO_4 powder from a W-enriched molten NaOH by controlling the water content in the melt, which enables the cyclic use of molten NaOH, as shown in Fig. 4 [24].

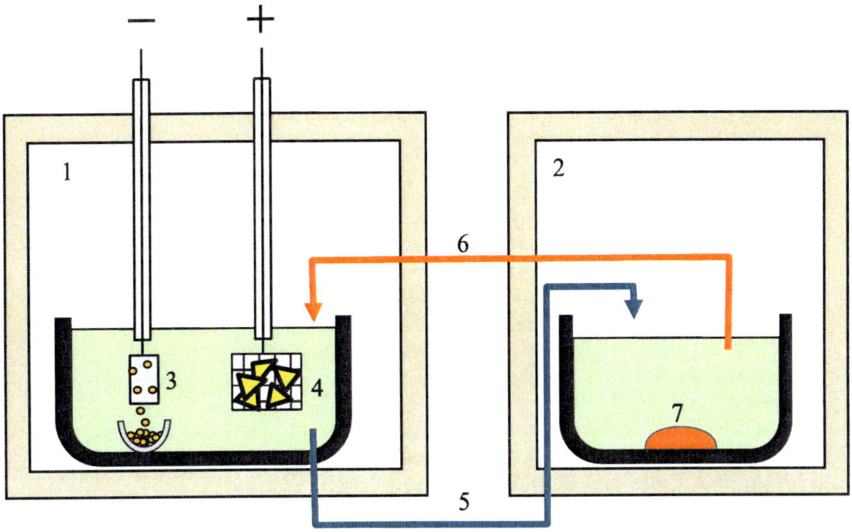

Fig. 4 Conceptual drawing of the new tungsten recycling process. 1. Anodic dissolution chamber kept at high partial pressure of water vapor and/or high temperature. 2. Precipitation chamber kept at low partial pressure of water vapour and/or low temperature. 3. Cathode of inert metals like Ni, on which H2 gas evolution and electrodeposition of powdery Co proceed. 4. Anode basket containing super hard alloy scraps, on which anodic dissolution of WC and binder metal of Co proceeds. 5. Flow path for tungstate ion-rich NaOH melt. 6. Flow path for tungstate ion-poor NaOH melt. 7. Precipitated Na2WO4

Conclusions

The molten salt process for tungsten recycling was reviewed. The molten salt process has been investigated as an alternative to the roasting and alkaline leaching process of super-hard alloy scrap in the conventional indirect method in which tungsten carbide is converted to water-soluble tungsten species. $NaNO_3$ is the most common oxidizing agent in this field, although the reaction is difficult to control. As a result of intensive investigation, one or more molten salt processes using $NaNO_3$ have been commercialized. The use of molten sulfates and carbonates is currently disadvantageous to the reaction rate, but this may be a candidate for a more economical and energy-saving process.

We proposed another molten salt process using molten hydroxide. In this process, the dissolution rate of super-hard alloy scrap was easily controllable, and Na_2WO_4 containing an aqueous solution was safely obtained. Further investigation is ongoing to simplify and reduce the cost of the process.

References

1. US Geological Survey (2016) Mineral commodity summaries
2. JOGMEC (2019) Material flow of mineral resources 2018, Tungsten
3. Oishi T (2017) Tungsten recycling technologies using Molten salts. Kinzoku 87:771–776 (in Japanese)
4. Lassner E (1995) From tungsten concentrates and scrap to highly pure ammonium paratungstate (APT). Int J Refract Met Hard Mater 13:35–44
5. Lassner E, Schubert WD (1999) Tungsten-properties, chemistry, technology of the element, alloys, and chemical compounds. Plenum Publishers, New York, pp 377–385
6. Shedd KB (2005) Tungsten recycling in the United States in 2000. US Geological Survey
7. Iguchi T, Ikegaya A (2008) Recycling of tungsten. In: Harada K, Nakamura T (eds) Technology of alternatives and recycling of rare metals. CMC Publishing Co, Ltd., Tokyo (in Japanese)
8. Nose K, Okabe TH (2011) Recycling technology of tungsten. Kinzoku 81:908–915 (in Japanese)
9. Itoh H (2013) Recent trend of rare metal recycling. J MMIJ 129:97–106 (in Japanese)
10. Bhosale SN, Mookherjee S, Pardeshi RM (1990) Current practices in tungsten extraction and recovery. High Temp Mater Processes 9:147–162
11. Ishida T, Itakura T, Moriguchi H, Ikegaya A (2012) Recycling technique for cemented carbide tools and development of tungsten-saving tools. SEI Tech Rev 181:33–39
12. Li KC, Wang CY (1955) Tungsten. Reinhold Publishing Corporation, New York, pp 175–176
13. Simon R (1981) DE3144295 (Germany patent)
14. Douglass AD, Reilly KT, Landmesser JE (1986) US4603043
15. Morishita M, Yamamoto H, Ikebe M, Yanagida H, Ueno T (2014) PCT. International patent, WO2014/045579
16. Yamamoto Y, Sasaya K, Fudo T, Nakano A, Yamanaka S, Iguchi T, Sato F, Ikegaya A (2010) PCT. International patent, WO2010/104009
17. Hayashi T, Sato F, Sasaya K, Ikegaya A (2016) Industrialization of tungsten recovering from used cemented carbide tools. SEI Tech Rev 189:8–14
18. Scott FH (1955) United Kingdom patent GB791925
19. Yasuda K, Nozaki F, Uehata R, Hagiwara R (2020) Oxidative dissolution of tungsten metal in Na_2CO_3 under $Ar–O_2–CO_2$ atmosphere. J Electrochem Soc 167:131501
20. Lohse M (1996) PCT. International patent, WO96/041768
21. Itakura T, Ikegaya A, Yamamoto Y (2016) Japanese patent JP6018958
22. Oishi T, Yaguchi M, Tanaka M (2013) Anodic dissolution of tungsten in a molten sodium hydroxide. J MMIJ 129:707–712
23. Oishi T, Yaguchi M (2017) Influence of partial pressure of water vapor on anodic dissolution of tungsten from super hard alloy tools in molten sodium hydroxide. Int J Refract Met Hard Mater 69:254–258
24. Oishi T, Yaguchi M (2018) Solubility of sodium tungstate in molten sodium hydroxide. Electrochemistry 86:61–65

Adaptability of the ISASMELT™ Technology for the Sustainable Treatment of Wastes

S. Nicol, D. Corrie, B. Barter, S. Nikolic, and B. Hogg

Abstract The recovery of resources from processing urban and industrial waste streams is essential to creating a sustainable society. The implementation of the ISASMELT™ Top Submerged Lance (TSL) technology for the retrieval of metals, and the capture of energy has been applied in the real world for over 20 years. The ISASMELT™ furnace is highly suited to treat these complex waste streams, with their varying compositions and dimensions, due to the flexibility and adaptability of the technology. The conditions in the furnace are tightly controlled to ensure the desired products and compositions are achieved. The turbulent, high temperature melt is capable of destroying hazardous chemicals. The slag generated can be used or stored without causing further pollution. This paper describes how existing and future ISASMELT™ operators can leverage the technology to supplement their feed supply and recover valuable materials from an increasing range of urban and industrial waste streams.

Keywords ISASMELT™ · Pyrometallurgy · Copper smelting · Sustainable recycling · Lead smelting · TSL · Urban waste · Circular economy · Scrap copper processing · Complex feed · Metallurgical process · Flowsheet design

Introduction

One of the major difficulties when treating and recycling end-of-life consumer products is obtaining a constant and consistent feed stream. Operating a recycling facility that treats different feed materials is typically challenging as process equipment is designed for specific feed materials and properties. The ISASMELT™ technology is unique in that a single furnace is able to smelt a wide range of feed materials, with a wide range of process conditions able to be operated without significant or expensive furnace modifications.

S. Nicol · D. Corrie · B. Barter · S. Nikolic · B. Hogg (✉)
Glencore Technology, 160 Ann St, Brisbane, QLD 4000, Australia
e-mail: ben.hogg@glencore.com.au

A. Lazou et al. (eds.), *REWAS 2022: Developing Tomorrow's Technical Cycles (Volume I)*, The Minerals, Metals & Materials Series,
https://doi.org/10.1007/978-3-030-92563-5_8

Glencore Technology developed the ISASMELT™ technology at its operations in Mount Isa in the 1980s and 1990s [1]. The technology was scaled up from bench scale tests, pilot tests, and demonstration plants and finally to full-scale industrial smelters. The extensive testing and development resulted in a furnace design widely regarded as one of the most robust and flexible smelting technologies in the industry. The history of this technology and its implementation in a variety of applications has been published in a number of papers [2–5].

The furnace technology developments resulted in a range of furnace features that enabled the ISASMELT™ furnace to be operated with almost any feed and under any smelting condition. The highly agitated bath is able to smelt materials up to 100 mm in size, under oxidising, reducing, or neutral conditions, in both batch and continuous operation modes.

This paper will outline the installation of the ISASMELT™ technology in recycling applications and the potential for the technology to smelt a range of materials.

The ISASMELT™ Furnace

The ISASMELT™ furnace is a bath smelting furnace, utilising a proprietary lance design. The feed material is typically charged into the furnace, via a feed device, through a port in the furnace roof. The feed system is straightforward, robust, and easy to operate, providing a simple installation and the ability to feed a wide range of materials without any modifications. The feed material enters the highly agitated bath and is rapidly digested, reacting quickly with the furnace melt.

The proprietary lance injects air and/or oxygen into the furnace, along with a trim fuel. The gases injected into the bath provide significant agitation and a direct contact with the bath. This ensures a rapid transfer of the gas species into the molten bath, resulting in fast reaction kinetics and efficient heat transfer (Fig. 1).

Another key aspect of the technology is the location of the bath agitation. As the lance is submerged in the centre of the bath and away from the furnace walls, the centre of the vessel is highly agitated, while the walls are almost stagnant. Coupled with accurate control of the slag composition and temperature, this results in unparalleled furnace campaigns without the use of furnace cooling elements. Campaigns of more than 6 years are possible with the ISASMELT™ technology, and 4-year campaigns have been demonstrated at a number of furnace installations around the world [6].

The furnace is typically tapped through a unique multi-phase taphole into a settling vessel. Coupled with the intense vessel agitation, this enables operating conditions to be targeted which are unachievable in other furnaces. Additionally, by performing phase separation in a separate vessel, the smelting and settling/separation processes can be optimised independently. Other furnaces perform this function in a single vessel, resulting in sub-optimal smelting and settling. Further developments in the ISASMELT™ furnace have enabled settling in the smelting vessel which, while less

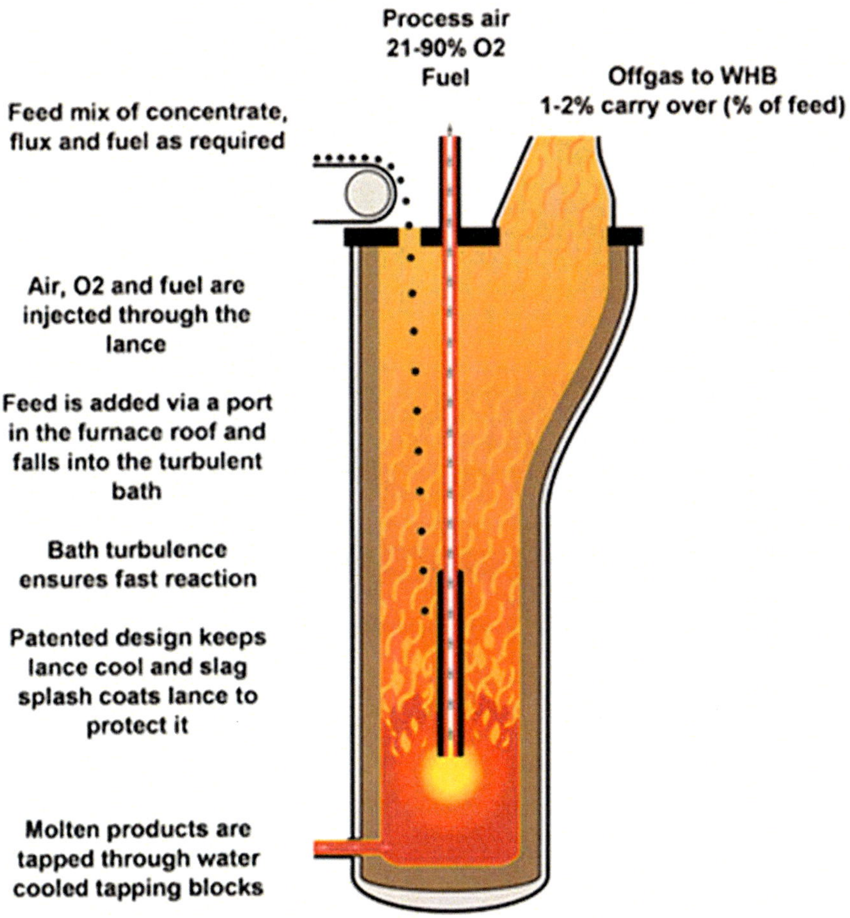

Fig. 1 ISASMELT™ furnace concept

efficient than that performed in a separate furnace, enables a single vessel to be installed to reduce smelter capital costs (Fig. 2).

Some key installations in recycling applications include.

1. 1991 Britannia Refined Metals, Northfleet, United Kingdom, Secondary Lead
2. 1997 Umicore Precious Metals, Hoboken, Belgium, Mixed Lead/Copper Secondaries
3. 2000 Metal Reclaim Industries, Pulau Indah, Malaysia, Secondary Lead
4. 2002 Aurubis (previously Hüttenwerke Kayser), Lünen, Germany, Secondary Copper.

The ISASMELT™ furnace design is ideal for recycling applications due to how the equipment operates. This furnace can be operated with fuels from renewable sources, leading to minimal energy consumption in smelting. Additionally, there is

Fig. 2 Multi-phase tapping of an ISASMELT™ furnace

minimal water use, and no hazardous liquids or solids are discarded from smelter. Emissions are easily able to be controlled with a gas cleaning plant.

Reductive Smelting–Secondary Lead

The first implementation of the ISASMELT™ technology for recycling was at Britannia Refined Metals (BRM) in 1991. At BRM, lead acid batteries were processed through a battery breaking and desulphurisation plant. The leaded materials, after desulphurisation, were fed, along with fluxes and reductant, into the ISASMELT™ furnace bath. The furnace was designed to produce 35 ktpa of lead bullion in a furnace with an internal diameter of 1.8 m. The furnace was operated to produce a lead bullion and a lead-rich slag, containing 58–66 wt% Pb in slag [7, 8].

The BRM installation is an example of reductive smelting in an ISASMELT™ furnace. Under reductive-smelting conditions, oxide materials are reduced to produce a metal product and a slag. Reductive smelting in an ISASMELT™ is typically selective such that some metals are intentionally pushed into the slag, and a relatively clean metal is produced. In the secondary lead smelting process, lead, tin, and antimony are selectively reduced to form two different metal products, hard lead (high Sb and Sn) and soft lead (low Sb and Sn).

The unique operating and design features of the ISASMELT™ furnace result in a secondary lead smelter with longer campaigns and with lower dusting when compared to other furnace technologies. Significant efforts and advancements were made with refractory design, enabling faster rebricks and preventing bullion penetration of the hearth. An additional taphole was installed at the top of the bath to remove a soda slag from the furnace, present as a result of the soda ash used in desulphurisation.

The lead-rich slags produced in secondary lead furnaces are typically extremely corrosive to refractory bricks. As such, without furnace cooling elements, campaigns as short as six months are common in the secondary lead industry [9]. This results in significant maintenance downtime, increased refractory costs, and lost production at sites. The ISASMELT™ technology at BRM, with its unique design features and operating parameters, was able to be operated with a furnace campaign of 1 year, without the use of water-cooling elements. This was achieved in the early 2000's, prior to the plant being closed. With current know-how, it is expected that the campaign life would be in the range of 2 years.

With the digestion of the feed directly in the molten bath and unique process parameters, a low dusting rate of the secondary lead ISASMELT™ furnace was observed. Due to the feed properties and chemistry of secondary lead smelters, a dust generation rate in the order of 15% is commonly observed [10]. With the ISASMELT™ technology, a dusting rate of less than 4% is able to be achieved. This resulted in a low recirculating load of dust and in turn simpler management of volatile elements such as cadmium.

The secondary lead ISASMELT™ furnace is able to be operated in batch or continuous mode. Both of these operating modes were used at BRM and have been proven on an industrial scale. In batch operation, the first stage produces a soft lead product and a lead-rich slag. In the second stage, the lead-rich slag is reduced to produce a hard lead product and a discardable slag. In a continuous operation, a soft lead product and a lead-rich slag is produced. The lead-rich slag is subsequently smelted in a separate furnace to recover a hard lead product and a discardable slag (Fig. 3).

Oxidative Smelting–Copper Scrap and E-Waste

The first installation of an ISASMELT™ furnace for the smelting of secondary copper, including materials such as E-Waste, was in 2002 at Hüttenwerke Kayser (now Aurubis, Lünen). The ISASMELT™ was installed to replace three blast furnaces and two Pierce-Smith converters operating the traditional Knudsen process for copper scrap recycling [11].

In the Knudsen process, the copper scrap, along with coke and fluxes, was smelted in the blast furnace to produce a black copper with appreciable quantities of iron [12]. The black copper was close to iron saturation to ensure that the discard slag was as low in base metals as possible. The black copper was subsequently oxidised in the converters to produce a slag and a dust rich in lead, tin, zinc, antimony, and other base metals. The blister copper produced by the converters was fire-refined in an anode

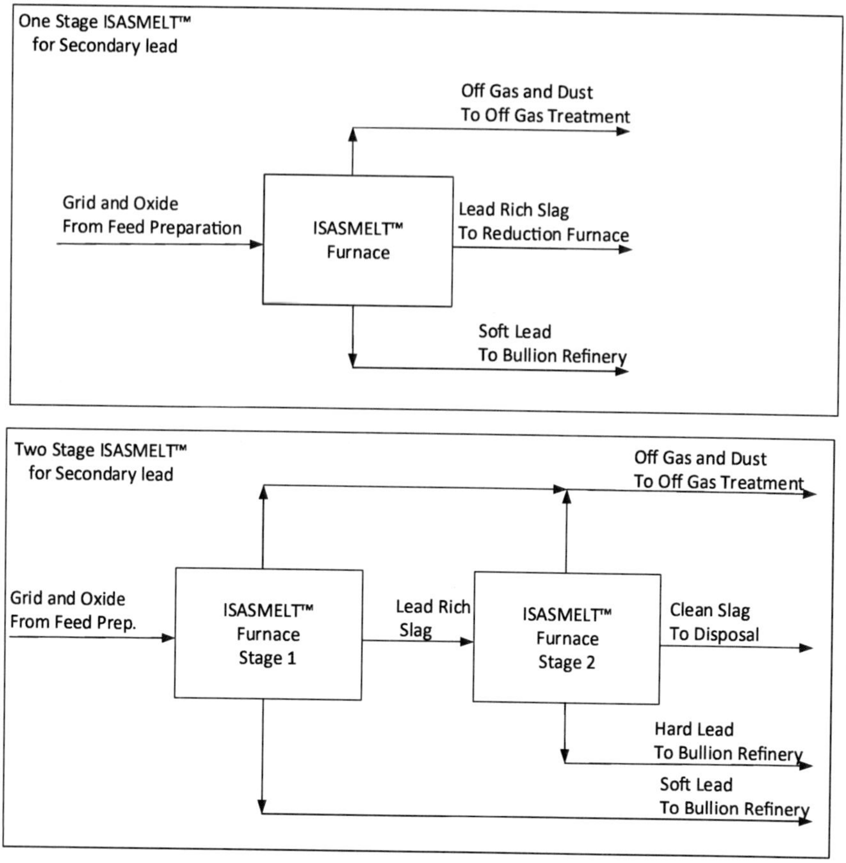

Fig. 3 ISASMELT™ smelting options for secondary lead

furnace, followed by casting and electro-refining to cathode copper. The secondary copper smelter operates the converters and anode furnace primarily to reject base metals from the black copper, with the anode copper that is sent to the electro-refinery higher in impurities than a typical primary copper smelter.

The ISASMELT™ furnace offers a number of significant advantages when compared to the traditional copper scrap recycling flowsheet. As the furnace is able to be operated in batch mode, under both reducing and oxidising conditions, just one furnace is required for both smelting and converting. The use of just one furnace for both operations significantly decreases the capital and operating costs for a secondary copper smelter. Unlike the blast furnace, the furnace is able to accept a wide range of feed material compositions and sizes without significant feed preparation, simplifying and decreasing the costs in this area of the smelter. The blast furnace has a counter current gas/solid flow, resulting in recirculating loads of volatiles in the furnace that inhibits the rejection of volatile elements to the dust phase. In the

ISASMELT™ furnace, the intense agitation and bath smelting conditions result in significant first-pass rejection of volatile elements into a fine dust [13].

The ISASMELT™ furnace installed at Aurubis Lünen was initially operated in a two-stage batch operation. The first stage produces a black copper rich in iron and the second stage a refined copper product for the anode furnace. The tin and lead-rich slag from the second stage was smelted in a specific furnace to produce a lead–tin alloy for sale, purging these elements from the smelter. The ISASMELT™ furnace fume was rich in zinc and lead, and the fine ISASMELT™ dust was sold for offsite processing to bleed these elements from the flowsheet.

The ISASMELT™ furnace at Aurubis Lünen was subsequently converted to a continuous operation, producing a black copper product and a discard slag. The black copper was then smelted in a separate furnace, followed by final refining in an anode furnace. The switch to continuous operation allowed steady process control of the initial reductive-smelting step and an increase in throughput of the whole plant by approximately 25% [14].

The experience at the Aurubis Lünen smelter with their ISASMELT™ furnace has proven on an industrial scale both the batch and continuous smelting of secondary copper materials. Both options have their merits and require assessment on a case-by-case basis.

Work is ongoing at Glencore Technology to further optimise the secondary copper flowsheet, as practiced by Aurubis Lünen. Significant developments have been made on slag and furnace chemistry to optimise the smelting of E-Waste feeds (Fig. 4).

Mixed-Feed Smelting–Copper and Lead Processing

With an increase in the intricacy of the end-of-life products that require recycling, the feed to a recycling facility is becoming increasingly complex and varied. When smelted, this leads to an extremely complicated chemistry and flowsheet. The ISASMELT™ furnace installed at the Hoboken Smelter for Umicore in 1997 provided a technical solution for the smelting of a wide range of extremely complex feed materials [15].

The combined smelting of copper and lead-rich feeds is a significant challenge. For any other furnace to operate successfully, the furnace must be operated tightly to a narrow operating window. The ISASMELT™ furnace flexibility enables the smelter to run within a comparatively large operating window.

Prior to the installation of the ISASMELT™ furnace, the Hoboken smelter utilised a sinter-blast furnace flowsheet to smelt a complex copper-lead feed material. The blast furnace produced a discard slag, copper matte, nickel speiss, and a lead bullion in a multi-phase forehearth. The operation of a 3+ phase blast furnace is extremely challenging, requiring careful control of the sulphur content of the feed to ensure that a matte phase is formed rather than a copper metal. If insufficient sulphur is present, there is the opportunity for copper to come out of solution as a solid phase, risking the formation of accretions and other issues inside the furnace.

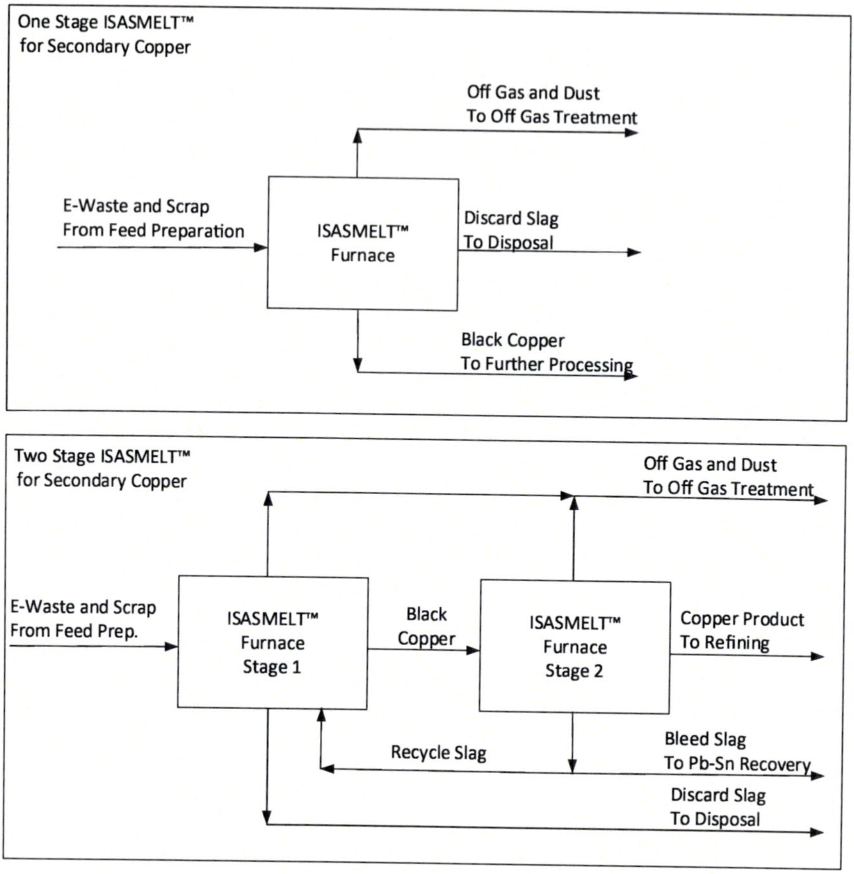

Fig. 4 ISASMELT™ smelting options for secondary copper

Significant research and piloting was performed by Umicore prior to the installation of the ISASMELT™ furnace to ensure that the full-scale furnace was ramped up rapidly and with as few issues as possible. Umicore has funded significant research in the field of extractive metallurgy to ensure that its operations have been successful and at the forefront of the secondary smelting industry. This work ensures that their ISASMELT™ furnace is operated as close to optimal conditions as possible.

The ISASMELT™ furnace installed at the Hoboken Smelter processes a wide range of complex feeds to produce a lead-rich slag and a copper metal product. The lead-rich slag is subsequently smelted in a blast furnace to produce a lead bullion and a clean discard slag. The copper metal product is refined to produce a clean copper metal and a precious metals anode slime for further processing. As demonstrated by Aurubis Lünen and BRM, the copper metal and lead-rich slag are both able to be processed within an ISASMELT™ furnace to produce a lead metal or refined copper and a discard slag [8, 16].

Replacing the sinter strand with the ISASMELT™ furnace at Hoboken resulted in a significant decrease in the impact of the smelter on the surrounding environment [17]. The furnace enabled a significant increase in off-gas capture and in turn a decrease in emission of dust and fugitive emissions to the surrounding area. The furnace was shown to be a cleaner and more efficient technical solution when compared with the older technology that was used onsite.

Future Applications of ISASMELT™ Technology to Recycling

As demonstrated by the installation of ISASMELT™ furnaces at recycling facilities around the world, this furnace provides a technical solution for the smelting of almost any feed material [16, 18]. The capability of the ISASMELT™ technology offers extreme flexibility; it is able to be operated in both batch and continuous modes with a wide range of feed materials and compositions. As demonstrated by the BRM, Aurubis Lünen, and Umicore Hoboken installations, the ISASMELT™ is a world leading technology for secondary smelting operations.

With the wide range of furnace operating conditions, the ISASMELT™ furnace is able to be installed as a single furnace solution for a recycling facility. In this application, the furnace would be operated in batch mode, performing both oxidation and reduction to produce a range of metal, dust, and slag products as either intermediates or final products. Due to the flexibility of the ISASMELT™ furnace, and with the use of advanced metallurgical calculations and tools, the furnace would be capable of smelting a wide range of feed materials on a campaign basis.

The ISASMELT™ furnace is able to be installed in existing flowsheets to transform smelters into advanced facilities. The transformation enables the smelters to process a wider range of feed materials with a lower environmental footprint. This was demonstrated at both Aurubis Lünen Kayser and Umicore Hoboken. In these applications, the ISASMELT™ furnace is typically operated in a continuous smelting operation. The molten products from the furnace are subsequently processed in the other existing furnaces at the smelter, to produce the valuable final products and a discard slag.

Furnace Options for Recycling with an ISASMELT™ Furnace

The ISASMELT™ is able to be designed and operated within a significant range of operating conditions. A summary of the operating conditions possible with the furnace is outlined in the table below.

Parameter	Value
Lance total flowrate	200–71,000 Nm^3/h proven
Furnace fuel/reductant	Coal, plastic, coke, and E-Scrap
Furnace trim fuel supply	Natural gas, diesel, pulverised coal, waste oil, and hydrogen
Lance oxygen enrichment	21 (air) to 95 Vol% O_2 proven
Furnace availability	92% (including re-brick and maintenance)
Furnace lining	Fully bricked, copper staves, and intensively cooled copper panels proven
Taphole types	Combined and separate metal/matte and slag tapholes proven
Furnace campaign	4+ years proven [6], 6+ years possible
Ramp-up to design capacity	3 months proven
Furnace operation	Batch and continuous proven, able to change back and forth during asset life
Feed size	<100 mm proven
Feed moisture	Up to 10 Wt% proven
Feed delivery system to furnace	Vibrating or belt-style for coarse feed Pneumatic injection for fine, volatile, or low-density feed

Conclusions

As demonstrated by numerous installations around the world, the ISASMELT™ furnace technology is a world leading technology for the recycling of many metals. The ISASMELT™ furnace is a proven solution for secondary lead, secondary copper, E-Waste, and mixed/complex feed materials. It is able to operate on both a batch and continuous basis. The furnace is able to be installed in both a single furnace smelter and in a multi-furnace smelter, with efficient gas capture and minimal environmental emissions demonstrated. With advanced metallurgical tools and calculations, the furnace is able to be operated on a campaign basis in a smelter with changing feed materials. With this technology, what was previously a waste and economically or technically unrecoverable is now able to be smelted into valuable products.

References

1. Arthur PS, Alvear Flores GRF (2011) Incremental scale up of ISASMELT™—the key to its success. In: Davis BR, Kapusta JPT (eds) 50th annual conference of metallurgists of CIM, pp 203–214
2. Bakker ML, Nikolic S, Mackey PJ (2011) ISASMELT™ TSL—applications for nickel. Miner Eng 24(7):610–619

3. Errington WJ, Arthur PS, Wang J, Dong Y (2005) The ISA-YMG lead smelting process. Proceedings of the international symposium on lead and zinc processing, Kyoto, Japan October, MMIJ, pp 581–599
4. Arthur PS, Hunt SP (2005) ISASMELT™—25 years of continuous evolution. In: Nilmani M, Rankin WJ (eds) Floyd international symposium on sustainable development in metals processing. NCS Associates (Australia), pp 73–94
5. Alvear Flores GRF, Nikolic S (2013) ISASMELT for recycling of valuable elements contributing to a more sustainable society. In: REWAS 2013: enabling materials resource sustainability, TMS annual meeting, San Antonio, Texas, U.S.A.
6. Nikolic S, Hogg B, Voigt P (2019) Freeze lining refractories in non-ferrous TSL smelting systems, TMS 2019, San Antonio, Texas, U.S.A., 10–14 Mar 2019
7. Port W, Errington B, Hawkins P, Lim A (2010) ISASMELT™ for lead recycling, COM2010, Vancouver, B.C., Canada, 3–6 Oct 2010
8. Ramus K, Hawkins P (1993) Lead/acid battery recycling and the new Isasmelt process. J Power Sources 42:299–313
9. Vondersaar M, Bulnes B (2000) Operation of a high output, one-pass smelting system recycling lead acid batteries. In: Fourth international symposium on recycling of metals and engineered materials, TMS 2000, Nashville, Tennessee, United States, 12–16 Mar 2000
10. Moenster J, Snakovitch M (2000) Operations at the Doe Run Companys buick resource recycling division. In: Fourth international symposium on recycling of metals and engineered materials, TMS 2000, Nashville, Tennessee, United States, 12–16 Mar 2000
11. Nolte A, Kreymann R (1999) Optimisation of the blast furnace process in a secondary copper smelter, Copper '99, Phoenix, Arizona, U.S.A., 10–13 Oct 1999
12. European Commission (2009) European dioxin inventory—secondary copper production [online]. https://ec.europa.eu/environment/archives/dioxin/pdf/stage1/seccopper.pdf. Accessed 09 Aug 2021
13. Player R (1996) Copper ISASMELT—process investigations. In: The Howard Warner international symposium on injection in pyrometallurgy, Melbourne, Victoria, Australia, 1–3 July 1996
14. Arubis Annual Report 2010/2011
15. Vanbellen F, Chintinee M (2007) Extreme makeover. UPMR's Hoboken Plant, EMC 2007
16. Alvear Flores GRF, Nikolic S, Mackey P (2014) ISASMELT™ for the recycling of E-scrap and copper in the U.S. case study example of a new compact recycling plant, JOM, vol 66, no 58, pp 823–832
17. Vanbellen F, Chintinne M (2006) The precious art of metals recycling. In: Sohn international symposium advanced processing of metals and materials, TMS 2006, San Antonio, Texas, U.S.A., 12–16 Mar 2006
18. Alvear Flores GRF, Nikolic S (2013) ISASMELT™ for recycling of valuable elements contributing to a more sustainable society. In: REWAS 2013, TMS 2013, San Antonio, Texas, U.S.A., 3–7 Mar 2013

Recovery of Precious Metal Silver from Scrap Computer Keyboards

Rekha Panda, Om Shankar Dinkar, Pankaj Kumar Choubey, Rukshana Parween, Manis Kumar Jha, and Devendra Deo Pathak

Abstract Silver (Ag) is extensively used in manufacturing of electronic goods due to its low cost and conductivity. In view of the escalating demand, stringent, environment rules, and limited sources of Ag, the present paper is focused on the development of hydrometallurgical process flow-sheet to extract Ag from scrap computer keyboards. These keyboards contain ~0.4% of Ag. Initially, keyboards were dismantled to separate the Mylar sheets scontaining Ag. The same were pyrolyzed at 300 °C for 2 h to get enriched metallic part. About 99.99% Ag was leached using 2 M HNO_3 at 60 °C within 30 min in close and proper condensed system. Separation techniques (precipitation/cementation) could be used to obtain pure Ag salt/metal. Based on the laboratory-scale experiments, the process flow-sheet developed is economical, eco-friendly, and has potential to be translated to industry for commercial exploitation after scale up/pilot trial.

Keywords Precious metal · Silver · Scrap keyboards · Recycling · Hydrometallurgy

Introduction

Silver has been recognized and appreciated by mankind for its astonishing luster as well as its numerous applications starting from coinage to modern technological use [1, 2]. The major industrial applications of silver are presented in Table 1. The unique physical properties of silver such as resistance to corrosion, excellent thermal and electrical conductance, and antibacterial nature have made it suitable for various industrial application, viz. electronics, solar cell, photographic, pharmaceutical, etc.

R. Panda · O. S. Dinkar · P. K. Choubey · R. Parween · M. K. Jha (✉)
Metal Extraction and Recycling Division, CSIR-National Metallurgical Laboratory, Jamshedpur 831007, India
e-mail: mkjha@nmlindia.org

R. Panda · D. D. Pathak
Department of Chemistry, Indian Institute of Technology (Indian School of Mines), Dhanbad 826004, India

A. Lazou et al. (eds.), *REWAS 2022: Developing Tomorrow's Technical Cycles (Volume I)*, The Minerals, Metals & Materials Series,
https://doi.org/10.1007/978-3-030-92563-5_9

Table 1 Major industrial applications of silver in the year 2018

S. No	Applications	Consumption of Ag (ton)
1	Electronic industries	9533.06
2	Jewelry	6609.38
3	Coins and bars	5635.86
4	Other industries	4569.03
5	Photovoltaic	2503.79
6	Silverware	1900.39
7	Photography	1222.34
8	Ethylene oxide	167.95

From Table 1, it is observed that electronic industries are the major consumer of silver compared to other sectors. About 9533 tons of silver was consumed by the electronic industries in the year 2018 [3].

Rapid advancement in the technological features, attractive designs, lucrative marketing, and compatibility of the electronic equipments have reduced the life spans of these products and consequently leads to the generation of huge quantity of waste electrical and electronic equipments (WEEEs). Around 53.5 million metric tons of WEEEs was produced in 2019, which is expected to reach 74.7 million metric tons by 2030 [4]. Table 2 shows the worldwide generation of WEEEs in 2019 where Asia tops the list followed by Europe, America, Africa, and Oceania [4]. According to Table 2, Asia was the largest producer of WEEEs (about 46.45% of the total contribution) in 2019, while the amount WEEEs collected for recycling was only 11.70%.

The WEEEs contain varieties of metals (non-ferrous, precious, rare and hazardous) and materials (plastics, ceramics, epoxies, rubbers, etc.). The illegal recycling practices (land filling and incineration) have led to the loss of such valuable metals and materials converting several billion into trash. The ever increasing demand for silver in the industrialized world has diminished its high grade resources and resulted in a huge gap between its demand and supply. In order to mitigate this gap, more emphasis is being laid on the exploitation of alternative resources (lean grade ores, metallurgical waste, industrial waste, electronic waste, etc.) of Ag.

Table 2 Worldwide WEEEs generation in 2019

Continents	Total WEEEs generated (MT)	Total contribution (%)	WEEEs recycled (%)
Asia	24.9	46.45	11.70
Europe	13.1	24.44	42.50
America	12.0	22.39	9.40
Africa	2.9	5.41	0.9
Oceania	0.7	1.30	8.80

Recently, R & D efforts have been made to recover silver from various secondary resources, viz. electronic waste, photographic waste, photovoltaic waste, and anode slime [1, 5, 6]. Xing and Lee, 2019 reported the hydrometallurgical extraction of silver (~77%) forms anode slime [6]. Samson and Muzenda, 2014 reported silver recovery from radiographic effluents and X-ray film waste [7]. Liu et al., 2019 reported leaching of silver (93%) and mercury (98%) from zinc refinery residues using 20 g/L thiourea and 4 g/L Fe^{3+} as oxidant at 40 °C in 2 h [8]. Li et al., 2012 reported a method to dissolve Au and Ag from the waste PCBs of mobile phones using thiourea [9]. Yoo et al., 2019 reported a precipitation method to recover silver from the leach liquor of electronic waste [10]. Gamez et al., 2019 reported a process of leaching using thiosulfate followed by ion-exchange using MTA 5011 resin to recover silver and gold from PCBs [11]. Gurung et al., 2013 reported acidothiourea leaching followed by adsorption using biosorbent prepared from persimmon tanin to recover gold and silver from spent mobile phones [12]. Among all secondary resources, recycling of WEEEs to recover silver is more attractive as the content of silver is quite high compared to its natural resources. Thus, keeping in view the limited high grade resources, recovery of silver from WEEEs will not only reduce the load on mining but also minimize the loss of valuables and protect the environment from the hazardous effect of WEEEs.

Therefore, in view of the limited natural resources of silver, increasing demand and stringent environmental regulations, present paper is focused on the recovery of silver from Mylar sheets of waste computer keyboards using pyro-hydrometallurgical route.

Experimental

Materials

The waste computer keyboards were manually dismantled to separate the Mylar sheets. The Mylar sheets containing silver were used as starting material for experimental purpose. Different mineral acids such as nitric acids (HNO_3), hydrochloric acid (HCl), and sulphuric acid (H_2SO_4) of laboratory grade were supplied by Merck, India. De-ionized water was used for preparing samples during the experiment.

Methodology

Sample Preparation

The keyboards were dismantled to liberate different parts like plastic casing material, PCBs, Mylar sheet, rubber, cables, etc. Among all parts, the Mylar sheet containing

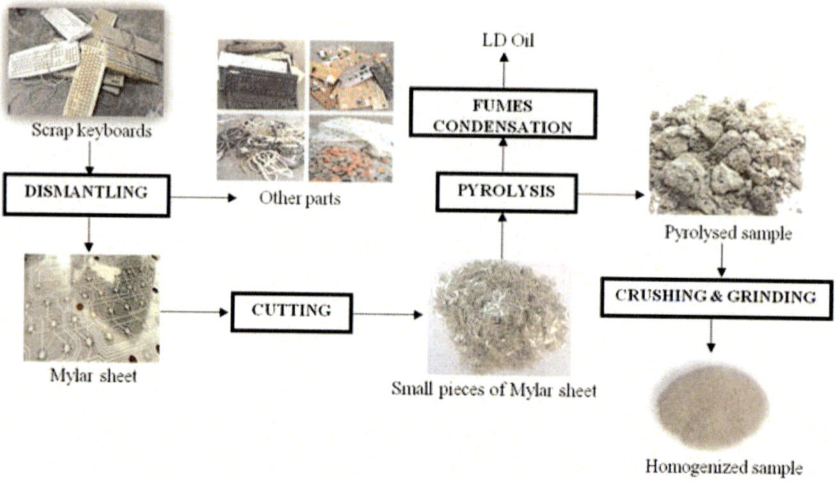

Fig. 1 Flow-sheet for sample preparation

silver was separated and cut into small pieces. The small pieces of Mylar sheet were pyrolysed at 300 °C for 2 h. The pyrolysed sample was crushed in mortar pestle, homogenized, and milled to reduce particle size (Fig. 1). The obtained pyrolysed sample was chemically analyzed and found to contain ~ 2% Ag.

Leaching Procedure

The pyrolysed sample was leached in a glass reactor which was well fitted with a reflux condenser to prevent the loss of gases/vapour evolved during the experiment. Hot plate with magnetic stirring facility and sensor to control the temperature was used for leaching experiment. Separation of leach liquor from the leached residue was carried out using a vacuum filter. The obtained leach liquor containing silver was processed through cementation/precipitation technique to recover silver as metal/salt. Based on our previous experience of drying, the leached residue was dried in vacuum oven for 2 h at 90 °C and will further be treated for the recovery of other non-ferrous metals.

Analytical Procedure

All samples prepared during the experiments were chemically analysed using atomic absorption spectrometer (AAS) (Perkin Elmer model, Analyst 200; USA). pH meter (Model: 797 Basic titrino; Make: Metrohm, Switzerland) with electrode and automatic temperature correction was used to measure the pH of solution during cementation studies.

Results and Discussion

In order to recover silver from Mylar sheet of waste keyboards, the leaching studies were carried out for the effective recovery of silver varying different process parameters, viz. selection of leachant, effect of acid concentration, temperature, time, and pulp density. The leach liquor obtained was processed for precipitation/cementation to recover silver as value added product (salt/metal). The obtained results are discussed below:

Leaching of Ag from Pyrolysed Mylar Sheet

Selection of Leachant

Metal dissolution mainly depends on nature of raw material and chemical properties of metals to be dissolved. Thus, selecting an appropriate leachant for effective leaching of metals is an important factor before proceeding for optimization of other experimental parameters. Initially, different mineral acids (HCl, H_2SO_4, and HNO_3) of 2 M each maintaining 90 °C, 60 min mixing time at 100 g/ L pulp density were used for the selective leaching of Ag from pyrolysed Mylar sheets obtained from scrap keyboards.

Leaching of Ag in HNO_3 under above experimental condition is very high compared to leaching with H_2SO_4 and HCl, where negligible dissolution of Ag was observed. Therefore, HNO_3 was considered to be the best leachant for maximum leaching of Ag and further used for optimizing various parameters. Ag leaching using HNO_3 is presented by Eq. 1 given below:

$$Ag + 2HNO_3 \rightarrow AgNO_3 + NO_2 + H_2O \quad (1)$$

Effect of Time

Effect of leaching time on the dissolution of Ag from pyrolysed Mylar sheet of scrap keyboards was investigated by varying leaching time from 5 to 60 min using 2 M HNO_3 at 90 °C and pulp density 100 g/L. With the increase in leaching time from 5 to 60 min, the leaching of Ag was found to increase from 31.6 to 99.99% as shown in Fig. 2. It was observed that equilibrium for complete dissolution of Ag (99.99%) was achieved in 30 min. Thus, 30 min was considered optimum for maximum dissolution of Ag from the pyrolysed Mylar sheet.

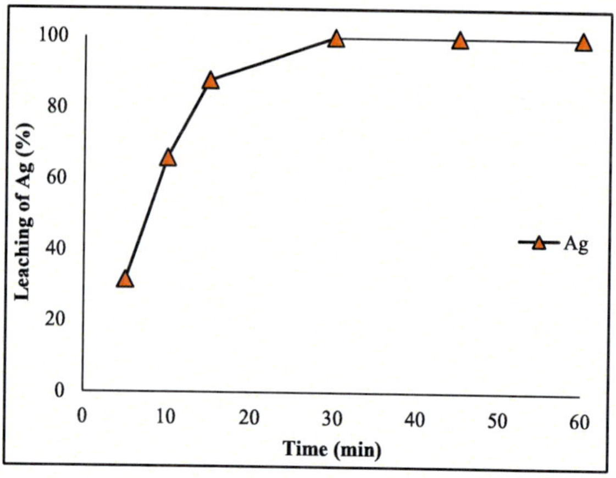

Fig. 2 Effect of leaching time [Acid: 2 M HNO_3; Temperature: 90 °C; Pulp density: 100 g/L]

Effect of Acid Concentration

In order to optimize the minimum amount of leachant required for complete dissolution of Ag from the Mylar sheet, experiments were carried out using different concentration of HNO_3. Concentration of HNO_3 was varied from 0.5 to 2 M at 80 °C and pulp density 100 g/L in 30 min mixing time. Leaching of Ag increased with the increase in HNO_3 concentration which might be due to the increase in flux of H^+ through particle boundaries for effective leaching [13]. Figure 3 shows leaching of Ag increased from 7.2 to 99.99% with the increase in HNO_3 concentration from 0.5 to 2 M. Thus, 2 M HNO_3 was optimized for dissolution of Ag from the pyrolysed sample.

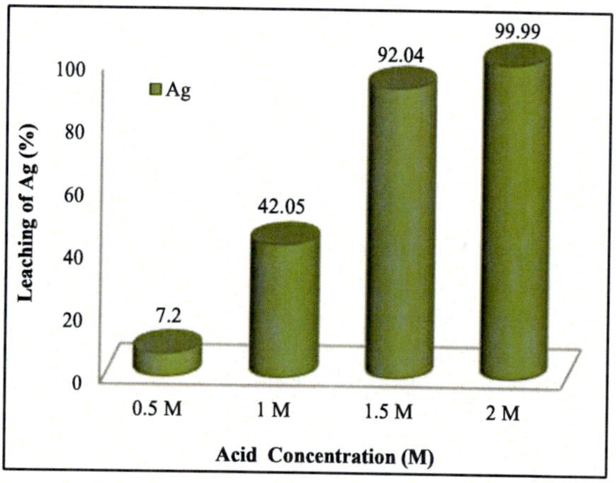

Fig. 3 Effect of acid concentration [Temperature: 90 °C; Time: 30 min; Pulp density: 100 g/L]

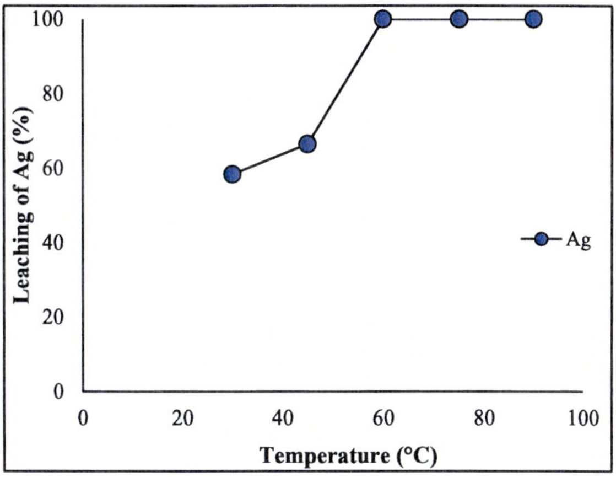

Fig. 4 Effect of temperature [Acid: 2 M HNO_3; Time: 30 min; Pulp density: 100 g/L]

Effect of Temperature

Temperature is a vital parameter to activate any kind of chemical reaction. Temperature can make any reaction vigorous by lowering its activation energy. In order to investigate the effect of temperature on the leaching of Ag from pyrolysed Mylar sheet sample, experiments were carried out at different temperatures from 30 to 90 °C using 2 M HNO_3 in 30 min at pulp density 100 g/L. Increase in temperature of the solution increases the reactivity, thereby increasing the dissolution of Ag. Results presented in Fig. 4 indicate increase in the leaching percentage of Ag from 58.33 to 99.99% with rise in temperature from 30 to 90 °C. Since maximum Ag dissolution was achieved at 60 °C, thus, it was chosen for further leaching experiments.

Effect of Pulp Density

Experiments were also carried out to optimize the pulp density for utmost dissolution of Ag. Leaching was conducted using 2 M HNO_3 at 60 °C for 30 min by varying pulp density between 50 to 200 g/L. Effect of pulp density on leaching of Ag is presented in Fig. 5. With the increase in pulp density, Ag dissolution was found to be decreased. An increasing trend in Ag dissolution was observed from 50 to 100 g/L, and then, sudden decrease in leaching trend was noticed. This decrease in the trend of Ag leaching at higher pulp density was mainly due to the increase in the surface area per unit volume of the leachant or may be due to the insufficient supply of H^+ that forbids thorough reactions between liquid and solid present in the system. Thus, 100 g/L pulp density was considered optimum for maximum leaching of Ag from the pyrolysed Mylar sheet.

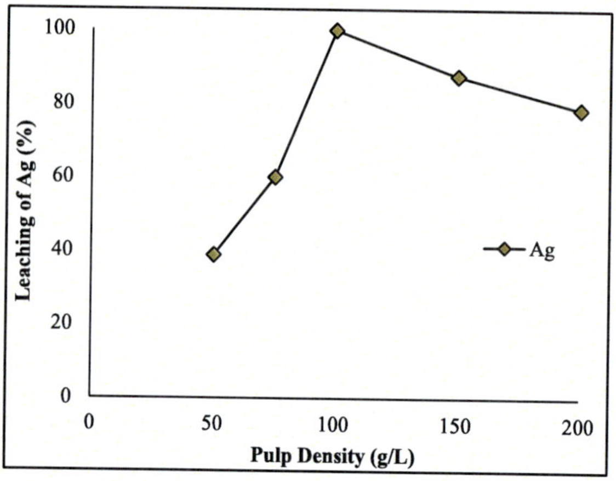

Fig. 5 Effect of pulp density [Acid: 2 M HNO_3; Time: 30 min; Temperature: 60 °C]

Recovery of Ag from Leach Liquor

The obtained leach liquor in nitrate medium could be further processed for Ag recovery using two different techniques i.e. precipitation/cementation to get value added products of Ag (salt/metal).

For precipitation, three different reagents, viz. HCl, KCl and NaCl, were used to precipitate Ag from the nitrate leach liquor. As per the solubility concept, it is well known that nitrate complexes are always soluble. On addition of chloride precipitants in the liquor containing $AgNO_3$, the strong bond between Ag^+ and NO_3^- breaks by the H_2O molecules due to ion–dipole attraction. The Ag^+, NO_3^-, Cl^-, and H_2O molecules exist in the solution. These ions freely interact with each other, but bonding of Ag^+ and Cl^- is so strong that it cannot be separated by ion–dipole force of H_2O molecules. This bonding suppresses the interaction of other ions and a white precipitate of AgCl formed. The obtained AgCl was characterized by dissolution using standard chemical procedure and found to be >95% pure.

In case of cementation, metallic Cu electro-won by PCBs recycling [14] was used to cement Ag present in the nitrate leach liquor. The cementation reaction between Ag^+ and Cu occurs as per the following reaction given below (Eq. 2):

$$Ag^+ + Cu^0 \rightarrow Ag^0 + Cu^{2+} \quad (2)$$

About 99% Ag cementation was achieved from the nitrate leach liquor at pH ~ 1, solution temperature 60 °C, and mixing time 15 min. The purity test of the obtained Ag powder was carried out and found to be 99% pure.

Conclusions

Based on the above research work carried out to recover Ag from the scrap keyboards of personal computers, the following conclusion could be drawn:

The Mylar sheets present in the scrap computer keyboards are potential secondary resource for silver recovery. The Mylar sheets were initially pyrolysed at 300 °C for 2 h to enrich the concentration of silver and collect plastic material as low density oil. The pyrolysed mass was further crushed to get a homogenized sample of size −100 mesh. Leaching of the pyrolysed sample was carried out at 60 °C for 30 min using 2 M HNO_3 maintaining 100 g/L pulp density to dissolve maximum silver (99.9%). The obtained leach liquor could be processed using precipitation/cementation techniques using chloride salts or Cu metal, respectively, to get pure silver product (salt/metal powder). The complete process flow-sheet developed to recover marketable product of Ag is presented in Fig. 6.

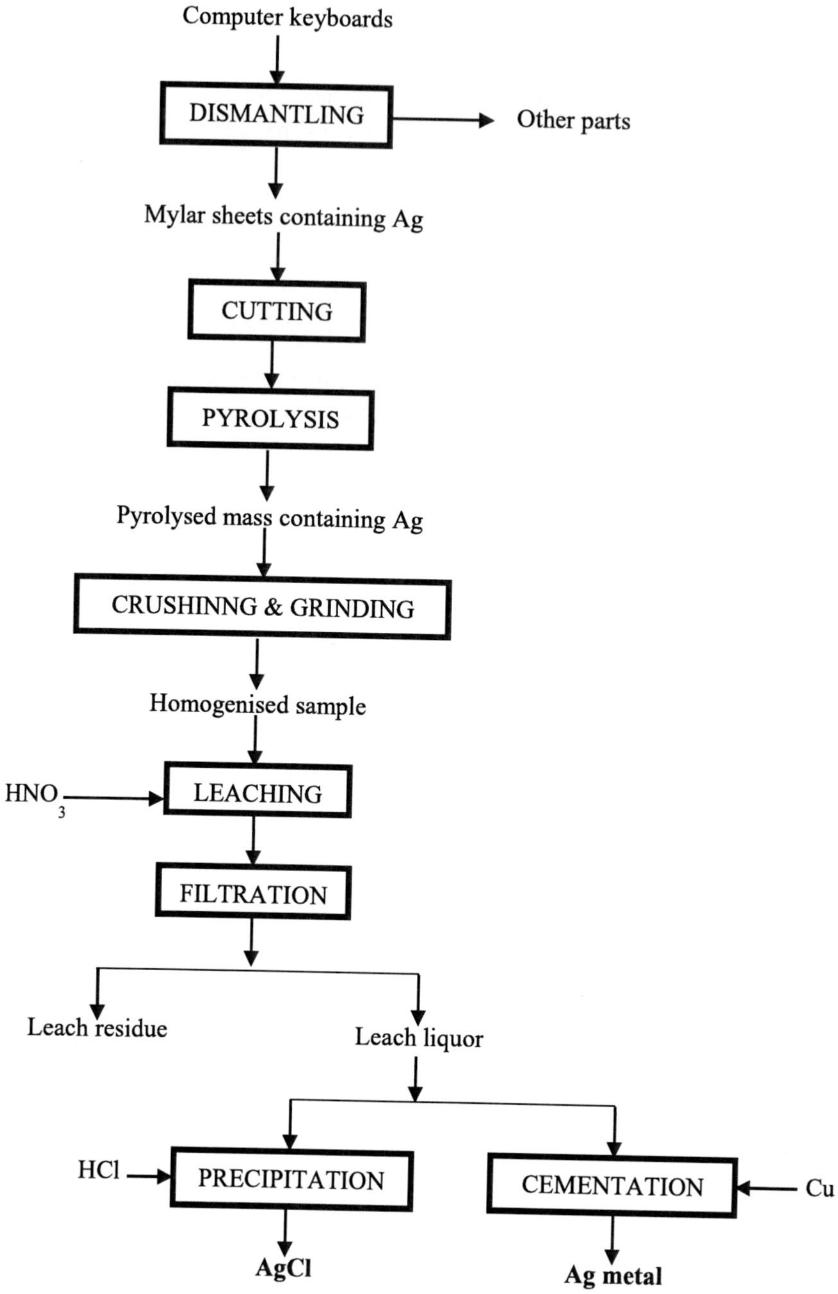

Fig. 6 Complete process flow-sheet developed to recover marketable product of Ag

Acknowledgements Authors are thankful to Director, CSIR-National Metallurgical Laboratory, Jamshedpur, India, for giving permission to publish this paper. One of the authors Ms. Rekha Panda would like to extend her sincere gratitude to CSIR, New Delhi, for providing Senior Research Fellowship (Grant: 31/10(64)/2017-EMR-I) to carry out this research work.

References

1. Syed S (2016) Silver recovery aqueous techniques from diverse sources: hydrometallurgy in recycling. Waste Manag 50:234–256
2. Tang B, Yu G, Fang J, Shi T (2010) Recovery of high-purity silver directly from dilute effluents by an emulsion liquid membrane-crystallization process. J Hazard Mater 177:377–383
3. Alexander C, Always B, Litosh S, Wiebe J, Yao W, Saha D, Norton K (2019) World silver survey 2019
4. Forti V, Baldé CP, Kuehr R, Bel G (2020) The global E-waste monitor 2020 quantities, flows, and the circular economy potential
5. Birloaga I, Vegliò F (2018) Overview on hydrometallurgicprocedures for silver recovery from various wastes. J Environ Chem Eng 6:2932–2938
6. Xing WD, Lee MS (2019) Development of a hydrometallurgical process for the recovery of gold and silver powders from anode slime containing copper, nickel, tin, and zinc. Gold Bull 52:69–77
7. Samson SO, Muzenda E (2014) Review of silver recovery techniques from radiographic effluent and x-ray film waste. In: Proceedings of the world congress on engineering and computer science 2014 Vol II WCECS 2014, 22–24 Oct 2014, San Francisco, USA
8. Liu F, Wang J, Peng C, Liu Z, Wilson BP, Lundström M (2019) Recovery and separation of silver and mercury from hazardous zinc refinery residues produced by zinc oxygen pressure leaching. Hydrometallurgy 185:38–45
9. Li JY, Xu XL, Liu WQ (2012) Thiourea leaching gold and silver from the printed circuit boards of waste mobile phones. Waste Manage 32(6):1209–1212
10. Yoo J, So H, Yang MH, Lee KJ (2019) Effect of chloride ion on synthesis of silver nanoparticle using retrieved silver chloride as a precursor from the electronic scrap. Appl Surf Sci 475:781–784
11. Gámez S, Garcés K, Torre EL, Guevara A (2019) Precious metals recovery from waste printed circuit boards using thiosulfate leaching and ion exchange resin
12. Gurung M, Adhikari BB, Kawakita H, Ohto K, Inoue K, Alam S (2013) Recovery of gold and silver from spent mobile phones by means of acidothiourea leaching followed by adsorption using biosorbent prepared from persimmon tanin. Hydrometallurgy 133:84–93
13. Kumari A, Parween R, Chakravarty S, Parmar K, Pathak DD, Lee J-C, Jha MK (2019) Novel approach to recover rare earth metals (REMs) from Indian coal bottom ash. Hydrometallurgy 187:1–7
14. Kumari A, Jha MK, Singh RP (2016) Recovery of metals from pyrolysed PCBs by hydrometallurgical techniques. Hydrometallurgy 165:97–105

Deoxidation of Titanium Using Cerium Metal and Its Oxyhalide Formation

Gen Kamimura, Takanari Ouchi, and Toru H. Okabe

Abstract The recycling of titanium (Ti) scraps requires the direct removal of oxygen (O) from the Ti scraps. Although several deoxidation techniques for Ti have been developed, the strong affinity between Ti and O limits their cost-effectiveness. In this study, we develop a new deoxidation process for Ti using cerium (Ce) metal, which is the most abundant and cost-effective rare earth element. Thermodynamic analysis suggests that the deoxidation through the formation of Ce oxyhalides in halide fluxes containing Ce ions enables the production of Ti with extremely low O concentrations. We experimentally demonstrate that the formation reaction of CeOCl with Ce metal can deoxidize Ti metal and produce highly pure Ti with 100 mass ppm O or below, which is lower than the O concentration of the virgin Ti produced by the Kroll process. This deoxidation process with Ce metal enables the recycling of Ti scraps contaminated with O.

Keywords Titanium · Recycling and secondary recovery · Pyrometallurgy

Introduction

Titanium (Ti) and its alloys are expected to be widely used as structural materials owing to their excellent properties, such as high specific strength, excellent corrosion resistance, and biocompatibility. Titanium resources are abundant worldwide; thus, there are no problems with its resource supply. However, at present, the applications of Ti are limited to high value-added materials such as aircraft because Ti has an

G. Kamimura (✉) · T. Ouchi · T. H. Okabe
Institute of Industrial Science, The University of Tokyo, 4-6-1 Komaba, Meguro-ku, Tokyo 153-8505, Japan
e-mail: gen-kami@iis.u-tokyo.ac.jp

T. Ouchi
e-mail: t-ouchi@iis.u-tokyo.ac.jp

T. H. Okabe
e-mail: okabe@iis.u-tokyo.ac.jp

A. Lazou et al. (eds.), *REWAS 2022: Developing Tomorrow's Technical Cycles (Volume I)*, The Minerals, Metals & Materials Series,
https://doi.org/10.1007/978-3-030-92563-5_10

extremely high affinity for oxygen (O). Furthermore, high-temperature processes such as smelting, melting, and casting of Ti require costly and complicated techniques to prevent O from contaminating the Ti materials.

The yield in the fabrication process of Ti ingots is low, resulting in the generation of numerous Ti scraps, which is another factor for the high cost of Ti products [1, 2]. These Ti scraps are recycled by remelting, and the high-temperature process drastically increases the O concentration in the Ti materials. Ti scraps with high O concentration, which cannot be used for Ti products, are recycled as additive materials and deoxidants for steelmaking (cascade recycling).

The demand for Ti is expected to increase significantly in the aircraft industry, which necessitates a more efficient utilization and recycling of the Ti scraps. Effective recycling of Ti scraps requires deoxidation technology that directly removes O from Ti. At present, there is no industrial process to deoxidize Ti, although several deoxidation techniques with calcium (Ca), magnesium (Mg), and rare earths (REs) have been developed [3–34]. Metallic calcium (Ca) has a high deoxidizing ability to Ti, and many researchers have investigated the deoxidation of Ti using metallic Ca as a deoxidant. However, the low efficiency in the production and regeneration of metallic Ca are the limiting factors for its industrialization.

Recently, Okabe et al. thermodynamically analyzed the deoxidation reactions of Ti using REs as deoxidants and predicted the effectiveness [21, 22]. The reliability of the thermodynamic data for RE compounds is relatively low. A series of experimental studies by Okabe et al. demonstrated the deoxidizing abilities of yttrium (Y), lanthanum (La), and holmium (Ho). Their experimental results roughly corresponded with the predictions of their thermodynamic calculations [23–34].

With an increasing demand and production of the REs for raw materials of permanent magnets, such as neodymium (Nd) and dysprosium (Dy), other REs such as Y, La, and cerium (Ce) are also increasingly produced simultaneously. However, compared to Nd and Dy, the demand of these REs (such as Y, La, and Ce) is expected to stagnate, resulting in oversupply in the future [35–39]. For resource utilization, developing the demand for these REs in oversupply has become more important. In this study, we aim to experimentally verify the deoxidation of Ti using Ce as a deoxidant for Ti recycling and RE resource utilization.

Experimental

The Ti samples used in the deoxidation experiments (~0.1 g each) were "Ti-1.2U" (~1000 mass ppm O, commercially pure Ti (CP–Ti) wire with a diameter of 1.2 mm, Shinkinzoku Industry Co., Ltd.), "Ti-2.0I" (~960 mass ppm O, CP-Ti wire with a diameter of 2.0 mm, Shinkinzoku Industry Co., Ltd.), "Ti-6 N" (~120 mass ppm O, Ti cube with side length of 2 mm, purified by electrolysis and electron beam melting), and "Ti64-1.2σ" (~1500 mass ppm O, Ti-6mass%Al-4mass%V alloy wire with a diameter of 1.2 mm, Shinkinzoku Industry Co., Ltd.). Prior to the experiments, the Ti samples were cleaned with a mixture solution of hydrofluoric acid (46.0–48.0%,

Kanto Chemical Co., Inc.), nitric acid (60–61%, Kanto Chemical Co., Inc.), and water (volumetric ratio of 1:4:10). They were subsequently placed in a Ti crucible (25.4 mm in outer diameter, 80 mm in height, 1 mm in thickness, Shinkinzoku Industry Co., Ltd.) with metallic Ce shot (99%, Santoku Co., Ltd.), CeO_2 powder (99.9%, High Purity Chemicals Co., Ltd.), KCl powder (99.5%, Kanto Chemical Co., Inc.), and $CeCl_3$ powder (95%, Fluorochem Ltd.). Several Ti crucibles containing the Ti samples were placed in a stainless steel container, the top of which was sealed with tungsten inert welding. The stainless steel chamber was heated at 1200 K (927 °C) for 173 ks (2 days) in a muffle furnace. After the experiments, the Ti samples were again cleaned with a mixture of hydrofluoric acid, nitric acid, and water (volumetric ratio of 1:4:10). The O concentrations in the Ti samples were determined by inert gas fusion and infrared absorption spectroscopy (ON836, LECO Corporation). The crystalline phases of the fluxes were identified using an X-ray diffractometer (D2 Phaser, Bruker).

Results and Discussion

The standard Gibbs energy change of the dissolution of oxygen (O_2) gas in β-Ti was determined by Okabe et al. as follows [40]:

$$\begin{aligned} 1/2\,\mathrm{O_2}\,(g) &= \mathrm{O}\,(1\ \text{mass\%}\ \text{in}\ \beta\text{-Ti}), \\ \Delta G^{\circ}_{1,\mathrm{Ti}} &= -583000 + 88.5\,T\ \mathrm{J \cdot mol^{-1}}(1173\text{–}1373\,\mathrm{K}), \end{aligned} \tag{1}$$

where $\Delta G^{\circ}{}_{1,\mathrm{Ti}}$ is the standard Gibbs energy change of the dissolution of O_2 gas in β-Ti, and T is the absolute temperature. The standard state of O in β-Ti is the Henrian standard state with respect to 1 mass% in β-Ti. The deoxidation of Ti by Ce metal in $CeCl_3$ flux proceeds through reaction (2) [21, 40–42]:

$$\begin{aligned} \mathrm{O}\,(1\ \text{mass\%}\ \text{in}\ \beta\text{-Ti}) + 2/3\ \mathrm{Ce}\ (l) + 1/3\ \mathrm{CeCl_3}\ (l) &= \mathrm{CeOCl}\ (s) \\ \Delta G^{\circ}_{\mathrm{deox},(2)} &= -48500\,\mathrm{J \cdot mol^{-1}}\text{at } 1200\ \mathrm{K}, \end{aligned} \tag{2}$$

where $\Delta G^{\circ}{}_{\mathrm{deox},(2)}$ is the standard Gibbs energy change of reaction (2). Notably, the deoxidation product in reaction (2) is cerium oxychloride (CeOCl) and not cerium oxide (Ce_2O_3). The deoxidation limit based on reaction (2) is calculated to be 77 mass ppm O in β-Ti at 1200 K.

Figure 1 shows the cross-section of the Ti crucible after the deoxidation of Ti samples using Ce metal in a $CeCl_3$ flux (Exp. no. 210731-lot4-4).

The O concentration in the Ti samples after deoxidation by Ce metal in $CeCl_3$ fluxes decreases to 60 ± 50 mass ppm O [43]. Thus, we experimentally demonstrate that deoxidation with Ce metal in a $CeCl_3$ flux can produce highly pure Ti with 100 mass ppm O or below.

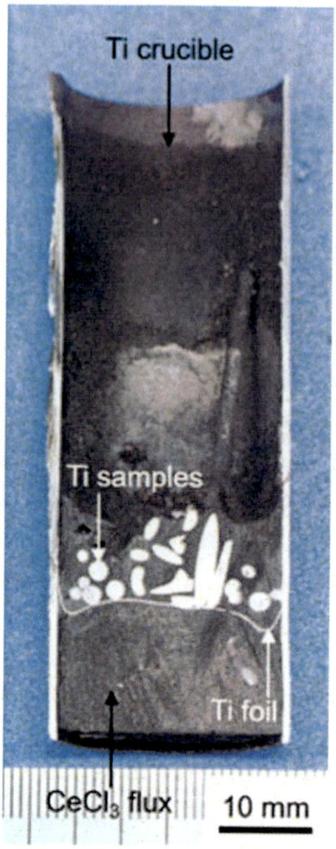

Fig. 1 Cross-section of the Ti crucible after the deoxidation experiment in the Ce/CeOCl/$CeCl_3$ equilibrium (Exp. no. 210731-lot4-4) [43]

Figure 2 shows the XRD pattern of the $CeCl_3$ flux at the bottom of the crucible after the deoxidation experiment using Ce metal (Exp. no. 210607-lot3-5). This XRD pattern includes the peaks attributed to CeOCl and indicates the co-existence of Ce metal, $CeCl_3$, and CeOCl in the system.

Thus, we experimentally demonstrate that the formation reaction of CeOCl with Ce metal can deoxidize Ti metal and produce highly pure Ti with O concentration of 100 mass ppm O or below. This O concentration is lower than that of virgin Ti produced by the Kroll process (~500 mass ppm O).

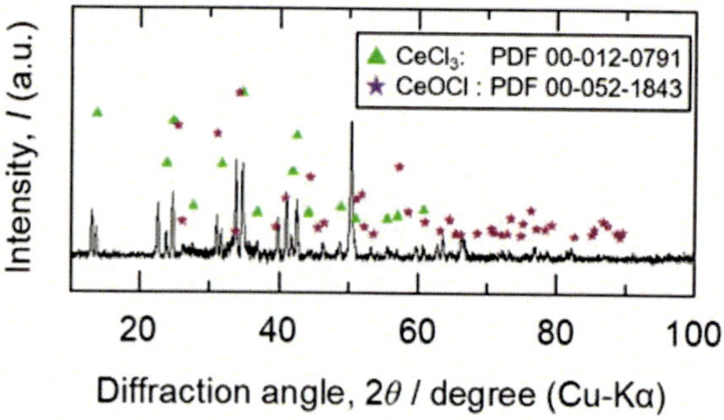

Fig. 2 XRD pattern of the $CeCl_3$ flux in the bottom of the Ti crucible after the deoxidation experiment (Exp. no. 210607-lot3-5) [43]

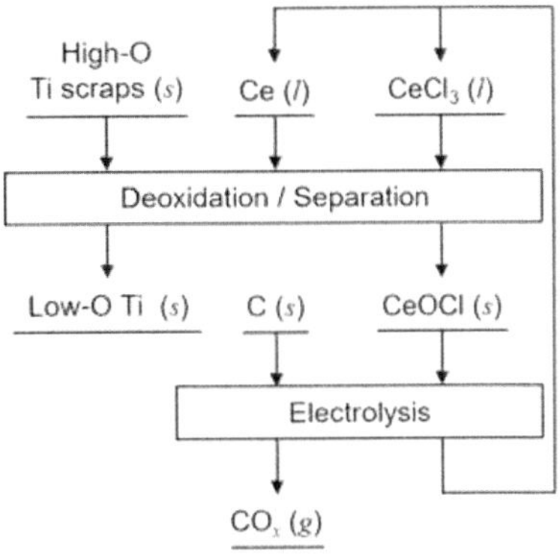

Fig. 3 Proposed recycling process for high-oxygen Ti scraps using Ce metal in a $CeCl_3$ flux

Conclusion

To develop a novel recycling process for Ti scraps that are heavily contaminated with O, we investigated the deoxidation of Ti using Ce metal in a $CeCl_3$ flux. The deoxidation reaction through the formation of CeOCl successfully produced high-purity Ti samples with extremely low O concentrations of 100 mass ppm O or below. To the best of our knowledge, this is the first study to demonstrate the deoxidizing ability of Ce to Ti. Cerium, which is oversupplied in RE production, is a promising deoxidant for Ti scraps. Ultimately, we aim to develop an overall process that consumes no Ce metal by recovering Ce metal and $CeCl_3$ from CeOCl with a low-cost process such as electrolysis, as shown in Fig. 3.

Acknowledgements We would like to thank Mr. Koichi Hirota at Shin-Etsu Chemical Co., Ltd. for providing us with rare earth metal samples. We are grateful to Mr. Akihiro Iizuka and Mr. Kenta Akaishi at the University of Tokyo for their experimental support and their comments on this manuscript. This work was financially supported by the Japan Society for the Promotion of Science (JSPS) through a Grant-in-Aid for Scientific Research (S) (KAKENHI Grant No. 19H05623). This paper is also based on results obtained from a project commissioned by the New Energy and Industrial Technology Development Organization (NEDO).

References

1. Takeda O, Okabe TH (2019) Current status of titanium recycling and related technologies. JOM 71(6):1981–1990
2. Takeda O, Ouchi T, Okabe TH (2020) Recent progress in titanium extraction and recycling. Metall Mater Trans B 51(4):1315–1328
3. Ono K, Miyazaki S (1985) Study on the limit of deoxidation of titanium and the reduction of titanium dioxide by saturated calcium vapors. J Jpn Inst Met 49(10):871–875
4. Fisher RL (1990) Deoxidation of titanium and similar metals using a deoxidant in a molten metal carrier. US Patent 4,923,531
5. Fisher RL (1991) Deoxidation of a refractory metal. US Patent 5,022,935

6. Okabe TH, Suzuki RO, Oishi T, Ono K (1991) Production of extra low oxygen titanium by calcium-halide flux deoxidation. Tetsu-to-Hagane 77(1):93–99
7. Okabe TH, Oishi T, Ono K (1992) Preparation and characterization of extra-low-oxygen titanium. J Alloys Compd 184(1):43–56
8. Okabe TH, Oishi T, Ono K (1992) Deoxidation of titanium aluminide by Ca-Al alloy under controlled aluminum activity. Metall Mater Trans B 23(5):583–590
9. Okabe TH, Nakamura M, Oishi T, Ono K (1993) Electrochemical deoxidation of titanium. Metall Mater Trans B 24(3):449–455
10. Fisher RL, Seagle SR (1993) Removal of oxide layers from titanium casting using an alkaline earth deoxidizing agent. US Patent 5,211,775
11. Fisher RL, Seagle SR (1993) DOSS, an industrial process for removing oxygen from titanium turnings scrap. In: Froes FH, Caplan IL (eds) Proceedings of titanium'92, science and technology 3. The Minerals, Metals & Materials Society, Warrendale, pp 2265–2272
12. Oh J-M, Lee B-K, Suh C-Y, Cho S-W, Lim J-W (2012) Preparation method of Ti powder with oxygen concentration of <1000 ppm using Ca. Powder Metall 55(5):402–404
13. Lim J-W, Oh J-M, Lee B-K, Suh C-Y, Cho S-W (2013) Deoxidation apparatus for preparing titanium powder with low oxygen concentration. US Patent 8,449,813 B1
14. Lim J-W, Oh J-M, Lee B-K, Suh C-Y, Cho S-W (2013) Method for preparing titanium powder with low oxygen concentration. US Patent 8,449,646 B1
15. Oh J-M, Kwon H, Kim W, Lim J-W (2014) Oxygen behavior during non-contact deoxidation of titanium powder using calcium vapor. Thin Solid Films 551:98–101
16. Oh J-M, Roh K-M, Lee B-K, Suh C-Y, Kim W, Kwon H, Lim J-W (2014) Preparation of low oxygen content alloy powder from Ti binary alloy scrap by hydrogenation–dehydrogenation and deoxidation process. J Alloys Compd 593:61–66
17. Oh J-M, Choi I-H, Suh C-Y, Kwon H, Lim J-W, Roh K-M (2016) Deoxidation and subsequent densification of sintered titanium by Ca deoxidizer. Met Mater Int 22(3):488–492
18. Kim S-J, Oh J-M, Lim J-W (2016) Thermodynamic evaluation of oxygen behavior in Ti powder deoxidized by Ca reductant. Met Mater Int 22(4):658–662
19. Xia Y, Fang ZZ, Fan D, Sun P, Zhang Y, Zhu J (2018) Hydrogen enhanced thermodynamic properties and kinetics of calciothermic deoxygenation of titanium–oxygen solid solutions. Int J Hydr Energy 43(27):11939–11951
20. Zhang Y, Fang ZZ, Sun P, Zhang T, Xia Y, Zhou C, Huang Z (2016) Thermodynamic destabilization of Ti-O solid solution by H_2 and deoxygenation of Ti using Mg. J Am Chem Soc 138(22):6916–6919
21. Okabe TH, Zheng C, Taninouchi Y (2018) Thermodynamic considerations of direct oxygen removal from titanium by utilizing the deoxidation capability of rare earth metals. Metall Mater Trans B 49(3):1056–1066
22. Okabe TH, Taninouchi Y, Zheng C (2018) Thermodynamic analysis of deoxidation of titanium through the formation of rare-earth oxyfluorides. Metall Mater Trans B 49(6):3107–3117
23. Okabe TH, Hirota K, Kasai E, Saito F, Waseda Y, Jacob KT (1998) Thermodynamic properties of oxygen in RE–O (RE=Gd, Tb, Dy, Er) solid solutions. J Alloys Compd 279(2):184–191
24. Okabe TH, Hirota K, Waseda Y, Jacob KT (1998) Thermodynamic properties of Ln-O (Ln = La, Pr, Nd) solid solutions and their deoxidation by molten salt electrolysis. Shigen-to-Sozai 114(11):813–818
25. Zheng C, Ouchi T, Iizuka A, Taninouchi Y, Okabe TH (2019) Deoxidation of titanium using Mg as deoxidant in $MgCl_2$-YCl_3 flux. Metall Mater Trans B 50(2):622–631
26. Zheng C, Ouchi T, Kong L, Taninouchi Y, Okabe TH (2019) Electrochemical deoxidation of titanium in molten $MgCl_2$–YCl_3. Metall Mater Trans B 50(4):1652–1661
27. Kong L, Ouchi T, Okabe TH (2019) Direct deoxidation of Ti by Mg in $MgCl_2$–$HoCl_3$ flux. Mater Trans 60(9):2059–2068
28. Kong L, Ouchi T, Zheng C, Okabe TH (2019) Electrochemical deoxidation of titanium scrap in $MgCl_2$-$HoCl_3$ system. J Electrochem Soc 166(13):E429–E437
29. Iizuka A, Ouchi T, Okabe TH (2020) Ultimate deoxidation method of titanium utilizing Y/YOCl/YCl_3 equilibrium. Metall Mater Trans B 51(2):433–442

30. Iizuka A, Ouchi T, Okabe TH (2020) Development of a new titanium powder sintering process with deoxidation reaction using yttrium metal. Mater Trans 61(4):758–765
31. Tanaka T, Ouchi T, Okabe TH (2020) Lanthanothermic reduction of TiO_2. Metall Mater Trans B 51(2):1485–1494
32. Tanaka T, Ouchi T, Okabe TH (2020) Yttriothermic reduction of TiO_2 in molten salts. Mater Trans 61(10):1967–1973
33. Tanaka T, Ouchi T, Okabe TH (2020) Magnesiothermic reduction of TiO_2 assisted by $LaCl_3$. J Sustain Metall 6(4):667–679
34. Kong L, Ouchi T, Okabe TH (2021) Deoxidation of Ti using Ho in $HoCl_3$ flux and determination of thermodynamic data of HoOCl. J Alloys Compd 863:156047
35. Alonso E, Sherman AM, Wallington TJ, Everson MP, Field FR, Roth R, Kirchain RE (2012) Evaluating rare earth element availability: a case with revolutionary demand from clean technologies. Environ Sci Technol 46(6):3406–3414
36. Binnemans K, Jones PT, Blanpain B, Gerven TV, Yang Y, Walton A, Buchert M (2013) Recycling of rare earths: a critical review. J Clean Prod 51:1–22
37. Binnemans K, Jones PT, Acker KV, Blanpain B, Mishra B, Apelian D (2013) Rare-earth economics: the balance problem. JOM 65(7):846–848
38. Zhou B, Li Z, Chen C (2017) Global potential of rare earth resources and rare earth demand from clean technologies. Minerals 7(11):203
39. Schulze R, Weidema BP, Schebek L, Buchert M (2018) Recycling and its effects on joint production systems and the environment—the case of rare earth magnet recycling—Part I—production model. Resour Conserv Recycl 134:336–346
40. Okabe TH, Suzuki RO, Oishi T, Ono K (1991) Thermodynamic properties of dilute titanium–oxygen solid solution in beta phase. Mater Trans JIM 32(5):485–488
41. Barin I (1995) Thermochemical data of pure substances, 3rd edn. Wiley-VCH, Weinheim
42. Wagman DD, Evans WH, Parker VB, Schumm RH, Halow I, Bailey SM, Churney KL, Nuttall RL (1982) The NBS tables of chemical thermodynamic properties: selected values for inorganic and C1 and C2 organic substances in SI units. National Bureau of Standards, Washington, DC
43. Kamimura G, Ouchi T, Okabe TH (2021) Deoxidation of titanium using cerium–chloride flux for upgrade recycling of titanium scraps. Mater Trans (submitted)

Estimation of the Generation and Value Recovery from E-waste Printed Circuit Boards: Bangladesh Case Study

Md. Khairul Islam, Nawshad Haque, Michael Somerville, Mark I. Pownceby, Suresh Bhargava, and James Tardio

Abstract This article analyses Bangladesh export–import data to quantify historically generated e-waste from four types of discarded electronic devices (mobile phones, TVs, tablets, and computers/laptops) and uses the trends to predict the generation of e-waste from these devices up to 2030. From this data, together with estimated redundancy rates, printed circuit board (PCB) masses, metal content, and value based on characterisation of indicative samples, the potential value of e-waste was evaluated. Through processing the PCBs in Bangladesh, metals including Cu, Ag, Au, Pd, and Sn worth more than US$2.4 billion till 2020 could be recovered. This value could reach US$7 billion when forecasted to 2030. The potential value varies mainly with the fluctuating metal prices in the international market.

Keywords E-waste · Recycling · Bangladesh · Precious metals

Introduction

Electronic waste (e-waste) encompasses the various units of antiquated electronic and electrical equipment (EEE) which have either lost their proper functionality or their appeal to the user with the arrival of new upgraded products. In 2019, worldwide e-waste generation was estimated to be 53.6 million metric tonnes (Mt), of which

Md. K. Islam (✉) · S. Bhargava · J. Tardio
Centre for Advanced Materials & Industrial Chemistry (CAMIC), School of Science, RMIT University, GPO Box 2476, Melbourne, VIC 3001, Australia
e-mail: s3737858@student.rmit.edu.au

Md. K. Islam · M. Somerville · M. I. Pownceby
CSIRO Mineral Resources, Private Bag 10, Clayton South, VIC 3169, Australia

Md. K. Islam
Bangladesh Council of Scientific and Industrial Research (BCSIR), IMMM, Joypurhat 5900, Bangladesh

N. Haque
CSIRO Energy, Private Bag 10, Clayton South, VIC 3169, Australia

A. Lazou et al. (eds.), *REWAS 2022: Developing Tomorrow's Technical Cycles (Volume I)*, The Minerals, Metals & Materials Series,
https://doi.org/10.1007/978-3-030-92563-5_11

only around 17% was adequately collected and recycled [1]. This is projected to rise to 70.4 Mt by the year 2030. An opportunity for urban mining through secondary resource recovery has been created since these electronic items are gradually and continuously being added to the e-waste stream. This could have a strong impact on the economy, especially of countries with no natural mineral resources and related industries to produce primary metals. E-waste contains several base metals (Cu, Zn, Sn, Ni, and Fe), precious metals (Ag, Au, Pt, and Pd), rare earths (Y, Ce, and Nd), and critical metals (Ga, In, Co, Li, and Ta) [2–5]. The sustainable recovery of these metals could provide economic benefits, decrease environmental hazards, and reduce the environmental footprint caused by mining by supplying metals for further manufacturing. A good understanding of the e-waste generation in different countries and regions requires proper estimation of the resources available and an assessment in terms of potential market value. This can lead to tailored processing routes designed to extract valuable materials considering the socio-economic situation of the country. Several articles have been published in recent years which estimated the e-waste generation for different countries and their potential value [6–8].

Bangladesh is a densely populated country in South Asia having around 170 million people. With its growing population and economy, in recent times, the use of EEE has increased rapidly. A few articles have been published on the Bangladesh e-waste scenario, surveys, policies, legal frameworks, and possible environmental impacts, etc. [9–11]; however, there is lack of quantitative data which could give a good estimate of the total e-waste generation within the country and the potential value that could be recovered. This kind of estimation encourages business entrepreneurs to invest and governments to formulate favourable policies for investors, stakeholders, and public–private partnership in this arena. Recently, the Department of Environment (DoE) in collaboration with the Centre for Environmental and Resource Management (CERM) published a report that shows the informal recycling of e-waste countrywide [12]. Another report for Bangladesh was previously published by Environment and Social Development Organization (ESDO). Although they reported a wide variety of e-waste categories and provided a sense of the huge stockpiling of e-waste, the data was not strong enough to support the conclusion of 2.7 Mt e-waste generation in Bangladesh annually [13]. Almost 90% of the total e-waste of Bangladesh was shown to come from ship breaking yards by assuming the number of obsolete ships coming into the shore every year and assuming they were full of e-waste. This is unlikely to be the actual scenario, and the data currently lacks a rigorous breakdown of the types of e-waste resources.

Methodology

This study uses trade data (export, import, re-import, and re-export) from the UN COMTRADE database [14]. The annual quantity of the selected electronic products was obtained from the database for harmonised system (HS) codes of the selected products and predicted until 2030 using a statistical extrapolation. The total amount of electronic products entering the Bangladesh market were assumed to be used internally, and the products would be gradually added to the e-waste stream at the end of their useful life. The Weibull distribution function was used to calculate the dynamic lifespan of individual electronic items and hence the proportion of generated e-waste [15]. To estimate the potential metals that are available for recovery from the e-waste, only the PCB portion of the e-waste was considered. The market value of the major and valuable metals was used to calculate potential revenue from the processing of these wastes. A more detailed description of the aforementioned is given in the subsequent sections.

Estimating Put-on-Market (POM) Products

Four types of electronics (mobile phones, TVs, tablets, and computers/laptops) that are used in Bangladesh and have a much higher growth in use were selected for this study. The import data was obtained from the UN COMTRADE database for individual products ranging from 1992 to 2019 [14]. The base data was chosen as 2000 to 2019 and then predicted until 2030. Estimation of the POM electronics, in tonnes, was calculated using the following equation:

$$\text{POM (t)} = \text{Domestic production(t)} + \text{Imports(t)} - \text{Exports(t)} \tag{1}$$

Bangladesh does not primarily manufacture electronics and hence does not export to other countries. In recent times however, some companies are assembling electronics such as mobile phones and televisions locally although accurate statistics for this kind of domestic production is limited. Therefore, to simplify the calculation, the domestic production and export were considered as zero. Thus, for Bangladesh, the equation becomes

$$\text{POM(t)} = \text{Imports(t)} \tag{2}$$

The HS codes for PC-Desktop used were HS-847141, HS-847149 and HS-847150. To get the sales data for Laptop/Tablets, it was HS-847130, and for TV sales data the HS codes were HS-852810, HS-852812, HS-852813, and HS-852820. However, the UN COMTRADE database does not report the mobile phone sales data before 2007 although Bangladesh started using mobile phones from 1997 when the first telecom company Grameenphone started business in Bangladesh [16]. Consequently, mobile phone subscriber data was taken from the Bangladesh Telecommunication Regulatory Commission (BTRC) website as the basis for mobile phone POM data. BTRC publishes monthly total subscriber data for the preceding 90 days period. In this regard, the unique subscriber penetration data was obtained as per the (Global System for Mobile Communications) GSMA report [17] to calculate the number of mobile phones (assuming every user has at least one mobile phone throughout their subscription period). To estimate the POM of the different electronic items, the number of units data obtained from the UN COMTRADE database was used. The weight data was also available but with less consistency. Therefore, the numbers were converted to weight by multiplying the data by the average unit weight of the individual items described in earlier research [6, 18].

Product Lifespan of Selected Electronics

The lifespan of electronic goods is considered as the period an electronic device/product remains in the service of its user. Generally, different electronic products have different lifetimes, and these are not always straightforward to predict since it depends on multiple factors including the product quality, type of electronics, consumer's choice, emergence of new technology/features, income of individuals, and the socio-economic situation of the country. For example, mobile phones have shorter lifespans compared to televisions or other large household appliances such as washing machines. There are arguments that the product lifespan should be higher in developed countries compared to the developing/underdeveloped countries and vice versa. One argument is that developed countries can afford and purchase higher quality products, and thus, the durability of the electronics should be higher compared to that of developing or lower income countries. Another argument is that in relatively lower income countries, consumers use products with much care and reuse, refurbishing to maximize the lifetime of a product, while consumers in developed nations usually do not reuse to such an extent. These scenarios can lead to an underestimation or overestimation of the total POM data; however in general, the method described in this article is relatively accurate for estimating POM and subsequent e-waste generation [6, 19].

The Weibull probability distribution function is a tool that has been used by many researchers to estimate the proportion of POM electronics being added to the waste stream within the broad lifespan of the product [6, 8]. This probability analysis tool is widely used in manufacturing lines to statistically measure probability of failure for a particular batch of product. It was also found to be effective in examining the behaviour of electronic products [18, 19]. The Weibull probability density function (PDF) is expressed by following:

$$f = \frac{\alpha}{\beta}\left(\frac{t}{\beta}\right)^{\alpha-1} \cdot e^{-\left(\frac{t}{\beta}\right)^{\alpha}} \tag{3}$$

where $f(t)$ is the probability density function indicating the failure rate. The shape factor (α) and scale factor (β) are the two parameters that can be empirically obtained for a particular product. This study adapts the shape factor and scale factor values derived for the four types of electronic products from a previously reported study [19] (Table 1). These values were empirically obtained for each product based on data from European and non-EU countries.

Table 1 Weibull function α, β values for individual electronic products

	Mobile phones	PC-Desktop	Laptop/Tablets	TVs[a]
α (shape)	1.52	1.80	1.94	2.19
β (scale)	5.62	10.33	8.76	11.51

[a] Average values for CRT and LCD/LED TVs combined are reported

Estimation of E-waste Generation from the Discarded Electronics

E-waste generated from the four electronic product types was calculated using the Weibull distribution function and the annual POM data using the following formula [19]:

$$\text{E-waste generated}(y) = \sum_{t=t_0}^{y} \text{POM}(t) \times L(t, y) \tag{4}$$

Here, E-waste generated (y) is the quantity of e-waste generated in the year; t_0 is the initial year of sale of the product; POM (t) refers to the electronics entered the market (in tonnes) in a historical year t, ahead of the year y, and $L(t, y)$ is the Weibull probability density function (Eq. 3) indicating the lifespan distribution for the batch of products sold per year (t).

Results and Discussion

POM Electronics and E-waste Generation

The mass of e-waste generation for each type was calculated by multiplying the number of units with the unit weight. The average weight of TVs was taken as 11.19 kg/unit and 15.2 kg/unit for Desktop-PCs, 3.2 kg/unit of Laptop/Tablets [19], and for mobile phones it was 0.1 kg/unit [6]. Figure 1 shows the annual POM for Mobile phones, Laptop/Tablets and Desktop-PCs follows a gradual incremental trend annually whereas the POM trend for TVs, while increasing overall from 2000 to 2019, between 2003 and 2013 there was a relative fall in the POM observed. The POM data was predicted until 2030 using the best fitted trend except for the TV data where it was difficult to fit a predictive trend. Hence, a 3% annual increment was considered to predict the future POM for TVs until 2030. This was based on the fact that e-waste has a growth rate of 3–5 wt.% for all over the world [2]. The lowest value of the range was therefore chosen as a conservative estimate.

The product lifespan profile of the electronics is shown in Fig. 2. This distribution function shows the gradual decay of a product after it enters the consumer market. The

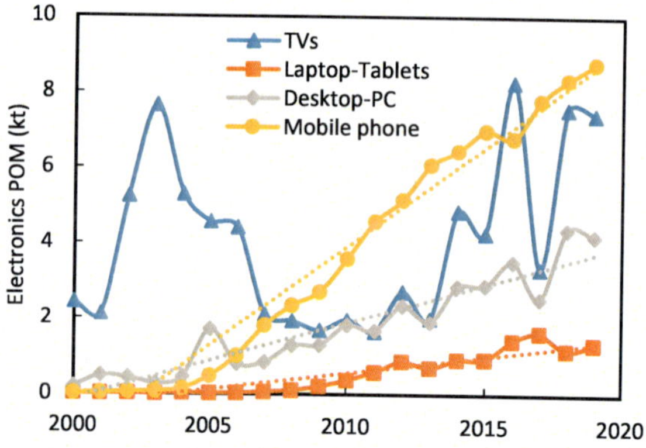

Fig. 1 Annual POM electronics in Bangladesh; data taken from Refs. [14, 20, 21]

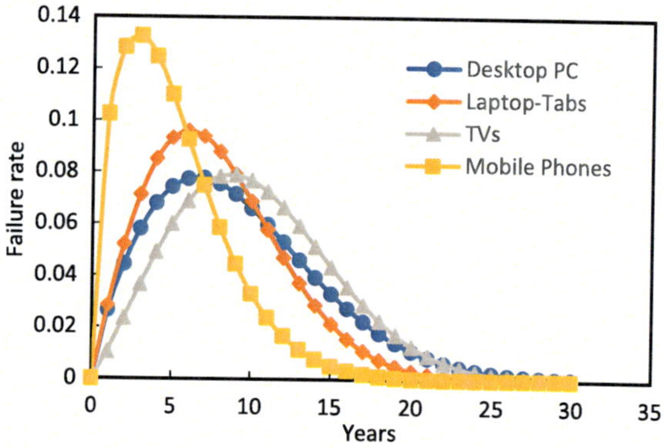

Fig. 2 Product lifespan profile based on Weibull probability distribution function; shape and scale factor data taken from [6, 19]

lifespan of mobile phones is the shortest, while TVs have the highest lifespan. Most of the mobile phones finish their usability within 4–5 years with some proportion having a higher lifespan. The sharp peak of the PDF for mobile phones at 4 years indicates the higher failure rate of mobile phones around this time, with very few having a higher lifetime. Laptop/Tablets have a relatively higher period of usability compared to mobile phones, and Desktop-PCs are more likely to serve for extended periods compared to mobile phones and Laptop/Tablets. Although having a relatively shorter lifespan from TVs, both distribution functions are similar having a relative blunt peak compared to mobile phones and or Laptop/Tablets. Note that while the calculation of e-waste generation depends on the Weibull function for the individual product type, the failure rate is less sensitive than the POM data of the electronics [19].

Generation of E-waste from Discarded Electronics

The generation of e-waste from the four different electronics from the year 2000 to 2030 is shown in Fig. 3a. Overall, the total annual e-waste generation increases with time from around 2.7 kt in 2000 to 25 kt in 2030. Although TVs dominate e-waste in the early years, mobile phones gradually become the highest contributor. The e-waste generation from Desktop-PCs and Laptops/Tablets increased with time as the accessibility of these devices increased. Nonetheless, the volume of e-waste coming from Desktop-PCs and Laptops/Tablets was smaller compared to that of mobile phones and TVs.

The mobile phone market has a saturation which is related to the number of subscribers and the total population, and for Bangladesh, it is near to this point as indicated by the subscriber data [20, 21]. Once this point is reached, the e-waste coming from mobile phones would become constant. However, the consumption of other electronics can continue to grow in coming years. In 2019, the generated e-waste by the four types of electronics was around 13 kt. The total e-waste generated in Bangladesh in 2019 was estimated as 199 kt with a per capita generation of 1.2 kg [1]. Clearly these four items make up only a very small proportion of the total e-waste generation in the country; however, they carry a significant value in terms of materials and their potential utility and price. Cumulative generation of e-waste from the selected EoL electronic devices is shown in Fig. 3b. While the current volume of e-waste resulting from these devices is more than 150 kt, it is predicted to rise to 350 kt by 2030.

For this study, the proportion of PCBs in the electronics was obtained from data reported in the literature. For mobile phones, the PCB was 30% of the total weight [3], while for Desktop-PCs, Laptops/Tablets, and TVs, the value was about 12% of total mass [18]. The total potential PCB contents from the individual electronics are shown in Fig. 4. The data indicates that by 2020 the annual waste PCB generation

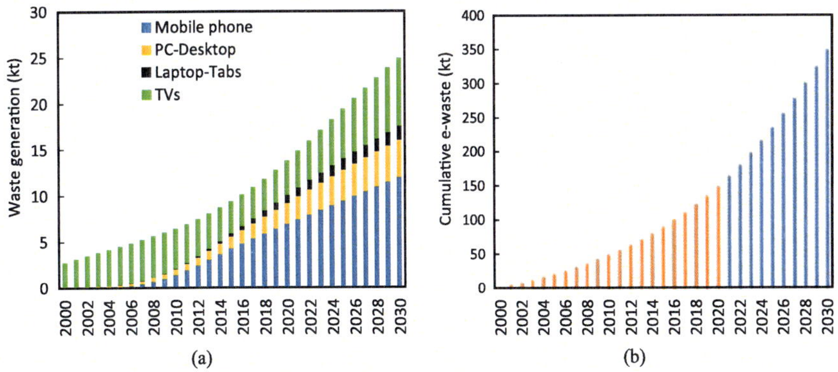

Fig. 3 **a** Annual e-waste generated from used electronics; **b** cumulative generation of e-waste from EoL mobile phones, TVs, Laptop/Tablets, and Desktop-PCs

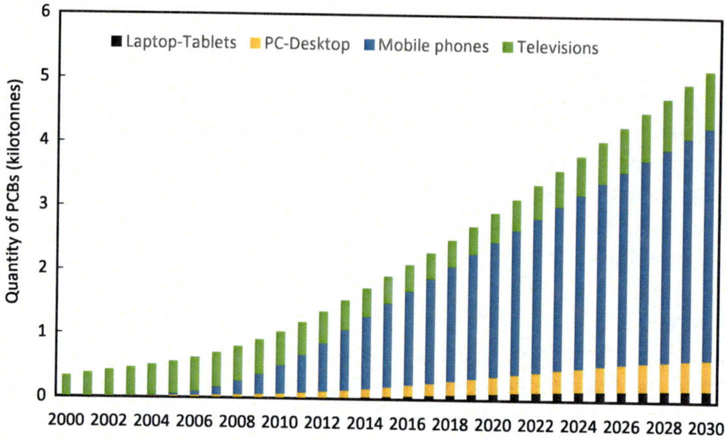

Fig. 4 Annual estimated amount of waste PCBs coming from the four types of EoL electronics

Table 2 Typical metal contents (wt.%) in PCBs; adapted from Ref. [3]

	Al	Cu	Fe	Sn	Zn	Ag	Au	Pd	Ba	Bi	Co	Sr	Ta
Mobile phone	1.5	33	1.8	3.5	0.5	0.38	0.15	0.03	1.9	0.044	0.028	0.043	0.26
Desktop-PC	1.8	20	1.3	1.8	0.27	0.057	0.024	0.015	0.19	0.005	0.0048	0.038	0.0007
Notebook PC	1.8	19	3.7	1.6	1.6	0.11	0.063	0.02	0.56	0.012	0.008	0.038	0.58
CRT TV	6.2	7.2	3.4	1.8	0.53	0.012	0.0005	0.002	0.24	0.028	0.0036	0.055	0.0
Plasma display TV	3.8	21	2.0	1.5	1.2	0.04	0.03	0.0	0.39	0.01	0.0	0.065	0.01
LCD TV	6.3	18	4.9	2.9	2.0	0.06	0.02	0.0	0.3	0.0	0.0	0.03	0.0

was close to 3 kt while it was predicted to rise to above 5 kt by 2030. Moreover, until 2006, TVs were the major electronics contributing to waste PCB generation. Later, waste generation from mobile phones and Desktop-PCs gradually increased with time. After 2010, the highest waste PCBs come from mobile phones followed by TVs and Desktop-PCs. The metal contents for PCBs from different electronic products were obtained from Oguchi et al. 2011 (Table 2).

Estimation of Potentially Recoverable Embedded Metals

Metals, plastics, and ceramics components of the electronic products could potentially be recovered from PCBs once they have finished their service by disassembly and dismantling. However, PCBs are complex composites consisting of various metals including precious metals, rare earths, and base metals [4, 5]. To recover the

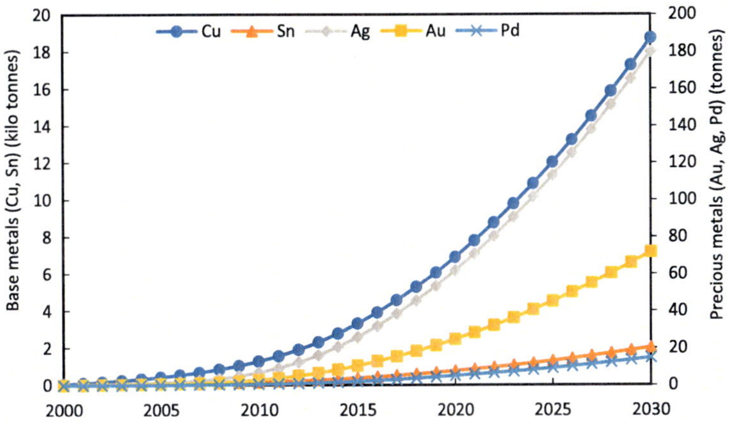

Fig. 5 Total quantity of base metals and precious metals embedded with the generated e-waste PCBs

materials within the PCBs, some physical and/or chemical processing is required. In this study, only the PCB portion of e-waste was considered to estimate the embedded metals that are either high in concentration or high in value. Among the base metals, Cu and Sn were estimated, while from the precious metals, Au, Ag, and Pd were estimated. Figure 5 shows the base metals and precious metal content of Bangladesh e-waste from 2000 till 2030. The data indicates that PCB e-waste generated in the year 2020 could provide around 7 kt of Cu, 764 t of Sn, 62 t of Ag, 25 t of Au, and 5 t of Pd. The amount will increase every year and is predicted to be close to 19 kt Cu, 2 kt Sn, 180 t Ag, 72 t Au, and 15 t of Pd in 2030.

Revenue Potential and Sensitivity

Reprocessing of the e-waste materials to recover valuable materials will generate revenue that could offset the processing cost. The potential value for the individual metals and the total potential revenue are plotted in Fig. 6. The current metal price was taken from the LME website and shown in Table 3 [22]. It is seen that the total potential of the e-waste for Bangladesh is close to US$3 billion dollars at present, while it is predicted to reach US$7 billion by 2030. Gold being the highest contributor could contribute around US$4 billion, while Pd is the second highest followed by Cu. This study shows that more than 82% of the total value comes from the precious metals while slightly above 7% comes from Cu. It was reported that precious metals (Au, Ag, and Pd) contribute almost 85% of waste PCBs and 93% in case of mobile phone PCBs [5]. Since Au is the major contributor and its price fluctuates with the volatility of world's economy, the potential revenue is highly sensitive to the gold price.

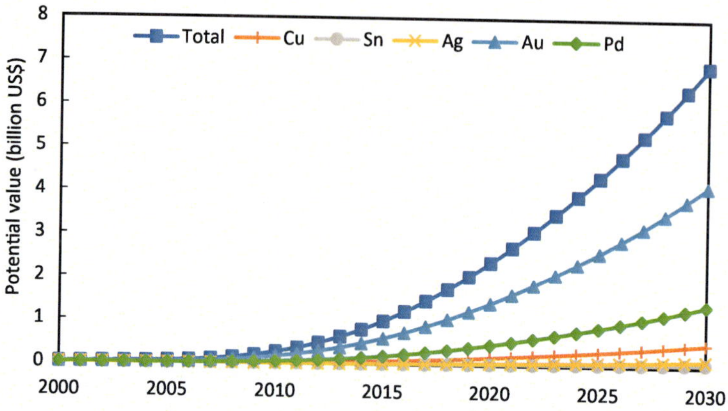

Fig. 6 Cumulative value of selected contributing metals within the waste PCBs of TVs, mobile phones, Desktop-PCs, and Laptop/Tablets

Table 3 Current metal price used in the value estimation of e-wastes

Metal	Cu	Sn	Au	Ag	Pd
LME price (US$/kg)	9.4	35.1	57,938	810	95,416

Limitation of the Present Study

This article described the POM and subsequent e-waste generation scenario of Bangladesh for the first time considering four types of electronics that are the fastest growing and most widely used in the country. There are, however, some limitations of the research with regard to the assumptions and estimation that were used.

- Most of the data was taken from UN COMTRADE database which is a reliable source of worldwide export–import data. Still, there may be some unreported electronics that entered Bangladesh market through informal routes. Some companies have recently started manufacturing and assembling mobile phones, televisions, and household electronics in Bangladesh. The data for locally manufactured products is not available and hence not included in this estimation.
- The number of individual electronics was multiplied by the average unit weight of the individual device to obtain the POM in weight. That could bring a variation in the estimated weight of the total POM and the generated e-waste.
- Based on the historical data, the future POM was predicted until 2030 using best fitted trends. These trends may not apply in coming years being highly dependent on the country's financial and political situation.
- To estimate the dynamic generation of e-waste from the selected electronics, the Weibull distribution function was used. For this, the α & β values were taken from the literature lifespan data based on a European-based study. The actual lifespan could be slightly different as the lifespan changes with evolution of the

technology and manufacturing year. For example, Mairizal et al. showed that the overall lifespan for mobile phones decreased with time compared to the earlier versions of mobile phones in Indonesia [8].

- The metal content of PCBs and the proportion of PCBs for the selected electronics was taken from the literature. These values could vary according to country, quality of the products, and usage periods.
- To calculate the potential value of embedded materials, the recent price of metals was taken into consideration. The market price of metals, especially precious metals, varies largely with time and other factors (economic and political stability). For example, the gold price increased recently in response to market volatility due to the COVID-19 global pandemic. The values in the current study are predicted until 2030; however, during the intervening, there may be a significant change in the metal market (for any or all of the metals examined). Thus, the expected revenue could be higher or lower.
- Complete recovery of the embedded metal portion of Cu, Sn, Ag, Au, and Pd was assumed in order to estimate the available metals and their potential economic value. However, it is unlikely that complete recovery of the metals could be achieved. Lower recoveries would decrease the metal yield and the subsequent revenue. On the other hand, some metals (Fe, Al, Zn, Pb, etc.) have been omitted from calculating the potential value. Depending on the recycling route, some of these metals could also be recovered, thereby increasing the total revenue.
- Moreover, this study considers only the PCB portion of the e-waste generated from the selected electronics which constitutes a small portion of the total mass. Other parts of the electronic devices such as body, casings, and wirings contain several materials that could also be recovered easily without any complex metallurgical/chemical processing and add value to the potential recovered materials.

Conclusion

This article analyses the export–import data of Bangladesh for TVs, Desktop-PCs, Laptop/Tablets, and mobile phone subscriber data to estimate the generation of total e-waste and the potential value of the embedded materials. The estimate indicates that about 7,000 t Cu, 764 t of Sn, 62 t of Ag, 25 t of Au, and 5 t of Pd would be available to recover from the generated waste PCBs by 2020. With a growth by 2030, this is predicted to be 19,000 t Cu, 2,000 t Sn, 180 t Ag, 72 t Au, and 15 t Pd. The potential value that could be recovered is currently close to US$2.4 billion and around US$7 billion by 2030. This value is sensitive to the market value of precious metals, especially gold and palladium. Despite the unavoidable limitations of this study, it gives a reasonable indication of the high potential of e-waste recycling industries in Bangladesh that could encourage entrepreneurs to consider investing in this sector.

References

1. Forti V, Balde CP, Kuehr R, Bel G (2020) The global e-waste monitor 2020: quantities, flows and the circular economy potential. https://collections.unu.edu/view/UNU:7737. Accessed 26 March 2021
2. Cucchiella F, D'Adamo I, Koh SL, Rosa P (2015) Recycling of WEEEs: an economic assessment of present and future e-waste streams. Renewable Sustain Energy Rev 51:263–272
3. Oguchi M, Murakami S, Sakanakura H, Kida A, Kameya T (2011) A preliminary categorization of end-of-life electrical and electronic equipment as secondary metal resources. Waste Manage 31:2150–2160
4. Islam MK, Haque N, Somerville MA (2021) Characterisation and techno-economics of a process to recover value from e-waste materials. In: TMS 2021 150th annual meeting & exhibition supplemental proceedings, pp 995–1006
5. Khaliq A, Rhamdhani MA, Brooks G, Masood S (2014) Metal extraction processes for electronic waste and existing industrial routes: a review and Australian perspective. Resources 3(1):152–179
6. Islam MT, Huda N (2019) E-waste in Australia: generation estimation and untapped material recovery and revenue potential. J Cleaner Prod 237:117787
7. Koshta N, Patra S, Singh SP (2021) Estimation of e-waste at micro level for reverse logistics: a case of Delhi. J Cleaner Prod 314:128063
8. Mairizal AQ, Sembada AY, Tse KM, Rhamdhani MA (2021) Electronic waste generation, economic values, distribution map, and possible recycling system in Indonesia. J Cleaner Prod 293:126096
9. Islam MT, Abdullah AB, Shahir SA, Kalam MA, Masjuki HH, Shumon R, Rashid MH (2016) A public survey on knowledge, awareness, attitude and willingness to pay for WEEE management: case study in Bangladesh. J Cleaner Prod 137:728–740
10. Rahman MA (2017) E-waste management: A study on legal framework and institutional preparedness in Bangladesh. Cost Manage 45:28–35
11. Alam M, Bahauddin K (2015) Electronic waste in Bangladesh: evaluating the situation, legislation and policy and way forward with strategy and approach. Present Environ Sustain Dev 1:81–102
12. Report on assessment of generation of e-waste, its impacts on environment and resource recovery potential in Bangladesh. https://doe.portal.gov.bd/publications/research. Accessed 09 July 2021
13. Hossain S, Sulatan S, Shahnaz F, Akram AB, Nesa M, Happell J (2010) E-waste: Bangladesh situation. Study report by environment and social development organization, Dhaka. Bangladesh
14. UN COMTRADE database 2021. https://COMTRADE.un.org/data/. Accessed 03 June 2021
15. Weibull W (1951) A statistical distribution function of wide applicability. J Appl Mech 18(3):293–297
16. Grameenphone. https://www.grameenphone.com/about/discover-gp/about-grameenphone/our-story. Accessed 12 July 2021
17. Country overview: Bangladesh Mobile industry driving growth and enabling digital inclusion. https://www.gsma.com/mobilefordevelopment/resources/bangladesh-mobile-industry-driving-growth-and-enabling-digital-inclusion/. Accessed 12 July 2021
18. Parajuly K, Habib K, Liu G (2017) Waste electrical and electronic equipment (WEEE) in Denmark: flows, quantities and management. Resour Conserv Recycl 123:85–92
19. Forti V, Baldé K, Kuehr R (2018) E-waste statistics: guidelines on classifications, reporting and indicators. http://collections.unu.edu/eserv/UNU:6477/RZ_EWaste_Guidelines_LoRes.pdf. Accessed 20 July 2021
20. BTRC 2021. http://www.btrc.gov.bd/telco/mobile. Accessed 07 July 2021
21. Statista 2021. https://www.statista.com/statistics/497091/number-of-mobile-cellular-subscriptions-in-bangladesh/. Accessed 07 July 2021
22. LME 2021. https://www.lme.com/en-GB. Accessed 24 July 2021

A Green Process to Acquire a High Purity Rare Earth Elements Leach Liquor from Nd–Fe–B Magnets by Caustic Digestion and Roasting Processes

Rina Kim, Kyeong Woo Chung, Ho-Sung Yoon, Chul-Joo Kim, and Yujin Park

Abstract To recover rare earth elements (REEs) from Nd–Fe–B magnets, a variety of hydrometallurgical processes have been developed, and among them, caustic digestion-acid leaching is the most promising. Through the caustic digestion, Nd and Fe in the magnet alloy could be converted into $Nd(OH)_3$ and Fe_3O_4, respectively, and they were easily recovered in the following acid leaching step. To remove iron from REE leach liquor, it was essential to precipitate iron after the acid leaching, and it consumed an amount of chemical like lime. So, to solve the aforementioned problem, a new green process was developed, lowering iron content in leach liquor; through an oxidative roasting of the digested product prior to the acid leaching, iron dissolution was considerably decreased. The digested powder was roasted at 350–450 °C, and the product was leached in 0.5 M HCl. The final leach liquor contained ca. 17 g/L Nd and ca. 330 mg/L Fe.

Keywords REE magnet · Recycling · Leaching · Caustic digestion · Roasting

Introduction

Rare earth permanent magnets have been widely used for small electronics, electric vehicles, wind turbines, and even for missile systems, because of the property of their high magnetism compared with other typical permanent magnets. The significance of the rare earth magnet is recently much more growing as many countries all over the world, including Korea, have declared to achieve carbon–neutral in 2050. Upon this circumstance, fighting global warming, the usage of the rare earth permanent magnets is being expanded to eco-friendly industries such as electric vehicles and renewable energy (mostly used for wind turbines). In electric vehicles and wind

R. Kim (✉) · K. W. Chung · H.-S. Yoon · C.-J. Kim · Y. Park
Korea Institute of Geoscience and Mineral Resources, Daejeon 34132, Korea
e-mail: rkim@kigam.re.kr

R. Kim · K. W. Chung
University of Science and Technology, Daejeon 34113, Korea

A. Lazou et al. (eds.), *REWAS 2022: Developing Tomorrow's Technical Cycles (Volume I)*, The Minerals, Metals & Materials Series,
https://doi.org/10.1007/978-3-030-92563-5_12

turbines, the rare earth magnet is a core component consisting of electric traction motors in electric vehicles and generators in wind turbines. According to a previous study, electric vehicles contain 1.25 kg of rare earth magnets per car, and 400 kg/MW of magnet is used for wind turbines [1]. Other than that, with expanding ICT industry, the amount of magnets used for hard disk drives is being expected to considerably increase.

With growing the size of those industries, it is expected that the number of waste magnets will be exponentially increased in the next 5–10 years, the time when the used vehicles or wind turbines start to be discarded in earnest. To achieve the carbon–neutral society, it is definitely important to use eco-friendly products, and end-of-life (EOL) products should be properly treated. In addition, it is well-known that EOL products are good resources to secure raw materials for manufacturing high-tech products. In this study, the authors focused on the recycling of rare earth magnet scraps to recover rare earth elements (REEs) contained in the magnet alloy.

Many hydrometallurgical recycling technologies of the rare earth magnet have been investigated and developed. The magnets can be directly leached in acidic or alkaline solution with or without pre-oxidation [2, 3]. In addition, the NaOH digestion-acid leaching process was also investigated in limited studies [4, 5]. However, from those processes, not only REEs were leached out, but also Fe was generally co-leached, so an additional step is essential to remove Fe which consumes lots of resources such as water and alkaline chemicals to control solution pH. Therefore, in this study, the alternative green hydrometallurgical process to treat the rare earth magnet scraps, NaOH digestion-oxidative roasting-acid leaching, was investigated.

Materials and Methods

For the test works, Nd–Fe–B magnet processing scrap was used as raw material, and it was obtained from a company manufacturing permanent magnets in Korea. It was analyzed that the magnet alloy was $Nd_2Fe_{14}B$ by XRD (Fig. 1).

The magnet scrap was first tested for caustic digestion. The caustic digestion was conducted at four different NaOH concentrations of 10, 20, 30, and 40% in a Teflon cell to study the digestion behavior depending on conditions. The digestion was conducted at 120 °C with 10 w/v% pulp density for 5 h. The cell was heated in an oil bath. The digested materials were analyzed by XRD to identify the transformation of the alloy into leachable forms.

As the following step, several selected digested materials were roasted in a muffle furnace under an oxidative condition for 3 h. The three different roasting temperatures of 250, 350, and 450 °C were tested, and the products were also analyzed by XRD.

To figure out the effect of oxidative roasting on the leaching behavior, both the digested only and the digested-roasted materials were used as samples for acid leaching. They were leached in 0.5 M acid solution with 10 w/v% pulp density

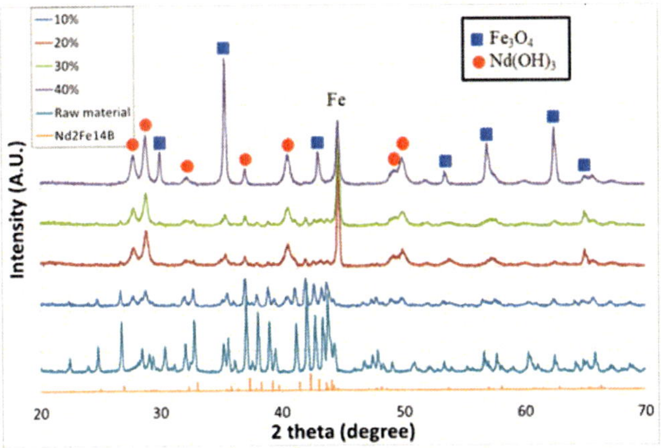

Fig. 1 Phase transformation of $Nd_2Fe_{14}B$ magnet alloy to $Nd(OH)_3$ and Fe_3O_4 at various NaOH concentrations

for 2 min (the digested only) or 20 min (the digested-roasted). The metal ion concentration in the leach liquors was analyzed by ICP-OES.

Results and Discussion

Caustic Digestion-Acid Leaching Process

To decompose NdFeB alloy and convert the alloy into leachable form, the magnet scraps were first digested in an alkaline solution. The scraps were treated at different NaOH concentrations from 10 to 40% to figure out the effect of NaOH concentration on the decomposition behavior. As shown in Fig. 1, the raw scrap material consisted of $Nd_2Fe_{14}B$ alloy, and its peaks from XRD analysis got weaker with increasing NaOH concentration, and $Nd(OH)_3$ and Fe_3O_4 peaks started to be evolved. Those two peaks were stronger at the elevated NaOH concentration. At 40% NaOH, $Nd_2Fe_{14}B$ peaks were completely disappeared. In addition, from 20% of NaOH concentration, some of Fe in the alloy was converted to metallic Fe, and its peak got weaker with increasing NaOH concentration. It can be noted that $Nd(OH)_3$ and Fe_3O_4 were formed through reactions shown in Eqs. 1 and 2.

$$Nd + 3H_2O = Nd(OH)_3 + 1.5H_2 \tag{1}$$

$$3Fe + 4H_2O = Fe_3O_4 + 4H_2 \tag{2}$$

Upon the result in Fig. 1, a higher NaOH concentration of 50% was applied for the caustic digestion process. In Fig. 2, $Nd(OH)_3$ and Fe_3O_4 peaks were more clearly observed from the digested product, and even the metallic Fe was also fully digested as its peak was disappeared.

After the caustic digestion, the digested product was leached under an acidic condition. As shown in Fig. 3, the digested products at different NaOH concentrations of 10, 20, 30, and 40% (sample No. 1, 2, 3, and 4, respectively) were leached by 0.5 M sulfuric acid, and the leaching efficiency of REEs (Nd and Dy) and Fe was calculated. To compare the leaching results of the digested products, the raw material was also

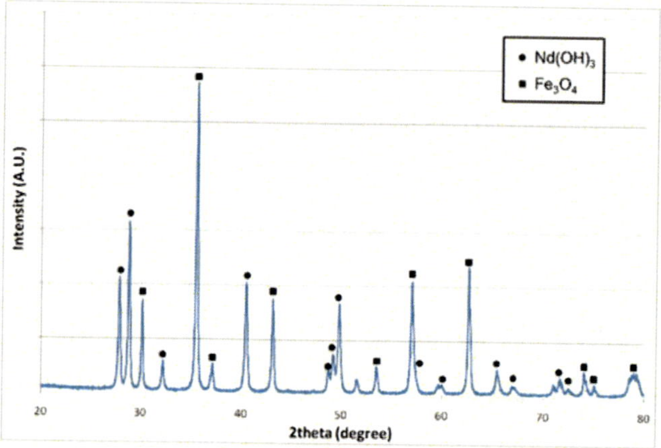

Fig. 2 Completed phase transformation of the magnet alloy to $Nd(OH)_3$ and Fe_3O_4 at 50% NaOH

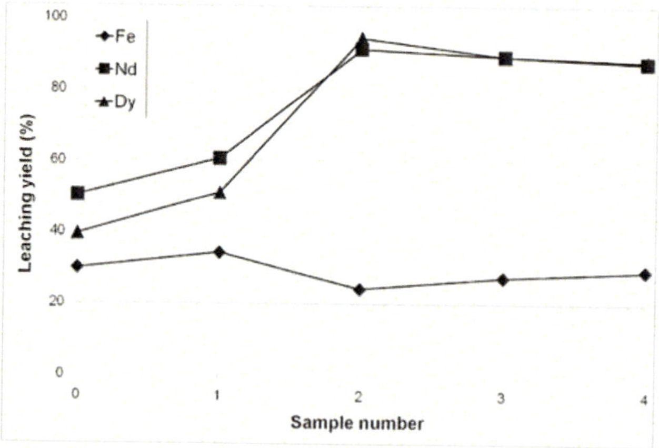

Fig. 3 Acid leaching results of the raw and NaOH digested materials (leaching conditions: 0.5 M H_2SO_4, 25 °C, 10 w/v% pulp density, 2 min; Sample information: 0—raw materials, 1 to 4—digested at 10, 20, 30, and 40% NaOH, respectively)

tested for acid leaching, which is depicted as sample No. 0. From the raw material, less than 50% of REEs were leached out, and the leaching efficiency of Fe was about 30%. At NaOH concentration of higher than 20%, No. 2 to 4, the leaching efficiency of REEs was considerably enhanced to approximately 90%, while that of Fe was in the same range or a little higher. This was because Nd and Fe were converted to hydroxide or oxide form, more leachable forms compared with the magnet alloy. Although the REEs leaching was improved by the caustic digestion, the Fe leaching was still too high. So, to lower the Fe co-leaching with REEs, the oxidative roasting was conducted between the caustic digestion and acid leaching. The detailed results are shown in the next section.

Caustic Digestion-Oxidative Roasting-Acid Leaching Process

In an attempt to decrease Fe concentration in a magnet leach liquor, oxidative roasting of the digested product was conducted. It was expected that the oxidative form of Fe_3O_4, i.e. Fe_2O_3, is less leachable in an acidic solution than Fe_3O_4 is. In Fig. 4, phases present in the digested product were changed depending on the roasting temperature. At 250 °C, Fe_3O_4 was oxidized to γ-Fe_2O_3 (maghemite), while $Nd(OH)_3$ was not changed. On the other hand, at the elevated temperatures of 350 and 450 °C, both α-Fe_2O_3 (hematite) and γ-Fe_2O_3 were formed through the oxidation of Fe_3O_4, and $Nd(OH)_3$ was converted to Nd_2O_3, losing water. At the higher temperature, the γ-Fe_2O_3 peak was weaker, which meant that it was further oxidized to α-Fe_2O_3. It is known that α-Fe_2O_3 starts to be formed at the higher temperature than γ-Fe_2O_3; generally, at the temperature higher than 300 °C, α-Fe_2O_3 can be formed. This Fe_3O_4 oxidation behavior was identical with other previous studies [6–9].

With the materials processed through the caustic digestion and the oxidative roasting, acid leaching using 0.5 M hydrochloric acid was conducted. As shown in Fig. 5, after the oxidation at 350 and 450 °C, 16–17 g/L Nd and 330–360 mg/L

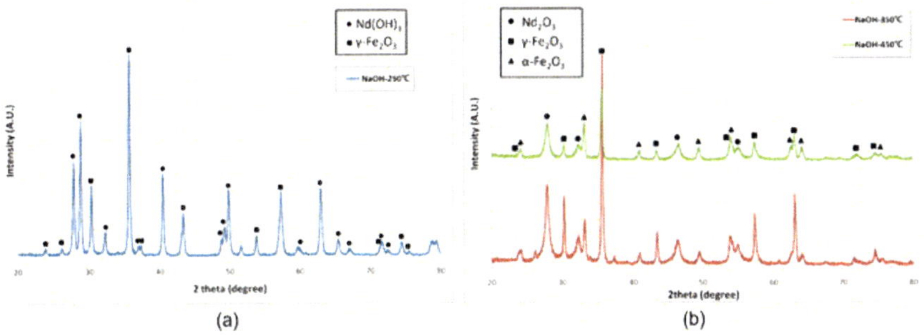

Fig. 4 Phase transformation of the digested magnet alloy after oxidative roasting at different temperatures: **a** 250 °C, **b** 350, and 450 °C

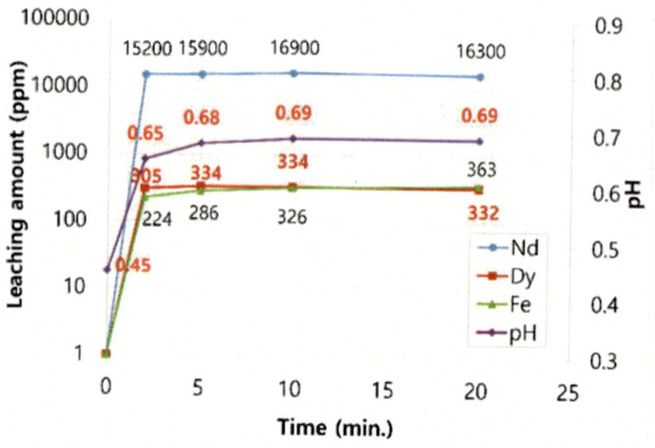

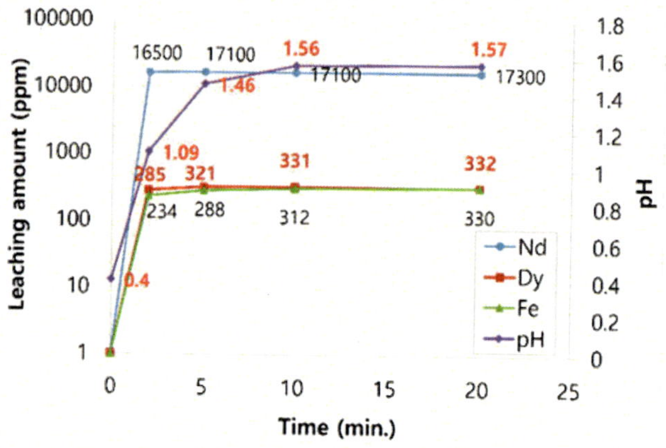

Fig. 5 Acid leaching results of the NaOH digested-oxidative roasted materials: roasted at **a** 350 and **b** 450 °C (leaching conditions: 0.5 M HCl, 10 w/v% pulp density)

Dy were leached out, which are equivalent to higher than 90% of leaching efficiency. Fe concentration in the leach liquor only reached 330 mg/L, and this represents less than 1% leaching efficiency.

By depressing Fe leaching from the magnet scrap, the leach liquor can be directly sent to solvent extraction process to separate the mixed REEs into the individual element. If the leach liquor contains a high concentration of Fe, the solution must be further treated to remove the impurity by precipitation. However, the precipitation is generally difficult to control, as REE loss can be happened during the precipitation due to surface adsorption onto the impurity precipitate or co-precipitation [10, 11]. Thus, with this new process, the whole process to recover REEs can be shortened, and at the same time, the REE loss can be prevented.

Conclusions

REEs can be effectively recovered from Nd–Fe–B magnets through caustic digestion-acid leaching process. However, the typical process has a problem which is producing an REE leach liquor containing a high concentration of Fe. To reduce the amount of Fe in the leach liquor, a new process was suggested in this study: caustic digestion-oxidative roasting-acid leaching. At the caustic digestion step, the Nd–Fe–B magnet alloy was first digested into $Nd(OH)_3$ and Fe_3O_4 in NaOH solution, and 50% NaOH was effective to fully digest the alloy. Then, the digested material was roasted at the temperature higher than 350 °C under the oxidative condition, and the product finally consisted of Nd_2O_3, α-Fe_2O_3 (hematite), and γ-Fe_2O_3 (maghemite). In the following acid leaching step, the REE leach liquor containing high REEs and low Fe was obtained; 17 g/L Nd and 330 mg/L Fe. It can be noted that the new green process was developed in this study, as it consumed less acid compared with the typical direct acid leaching. In addition, the final leach liquor is acceptable for direct solvent extraction to separate REEs into the individual element.

Acknowledgements This work was supported by the Energy Resource Technology Developing Project of the Korea Institute of Energy Technology Evaluation and Planning (KETEP) granted financial resource from the Ministry of Trade, Industry and Energy (MOTIE) of the Korean government (No. 20217510100070/KIGAM NP2021-037).

References

1. Yang Y, Walton A, Sheridan R, Güth K, Gauß R, Gutfleisch O, Buchert M, Steenari B-M, Gerven TV, Jones PT, Binnemans K (2017) REE recovery from end-of-life NdFeB permanent magnet scrap: a critical review. J Sustain Metall 3:122–149
2. Tunsu C, Petranikova M, Gergorić M, Ekberg C, Retegan T (2015) Reclaiming rare earth elements from end-of-life products: A review of the perspectives for urban mining using hydrometallurgical unit operations. Hydrometallurgy 156:239–258
3. Jha MK, Kumari A, Panda R, Kumar JR, Yoo K, Lee JY (2016) Review on hydrometallurgical recovery of rare earth metals. Hydrometallurgy 165:2–26
4. Yoon H-S, Kim C-J, Chung KW, Jeon S, Park I, Yoo K, Jha MK (2015) The effect of grinding and roasting conditions on the selective leaching of Nd and Dy from NdFeB magnet scraps. Minerals 5(3):1306–1314
5. Önal MAR, Riaño S, Binnemans K (2020) Alkali baking and solvometallurgical leaching of NdFeB magnets. Hydrometallurgy 191:105213
6. Colombo U, Gazzarrini F, Lanzavecchia G, Sironi G (1965) Magnetite oxidation: a proposed mechanism. Science 147(3661):1033
7. McCarty KF, Monti M, Nie S, Siegel DA, Starodub E, Gabaly FE, McDaniel AH, Shavorskiy A, Tyliszczak T, Bluhm H, Bartelt NC, Figuera J (2014) Oxidation of magnetite(100) to hematite observed by in situ spectroscopy and microscopy. J Phys Chem C 118:19768–19777
8. Kumar TKS, Viswanathan NN, Ahmed H, Dahlin A, Andersson C, Bjorkman B (2019) Investigation of magnetite oxidation kinetics at the particle scale. Metall Mater Trans B 50B:150–161
9. Li Z, Chaneac C, Berger G, Delaunay S, Graff A, Lefevre G (2019) Mechanism and kinetics of magnetite oxidation under hydrothermal conditions. RSC Adv 9:33633–33642

10. Sadri F, Kim R, Yang Z, Ghahreman A (2018) The effect of calcium sulfate crystallization and the crystal modification on aqueous REE stability in Ca saturated REE-Ca-SO_4-H_2O systems. Hydrometallurgy 182:82–96
11. Sadri F, Kim R, Ghahreman A (2021) Behavior of light and heavy rare-earth elements in a two-step Fe and Al removal process from rare-earth pregnant leach solutions. J Sustain Metall 7:1327–1342

Physicochemistry of Lithium-Ion Battery Recycling Processes

Alexandre Chagnes

Abstract The electric mobility and the energy transition rely on the development of performant energy storage devices such as fuel cells and Lithium-ion batteries. It is expected a huge increase of Lithium-ion battery production in the next years due to the increase of electric vehicles on the market. These batteries will have to be recycled in the next ten years. It is therefore of great importance to develop the recycling sector of Lithium-ion batteries. Among other, the search for efficient, cheap, and environmentally friendly processes for recycling Lithium-ion batteries must be prioritized under the impulsion of governmental regulations. Hydrometallurgy will replace the pyrometallurgical processes in a closed loop recycling strategy to produce metallic salts from spent Lithium-ion batteries that could be reused to manufacture new batteries. This paper gives a brief overview of the key elements for designing appropriate Lithium-ion battery recycling processes.

Keywords Lithium-ion battery · Recycling · Hydrometallurgy

Introduction

The energy transition relies on the development of technologies that make it possible to produce energy in a sustainable manner from resources such as wind, sun, and potential energy. The energy produced as part of the energy transition is intermittent, and it is therefore necessary to be able to store it. Electric mobility is also a major contributor in reducing the impacts of human activity on the environment and the climate since it contributes to reducing greenhouse gas emissions. Lithium-ion batteries (LiBs) are at the heart of energy storage for stationary applications or electric mobility. After a decade of rapid growth, in 2020, the global electric car stock hit the 10 million mark, a 43% increase over 2019, and representing a 1% stock share. Battery electric vehicles accounted for two-thirds of new electric car registrations and two-thirds of the stock in 2020. While only a small number of EV batteries have

A. Chagnes (✉)
Université de Lorraine, CNRS, GeoRessources, Nancy, France
e-mail: alexandre.chagnes@univ-lorraine.fr

A. Lazou et al. (eds.), *REWAS 2022: Developing Tomorrow's Technical Cycles (Volume I)*, The Minerals, Metals & Materials Series,
https://doi.org/10.1007/978-3-030-92563-5_13

aged off the streets already, millions of tons of batteries are expected to be decommissioned over the coming decades. Those batteries could supply a significant fraction of the EV industry's future mineral demand—but better recycling methods and government policies to support them are needed to ensure that batteries do not wind up in landfills instead. Spent Lithium-ion batteries from electric vehicles which cannot be reused for other applications will be treated by appropriate recycling processes to extract valuable metals that could be transformed in metallic salts needed to produce Lithium-ion batteries.

A few companies like Umicore are pioneers in recycling Lithium-ion batteries. With the emergence of electric vehicles, new consortia are appearing to develop the recycling sector, such as the recent partnership between Véolia and Solvay. The recycling of electric vehicle batteries is now based on two routes which are to be developed in parallel: the reuse of spent batteries whose performance remains sufficient for other applications or the complete recycling that involves chemical processes, i.e. pyrometallurgy and hydrometallurgy.

This paper aims to giving a short overview of the different operations involved in extractive metallurgy to recover these metals from spent Lithium-ion batteries with a special attention to hydrometallurgy.

The Lithium-Ion Battery Technology

The Lithium-ion battery relies on the reversible exchange of lithium ion between a positive electrode, most often a lithiated transition metal oxide denoted LiMeO, and a negative graphite electrode as illustrated in Fig. 1 and according to the following reactions:

$$\mathrm{LiMeO} = \mathrm{Li}_{1-x}\mathrm{MeO} + x\,\mathrm{Li}^{+} + x\,\mathrm{e}^{-} \quad (1)$$

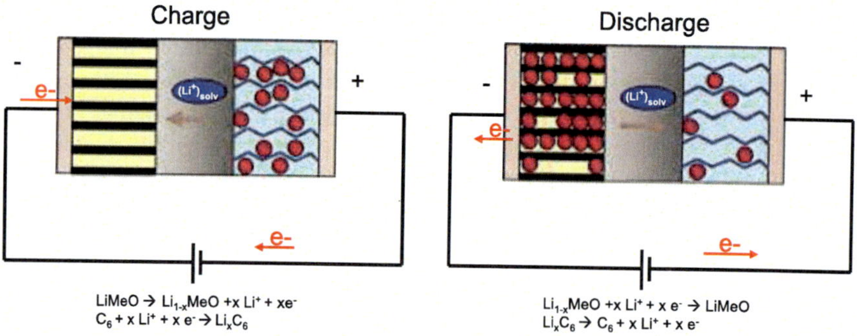

Fig. 1 Principle of Lithium-ion battery [4]

$$C_6 + x\,Li^+ + x\,e^- = LiC_6 \quad (2)$$

In these equations, the charge corresponds to the reaction going from the left to the right, and the discharge corresponds to the reverse reaction going from the right to the left.

Lithium-ion batteries are composed of lithium salt usually at 1 mol L^{-1} (usually $LiPF_6$, $LiBF_4$, or LiTFSI) dissolved in a mixture of aprotic dipolar organic solvents (usually alkyl carbonates) [1]. The electrolyte is soaked into a porous polypropylene–polyethylene separator which separates the positive electrode and the negative electrode [2]. In the future, the liquid electrolytes will most likely be replaced by solid electrolytes such as garnet, Nasicon, Perovskite–structured electrolytes, and polymeric electrolytes to produce safer batteries [3]. Less industrial development was devoted to the negative electrode than for the positive electrode.

The negative electrode has not changed since the first Lithium-ion battery manufactured by Sony in 1991. It is still graphite even many research works were performed to find alternative electrode exhibiting more capacity such as silicon. Much research works were focused on the development of positive electrodes like $LiCoO_2$, $LiNiO_2$, and $LiMn_2O_4$. Today, polymetallic electrodes such as NMC electrodes which are lithiated cobalt, nickel, and manganese oxides of various stoichiometries are the most used, especially for laptops but also for electric vehicles.[1]

Cobalt, nickel, manganese, and lithium are therefore key elements for the production of Lithium-ion batteries today. They are also essential for the development of the electric vehicle since Lithium-ion batteries are now considered one of the most suitable technologies for electric mobility, even if other technologies appear to be potentially interesting such as fuel cells. Given the large market for electric vehicles, we can therefore expect a significant growth in demand for these metals, but also a surge in the price and an increase in tensions. Demand is thus expected to increase by thirty-one for cobalt, sixty-nine for nickel, and forty-six for lithium by 2030 compared to 2017 according to Avicenne [5].

The supply of cobalt, nickel, manganese, and lithium (but also graphite and fluorine, which are the other key elements for the production of Lithium-ion batteries) requires the development of efficient, economical, and sustainable processes capable of recovering these metals contained in current resources or those that will be exploited in the near future, including spent Lithium-ion batteries.

The Hydrometallurgical Route to Recycle Lithium-Ion Batteries

Lithium-ion batteries are complex because of their assembly and because they must be handled with care due to the presence of toxic and flammable compounds in

[1] A. Chagnes, J. Swiatowska (Editeurs), "Lithium Process Chemistry: Resources, Extractions, Batteries and Recycling", Elsevier, 2015, 313 pages (ISBN: 978-0-12-801417-2).

Lithium-ion batteries. The development of recycling processes is also tricky because of the diversity of existing technologies and the rapid evolution of these technologies over time. It is therefore necessary to think about developing flexible recycling processes. There are several hundreds of strategies reported in the literature based on different chemistries to recover the metal species contained in spent Lithium-ion batteries. Most of these works focus on the recovery of the metals contained in the positive electrodes because cathodic materials represent a significant part of the battery cost. However, recent works started focusing on the recovery of graphite because it is one of the chemical elements identified as strategic by the European Commission due to the Chinese monopoly in spite of efforts to start new graphite mining in the world. Likewise, the recovery of lithium (and of the PF_6^- anion) contained in the electrolyte is starting to be the subject of research because lithium is also a key element for the development of the Lithium-ion batteries technology (pyrometallurgical processes implemented today at the industrial scale for recycling Lithium-ion batteries do not recover lithium). The following figure gives an idea of the different stages generally implemented for the recycling of spent Lithium-ion batteries by an hydrometallurgical process (Fig. 2).

For example, the Umicore process was based at its beginnings on a pyrometallurgical steps. The spent Lithium-ion batteries were sent into furnaces at 1500 °C to produce an alloy containing copper, cobalt, and nickel, while lithium and aluminum (from the current collector of the positive electrode) were lost in the slags. Selective leaching and then liquid–liquid extraction steps were implemented to separate and extract the cobalt and nickel from the positive electrode and the copper from the current collector of the negative electrode. Today, hydrometallurgical processes (with or without thermal pretreatment) are favored in order to be able to finely separate the different metallic species and enhance them. The electrolyte was lost in the Umicore process or more generally in pyrometallurgical routes.

With the new European regulations regarding Lithium-ion batteries recycling, it will be mandatory to recycle at least 80% of the weight of the batteries. Likewise, the European commission expect that 95% of cobalt, copper, and nickel as well as 70% of lithium will be recovered from spent Lithium-ion batteries by the recycling

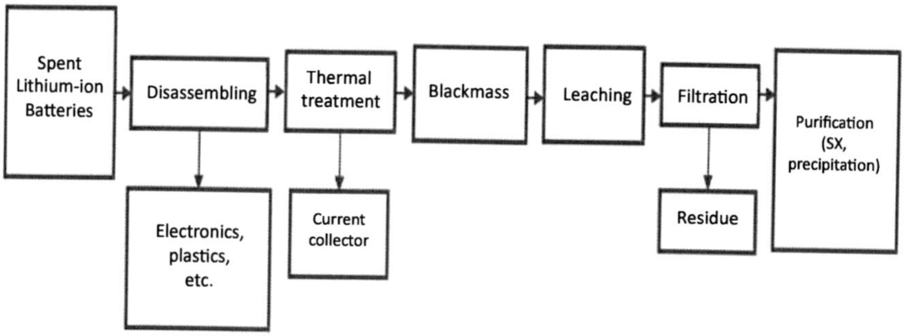

Fig. 2 General scheme for Lithium-ion battery recycling

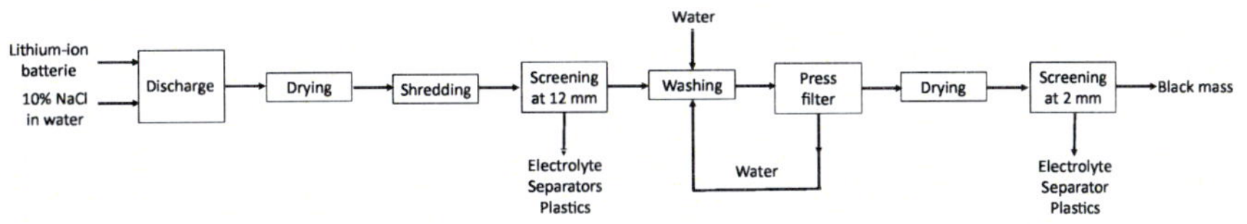

Fig. 3 General flowsheet of typical pretreatment of spent Lithium-ion battery before the hydrometallurgical stage [7]

industry by the end of 2030. The European commission also expects that 12% of the cobalt, 4% of the lithium, and 4% of the nickel needed to make a Lithium-ion battery come from recycling by 2030 [6]. Only hydrometallurgical processes can allow reaching these goals. However, a pretreatment must be implemented before the hydrometallurgical treatment to produce the blackmass, which could be processed by hydrometallurgy [7] (Fig. 3).

The battery is first discharged in a conductive sodium chloride solution for 6 h to reduce the danger of the battery. The plastics from the separator and the coil frame are removed by sieving. After this step, about 5% of plastic remains in the flow to be treated and must therefore be removed subsequently by a finer sieving step (2 mm against 12 mm). The separation of the current collector (copper for the negative electrode and aluminum for the positive electrode) may require the use of organic solvents to dissolve the binder or using an ultrasound bath. According to Jinhui et al. (2009), all of these operations lead to the production of a powder containing 0.3% copper, 0.8% aluminum, and 1.4% iron (mass percentages). This pretreatment would thus make it possible to reduce the quantities of copper, aluminum, and iron by 97, 73, and 93%, while 8% of the powdery material containing lithium, nickel, cobalt, and manganese is lost.

The resulting concentrate must then undergo a leaching step to dissolve the metals and then extraction-separation to selectively recover the transition metals and lithium. The extracted metals can then be converted into salts by appropriate crystallization/precipitation stages. Different inorganic and organic acids were tested in the literature under various experimental conditions to leach the cathodic materials [8]. For the sake of illustrations, several reactions corresponding to $LiCoO_2$ leaching by hydrochlorhydric acid, sulfuric acid, and nitric acid are reported below:

$$2LiCoO_2 + 6HCl + H_2O_2 \rightarrow 2CoCl_2 + 2LiCl + O_2 + 4H_2O \tag{3}$$

$$LiCoO_2 + 1.5H_2SO_4 + 1.5H_2O_2 + CoSO_4 + 0.5Li_2SO_4 + O_2 + 3H_2O \tag{4}$$

$$2LiCoO_2 + 6HNO_3 + H_2O_2 \rightarrow 2C(NO_3)_2 + 2LiNO_3 + O_2 + 4H_2O \tag{5}$$

More recent studies concern the investigation of NMC leaching mechanisms in sulfuric acid or hydrochloric acid [9–11]. These works showed that NMC dissolution is always a two-step process. In the first step, NMC is transformed very quickly into

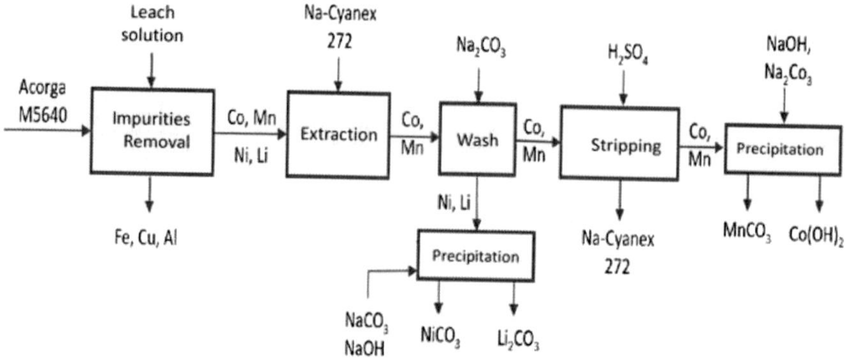

Fig. 4 Example of flowsheet to recover metals from spent Lithium-ion batteries [12–14]

a new lithium-deficient phase, and this phase is subsequently dissolved during a second-rate limiting step.

Sulfuric acid in the presence of 1–10% (vol.) of hydrogen peroxide is more efficient at high temperatures, i.e. 50–60 °C (H_2O_2 is necessary to increase the cobalt solubility by reducing cobalt(III) into cobalt(II)). This acid is efficient both for dissolving $LiCoO_2$ and NMC electrodes. Figure 4 shows an example among others that can potentially be implemented to recover lithium, cobalt, nickel, and manganese from spent Lithium-ion batteries containing NMC technology. In order to remove the graphite coming from the negative electrode, it is possible to set up a heat treatment of the concentrate before the leaching step.

The iron, copper, and aluminum contained at low concentrations in the leach solution are first removed by liquid–liquid extraction using Acorga®M5640 diluted in kerosene at pH between 0 and 2. The raffinate then contains cobalt(II), nickel(II), manganese(II), and lithium(I) at concentrations equal to 32.7, 0.07, 18.6, and 2.26 gL^{-1}. The liquid–liquid extraction step is then implemented to extract the cobalt(II), nickel(II), manganese(II), and lithium(I) from the leach solution by means of saponified Cyanex®272 diluted in kerosene.

The extraction of manganese(II), cobalt(II), and nickel(II) by saponified Cyanex®272 takes place, respectively, at pH values of 3.5, 5, and 8, while the extraction of lithium(I) varies very little with the pH and remains relatively low. The cobalt(II), nickel(II), and manganese(II) extraction yields are 89.3%, 95.6%, and 91.2%, respectively. After the liquid–liquid extraction step, a 0.1 mol L^{-1} Na_2CO_3 solution is used to remove the nickel(II) and lithium(I) slightly co-extracted in the organic phase. Nickel(II) and lithium(I) contained in the sodium carbonate solution are then precipitated at pH 9 and 11–12, respectively. A back-extraction of the organic phase loaded with cobalt(II) and manganese(II) after the extraction step with Acorga®M5640 is carried out using sulfuric acid at 0.1 mol L^{-1}. The manganese(II) and cobalt(II) are separated by precipitation by adjusting the pH of the solution to 7.5 to recover manganese(II) and to pH 11 to recover cobalt(II). This process allows to produce salts with a purity greater than 99%.

Conclusion

Lithium-ion batteries are complex objects to fully recycle as they contain cathode, electrode, electrolyte, and polymeric separators which the technologies change throughout the time. Furthermore, safety issues can occur during recycling, which increase the cost of recycling processes. Next generations of recycling processes will have to recover all components contained in Lithium-ion batteries and generate nearly zero-waste. The main challenge of Lithium-ion battery recycling is therefore to find a cheap process (given that LiBs prices will continue decreasing in the next decades), which is environmentally friendly and sufficiently flexible to face up technological changes. The literature review shows that technical solutions exist to recycle batteries, but the recycling cost and the flexibility of these processes must be improved. After blackmass production by physical concentration, the hydrometallurgical route appears to be the best route to recycle Lithium-ion batteries. Today, the sulfate route is the most developed at pilot scale, whereas other chemistries could find their place in recycling processes such as the use of organic acid, i.e. citric acid, and other inorganic acid such as hydrochloric acid in which different selectivity could be achieved.

References

1. Chagnes A (2015) Lithium battery technologies: electrolytes. In: Chagnes A, Swiatowska J (eds) Lithium process chemistry: resources, extractions, batteries and recycling, 1st edn. Elsevier, Amsterdam, pp 167–189
2. Chagnes A (2012) La technologie lithium-ion. La Revue 3EI 69:69–74
3. Zheng F, Kotobuki M, Song MOL, Lu L (2018) Review on solid electrolytes for all-solid-state Lithium-ion batteries. J Power Sources 389:198–213
4. Chagnes A, Swiatowska J (eds) (2015) Lithium process chemistry: resources, extractions, batteries and recycling. Elsevier, Amsterdam
5. Pillot C (2018) The rechargeable battery market and main trends 2017–2025. Avicenne Energy 2018. www.avicenne.com
6. New EU regulatory framework for batteries Setting sustainability requirements. https://www.europarl.europa.eu/RegData/etudes/BRIE/2021/689337/EPRS_BRI(2021)689337_EN.pdf. Accessed 20 Oct 2021
7. Jinhui LX, Hu Q, Wang Z, Zheng J, Wu L, Zhang L (2009) Study of extraction and purification of Ni, Co and Mn from spent battery material. Hydrometallurgy 99(1–2):7–12
8. Chagnes A, Pospiech B (2013) A brief review on hydrometallurgical technologies for recycling spent lithium-ion batteries. J Chem Technol Biotechnol 88:1191–1199
9. Xuan W, Otsuki A, Chagnes A (2019) Investigation of leaching mechanism of NMC 811 ($LiNi_{0.8}Mn_{0.1}Co_{0.1}O_2$) by hydrochloric acid for recycling cathodes in lithium-ion batteries. RSC Adv 9:38612–38618
10. Xuan W, De Souza Braga A, Korbel C, Chagnes A (2021) New insights in the leaching kinetics of cathodic materials in acidic chloride media for Lithium-ion battery recycling. Hydrometallurgy 204:105705
11. Chan KH, Anawati J, Malik M, Azimi G (2021) Closed-loop recycling of lithium, cobalt, nickel, and manganese from waste lithium-ion batteries of electric vehicles. ACS Sustain Chem Eng 9(12):4398–4410

12. Nayl AA (2010) Extraction and separation of Co(II) and Ni(II) from Acidic sulfate solutions using Aliquat 336. J Hazard Mater 173(1–3):223–230
13. Nayl AA, Ahmed IM, Aly MI (2014) Liquid-liquid extraction and separation of divalent manganese and zinc by Na-CYANEX 272 from sulfate solution. Sep Sci Technol 49(2):290–297
14. Nayl AA, Hamed Mostafa M, Rizk SE (2015) Selective extraction and separation of metal values from leach liquor of mixed spent Li-ion batteries. J Taiwan Inst Chem Eng 55:119–125

Shifting the Burden of Selectivity from Chemical to Physical Separation Processes via Selective Sulfidation

Caspar Stinn and Antoine Allanore

Abstract Increasing demand for critical metallic elements for sustainability applications motivates new approaches in primary and secondary production to handle falling ore grades and increasingly convoluted recycling streams. Separation of elements in distinct phases is generally less energy intensive than separation of elements substituted within a single phase, a phenomenon referred to in primary extraction as the "mineralogical barrier". Engineered materials leverage element substitution within single phase solutions to achieve target material performance. This results in large energy requirements during end-of-life recycling to selectively recover, via chemical separation, the target elements contained within a single phase. Herein, we present selective sulfidation as a novel, pyrometallurgical pretreatment to selectively partition target elements from a single phase into distinct, separate phases. We find such approach may support more competitive physical separation of difficult to isolate elements that previously required separation via complete hydrometallurgical dissolution and aqueous-organic liquid–liquid solvent extraction. We demonstrate selective sulfidation as applied to end-of-life magnet, battery, and copper slag recycling as a means to shift the burden of selective separation from chemical to physical processes.

Keywords Sulfidation · Recycling · Lithium-ion batteries · Rare-earth magnets · Slag · Copper smelting · Physical separation · Pyrometallurgy · Mineralogical barrier

Introduction

The electrification of sectors ranging from transportation [1] to heavy industry [2] stimulates growing demand for the materials needed to establish infrastructure based on renewable technologies, spanning from base elements such as copper and nickel

---

C. Stinn · A. Allanore (✉)
Department of Materials Science and Engineering, Massachusetts Institute of Technology, 77 Massachusetts Avenue, Cambridge, MA 02139, USA
e-mail: allanore@mit.edu

A. Lazou et al. (eds.), *REWAS 2022: Developing Tomorrow's Technical Cycles (Volume I)*, The Minerals, Metals & Materials Series,
https://doi.org/10.1007/978-3-030-92563-5_14

[3] to specialty elements such as the rare earths [4]. Meeting growing demand for these critical metallic elements motivates strategies for expanding both primary and secondary production [5, 6]. Addressing growing demand in primary production requires designing processes to handle lower mineral and concentrate grades [7], while continuing to tackle the high emissions and energy usage characteristic of both conventional pyrometallurgy and hydrometallurgy [8]. Meanwhile, recycling of materials from secondary feedstocks requires the reprocessing of engineered materials designed with mixed-metal compounds and solid solutions [9–11]. In practice, primary and secondary productions are often intertwined due to the role primary smelters play in recycling both base and rare elements [12, 13], and the similar thermodynamic and energetic challenges faced in optimizing physical and chemical separations for recovery of elements from complicated, mixed-element feedstocks [12–14].

In primary production, metallic element sources can be grouped into two categories: those that exist as the dominant cationic species in a single mineral or phase, and those that exist as minor components in phases or minerals via atomic substitution. This distinction between physical and chemical mixtures of natural minerals is termed the "mineralogical barrier" [15]. Across the mineralogical barrier, energy requirements and costs are higher for chemical separation of elements versus physical separation [16]. Secondary sources of materials likewise exhibit their own "mineralogical barrier" between systems where target elements exist as physically separable entities—such as cathode separation from casing materials in batteries—and systems where target elements are engineered to be in solid solution or mixed-metal compounds. The latter requires chemical separations—such as cobalt separation from manganese in nickel–manganese–cobalt (NMC) oxide battery cathode chemistries [17]. While study of the mineralogical barrier does not replace detailed life cycle assessment with well-defined system boundaries tailored to individual processes or materials, it serves as a useful generalization of energy use trends in materials extraction.

Theoretical and practical mineralogical barriers are compared in Fig. 1 as a function of product grade in the material feedstock, with the grades of critical elements from recycled magnet, battery, and slag sources noted [18–20]. The theoretical mineralogical barrier for a target element grade in a material feedstock may be determined by the difference in the minimum work required to chemically extract that component from a single-phase mixture, derived from the ideal Gibbs energy of mixing at a temperature of 25 °C, and the minimum work required for physical liberation. This can be assessed for one spherical grouping of particles from the bulk material, derived using crack resistance energy [21] and the King liberation model [22, 23]. Meanwhile, a practical mineralogical barrier may be determined by comparing the actual energy input for comminution and physical separations in mineral concentrate production to that of leaching and solvent extraction in chemical separations. For physical separation processes, energy for grinding using high-intensity stirred mills and physical separation via froth flotation as a function of liberated particle size is sometimes reported [24]. Here, it is taken from zinc sulfide concentrate production

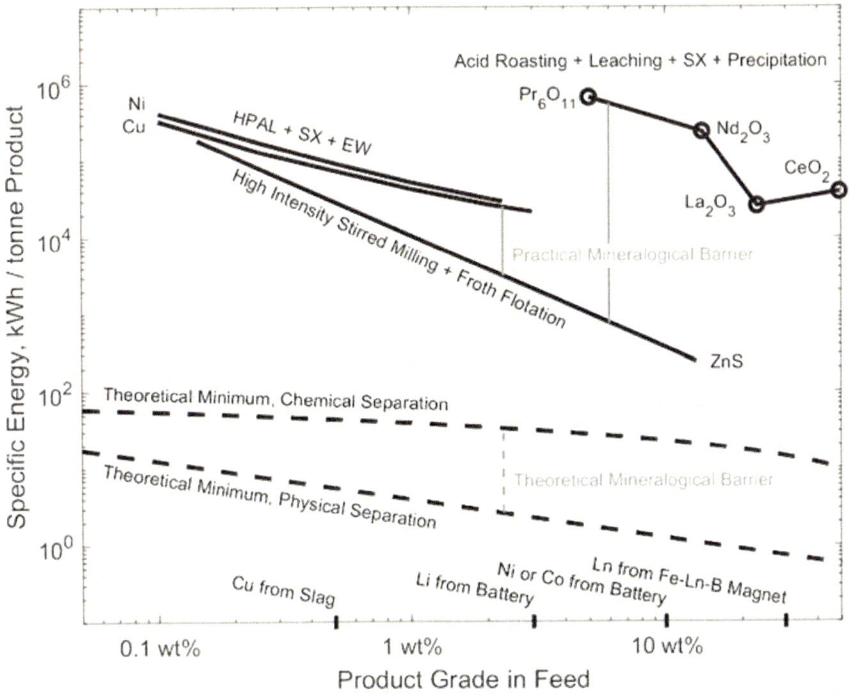

Fig. 1 Calculated theoretical and practical mineralogical barriers suggests that physical separations are generally less energy intensive than chemical separations at sufficiently high target element grades in feedstocks. If the burden of selectivity can be shifted from chemical to physical separation processes via a suitable, energy efficient pretreatment, significant energy savings in materials separation are possible at grades relevant for recovery of critical elements from recycled magnet, battery, and slag sources (HPAL: high pressure acid leaching. SX: solvent extraction. EW: electrowinning. Ln: lanthanide)

data [25], chosen as a model system due to the wide range of liberation sizes practiced in the industry. Energy requirements for high-pressure acid leaching (HPAL), solvent extraction (SX), and electrowinning (EW) for copper and nickel [8] are used as model systems for simple hydrometallurgical chemical separations. They provide a fair proxy due to the limited number of solvent extraction stages required for separation [12, 13]. Energy requirements for rare-earth element acid roasting, leaching, solvent extraction (SX), and compound precipitation are used as a model system for chemical separations requiring numerous hydrometallurgical separation stages [14].

The practical mineralogical barrier between physical and simple chemical separations is observed to decrease in relative magnitude with decreasing material grade. At a grade of approximately 0.1 wt%, the difference in energy requirements for material separation via physical and simple hydrometallurgical methods becomes negligible. This correlation is consistent with the conventional wisdom that when minerals become too fine-grained or low grade, leaching and hydrometallurgical processing are necessary for product recovery due to decreasing comminution efficiency at decreasing liberation sizes. Meanwhile, with increasing material grade, the energy reduction of physical separations over chemical separations is readily apparent. For example, at feedstock grades of 0.5 wt%, 3 wt%, 10 wt%, and 30 wt%, corresponding to copper recycling from slag, lithium recycling from lithium-ion batteries, nickel or cobalt recycling from lithium-ion batteries, and lanthanide recycling from iron–lanthanide–boron magnets, respectively, the practical barrier is approximately 45,000 kWh, 19,000 kWh, 11,000 kWh, and 6,000 kWh, respectively, per tonne of product, or 225 kWh, 570 kWh, 1100 kWh, and 1800 kWh, respectively, per tonne of feed. For pretreatment processes that enable physical separation, this can be viewed as an energy budget, effectively a budget for conversion cost since energy use remains one of the largest contributors to operating cost in minerals processing [26]. For reference, the energy burden of copper slag cleaning in a flash furnace is approximately 100 kWh per tonne of feed [13], and nickel smelting in an electric arc furnace is approximately 500 kWh per tonne of feed [12]. Therefore, for high-grade feedstocks where target elements are trapped in solid solutions, such as many secondary sources of critical elements, a simple pyrometallurgical pretreatment is attractive. It must facilitate cracking of solid solutions into physically separable phases, so that the burden of selectivity for the product elements is shifted from chemical to physical separation processes.

Pyrometallurgical roasting processes based on sulfide chemistry are promising pretreatments for physical separation [27]. They offer to selectively partition target metallic elements from mixed oxide phases into sulfide phases [27–33], allowing for liberation and physical separation based on the numerous differences in the physical properties between oxides and sulfides [13]. Sulfidation pretreatments for primary copper and nickel processing have been successfully performed at industrial scales [12, 34] and are predicted to require low energy inputs due to the exothermic nature of oxide sulfidation reactions. Herein, we present a thermodynamic framework for predicting selectivity in sulfidation. We then demonstrate selective sulfidation as a pretreatment to promote metallic element recovery from rare-earth magnets, lithium-ion batteries, and copper slag, as an avenue to shift the energy burden of selectivity in metal recovery from chemical to physical separation methods.

Thermodynamic Framework for Selective Sulfidation

Formation of a sulfide from an oxide, oxysulfide, or sulfate may be described by the following pyrometallurgical anion-exchange reaction, where M is a metallic element and ω, ϕ, υ, τ, and σ define the reaction stoichiometry:

$$\frac{4\tau}{2\omega\sigma + \phi\tau - 2\upsilon\tau} M_\omega O_\phi S_\upsilon + S_2 = \frac{4\omega}{2\omega\sigma + \phi\tau - 2\upsilon\tau} M_\tau S_\sigma + \frac{2\phi\tau}{2\omega\sigma + \phi\tau - 2\upsilon\tau} SO_2 \quad (1)$$

Elemental sulfur is chosen as the gaseous sulfidizing agent due to its low cost [35] and the fact that the other commonly used gaseous sulfidizing agents, hydrogen sulfide and carbon disulfide, are known in the catalyst industry to be non-selective in their sulfidation of oxides [36]. At thermodynamic equilibrium, the activities of the oxide, oxysulfide, or sulfate reactant ($a_{M_\omega O_\phi S_\upsilon}$) and the sulfide product ($a_{M_\tau S_\sigma}$) are related to a stoichiometry-dependent ratio (ψ) of the sulfur and sulfur dioxide partial pressures (P_{S_2} and P_{SO_2}, respectively) via the standard Gibbs energy of reaction ($\Delta_r G^\circ$), gas constant (R), and absolute temperature (T) [27]:

$$\log_{10}\left(e^{\frac{\Delta_r G^\circ}{RT}}\right) + \log_{10}\left(\frac{a_{M_\tau S_\sigma}^{\frac{4\omega}{2\omega\sigma+\phi\tau-2\upsilon\tau}}}{a_{M_\omega O_\phi S_\upsilon}^{\frac{4\tau}{2\omega\sigma+\phi\tau-2\upsilon\tau}}}\right) = \log_{10}\left(\frac{P_{S_2}}{P_{S_2}^{\frac{2\phi\tau}{2\omega\sigma+\phi\tau-2\upsilon\tau}}}\right) = \psi \tag{2}$$

In turn, ψ describes contributions from both pure compound behavior (ψ_{rxn}) and solution behavior (ψ_{sol}) [27], where activities raised to their stoichiometric power are abbreviated by $A_{M_\omega O_\phi S_\upsilon}$ and $A_{M_\tau S_\sigma}$.

$$\log_{10}\left(e^{\frac{\Delta_r G^\circ}{RT}}\right) = \psi_{rxn} \tag{3}$$

$$\log_{10}\left(\frac{a_{M_\tau S_\sigma}^{\frac{4\omega}{2\omega\sigma+\phi\tau-2\upsilon\tau}}}{a_{M_\omega O_\phi S_\upsilon}^{\frac{4\tau}{2\omega\sigma+\phi\tau-2\upsilon\tau}}}\right) = \log_{10}\left(\frac{A_{M_\tau S_\sigma}}{A_{M_\omega O_\phi S_\upsilon}}\right) = \psi_{sol} \tag{4}$$

Equations 1–4 can be compiled for all metal elements, assuming the data for and identity of $M_\omega O_\phi S_\upsilon$ and $M_\tau S_\sigma$ are known. ψ_{rxn} for chemistries studied herein, calculated using FactSage 8.0 supplemented with literature data [37], and the relative scale of ψ_{sol}, are depicted in Fig. 2. ψ, ψ_{rxn}, and ψ_{sol} may be related to stoichiometric-independent ratios of the sulfur and sulfur dioxide partial pressures via the Gibbs phase rule by employing the formalism of Pourbaix [38] or Kellogg [39]. Selective sulfidation of target elements from a single phase can lead to the precipitation of distinct, physically separable sulfide phases ($M_\tau S_\sigma$) from the surrounding matrix, as a result of the natural immiscibility of oxides, oxysulfides, sulfates, and sulfides [27, 40–42]. When differences in ψ_{rxn} outweigh differences in ψ_{sol} between target elements, selectivity in sulfidation is well-captured by the behavior of the pure compounds [27]. When differences in ψ_{rxn} are comparable to or outweighed by differences in ψ_{sol}, knowledge of solution behavior is required to describe sulfidation selectivity. Unsurprisingly, the relative ease of sulfidation of oxides roughly follows the Goldschmidt geochemical classification of elements [43], with the oxides of chalcophile and siderophile elements generally sulfidizing at lower ψ_{rxn} than the oxides of lithophiles. A notable exception are the alkali and heavy alkaline earth oxides of calcium, strontium, and barium, which exhibit more moderate ψ_{rxn} than other lithophiles due to their natural propensity for sulfate formation [44], occurring at ψ_{rxn} intermediate to oxide or sulfide stabilization.

For many challenging materials separations, elements requiring energy-intensive chemical separation exhibit stark differences in ψ_{rxn}, such as iron-neodymium in

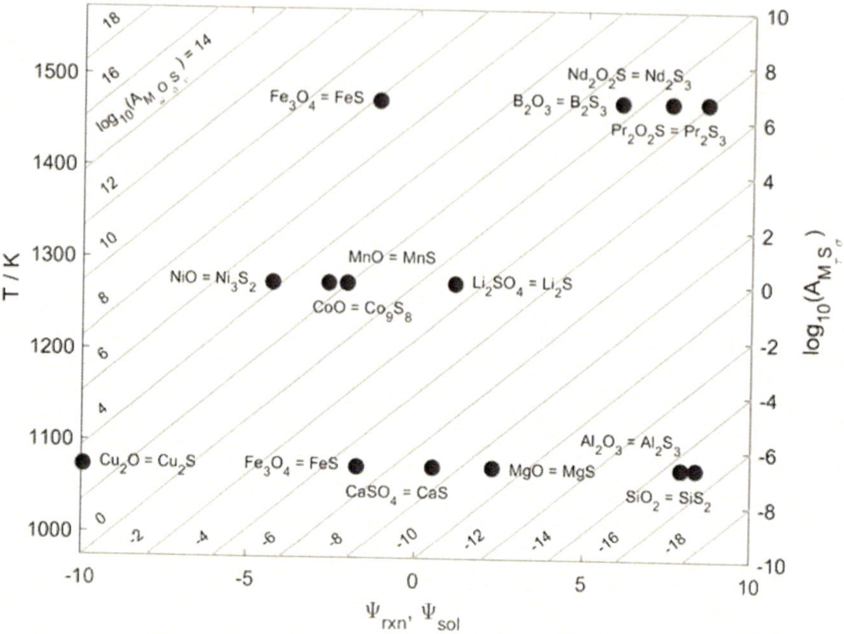

Fig. 2 Sulfidation series for relevant chemistries considered herein, with the relative contributions from pure compound (ψ_{rxn}) and solution effects (ψ_{sol}) to the effective sulfur to sulfur dioxide ratio required for sulfidation to occur (ψ), as described in Eqs. 1–4

magnet recycling, nickel–cobalt in battery recycling, and copper–silicon in slag recycling (Fig. 2). Therefore, sulfide chemistry is a promising approach to shift the burden of selectivity in separation from expensive chemical methods to less energy-intensive physical methods. In the following sections, we demonstrate selective sulfidation as applied to rare-earth magnet recycling, lithium-ion battery recycling, and metal recovery from copper slags to overcome the practical mineralogical barrier (Fig. 1) to chemical and dilute physical separations for recycling of high-grade elements from engineered materials.

Selective Sulfidation for Rare-Earth Magnet Recycling

Rare earth element magnets based on iron-lanthanide-boron chemistry are essential for advanced sustainability and electronic applications, ranging from renewable power generation to electric vehicles [4]. The supply of critical and strategically significant lanthanide elements remains problematic due to environmentally unsustainable production [45], a lack of geographic diversity in processing infrastructure that is further confounded by illegal mining operations [46], and the rare-earth element balance problem [6]—a mismatched supply and demand of different lanthanide coproducts with respect to their natural abundances. Recycling of high-demand lanthanides, such as neodymium, can lower the environmental impact of processing, decentralize rare-earth element production, and selectively supplement the supply of disproportionately demanded lanthanide elements [47]. As shown in Fig. 1, end-of-life rare-earth magnets contains approximately 30 wt% lanthanides, with a neodymium grade higher than that of natural concentrate. Due to the similarity in electronic structure between lanthanides and the resulting difficulty in separation [14], the mineralogical barrier between chemical and physical separation processes is higher than that of base metals, with the energy burden of conventional chemical separation predicted to be on the order of 100× larger than an equivalent physical separation process at a lanthanide grade of 30 wt%. Numerous hydrometallurgical [18], pyrometallurgical [48], and liquid metal [49] processes have been explored for the chemical separation of elements from rare-earth magnet materials, yet current recycling processes of both end-of-life magnets and production waste is complicated by the presence of oxide impurities [9, 48]. Selective anion exchange pretreatments utilizing boron chemistry have been shown to facilitate physical separation of rare-earth elements from one another for primary mineral processing [50], but such an approach has never been attempted with sulfide chemistry until recently [27]. As illustrated by ψ_{rxn} in Fig. 2, at a sulfur to sulfur dioxide ratio on the order of 10:1, iron, rare-earth elements, and boron are predicted to be stable in distinct sulfide, oxysulfide, and oxide phases, respectively. Indeed, oxide–sulfide anion exchange chemistry is a thermodynamically promising pretreatment for facilitating selective recovery of lanthanides from rare-earth magnet waste via physical separation.

To demonstrate selective sulfidation as a pretreatment to separate iron from lanthanides in rare-earth magnet waste, nickel-plated iron–lanthanide-boron magnets (Ni-plated Fe-Nd-Pr-Dy-B, Grade N45, 6.25 mm × 6.25 mm × 6.25 mm, McMaster-Carr) were heated under air at 500 °C to demagnetize, crushed to a particle size of 90–212 μm, then calcined under air at 1000 °C for 5 h in a boron nitride crucible [27]. The brittle iron-lanthanide-boron magnet was separated from the ductile nickel coating during comminution. The calcined rare-earth oxide was reground to a particle size of 90–212 μm, then sulfidized at a scale of 2 g at 1200 °C for 1 h using vaporized elemental sulfur (S_x, 99.5% purity, Acros Organics) as a sulfur source, at a sulfur partial pressure of approximately 0.1 atm. The sulfur to sulfur dioxide ratio was set at approximately 10:1 following methods and reactor design described previously [27].

Figure 3 illustrates optical (dark field) microscopy and SEM/EDS (SEM: JEOL JSM-6610LV, JEOL Ltd., EDS: Sirius SD detector, SGX Sensortech Ltd.) element maps detailing oxygen, sulfur, iron, neodymium, and praseodymium distribution in the sulfidized product. As shown in Fig. 3, iron partitioned to a sulfide phase, whereas neodymium and praseodymium remained as oxides (Fig. 3), with product phases on the order of 20–100 μm in size and large enough for effective liberation [22, 23] and physical separation using standard industrial mineral processing methods [27, 51]. Boron, not shown in Fig. 3, was also observed to partitioned to the oxide phase. The suppression of thermodynamically predicted rare-earth oxysulfide formation may be due to sluggish sulfidation kinetics for rare-earth oxides [27], or due to solution

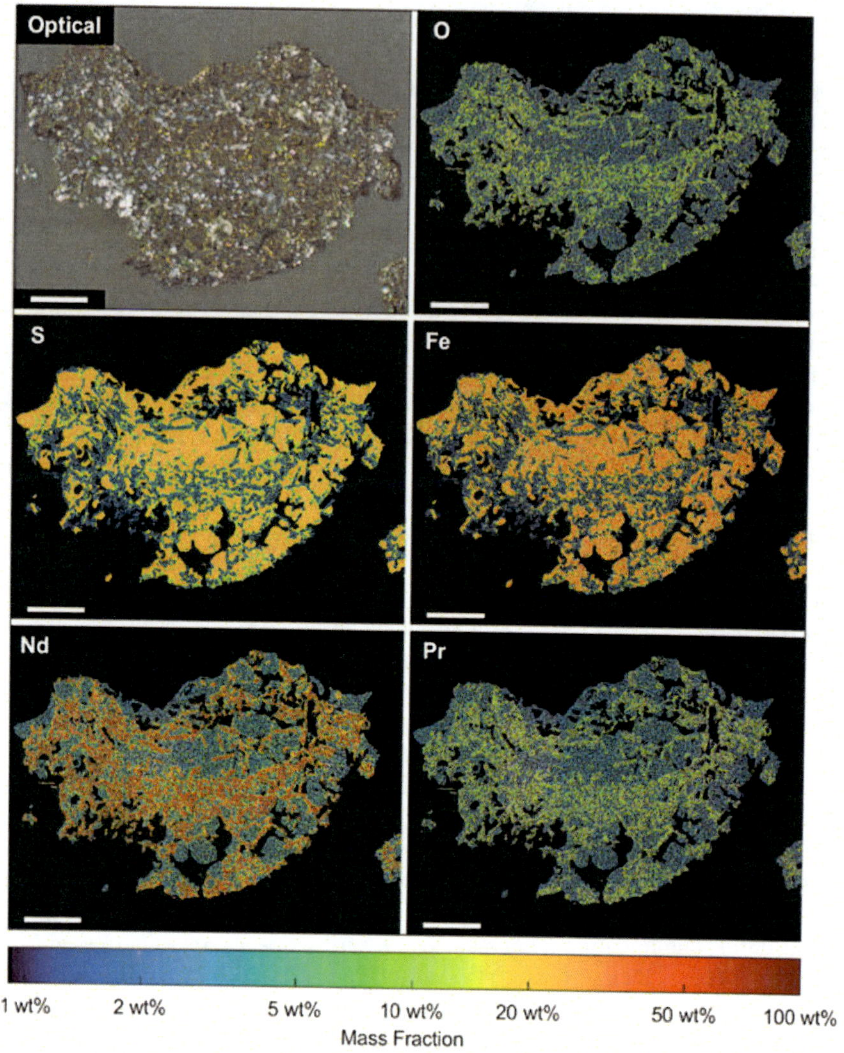

Fig. 3 Selective sulfidation of calcined, de-coated iron-neodymium-boron magnet nucleates distinct iron sulfide and rare earth oxide phases on the order of 20–100 μm in size, as illustrated via EDS mapping for the region shown in the optical image, large enough for liberation of Fe and Nd-Pr via comminution and physical separation with optimization of phase growth and coarsening behavior. Scale bars correspond to 200 μm

effects (ψ_{sol}) stemming from compound-forming interactions between lanthanide and boron oxides [50, 52]. Further research on lanthanide oxide sulfidation kinetics and iron–lanthanide–boron–oxygen–sulfide solution thermodynamics is necessary to design sulfidation pretreatments for physical separation of individual lanthanides. Our preliminary results demonstrate that selective sulfidation is a technically viable pretreatment to facilitate the physical separation of iron from lanthanide elements in rare-earth magnet waste. The burden of selectivity for iron and lanthanides in rare-earth magnet recycling may therefore by shifted across the mineralogical barrier from chemical separations to physical separations using a sulfidation pretreatment. In the following section, we apply a similar selective sulfidation process to overcome the mineralogical barrier in lithium-ion battery recycling.

Selective Sulfidation for Lithium-Ion Battery Recycling

As electrification of the transportation sector continues, effective recycling of lithium-ion battery materials from electric vehicles is critical to minimizing energy use in their production [53]. Furthermore, the rapidly increasing demand for battery elements runs the risk of overwhelming geopolitically uncertain supply chains for critical components [5]. In their current form, lithium-ion batteries largely rely on cathodes engineered from solid solutions or mixed-metal compounds of lithium, nickel, manganese, and cobalt (NMC) oxide [10, 11]. While recycling methods based on physical separation for direct reuse have been explored, the constantly evolving state of the art for battery chemistries motivates hydrometallurgical and pyrometallurgical approaches that allow for recovery of pure components [1, 17], in particular lithium, nickel, and cobalt compounds. In end-of-life lithium-ion batteries, the average lithium grade is on the order of 3 wt%, whereas nickel and cobalt grades are on the order of 5–15 wt% depending on the cathode chemistry [19]. At these grades, a practical mineralogical barrier likely exists between physical and chemical methods of element recovery, with chemical separation predicted to require 10–20× more energy than an equivalent physical separation process (Fig. 1). Significant energy savings in recycling may therefore be possible thanks to the use of a low-energy pretreatment, such as sulfidation, to shift the burden of selectivity for lithium, nickel, and cobalt from chemical to physical separation processes.

Sulfidation chemistry has previously been considered to facilitate selective metallic element recovery from end-of-life lithium-ion batteries via leaching, molten sulfide electrolysis, and physical separation [27, 33, 54, 55]. From ψ_{rxn} in Fig. 2, at a 10:1 sulfur to sulfur dioxide ratio lithium is predicted to be stable as a sulfate, with nickel, manganese, and cobalt stable as sulfides. The relative contributions of ψ_{rxn} and ψ_{sol} to ψ are unknown however due to mixed-metal compound formation between lithium and NMC oxides [10, 11] and the unexplored thermodynamics of the lithium–nickel–manganese–cobalt–oxygen–sulfur system. Herein, we explored selective sulfidation for 2 g of simulated NMC oxide cathode material ($LiNi_{1/3}Mn_{1/3}Co_{1/3}O_2$ solid solution, 98% purity, Sigma Aldrich) at 1000 °C for one hour using vaporized elemental sulfur (S_x, 99.5% purity, Acros Organics) as a sulfur source. The sulfur partial pressure was approximately 0.1 atm with a 10:1 sulfur to sulfur dioxide molar ratio, following methods and reactor design described previously [27, 33].

Figure 4 shows the SEM/EDS (SEM: JEOL JSM-6610LV, JEOL Ltd., EDS: Sirius SD detector, SGX Sensortech Ltd.) backscattered electron composition (BEC) image

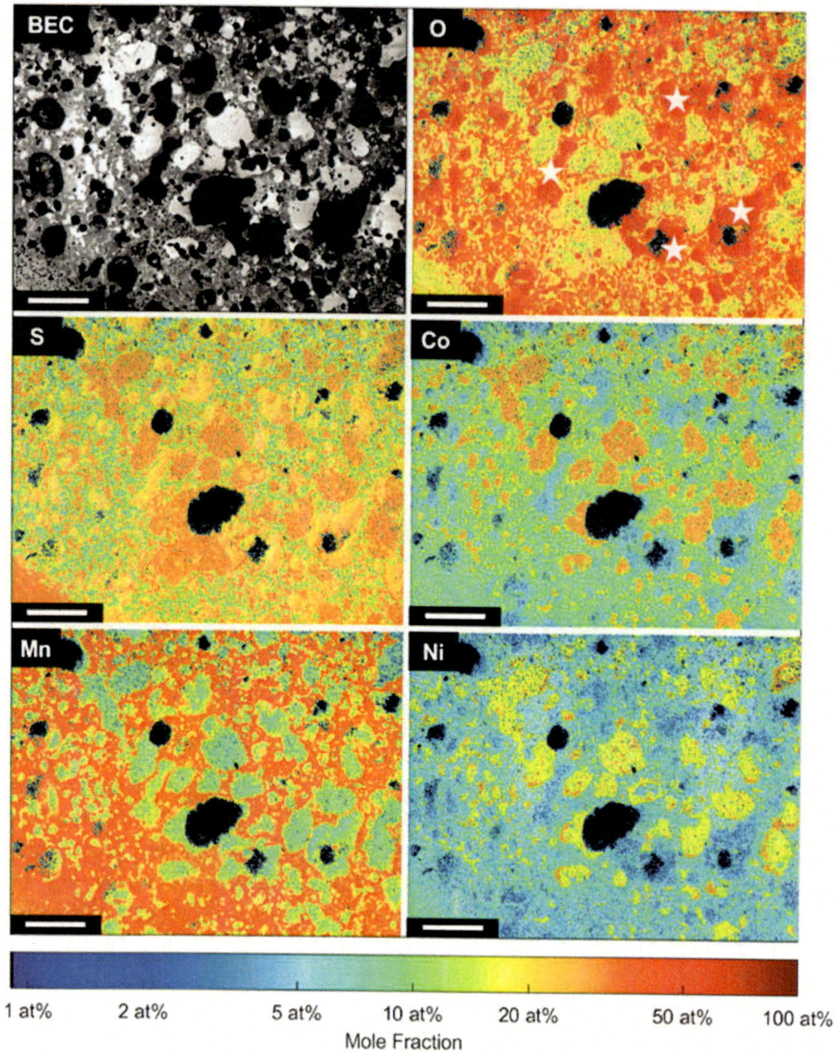

Fig. 4 Selective sulfidation of synthetic nickel-manganese-cobalt oxide (NMC111) lithium ion battery cathode nucleates distinct nickel-rich, cobalt-rich, and manganese-rich phases on the order of 100–200 μm in size, as illustrated via EDS mapping for the region shown in the SEM/BEC image, large enough for liberation of Ni, Co, and Mn via comminution and physical separation. XRD analysis reveals the presence of lithium as a sulfate, possibly corresponding to oxygen-sulfur-rich, metal-deficient regions in the element maps, denoted by (☆) in the oxygen panel. Scale bars correspond to 200 μm

of and element maps detailing oxygen, sulfur, cobalt, manganese, and nickel distribution in the sulfidized lithium-ion battery material. As depicted in Fig. 4, sulfidation was indeed performed selectively, with the formation of distinct nickel-rich and cobalt-rich sulfide phases, with some solubility for oxygen, on the order of 100–200 μm in size, large enough for effective liberation [22, 23] and physical separation using standard industrial mineral processing methods [25, 51]. Meanwhile, instead of manganese sulfide as predicted from Fig. 2, an under-sulfidized manganese-rich oxysulfide phase is formed; whether this phase is in fact an oxysulfide [56], oxysulfate [56], sulfatosulfide [56], or mixed oxide/sulfate/sulfide [56] is indiscernible from SEM/EDS element mapping in Fig. 4. XRD analysis of the sulfidized product suggests the presence of some manganese sulfide; however, oxysulfides of manganese may show poor crystallinity [57], so the precise nature of the manganese oxysulfide product phase remains unclear. While lithium is not visible via SEM/EDS, the presence in Fig. 4 of oxygen–sulfur-rich regions lacking significant amounts of nickel, manganese, or cobalt, denoted by (☆) in the oxygen panel, suggests the existence of lithium-rich sulfate phases. The presence of lithium sulfate is confirmed by XRD analysis. Overall, the formation of distinct phases of nickel, manganese, cobalt, and lithium that are physically separable suggests that selective sulfidation is a technically viable pretreatment. It offers to shift the burden of selectivity in lithium-ion battery recycling from energy intensive pyrometallurgical or hydrometallurgical chemical separations to more benign chemical pretreatments for physical methods [1, 17]. In the following section, we explore selective sulfidation applied to copper recovery from copper smelting slags as a means to overcome the mineralogical barrier at low product grades by increasing the grade of recoverable sulfide phases.

Selective Sulfidation for Copper Recovery from Slags

While electrification of the transportation sector continues to increase demand for battery metals [3], electrification of society is predicted to more broadly stress copper supplies [58], motivating efforts to increase recovery of copper from both primary and secondary sources. Presently, approximately 80% of the world's copper ore is produced from smelting of copper–iron–sulfide minerals [13]. For separation of iron from copper via matte smelting, the molten copper–iron–sulfide matte is contacted with a silica-rich slag phase. Following matte smelting, the slag contains approximately 75–90% fayalite, <10% alumina, <10% calcia, <4% magnesia, and 0.5–2.1% copper in the form of entrained sulfide matte particles and copper dissolved in fayalite [20]. Considering that smelters pay for about 96% of the copper value when purchasing concentrate [59], there is a strong economic incentive to minimize and recover copper lost to the slag phase during matte smelting. Copper in the form of entrained matte particles is presently liberated from slags via comminution and physically separated [13], yet at copper grades less than 1%, the practical mineralogical barrier between chemical and physical separation begins to diminish. As shown in Fig. 1 for copper recovery from slag at a grade of 0.5 wt%, energy needs for physical and chemical separation are predicted to be on the same order of magnitude. This calls into question the relative energetic effectiveness of selectivity in current physical separation methods. Meanwhile, the solubility of copper oxide in fayalite is approximately 1% [20], representing a significant fraction of copper in the slag that is inaccessible to recovery via physical separation without pretreatment.

Numerous pretreatments based on sulfidation have been explored to recover copper dissolved in fayalite phases of slag, generally utilizing roasting the slag in the presence of pyrite [60, 61] or iron sulfate [62] to form leachable or floatable copper phases. As shown by ψ_{rxn} in Fig. 2, the pure oxides of copper and iron are thermodynamically predicted to sulfidize at far lower sulfur to sulfur dioxide gas ratios than pure magnesia, alumina, and silica slag-formers. Under conditions where pure iron oxide sulfidizes, calcium is predicted to be stable as a sulfate. However, even if copper can be selectively precipitated as a sulfide via a selective sulfidation pretreatment, closing of the practical mineralogical barrier (Fig. 1) at grades relevant to recover copper from fayalite slag phases suggests that the energy difference may be small between chemical separation and physical separation with pretreatment. To overcome the practical mineralogical barrier, a higher-grade phase may be used to serve as a collector for copper upon sulfidation pretreatment. This will allow copper to be liberated with this higher grade "collector" phase and thereby lower energy input. From the fayalite phase, iron may be co-sulfidized with copper, predicted to increase the total sulfide product grade in the slag for liberation and physical separation from approximately 0.5 wt% up to 40 wt% and reduce the energy burden from chemical separation at a grade of 0.5 wt% by three orders of magnitude (Fig. 1).

We conducted the selective sulfidation at 800 °C of copper and iron from copper smelter slag ground to a particle size of 90–212 μm, at a scale of 2 g, using vaporized elemental sulfur (S_x, 99.5% purity, Acros Organics) at a partial pressure of approximately 0.1 atm. The sulfur to sulfur dioxide gas ratio was approximately 10:1, following methods and reactor design described previously [27]. Here, the alumina reactor bed previously employed [27] was substituted for graphite to minimize interactions between slag phases and the reactor. Figure 5 show the SEM/EDS (SEM: JEOL JSM-6610LV, JEOL Ltd., EDS: Sirius SD detector, SGX Sensortech Ltd.) element maps detailing ratios of sulfur to oxygen, iron to silicon and aluminum, and copper to silicon and aluminum in the untreated and sulfidized copper smelter slag.

As shown in Fig. 5, iron and copper are observed to preferentially segregate over silicon and aluminum to the sulfide phase, resulting in iron–copper–sulfide and aluminum–silicon oxide product phases. In the untreated slag, sulfide phases are observed to constitute a grade of approximately 1 wt%, increasing to approximately 16 wt% upon sulfidation. Product sulfide phases were observed to be on the order of 50 μm, large enough for effective liberation [22, 23] and physical separation using standard industrial mineral processing methods [25, 51]. The depletion of copper content from fayalite slag phases was difficult to quantify via SEM/EDS due to an abundance near the detection limit of the equipment. Further analysis will be needed to determine the conversion of solubilized copper in fayalite to sulfide and the

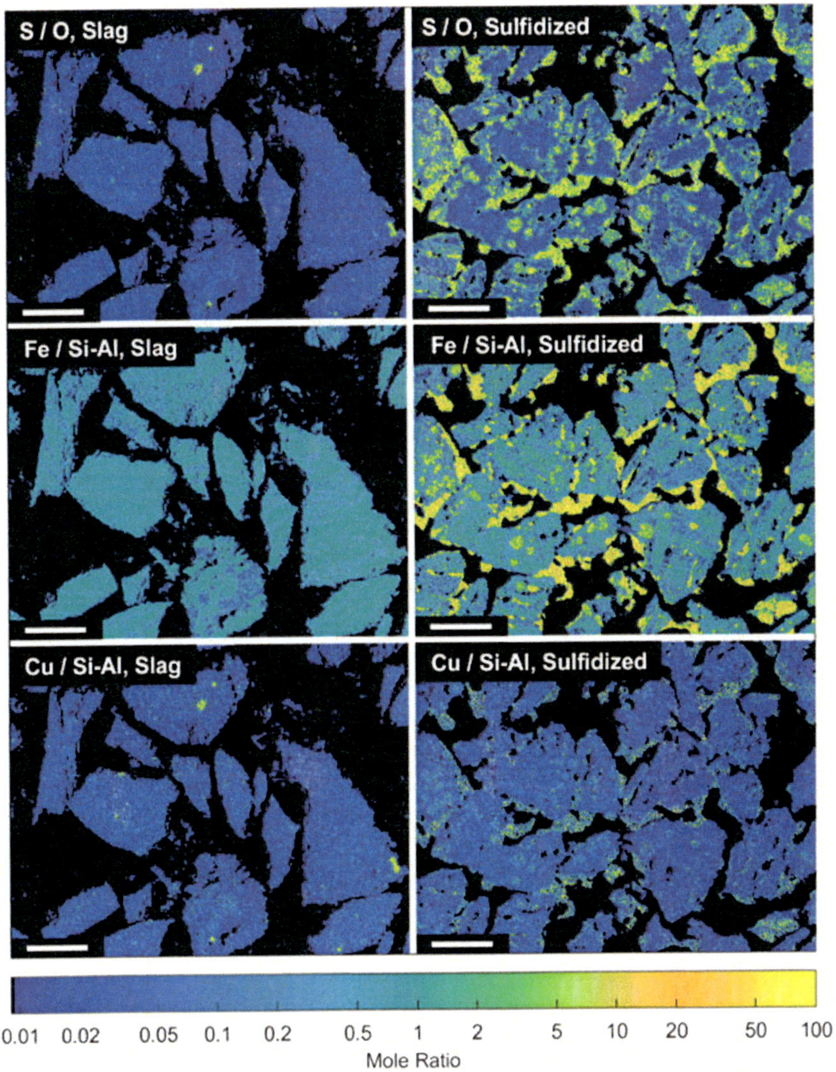

Fig. 5 Selective sulfidation of copper smelter slag results in the nucleation of iron-rich sulfide phases, with copper preferentially partitioning into these new sulfide phases versus silicon and aluminum. Iron sulfide formed via selective sulfidation serves as a "collector" phase for copper, potentially supporting increased recovery of copper from slags via comminution and physical separation. Scale bars correspond to 100 μm

partitioning of valuable trace elements such as cobalt and platinum group elements between oxide and sulfide phases. The use of an iron sulfide collector phase, formed in situ from sulfidation of iron from fayalite, allows the energetic burden of selectivity to be shifted from low product grades (~1 wt%) to higher grades (~16 wt%) via the selective sulfidation pretreatment. This indicates that the practical mineralogical barrier is widened, now in favor of lower-energy physical separation methods. The sulfide grade may be further increased toward 40% by increasing the sulfur to sulfur dioxide ratio in the reactor to improve the conversion of fayalite to iron sulfide. Meanwhile, excess heat produced from smelting of the iron-rich recovered copper sulfide may in turn be used to melt higher concentrations of gangue impurities in copper concentrate during smelting. This would facilitate the use of lower grade concentrates at a time of falling ore grades, while simultaneously increasing copper recovery [63].

Conclusions

Electrification-induced growth in demand for sustainably sourced metallic elements drives efforts to reduce the energy use of materials separations in both primary and secondary production. A mineralogical barrier exists between the lower energy burden for physical separation of elements concentrated in distinct phases and the higher energy burden for chemical separation of elements mixed at the atomic scale. Through comparing the energy usage of industrial-scale materials separation processes, this difference in energy is generally found to range from approximately 200–2,000 kWh per tonne of feed at product grades ranging between 1 and 30 wt%. While this analysis does not replace detailed thermodynamic and life cycle assessments of individual materials systems, it illustrates that in general significant energy savings in material separation can be realized through the development of energy-efficient pretreatments that facilitate physical separation in place of chemical separation. Selective sulfidation is a possible pretreatment to partition metallic elements from multi-metal compounds and solid solutions into separate oxide and sulfide phases, allowing the burden of selectivity to be shifted from chemical to physical methods of separation. We demonstrate selective sulfidation of calcined rare-earth magnets to partition iron and lanthanide elements into distinct, physically separable phases for recycling by physical separation. Similarly, we partition nickel, manganese, cobalt, and lithium from lithium-ion battery cathodes into distinct, physically separable phases for lithium-ion battery recycling by physical separation. Finally, we use selective sulfidation to form a dedicated sulfide collector phase for copper recovery from copper smelter slag, increasing the grade of copper-containing sulfide phases from approximately 1 wt% to approximately 16 wt%. For these recycling challenges, the selective sulfidation pretreatment facilitates the use of physical separations that are predicted to require 10–100× less energy than comparable chemical separations.

Acknowledgements The authors wish to thank Marko Yakasovic for providing samples of copper smelter slag and Dr. Mary Elizabeth Wagner for insight pertaining to rare-earth magnet thermodynamics and recycling. Part of this work was funded from the DOE-AMO-EERE Office under project DE-EE0008316.

References

1. Harper G, Sommerville R, Kendrick E, Driscoll L, Slater P, Stolkin R, Walton A, Christensen P, Heidrich O, Lambert S, Abbott A, Ryder K, Gaines L, Anderson P (2019) Recycling lithium-ion batteries from electric vehicles. Nature 575(7781):75–86. https://doi.org/10.1038/s41586-019-1682-5
2. Allanore A (2017) Electrocehmical engineering for commodity metals extraction. Electrochem Soc Interface 26(2):63–68. https://doi.org/10.1149/2.F05172if
3. Nguyen RT, Eggert RG, Severson MH, Anderson CG (2021) Global electrification of vehicles and intertwined material supply chains of Cobalt, Copper and Nickel. Resour Conserv Recycl 167:105198. https://doi.org/10.1016/J.RESCONREC.2020.105198
4. Cheisson T, Schelter EJ (2019) Rare earth elements: mendeleev's bane, modern marvels. Science 363(6426):489–493. https://doi.org/10.1126/science.aau7628
5. Olivetti EA, Ceder G, Gaustad GG, Fu X (2017) Lithium-Ion battery supply chain considerations: analysis of potential bottlenecks in critical metals. Joule 1(2):229–243. https://doi.org/10.1016/j.joule.2017.08.019
6. Binnemans K, Jones P T, Müller T, Yurramendi L (2018) Rare earths and the balance problem: how to deal with changing markets? J Sustain Metall 4(1):126–146. https://doi.org/10.1007/s40831-018-0162-8
7. Rötzer N, Schmidt M (2018) Decreasing metal ore grades—is the fear of resource depletion justified? Resources 7(4):88. https://doi.org/10.3390/RESOURCES7040088
8. Norgate T, Jahanshahi S (2010) Low grade ores—Smelt, leach or concentrate? Miner Eng 23(2):65–73. https://doi.org/10.1016/j.mineng.2009.10.002
9. Wagner M-E, Allanore A (2020) Chemical thermodynamic insights on rare-earth magnet sludge recycling. ISIJ Int 60(11):2339–2349. https://doi.org/10.2355/isijinternational.ISIJINT-2020-320
10. Wang M (2004) Enthalpy of formation of $LiNiO_2$, $LiCoO_2$ and their solid solution, LiNi1−xCoxO2. Solid State Ionics 166(1–2):167–173. https://doi.org/10.1016/j.ssi.2003.11.004
11. Chang K, Hallstedt B, Music D (2011) Thermodynamic and electrochemical properties of the Li–Co–O and Li–Ni–O systems. Chem Mater 24:97–105. https://doi.org/10.1021/cm201964r
12. Crundwell FK, Moats MS, Ramachandran V, Robinson TG, Davenport WG (2011) Extractive metallurgy of Nickel, Cobalt, and Platinum-group metals. Elsevier
13. Schlesinger ME, King MJ, Sole KC, Davenport WG (2011) Extractive metallurgy of copper. Elsevier
14. Zhao B, Zhang J, Schreiner B (2016) Separation hydrometallurgy of rare earth elements. Springer International Publishing AG Switzerland, Cham
15. Skinner BJ (1976) Second Iron Age Ahead? The distribution of chemical elements in the earth's crust sets natural limits to man's supply of metals that are much more important to the future of society than limits on energy. Sci Am 64(3):258–269
16. Skinner BJ (1979) Earth resources (minerals/metals/ores/geochemistry/mining). Proc Natl Acad Sci USA 76(9):4212–4217
17. Ciez RE, Whitacre JF (2019) Examining different recycling processes for Lithium-Ion batteries. Nat Sustain 2:148–156. https://doi.org/10.1038/s41893-019-0222-5

18. Zhang Y, Gu F, Su Z, Liu S, Anderson C, Jiang T (2020) Hydrometallurgical recovery of rare earth elements from NdFeB permanent magnet scrap: a review. Met 10(6):841. https://doi.org/10.3390/MET10060841
19. Ahmed S, Nelson PA, Gallagher KG, Susarla N, Dees DW (2017) Cost and energy demand of producing nickel manganese cobalt cathode material for lithium ion batteries. J Power Sources 342:733–740. https://doi.org/10.1016/J.JPOWSOUR.2016.12.069
20. Gorai B, Jana RK (2003) and Premchand. Characteristics and utilisation of copper slag—A review. Resour Conserv Recycl 39(4):299–313. https://doi.org/10.1016/S0921-3449(02)00171-4
21. Fuerstenau D, Abouzeid A-Z (2002) The energy efficiency of ball milling in comminution. Int J Miner Process 67(1):161–185. https://doi.org/10.1080/10643389.2012.728825
22. King RP (1979) A model for the quantitative estimation of mineral liberation by grinding. Int J Miner Process 6(3):207–220. https://doi.org/10.1016/0301-7516(79)90037-1
23. Finch JA, Petruk W (1984) Testing a solution to the King liberation model. Int J Miner Process 12(4):305–311. https://doi.org/10.1016/0301-7516(84)90036-X
24. Taylor L, Skuse D, Blackburn S, Greenwood R (2020) Stirred media mills in the mining industry: material grindability, energy-size relationships, and operating conditions. Powder Technol 369:1–16
25. de Bakker J (2014) Energy use of fine grinding in mineral processing. Metall Mater Trans E 1(1):8–19. https://doi.org/10.1007/s40553-013-0001-6
26. Curry JA, Ismay MJL, Jameson GJ (2014) Mine operating costs and the potential impacts of energy and grinding. Miner Eng 56:70–80
27. Stinn C, Allanore A (2021) Selective sulfidation of metal compounds. Nature 2021:1–9. https://doi.org/10.1038/s41586-021-04321-5
28. Harris CT, Peacey JG, Pickles CA (2013) Selective sulphidation and flotation of nickel from a nickeliferous laterite ore. Miner Eng 54:21–31
29. Han J, Liu W, Wang D, Jiao F, Qin W (2015) Selective sulfidation of lead smelter slag with sulfur. Metall Mater Trans B 47:344–354. https://doi.org/10.1007/s11663-015-0526-4
30. Han J, Liu W, Wang D, Jiao F, Zhang T, Qin W (2016) Selective sulfidation of lead smelter slag with pyrite and flotation behavior of synthetic ZnS. Metall Mater Trans B 47B:2400–2410. https://doi.org/10.1007/s11663-016-0693-y
31. Ke Y, Peng N, Xue K, Min X, Chai L, Pan Q, Liang Y, Xiao R, Wang Y, Tang C, Liu H (2018) Sulfidation behavior and mechanism of zinc silicate roasted with pyrite. Appl Surf Sci 435:1011–1019. https://doi.org/10.1016/j.apsusc.2017.11.202
32. Zhuo-yue L, Zhen-ning L, Yong-xing Z, Jin-fang L, Jie P, Ji-lai N, Cent J (2020) Recovery of Zn, Pb, Fe and Si from a low-grade mining ore by sulfidation roasting-beneficiation-leaching processes. J Cent South Univ 27(1):37–51. https://doi.org/10.1007/s11771-020-4276-3
33. Stinn C, Allanore A (2021) In Anderson C, Goodall G, Gostu S, Gregurek D, Lundström M, Meskers C, Nicol S, Peuraniemi S, Tesfaye F, Tripathy PK, Wang S, Zhang Y (eds) Selective sulfidation and electrowinning of Nickel and Cobalt for Lithium Ion Battery Recycling. In: Anderson C (ed) Ni-Co 2021: the 5th international symposium on Nickel and Cobalt. Springer Nature Switzerland AG, Cham, p 99–110
34. Adham K, Lee C (2012) Fluid bed roasting of metal ores and concentrates for arsenic removal. Towards clean metallurgical processing for procfit, social, and environmental stewardship. 109–116
35. U.S. Geological Survey (2021) Mineral commodity summaries
36. Afanasiev P, Fischer L, Beauchesne F, Danot M, Gaborit V, Breysse M (2000) Preparation of the mixed sulfide $Nb_2Mo_3S_{10}$ catalyst from the mixed oxide precursor. Catal Letters 64(1):59–63. https://doi.org/10.1023/A:1019058113931
37. Akila R, Jacob KT, Shukla AK (1987) Gibbs energies of formation of rare earth oxysulfides. Metall Trans B 18B(March):163–168
38. Pourbaix MJN, Rorive-Bouté MCM (1948) Graphical study of metallurgical equilibria. Discuss Faraday Soc 4(0):139–154. https://doi.org/10.1039/DF9480400139

39. Kellogg HH (1964) A critical review of sulfation equilibria. Trans Metall Soc AIME 230:1622–1634
40. Fleet ME, MacRae ND (1987) Sulfidation of Mg-rich olivine and the stability of niningerite in enstatite chondrites. Geochim Cosmochim Acta 51(6):1511–1521. https://doi.org/10.1016/0016-7037(87)90333-4
41. Fleet ME, Tronnes RG, Stone WE (1991) Partitioning of platinum group elements in the Fe-O-S System to 11 GPa and their fractionation in the mantle and meteorites. J Geophys Res Solid Earth 96(B13):21949–21958. https://doi.org/10.1029/91JB02172
42. Bataleva YV, Palyanov YN, Borzdov YM, Sobolev NV (2016) Sulfidation of silicate mantle by reduced S-bearing metasomatic fluids and melts. Geology 44(4):271–274. https://doi.org/10.1130/G37477.1
43. Goldschmidt VM (1937) The principles of distribution of chemical elements in minerals and rocks. The seventh Hugo Müller Lecture, delivered before the Chemical Society on March 17th, 1937. J Chem Soc (0):655–673. https://doi.org/10.1039/JR9370000655
44. Holser WT, Kaplan IR (1966) Isotope geochemistry of sedimentary sulfates. Chem Geol 1(C):93–135. https://doi.org/10.1016/0009-2541(66)90011-8
45. Bailey G, Joyce PJ, Schrijvers D, Schulze R, Sylvestre AM, Sprecher B, Vahidi E, Dewulf W, Van Acker K (2020) Review and new life cycle assessment for rare earth production from bastnäsite, ion adsorption clays and lateritic monazite. Resour Conserv Recycl 155:104675. https://doi.org/10.1016/j.resconrec.2019.104675
46. Lee JCK, Wen Z (2018) Pathways for greening the supply of rare earth elements in China. Nat Sustain 1:598–605. https://doi.org/10.1038/s41893-018-0154-5
47. Jowitt SM, Werner TT, Weng Z, Mudd GM (2018) Recycling of the rare earth elements. Curr Opin Green Sustain Chem 13:1–7. https://doi.org/10.1016/J.COGSC.2018.02.008
48. Firdaus M, Rhamdhani MA, Durandet Y, Rankin WJ, McGregor K (2016) Review of high-temperature recovery of rare earth (Nd/Dy) from magnet waste. J Sustain Metall 2016 24 2(4):276–295. https://doi.org/10.1007/S40831-016-0045-9
49. Rasheed MZ, Nam S-W, Cho J-Y, Park K-T, Kim B-S, Kim T-S (2021) Review of the liquid metal extraction process for the recovery of Nd and Dy from permanent magnets. Metall Mater Trans B 52(3):1213–1227. https://doi.org/10.1007/S11663-021-02102-Z
50. Yin X, Wang Y, Bai X, Wang Y, Chen L, Xiao C, Diwu J, Du S, Chai Z, Albrecht-Schmitt TE, Wang S (2017) Rare earth separations by selective borate crystallization. Nat Commun 8. https://doi.org/10.1038/ncomms14438
51. Corin KC, McFadzean BJ, Shackleton NJ, O'Connor CT (2021) Challenges Related to the Processing of Fines in the Recovery of Platinum Group Minerals (PGMs). Miner 11(5):533. https://doi.org/10.3390/MIN11050533
52. Cohen-Adad MT, Aloui-Lebbou O, Goutaudier C, Panczer G, Dujardin C, Pedrini C, Florian P, Massiot D, Gerard F, Kappenstein C (2000) Gadolinium and Yttrium Borates: Thermal Behavior and Structural Considerations. J Solid State Chem 154(1):204–213. https://doi.org/10.1006/JSSC.2000.8837
53. Dunn JB, Gaines L, Sullivan J, Wang MQ (2012) Impact of recycling on cradle-to-gate energy consumption and greenhouse gas emissions of automotive lithium-ion batteries. Environ Sci Technol 46(22):12704–12710. https://doi.org/10.1021/ES302420Z
54. Korkmaz K, Alemrajabi M, Rasmuson Å, Forsberg K (2018) Recoveries of valuable metals from spent nickel metal hydride vehicle batteries via sulfation, selective roasting, and water leaching. J Sustain Metall 4(3):313–325. https://doi.org/10.1007/s40831-018-0169-1
55. Shi J, Peng C, Chen M, Li Y, Eric H, Klemettinen L, Lundstrom M, Taskinen P, Jokilaakso A (2019) Sulfation roasting mechanism for spent Lithium-Ion battery metal oxides under SO_2-O_2-Ar Atmosphere. JOM 71(12):4473–4481
56. Bastanov SS, Ryabinina OI, Obzherina KF, Derbeneva SS (1968) On the chemical structure of manganese oxysulfides. Bull Acad Sci USSR Div Chem Sci 17:6–11
57. Liu M, Fu Y, Ma H, Wang T, Guan C, Hu K (2016) Flower-like manganese-cobalt oxysulfide supported on Ni foam as a novel faradaic electrode with commendable performance. Electrochim Acta 191:916–922. https://doi.org/10.1016/J.ELECTACTA.2016.01.156

58. Schipper BW, Lin HC, Meloni MA, Wansleeben K, Heijungs R, van der Voet E (2018) Estimating global copper demand until 2100 with regression and stock dynamics. Resour Conserv Recycl 132:28–36. https://doi.org/10.1016/J.RESCONREC.2018.01.004
59. Bustillo M (2018) Mineral resources: from exploration to sustainable assessment
60. Guo Z, Zhu D, Pan J, Zhang F (2016) Mechanism of mineral phase reconstruction for improving the beneficiation of copper and iron from copper Slag. JOM 68(9):2341–2348. https://doi.org/10.1007/S11837-016-2048-1
61. Tümen F, Bailey NT (1990) Recovery of metal values from copper smelter slags by roasting with pyrite. Hydrometallurgy 25(3):317–328. https://doi.org/10.1016/0304-386X(90)90047-6
62. Altundoğan HS, Tümen F (1997) Metal recovery from copper converter slag by roasting with ferric sulphate. Hydrometallurgy 44(1–2):261–267. https://doi.org/10.1016/S0304-386X(96)00038-2
63. Rush LT (2021) Integrative approach to metal extraction and electrification, Massachusetts Institute of Technology

Pre-study of the Dissolution Behavior of Silicon Kerf Residue in Steel

Adamantia Lazou, David Nilssen, Mertol Gökelma, Maria Wallin, and Gabriella Tranell

Abstract Silicon kerf residue is generated during the wafering process of pure silicon in the photovoltaic value chain. The generated by-product has a high volume, and the particle size is typically below 1 μm. Although the fine particles are partly oxidized, the material may be beneficial in different metallurgical applications such as grain refining and alloy composition adjustments. This work studies the dissolution behavior of silicon kerf in low alloy steel melts with the aim to upcycle the kerf material in the steel industry for different purposes. In this study, a steel alloy and the kerf residue were melted (at 1580 °C) in an alumina crucible placed in an induction furnace. The amount of added kerf residue was varied. The behavior of the particles in the solidified alloy was characterized by using an optical microscope, electron probe microscope (EPMA), and wavelength-dispersive X-ray spectroscopy (WDS) in order to study the dissolution behavior of the Si-kerf residue in the steel.

Keywords Silicon kerf residue · Utilization of by-products · Low alloy steels · Melting and dissolution of silicon kerf

Introduction

Solar energy is a clean renewable resource, allowing a rapid increase in worldwide crystalline Si (cSi)-based photovoltaic (PV) distribution over the past couple decades [1]. In the photovoltaic value chain, illustrated in Fig. 1, metallurgical grade silicon (MG-Si) is produced from quartz and carbon reductants. The MG-Si is further purified to solar grade silicon (SoG-Si) through chemical or metallurgical methods. For photovoltaic applications, a typical purity of the SoG-Si is >6 N [2]. Single-

A. Lazou (✉) · D. Nilssen · M. Wallin · G. Tranell
Department of Materials Science and Engineering, Norwegian University of Science and Technology, NTNU, 7034 Trondheim, Norway
e-mail: adamantia.lazou@ntnu.no

M. Gökelma
Department of Materials Science and Engineering, İzmir Institute of Technology, 35430 İzmir, Turkey

A. Lazou et al. (eds.), *REWAS 2022: Developing Tomorrow's Technical Cycles (Volume I)*, The Minerals, Metals & Materials Series,
https://doi.org/10.1007/978-3-030-92563-5_15

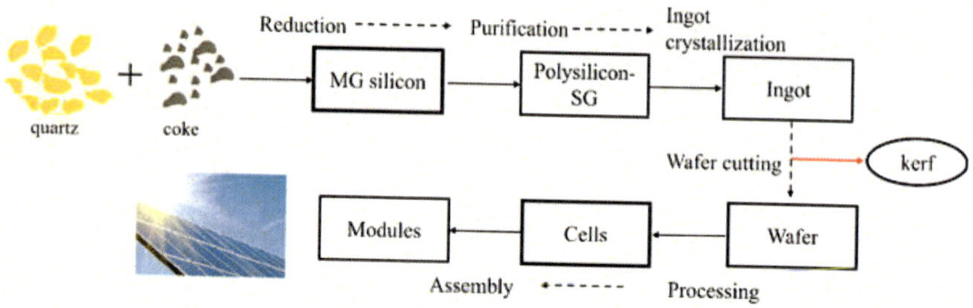

Fig. 1 PV value chain

or multi-crystalline ingots are then produced through crystal pulling or directional solidification. The ingots are cut into blocks which are sliced into wafers, mainly by diamond wire sawing [3, 4]. During this process, approximately 35–40% of crystalline silicon, known as silicon kerf residue, is lost. The material is complex and is usually composed of fine (typically in the order of 1 μm, due to the brittle structure of silicon) partly oxidized silicon particles with some metallic contamination from the sawing wire and lubricants [1, 2, 5]. The disposal of this material represents both a resource loss and creates environmental challenges. The global installed capacity of PV is estimated to reach 1270.5 GW by the end of 2022, mainly through cSi technologies, which is more than 200 times the capacity that was estimated by the end of 2017 [6]. The annual production of Si-kerf is estimated around 168.000 tonnes/year [5]. By effectively valorising Si-kerf residue, the specific process emissions and energy consumption will be reduced in the primary Si production, and less waste will be landfilled.

Research for the recycling of Si-kerf refers among others to use in silicon-containing alloys [7], refining by centrifugation [8], and directional solidification techniques. The recycling of silicon from the Si-kerf is, however, challenging due to the small particle size and impurity content [1, 9]. Hence, the reuse of Si-kerf into other applications reduces the mentioned challenges [1], and the most studied uses are the production of silicon carbide powder, silicon nitride [10], in ceramics, etc. [1].

Perspectives of the Present Study

The use of inoculant agents for grain refining in metal casting is well known. The use of such agents, in the wire and arc additive manufacturing (WAAM), is, however, still under investigation [11]. Recently, the WAAM technology has drawn attention due to the benefits that it can offer to create, repair, and extend the life time of large industrial equipment and in this way reduce the cost and material waste as well as increase the materials efficiency [12]. Nevertheless, challenges exist that are related to the manufacturing accuracy, quality assurance, and automation of the WAAM system

[13]. The formation of large grain size microstructures during the production reduces the mechanical properties of the product which might be detrimental for the material's performance. Considering this, the introduction of inoculating agents during the WAAM process can promote the grain refinement and modify the existing phases. Typical inoculation agents for low alloyed steels are the rare earth metal oxides, NbC-containing ferroalloy, or TiN-containing ferro alloy [14, 15]. One potential inoculant agent for low alloyed steel is SiC particles due to their properties such as high hardness and high melting point [11], opening possibilities for use of carburized kerf residue. The Si-kerf residue could also be used to fine-tune the composition of a steel alloy. Microalloying or modifying the composition of an alloy is a tool to reduce the cost related to post-processing and enhance the properties of alloys [16].

The present study is a preliminary experimental investigation of the dissolution behavior of the Si-kerf residue introduced into a steel melt, which offers the basis for further research and experimental investigations into the potential for recycling and reuse of the kerf in the steel industry. In the current work, the main objectives are being to observe and evaluate the physical interaction and dissolution behavior of the Si particles in steel.

Materials and Methods

The steel used was a commercial low alloy steel (ASTM A182 black rolled bar, EN10204) with chemical composition as listed in Table 1. The steel was received as a rod (Ø76.2 × 2000 mm) which was cut into small pieces prior to the experimental trials. The Si-kerf was received as dried powder material (thermally treated to remove the lubricant), and its chemical composition is presented in Table 1. The steel and Si-kerf samples were dried at 100 °C for 24 h prior to analysis and experimental trial. Approximately 98 g steel was used in each experiment, while the amount of Si-kerf was varied. The Si-kerf particles were introduced and melted together with the steel (before melting) for the experiments presented in the current work (two in total). The amount of kerf was varied between the trials, while the temperature and time were kept constant. The additions of Si-kerf were aimed to achieve, 25 (sample I), and 50 (sample II) times more (wt%) Si in the final alloy that corresponds to approximately 5 and 10 wt%, respectively. A steel sample was melted without the addition of Si-kerf residue particles, to provide the reference comparison for the samples I and II.

Table 1 Chemical composition of steel in weight % and Si-kerf in ppm

Steel	Fe	Cr	Mo	Ni	Si	C	Mn	P	Cu	–	
	Balance	2.34	1.03	0.20	0.21	0.13	0.53	0.01	0.17	–	
Si-kerf	Al	Co	Cr	Cu	Fe	Mg	Ni	Pb	Ti	Zn	Si
	81.4	0.218	2.05	4.45	352	6.74	560	1.49	25.4	7.86	94×10^4

The experiments were conducted in a closed 30 kW induction furnace, in which 5–6 kW corresponds to the temperature range of 1400–1600 °C in the present study (Fig. 2a). The furnace can operate either under vacuum or in controlled atmosphere (Ar in the present study with purity of 99.9% and approximately 1 l/min flow rate). Prior to the experiments, the oxygen was removed from the furnace chamber (0.5–0.6 m^3) atmosphere through one-time evacuation and argon flashing up to 1030 mbar. During the experiments, Ar was flowing to the chamber keeping the pressure around 1030–1050 mbar. The experiments were conducted in alumina crucibles (A5, Ø50/44 × 75 mm) that were placed inside a graphite crucible (IG-15, Ø60/52 × 100 mm) (Fig. 2b). The temperature was recorded with the use of a thermocouple type C that was placed in the sample. The heating was controlled with a manual increase of the supplied power, and the target experimental temperature was 1580 °C (±20), to ensure complete melting of the steel (Fig. 2c). After reaching the target temperature, the samples were immediately left to cool to room temperature. The samples were then prepared for electron probe microscope analysis (EPMA) to examine the microstructure, the chemical composition of the product phases, and the distribution of the elements between phases. The EPMA analysis was carried out in a fixed working distance of 11 mm, different magnifications, and an acceleration voltage of 15 kV was applied. The microstructure of the reference sample was analyzed by using scanning electron microscope (SEM). Images were taken in different magnifications, and the composition was measured with energy dispersive spectroscopy (EDS) analysis. The SEM analysis was executed with an acceleration voltage of 15 kV in 10 mm working distance.

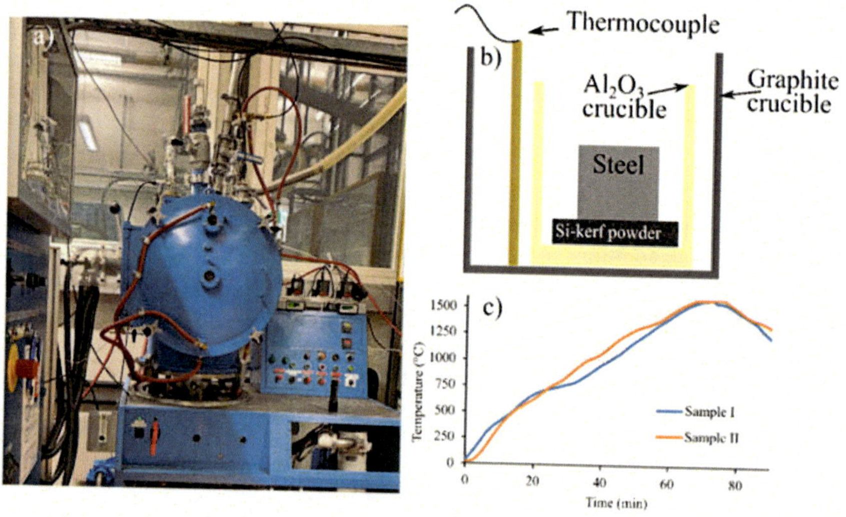

Fig. 2 **a** Induction furnace used in this work, **b** sample set up (before melting) in the crucible, and **c** temperature profile for the samples I and II

Results and Discussion

The chemical composition of the steel can be seen in Table 1. The steel is a Cr-rich low alloy steel which is commonly used in forged components due to its properties [17, 18]. The initial Si content of the steel was approximately 0.21 wt%. In Table 1, the chemical analysis of the Si-kerf is also presented (in ppm), as measured with ICP-MS analysis. The Si-kerf is mainly composed of Si, while the rest of the elements that were detected are attributed to the sawing process [4]. The balance is attributed mainly to atmospheric impurities such as oxygen and carbon [4].

The SEM analysis of the reference sample (steel melted without additions of Si-kerf residue) can be seen in Fig. 3, and the quantification of the marked points is provided in Table 2. Figure 3a and b presents an overview of the sample in low magnification in secondary and backscatter electron images, respectively. Figure 3c provides an image in higher magnification. The sample is mainly composed of a

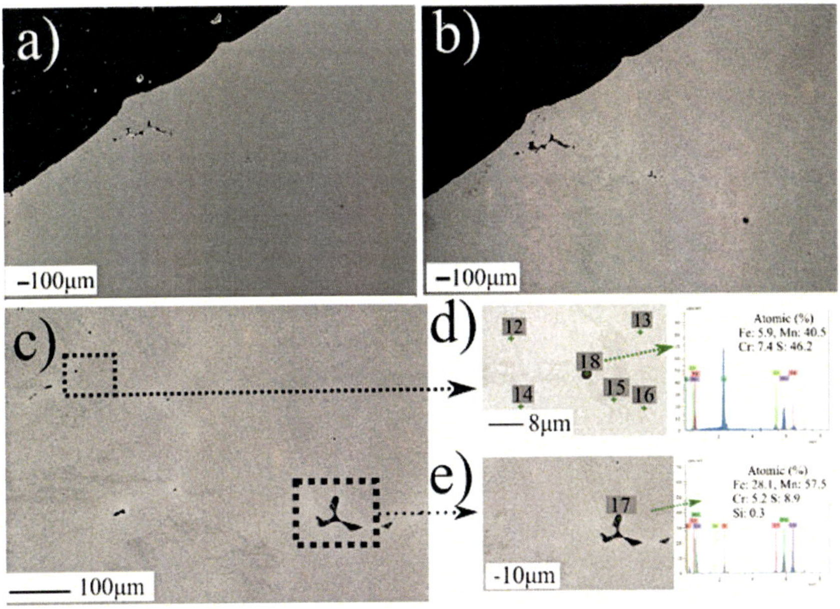

Fig. 3 Micrographs of reference sample **a** secondary, **b**, **c**, **d** and **e** backscattered electron images in 100×, 100×, 500×, 5000×, and 1000× magnification, respectively. The magnified areas are marked in the figure. The marked points are quantified in the figure and in Table 2

Table 2 Composition (atomic %) of the reference sample as measured with EDS

Points	Fe	Cr	S	Si	Mn
12–14	94.3	3.1	1.4	0.7	0.8
15–16	95.3	2.7	0.9	0.6	0.7

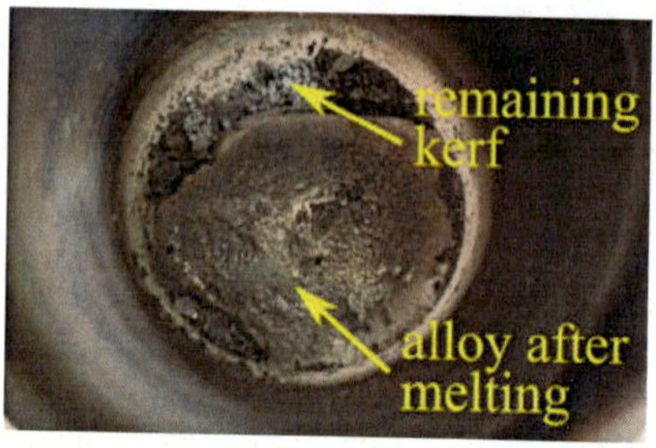

Fig. 4 Sample II after melting at 1580 °C with immediate cooling (kerf and steel were melted together)

light grey phase (points 12–14), while a slight darker phase (points 15–16) presents in less extend. The difference in their appearance could be attributed to the slight difference in their composition (lower Cr and S for points 15–16).

Figure 3d, e present the analysis of the dark phase which contains a higher Mn, Cr, and S content in comparison with the rest of the analysed points (Table 2).

As mentioned, the samples after melting were prepared for EPMA analysis. Remaining powder was observed after the melting process, attributed to undissolved Si-kerf residue (Fig. 4). This might indicate that the given time at the target temperature was not enough to achieve complete melting of the Si-kerf residue, considering also the added amount of Si-kerf.

The micrographs for sample I can be seen in Fig. 5, and the chemical composition of the different phases is provided in Table 3. Figure 5a, b presents the cross-section of the sample from the upper and inner part, as shown with the arrow. According to Fig. 5a, it seems that in the upper part of the sample there is a dark grey phase that was mainly composed of Si, indicating the presence of remaining Si-kerf. In general, the sample seems to be composed mainly of a bright grey phase A, while a slight darker grey phase B was observed in a random distribution (Fig. 5b). Based on the analysis of the marked phases (Table 3), the major component is Fe, while it seems

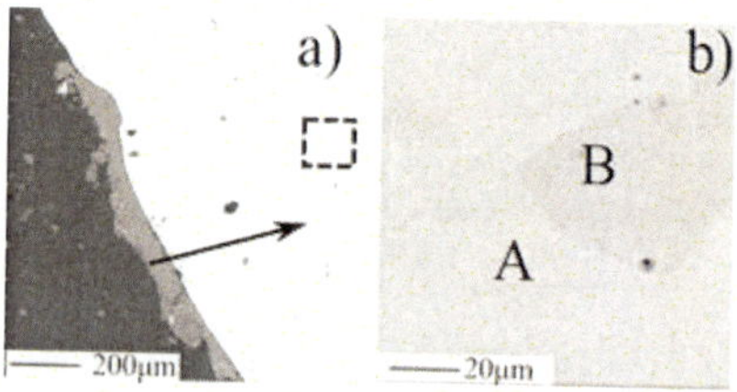

Fig. 5 Micrographs of sample I from **a** upper and **b** inner part, in 100× and 1000× magnification, respectively. The magnified area is marked in the figure. Marked phases are quantified in Table 3

Table 3 Chemical analysis (wt%) of the phases for the sample I, as measured with EPMA

Points	Fe	Cr	Mo	Si	Mn	S
A[a]	95.07	2.39	0.09	2.25	0.53	0.02
B[a]	95.75	2.29	0.13	2.60	0.47	0.03

[a] Average from 5 points

that there is a slight difference in the Si content. Phase B seems to have slightly lower Si concentration.

A mapping analysis might indicate more clear the distribution of elements between these phases. The Si content of these phases is approximately 2% that is lower than the expected (around 5%), confirming that less Si than added was dissolved. Nevertheless, the Si content of the final alloy has increased (around 11% more Si) as compared to the initial alloy. The composition is close to the bcc phase, which indicates that silicon is dissolved in iron rather than forming intermetallic phases. Si seems to be homogenously dissolved in the sample, although it should be emphasized that the size of the Si-kerf particles is small which makes the detection of the particles challenging.

The mapping analysis of the sample I can be seen in Fig. 6. The analysis confirms the observations mentioned earlier as the distribution of Si and Fe seems even between phases A and B. This might indicate further that the difference in the appearance of phase A and B could be attributed on the density difference, possibly due to the higher content of Mo in the B phase. The micrograph of phase A suggest a similar morphology with the reference sample morphology of phase A, although the chemistry of the different phases is changing due to the dissolution of Si-kerf.

The micrographs for sample II can be seen in Fig. 7. The analysis of the marked phases is provided in Table 4. Figure 7a and b presents the cross-section of the

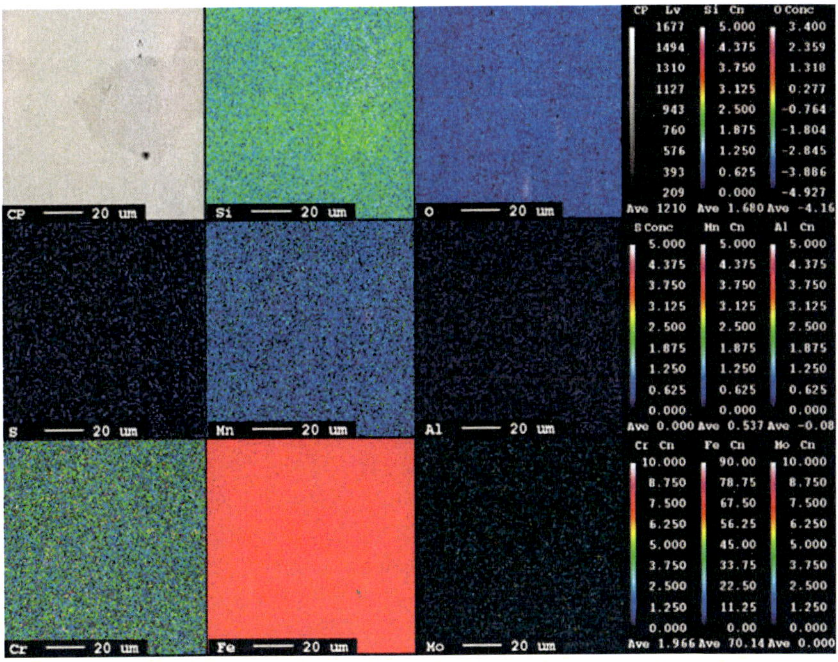

Fig. 6 EPMA map of silicon, oxygen, sulfur, manganese, aluminium, chromium, iron, and molybdenum, for sample I in 1000× magnification

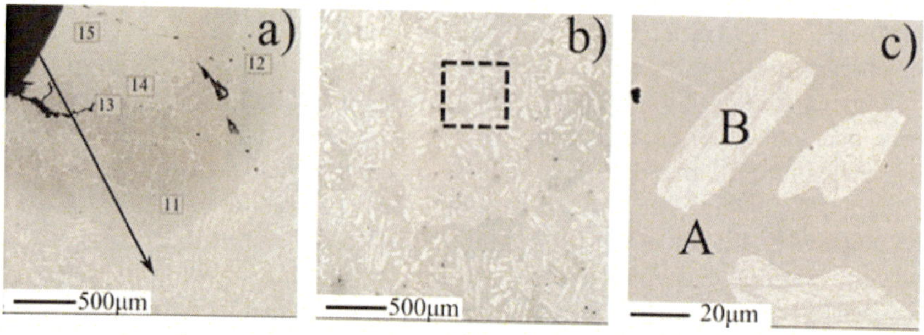

Fig. 7 Micrographs for sample II **a** on the upper part, **b** and **c** in the inner of the sample in 40×, 40×, and 1000× magnification, respectively. The magnified area is marked in the figure. Marked phases are quantified in Table 4

Table 4 Chemical analysis (wt%) of the phases for the sample II, as measured with EPMA

Points	Fe	Cr	Mo	Si	Mn	S
A[a]	94.34	2.35	0.06	3.64	0.53	0.03
B[a]	94.77	2.36	0.04	3.25	0.57	0.01
11–12	90.5	2.08	0.06	8.06	0.48	0.01
13	88.22	1.85	0.01	10.37	0.47	0.04
14	88.06	1.79	–	9.36	0.57	0.01
15	90.3	1.81	0.06	6.48	0.43	0

[a] Average from 5 points

sample from the outer (upper part, not in contact with crucible) and inner parts of the sample, respectively, as shown by the arrow. Figure 7c presents the micrograph from the inner part of the sample in higher magnification than Fig. 7b. Based on Fig. 7a, it seems that in the upper part of the sample, there is a region where the microstructure appears to more homogeneous (dark grey phase) than the inner part. Moreover, this region seems to have clear boundaries with the inner part of the sample.

In this region (points 11–15 in Table 4), there is significantly higher Si content. This might correlate with the fact that in this sample there was remaining kerf, indicating that part of the added Si-kerf remained undissolved. On the other hand, on the inner part of the sample, there are mainly two phases with clear boundaries, the dark grey phase A and brighter grey phase B.

The analysis of these two phases (in Table 4) indicates that their chemical characteristics are similar although the Si content in phase B is slightly lower. Individual SiO_2 particles could, however, not be observed. Moreover, the Si content was higher than in sample I, indicating that more kerf was dissolved in this case. The measured compositions can be seen in Table 4. For the rest of the analyzed points, Si is significantly increased. The morphology and composition of sample II are different than both the reference sample and sample I. This indicates that the addition of more Si-kerf residue changes the microstructural appearance. However, the effects of dissolution of the Si-kerf residue on the microstructure of the steel need to be addressed in future work.

The particles of Si-kerf residue are believed to contain some amount of SiO_2 on their surface due to oxidation, considering the stability of SiO_2 over Si. As an example, approximately 2 wt% of O_2 was detected in the analysis of Si-kerf residue with an average calculated thickness of 0.01 μm [4]. The fact that a remaining powder was observed after melting (Fig. 4) could be explained by the presence of SiO_2 that might prevent complete melting of the Si-kerf residue at 1580 °C. The region being rich in Si for the sample II (Fig. 7) might be attributed to SiO_2 in the surface of the alloy due to the density difference. If the SiO_2 particles remain unchanged during melting, then they could potentially act as inoculant agents for the steel. Nevertheless, such behavior was not clearly observed in the current study. If the oxide layer on the particles surface breaks during the melting process, the Si can be dissolved in liquid iron and alloy the steel. As such, parameters like the density and stability of the added particles in the melt should be explored further. The addition of the particles prior the liquid state might promote the wetting of the particles with the steel and as such achieve a better dissolution. As a result, the Si being evenly dispersed in the analyzed samples could be explained accordingly.

Conclusions and Suggestions for Further Work

The current work is a preliminary study into the valorisation of Si-kerf residue in steel applications. The dissolution behavior of the Si-kerf residue into low alloyed steel was investigated, and the preliminary experiments indicate that

- Si particles dissolve rapidly in the steel alloy, and the addition of more Si-kerf typically results in a higher Si content in the final alloy.
- The microstructure of the samples is changing with the addition of the Si-kerf residue particles.
- Si is detected in all the analysed phases of the samples, while SiO_2 individual particles were not seen.
- Regions with higher Si content were observed on the outside surface of the alloy, confirming that part of the added Si-kerf remained undissolved.

Based on the results obtained from the present work, it is suggested that the addition of Si-kerf residue in different amounts should be studied in a wider range for the purpose of micro-alloying or deoxidation. Furthermore, the undissolved Si-kerf residue particles should be analyzed to examine the chemical composition and appeared phases. The effect of holding time and temperature on the behavior of Si particles in the steel at liquid state should also be addressed, since these parameters will affect the dissolution behavior. Finally, the introduction of Si-kerf residue particles after the melting of steel should be investigated further, in terms of its dissolution behavior. Further research will also include a more in-depth metallographic analysis with the use of proper etching agents to examine whether remaining SiO_2 particles were left in the grains and in this way address whether Si-kerf could be used as inoculant agent. In addition, further experiments will try to address whether carbonized kerf residue or other type of additions can be used as inoculant agents.

Acknowledgements This project has received funding from the European Institute of Innovation and Technology (EIT), a body of the European Union, under the Horizon Europe, research and innovation program under Project Agreement 20144.

References

1. Guo J, Liu X, Yu J, Xu C, Wu Y, Pan D et al (2021) An overview of the comprehensive utilization of silicon-based solid waste related to PV industry. Resour Conserv Recycl (Elsevier) 169:105450. https://doi.org/10.1016/J.RESCONREC.2021.105450
2. Yang S, Ma W, Wei K, Xie K, Wang Z (2019) Thermodynamic analysis and experimental verification for silicon recovery from the diamond wire saw silicon powder by vacuum carbothermal reduction. Sep Purif Technol (Elsevier) 228:115754. https://doi.org/10.1016/J.SEPPUR.2019.115754
3. Tomono K, Miyamoto S, Ogawa T, Furuya H, Okamura Y, Yoshimoto M et al (2013) Recycling of kerf loss silicon derived from diamond-wire saw cutting process by chemical approach. Sep Purif Technol (Elsevier) 120:304–309. https://doi.org/10.1016/J.SEPPUR.2013.10.014
4. Lim K (2018) Recovery and utilisation of Kerf Waste from silicon wafering process. Master's thesis, NTNU, Norwegian University of Science and Technology
5. Yang HL, Liu IT, Liu CE, Hsu HP, Lan CW (2019) Recycling and reuse of kerf-loss silicon from diamond wire sawing for photovoltaic industry. Waste Manag (Pergamon) 84:204–210. https://doi.org/10.1016/J.WASMAN.2018.11.045
6. Lu X, Miki T, Takeda O, Zhu H, Nagasaka T (2019) Thermodynamic criteria of the end-of-life silicon wafers refining for closing the recycling loop of photovoltaic panels. (Taylor & Francis) 20:813–825. https://doi.org/10.1080/14686996.2019.1641429. http://Www TandfonlineCom/Action/JournalInformation?Show=aimsScope&journalCode=tsta20#VmBm uzZFCUk
7. Wei D, Gao S, Kong J, Jin X, Jiang S, Zhou S et al (2020) Recycling silicon from silicon cutting waste by Al–Si alloying. J Clean Prod (Elsevier) 251:119647. https://doi.org/10.1016/J.JCLEPRO.2019.119647
8. Lin YC, Wang TY, Lan CW, Tai CY (2010) Recovery of silicon powder from kerf loss slurry by centrifugation. Powder Technol (Elsevier) 200:216–223. https://doi.org/10.1016/J.POWTEC.2010.02.028
9. Hachichi K, Lami A, Zemmouri H, Cuellar P, Soni R, Ait-Amar H et al (2017) Silicon recovery from Kerf Slurry waste: a review of current status and perspective. Silicon 10:4. (Springer) 10:1579–1589. https://doi.org/10.1007/S12633-017-9642-X
10. Jin X, Kong J, Zhou X, Xing P, Zhuang Y (2020) Recycling of silicon kerf loss derived from diamond-wire saw cutting process to prepare silicon nitride. J Clean Prod (Elsevier) 247:119163. https://doi.org/10.1016/J.JCLEPRO.2019.119163
11. Rodrigues TA, Duarte VR, Tomás D, Avila JA, Escobar JD, Rossinyol E et al (2020) In-situ strengthening of a high strength low alloy steel during wire and arc additive manufacturing (WAAM). Addit Manuf (Elsevier) 34:101200. https://doi.org/10.1016/J.ADDMA.2020.101200

12. Wu B, Pan Z, Ding D, Cuiuri D, Li H, Xu J et al (2018) A review of the wire arc additive manufacturing of metals: properties, defects and quality improvement. J Manuf Process (Elsevier) 35:127–139. https://doi.org/10.1016/J.JMAPRO.2018.08.001
13. Xia C, Pan Z, Polden J, Li H, Xu Y, Chen S et al (2020) A review on wire arc additive manufacturing: monitoring, control and a framework of automated system. J Manuf Syst (Elsevier) 57:31–45. https://doi.org/10.1016/J.JMSY.2020.08.008
14. Gennesson M, Daloz D, Zollinger J, Rouat B, Demurger J, Poirier D et al (2018) Inoculation in lab scale low alloyed steel castings. Minerals, metals and materials series. Springer International Publishing, Part F12, pp 577–585. https://doi.org/10.1007/978-3-319-72526-0_54
15. Arvola DA (2018) Grain refinement of high alloy stainless steels in sand and directionally solidified castings. Missouri University of Science and Technology
16. Ollilainen V, Kasprzak W, Holappa L (2003) The effect of silicon, vanadium and nitrogen on the microstructure and hardness of air cooled medium carbon low alloy steels. J Mater Process Technol (Elsevier) 134:405–412. https://doi.org/10.1016/S0924-0136(02)01131-7
17. Escrivà-Cerdán C, Ooi SW, Joshi GR, Morana R, Bhadeshia HKDH, Akid R (2019) Effect of tempering heat treatment on the CO_2 corrosion resistance of quench-hardened Cr-Mo low-alloy steels for oil and gas applications. Corros Sci (Pergamon) 154:36–48. https://doi.org/10.1016/J.CORSCI.2019.03.036
18. Fallahmohammadi E, Bolzoni F, Fumagalli G, Re G, Benassi G, Lazzari L (2014) Hydrogen diffusion into three metallurgical microstructures of a C-Mn X65 and low alloy F22 sour service steel pipelines. Int J Hydrogen Energy (Pergamon) 39:13300–13313. https://doi.org/10.1016/J.IJHYDENE.2014.06.122

Characterisation of Hyperaccumulators for Lithium Recovery from Ancient Mine Soils

Lorna Anguilano, Uchechukwu Onwukwe, Danny Aryani, Jesus J. Ojeda, Guido Lingua, Valentina Gianotti, and Alessandra Devoto

Abstract The importance of lithium in modern industry is proven by a staggering triplication of the market for Li-based batteries, valued at $30b in 2017 and expected to reach $100b by 2025. Lithium is used as nanoparticles, particularly for batteries and electronics applications. Presently, lithium nanoparticles are manufactured using induction thermal plasma and other energy-demanding technologies. Furthermore, lithium is mined using significant volumes of water in areas such as South America, where aquifers are facing an ever-growing pollution from over-mining and agriculture, affecting the provision of clean and safe drinking water. We propose a switch in mining and manufacturing methods through the use of phytomining in ancient mine locations, to foster economic sustainability in areas affected by unemployment while maintaining the historic splendour of these sites. This paper will focus on the report of preliminary experiments using Agrostis Tenuis as hyperaccumulator for lithium and the characterisation of the biomass to assess its metal collecting behaviour. Our experiments have proven that in such areas, the amount of lithium (~1000 ppm) present in the sludges derived from mine adits can be recovered by autochthonous grasses (~20% per harvest) and transformed into re-usable nanoparticles using low-energy bio-synthesis.

Keywords Recycling and secondary recovery · Sustainability · Characterization

L. Anguilano (✉) · U. Onwukwe · D. Aryani
Brunel University London, London, United Kingdom
e-mail: Lorna.Anguilano@brunel.ac.uk

J. J. Ojeda
Swansea University, Swansea, United Kingdom

G. Lingua · V. Gianotti
Universita' del Piemonte Orientale, Vercelli, Italy

A. Devoto
Royal Holloway University London, London, United Kingdom

A. Lazou et al. (eds.), *REWAS 2022: Developing Tomorrow's Technical Cycles (Volume I)*, The Minerals, Metals & Materials Series,
https://doi.org/10.1007/978-3-030-92563-5_16

Introduction

The importance of lithium in modern industry is proven by a staggering triplication of the market for Li-based batteries, valued at $30b in 2017 and expected to reach $100b by 2025. Lithium is used as nanoparticles, particularly for batteries and electronics applications. Presently, lithium nanoparticles are manufactured using induction thermal plasma and other energy-demanding technologies. Furthermore, lithium is mined using significant volumes of water in areas such as South America, where aquifers are facing an ever-growing pollution from over-mining and agriculture, affecting the provision of clean and safe drinking water.

We propose a switch in mining and manufacturing methods through the use of phytomining in ancient mine locations, to foster economic sustainability in areas affected by unemployment while maintaining the historic splendour of these sites. This paper will focus on the report of preliminary experiments using Agrostis Tenuis as hyperaccumulator for Lithium and the characterisation of the biomass to assess its metal collecting behaviour.

Our experiments have proven that in such areas, the amount of lithium (~1000 ppm) present in the sludges derived from mine adits can be recovered by autochthonous grasses (~20% per harvest) and transformed into re-usable nanoparticles using low-energy bio-synthesis.

Hyperaccumulator plants can accumulate high concentration of contaminants (especially metals) in their shoots without showing significant signs of toxicity [3, 10]; the term was coined by Brooks and co (1977) to indicate plants that uptake large amounts of heavy metals from the soil, a behaviour contrary to that of excluder plants. In quantitative terms, the level of metal concentration achievable by hyperaccumulators is metal-dependent. A plant is classified as hyperaccumulator if it achieves: 10,000 mg/kg dry weight for Mn and Zn concentrations, 100 mg/kg dry weight for Cd, and 1,000 mg/kg dry weight for As, Cu, Co and Ni [1]. Aside from the "high" metal concentration that these plants can achieve, another intrinsic feature is the ability to translocate the metals from their roots system to their shoots [1]. While the hyperaccumulation of nickel, copper, and noble metals such as gold has received strong attention in the past decades, other metals such as lithium have not been widely researched, and the literature available up to date is sparse. [4] indicate the ability of *Brassica juncea* to uptake lithium; Kavanagh and co-authors (2017) present the accumulation of lithium in edible plants such as cabbage and tomatoes and indicate the capability of cress of accumulating up to 4 times the amount of lithium, and its consequent extraction using incineration. In 2018, the same authors present a paper showing 16 lithium-tolerant plants showing an accumulation between 1000 mg/kg (*Distichlis spicate*) and 8000 mg/kg (*Brassica cartinata*) and assessing 5 more hyperaccumulators highlighting excellent results from *Cardamine hirsute*, *Brassica oleracea* and *Helianthus annus*. Liang et al. (2018) use *Apocynum pictum,* while Henschel and co-authors (2020) trialled the nickel hyperaccumulator *Alyssum murale* for the accumulation of exhaust batteries elements, nickel, cobalt, and lithium and show the ability of these plants to accumulate 1% of the Li available in the soil within one harvest. These results demonstrate that accumulators known for other metals such as nickel, vanadium, or cobalt can play a fundamental role also in the extraction of lithium from contaminated soils/industrial e-waste, potentially indicating a route for green mining of low grade ores.

Methods and Materials

This project evaluates the potential of lithium extraction and upcycling from contaminated soils in the vicinity of ancient mines in Wales (UK). The process was divided in three stages: (i) mining: the capacity of hyperaccumulator plants to accumulate lithium and the speed of the plant growth and consequently volume of harvestable biomass per year (biomass yield); (ii) extraction: methods to extract the metal from the biomass; (iii) manufacturing: nanosynthesis methods using the bio-extracts.

The work focuses on the concentration of lithium (Table 1) measured in soils samples recovered in the vicinity of an adit discharge (Pugh adit—https://cdn.cyfoethnaturiol.cymru/media/679801/cwmystwyth-mine-case-study_2016_06.pdf) at the mine of Cwmystwyth, Ceredigion, Wales (Fig. 1). This adit was initially selected as it showed the highest discharge of zinc in the water basin within this region (zinc being very much related with the main ore worked in this mine which was galena); however, a variety of other metal-containing minerals are present in the area; hence, they are present within the leachate portion that percolates in the surrounding soils. Five samples of soils surrounding the adit discharge (Fig. 1) were selected, 4 within the discoloured area (orange mud) and one downstream (point 4). 8A and C were selected 1 m above the river level, 1E, F, and 4 20 cm above the river level. As can be noted in Table 1, the 2 points downstream 20 cm above the river level show a concentration of Li > 1000 ppm.

Agrostis tenuis was selected as hyperaccumulator due to its presence in close proximity to the adit and in other mining settings of the Welsh area. The plant is a well-known hyperaccumulator of a variety of metals including Cd, Cu, Mn, Ni,

Fig. 1 Pugh adit at Cwmystwyth mine and location of the soils samples retrieved for analyses

Fig. 2 Lithium accumulation (green spots) in Agrostis leaves (TOF–SIMS) (Color figure online)

Pb, Zn, and Sb and as an autochthonous species will not affect the ecosystem. Two different experimental set-ups were selected to evaluate the ability of Agrostis to hyperaccumulate lithium: (1) high concentration method, where the same concentration of lithium detected in point 1F (Table 1) was added to the soil as of lithium nitrate diluted in water to achieve an acceptable pH for plant growth (pH 5). After stabilisation (2 days), the seeds were sawn and the soil was watered every 2 days. (2) Seeds were sawn in clean commercial compost and then lithium nitrate was added in aliquots every 2 days (during watering) to achieve the concentration of point 1F (Table 1). The plants were let to grow for generation generations to assess the level of adaptability to high lithium concentration and the changes in accumulation rate. The plants were harvested every 6 weeks. The biomass harvested was analysed using ICP-OES to evaluate the concentration of lithium and by TOF-SIMS to ascertain where the lithium was localised in the grass leaf. The analysis showed that Agrostis is able to remove the lithium from the soil and store it in the leaves (Fig. 2).

Conclusion

The preliminary data obtained so far demonstrate the ability of Agrostis tenuis to uptake Lithium and move it to its aerial part.

Work on metal extraction and nanosynthesis is ongoing utilising mild acids (acetic and citric) to evaluate the potential to add by-products of other industrial processes supporting the circular economy targets.

References

1. Asad SA, Farooq M, Afzal A, West H (2019) Integrated phytobial heavy metal remediation strategies for a sustainable clean environment—a review. Chemosphere 217:925–941
2. Brooks R, Lee J, Reeves R, Jaffre T (1977) Detection of nickeliferous rocks by analysis of herbarium specimens of indicator plants. J Geochem Explor 7:49–57
3. Ebbs SD, Lasat MM, Brady DJ, Cornish J, Gordon R, Kochian LV (1997) Phytoextraction of cadmium and zinc from a contaminated soil. J Environ Qual 26(5):1424–1430
4. Elektorowicz M, Keropian Z (2015) Lithium, Vanadium and Chromium Uptake Ability of Brassica juncea from Lithium Mine Tailings. Int J Phytoremed 17(1–6):521–528
5. Ernst WHO (2005) Phytoextraction of mine wastes—options and impossibilities. Geochemistry 65:29–42
6. Henschel, J. Mense, M. Harte, P. Diehl, M. Buchmann, J. Kux, F. Schlatt, L., Karst, U, Hensel, A. Winter, M. Nowak, S. Phytoremediation of soil contaminated with lithium ion battery active materials—a proof-of concept study. Recycling 2020(5):26
7. Jiang L, Wang L, Zhang L, Tian C (2018) Tolerance and accumulation of lithium in Apocynum pictum Schrenk. PeerJ 6:e5559. https://doi.org/10.7717/peerj.5559
8. Kavanagh, L. Keohane, J. Cabellos, G. G. Lloyd A. T. Cleary, J. (2017) The viability of a phyto-mining protocol for extracting lithium from low grade ore. In: 3rd International Conference on Environmental Science and Technology (ICOEST) held in Budapest, Hungary on 19–23 October 2017
9. Kavanagh L, Keohane J, Cabellos GG, Lloyd AT, Cleary J (2018) (2018) Induced Plant Accumulation of Lithium. Geosciences 8:56
10. Thijs, S., Langill, T. and Vangronsveld, J. (2017) The Bacterial and Fungal Microbiota of Hyperaccumulator Plants. Adv Bot Res 43–86

An Innovative Separation Process for Spent Lithium-Ion Battery Using Three-Stage Electrodialysis

Ka Ho Chan, Monu Malik, and Gisele Azimi

Abstract An innovative electrochemical process to separate lithium, nickel, manganese, and cobalt using a three-stage electrodialysis with the addition of ethylenediaminetetraacetic acid (EDTA) is proposed. Before the electrodialysis experiment, ultraviolet–visible spectroscopy is used to study the complexation behavior of EDTA with four different metals. During the electrodialysis experiment, a synthetic solution mixed with a stoichiometric ratio of EDTA to the target metal is supplied to a lab-scale electrodialysis stack, in which nickel is separated in the first stage, cobalt is separated in the second stage, and manganese is separated from lithium in the third stage. The results reveal that the concept of separating four different metals in a multi-metallic system using the three-stage electrodialysis is feasible, which could be potentially implemented in other industries that involve the waste valorization of critical metals.

Keywords Lithium-ion batteries (LIBs) · Electrodialysis · Separation · Ultraviolet–visible spectroscopy · Ethylenediaminetetraacetic acid (EDTA)

Introduction

In the past two decades, lithium-ion batteries (LIBs) have been considered as the most dominant energy storage device in electric vehicles due to their high energy and power densities compared with other commercialized batteries [1]. With the increasing popularity of electric vehicles and significant environmental issues, a suitable treatment for end-of-life LIBs has become important. Traditional separation

K. H. Chan · M. Malik · G. Azimi (✉)
Laboratory for Strategic Materials, Department of Chemical Engineering and Applied Chemistry, 200 College Street, Toronto, ON M5S 3E5, Canada
e-mail: g.azimi@utoronto.ca

G. Azimi
Department of Materials Science and Engineering, University of Toronto, 184 College Street, Toronto, ON M5S 3E4, Canada

A. Lazou et al. (eds.), *REWAS 2022: Developing Tomorrow's Technical Cycles (Volume I)*, The Minerals, Metals & Materials Series,
https://doi.org/10.1007/978-3-030-92563-5_17

and recovery processes in the LIB recycling industry comprise chemical precipitation and solvent extraction, which are complicated processes with high chemical consumption and high environmental pollution [2]. Accordingly, the LIB recycling route needs to be improved to reduce the environmental impacts and enhance the sustainable development.

Electrodialysis, as a modern progressive separation technique, has recently gained significant interest in various industries. Compared with the traditional separation and recovery processes, the advantages of electrodialysis are low energy consumption, less waste generation, and long membrane lifetime [3]. The ion-exchange membranes are the most important components in electrodialysis. Under the influence of an electric potential, the ions which have a similar charge with the functional group of the membranes are stopped by the membrane, while the ions with opposite charge to the functional group of the membranes can pass through the membrane.

Most previous studies have focused on using electrodialysis for the separation of two metals in a binary system, such as separation of Li and Co [4] and separation of Ni from Co [5] with the addition of ethylenediaminetetraacetic acid (EDTA). However, the separation of various metals in a multi-metallic system using electrodialysis has not been reported yet. The objective of this study was to separate Li, Ni, Mn, and Co using a three-stage electrodialysis with the addition of EDTA. Before the electrodialysis experiment, the complexation behavior of EDTA with four different metals was studied using ultraviolet–visible spectroscopy. During the electrodialysis experiment, a synthetic solution mixed with a stoichiometric ratio of EDTA to the target metal was supplied to a lab-scale electrodialysis stack and the feasibility of separating four different metals in a multi-metallic system using the three-stage electrodialysis was examined.

Experimental

Characterization of EDTA Complexation

The main drawback of electrodialysis is the non-ideal selectivity of the ion-exchange membranes. Thus, complexing agents such as EDTA have been commonly used to increase the membrane selectivity. When EDTA completely dissociates, the species Y^{4-} combines with the metal ions to form metal-EDTA complexes at certain pH. According to the formation constants of metal-EDTA complexes (Table 1), Ni^{2+} forms complexes with EDTA first and then Co^{2+}, Mn^{2+}, and Li^{+} [6]. To investigate the EDTA complexation behavior with four different metals, EDTA was mixed with metal ions in a stoichiometric ratio at different pH and the absorbance of the metal-EDTA complexes was measured using an ultraviolet–visible spectrometer (Lambda 365).

Table 1 Formation constants of metal-EDTA complexes

Metal ion	K_f
Ni^{2+}	$10^{18.4}$
Co^{2+}	$10^{16.5}$
Mn^{2+}	$10^{13.9}$
Li^{+}	$10^{2.95}$

Synthetic Solution Preparation and Electrodialysis Experimental Procedures

Our previous study shows that ~99% of Li, Ni, Mn, and Co was extracted when the cathode material of spent LIBs was leached in $H_2SO_4 + H_2O_2$ leachant under the optimal operating conditions (temperature = 51 °C, H_2SO_4 concentration = 1.0 M, H_2O_2 concentration = 0.62 wt%, liquid-to-solid ratio = 25.8 mL g^{-1}) [7]. To prepare the synthetic solution, four different metal sulfate salts were dissolved in deionized water and the concentration of the synthetic solution was equivalent to the concentration of the extracted metals in ten times dilution. Figure 1 shows the electrodialysis experimental setup. The lab-scale electrodialysis stack consists of five compartments separated by two anion-exchange membranes (AEMs) and two cation-exchange membranes (CEMs). The electrodes of the electrodialysis stack were connected to a power supply. The feed compartment was filled with the synthetic solution, whereas the other compartments were filled with dilute sulfuric acid. The ion concentrations were determined by inductively coupled plasma optical emission spectrometry (ICP-OES, PerkinElmer Optima 8000) to compute the separation efficiency using the following equation.

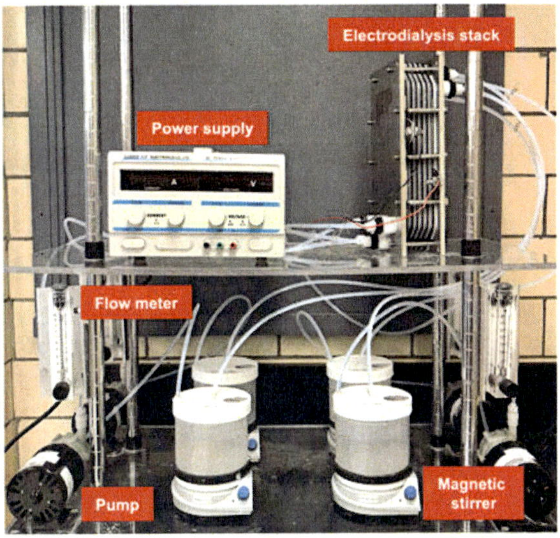

Fig. 1 Electrodialysis experimental setup

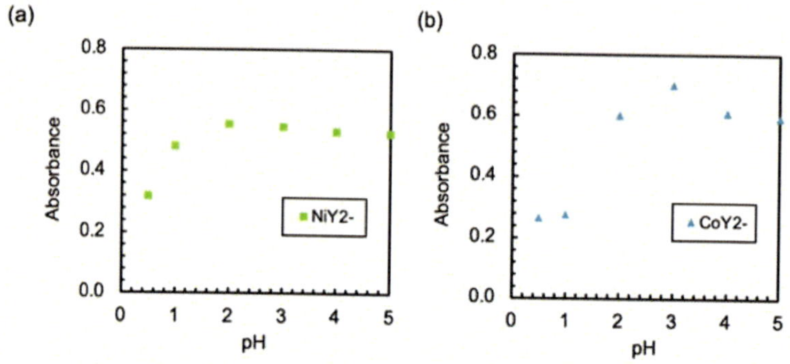

Fig. 2 Absorbance of NiY^{2-} and CoY^{2-} complexes as a function of pH

$$S_i = \frac{C_i - C_f}{C_i} \times 100\% \tag{1}$$

where C_i and C_f are the initial and final concentrations of the target ion i in the feed compartment, respectively.

Results

Complexation Behavior of EDTA

It was observed that only Ni and Co had a color change for the formation of metal-EDTA complexes among all four metals. Figure 2 shows the absorbance of NiY^{2-} and CoY^{2-} complexes at the maximum wavelength as a function of pH measured by ultraviolet–visible spectroscopy. As shown in this figure, the absorbance of NiY^{2-} complexes is the strongest at pH 2 and that of CoY^{2-} complexes is the strongest at pH 3, suggesting that a dominant amount of NiY^{2-} and CoY^{2-} complexes forms at pH 2 and pH 3, respectively.

Working Principle of Three-Stage Electrodialysis

Figure 3 shows the working principle of metal separation using electrodialysis with the addition of EDTA. The electrodialysis stack consists of five compartments separated by two anion-exchange membranes and two cation-exchange membranes. When the electrodes of the electrodialysis stack are connected to the power supply, metal-EDTA complexes pass through the anion-exchange membrane to the metal-EDTA complex recovery compartment, whereas free metal ions pass through the

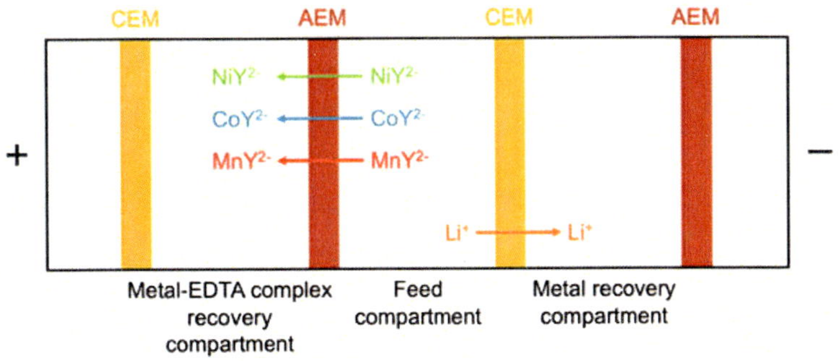

Fig. 3 Working principle of metal separation using electrodialysis with the addition of EDTA

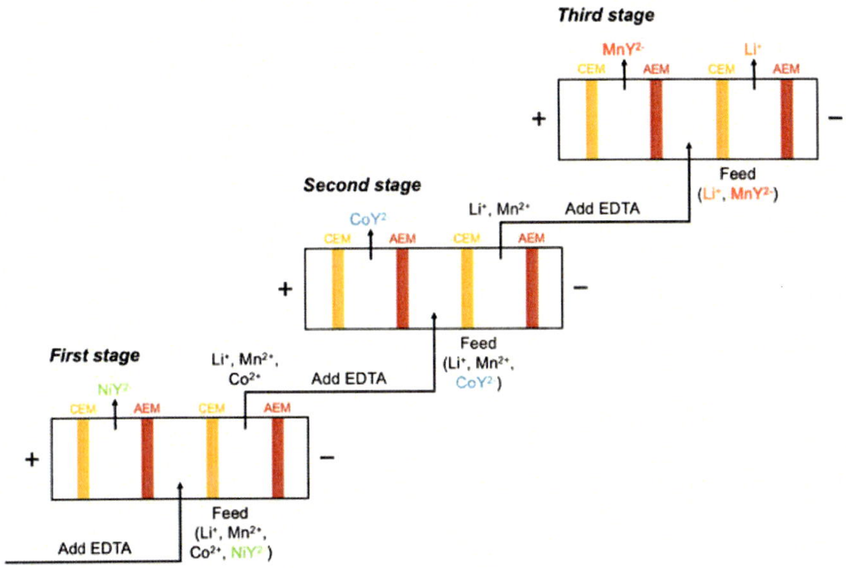

Fig. 4 Process flow diagram of the three-stage electrodialysis

cation-exchange membrane to the metal recovery compartment. Therefore, ions with different charges are separated during electrodialysis.

The proposed method for separating four different metals in the multi-metallic system using electrodialysis consists of three stages as shown in Fig. 4. In the first stage, a stoichiometric ratio of EDTA to Ni is added to the synthetic solution and NiY^{2-} complexes form after adjusting the pH; hence, during electrodialysis, Ni is separated from other ions. After the first stage, the solution from the metal recovery compartment mixed with a stoichiometric ratio of EDTA to the target metal is supplied to the feed compartment in the next stage. Using the same strategy, Co

Table 2 Results of the electrodialysis experiments

Stage	Electric potential (V)	Operating time (h)	Separation efficiency (%)
First	18	3	95
Second			81
Third			25

is separated from other ions in the second stage and Mn is separated from Li in the third stage.

Electrodialysis Results

Table 2 summarizes the results of the electrodialysis experiments. In all three stages of electrodialysis, the electrodialysis stack was operated at 18 V and the experimental operating time was set at 3 h. In the first stage, the synthetic solution mixed with a stoichiometric ratio of EDTA to Ni was supplied to the feed compartment, 95% of NiY^{2-} complexes was separated from other ions. Similarly, in the second stage, 81% of CoY^{2-} complexes was separated from other ions. However, in the third stage, only 25% of MnY^{2-} complexes was separated from Li. Future study is currently underway to further improve the separation efficiency of all four metals.

Conclusions

A three-stage electrodialysis process for separating Li, Ni, Mn, and Co was proposed. Before the electrodialysis experiment, the complexation behavior of EDTA with four different metals was investigated using ultraviolet–visible spectroscopy. During the electrodialysis experiment, the synthetic solution mixed with a stoichiometric ratio of EDTA to the target metal was supplied to the lab-scale electrodialysis stack, in which 95% of Ni was separated in the first stage, 81% of Co was separated in the second stage, and only 25% of Mn was separated from Li in the third stage. The results show that using the three-stage electrodialysis to separate four different metals in the multi-metallic system is feasible. It is believed that this electrodialysis process can improve the LIB recycling route by reducing the environmental impacts and enhancing the sustainable development.

References

1. Tan J, Wang Q, Chen S et al (2021) Recycling-oriented cathode materials design for lithium-ion batteries: Elegant structures versus complicated compositions. Energy Storage Mater 41:380–394. https://doi.org/10.1016/j.ensm.2021.06.021
2. Chan KH, Malik M, Azimi G (2021) Recovery of valuable metals from end-of-life lithium-ion battery using electrodialysis. In: Azimi G, Ouchi T, Forsberg K et al (eds) Rare Metal Technology 2021. Springer International Publishing, Cham, pp 11–17
3. Mosadeghsedghi S, Mortazavi S, Sauber ME (2018) The feasibility of separation of rare earth elements by use of electrodialysis. Extraction 2018. Springer International Publishing, Cham, pp 1831–1838
4. Iizuka A, Yamashita Y, Nagasawa H et al (2013) Separation of lithium and cobalt from waste lithium-ion batteries via bipolar membrane electrodialysis coupled with chelation. Sep Purif Technol 113:33–41. https://doi.org/10.1016/j.seppur.2013.04.014
5. Chaudhary AJ, Donaldson JD, Grimes SM, Yasri NG (2000) Separation of nickel from cobalt using electrodialysis in the presence of EDTA. J Appl Electrochem 30:439–445
6. Harris DC (2010) Quantitative Chemical Analysis. Freeman, W. H
7. Chan KH, Anawati J, Malik M, Azimi G (2021) Closed-loop recycling of lithium, cobalt, nickel, and manganese from waste lithium-ion batteries of electric vehicles. ACS Sustain Chem Eng 9:4398–4410. https://doi.org/10.1021/acssuschemeng.0c06869

Extraction of Nickel from Recycled Lithium-Ion Batteries

Meng Shi, Sabrina M. Reich, Ankit Verma, John R. Klaehn, Luis A. Diaz, and Tedd E. Lister

Abstract A series of operations have been developed to separate and recover individual critical metals from the end-of life lithium-ion batteries (LIB) based on their electrochemical and chemical properties. The black mass from waste LIBs contained Ni, Co, Li, and Mn, as well as contaminates such as Al, Fe and Cu. This paper highlights the leaching of metals and the recovery of Ni as part of a comprehensive recovery scheme. The electrochemical leach successfully dissolved over 97% of these metals into leachate and deposited over 98% of Cu onto cathode. The produced leachate was mildly acidic which could be used directly for Ni extraction through ion-exchange. Purity of the Ni-rich product was over 99%, and the precipitated $NiSO_4$ powder possessed a higher Ni purity at 99.8%. Through these operations, we have successfully developed a feasible and flexible approach in recovery of critical materials and high purity metal salts from recycled LIBs.

Keywords Waste lithium-ion batteries · Electrochemical leaching · Ion-exchange · Ni extraction

---

M. Shi · S. M. Reich · A. Verma · J. R. Klaehn · L. A. Diaz · T. E. Lister (✉)
Idaho National Laboratory, 1955 Fremont Ave., 83415 Idaho Falls, ID, USA
e-mail: Tedd.Lister@inl.gov

M. Shi
e-mail: Meng.Shi@inl.gov

S. M. Reich
e-mail: Sabrina.Reich@inl.gov

A. Verma
e-mail: Ankit.Verma@inl.gov

J. R. Klaehn
e-mail: John.Klaehn@inl.gov

L. A. Diaz
e-mail: Luis.DiazAldana@inl.gov

A. Lazou et al. (eds.), *REWAS 2022: Developing Tomorrow's Technical Cycles (Volume I)*, The Minerals, Metals & Materials Series,
https://doi.org/10.1007/978-3-030-92563-5_18

Introduction

Since 1991 when the LIBs became commercially available, the demand for LIBs has been growing due to its wide application in areas such as power backups, portable electronic goods, electric mobility, and energy storage systems [1]. Facing a possible shortage of Li, Co, and Ni in the near future for manufacturing LIBs, the Department of Energy made investments in advanced battery material supply chains and technologies [2]. Due to the growing costs in mining and problems such as human rights, labor, and environment abuses, producers of goods using critical metals started seeking solutions to reduce reliance on the existing supply chains. For example, Tesla announced its strategy in Co-free batteries in 2020 [3]. This announcement on the Tesla Battery Day encouraged a stronger Ni supply chain for batteries used in electronic vehicle.

While mining is the primary source of Ni products, recycling from used LIB materials is becoming an important alternative source. In June 2021, the National Blueprint for Lithium Batteries (2021–2030) states that recycling LIB materials at a competitive price as one of the five goals to establish a secured battery materials and technology supply chain [4]. Encouraged by economic markets, limited storage, and national strategy, studies on battery materials recycle and reuse become more important and attractive to research groups in academic and industrial institutes.

There are international companies and research institutes working on LIB recycling, such as Retriev Technologies (USA) and Li-Cycle (Canada). Methods in recycling battery materials can be classified into three types: (1) direct recycling, (2) pyrometallurgy, and (3) hydrometallurgy [5–7]. Direct recycling recovers useful battery components through mostly physical recycling processes with minimum chemical treatments [6, 7]. This method has a short recovery route and a high recovery rate. However, various battery types and materials make the direct recycling a challenge due to technological change and uncertainty in long-term performance. Therefore, direct battery recycling is not economically favorable [8]. It can also be assumed that direct recycling is limited in cycles such that more comprehensive recovery of a material is eventually required. Pyrometallurgy and hydrometallurgy are chemical approaches to recovering individual materials from spent batteries. Pyrometallurgy processes employ high-temperature smelting where graphite is consumed and primarily Ni, Co, and Cu recovered with Li reporting to the slag [6]. To separate individual metals and recycle Li, hydrometallurgy processes are generally required at the back end of pyrometallurgy process. However, owing to the energy consumption and carbon footprint, safety concerns at high temperatures, and growing interests in Co-free LIBs, pyrometallurgy processes have less development prospects. Compared with the traditional pyrometallurgy method, hydrometallurgy process offers several advantages in terms of lower energy consumption and capital cost, along with higher recovery rate, product purity, and selectivity [9]. Leaching, separation, and extraction are involved in the process under mild working conditions. Nevertheless, improvements can be made to reduce chemical usage, reduce process steps, and minimize wastewater generation [10].

This paper focuses on the recovery of high-purity Ni as part of a project to recover metals from waste LIBs. Our research group is attempting to develop scalable processes with minimum chemical consumption. The scheme starts with black mass (BM), an intermediate product obtained from the physical beneficiation of LIBs. BM is a mixture of both anode and cathode chemistry containing graphite, metal oxides (Li, Mn, Co, Ni, Al) and impurities such as Fe and Cu [5]. An electrochemical leaching (EC-Leach) process effectively extracts cathode metal oxides into the liquid phase while recovering graphite and Cu metal [5]. Compared with our previous work, the leaching process has been scaled to process 500 g black mass and generates 2 L leachate per day. The mild acid leachate (pH > 1) is rich in Ni, Co, Li, and Mn. Ni was separated from the leachate in the form of sulfate salts through ion-exchange (IX) steps [11, 12]. The current IX system can process 300 mL leachate and recovery around 4 g Ni per day. Advantages of employing IX method included simple operation, lack of volatile organic chemicals, good selectivity, wide operational pH range, and simple contaminant removal steps. After producing rich eluates, high purity $NiSO_4$ powder was produced by evaporation and crystallization.

Experimental

Materials

Potassium hydroxide (KOH, Fisher Scientific, > 85%), iron (II) sulfate ($FeSO_4$, Sigma-Aldrich, $\geq$ 99.0%), hydrochloric acid (HCl, Fisher Scientific, 36.5% ~ 38%) and sulfuric acid (H_2SO_4, Fisher Scientific, 95–98 wt.%) were used as received. Nano-pure water used in this study was obtained from the Milli-Q system (Millipore, 18.2 MΩ). LIB black mass, a mixture of shredded anode and cathode materials, was provided by Retriev Technologies. The metal composition of BM was determined by analyzing liquid phase of an aqua regia digestion (Table 1).

Table 1 Weight of metals in BM (in wt.%), metal compositions in leachate (1110 mL), and recovery efficiency of leaching process

	Li	Cu	Al	Mn	Fe	Co	Ni	Zn
Metals in BM /wt.%	2.8	0.4	0.6	2.4	0.4	9.4	12.0	0.0
Leachate composition /mg/L	6124.1	9.8	1252.6	5690.4	1107.9	24573.8	27573.9	15.5
Metals in leachate /g	6.80	0.01	1.4	6.3	1.2	27.3	30.6	0.02
Leaching recovery efficiency (%)	97.4	1.2	98.8	105.5	88.3	115.5	102.1	74.9

EC-Leach

The electrochemical leaching system was designed and build in house [5]. The leaching setup consists of two identical electrochemical cells connected in series. A typical EC-Leach reactor was a 1500 mL cell divided into two compartments (anode and cathode) by a bipolar membrane Fumasep FBM-PK. The anolyte and catholyte compartments had a capacity of 400 mL and 1000 mL, respectively. The anode was a double-layered Ni mesh with a total immersed surface area of 450 cm^2, and the cathode was a stainless-steel mesh with an immersed surface area of 270 cm^2. The anolyte was 350 mL of 1 M KOH, and loaded catholyte was 800 mL mixture of 2 M H_2SO_4 and 10 mM $FeSO_4$. Agitation was applied in the cathodic bath with an overhead mixer (XZB ELEC, DX-120D) at a rotating speed of 450 rpm. Potential was provided by a DC power supply (BK Precision, 1735A) after the first 50 g BM was mixed with the catholyte. Then, another 50 g of BM was added every other 10 min until a total of 250 g BM was reached. The pH of catholyte was monitored by a Black Stone BL 931700 pH controller. The controller operated a microtube peristaltic pump (Masterflex® C/L EW-77120) which delivered 4 M H_2SO_4 to maintain the pH of catholyte at 1.00. A constant potential of 4.2 V was applied to a series of two identical cells (2.1 V per cell) for 24 h. Catholyte mixture was filtered to remove undissolved materials (primarily graphite) from leachate. All experimental operations were performed at room temperature on a bench top. Table 1 shows the composition of the obtained leachate and the efficiencies of the EC-Leach process.

Ion-Exchange

Ni was absorbed from the leachate using AmberSep™ M4195 chelating anion exchange resin (previously known as DOWEX® M4195) loaded into PVC pipes (inner diameter: 0.75 in, length: 3 in). The bed volume was 20 mL. Newly packed columns were washed with nano-pure water before use. Leachate and IX eluent were injected in the system with a microtube peristaltic pump at a flow rate 0.5 mL/min.

Previous work developed the recovery of Ni and Co from LIB leachate using IX [12]. A series of three packed columns were used for each loading and elution cycle (Fig. 1). A 20 mL volume of leachate was loaded through the bottom of the first column in series (Fig. 1a). After the leachate was fed, the first column was washed with 30 mL 5 g/L H_2SO_4 from the top followed by 5 mL 2 M H_2SO_4 (Fig. 1b). The leachate coming from the bottom of column #1 passed through columns #2 and #3 until 20 mL of effluent were collected (same as loaded). Then, the first column was eluted with 15 mL 2 M H_2SO_4 followed by 20 mL nano-pure water (Fig. 1c). The loading column was alternated among three columns. Cycles of loading and elution were continued until 500 mL of leachate was processed.

Around 875 mL of Ni-rich eluate were obtained on each 500 mL cycle. Then, the eluate was evaporated on a hot plate until around 1/10 of the initial volume. The

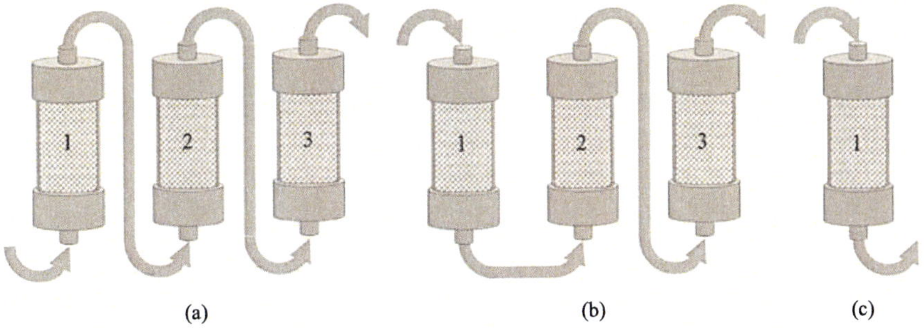

Fig. 1 IX setup for Ni extraction. **a** Leachate loading, **b** column washing, and **c** Ni elution

concentrated $NiSO_4$ solution was allowed to cool overnight in a covered container. A layer of light green solids grew at the bottom of the container. The generated solids were separated through vacuum filtration and dried in an oven at 75 °C.

Analysis

Analysis of metal concentration in the different streams and feedstock was performed with the atomic absorption spectroscopy (AAS, Agilent, 240FS AA). Calibration standards were prepared for each metal from commercially available standard solutions. Prepared standard solution was 1 mg/L for Zn, 40 mg/L for Al, and 5 mg/L for all other metals. Liquid samples were diluted with 3% HCl. Solid samples were firstly dissolved in either aqua regia or concentrated HCl, and then diluted with 3% HCl. A previous study on the AAS analysis proved that there was error in the AAS measurements [13]. This study determined the maximum errors in metal concentration measurements. For example, the maximum measurement errors for Ni and Co were both at 10%.

Results and Discussion

EC-Leach of Waste LIB Materials

In the EC-Leach process, the bipolar membrane splits water, allowing H^+ to transfer to the cathode and OH^- to the anode (Eq. 1). Therefore, the anolyte and catholyte maintain different pH environments while avoiding electrolyte crossover. At the anode, O_2 is evolved from the transferred OH^- (Eq. 2).

$$H_2O \rightarrow H^+ + OH^- \quad (1)$$

$$4OH^-(aq) \rightarrow O_2(g) + 2H_2O + 4e^- \quad (2)$$

Fe^{2+} was employed as a reducing agent to convert metals in spent LIB oxide cathode material to their soluble form (Eq. 3) [5]. As Fe^{3+} was generated via Eq. 3, it then acted as an oxidation agent to dissolve small amounts of Cu foil in the waste battery materials (Eq. 4). Unreacted Fe^{3+} is then reduced to Fe^{2+} at the cathode (Eq. 5) and the cycle continues until all the metals have been leached. The cathode also reduced dissolved Cu^{2+} forming a deposit on the cathode mesh (Eq. 6).

$$\begin{aligned} &LiNi_xMn_yCo_zO_2(s) + Fe^{2+}(aq) + 4H^+(aq) \rightarrow \\ &Li^+(aq) + xNi^{2+}(aq) + yMn^{2+}(aq) + zCo^{2+}(aq) + Fe^{3+}(aq) + 2H_2O \end{aligned} \quad (3)$$

$$2Fe^{3+}(aq) + Cu(s) \rightarrow Cu^{2+} + 2Fe^{2+}(aq) \quad (4)$$

$$Fe^{3+}(aq) + e^- \rightarrow Fe^{2+}(aq) \quad (5)$$

$$Cu^{2+}(aq) + 2e^- \rightarrow Cu(s) \quad (6)$$

During the leaching process, the overall reaction (Eqs. 1–6) can be condensed as shown in (Eq. 7) where the EC-Leach allows a decrease in acid consumption. However, as protons were consumed, 4 mL 4 M H_2SO_4 was pumped to catholyte to maintain the pH value < 2 and avoid Fe precipitation.

$$\begin{aligned} &LiNi_xMn_yCo_zO_2(s) + 3H^+(aq) \rightarrow \\ &Li^+(aq) + xNi^{(2+)}(aq) + yMn^{(2+)}(aq) + zCo^{(2+)}(aq) + 3/2H_2O + 1/4O_2(g) \end{aligned} \quad (7)$$

After filtration and washing of the remaining solids, 1110 mL and 1000 mL of leachates were collected from each cell. The obtained leachate was a dark burgundy solution at pH 1.45. In this manuscript, discussions and product processes were based on the 1110 mL leachate product. Efficiency of the leaching process is provided in Table 1. Leaching efficiencies of metals of interests such as Li, Mn, Co and Ni were high. Efficiency results higher than 100% are due to the errors in instrument and operations. Over 97 wt.% Li, Mn, Co, and Ni were recovered in the leachate. Ni and Co concentrations were 24.6 g/L and 27.6 g/L, respectively. Li and Mn compositions were 6.1 g/L and 5.7 g/L, respectively. Impurity metals Al and Fe were greater than 1100 mg/L. 88.3 wt.% of Fe was extracted in leachate, taking account of the amount of Fe initially added to the catholyte. During EC-leach, the cathode was plated with a Cu layer, leaving only 1.2% of original Cu in leachate, removing 98% of the leached Cu. This effectively eliminates a process step to remove Cu. In conclusion, it proved that the EC-leach process could effectively extract Li, Mn, Co, and Ni to leachate, recover Cu on the cathode, and leave graphite as a solid which could be recycled after leaching.

Ni Extraction Through Ion-Exchange

500 mL from the leachate solution generated in Sect. 3.1 was tested for Ni separation through IX without adjusting pH. The flowrate was controlled at 0.5 mL/min which was the optimized flow rate based on the column bed volume [14]. After Cu, which was removed during the leaching process, Ni possesses the strongest affinity to the resins, compared with other metals in leachate. AmberSep™ M4195 resins have been widely employed to process leachates different that LIB leachates for the separation of individual metals Ni, Co, Li, Pb, Fe, and Mn using various processing mediates [14–16].

Following the operations described in Sect. 2.3, three liquid products were obtained: effluent, wash, and eluate (Fig. 2). Under mild acid conditions, Ni favorably bonded to the resins while other metals passed through the columns allowing the selective loading of the resin. The effluent contains the remainder of metals with 76% of Ni removed. Then, the binding of Ni to the resins was relieved by elution with 2 M H_2SO_4. Between loading and elution, the loaded column needs to be washed to remove retained leachate and avoid contamination of the eluate. Table 2 summarized the metal compositions in products after loading 500 mL leachate. Metals were not perfectly balanced due to errors in analysis and metal retentions in columns. The effluent was a Co-rich solution containing 66.5% of Co and 23.8% of Ni of the

Fig. 2 Pictures of leachate IX products, including eluate, effluent, and wash (from left to right)

Table 2 Metal compositions in leachate IX separation products (unit: mg)

Product	Li	Cu	Al	Mn	Fe	Co	Ni	Zn
Effluent	2048.4	0.0	377.3	1596.4	403.8	8171.1	3282.5	3.9
Wash	320.5	0.0	56.4	317.9	60.6	1127.2	2782.3	0.3
Eluate	3.2	3.0	0.0	3.8	0.7	40.9	6553.7	0.0

loaded leachate, which could be processed in Co extraction in future experiments. Eluate was the Ni extraction product containing 47.5% of Ni in leachate. This Ni-rich solution had a Ni purity of 99.2% (metal basis). The wash solution had a pH of 0.93 and thus could be reused as part of catholyte in future EC-leach operation.

Eluate products from similarly processed leachate solutions were mixed with this batch and processed for $NiSO_4$ crystal generation (Table 3). After solid $NiSO_4$ was separated, it left a saturated $NiSO_4$ solution. Green $NiSO_4$ powders (Fig. 3) were formed after dried in oven at 75 °C overnight. The $NiSO_4$ powder had a Ni purity

Table 3 Compositions of Ni-rich solution mixture and saturated Ni solution and $NiSO_4$ powder after evaporation

Product	Li	Cu	Al	Mn	Fe	Co	Ni	Zn
Ni-rich solution/mg	6.7	9.7	1.3	6.3	6.4	89.1	16459.8	0.1
Saturated solution/mg	0.1	0.0	0.0	0.3	0.0	0.0	1273.2	0.0
$NiSO_4$ powder/mg	3.1	0.0	0.00	9. 6	0.00	16.1	13625.1	0.0
$NiSO_4$ powder (Metal basis) (%)	0.02	0.00	0.00	0.07	0.00	0.12	99.79	0.00

Fig. 3 $NiSO_4$ produced from Ni-rich solutions after evaporation

of 99.8% (metal basis). Ni in the powders was 31.93 wt.% of the total $NiSO_4$ solids generated, which was higher than the battery-grade Ni purity requirement (22.3 wt.%, CoreMax Nickel Sulfate-Battery Grade). The weight percentage of Ni in the $NiSO_4$ products was highly dependent on the amount of residual water retaining in the powder.

Conclusions

A multi-step process including electrochemical leaching and IX separation with low chemical consumption was demonstrated in this work. Starting from BM, the electrochemical leaching could effectively dissolve Li, Mn, Co, and Ni into leachate and deposit Cu on electrode. The leaching method works well with BM from a wide variety of battery types (Ni or Co rich). The acidic leachate could be utilized directly in the IX columns to separate Ni component. IX products included a Ni-rich solution with Ni purity > 99%, a Co-rich phase for future Co extraction, and a wash solution which could be recycled in future leaching operations. After evaporation, high-quality $NiSO_4$ powder could be produced. Ni component in the received $NiSO_4$ powder had a higher concentration than that in battery-grade $NiSO_4$.

Acknowledgements This manuscript has been authored by Battelle Energy Alliance, LLC under contract no. DE-AC07-05ID14517 with the U.S. Department of Energy. This work was supported by the Critical Materials Institute (CMI), an Energy Innovation Hub funded by the U.S. Department of Energy, Office of Energy Efficiency and Renewable Energy, Advanced Manufacturing Office. The authors would like to acknowledge Retriev Technologies for providing the black mass used in this study as well as technical guidance.

References

1. Lithium Ion Battery Applications and Advantages. https://www.eletimes.com/lithium-ion-battery-applications-advantages
2. DOE Announces Actions to Bolster Domestic Supply Chain of Advanced Batteries. https://www.energy.gov/articles/doe-announces-actions-bolster-domestic-supply-chain-advanced-batteries, Accessed 8 June 2018
3. Morris J, Tesla Battery Day 2020: Big Elon Musk EV Announcements Expected On September 22. https://www.forbes.com/sites/jamesmorris/2020/09/06/tesla-battery-day-2020-big-elon-musk-ev-announcements-expected-on-september-22/?sh=721a31c63638
4. National Blueprint for Lithium Batteries. https://www.energy.gov/eere/vehicles/articles/national-blueprint-lithium-batteries
5. Diaz LA, Strauss ML, Adhikari B, Klaehn JR, McNally JS, Lister TE (2020) Electrochemical-assisted leaching of active materials from lithium ion batteries. Res Conserv Recycl 161:104900. https://doi.org/10.1016/j.resconrec.2020.104900
6. Zhou L-F, Yang D, Du T, Gong H, Luo W-B (2020) The current process for the recycling of spent lithium ion batteries. Front Chem 8:1027. https://doi.org/10.3389/fchem.2020.578044

7. Ciez RE, Whitacre JF (2019) Examining different recycling processes for lithium-ion batteries. Nat Sustain 2(2):148–156. https://doi.org/10.1038/s41893-019-0222-5
8. Harper G et al. (2019) Recycling lithium-ion batteries from electric vehicles. Nature 575(7781):75–86. https://doi.org/10.1038/s41586-019-1682-5
9. Chen M et al. (2019) Recycling end-of-life electric vehicle lithium-ion batteries. Joule 3(11):2622–2646. https://doi.org/10.1016/j.joule.2019.09.014
10. Meshram P, Abhilash, Pandey BD, Mankhand TR, Deveci H (2016) Comparision of different reductants in leaching of spent lithium ion batteries. JOM 68(10):2613–2623. https://doi.org/10.1007/s11837-016-2032-9
11. Strauss ML (2020) Improved methods for separation of Cobalt, Manganese, Nickel, Lithium from Waste Lithium-ion Battery Scrap (Conference: Society of Mining, Metallurgy and Exploration (SME) 2020, Phoenix Convention Center, 02/23/2020 - 02/26/2020). Idaho National Lab. (INL), Idaho Falls, ID (United States), 2020, p. Medium: ED
12. Strauss ML, Diaz LA, McNally J, Klaehn J, Lister (in press) Cobalt, Nickel, and Manganese Separation from Waste Lithium-Ion Batteries Using Ion Exchange. Hydrometallurgy
13. Govett GJS, Whitehead RE (1973) Errors in atomic absorption spectrophotometric determination of Pb, Zn, Ni and Co in geologic materials. J Geochem Explor 2(2): 121–131. https://doi.org/10.1016/0375-6742(73)90011-3
14. Rosato L, Harris GB, Stanley RW (1984) Separation of nickel from cobalt in sulphate medium by ion exchange. Hydrometallurgy 13(1):33–44. https://doi.org/10.1016/0304-386X(84)90015-X
15. Nguyen TH, Lee MS (2018) A review on the separation of lithium ion from leach liquors of primary and secondary resources by solvent extraction with commercial extractants. Processes 6(5):55. https://www.mdpi.com/2227-9717/6/5/55.
16. Diniz CV, Ciminelli VST, Doyle FM (2005) The use of the chelating resin Dowex M-4195 in the adsorption of selected heavy metal ions from manganese solutions. Hydrometallurgy 78(3–4):147–155. https://doi.org/10.1016/j.hydromet.2004.12.007

Development of Technology for Recycling Large-Capacity Lithium-Ion Batteries for EV, ESS

Hong-In Kim, Jeong-Soo Sohn, Soo-Kyung Kim, Dong-Hyo Yang, and Suk-hyun Byun

Abstract The use of lithium secondary batteries for electric vehicles is increasing rapidly. In order to recycle batteries for electric vehicles, the discharge process is essential. The lithium-ion battery pack that is 227 kg and is series with 12 modules is composed of 20% cover materials, 14.4% plastics, and 59% battery cells. Each module which is series with 11 unit cells is composed of 18% frame body, 12% heat-reduce plate, and 70% battery cells. The waste LIBs pack was prepared by physical treatment, including hand disassembly, electric discharging, drying, and crushing. In the leaching process, 1mol/L H_2SO_4 and 10vol% H_2O_2 are added to the sulfuric acid reduction leaching and waste water reduction leaching process, and Al, Cu, Fe, etc. contained as impurities are controlled to be less than 10mg/L by precipitation as a pH adjustment. The solution with controlled impurities is applied to the solvent extraction system process with PC88A to selectively separate and extract manganeses, cobalt, and nickel. Finally, lithium left in the solution was concentrated to 18g/L and then, was precipitated as Li_2CO_3 with efficiency over 88%.

Keywords Lithium-ion battery · Electric vehicle · Recycling · Discharge · Dismantling

Introduction

The lithium-ion batteries used in electric vehicle are composed of one pack. This pack is composed of several modules, and each module is composed of several cells. The weight of battery pack is above 400 kg, but the weight percent of cell is only 60%. The other 40% is not lithium-ion cells but another components such as battery management system (BMS), power relay assembly (PRA), safety plug,

---

H.-I. Kim (✉) · J.-S. Sohn · S.-K. Kim · D.-H. Yang
Korea Institute of Geosciences and Mineral Resources (KIGAM), Daejeon 34132, Korea
e-mail: hi@kigam.re.kr

S. Byun
SungEelHitech. 143-12, Gunsansandan-ro, Gunsan-si 54002, Jeollabuk-do, Korea

A. Lazou et al. (eds.), *REWAS 2022: Developing Tomorrow's Technical Cycles (Volume I)*, The Minerals, Metals & Materials Series,
https://doi.org/10.1007/978-3-030-92563-5_19

cable, cooling system and pack case. Recycling technologies for valuable metals such as cobalt, nickel, manganese and lithium are now commercialized, but non-battery components in battery pack of electric vehicles are not fully recycled [1]. In order to complete the recycling of spent battery pack in electric vehicles, recycling of non-battery components is necessary in addition to lithium-ion cell recycling. So in this study, we introduced the physical treatment processes such as dismantlement, crushing, grinding, magnetic separation, and size separation for non-battery components recycling.

Unlike the currently commercialized small lithium-ion battery recycling process, large-capacity lithium-ion battery pack dismantling and sorting process takes a lot of time and high process cost, so the interest of domestic recycling companies is low. Due to the composition change to secure stability due to the enlargement of lithium-ion batteries, Mn and Ni components are predominant rather than Co components, so the economic feasibility of recycling is low. There is room to gain an edge in the competition. Belgium, China, and Korea are currently competing for recycling of small waste lithium-ion batteries, and in the case of large-capacity lithium-ion batteries, the amount of rare metals used will increase. It will be possible to increase the competitiveness of the related industry by contributing to securing it. As the amount of waste lithium-ion batteries generated by electric vehicles is expected to gradually increase in the future, there is a limit to the existing small-sized lithium-ion battery treatment methods. It is necessary to establish recycling system through classification by composition and characteristics of waste lithium-ion batteries. Currently, the composition of waste lithium-ion batteries is diversely distributed into LCO-based, LMO-based, NCM-based, etc., but a recycling system for each composition has not been established [2]. Therefore, it is difficult to develop a recycling process. In order to solve the problem of reliance on imports of cathode active material while the secondary battery industry is developed like Korea, an efficient large-sized lithium secondary battery recycling technology must be established. Active exchanges between companies that produce batteries and research institutes participating in process development are required. The cathode material used in secondary batteries for automobiles has to be used with materials that can secure the long-term reliability and safety of the battery among various characteristics [3].

However, in the long term, in order for electric vehicles to realize the performance of an internal combustion engine vehicle with a driving distance of 500 km or more per charge, it is necessary to develop a new concept battery that is higher than the current lithium secondary battery. Further development of core component materials such as cathode materials is required.

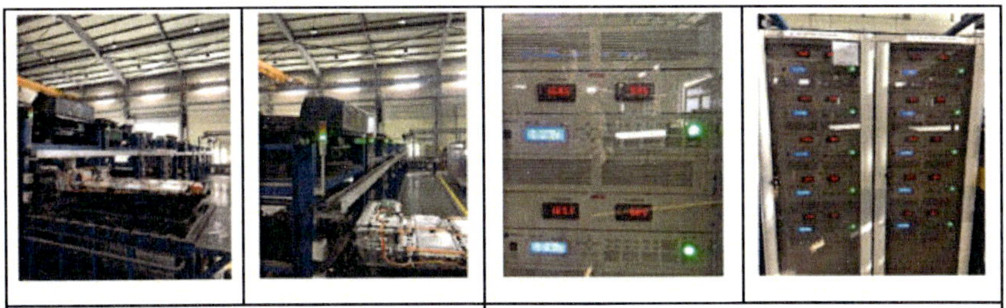

Discharge system and storage rack	10set Discharger System(163V, 10A)

Fig. 1 Lithium-ion battery storage and discharge system for electric vehicle

Discharge System

In order to recycle batteries for electric vehicles, the discharge process is essential. This is because batteries for electric vehicles require high-output power, so high-voltage, high-capacity batteries are used. In order to dismantle or crush it, a discharge process is required.

The initial voltage before discharging of the electric vehicle module was 35.45 V, and the voltage was lowered to 5.77 V after 6 h of starting discharging by connecting this pack to the discharger. After that, the discharger was removed and the voltage measured again the next day was 6.55 V, and the discharger was connected to perform discharging again. After 30 min of discharging, the voltage measured was 1.078 V, and it was continuously discharged, but it was confirmed that it did not drop significantly to 0.941 V even after 1 h. As a result, it was found that an additional discharge is required for a complete discharge.

The installed EV discharge system can discharge 10 units at the same time. The display and buzzer can be used during discharging when discharging is complete. By applying 10 dischargers, the EV pack can be discharged at the same time, so that 50 units/day discharge system is possible (Fig. 1).

Dismantling Process

The spent lithium-ion battery packs were obtained from the domestic car manufacturer as a prototype that will be sold to the automobile market. The lithium-ion battery pack that is 227 kg and is series with 12 modules is composed of 20% cover materials, 14.4% plastics, 59% battery cells. The dismantling of lithium-ion batteries for electric vehicles takes the most time in the process of separating them into cells.

Each module which is series with 11 unit cells is composed of 18% frame body, 12% heat-reduce plate, 70% battery cells. The waste LIBs pack was prepared by physical treatment, including hand disassembly, electric discharging, drying, and crushing. Spent LIBs module was first treated manually disassembly to frame, each

Table 1 Dismantling time and parts for battery packs for electric vehicles

No	Before dismantling	Tool	Tmie (sec)	After dismantling
1		10 mm bolt 8ea	60	
2		10 mm bolt 12ea cross-bolt 5ea	30	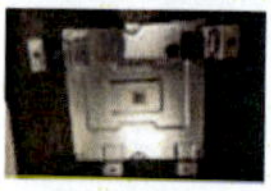
3		10 mm bolt 14ea 13 mm bolt 2ea cross-bolt18ea	660	
4		10 mm bolt 34ea	480	
5		10 mm bolt 54ea cross-bolt 32ea insulated scissors	3,080	

battery cell, and Al cooling plate. The frame and Al cooling plate can be reused at manufacturing step. Battery cell was treated manually disassembly with press machine, of which it was separated with Al laminate casing, cathode and anode materials, and separator. The cathode and anode materials were soaked in distilled water for electric discharging for 1 h at 25 °C. After dehydration and drying, electric discharged cathode and anode materials were subjected to crushing and screening.

After separating the waste lithium-ion battery for ESS, the final product, the cell, was dismantled. As for the main material constituting the waste lithium-ion battery for ESS, the cell accounted for 59.0%, similar to the car pack, followed by the case with iron, accounting for 21.3%. The weight and composition ratio of the anode electrode, cathode electrode, and separator contained in the cell, which occupies 59.0%, are shown in Table 2. This is similar to the previously analyzed electric vehicle battery, and it can be seen from the analysis result of the cathode active material that an NCM ternary cathode electrode material having a composition of 1:1:1 was used.

For the effective recycling of waste lithium-ion batteries for ESS, it is necessary to develop pretreatment technology that considers risk factors such as explosion and fire that may occur when separating/disassembling unit cell into components such as can, electrodes, electrolytes, and separators.

The following results were obtained by disassembling the lithium-ion battery cell for electric vehicle (Table 2).

Table 2 Dismantling rate of battery packs for electric vehicles

Classification	Weight (kg)	Ratio (%)
Cell	160.99	59.0
Al	7.04	2.6
Cu	7.15	2.6
Fe	58.06	21.3
PCB	0.35	0.1
Valuable resources (%)	233.59	85.6
Plastic	39.25	14.4
Etc	0.08	0.0
Total waste (%)	39.33	14.4
Total	272.92	100.0

Table 3 Cell decomposition results for electric vehicles

Classification	Cathode	Cathode active material	Al foil

	358 g	286.4 g	71.6 g
	Anode	Graphite	Cu foil
Cell (639.6 g)			
	218 g	174.4 g	43.6 g
	Case	separator	Terminals

	63.5 g	56.8 g	48.5 g

Spent EV/ESS-Batteries Recycling

Lithium-ion batteries for electric vehicles and lithium-ion batteries for ESS can be implemented by designing the recycling process in the same way as in Fig. 2.

This research aims to introduce the results of a commercialization process study on the stable dismantling and heat treatment of waste mixed batteries by mixing lithium-ion batteries, lithium primary batteries, manganese alkaline batteries, nickel hydride batteries, etc. First, in the case of lithium-based mixed batteries, the ratio of lithium-ion batteries and lithium primary batteries is mixed at 4:1, a stable heat treatment furnace is used to explode at a temperature higher than 400°C, and then the crushing and grinding process is conducted. This heat treatment furnace allows disposal of 200 kg/batch waste cells. Also, through a stable heat treatment process, electrolytes in the battery that are harmful to the human body can be evaporated and removed, while separating separators and plastics that adversely affect crushing and grinding can be removed.

The obtained powder is injected into the leaching equipment system. In the leaching process, 1 mol/L H_2SO_4 and 10 vol% H_2O_2 are added to the sulfuric acid reduction leaching and wastewater reduction leaching process, and Al, Cu, Fe, etc., contained as impurities are controlled to be less than 10 mg/L by precipitation as a pH adjustment. The solution with controlled impurities is applied to the solvent extraction system process with PC88A to selectively separate and extract manganeses, cobalt, and nickel. After recovering cobalt and manganese, the nickel was recovered as $Ni(OH)_2$ by precipitation process or nickel was extracted by solvent extraction process at pH higher than 6.4. Finally, lithium left in the solution was concentrated to

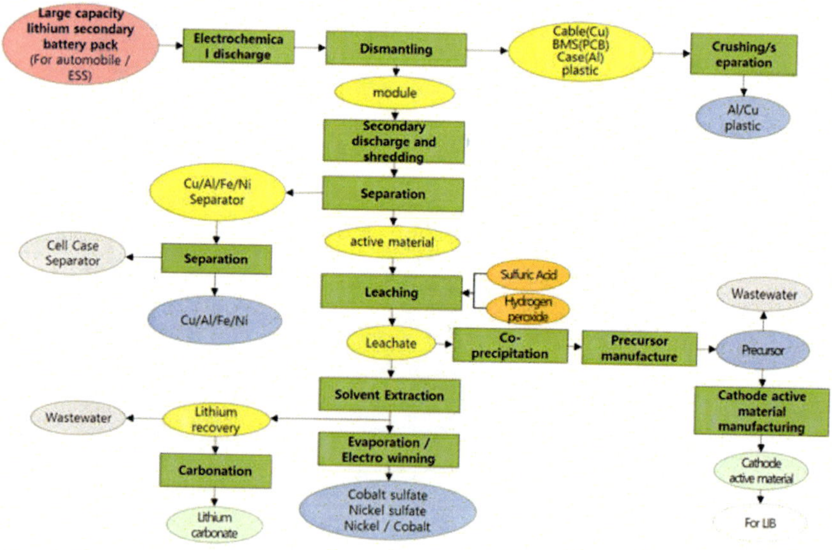

Fig. 2 Process diagram of lithium-ion battery pack for electric vehicles

18 g/L and then was precipitated as Li_2CO_3 with efficiency over 88%. The remaining 12% of lithium was precipitated as Li_3PO_4 at room temperature with efficiency over 90%. Therefore, the recovery efficiency of lithium was more than 96%. For non-lithium-based batteries, a solution with controlled impurities is obtained through the wastewater reduction process and then subjected to a solvent extraction process. Zinc metal and manganese zinc sulfate can be obtained by solvent extraction using D2EHPA as the extractant. In this new development process approximately 90% of the valuable metals such as cobalt, manganese, nickel, lithium, and zinc was recovered.

Copper, aluminum, and plastic separated during crushing become reusable through a physical sorting process, and a mechanical sorting system is established to ensure the complete separation by material.

To separate the active material from the lithium-ion battery used in electric vehicles, crushing, pulverization, and sorting are employed. Then, a high-purity nickel and cobalt solution is produced through leaching, impurity removal and electrolytic extraction. The recycled nickel and cobalt sulfate are used as raw materials to manufacture new lithium secondary batteries. Based on the hydrometallurgical process, we have obtained a complete recycling and commercialization technology for waste lithium secondary batteries from electric vehicles (Fig. 2).

By applying the discharging, dismantling, crushing, and separation processes of lithium-ion battery packs for electric vehicles, we built an industrial facility with a capacity of 50 units/day yielding a process that can dismantle 15,000 units/year. The technology can be applied for large-capacity industries treating waste lithium-ion battery packs from electric vehicle.

Conclusions

The use of lithium secondary batteries for electric vehicle is increasing rapidly. Discharge process is essential for recycling large-capacity, high-capacity lithium secondary batteries for automobiles. Recycling of lithium-ion batteries for electric vehicles requires pretreatment processes such as dismantling. After dismantling and crushing the lithium-ion battery, it can be treated in a process similar to the small battery treatment process. In general, a leaching process using sulfuric acid and hydrogen peroxide is used for the leaching process of the lithium-ion battery active material. In the separation process of the leaching solution, a solvent extraction process is used to recover manganese, cobalt, nickel and lithium. After the solvent extraction process, it is made of cobalt, nickel sulfate and metals through a crystallization process and an electrowinning process to recover cobalt and nickel metals.

References

1. Kotak Y, Fernández CM, Casals LC, Kotak BS, Koch D, Geisbauer C, Trilla L, Gómez-Núñez A, Schweiger H-G (2021) End of electric vehicle batteries: reuse vs. recycle. Energies 14:2217
2. Beaudet A, Larouche F, Amouzegar K, Bouchard P, Zaghib K (2020) Key challenges and opportunities for recycling electric vehicle battery materials. Sustainability 12:5837
3. Global EV (2021) Outlook. https://iea.blob.core.windows.net/assets/ed5f4484-f556-4110-8c5c-4ede8bcba637/GlobalEVOutlook2021.pdf

Setting New Standards for Circular Economy in the Cement Industry

Michael Klitzsch and Martin Geith

Abstract In the past, recycling of refractory linings from the cement industry has not been possible due to contamination impregnating the linings during the cement production process. These contaminations lead to production issues and negative effects on the refractoriness when spent material is reused as raw material. RHIM has developed and patented a treatment method to reduce the contamination of spent refractories which technically enables their reuse. Linings based on this technology have a reduced CO_2 footprint of up to 20%. However, a business model based on a circular economy requires additional developments to be implemented in the organization. As spent refractory linings are classified as waste, the legal framework of the supply chain must be adapted. A changed supply chain setup enables the sourcing and cross-border shipment of spent refractories. With the introduction of products containing recycled material to the market, the perception of these "waste" products must be positively changed.

Keywords Recycling · Cement industry · Refractory industry · Sustainability · Circular economy · Washing treatment · Recycling · Cement industry · Refractory industry · Sustainability · Circular economy · Washing treatment

Introduction

Cement, the most important inorganic binder for the construction industry, is produced in large cement plants with a capacity of up to 10,000 tons/day. During operation of cement plants, volatile compounds evaporate in the hottest area of the kiln and are transported into cooler areas. This effect leads to a formation of so-called

M. Klitzsch (✉) · M. Geith
RHI Magnesita GmbH, Magnesitstrassse 2, 8700 Leoben, Austria
e-mail: Michael.Klitzsch@rhimagnesita.com

M. Geith
e-mail: Martin.Geith@rhimagnestia.com

A. Lazou et al. (eds.), *REWAS 2022: Developing Tomorrow's Technical Cycles (Volume I)*, The Minerals, Metals & Materials Series,
https://doi.org/10.1007/978-3-030-92563-5_20

alkali-/sulfate-circles in the cement kiln atmosphere. This results in a strong penetration of the volatile compounds into the open pores of the brick structure, followed by condensation and in the worst case corrosion of the refractory material. Additionally, in the last decades, the common fuels as coal-dust or gas were replaced more and more often by secondary fuels such as crushed plastics, used car-tires or even animal meal, resulting in a further increase of volatile compounds and the thermochemical load on refractory linings.

Recently, the idea to reuse the broken-out spent linings of cement rotary kilns was developed. Through this measure, the replacement of virgin magnesia with spent cement rotary kiln material a positive environmental effect would result. To ensure a stable refractory recycling business in the cement industry, both the production method and the final properties of the bricks containing recycled material must be carefully evaluated.

Recycling of Material from Cement Rotary Kilns

To evaluate the chemical load, a full chemical analysis was carried out by XRF acc. to ISO 12677 of a basic magnesia–spinel refractory material, retrieved from a rotary kiln from a European cement plant. The alkalis Na and K, the main elements penetrating the pores of the refractory lining during cement plant operation, were measured quantitatively by ICP-OES acc. to ISO 26845, SO_3 was measured quantitatively by elementary analysis acc. to DIN 51,085, and Cl was measured quantitatively by titration. To obtain a more detailed analysis, the used refractory material was crushed into different grain sizes and each grain size fraction was analyzed in detail individually. Through these procedures, the accumulation effects can be easily identified. Table 1 shows the result of the analysis for the volatile components. The overall amount of alkali salts of 3.79% was quite high, and the highest contamination was measured in the coarsest and finest fractions. As main elements K, Cl, and SO_3 could be found, while the Na content was quite low.

Table 1 Analysis of a spent material from cement rotary kilns with different grain fractions

Grain size (mm)	m (%)	m salts (%)	g Na (%)	g K (%)	g Cl (%)	g SO_4 (%)
3.15–5.00	27.20	4.48	0.06	0.88	0.70	2.85
1.00–3.15	36.90	2.72	0.03	0.49	0.61	1.59
0.50–1.00	10.35	2.79	0.03	0.42	0.70	1.65
0.315–0.50	5.20	3.30	0.06	0.76	0.77	1.71
0.10–0.315	10.30	4.31	0.08	1.00	1.16	2.07
<0.10	10.05	6.63	0.13	1.68	2.22	2.61
sum	100.0	3.79	0.05	0.77	0.87	2.10

m… weight amount of specific grain size, m salts… cumulated weight content of volatile compounds ($Na + K + SO_3 + Cl$), g Na, K, Cl, SO_4… weight content of volatile substances

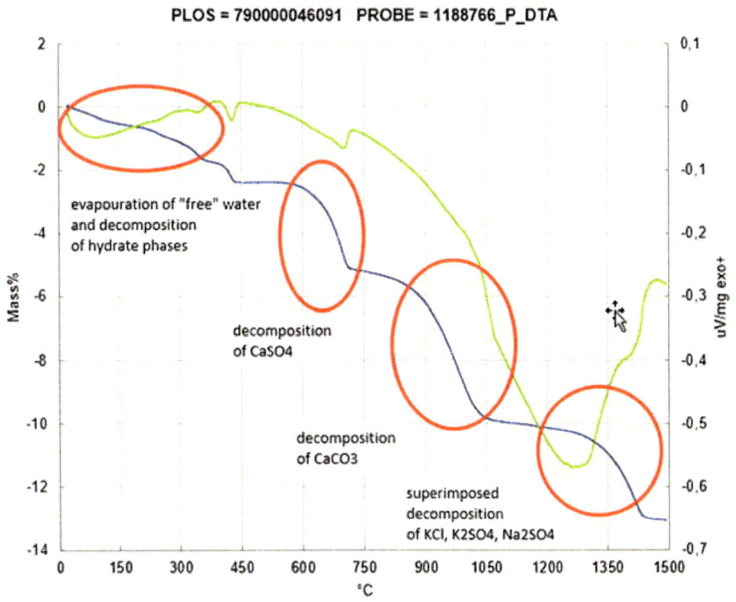

Fig. 1 TG/DTA analysis of a cement rotary kiln material

To obtain a more detailed understanding of the contained mineralogical compounds, a DTA/TG analysis of the spent material was carried out and is shown in Fig. 1. The weight loss indicates that the penetrating volatiles form mainly $CaSO_4$. $CaCO_3$. Na_2SO_4. K_2SO_4 and KCl. In an X-ray diffraction analysis, this result was verified, showing also $CaSO_4$. KCL, and additionally K-Langbeinite ($K_2Mg(SO_4)2.6H_2O$ as the main phases.

To evaluate the usability of such highly contaminated spent material in recycled brick production, refractory bricks for the lining of cement rotary kilns were produced. The produced bricks were comprised of between 10 and 50% of recycled material and were produced and burned in a tunnel kiln at 1600 °C. At this temperature, it was expected that the volatiles, contained in the recycled material, would react with the brick components or evaporate into the kiln atmosphere during heat up.

In this trial series, it was observed that the volatile substances were melting prior to evaporation. This resulted in a high tendency for the individual bricks to stick together, which disabled the automatic unloading procedure, which is required in state-of-the-art refractory production plants. Due to this effect, the amount of recycled material was limited to a maximum content of 20%, thereby reducing the occurrence of stuck bricks and ensuring the essential automatic operation in production.

Bricks containing 20% recycled material were pressed on a hydraulic press and fired at 1600 °C in a tunnel kiln and then investigated in detail. To evaluate the brick, the following properties were measured, bulk density and open porosity acc. to ISO 5017, cold crushing strength acc. to ISO 10059–1 and refractoriness under load acc.

Table 2 Physical properties of bricks using recycled material versus bricks containing virgin magnesia

Recycled material content	BD (g/cm^3)	Por (%)	CCS (MPa)	RUL $T_{0.5}$(°C)
0%	2.95	15.3	68	1520
20%	2.90	17.6	65	1396

BD, por… bulk density, porosity acc. to ISO 5017, CCS … cold crushing strength acc. to ISO10059-1, RUL /0,5… refractoriness under load $T_{0.5}$ acc. To ISO 1893

to ISO 1893. Test results are shown in Table 2. It was shown that the bricks containing recycled material had a decrease in the bulk density and cold crushing strength, while the open porosity increased. Furthermore, the refractoriness under load $T_{0.5}$ value decreased significantly.

Due to the evaporation of volatile substances, during production of bricks containing recycled material an impact on the environment was expected. Especially during burning, the formation of a typical odor in the exhaust gas was observed, volatile components could be proven analytically. The observations during brick production caused by evaporation of the volatile substance require further measures to enable a safe use of basic magnesia–spinel recycled material from cement rotary kilns. To enable this from H&S, environmental, and quality perspective, a method to "clean" the used cement rotary kiln lining needed to be developed.

Development of a Treatment Method for Recycling Material

To "clean" the material, various treatment methods were tested. While most of the contaminations show a good solubility in water, a washing treatment was identified as most efficient. To develop the washing treatment, five used linings from cement rotary kilns across Europe were chosen. The contamination (overall salt content) of these five materials varied between 2.4% and 7.1%. Initially, it was investigated with one used lining, which grain size of the bricks should be treated. Therefore, the received used linings were crushed into 2 grain sizes and washed 5 times with fresh water for 10 min in a concrete mixing machine. Table 3 shows that the cleaning efficiency increases with a reduction of the grain size. The content of volatile substances of the washed grain size 0–5 mm was significantly lower compared to 0–40 mm. Thus, the following investigations took place with 0–5 mm grain size.

For industrial purposes, washing the material 5 times with fresh water is not very efficient, a parameter study considered washing time, water to recycling material ratio, and the number of washing steps was carried out. It was determined that a washing process which was comprised of two steps with a water to recycling material ratio of 1 and a washing time of 20 min was identified as best compromise between washing effort and washing yield. These parameters were used to treat the five European used linings. Figure 2 shows the achieved reduction of contaminations. One box represents the variation of the reduction within the five different materials

Table 3 XRF analysis of untreated and washed recycling material, (1) after determination of LOI, (2) tested on the unannealed sample

	Untreated material	Washed material 0–40 mm	Washed material 0–5 mm
LOI (%)	2.48	2.44	0.71
MgO (%)$^{(1)}$	83.65	83.47	87.36
Al_2O_3 (%)$^{(1)}$	11.73	12.44	10.22
CaO (%)$^{(1)}$	1.41	1.15	0.99
SiO_2 (%)$^{(1)}$	0.99	1.15	0.36
K2O (%)$^{(2)}$	1.40	0.80	0.11
Na_2O (%)$^{(2)}$	0.19	0.02	0.00
SO_3 (%)$^{(2)}$	1.82	1.39	0.69
Cl (%)$^{(2)}$	1.05	0.52	0.04

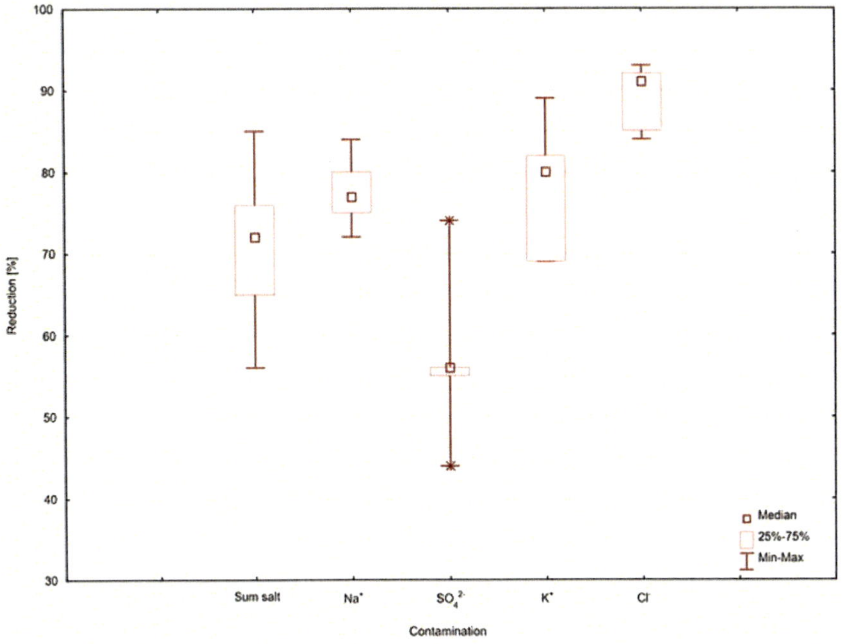

Fig. 2 Reduction of salt contents received by washing treatment of five different cement rotary kiln recycling materials

which were investigated. The average reduction of contained salts was approximately 70%, and specific contaminations were reduced by between 55 to 85%. It could also be shown that the washing process was most efficient for Cl^- (90% reduction). The lowest washing efficiency was observed for sulfate.

Table 4 Physical properties of bricks using washed material versus bricks containing virgin magnesia

Recycled content	BD (g/cm^3)	Por (%)	CCS (MPa)	RUL $T_{0.5}$(°C)
0%	2.95	15.3	68	1520
20%	2.93	16.2	77	1502

BD, Por… bulk density, Porosity acc. to ISO 5017, CCS … Cold crushing strength acc. to ISO10059-1, RUL /0.5… Refractoriness under load $T_{0.5}$ acc. To ISO 1893

Properties of Bricks Containing Recycled Material for the Lining of Cement Rotary Kilns

The washed spent material, containing reduced amounts of volatiles, was used to produce refractory bricks. To enable a direct comparison, the same recipes and process parameters that were used with untreated spent material were processed. Table 4 provides an overview of the properties obtained. By reducing the contained volatile substances through washing the spent material, an improvement of the brick properties was obtained. Both, bulk density, and open porosity were significantly improved. A significant improvement was observed in the refractoriness under load, and with a $T_{0.5}$ value of 1502 °C, the property values of a virgin refractory magnesia–spinel brick were achieved. In the trial's series, it was proven that a pretreatment of spent refractory material leads to a significant improvement in the properties of the newly produced refractory bricks. Furthermore, the production of the bricks was improved as the tendency of the bricks to stick together during firing was reduced and a positive environmental effect was observed, with a reduction in pollution from volatile substances. Nevertheless, to ensure a "green" brick production, manufacturing plants were equipped with desulfurization units to clean the exhaust gas.

As the washed material substitutes traditional raw materials (magnesia), the product carbon footprint (PCF) of bricks containing recycled material decreases significantly. Table 5 shows the PCF according to ISO 14044 of a "traditional" and a brick containing 20% recycled material. It shows that the burning and dissociation of magnesite ($MgCO_3$) to magnesia and CO_2 has the highest impact on PCF. The use of washed material avoids the CO_2 emissions related to the burning and dissociation of magnesite. The reduction rate depends on the substitution rate. A substitution rate of 20% reduced PCF of the final product (brick) by 12%. The calculations on PCF are assured by Denkstatt, a consulting company for sustainability and environment.

Development of a New Business Model

Previously, bricks containing recycled material were considered as cheap and low-quality products in the cement industry. To market this product, a new business model

Table 5 PCF (in tons of CO_2 per ton of finished product) of traditional refractory brick for the cement industry and bricks containing recycled material

Sources of product carbon footprint	PCF of traditional product	PCF of recycling containing product	CO_2 reduction (%)
Burning and dissociation of $MgCO_3$ -> $MgO + CO_2$	1.2	1.0	−17
Production of finished product	0.1	0.1	0
Others (PCF of external raw materials, transportation)	0.4	0.4	0
SUM = PCF of finished product	1.7	1.5	−12

and branding was developed. The new branding of the bricks containing recycled material refers to the low carbon footprint of the product. The new business model is based on circular economy, shown in compare Fig. 3.

In general, a circular economy is based on the principles of minimizing waste, pollution, carbon emission, and usage of natural resources, keeping products and materials in use, and regenerating natural systems. That means, that this business model approach requires a collaboration with the customer. When breaking out the spent lining, the cement plant is required to take the initial steps to support the quality of the recycled product. The spent lining is separated into different recycling grades according to the mineralogical composition of the bricks, shown in Fig. 3

Fig. 3 Circular economy business model

process step "sorting and declaration". In addition, the circular economy approach requires that the refractory producer takes possession of the pre-sorted material for additional treatments and recycling purposes. This collaboration ensures a high-quality recycling product. The promotion of the business model is based on two pillars. The management of cement producers is incentivized by the possibility to reduce the carbon footprint of the supply chain, and cement plants have the advantage of a new disposal route for their spent linings.

Development of the Organization

The implementation of circular economy to an organization requires changes on many levels inside the organization. The most important change comes with the desired use of spent linings. Spent lining bricks are legally classified as waste. Thus, permits for the shipment, collection, and treatment of waste must be acquired. These permits are also associated with many constraints by the authorities. This effects the supply chain and operational setup. Employees must be trained on the legal aspects and organizational changes are required to deal with waste efficiently. On the other hand, the treatment and the products are also accompanied with challenges. The sorting and cleaning, and the washing processes also generate byproducts which cannot be used in the production of cement rotary kiln bricks. Furthermore, the disposal of these byproducts would disrupt the whole business model. Thus, ways to utilize these by-products must be developed. This results in a new playground for R&D within an organization.

From Aluminium Industry Waste to Soil Amendment for Cash Crops and Fertiliser for Lowbush Blueberry

Julie Élize Guérin, Lotfi Khiari, Abdelkarim Lajili, Claude Villeneuve, Patrick Faubert, Marie-Christine Simard, Marc-André Séguin, Jean Lavoie, and Stéphane Poirier

Abstract Rio Tinto's Aluminium division operates smelters in Canada, primarily in the Saguenay–Lac-Saint-Jean region in Quebec, where its entire aluminium production line generates two specific calcium sulphate ($CaSO_4$) by-products. The aqua-catalysed hydrated lime (CHAC) is the by-product of the sulphur scrubber at the coke calciner plant, and the neutralised synthetic anhydrite (SA) is derived from the industrial processes of the chemical transformation of calcium fluoride into aluminium fluoride, which is used in the manufacture of electrolysis bath for aluminium production. Since 2015, several research projects have been developed at the Rio Tinto-Arvida Research and Development Center, in collaboration with local universities, to evaluate the potential of these by-products as a liming agent for various agroecosystems and as a certified calcium-rich fertiliser for lowbush blueberries, our regional flagship. The aim of this paper is to present the research and development programme that has led to the agricultural recycling of these two by-products.

Keywords Calcium sulphate · Flue gas desulphurisation · Gypsum · Lowbush blueberry · Agriculture

J. É. Guérin (✉) · M.-C. Simard
Rio Tinto, Arvida Research and Development Center (ARDC), 1955 Mellon Boulevard, Building 110, Jonquiere, QC G7S 4K8, Canada
e-mail: julieelize.guerin@riotinto.com

L. Khiari · A. Lajili
Soil and Food Engineering Department, Laval University, 2425 Agriculture Street, Quebec, QC G1V 06, Canada

C. Villeneuve · P. Faubert
Département des sciences fondamentales, Université du Québec à Chicoutimi, 555, de l'Université Boulevard, Chicoutimi, QC G72B1, Canada

M.-A. Séguin · J. Lavoie · S. Poirier
Rio Tinto, By-Products Valorisation and Commercialisation, 1955 Mellon Boulevard, Building 100, Jonquiere, QC G7S 4K8, Canada

A. Lazou et al. (eds.), *REWAS 2022: Developing Tomorrow's Technical Cycles (Volume I)*, The Minerals, Metals & Materials Series,
https://doi.org/10.1007/978-3-030-92563-5_21

Introduction

Rio Tinto is committed to being part of the solution to address the challenges of climate change. Thus, many efforts must be made to minimise industrial environmental impacts and provide major benefits to the surrounding communities. One of the key elements in achieving these objectives is, of course, the complete management of industrial residues. The Saguenay-Lac-Saint-Jean (SLSJ) region, in Quebec, Canada, is an important hub for Rio Tinto's aluminium production. The Jonquiere smelter site (SLSJ) generates annually more than 400 kt (kilotonnes) of by-products from its operations, of which 80 kt of synthetic anhydrite (SA) are produced by the fluoride plant and 17 kt of by-product from the SO_2 scrubber system of the coke calciner kiln (CCK) plant.

By-product recycling in agriculture is a promising avenue to increase the company's environmental performance, reduce greenhouse gas (GHG) emissions, and participate in the regional circular economy. In addition to aluminium (Al) production, lowbush blueberry production is also a characteristic flagship of the SLSJ region, accounting for nearly 50% of the marketed volumes in Canada [1]. However, lowbush blueberry production costs are increasing while selling prices are decreasing. Soil conditions in which lowbush blueberries are grown often have calcium (Ca) deficiencies [2], and gypsum is commonly applied to ensure good Ca nutrition of lowbush blueberries [3].

Cash crop producers in our region are also reluctant to lime their soils, as the calcic materials available on the market are often found outside the region, making these latter very expensive. Flue gas desulphurisation gypsum (FGDG) is well known in countries where electricity generation is obtained from coal combustion and hundreds of thousands of tonnes of FGDG are applied each year in agriculture [4]. In addition to being considered as an amendment to correct soil acidity, several authors have raised the agri-environmental benefits of using FGDG. Indeed, FGDG is an excellent source of calcium (Ca) and sulphur (S) for plants, improves soil physical properties, and reduces soil erosion and phosphorus (P) losses by leaching [5, 6]. Thus, the use of neutralised synthetic anhydrite (SA), as a Ca source in blueberry fields, and CCK scrubber by-product, as a liming agent for cash crops, could therefore be an excellent way to reduce production costs since these products are available locally and at a lower cost compared to other calcic fertilisers and amendment sources. Subsequently, developing a regional symbiosis between these important industries in the SLSJ region, regarding the supply of agricultural inputs, seemed obvious!

Although elsewhere in the world, the use of SA and FGDG is well documented, it was nevertheless, necessary to convince local authorities that these Al industry by-products were environmentally safe, comparable to other calcic products already in use and, above all, had interesting agronomic attributes. Therefore, to determine the agri-environmental impacts of the use of our by-products in agricultural settings, and to define the use conditions, the Arvida Research and Development Center (ARDC) has implemented a major research and development programme in collaboration with Laval University, Université du Québec à Chicoutimi (UQAC) and several local private companies. The aim of this paper is to briefly present the research and development programme that has led to recovering the unrecoverable in agronomic products.

By-Product Descriptions

Calcium sulphate ($CaSO_4$) is mainly stable in the form of three elements which differ according to their degree of hydration: gypsum ($CaSO_4 \cdot 2H_2O$), basanite ($CaSO_4 \cdot ½\ H_2O$), and anhydrite ($CaSO_4$). The most abundant form of gypsum is found in its natural state (geological deposit), and anhydrite is mainly produced by the dehydration of gypsum or basanite. Gypsum is also one of the first fertilisers used and applied in agriculture for over 250 years [4]. Calcium sulphate can also be of synthetic origin and are generally residues resulting from industrial processes such as the AlF_3 production process and residues from gas desulphurisation processes. This is the case for the fluoride plant and the CCK at Rio Tinto's Jonquière smelter site.

Synthetic Anhydrite

The fluoride plant has been in operation since 1985 and has a production capacity reaching 110 kt of aluminium fluoride (AlF_3) per year. The latter is an important additive used by aluminium smelters. Synthetic anhydrite ($CaSO_4$) is the solid by-product of this plant and originates from the chemical transformation of fluorite or fluorspar (CaF_2), into AlF_3. Anhydrite is produced in the gasifier where spathfluor and sulphuric acid $(H_2SO_4)_{(l)}$ react to produce hydrofluoric acid $(HF)_{(g)}$, according to the following reaction.

$$CaF_{2(s)} + H_2SO_{4(l)} \rightarrow 2HF_{(g)} + CaSO_{4(s)}$$

At the gasifier exit, SA is neutralised with lime (CaO) [7]. For every tonne of AlF_3 produced, 2.4 t of SA are generated. It consists of more than 95% $CaSO_4$ and its purity, well above 90%, is comparable to that of a geological deposit [8]. Calcium and sulphur concentrations are 28% and 23%, respectively. These are characterised by a white colour, a bulk density of 1.4 g cm^{-3}, and a fine particle size where 80% of the particles are smaller than 6.4 mm (Table 1).

Table 1 Main features of synthetic anhydrite (SA) and aqua-catalysed hydrated lime (CHAC)

Composition	SA	CHAC
Chemical properties (%)		
Anhydrous calcium sulphate ($CaSO_4$)	>95	45–50
Calcium fluoride (CaF_2)	1–4%	
Calcium dihydroxide ($Ca(OH)_2$)		35–40
Calcium oxide (CaO)	<1	
Calcium carbonate ($CaCO_3$)		2–3
Silica dioxide (SiO_2)	<0.3	
pH	5–11	12–12.5
Physical properties (%)		
Bulk density (g cm^{-3})	1.4	0.42
Particle size passing 12.5 mm (% d.b.)	>95	>95
Moisture content (%)	0.05	0.05
Agronomic properties		
Calcium (Ca)	28	35
Sulphur (S)	23	5
Sulphate (SO_4)	67	35
Neutralising power (eq$CaCO_3$)		87
Metallic trace elements (%)		
Aluminium (Al)	388	1.314
Arsenic (As)	5	7
Cadmium (Cd)	0.5	0.5
Cobalt (Co)	2	3
Chromium (Cr)	5	4
Copper (Cu)	6	2
Iron (Fe)	558	1 036
Mercury (Hg)	0.42	0.02
Lead (Pb)	5	6
Zinc (Zn)	10	48

Aqua-Catalysed Hydrated Lime

The Rio Tinto-Arvida Research and Development Center has developed a new SO_2 scrubbing process called the aqua-catalysed hydrated lime (CHAC) process. Sulphur capture is achieved by injecting hydrated lime into the gas channel between the furnaces and CCK scrubbers. The reagent chemical reaction with the SO_2 gas released during coke calcination allows the neutralisation of SO_2 when the reagent is injected into the process gases. The CHAC residue consists of $CaSO_4$, calcium sulphite ($CaSO_3$), and hydrated lime ($Ca(OH)_2$) captured by the bag filters together with the coke dust and is found in approximate proportions of 45–50%, 0.5–1%, 35–40%, and 5–10%, respectively (Table 1). CHAC is a grey powder with a very fine grain size, high dryness (99.5% dry matter), and 98% of the material passes through a sieve mesh of <0.150 mm, thus classifying the material as a fine hydrated lime. The Ca, S, and residual lime contents give this material very interesting properties for crop nutrition, but also for soil acidity correction. The level of trace elements, polycyclic aromatic hydrocarbons, volatile organic compounds, and dioxins and furans, in our two calcium by-products, complies with the regulator requirements [9] (BNQ 2015) (data not shown).

Research and Development Programme Description

Since 2017, incubation, greenhouse, experimental field plot, and industrial-scale trials have been conducted in several local agroecosystems where varying doses of SA and CHAC have been applied. Soil, water, air, and crop quality parameters have been measured at different crop development stages. However, only experimental field and industrial-scale plot trials of SA application in lowbush blueberry fields and CHAC recycling in cash crops will be presented in this paper.

Material and Methods

Field Scale Trials

Although SA has already been used as a fertiliser for potato crops, and CHAC as a liming agent in cash crops, their actual requirements and agricultural value for crops still need to be determined. To establish standardised reference grids for these calcic products, field scale trials were conducted by Laval University, using different crops and soils in Canada. A total of six experimental field sites have been established since 2020 to test both limed and non-limed products in different crops and soils (alfalfa, corn, broccoli, and millet crops). Rio Tinto participated in these research activities with both CHAC and SA products using doses of 3 and 9 t ha $^{-1}$ in alfalfa and corn fields and in broccoli and millet fields, respectively. Soil and biomass samples were taken for all crops during the different agronomic cycles in each experimental unit. The full results of this experiment will be presented in another scientific paper, as all the data have not been collected and analysed yet.

Since 2017, a total of three experimental sites have been established, two in commercial lowbush blueberry fields (2017, 2020) and one at the UQAC Blueberry Research and Teaching Station (BRTS) (2021). All of these sites are located in the SLSJ region. The lowbush blueberry is produced according to a two-year production system, one year of vegetative growth and one year of harvest. A first experimental site was established in 2017 where three increasing doses of SA (0, 1.5, 3, 4.5 t

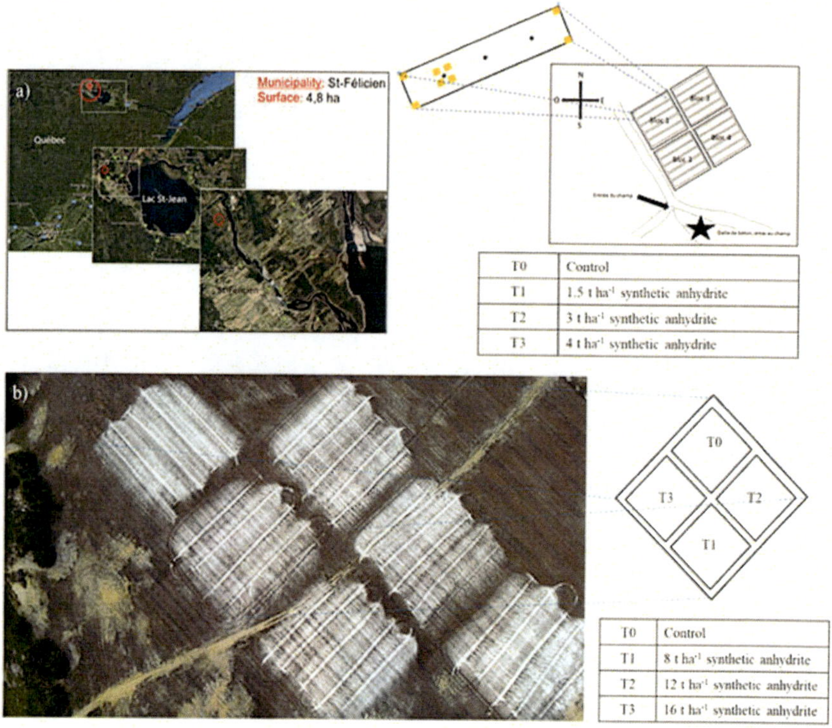

Fig. 1 Lowbush blueberry experimental designs: **a** 2017 field scale trial; **b** 2021 industrial-scale trial

ha^{-1}) were applied on a field, which was divided into four distinct experimental blocks (Fig. 1a). The same process was done in 2020 and 2021, but where seven increasing doses of SA (0, 3, 6, 9, 12, 15, 18, and 21 t ha$^{-1)}$ were applied. For each site, the SA applications were carried out in the spring of the vegetation year and following an experimental device where each treatment was repeated three times. Each experimental block also had a control treatment that had not received any application of anhydrite. Soil and biomass samples were randomly collected from each experimental field plot, before and after application. Fresh lowbush blueberry yields were determined in each experimental plot by harvesting a 1-m^2 area. The 2021 site will be harvested in summer 2022.

Part of this project is also being conducted by the UQAC team (BRTS), whose objective is to determine the effects of SA application on GHG emissions, soil and plant nutrient status, blueberry yields, and plant biomass. The experimental design was installed in May 2021 on 10-m^2 plots and consisted of the application of two increasing doses of SA (0, 6, 12 t ha^{-1}), replicated in four blocks on two experimental sites. On site 1, soil-surface GHG emissions were monitored using non-flow-through, non-steady chambers (area: 55 × 55 cm) (Fig. 2) [10]; nitrous oxide (N_2O), carbon dioxide (CO_2), and methane (CH_4) fluxes as well as cumulative area-based emissions

Fig. 2 Non-flow-through, non-steady chamber for the greenhouse gas emission monitoring

were compared following a regular sampling frequency over the 2021 and 2022 growing seasons (May to October).

The GHG concentrations in air samples were analysed using a gas chromatograph equipped with an electron capture detector for N_2O and a flame ionisation detector, for CH_4 and CO_2. The soil water quality was also monitored using lysimeters inserted at 60-cm depths. On sites 1 and 2, a complete soil chemical characterisation (exchangeable elements) was performed. Soil nitrate and ammonium exposures were monitored using anion/cation exchange membranes during the 2021 and 2022 growing seasons [11, 12]. Plant biomass was estimated using the point intercept method [13]. Plant foliar nutrient status, blueberry fresh fruit yields, and plant biomass were monitored in a 1-m^2 area during the vegetative growth year (2021), harvest year (2022) and both vegetative and harvest years, respectively. Results will be available at the end of this experimentation in 2022.

Industrial-Scale Trial

To evaluate the transport, handling, spreading, and required equipment for the application of CHAC and SA doses, industrial-scale tests were implemented in larger plots. To test CHAC, applications have been conducted in more than 15 cash crop fields (meadow, cereals, canola, soya, and alfalfa) since 2018. The different crop lime requirements were determined according to the water pH_{water} and pH_{buffer}, where a reduction factor was calculated from the material neutralising power in comparison with agricultural lime ($CaCO_3$) [14]. Doses between 1.4 and 2 t ha^{-1} of CHAC were applied. The effect of CHAC applications on soil properties, crop quality and application rate have been monitored and compared to a controlled area not subjected to CHAC application. Large-plot tests were also conducted with SA in a blueberry field, in the Lac-St-Jean region, in the spring of 2021 during the vegetative season. Four treatments representing increasing doses of SA (0, 8, 12 and 16 t ha^{-1}) were applied according to an experimental design in three blocks (Fig. 1b). The large plots were 1-ha in size and separated from each other by 10-m borders. Soil and biomass samples were randomly collected, from each large plot, before and after SA application. Lowbush blueberry fresh fruit yields, soil, and biomass will be measured during the summer of 2022.

Legal Authorisation

Under Canadian agricultural laws and regulations, industrial residues can be valorised in the agricultural sector through direct application as fertilisers (fertilisers, soil conditioners, or amendments) or through a formulated fertiliser. Therefore, to be recognised for agricultural use, a residue must meet the requirements of the *Fertilisers Act* (R.S.C., 1985, c. F-10) or the BNQ Standard [9]. In 2019, an application for registration was therefore filed to have SA recognised as a commercial fertiliser, in accordance with the specific requirements of the Canadian Food Inspection Agency (CFIA) safety standards, guaranteed minimum analyses, and labelling standard. Thus, the quality of our product and the control of our production are ruled according to a very rigorous protocol. An application is underway for the registration of CHAC with the BNQ standard. In addition, an environmental authorisation has already been obtained from local regulators for all CHAC experimentations.

Preliminary Results

Studies on small plots in the lowbush blueberry experimental site established in 2017 (Fig. 1a) showed significant results in some foliar nutrient concentration following application of different SA doses (0, 1.5, 3, 4.5 t ha^{-1}). The increase in SA doses leads to an increase in foliar concentrations of P, K, and S with a p value <0.05 (Table 2). Foliar Ca concentration, chlorophyll index, and fresh fruit yield suggest a positive response to the amendment following increased SA application, but statistically, the results did not show significant differences (Table 2).

Table 2 Effect of increasing the dose of SA on quantitative diagnosis of the foliar concentration of P, K, Ca, and S, the chlorophyll index, and fresh fruit yields in small plots scales

SA dose (t ha^{-1})	0	1.5	3	4.5	F values (*P*)[a]
Foliar concentration in P (%)	0.114[a]	0.125[a]	0.133[a]	0.131[a]	5.02 (0.017)
Foliar concentration in K (%)	0.494[a]	0.572[a]	0.662[a]	0.703[a]	8.96 (0.002)
Foliar concentration in Ca (%)	0.473	0.440	0.517	0.529	2.42 (0.116)
Foliar concentration in S (%)	0.112[a]	0.183[a]	0.217[a]	0.249[a]	16.73 (0.0001)
Chlorophyll index	36.00	37.13	37.28	37.83	0.36 (0.778)
Fresh fruit yield (kg ha^{-1})	4 971	3 588	4 737	6 300	0.5 (0.754)

[a]Treatment effect are significant at $\alpha = 0.05$

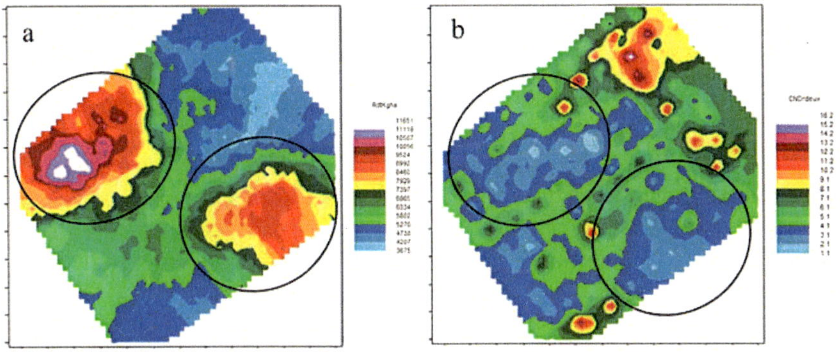

Fig. 3 Spatial correspondence between yield **a** and nutrient balance **b** in large plot scales

The SA application test conducted on a large-scale research plot revealed an improvement in plant nutrient balance in certain high-yielding areas of the field. The two georeferenced maps of yield and compositional nutrient diagnosis (CND) show a spatial correspondence (Fig. 3). The better the nutrient balance (R^2 of CND is low), the higher the yield.

Finally, industrial-scale trials showed that CHAC application had a positive effect on soil pH_{water} and pH_{buffer} under different crop ecosystems, and this will be confirmed in 2024 at the end of the field scale trials led by Laval University. An increase in soil pH between spring (before application) and fall (following application) was observed in all crop soils studied without causing crop damage. However, the main objective of these trials, carried out with the help of local farmers, was to allow them to validate whether the CHAC and SA were suitable for their machinery, to calibrate their equipment according to needs and to verify the effectiveness of these fertilisers on their own crops.

Summary

Rio Tinto's synthetic anhydrite (SA), composed of more than 95% of $CaSO_4$, was developed according to a very rigorous research programme and quality control protocol that led to its approval by the CFIA. This fertilising by-product was also awarded by the 2020 Quebec Circular Contest in the Large Business category, standing out among 40 applications across Quebec. Preliminary results from the research programme showed a measurable improvement in the nutrient status of blueberry plants, as shown by an improved P, K, and S uptake and increased chlorophyll index, and a tendency to increase blueberry yields when using doses of 1.5 and 4.5 t ha^{-1}. Aqua-catalysed hydrated lime (CHAC) has also demonstrated that it could be used to correct soil acidity and ensure good plant growth. The industrial-scale trials allowed farmers to validate whether our agronomic products could be used, in the right dose, with their equipment and thus offer an alternative quality product at a lower cost. Aluminium is the third most abundant metal on earth. This silvery metal is not only practical and popular for making products from beverage cans to Tesla Roadsters, it plays a central role in soil acidity. In addition to being a Ca and S rich fertiliser and soil amendment, RTA's calcic products could help mitigate phytotoxic conditions due to an excess of soluble Al in acidic soil and promote deeper plant rooting. Finally, once considered as buried residues representing a heavy environmental legacy and sources of considerable management costs, these RTA by-products (SA and CHAC) are now reused in agricultural markets and circulating in the Quebec economy.

Acknowledgements The authors would like to thank Produits BCC, Viridis Environment, field crop producers from the city of Saguenay, the Club Conseil Bleuet, the Coop Lac-St-Jean, the Corporation d'aménagement forêt Normandin, UQAC Blueberry Research and Teaching Station (UQAC project's part is partly funded by Consortium de recherche et innovations en bioprocédés industriels au Québec and Centre québécois de recherche et de développement de l'aluminium) for their participation in this research program, and Natural Sciences and Engineering Research Council of Canada for financial funds for research activities at Laval University. Thanks also to Ms. Sandras de Moor, from ARDC, for linguistic revision and critical reading of the manuscript.

References

1. Québec (2019) Ministère de l'Agriculture, des Pêcheries et de l'Alimentation du Québec (MAPAQ). Culture du bleuet
2. Lafond J (2014) Calcium and magnesium fertilisation in wild lowbush blueberry in Quebec. Can J Soil Sc 94:67–76
3. Sanderson KR, Carter MR, Ivany JA (1996) Effects of gypsum on yield and nutrient status of native lowbush blueberry. Can J Plant Sci 76:361–366
4. Chen L, Dick WA (2011) Gypsum as an agricultural amendment: general use guideline. The Ohio State University. Bulletin 945, 35 p

5. Clark RB, Ritchey KD, Baligar VC (1999) Benefits and constraints for use of FGD products on agricultural land. 1999 international Ash Utilization symposium. Center for Applied Energy Research, University of Kentucky, Paper # 22
6. Baligar VC, Clark RB, Korcak RF, Wright RJ (2011) Chapter two—flue gas desulfurization product use on agricultural land. Adv Agron 111:51–86
7. US Patent number: 6517790. Converting fluorgyp to calcium sulfate
8. Maubert F (1987) Le gypse et l'anhydrite. Bureau de recherches géologiques et minières. 87 SGN 587 GEO. 34 p
9. Bureau de normalisation du Québec (BNQ) (2015) Norme BNQ 0419-090/2015 Amendements minéraux—Amendements calciques et magnésiens provenant de procédés industriels, Québec. 3e édition. ISBN 978-2-551-25683-9, 48 p
10. Rochette P, Bertrand N (2008) Soil-surface gas emissions. In: Carter MR, Gregorich EG (eds) Soil sampling and methods of analysis. CRC Press, Boca Raton, FL, USA, pp 851–861
11. Lafond J, Ziadi N (2013) Biodisponibilité de l'azote et du phosphore dans les sols de bleuetières du Québec. Can J Soil Sc 93(1):33–44. https://doi.org/10.4141/cjss2011-106
12. Ziadi N, Simard RR, Allard G, Lafond J (1999) Field evaluation of anion exchange membranes as a soil testing method for grasslands. Can J Soil Sc 79(2):281–294. https://doi.org/10.4141/s98-062
13. Lévesque JA, Bradley RL, Bellemare M, Lafond J, Paré MC (2018) Predicting weed and lowbush blueberry biomass using the point intercept method. Can J Plant Sc 98(4):967–970. https://doi.org/10.1139/cjps-2017-0201
14. Centre de référence en agriculture et agroalimentaire du Québec (CRRAQ) (2010) Guide de référence en fertilisation – 2e édition. Québec. ISBN 978-2-7649-0231-8, 473 p

Maximizing the Efficiency of By-Product Treatment by Multi-metal Recovery and Slag Valorization

Gustav Hanke, Jürgen Antrekowitsch, Fernando Castro, and Helmut Krug

Abstract Complete utilization of every material mined seems to be a highly obvious strategy to save natural resources and avoid residues to be dumped. In fact, examples where this is practiced are very limited. Often residues out of metal production are not fully exploited and contain still considerable amounts of various valuable elements when being dumped. Using the full potential of mined material would significantly support the metal supply, while optimizing the CO_2 balance, as much of the afford of winning a metal is already done, especially the mining. Not only the CO_2 footprint, but also the avoidance of residues is an omnipresent topic nowadays. Pyrometallurgical treatment is unavoidably linked to slags. Regarding a zero-waste solution, these slags have high potential to be used as construction materials. Referring to this, zero-waste solutions for residues from primary zinc and lead production were tested, emphasizing the potential of this strategy.

Keywords Sustainability · Pyrometallurgy · Recycling and secondary recovery

---

G. Hanke (✉) · J. Antrekowitsch
Chair of Nonferrous Metallurgy, Montanuniversität Leoben, Franz-Josef Straße 18, 8700 Leoben, Austria
e-mail: gustav.hanke@unileoben.ac.at

F. Castro
University of Minho, Research Centre Metrics, Campus de Azurém, 4800-058 Guimarães, Portugal

H. Krug
R+M Ressourcen + Management GmbH, Marktgasse 34, 7434 Bernstein, Austria

A. Lazou et al. (eds.), *REWAS 2022: Developing Tomorrow's Technical Cycles (Volume I)*, The Minerals, Metals & Materials Series,
https://doi.org/10.1007/978-3-030-92563-5_22

Introduction

Treating residues from metal producing industry for metal extraction is getting more and more attention [1]. However, such residues are not only of interest as secondary resources for metal winning, but also because of increasing challenges of landfilling due to the ever-growing environmental awareness. With this, costs for dumping are increasing and requirements are getting stricter.

Residues from nonferrous metal production often contain considerable amounts of various metals [2]. This is because the corresponding ores, in opposite to iron ores for example, usually contain many different metals. Whilst the main metal and in some cases a second or third one is won, many of the present metals are not recovered and end up in the waste material. The metal contents within these materials can be even higher than in primary ores, but can cause major problems for methods of mineral processing due to grain size, phase distribution, etc. However, a big part of the effort of winning metals, namely the mining, is already done for these materials, making it even more obvious to use the materials full potential and with this, to minimize the overall CO_2 footprint.

Modern efforts of using metallurgical residues as secondary resources often face the problem of causing again large amounts of new residues. Even if a residue is treated successfully for metal winning, the handling of the non-metal fraction causes challenges again. The only possibility to avoid dumps is to use every fraction produced during a process. For some residues, like blast furnace slag, the use as construction material is very common, whereas in many cases, especially in nonferrous metallurgy, there is no possibility of valorization so far.

Natural sand is an often underestimated resource. It must fulfill different criteria such as grain size distribution and grain shape to be used in construction. Due to the low price of sand, gravel, etc., transport distances also have to be short. Sand is not yet being classified as critical resource, but following the exponentially growing building activities, it will most likely become critical in future. This, together with the avoidance of dumps in general, makes the use of all kinds of technical materials for construction a highly reasonable strategy.

Materials

The main part of research to date and presented here has been done on residues from hydrometallurgical zinc winning process. Main component of these residues is jarosite, a sulfate mineral which is being formed in order to remove iron out of sulphuric acid solution used to leach zinc from the roasted concentrate [3, 4]. Commonly, the whole residue is also called jarosite, even though it contains not only the precipitation product (jarosite phase), but also other residues formed during the overall metal winning process. Treatment of the jarosite residue is of interest due to different reasons. Because of its hazardous properties, dumping is an increasing problem for many plants. As existing dumps will be full in near future and there will be no permissions for new ones. From another point of view, the jarosite residue bears several valuable metals, especially zinc, lead, and silver, in some cases also copper, gold, indium, and germanium. Recovery of these elements using a roast-reduction process was tested extensively with promising success up to 1000 kg scale [5, 6]. Further trials in smaller scale were performed not only to win the valuable metals, but also to evaluate the usability of the produced slag as construction material. Initially, its suitability as substitute for natural sand in Portland cement-based concrete was tested.

A second residue, whose treatment for multi-metal recovery has been studied, is slag out of lead production. The slag was taken from a historical dump of a former lead and zinc smelter, which used the imperial smelter furnace technology [2]. The dump located in Macedonia is considered as potential source for pollution, mainly because of its lead and arsenic content [7]. To date, the work on this material focused on the pyrometallurgical treatment of this specific slag to recover zinc and minor amounts of lead. The process is based on selective reduction on a lead bath, producing a liquid slag which can be modified to meet requirements for further utilization. Trials on its applicability of the remaining slag as construction material will be part of upcoming research in near future. Also, in this case, the aim is to realize a zero-waste solution and to minimize the overall CO_2 footprint of lead and zinc production.

Characterization of Jarosite Material

As the main part of the material is a precipitation product, the grain size is generally very small, with about 90 wt% passing 25 μm and 60 wt% passing 10 μm. However, due to agglomerations and the measurement by sieve analysis, the grains size distribution is suspected to be even smaller. Characterization of such small-grained material quickly reveals the limits of common methods such as light microscopy and even scanning electron microscopy [8]. Table 1 shows the main elements of interest measured by X-ray fluorescence analysis. Zinc, lead, and silver are the most valuable phases in jarosite residues. Copper is often present, but of lesser importance due to the low amount. Indium can be of interest in some specific jarosite residues.

Table 1 Chemical analysis (XRF) of a typical jarosite material [8]

Component	Fe	Zn	Pb	Cu	Si	Ca	S	Ag	In
Wt%	25.3	5.4	5.8	0.8	2.7	3.2	11.3	0,026	0,012

The main components, identified mainly by scanning electron microscopy and X-ray diffraction analysis, are obviously jarosite phases. The typical endmember formed in the process is natrojarosite ($NaFe_3(SO_4)_2(OH)_6$) or ammoniojarosite ($(NH_4)Fe_3(SO_4)_2(OH)_6$). To a minor extend also, hydroniumjarosite $(H_3O)Fe_3(SO_4)_2(OH)_6$) and plumbojarosite ($PbFe_6(SO_4)_4(OH)_{12}$) can be present. Contrary to expectations out of the overall hydrometallurgical zinc winning process, the jarosite phase proved also to bear considerable amounts of zinc and lead, which are the two most important valuable elements. Silver can theoretically also be present, even though it was not clearly identified [9]. Other phases bearing these elements are zinc sulfide, zinc ferrite, lead oxide, sulfate, and sulfide. Silver is often present in inclusions together with copper and lead in various phases, especially in silicates (quartz and feldspar) [6].

Characterization of Lead Slag

The granulated slag is fine grained, with about 95 wt% passing 3 mm. Grains of agglomerated slag with diameters up to 20 mm can be found. The glassy and homogenous matrix consists of a CaO-SiO_2-FeO phase. Main valuable elements are zinc (7.9 wt%) and lead (2.4 wt%) (Table 2). Beside the amorphous matrix, zinc oxide, iron zinc oxide, ferrous sphalerite, iron silicate, calcium arsenate, and pyrite were identified using X-ray diffraction analysis and scanning electron microscopy. These phases are present as irregular-shaped inclusions. Lead appears mainly as oxide or sulfide as droplets with diameters of up to 75 μm. Metallic lead is rare. Zinc is mainly bound to the matrix, where it makes up to 7 wt%. Other zinc phases, such as zinc sulfide may be found, but are very rare [2, 10, 11].

Table 2 XRF analysis of the investigated lead slag [10]

Component	MgO	Al_2O_3	SiO_2	CaO	MnO	Fe	Zn	Pb	Cu	S	As
Wt%	1.8	5.7	18.6	12.8	1.5	38.1	7.9	2.4	0.7	2.7	0.2

Pyrometallurgical Treatment of Jarosite Material

Due to the dominance of the jarosite phase, and very limited amount of other phases, results gathered by the detailed characterization do not allow any scope for concentration by means of mineral processing. Nevertheless, small scale trials in grain size and magnetic separation and flotation were performed in parallel. Despite smaller concentrations of single elements, the findings from the characterization were confirmed and no noteworthy concentration was achieved [8].

Prior to metallurgical treatment, the material was pelletized with a disc pelletizer. This was necessary, mainly for making sample handling easier and to avoid carryover of material with the off gas, but also to allow a better gas flow. Besides water, no additives were necessary to form stable pellets with a diameter of about 3–10 mm.

The investigated pyrometallurgical process for multi-metal recovery consists of two steps. First the material is calcinated, to remove volatile compounds, especially hydroxides and sulfur. This could be done in a top-blown-rotary-converter or short-drum-furnace. The second step is the reduction in an electric furnace where all valuable metals are won. Zinc evaporates and can be caught in the off gas after reoxidation as ZnO. Lead and silver (copper, gold) are collected in the liquid lead bath at the bottom. The slag is theoretically free of heavy metals and has a fayalitic composition, meaning that iron is in this case not part of the metal product but of the non-metal product (slag).

Use of Slag from Jarosite Treatment

The slag from pyrometallurgical jarosite treatment was tested for its potential in being used as sand replacer in concrete. During the reduction step, after the successful separation of slag and metal (Zn, Pb, Ag, Cu...), the fayalitic slag was quenched in water and sieved to three grain sizes, <1 mm, 1–2 mm, and 2–4 mm. Every grainsize was introduced to concrete (with Portland cement 32.5 and normalized sand EN196) in three different ratios, replacing natural, normalized sand. As reference, one concrete sample was produced only with Portland cement and sand (Table 3). Water was added until the result of the flow table test reached a value between 160 and 180 mm.

For the utilization as construction material, the mechanical strength (compression and flexural) and chemical resistance (leachability) were tested.

Compressive resistance determination was done on 12 cubic samples (5 × 5 cm) for each condition and flexural strength evaluation on 6 samples 16 × 4 × 4 cm per condition.

After 7 and 28 days of curing at 20–23 °C, the tests on flexural and compression strength were performed on the produced samples. For eluate testing, samples after 28 days of curing time were used.

Table 3 Mixtures prepared for mechanical and chemical testing

Reference	% Cement	% Sand	% slag < 1 mm	% slag 1–2 mm	% slag 2–4 mm
1	25	75	0	0	0
2	25	72	5	0	0
3	25	65	10	0	0
4	25	55	20	0	0
5	25	72	0	5	0
6	25	65	0	10	0
7	25	55	0	20	0
8	25	72	0	0	5
9	25	65	0	0	10
10	25	55	0	0	20

Pyrometallurgical Treatment of Lead Slag

The proposed technology for multi-metal recovery is very similar to the treatment of jarosite material concerning the reduction step. However, due to the composition of the lead slag, calcination is not necessary. Often iron is already present as fayalite out of primary lead winning, forming ideal conditions for the re-melting on a lead bath [11].

After removing zinc and lead, the liquid slag can be optimized by additives to fulfil the requirements of a sand substitute.

Results

Results from Jarosite Material Treatment

The pyrometallurgical two-step process proved its applicability for jarosite material successfully in different scales. Using only the second step (reduction) is possible, but leads to a big loss of metal to the slag phase due to the sulfur content. Temperature is a critical parameter in both process steps. For calcination, it must be high enough for volatilizing SO_2, water, and hydroxides but must not be too high to avoid the loss of valuable components, such as PbO.

For the reduction step, the temperature must be high enough for melting all the material, but again, unwanted vaporization of components must be avoided. Additionally, due to energetic and economic reasons, the temperature should be as low as possible. The success of the trials was measured by analyzing remaining contents of valuables, especially lead and zinc, in the slag. With remaining 2–3 wt% each, the values did not get as low as expected according to other similar trials (Table 4).

Table 4 Chemical composition of the produced slag from jarosite treatment (X-ray fluorescence analysis)

	wt%
FeO	59.2
SiO_2	25.1
CaO	2.07
ZnO	2.73
PbO	2.12

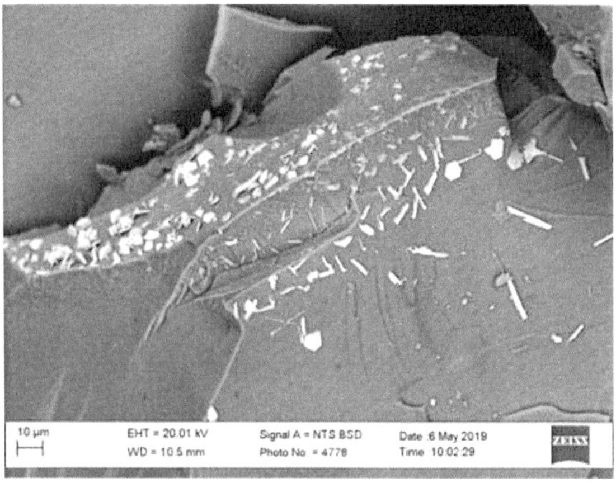

Fig. 1 SEM-BSE image of slag from jarosite treatment. Bright spots are lead and iron containing phases

After detailed investigations of the slag, using a scanning electron microscope, it becomes clear that especially the remaining lead is present as metallic droplets in the slag (Fig. 1). This indicates problems in the settling behavior of the liquid metal. Therefore, the problem does not relate to the chemical process of separation, but to the technical implementation, respectively the furnace. Due to turbulent conditions, the separation of slag and metal was obviously not as clear as expected. In further optimized trials, it has been shown that remaining values of lead and zinc distinctly lower than 1 wt% each are feasible.

Use of Slag from Jarosite Treatment

The tests on slag–concrete samples showed very promising results, especially in mechanical properties. The samples containing slag were at least comparable, but in many cases even higher in their flexural and compressive strength compared to the slag-free reference sample. This touches especially the samples containing slag with grain sizes 1–2 mm and 2–4 mm. Also, the differences were higher after 7 days of curing time. Interestingly, the amount of slag in each sample did not show big

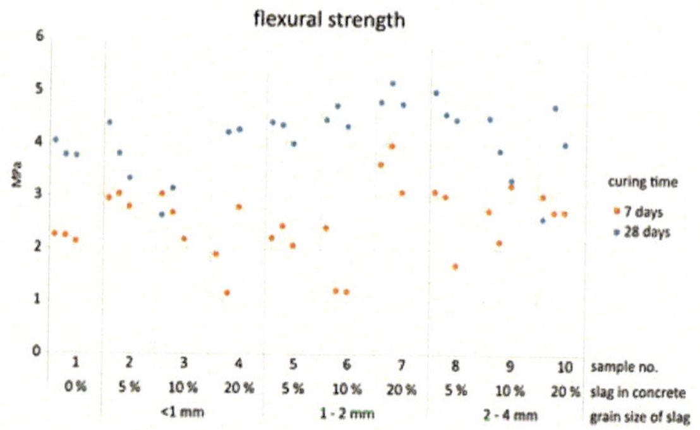

Fig. 2 Results from flexural strength testing of cement–slag concrete

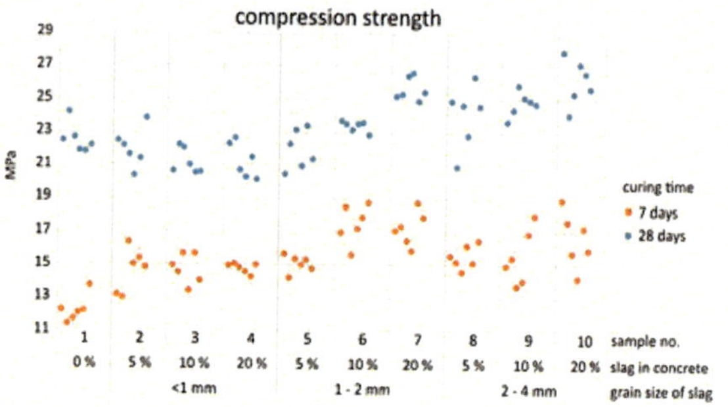

Fig. 3 Results from compression strength testing of cement–slag concrete

differences for grainsizes < 1 mm and 2–4 mm, whereas the strength of samples with grainsizes 1–2 mm increased with higher amount of slag (Figs. 2 and 3).

For the leachability testing, one sample of each slag grain size containing 20 wt% of slag was crushed (<10 mm) and mixed with distilled water in a proportion solid:liquid 1:10. Referring to international standards for inert landfilling, only the leachability of lead did not match the limits, with 0.93 mg/kg exceeding the limit of 0.5 mg/kg nearly twice (Table 5: Results from leachability testing.). However, as mentioned above, the slag contained comparably high amounts of remaining lead present as metallic inclusions, which is due to technical reasons during the trial. However, this value does not exceed the limits for deposition of residues in non-hazardous wastes landfills. For some other elements, there is some limited leaching but not exceeding the respective limits.

Table 5 Results from leachability testing

Parameters	Analytical methods	Specification [mg/kg]	Result [mg/kg]
As	US EPA 200.8, CSN EN ISO 17294–2	0.5	<0.01
Ba		20	4.4
Cd		0.04	0.003
Cr total		0.5	<0.01
Cu		2	<0.05
Hg		0.01	<0.002
Mo		0.5	0.38
Ni		0.4	<0.01
Pb		0.5	0.93
Sb		0.06	0.01
Se		0.1	<0.01
Zn		4	<0.1
Chloride	CSN EN ISO 10304–1	800	30
Fluoride		10	3
Sulphate		1000	48
Phenol index	CSN ISO 6439	1	<0.10
Dissolved organic carbon	CSN EN 1484	500	36

Results from Lead Slag Treatment

Small-scale trials proved the pyrometallurgical technique being applicable for treatment of lead slag. Yields of zinc and lead were >85 wt% and >95 wt%. The slag shows more the 60 wt% of the material present as fayalite. Some optimizations regarding basicity have been done. In one case, sand was added to reach a fayalite percentage of more than 80 wt%. Minimizing the lead content in the slag allowed values after standard leaching which are below the official regulations. Further investigations will focus on optimizing the slag composition and with this the fayalitic part as well as the influence of cooling conditions after slag tapping.

The usability of this slag as construction material was not yet evaluated, but this will be a focus of further research, and investigations similar to those for the slags from jarosite treatment will follow.

Summary

The potential of jarosite material was clearly shown. Not only for the winning of valuable metals, but also in producing a slag, applicable as construction material. In both cases, it can contribute to save natural resources and to avoid waste material. Furthermore, also historical dumps could be treated to remove the dump or to make it available for other, even more problematic materials, which still cannot be recycled.

The applicability of the pyrometallurgical process is out of question. Critical in this case is of course the energy required for the process. However, together with the global efforts of developing and optimizing sources for green energy, electricity will become greener and get ever closer to CO_2 neutrality. With this, technologies relying on electrical energy will do so as well.

Tests on using slag out of jarosite treatment as replacer for sand in concrete showed very satisfying results. In terms of mechanical properties, it is even slightly better than concrete using natural sand only. Leaching tests were not as successful in the first run. However, lower amounts of sand being replaced by slag will most likely also reduce the leaching of lead. Furthermore, in a second analysis with slag from another but similar test, only with a lower remaining lead content, the necessary requirements were fulfilled, even when replacing higher amounts of natural sand.

Further research will focus on increasing the lead and zinc yield of the pyrometallurgical process (and with this, smaller contents in the slag) and additional tests on the leachability of the produced concrete.

Multi metal recovery of lead slag showed also very promising results. However, concerning a zero-waste solution, the slag must also become a usable product. For this, test work on its usability as construction material will take place.

Acknowledgements COMMBY is a FFG Austria COMET project. Thank you to the funders Österreichische Forschungsförderungsgesellschaft (FFG), Steirische Wirtschaftsförderungsgesellschaft (SFG), Wirtschaft Burgenland GmbH, Land Steiermark and Land Burgenland.

References

1. Höber L, Steinlechner S (2021) A comprehensive review of processing strategies for iron precipitation. Cleaner Eng Technol
2. Schatzmann W, Antrekowitsch J (2019) Assessment of by-products—from waste to values. Wastes: Solutions. Treat Opport III:506–511
3. Sinclair R (2005) The extractive metallurgy of zinc. Victoria: Australasian Institute of Mining and Metallurgy
4. Svens K, Kerstiens B, Runkel M (2003) recent experiences with modern zinc processing technology. Erzmetall 56(2):94–103
5. Steinlechner S, Antrekowitsch J (2018) Thermodynamic considerations for a pyrometallurgical extraction of indium and silver from a jarosite residue. Metals 8(5)
6. Antrekowitsch J, Hanke G (2020) Characterization and processing of residues from hydrometallurgical zinc smelters. In: PbZn 2020: 9th international symposium on lead and zinc processing, pp 437–445

7. Stafilov T, Sajn R, Pancevski Z, Boev B, Frontasyeva M, Strelkova L (2010) Heavy metal contamination of topsoils around a lead and zinc smelter in the Republic of Macedonia. J Hazard Mater 175:896–914
8. Hanke G, Bernhart W, Antrekowitsch J (2019) Treatment of precipitation residues from. World of metallurgy-Erzmetall: internationale Fachzeitschrift für Metallurgie 72(6)
9. Pappu A, Saxena M, Asolekar S (2006) Jarosite characteristics and its utilisation potentials. Sci Total Environ 359:232–243
10. Leuchtenmüller M, Schatzmann W, Steinlechner S (2020) A kinetic study to recover valuables from hazardous ISF slag. J Environ Chem Eng 8(4)
11. Unger A (2015) Entwicklung eines Recyclingprozesses zur simultanen Rückgewinnung von Wertmetallen aus Reststoffen der Blei- und Zinkindustrie. Ph.D. thesis, Montanuniversität Leoben

Characterization and Thermal Treatment of Eggshell and Olive Stones for Heavy Metals Removal in Mining Environmental Liabilities Sites

Mery C. Gómez-Marroquín, Henry Colorado, Dalia E. Carbonel-Ramos, Jhony F. Huarcaya-Nina, Stephany J. Esquivel-Lorenzo, Alfredo Ceroni-Galloso, and Hugo D. Chirinos-Collantes

Abstract This paper presents the results of characterization, thermal, and chemical treatments made over natural eggshells (ES) and olive stones (OS) samples. The main goal was to perform an analysis to assess the potential of treated ES and OS to remediate heavy metals from polluted soils with mining tailings (PSMT). Sample powders of ES, OS, and PSMT were mechanically treated and characterized with chemical analysis, X-ray diffraction, Fourier transform infrared spectroscopy (FT-IR), thermogravimetric, and differential thermogravimetric techniques (TGA & DTG). ES was subjected to a calcination test. OS samples were thermally treated and chemically modified with thiol. PSMT was characterized using specific gravity, average particle size, and environmental indicators. FT-IR analysis indicated the presence of $CaCO_3$, Li_2CO_3, K_2CO_3, and KNO_3 functional groups in ES and OS. The XRD analysis of ES samples showed the presence of one mineralogical phase ($CaCO_3$). Diffraction patterns in the OS could not be identified due the amorphous condition of the sample. ES calcination tests presented a mass loss near 47% (w/w). The process to obtain chemically modified pre-oxidized biochar from OS reported pyrolysis mass loss of 80.59%, pre-oxidized mass loss of 35.43%, and an addition of 0.14 g during the chemical treatment. Equilibrium diagrams showed that Cd^{+2} and Hg^{+2} are most likely in the PSMT soil solution (aqueous phase), whose cations could be removed via cation exchange of the calcinated ES. The mercaptoethanol content in the chemically modified pre-oxidized biochar could be effective in removing the Hg content in PSMT.

Keywords Eggshell · Olive stones · Calcination · Thermal treatment · Chemical modified · Agricultural recycling · Environmental effects · Processing extraction · Secondary recovery

---

M. C. Gómez-Marroquín (✉) · D. E. Carbonel-Ramos · J. F. Huarcaya-Nina · S. J. Esquivel-Lorenzo · A. Ceroni-Galloso · H. D. Chirinos-Collantes
National University of Engineering, 210 Túpac Amaru Ave, Rímac, Lima 25, Peru
e-mail: mgomezm@uni.edu.pe

H. Colorado
CCComposites Laboratory, Universidad de Antioquia UdeA, Calle 70 No. 52-21, Medellin, Colombia

A. Lazou et al. (eds.), *REWAS 2022: Developing Tomorrow's Technical Cycles (Volume I)*, The Minerals, Metals & Materials Series,
https://doi.org/10.1007/978-3-030-92563-5_23

Introduction

Mining is one of the most important economic activities around the world, involving a great investment and natural resources demand [1]. In the areas where they are developed, mining activities can have large economic, environmental, and social impacts [2], and therefore, if not managed properly, mining can have a detrimental effect on human health and wild life [3]. Latin America has one of the largest metal land reserves. In recent years, Peru has become an emerging economy, amongst the strongest in South America [4]. According to the National Society of Mining, Oil and Energy, the mining sector in Peru is responsible for 10% of the gross domestic product, 60% of total exports, 16% of private investment, and 19% of taxes paid by companies [5]. Despite its main contribution to the economy, the mining sector represents a considerable toll on the environment. In 2019, the number of Mining Environmental Liabilities (MEL) sites in Peru was of at least 8448, most of them without an identified responsible person. The regions of Ancash, Apurímac, Arequipa, Cajamarca, Huancavelica, Junín, Pasco, and Puno, concentrate the greatest number of MEL sites in the country [6]. Mining activities have a distinct impact on heavy metal soil contamination, which is an important topic to be addressed since can cause many physical and biochemical changes in the structure of soil microorganisms [7]. The use of chicken eggshell is an effective and economically feasible technique in soil remediation mainly due to its large specific surface area that allows cation exchange and increases soil pH [8]. Biochar-based olive stones also represent a promising material for soil remediation. Olive stones are generated in large quantities in the olive oil industry as waste products. Nowadays, the use of these biomasses can signify a considerable cost reduction due to their easy application in remediating polluted soils with mine tailing (PSMT) [9]. Olive stones can be subject to carbonization processes (i.e., pyrolysis and pre-oxidation) to obtain biochar. This biochar can later be chemically modified with thiol to increase its capacity to remove mercury from PSMT. The thiol promotes the formation of mercaptans into the biochar [10]. In this research, we characterized and thermally treated natural eggshell and olive stones. In addition, olive stones were chemically treated. The main objective was to assess the potential of these natural materials for heavy metal remediation in PSMT. Eggshells were treated via calcination, while olive stones were treated in two ranges of particle size via pyrolysis and pre-oxidation. The added value in different environmental applications to these huge quantities of agriculture wastes still represents a great challenge for the researchers; however, the use of biomasses such as eggshell and olive stones (natural, calcinated, or chemically modified) can be regarded as a feasible alternative for environmental remediation in comparison with other conventional methods.

Experimental Procedures

In this research, we performed the mechanical preparation, characterization, and thermal treatment of two types of biomasses: Peruvian Red Chicken Eggshell (ES) and Peruvian Black Olive Stones (OS), both collected from domestic consumption. ES samples were separated of their eggshell membranes, cleaned with distilled water, dried at 100 °C for 2 h, and ground in ceramic mortars to obtain powders of 100%-100 mesh. OS samples were cleaned with a detergent dissolution and potassium hydroxide to remove organic impurities. Then, they were rinsed with distilled water, dried at room temperature, ground in a grain mill, and ceramic mortars to obtain particles of 25%-100 mesh. PSMT was obtained by grid sampling from a site called "FIGMM UNI Zone A"; located in the campus of the Geological, Mining and Metallurgical Engineering College of the National University of Engineering in Lima, Perú. Sample powders of ES, OS, and PSMT were mechanically treated and characterized with chemical analysis, X-ray diffraction (XRD), Fourier transform infrared spectroscopy (FT-IR), thermogravimetric (TGA), and differential thermogravimetric (DTG) techniques. ES was subjected to a calcination test. OS samples were thermally treated and chemically modified with thiol. Finally, a series of test to characterize the PSMT were conducted: specific gravity, average particle size, and environmental indicators.

Chemical Compositions

To determinate the chemical composition of the ES, OS, and PSMT, samples were analyzed with the following techniques: energy dispersive X-ray fluorescence spectrometry with an EDX 800HS spectrometer, SHIMADZU; flame atomic Absorption spectrophotometry with an AA 7000 spectrometer, SHIMADZU; UV–visible spectrophotometry with a UV-1800 spectrometer, SHIMADZU; and elemental analyzer. All experiments were made at room temperature (22.2 °C) and 65% of relative humidity. Elemental analysis (from Na to U) was performed with X-ray fluorescence spectrometry using a semi-quantitative analysis under vacuum atmosphere, with a 100 g sample. OS sulfur analysis was conducted in a LECO carbon analysis with organic application for sulfur and carbon in plant, feed, grain, and flour.

XRD Studies

XRD of raw materials (ES, OS, and PSMT) was performed in an ADVANCE D8 X-ray diffractometer, BRUKER. This equipment scanned with a goniometer (radius 240 mm), and the patterns were collected in the $2\ \theta$ range of 5° to 90° with 0.02° increment and a divergence slit size of 0.6°. The diffractometer has copper anodes

($\lambda = 1.5406$ Å, Cu K-α; $\lambda = 1.39\ 225$ Å, Cu K-β) and operates with 40 kV voltage and 30 mA current. We analyze the crystalline phases and/or inorganic compounds in the tested samples qualitatively (by comparison with the ICSD database) and quantitatively (using the Rietveld method).

FT-IR Spectra

FT-IR was conducted to ES and OS samples. All spectra were obtained on a Shimadzu IR-affinity FT-IR spectrophotometer with a Pike Technologies MIRacle ATR. IR spectra was performed on a wave number range from 500 cm^{-1} to 4000 cm^{-1} with a 4 cm^{-1} resolution and 45 scans. We use a pure $CaCO_3$ standard from the FT-IR spectrometer database to compare the $CaCO_3$ contained in the ES sample.

TGA and DTG Evaluations

For TGA and DTG analysis, a Perkin Elmer TGA 4000 device was used. ES and OS samples were grounded and mechanically homogenized in mortars until reaching a granulometry of approximately 100%–100 mesh (about 0.149 mm). ES and OS samples were placed in a desiccator for 6 h prior the TGA and DTG evaluations. Then, 32.971 mg and 16.723 mg of ES and OS samples were analyzed. TGA analysis and DTG analysis were programmed according to the following conditions: 3–10 °C/min heating rate, initial and final temperature of 30° and 900° C, isothermal period of 5 min at 900 °C, and inert gas flow of nitrogen (N_2) at 50 mL/min. To compare the estimation of the decomposition temperature, samples were heated from 30 °C and 900 °C at a heating speed of 10 °C/min for 5 min. During the test were measured the derivative weight %, (%/min) and weight (%) of the DTG and TGA. Pure $CaCO_3$ standard from the laboratory database was used to compare the contained $CaCO_3$ in the ES sample.

Thermal Treatment of Biomasses

The thermal treatment of ES and OS samples was carried out in an electric muffle furnace to a maximum temperature of 1000 °C. Several controls were made for ensuring the correct performance of the electric furnace: (a) preparing a heating speed profile of the reactor to be used for the tests, (b) recording the temperatures and times reached since room temperature until the maximum test temperature (900 °C), and (c) preheat crucibles to 200 °C for about 3 h. Samples of 100 g ES and OS were grounded and mechanically homogenized in mortars, until reaching a granulometry of approximately 100%–100 mesh (about 0.149 mm). A proximity analysis of OS was performed to determine moisture, volatile matter, ash, and fixed carbon percentage according to ASTM standards (7582, D2013, and D346). For the ES calcination, 1 g was heated in triplicate (three samples of 1 g ES, each in their respective crucible) at 900 °C for 5 h in the electric muffle furnace. The pyrolysis of 1328.2 g of OS was carried out in a vacuum pyrolysis reactor (67,461 kPa abs), at a heating rate of 16 °C/min and target temperature of 500 °C during a 3-h period. The biochar pre-oxidation was conducted in a muffle furnace under air atmosphere, with a heating ramp of 2 °C/min, from room temperature to 250 °C, later followed by an isothermal ramp at 250 °C for 1 h, and then cooling to room temperature. We reserve 20 g of the resulting biochar. The pre-oxidized biochar was homogenized into two particle sizes ranges, one less than 250 μm and the other between 500 μm and 250 μm. To chemically modify the biochar, we placed 1 g of the pre-oxidized sample in a flask with 4 ml of 2-mercaptoethanol (99%), 2.8 ml of acetic anhydride (98.5%), and 0.2 ml of sulfuric acid (95%) for each gram of pre-oxidized biochar. The flask was stoppered and stirred for 18 h at 80 °C. The obtained product was filtered and washed with distilled water up to a pH between 5.5 to 6.5 and then dried in an oven for 12 h at 35 °C. The entire procedure was realized in a fume hood.

Sampling, Physical Properties, and Environmental Indicators of PSMT

PSMT was sampled using the sampling grid technique. Each delimited gride had about 1 m side. The total removed volume of PSMT was 48 m^3. The specific gravity was calculated using the pycnometer method. For the average particle size estimating, a series of ASTM meshes (m14, m20, m30, m40, m50, m70, m100, m140, m200, m270, m400) and a background of -m400 were used. To measure some environmental indicators in the PSMT, a multiparameter (HI 98,194, HANNA) was calibrated. PSMT solution (100 g/L) was stirred for 30 min at room temperature (22.2 °C) in a magnetic hotplate stirrer (MSH 20D, Daihan). These measurements were pH, voltage (mV), oxidation–reduction potential (ORP), electrical conductivity (EC), total dissolved solids (TDS), resistivity, salinity, σ seawater, dissolve oxygen (DO), atmospheric pressure (AP), and temperature (T).

Results and Discussion

Chemical Compositions

Chemical compositions of ES, OS, and PSMT are shown in Table 1, Table 2, and Table 3, respectively.

Table 1 Chemical composition of ES

Elements	%	ppm (mg/kg)	Reference methods
Ca	35.52		Sample treatment: Nitric acid digestion Quantitative analysis: APHA 3111B
Na	0.28		
K		<0.00	
Al		<0.40	
Si	<0.20		

Table 2 Chemical composition of OS

Elements	%	ppm (mg/kg)	Reference methods
Ca	1.59		AOAC 999.11 (Sample treatment) APHA 3111B (Quantitative analysis)
Na	0.32		
K	0.16		
Mg		103.55	
Fe		709.43	
Mn		<2.00	
Cr		<2.00	
Cu		6.66	
Hg		<500.00	
Ni		<6.00	
Zn		<0.80	
Al		<40.00	
Ti		<10.00	
P		26.60	Nitric acid digestion (sample treatment) APHA 4500-p (quantitative analysis)
S	0.039%		Note. Form No 203–821-321

Values <2 ppm, <500 ppm, <6 ppm, <0.8 ppm, <40 ppm and <10 ppm represent the detection limits for Mn, Cr, Hg, Ni, Zn, Al, and Ti, respectively

ES samples mainly contain Ca (Table 1), which is lower than the stoichiometric amount (40.00%). In 100 g (molecular weight of $CaCO_3$) of ES, the $CaCO_3$ content is near 88.80%. This value is similar to the usual content of $CaCO_3$ in ES (85%–95% $CaCO_3$) [11].

In comparison with the other trace metals, OS contains a high grade of Ca (1.59%). Ca, Na, and K are common metals in ES and O. However, the K and Na contents in the ES samples are higher than in the OS samples. Therefore, ES could be used directly as an acidity corrector in the PSMT. The S content in OS (0.039%), although low, could provide them with a similar behavior to mercaptans (see the curve area of the sulfur percentage in the OS sample in Fig. 3b).

Table 3 Chemical composition of PSMT

Elements	%	ppm (mg/kg)	ECAs (1) (mg/kg)	Reference methods
Sb		559.37		Sample treatment: Nitric acid digestion Quantitative analysis: APHA 3111B(1)
As	0.83		50	
Cd		116.88	10	
Cu	0.26			
Cr		<2.00	400	
Sn		<200.00		
Ag		170.27		
Pb	3.13		140	
Se		<60.00		
Zn		<0.80		
Hg		<500.00	6.6	
Mo	< 0.20			

Values <2 mg/kg, <200 mg/kg, <60 mg/kg, <0.8 mg/kg and 500 mg/kg represent the detection limits of the methods for Cr, Sn, Se, Zn, and Hg, respectively
(1) Peruvian soil environmental quality standards for residential soils and parks (D.S N° 011–2017-MINAM) (*Estándar de Calidad Ambiental para suelo*, by its acronym in Spanish)

In Table 3, PSMT exhibits a multi-elemental chemical composition with high Pb percentage (3.13%). According to the Peruvian soil environmental quality standards, for residential soil and parks, the cadmium content is higher than the regulation. Hg content, however, cannot be compared since the detection limit is higher that the value in the Peruvian regulation. As, Cr and Pb did not exceed the quality standards. The Peruvian legislation has not yet established environmental quality standards for Sb, Cu, Sn, Ag, Se, or Zn.

FT-IR Spectra

FT-IR ES and OS spectra are shown in Fig. 1, where the ES, pure $CaCO_3$, and OS FT-IR spectra shows compounds such as calcium, lithium, potassium carbonates, and potassium nitrate. Table 4 shows a list of coincident absorbance bands of ES (gray line in Fig. 1a) and $CaCO_3$ (blue line in Fig. 1a) at 879, 824, 798, and 796 cm^{-1} that indicate the presence of $CaCO_3$, Li_2CO_3, K_2CO_3, and KNO_3 functional groups, respectively. There is only one coincident band at 807 cm^{-1} that corresponds to the Polyimide K group in FT-IR spectrum of ES and pure $CaCO_3$ [12]. In Fig. 1b, the ES and OS FT-IR spectra have two absorbance bands that are coincident at 879 and 824 cm^{-1} according to the Shimadzu spectrum library. In the three analyses, the FT-IR spectrometer cannot identify the spectrum reflections in the bands of their functional groups, specifically in the range of 1000 to 2000 cm^{-1}.

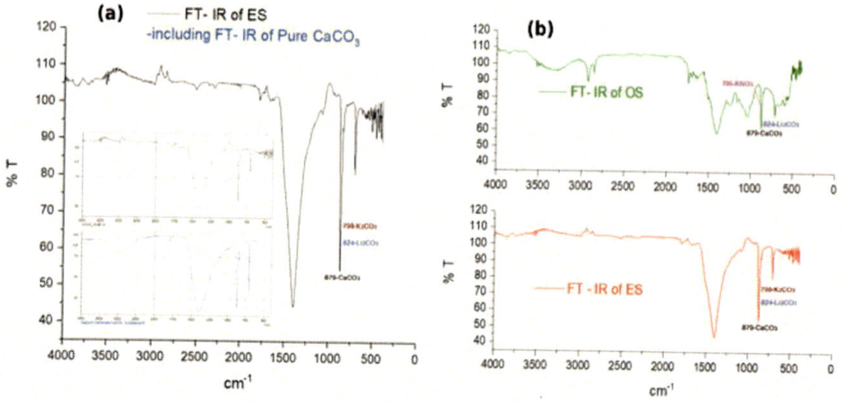

Fig. 1 FT-IR spectra of **a** ES and pure $CaCO_3$, and **b** OS and ES samples

Table 4 Information about FT-IR bands of main functional groups of ES, pure $CaCO_3$, and OS

Samples	Score	Shimadzu Library	Note found/title
OS, ES and	879	7—inorganic	D_ $CaCO_3$/Calcite
Pure $CaCO_3$			
OS, ES and	824	10—inorganic	D_ Li_2CO_3/Lithium
Pure $CaCO_3$			
Eggshell-ES	807	45—IRs Polymer1	PL1,/Polyimide K
Pure $CaCO_3$			
Eggshell-ES	798	9—inorganic	D_K_2CO_3/Potassium
Pure $CaCO_3$			
Olive stones-OS	796	16—Inorganic	D_KNO_3,/Potassium
Pure $CaCO_3$			
Eggshell-ES	788	205—IRs Agrichemicals	Methyldymron/M
Pure $CaCO_3$	784	173—IRs Agrichemicals	Bethrodine/Bethrodine
Eggshell-ES	778	69—Polymer	D_Additive 7/Polymer
Pure $CaCO_3$			
Eggshell-ES	776	103—IRs Agrichemicals	Cyprodinil/Cyprodinil
Pure $CaCO_3$	775	179—IRs Agrichemicals	Pyributicarb/Pyrite

XRD Studies

The XRD analysis results of ES and OS samples are shown in Figs. 2 and 3.

Figure 2 shows the presence of only one mineralogical phase in ES, $CaCO_3$. From Fig. 3a is difficult to identify any diffraction patterns due the amorphous condition of the OS sample. The latter can be related to the presence of organic impurities that

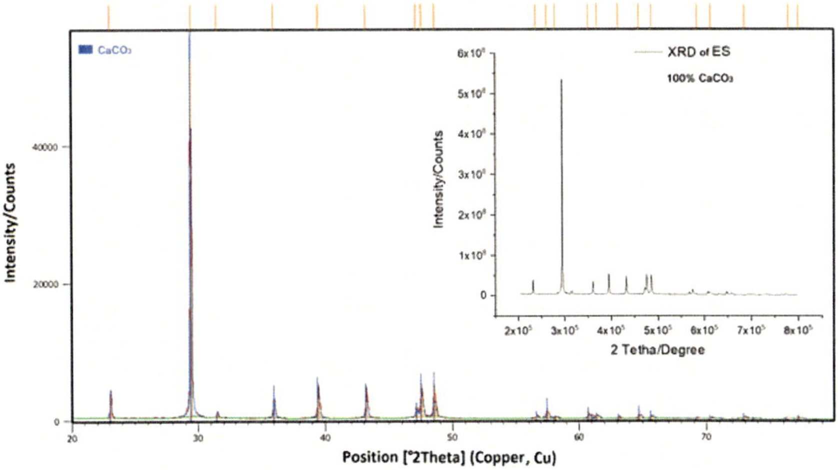

Fig. 2 XRD analysis of ES sample

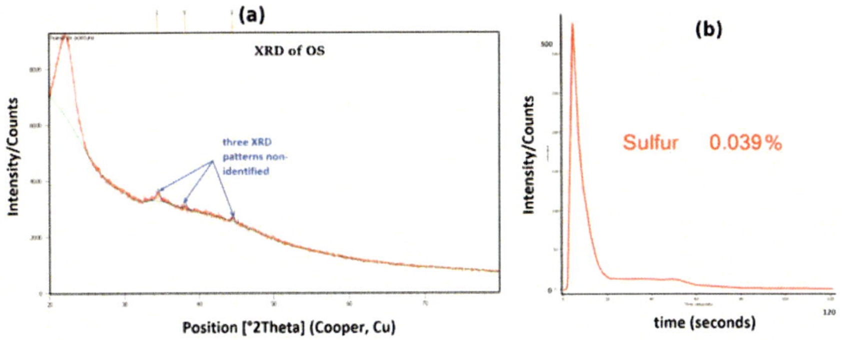

Fig. 3 XRD analysis of **a** OS sample and **b** sulfur analysis

were not separated during the cleaning stage with potassium hydroxide and distillated water. The XRD analysis results of the PSMT samples are shown in Fig. 4 and Table 5.

TGA and DTG Evaluations

Results of TGA and DTG analysis from the ES and OS samples are shown in Fig. 5 and Table 6.

In Fig. 5a, DTG of ES (blue line) and pure $CaCO_3$ (green line) exhibit peaks at 834.48 and 805.96 °C. The decomposition temperature of the ES sample (88.80%

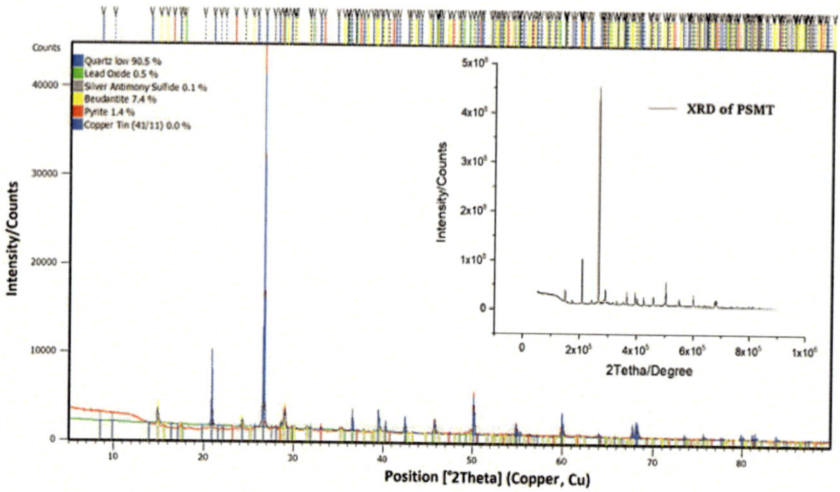

Fig. 4 XRD analysis of the PSMT sample

Table 5 Diffraction patterns used to quantify the crystal phases in the PSMT sample

Ref. code	Score	Compound name	Scale factor	%	Chemical formula
01–087-2096	73	Silicon oxide	0.890	90.50	SiO_2
01–085-1289	34	Lead oxide	0.020	0.50	PbO
98–060-5711	23	Silver antimony sulfide	0.018	0.10	Ag S_2 Sb
01–079-2206	38	Lead iron aluminum arsenate sulfate hydroxide-beudantite	0.066	–7.40	Pb ($Fe_{2.54}$ $Al_{0.46}$) ($As_{1.07}$ O_4) ($S_{0.93}$ O_4) ($OH)_6$ $PbFe_3(AsO_4)(SO_4)(OH)_6$
98–001-0422	31	Pyrite	0.015	1.40	Fe1 S_2
98–000-0439	21	Copper tin (41/11)	0.008	0.10	Cu_{41} Sn_{11}

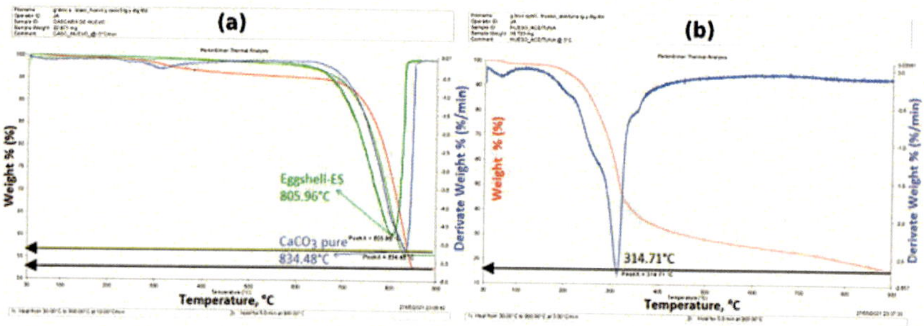

Fig. 5 TGA and DTG of **a** ES and pure $CaCO_3$ and **b** OS samples

Table 6 TGA and DTG results from ES and OS samples and pure $CaCO_3$

			TGA		DTG
Samples	Initial mass, mg	Heating rate, °C/min	Volatile matter, %	Weight/mass loss, %	Decomposition temperature, °C
ES	32.971	10	47.00	53.00	834.48
pure $CaCO_3$	32.971	10	44.00	56.00	805.96
OS	16.723	3	84.00	16.00	314.71

$CaCO_3$) is higher than for the pure $CaCO_3$ (decomposition temperature about 700 °C to 750 °C). Moreover, TGA of both samples observed a little deflection between the red lines in relation to the green lines (Fig. 5a, b), taken place near 250 °C due to the impurities present in the ES sample. TGA and DTG of OS presented a mass loss near 84% (w/w) and average moisture of 9.29%. In Fig. 5b, the OS sample evidenced a peak at 314.71 °C (maximum decomposition temperature) after a heating rate of 3 °C/min for about 5 min [13].

Calcination of the ES Sample

According to the reaction: $CaCO_3(s) = CaO(s) + CO_2(g)$, the calcination of the ES sample confirmed a generation of 53% CaO. Both ES TGA analysis and calcination tests observed a mass loss near 47% (w/w).

Thermal Treatment of the OS Sample (Proximity Analysis)

Proximity analysis presents 9.01% moisture, 84.06% volatile matter, 4.90% ash and 2.03% fixed carbon. As Na (1.59%) and fixed carbon (2.03%) quantities are low, it will be necessary to active the OS carbon content using a chemical route [14].

Route for Obtaining Chemical Modified Pre-Oxidized Biochar from OS Sample

Table 7 depicts a summary of results for the pyrolysis, pre-oxidation, and chemical modification processes.

Chemically modified pre-oxidized biochar (w/w) was constituted of 50 g of average particle size $\leq$250 μm, 75.40 g of an average particle size range between $\geq$500 μm and $\leq$250 μm, and 40 g of average particle size $\geq$500 μm. The chemical treatment to the pre-oxidized biochar, to form chemically modified pre-oxidized

Table 7 Route for obtaining chemically modified biochar from OS

Pyrolyzed mass, g	Biochar mass, g	% Biochar (w/w)	Pre-oxidized biochar mass, g	% Pre-oxidized biochar (w/w)	Chemically modified pre-oxidized biochar mass, g	% Chemically modified pre-oxidized biochar (w/w)
1328.20	257.80	19.41	164.54	64.57	165.40	100.00

biochar, increased the mass content in 0.14 g. Results suggest that the presence of mercaptoethanol in the chemically modified pre-oxidized biochar could be effective method for removing the Hg content in the PSMT sample.

Physical Properties and Environmental Indicators of the PSMT

PSMT samples had an average particle size of 96.24 μm and specific gravity of 2890 kg/m^3. The environmental indicators values are: 158 mV/pH, 4.37 pH, 306.9 mV (ORP), 88.5% DO, 7.57 ppm DO, 5906 μS/cm, 5526 μS/cm^4, 0.0002 MΩ.cm, 2953 ppm TDS, 3.21 PSU, 0.3σt, and 21.06 °C. Equilibrium diagrams and Eh and pH values are very useful to identify the chemical speciation of inorganic compounds. Looking for the intersection of the Eh (158 mV) and pH (4.37) values in the Cd–O-H and Hg-O–H equilibrium diagrams, we could find that Cd^{+2} or Hg^{+2} are in the soil solution (aqueous phase). Natural and Calcinated ES could remove this Cd^{+2} and Hg^{+2} through a dissociation of the Ca^{+2} contents, promoting a basic medium through cation exchange with other metal ions [15].

Conclusions

In the FT-IR spectra, the ES and pure $CaCO_3$, and OS absorbance bands at 879, 824, 798 and 796 cm^{-1} indicate the presence of $CaCO_3$, Li_2CO_3, K_2CO_3, and KNO_3 functional groups, respectively. The XRD analysis of ES samples showed the presence of one mineralogical phase ($CaCO_3$). Diffraction patterns in the OS could not be identified due the amorphous condition of the sample. DTG analysis showed ES exhibited peaks at 834.48 °C. Decomposition temperature of the ES sample was higher than that of pure $CaCO_3$, (decomposition temperature of 700 to 750 °C). TGA and DTG analysis of OS presented a mass loss near 84% (w/w), with an average moisture of 9.29%. TGA and ES calcination tests presented a mass loss near 47% (w/w). The proximity analysis showed a low content (2.09%) of fixed carbon in the OS, which indicate that it will be necessary to activate the contained carbon in the OS via chemical route. The process to obtain chemically modified pre-oxidized biochar from OS reported a pyrolysis mass loss of 80.59%, pre-oxidized mass loss of 35.43%, and an addition of 0.14 g during the chemical treatment. According to the Peruvian soil environmental quality standards for residential soils and parks, the PSMT sampled has cadmium content higher than the regulation limits, while As, Cr, and Pb did not exceed the quality standards. Hg content could not be compared since the detection limit is higher that the value in the Peruvian regulation. The PSMT sample had an average particle size of 96.24 μm and a specific gravity of 2890 kg/m^3. Equilibrium diagrams showed that Cd^{+2} cations are most likely in the PSMT soil solution (aqueous phase). This compound could be removed by the cation exchange properties of calcinated ES. The mercaptoethanol content in the chemically modified pre-oxidized biochar could be effective in removing the Hg content in PSMT.

Acknowledgements Authors would like to thank the Research Vice Rectorate and the College of Geological Mining and Metallurgical Engineering Research Institute of the National University of Engineering for the financial assistance granted, without which these programmed experiences could not have been carried out.

References

1. Michael S (2019) Mining capital. Methods, best-practices and case studies for financing mining projects. Springer Link, 2019. https://www.springer.com/gp/book/9783030312244
2. Worlanyo AS, Jiangfeng L (2020) Evaluating the environmental and economic impact of mining for post-mined land restoration and land-use: a review. J Environ Manag, 111623
3. Amirshenava S, Osanloo M (2019). A hybrid semi-quantitative approach for impact assessment of mining activities on sustainable development indexes. J Clean Prod 218:823–834
4. Mineral Commodity Summaries (2018) U.S. Geological Survey, U. S. Geological Survey
5. Gestión D (2021) Minería puede ser el salvavidas para economía peruana en recesión. https://gestion.pe/economia/mineria-puede-ser-el-salvavidas-para-economia-peruana-en-recesion-noticia/?ref=gesr
6. Cruzado-Tafur E, Torró L, Bierla K, Szpunar J, Tauler E (2021) Heavy metal contents in soils and native flora inventory at mining environmental liabilities in the Peruvian Andes. J South American Earth Sci 106:103107
7. Zhang X, Yang H, Cui Z (2018) Evaluation and analysis of soil migration and distribution characteristics of heavy metals in iron tailings. J Clean Prod 172:475–480
8. Al-Ghouti MA, Salih NR (2018) Application of eggshell wastes for boron remediation from water. J Mol Liq 256:599–610
9. Pardo T, Bernal P, Clemente R (2017) The use of olive mill waste to promote phytoremediation. Olive Mill Waste, 183–204
10. Delgado-Moreno L, Bazhari S, Gasco G, Méndez A, El Azzouzi M, Romero E (2021). New insights into the efficient removal of emerging contaminants by biochars and hydrochars derived from olive oil wastes. Sci Total Environ 752, 141838
11. Awogbemi O, Inambao F, Onuh EI (2020) Modification and characterization of chicken eggshell for possible catalytic applications. Heliyon 6(10):e05283
12. Umesh M, Choudhury DD, Shanmugam S, Ganesan S, Alsehli M, Elfasakhany A, Pugazhendhi A (2021). Eggshells biowaste for hydroxyapatite green synthesis using extract piper betel leaf-Evaluation of antibacterial and antibiofilm activity. Environ Res, 111493

13. Gomez-Martin A, Chacartegui R, Ramirez-Rico J, Martinez-Fernandez J (2018) Performance improvement in olive stone's combustion from a previous carbonization transformation. Fuel 228:254–262
14. Puig-Gamero M, Esteban-Arranz A, Sanchez-Silva L, Sánchez P (2021) Obtaining activated biochar from olive stone using a bench scale high-pressure thermobalance. J Environ Chem Eng 9(4):105374
15. Gholizadeh M, Hu X (2021) Removal of heavy metals from soil with biochar composite: a critical review of the mechanism. J Environ Chem Eng, 105830

Utilization of Copper Nickel Sulfide Mine Tailings for CO_2 Sequestration and Enhanced Nickel Sulfidization

Fei Wang, David Dreisinger, Glenn Barr, and Chris Martin

Abstract Global warming mitigation strategies include reducing CO_2 emissions and encouraging the use of battery-powered electrical vehicles. Nickel in olivine which is an important mineral of many nickel sulfide mines and suitable for permanent CO_2 sequestration is regarded as non-recoverable and discarded in mine tailings as wastes. Enhancing nickel sulfide production from hitherto non-recoverable resources could help increase battery supply. This work shows that a copper nickel sulfide flotation mine tailing in Minnesota could be used for directly accelerated mineral carbonation and concurrent nickel sulfidization. In addition to converting magnesium and iron silicates to mineral carbonates for permanent CO_2 storage, nickel in olivine is concurrently converted to nickel sulfide for potential recovery. The mineral carbonation of olivine is the dominant chemical reaction process and provides the precondition for nickel sulfidization. A pre-concentration step of olivine from tailings is recommended to enrich olivine and nickel contents for potential application.

Keywords CO_2 sequestration · Nickel sulfidization · Mine tailings · Olivine · Mineral carbonation

F. Wang (✉) · D. Dreisinger
Department of Materials Engineering, The University of British Columbia, Vancouver, BC V6T 1Z4, Canada
e-mail: fei.wang@alumni.ubc.ca

G. Barr
Twin Metals Minnesota LLC, St. Paul, MN 55102, USA

C. Martin
Blue Coast Research Ltd., Parksville, BC V9P 1R2, Canada

A. Lazou et al. (eds.), *REWAS 2022: Developing Tomorrow's Technical Cycles (Volume I)*, The Minerals, Metals & Materials Series,
https://doi.org/10.1007/978-3-030-92563-5_24

Introduction

The current world is working to mitigate global warming. The United States expects to achieve carbon neutrality by 2050 [1]. Accelerated mineral carbonation could be an important method in achieving the target of reducing CO_2 emissions and to permanently store CO_2 as stable mineral carbonates [2]. The Duluth Complex is the largest single resource of nickel, copper, cobalt, and precious metals in North America. The over 9.5 billion tonnes of mineralised material occur in mafic and ultramafic rocks. The potential flotation tailings of the complex contain approximately 25% olivine in average, i.e., 2.4 billion tonnes, and thus can make significant contributions to reduce CO_2 emissions via mineral carbonation.

In addition to reducing CO_2 emissions, encouraging battery-powered electrical vehicles instead of fossil fuel-powered traditional vehicles is also important to mitigate global warming. Metals including nickel and cobalt are significant elements for batteries. Most of the residual nickel and cobalt of the Duluth Complex tailings are present as replacement ions in the basic silicate mineral olivine and thus are considered as non-recoverable by flotation. This situation is quite common for many Ni-hosted sulfide deposits. Thus, enhancing recovery of the non-recoverable nickel and cobalt could play a significant role in both reducing CO_2 emissions and improving battery supply. The previous work [3] on high-purity olivine (forsterite) has confirmed the possibility of converting nickel contained in olivine to nickel sulfides by utilizing mineral carbonation, as shown in Fig. 1. Mineral carbonation provides the precondition to release nickel ions for sulfidization and magnesium

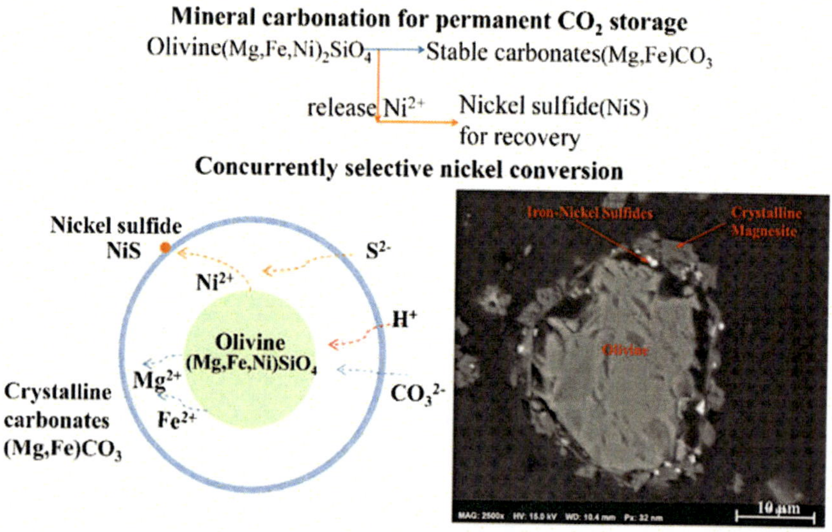

Fig. 1 Schematic diagram of mineral carbonation of olivine and concurrent nickel sulfidization [3]

and ferrous ions for concurrently permanent CO_2 sequestration. The supply of a gas mixture containing 95% CO_2 and 5% H_2S can be used to achieve the selectivity of nickel conversion and mineral carbonation.

However, the effect of the high iron content of the Duluth Complex tailings on mineral carbonation and nickel conversion is not known and tailings have not been tested. This work is to characterize the Twin Metals Minnesota LLC Duluth Complex-hosted Maturi deposit tailings for carbon sequestration capacity and to test the sulfidization during carbon sequestration in order to provide a pathway to enhanced metal recovery.

Materials and Methods

Materials

The tests refer to two types of Maturi tailings: one contains the life of mine average olivine content of the deposit, described in the following context as low olivine group (LOG); the other contains a relatively high olivine content corresponding to several specific sizeable zones in the deposit, described in the following context as high olivine group (HOG). The chemical compositions of HOG and LOG samples together with high-purity forsterite as comparison are shown in Table 1, analyzed via inductively coupled plasma-atomic emission spectroscopy (ICP-AES). Both the HOG and LOG tailings contain nickel in 0.08% and copper in 0.03%, respectively. As can be seen, although the nickel content in the tailings is lower than nickel of the high-purity forsterite in previous work, the nickel in olivine of the tailings is similar to the nickel content in the forsterite. The iron content is higher than 11% compared to 7% of the high-purity forsterite. The theoretical mineral carbonation capacity, calculated based on total contents of magnesium, iron, and calcium, is 0.29 tonne CO_2 per tonne HOG and 0.26 tonne CO_2 per tonne LOG tailing, respectively.

The mineral composition through the quantitative X-ray diffraction (QXRD) method is shown in Table 2. The dominant mineral of both HOG and LOG samples is plagioclase, accounting for 48% and 55%, respectively, followed by olivine occupying 35% and 26%, respectively. The third gangue mineral is pyroxene (diopside) accounting for 7% and 8% for HOG and LOG samples, respectively. Each of the rest gangue minerals is less than 2% of the tailings. The sulfur content shown in Table 1

Table 1 Chemical compositions of the high olivine group (HOG) and low olivine group (LOG) samples and high-purity forsterite, wt%

Sample	Ni	Cu	Fe	S	C	Mg	Si	Co	Al	Cr	Mn	Ca	Ti
HOG	0.082	0.030	12.34	0.063	0.069	7.6	19.4	0.0087	7.9	0.047	0.14	5.1	0.55
LOG	0.083	0.031	11.12	0.068	0.11	6.0	20.0	0.0072	8.6	0.038	0.14	5.6	0.67
Forsterite	0.27	–	7.09	0.00	0.00	27.4	20.8	–	0.09	0.21	0.08	0.11	–

Table 2 Mineral compositions of the HOG and LOG samples through the quantitative X-ray diffraction (QXRD) analysis

Mineral	Ideal formula	HOG	LOG
Ankerite-Dolomite	$Ca(Fe^{2+},Mg,Mn)(CO_3)_2$–$CaMg(CO_3)_2$	0.4	0.8
Biotite	$K(Mg,Fe^{2+})_3AlSi_3O_{10}(OH)_2$	1.7	1.2
Clinochlore	$(Mg,Fe^{2+})_5Al(AlSi_3O_{10})(OH)_8$	1.3	1.3
Pyroxene (Diopside)	$CaMgSi_2O_6$	7.0	7.9
Ferroactinolite	$Ca_2Fe_5{}^{2+}Si_8O_{22}(OH)_2$	0.5	1.0
Olivine	$(Mg,Fe)_2SiO_4$	35.2	25.7
Ilmenite	$Fe^{2+}TiO_3$	1.6	2.2
Iron-alpha	α-Fe	2.2	2.1
Magnetite	$Fe^{2+}Fe^{3+}{}_2O_4$	1.4	1.5
Plagioclase	$NaAlSi_3O_8$–$CaAl_2Si_2O_8$	48.2	54.9
Quartz	SiO_2	0.5	0.7
Talc	$Mg_3Si_4O_{10}(OH)_2$	–	0.7
Total		100.0	100.0

is in forms of chalcopyrite and cubanite, followed by pentlandite and pyrrhotite. This may result in an initially high molar ratio of Fe/Ni for the tailings, higher than 30 for the high-purity forsterite, regarding nickel sulfide chemical analysis of selective sulfidization of nickel over iron.

It is expected that olivine is the major mineral involved in mineral carbonation based on the previous research on raw natural silicate samples [4]. Correspondingly, the theoretical maximum mineral carbonation capacity needs to be re-calculated based on olivine content where each mole olivine containing two moles of bivalent metal elements can sequester two moles of CO_2 gas. The corrected mineral carbonation capacities for HOG and LOG are 0.17 and 0.12 tonne CO_2 per tonne sample, respectively. The scanning electron microscope with an energy dispersive X-ray detector (SEM–EDX) analysis further confirms that the Fe/Mg molar ratio of the olivine is approximately 2.6 ~ 3.1, much higher than the ratio in high-purity forsterite (0.075). According to the peer work [5], mineral carbonation of olivine containing high Fe/Mg molar ratio may need slightly reductive atmosphere to be accelerated.

Methods

The samples were the tailings after copper sulfide and nickel sulfide flotation. Each of the HOG and LOG samples was homogeneously divided into 3 different charges with different particle size distribution, through grinding and passing sieves of 200 mesh, 325 mesh, and 400 mesh, respectively. As a result, six sub-samples were obtained

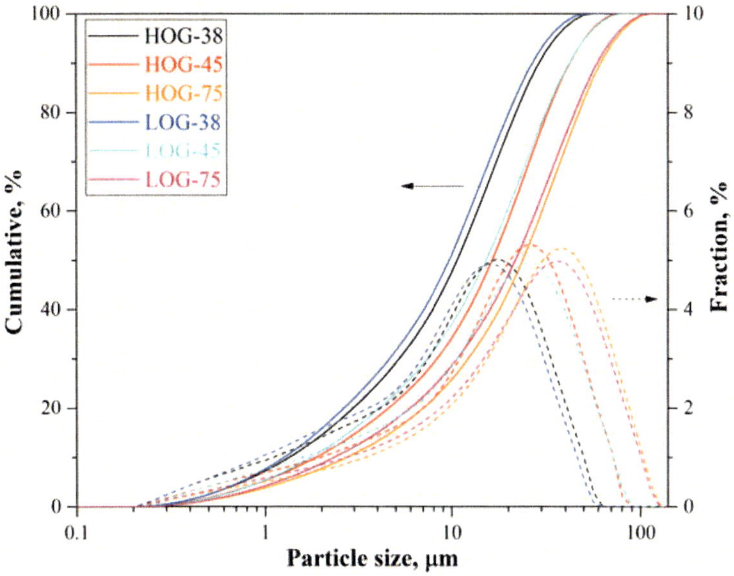

Fig. 2 Particle size distribution of HOG and LOG samples

and identified as HOG-38, HOG-45, and HOG-75 for the HOG, and LOG-38, LOG-45, and LOG-75 for the LOG, respectively. The particle size distribution (PSD) of the six samples is shown in Fig. 2. There was no obvious difference in PSD overall trend between HOG and LOG samples, although LOG samples were slightly finer than the corresponding HOG samples. The P_{80} values were 23–25 μm for HOG-38 and LOG-38, 36 μm for HOG-45 and LOG-45, and 52–53 μm for HOG-75 and LOG-75, respectively. The mineral carbonation and sulfidization tests were carried out in a 600 mL stainless steel autoclave with a sampling kit. The mineral carbonation tests were conducted under the conditions of 175 °C, $PCO_2 = 34.5$ bar, 10% pulp density, addition of 1.5 m $NaHCO_3$ with a reaction time up to 8 h. The mineral carbonation tests with concurrent nickel sulfidization were carried out under the identical conditions except for temperature at 150 °C and gas mixture supply of 5% H_2S and 95% CO_2 ($PH_2S = 1.4$ bar + $PCO_2 = 26$ bar) because of maximum pressure limitation of H_2S regulator. The calculations of mineral carbonation efficiency and nickel conversion efficiency are based on Eqs. (1) and (2).

$$\alpha = \frac{(\theta_2 - \theta_1)}{m \times (\frac{12.01}{44.01} - \theta_2)} \times 100\% \tag{1}$$

$$\beta = \frac{(m_2 \times \delta_2 - m_1 \times \delta_1)}{m_1 \times (1 - \delta_1)} \times 100\% \tag{2}$$

where α is the mineral carbonation efficiency; m_1 and m_2 are the weights of solid sample before and after reaction, respectively; θ_1 and θ_2 are the total carbon content of solid sample before and after reaction, respectively; m is the theoretically maximum mineral carbonation capacity; β is the nickel conversion efficiency; δ_1 and δ_2 are the nickel in sulfides content of solid sample before and after reaction and analyzed by bromine–methanol digestion-ICP AES [6, 7], respectively.

The chemical compositions of the mineral samples and solutions were analyzed by ICP-AES. QXRD analysis was used to determine the mineral composition of the HOG and LOG samples. Detailed PSD analyses were obtained through laser sizing using a Malvern Mastersizer 2000 instrument. The changes in mineral compositions of the samples were analyzed by regular XRD analysis using a Rigaku MultiFlex XRD instrument with Cu-Kα radiation over the range of 3°–90° of 2θ. The total carbon and sulfur contents were determined by LECO CS3200 instrument. The nickel sulfide analysis was analyzed by bromine–methanol digestion-AAS. SEM–EDX analysis using an FEI Quanta 650 instrument was used to investigate the process of the mineral carbonation and nickel sulfidization.

Results and Discussion

Mineral Carbonation

As shown in Fig. 3, the effect of particle size on the mineral carbonation efficiency is significant. The finer the particles of the tailing samples, the easier the mineral carbonation process owing to the larger specific surface area supporting the chemical reaction. Besides, when the mineral carbonation capacity was based on the total contents of magnesium, iron, and calcium as the traditional way shown in Fig. 1a1, b1, the mineral carbonation efficiency of HOG samples was higher than the results of LOG samples, 45% for HOG-38 compared to 32% for LOG-38 at 8 h. The difference in carbonation efficiency between HOG and LOG samples was gradually reduced for the coarser particles, 32% for HOG-45 vs. 26% for LOG-45, followed by 21% HOG-75 vs. 20% for LOG-75 at 8 h. In fact, the mineral carbonation efficiency should be further optimized based on the olivine content in tailings, since previous research shows that carbonation of olivine is the dominant chemical reaction and that the mineral carbonation capacity should be corrected based on the olivine content for natural silicate mixtures [4]. As shown in Fig. 3a2, b2, the effect of particle size on the corrected mineral carbonation efficiency was further magnified; 79% of olivine for HOG-38 has been reacted at 8 h while only 37% of olivine for HOG-75 reacted; similarly for LOG samples, 68% for LOG-38 compared to 43% for LOG-75. In addition, the mineral carbonation efficiency based on olivine was much higher than based on the traditional way. This indicates that olivine was the dominant mineral

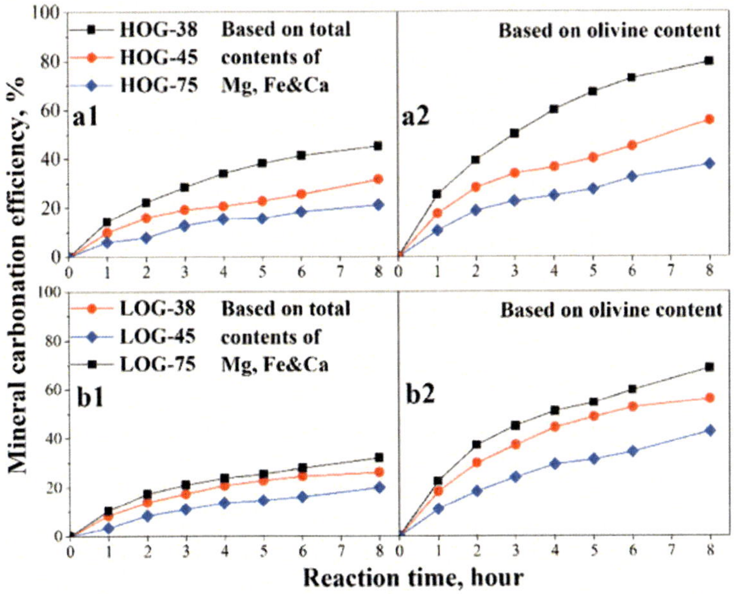

Fig. 3 Effect of particle size of the HOG **a** and LOG **b** samples on mineral carbonation efficiency and role of olivine on mineral carbonation: theoretical carbonation capacity based on the total contents of Mg, Fe and Ca (a1 and b1) and based on olivine content (a2 and b2)

involving mineral carbonation, which supports the previous research. The concentration of magnesium, iron and the other bivalent metals in aqueous solution was less than 0.8 mg/L. All the released bivalent elements from olivine during carbonation precipitated as mineral carbonates.

The results of XRD analysis (Fig. 4) further support the role of olivine during the mineral carbonation. With an increase in reaction time, the peak intensity of olivine gradually decreased and correspondingly the peak intensity of the produced mineral carbonates (magnesite–siderite) gradually increased. Together with formation of mineral carbonates, the amount of produced quartz also gradually increased owing to the reaction of olivine. There was no obvious change in the other minerals including plagioclase and pyroxene. SEM images of carbonated HOG-38 can confirm the experimental and XRD results. The outer surface (Fig. 4a, b) shows the formation of massive crystalline mineral carbonates, while plagioclase and ilmenite were untouched. The transverse surface (Fig. 4c, d) reveals that the unreacted olivine was surrounded by crystalline carbonates with a porous void between the carbonates layer and the olivine. There was no obvious formation of carbonates outside of plagioclase and pyroxene. It is also noted that the olivine in Maturi tailings was slightly more resistant to carbonation compared to the high-purity forsterite because of the high Fe/Mg ratio of the olivine in tailings (average 2.84 vs. 0.075 for the high-purity forsterite). The mineral carbonation efficiency of HOG-38 olivine was 79% compared with the complete reaction for the high-purity forsterite under the

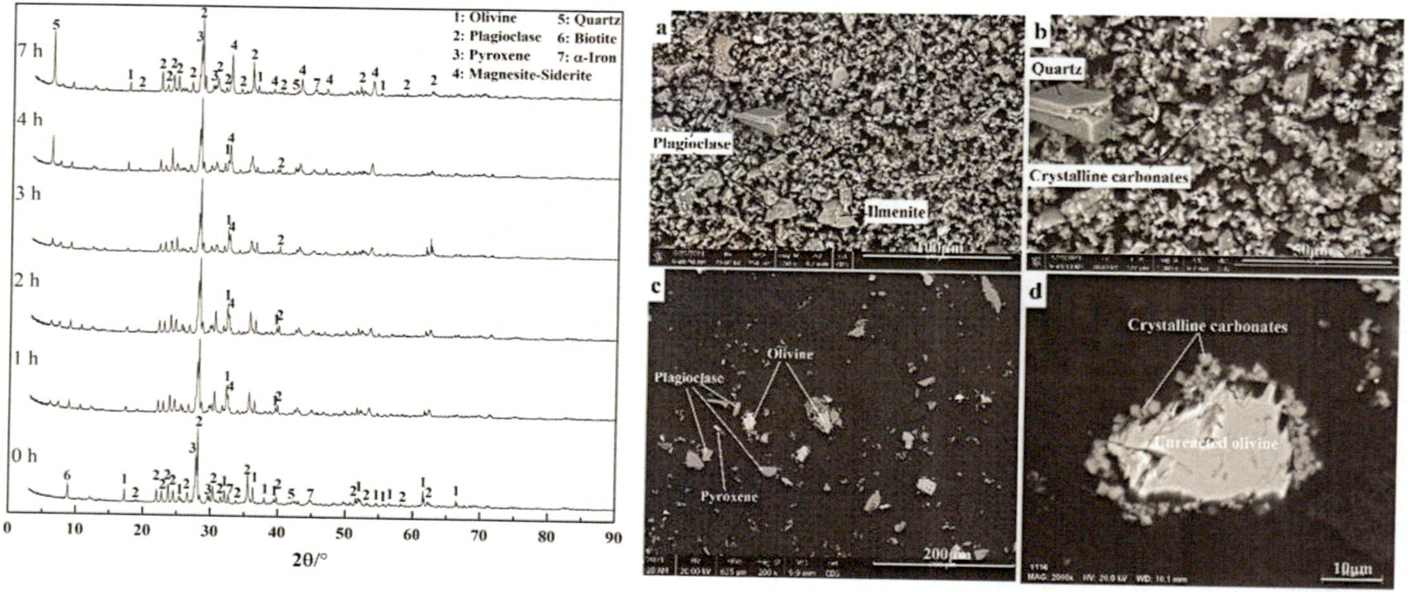

Fig. 4 Changes in mineral compositions (left) and SEM images (right) of HOG-38 tailings during mineral carbonation through XRD patterns: **a**, **b** outer surface of particles; **c**, **d** transverse surface of particles

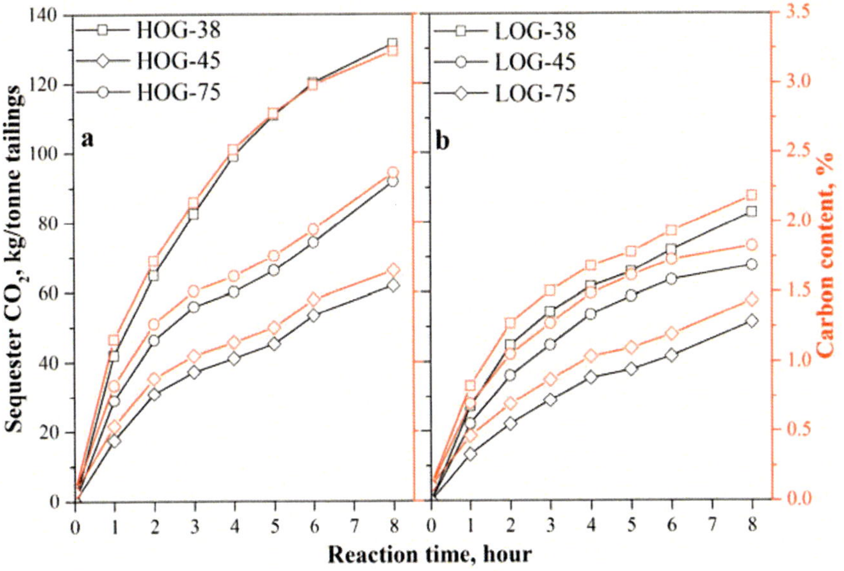

Fig. 5 Amount of sequestered CO_2 gas through mineral carbonation and the increase of carbon content in residues

identical conditions: 175 °C, 34 bar CO_2 partial pressure, 1.57 m^2/g surface area (1.53 for the high-purity forsterite), and 1.5 M sodium bicarbonate for 8 h reaction. A slight reductive atmosphere might be needed to accelerate the mineral carbonation of the Maturi tailings.

Nevertheless, the mineral carbonation progress can be also described as the amounts of sequestered CO_2 gas per tonne of tailings as shown in Fig. 5. With the increase in reaction time, the carbon content gradually increased from less than 0.1% for raw tailings to 3.23% for HOG-38 and 2.18% for LOG-38 at 8 h. All the increased carbon was the sequestered CO_2 in mineral carbonates. Correspondingly, the amount of sequestered CO_2 gas gradually increased up to 131 kg per tonne HOG-38 tailings and 83 kg per tonne LOG-38 tailings.

Mineral Carbonation with Concurrent Nickel Sulfidization

In addition to CO_2 sequestration, the mineral carbonation process of the tailings can be utilized for nickel conversion from silicates to sulfides for potential enhanced recovery. The mineral carbonation efficiency and the corresponding sequestered CO_2 gas are shown in Fig. 6. Compared with Fig. 3 and Fig. 5, the mineral carbonation efficiency based on the olivine content and the amount of sequestered CO_2 gas were markedly decreased from 79% and 131 kg CO_2 per tonne tailing to 34% and 57 kg CO_2 per tonne tailing for HOG-38 samples. The coarse samples had even

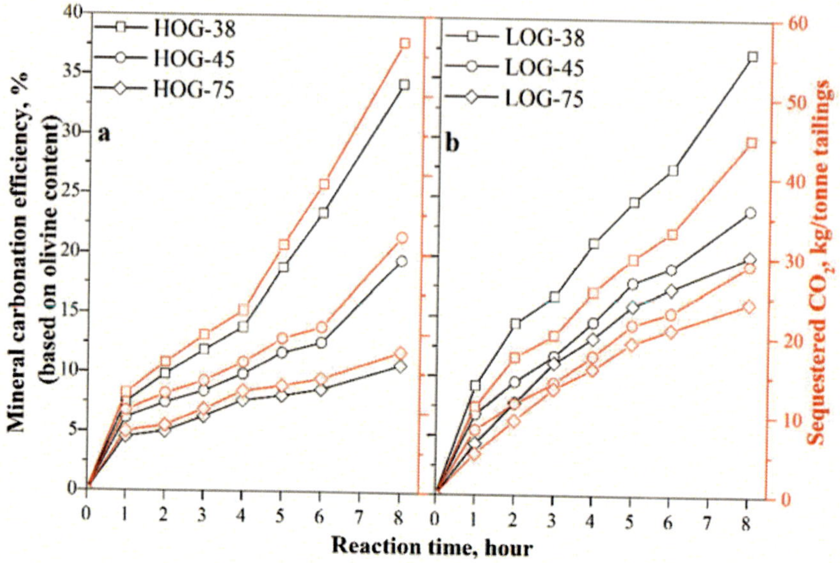

Fig. 6 Mineral carbonation efficiency and the sequestered CO_2 amount of HOG **a** and LOG **b** samples during carbonation and concurrent sulfidization

lower carbonation progress, e.g., 11% and 18 kg CO_2 per tonne tailing for HOG-75 samples. Similar results were shown for LOG samples. It may be because of the high molar ratio of Fe/Mg in olivine that increased the sensitivity of mineral carbonation to temperature and CO_2 partial pressure. As shown in Fig. 7, the decrease of temperature

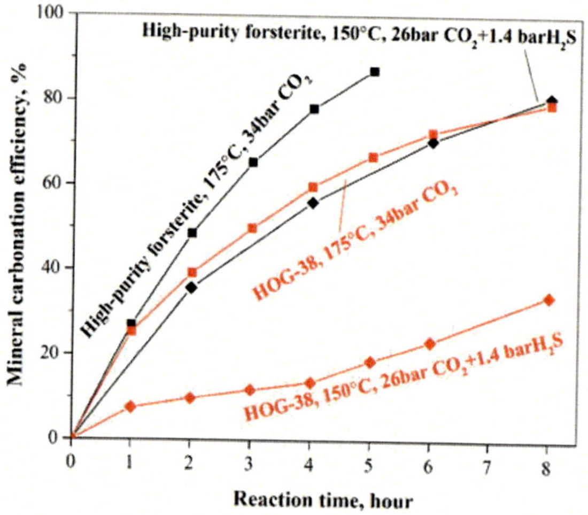

Fig. 7 Comparison of mineral carbonation and sulfidization between high-purity forsterite and HOG-38 tailings

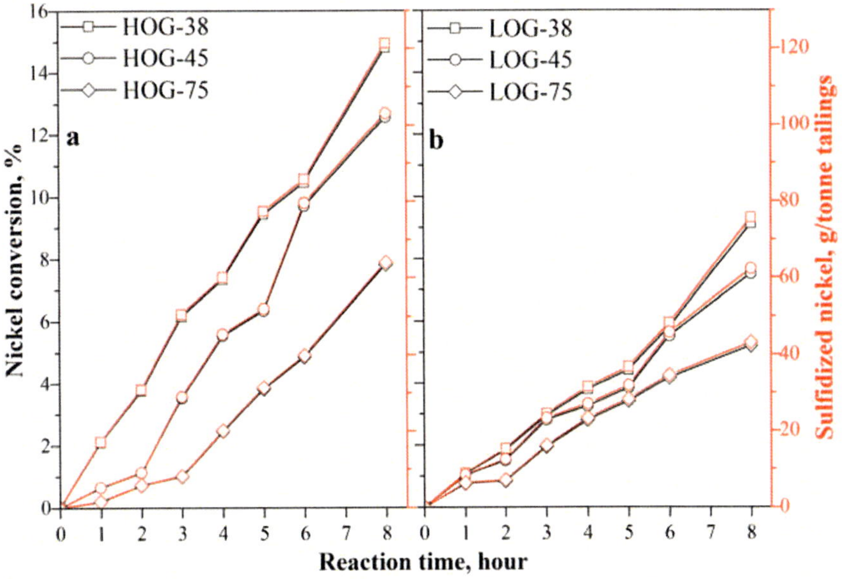

Fig. 8 Nickel conversion efficiency and sulfidized nickel amount during mineral carbonation and concurrent sulfidization process for HOG **a** and LOG **b** tailings

and CO_2 pressure resulted in decrease by 49% of mineral carbonation efficiency for HOG-38 samples whereas only up to 22% of mineral carbonation efficiency was reduced for high-purity forsterite. It is important to maintain the carbonation and sulfidization at high temperature and CO_2 pressure.

Together with mineral carbonation where CO_2 converted to mineral carbonates, some nickel in silicates also converted to nickel sulfides as shown in Fig. 8. The nickel conversion efficiency gradually increased with reaction time and with increase in mineral carbonation efficiency. The highest nickel conversion efficiency for HOG and LOG samples was 15% and 9%. Correspondingly, up to 121 g and 76 g nickel per tonne tailings was converted from silicates to sulfides for HOG and LOG samples, respectively. As for selectivity of sulfidization, since there were iron sulfides and high iron content in raw tailings, the initial molar ratio of Fe/Ni was very high at average 100; the mineral carbonation and sulfidization process gradually lowered down the molar ratio of Fe/Ni in sulfidized residues to around 20. It means that the amount of nickel converted to nickel sulfide was increasing with sulfidization process and was higher than the amount of iron converted from olivine. In other words, the sulfidization process was selective of nickel over iron.

Through the comparison of mineral compositions among raw tailings, carbonated residue and sulfidized residue for HOG-38 as shown in Fig. 9, it can be seen that there was obvious formation of mineral carbonates and quartz during sulfidization but the extent of reaction was lower than that in the carbonation only. The XRD results are

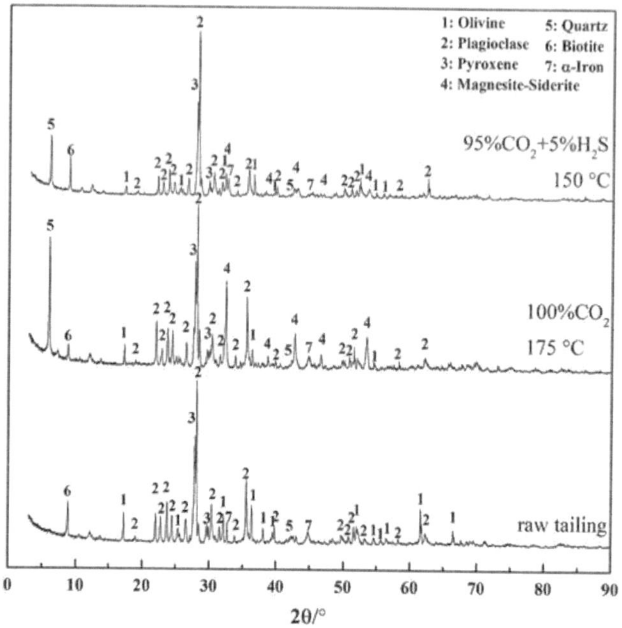

Fig. 9 Comparison of changes in mineral compositions of HOG-38 tailings during mineral carbonation and nickel sulfidization through XRD patterns

identical with the experimental data for sulfidization tests. In addition, the SEM–EDX analysis further proved the decrease in the extent of reaction with smaller amounts of mineral carbonates. The carbonated olivine surrounded by crystalline carbonates with a porous void was not found. The mineral carbonation efficiency may be dominantly owing to the reaction of superfine particles.

With considering that olivine was the dominant mineral for mineral carbonation of tailings and the other minerals were not reacted, it would economically to pre-concentrate olivine from the tailings to reduce the capital costs. The pre-concentration step would also benefit the effective nickel sulfidization.

Conclusion

The overall finding of the study performed on the Maturi flotation tailings is encouraging. The work confirms that the tailings can be utilized to sequester CO_2 gas through accelerated mineral carbonation and to concurrently convert nickel silicate to nickel sulfide for potential recovery. The mineral carbonation of olivine is the dominant chemical reaction process and provides the precondition for nickel sulfidization. The other minerals are not significant for mineral carbonation. Although the olivine in the tailings contains high Fe(II) content, each tonne tailings can still sequester up to

131 kg CO_2 at pure compressed CO_2 pressure supply; at gas mixture of 95% CO_2 and 5% H_2S and lower temperature, up to 15% of nickel from tailings can convert to nickel sulfide, up to 121 g nickel per tonne tailings, together with sequestered 57 kg CO_2 per tonne tailings. Further studies are still needed including pre-concentrating olivine from the tailings and using gas mixture containing reductive gas during mineral carbonation to facilitate the reaction process.

Acknowledgements The authors thank Twin Metals Minnesota LLC for financial support. Blue Coast Research Ltd is also thanked for preparing the samples for this study.

References

1. The White House. Fact Sheet: President Biden Sets 2030 Greenhouse Gas Pollution Reduction Target Aimed at Creating Good-Paying Union Jobs and Securing U.S. Leadership on Clean Energy Technologies. *Whitehouse.Gov* 6 (2021). Accessed 2nd September 2021
2. Wang F, Dreisinger DB, Jarvis M, Hitchins T (2018) The technology of CO_2 sequestration by mineral carbonation: current status and future prospects. Can Metall Q 57:46–58
3. Wang F, Dreisinger D, Jarvis M, Hitchins T, Trytten L (2021) CO_2 mineralization and concurrent utilization for nickel conversion from nickel silicates to nickel sulfides. Chem Eng J 406:126761
4. Wang F, Dreisinger D, Jarvis M, Hitchins T (2021) Kinetic evaluation of mineral carbonation of natural silicate samples. Chem Eng J 404:126522
5. Wood CE, Qafoku O, Loring JS, Chaka AM (2019) Role of Fe(II) content in olivine carbonation in wet supercritical CO_2. Environ Sci Technol Lett 6:592–599
6. Penttinen U, Palosaari V, Siura T (1977) Selective dissolution and determination of sulphides in nickel ores by the bromine-methanol method. Bull Geol Soc Finl 49:79–84
7. Klock PR, Czamanske GK, Foose M, Pesek J (1986) Selective chemical dissolution of sulfides: an evaluation of six methods applicable to assaying sulfide-bound nickel. Chem Geol, 157–163

Removal Behavior of Benzohydroxamic Acid from Beneficiation Wastewater via Fe(III) Chelation and Precipitation Flotation Process

Guihong Han, Yifan Du, Yanfang Huang, Wenjuan Wang, Shengpeng Su, and Bingbing Liu

Abstract Benzoximic acid (BHA) is a commonly used collector in mineral processing industry. The accumulation of BHA in water will cause serious damage to environment. Therefore, how to efficiently treat the residual BHA in mineral processing wastewater is an important issue in the mineral processing industry. In this study, Fe(III) was used as chelating agent to separate BHA from wastewater by precipitation flotation. The optimum flotation conditions of BHA and Fe(III) were obtained by optimizing the chelation and flotation processes. The results showed that the chelating efficiency was the highest when the chelating pH was 7 and Fe(III) was 40 mg/L. When the addition amount of surfactant CTAB is 4 mg/L and the gas flow rate is 40 mL/min, the flotation removal rate of BHA reaches 90%, and the COD removal rate of the solution is 73%. The treated solution meets the first-class standard of industrial waste water discharge.

Keywords Benzohydroxamic acid · Organic contaminant · Precipitate flotation

Introduction

The rapid development of beneficiation mining produces a large amount of beneficiation wastewater. Mineral processing wastewater contains a large amount of mineral processing agents, leading to the increase of its chemical oxygen demand (COD) [7]. Benzohydroxamic acid (BHA) is a highly efficient collector for zinc ore and tungsten ore, which is widely used in metallurgical industry [6, 11]. Generally, when these agents such as BHA enter the water environment, it may even contain biotoxicity endangering ecosystems, fishery, and human health [1]. Thus, tremendous hazardous beneficiation wastewaters containing low concentration of BHA organic contaminant

G. Han · Y. Du · Y. Huang (✉) · W. Wang · S. Su · B. Liu
School of Chemical Engineering, Chemical Building, Zhengzhou University, Room No. 216, 450001 Henan, People's Republic of China
e-mail: huangyf@zzu.edu.cn

B. Liu
e-mail: liubingbing@zzu.edu.cn

A. Lazou et al. (eds.), *REWAS 2022: Developing Tomorrow's Technical Cycles (Volume I)*, The Minerals, Metals & Materials Series,
https://doi.org/10.1007/978-3-030-92563-5_25

needs urgent treatment [12]. However, most of the treatment methods for BHA, such as adsorption and catalytic degradation, are faced with the characteristics of high cost, high energy consumption and low efficiency, which urges us to look for new treatment methods for wastewater containing BHA.

Precipitation flotation is a commonly used method in wastewater treatment in recent years, which originated from the mineral processing industry [3, 8]. Precipitation flotation technology is to produce fine precipitate particles by adding precipitator or chelating agent through chemical reaction [13]. The particle surface properties of the target substance and the surface tension of the liquid are changed by the action of the surfactant, and then bubbles are introduced to separate the target substance from the solution [14, 15]. Compared with other methods, precipitation flotation requires fewer flotation agents, has a wider application range, and has a more stable flotation effect, which has a satisfactory effect on the treatment of low concentration wastewater [10]. However, the selection of appropriate chelating precipitator for BHA becomes an important step to precipitate flotation. Fortunately, it is reported that soluble BHA can form insoluble metal chelates with the addition of metal ions under weakly basic conditions [2, 4]. This makes it possible to remove soluble BHA from wastewater by mature and simple flotation technology.

In this paper, the efficient removal of BHA organic pollutants from industrial mineral processing wastewater by chelating, flocculation, and precipitation flotation process was studied. The dissolved BHA was firstly chelated by Fe(III) to form a hydrophobic precipitate, and then, the precipitate was removed by flotation in a microbubble column. The process flow of precipitation flotation was optimized through experiments, and the mechanism of chelation was also briefly explored.

Experimental

Materials

In this work, the simulated BHA-bearing wastewaters were prepared using the analytically pure BHA (Aladdin Biological Technology Co., Ltd., China) as raw material. Chelating reagent of Fe(III) solutions was prepared from iron chloride hexahydrate ($FeCl_3 \cdot 6H_2O$, Tianjin Kemiou Chemical Reagent Co., Ltd., China). Sodium hydroxide (NaOH, Sinopharm Chemical Reagent Co., Ltd. China) and sulfuric acid (H_2SO4, Sinopharm Chemical Reagent Co., Ltd. China) were used to adjust the solution pH. Cetyl trimethyl ammonium bromide (CTAB, Aladdin Biological Technology Co., Ltd., China) was employed as a collector, and nonylphenol polyoxyethylene ether (NP-40, Aladdin Biological Technology Co., Ltd., China) was used as frother during the flotation process of BHA-Cu sediments. All the chemical reagents were in analytical grade, and deionized water was used to prepare the simulated wastewaters and stock solutions.

Experimental Procedures

Figure 1 shows the schematic diagram of the chelation and precipitation flotation process. Precipitation reagent of Fe(III) was added into 500 ml simulated BHA wastewater in a beaker, then the mixed solutions were blended by a magnetic stirrer with 150 rpm (Shanghai Rex Instrument Factory, China) for 30 min at 20 °C. CTAB surfactant was applied to enhance the hydrophobic property of the BHA-bearing particles in the solution. Afterward, the suspension liquid was transferred to the microbubble flotation column, and a certain amount of NP-40 foaming agent was further added to promote foam production and the adhesion of particles to air bubbles. The organic contaminant BHA and additive Fe(III) were concentered in the foams.

The turbidity and COD of treated solution were measured by Turbidity meter (WGZ 2000, Shanghai Xinrui Instruments & meters Co. Ltd. China) and potassium dichromate method (Chinese standard of HJ 828–2017). The removal efficiency of COD of the aqueous solution is defined as the following Eq. (1):

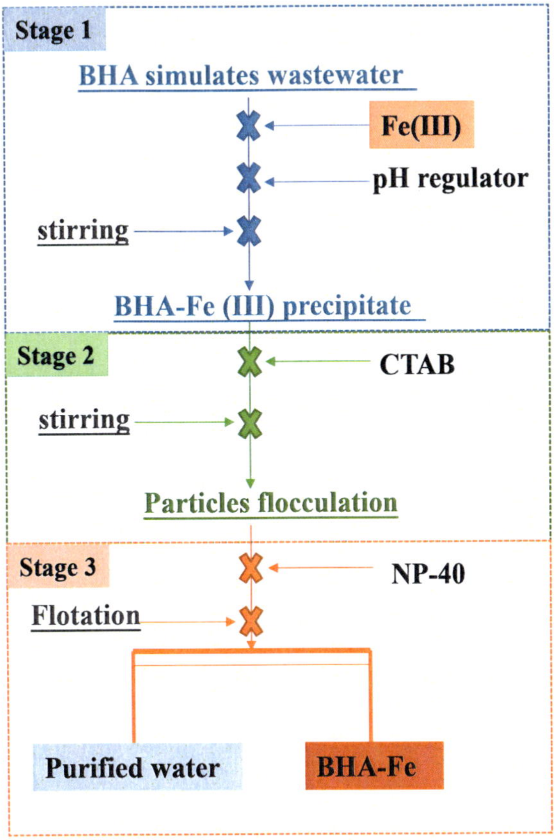

Fig. 1 Schematic diagram of chelation and precipitation flotation process

$$R_{\mathrm{COD}} = (1 - \frac{V_P COD_p}{V_0 COD_0}) \times 100\% \tag{1}$$

where R_{COD} is the removal efficiency of COD, %; COD_p and COD_0, respectively, represent the value of COD before and after chelation precipitation, mg/L; V_0 and V_p, respectively, represent the volume of the solution before and after chelation precipitation or flotation, mL.

The UV–vis analysis (TU-1901, Beijing Purkinje General Instrument Co., Ltd., China) was used to determine the BHA at the wavelength of 330 nm. The flotation products and residual solution were sampled and analyzed, and the removal efficiency of BHA in the precipitation flotation process was determined according to Eq. (2).

$$R_f = \left(1 - \frac{V_f C_f}{V_0 C_0}\right) \times 100\% \tag{2}$$

where R_f represents the flotation removal efficiency of BHA, %; C_0 and C_f, respectively, represent the concentrations of BHA before and after flotation, mg/L; V_0 and V_f, respectively, represent the volume of the solution before and after flotation, mL.

Results and Discussion

Chelation Precipitation of BHA by Fe(III) Addition

The first task to precipitate BHA flotation is to form chelating precipitation of free BHA in solution under a complex solution environment. Therefore, a series of initial pH values and Fe(III) dosages were investigated for the solution to explore the chelating process and determine the best conditions for the chelation. The corresponding results are shown in Fig. 2.

The chelating efficiency of BHA was significantly affected by the pH and Fe(III) content added in the solution. As shown in Fig. 2a, with the increase of pH, the chelating efficiency of BHA and Fe(III) firstly increased and then gradually decreased. At pH of 7, the chelating efficiency of BHA by Fe(III) is the highest. With the increase of Fe(III) concentration in the solution, the collision between Fe(III) and BHA molecules in the solution is strengthened, and the chelating efficiency also increases. When the initial concentration of Fe(III) in the solution is increased to 30 mg/L, the chelation of BHA by Fe(III) reaches saturation. As the concentration of Fe(III) in the solution continues to increase, the combination of OH- with excessive Fe(III) in the solution is strengthened and generates Fe(OH)3. The pH of the solution changes, resulting in the reduction of chelating efficiency.

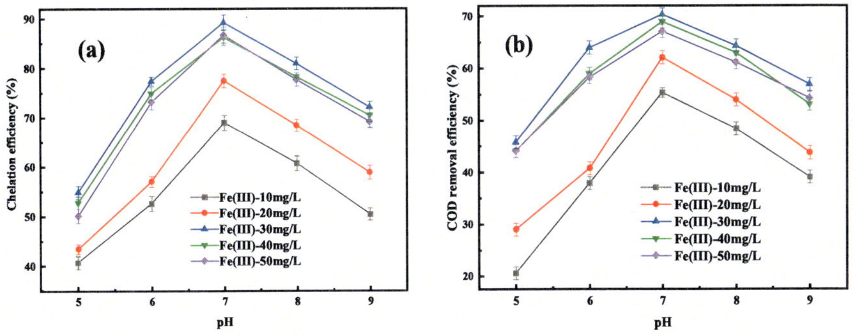

Fig. 2 **a** Effect of Fe(III) concentrations on the chelation efficiency of BHA; **b** removal efficiency of COD after chelation

Resultantly, the optimal chelating procedure was conducted at pH of 7 with Fe(III) content of 30 mg/L. At this condition, the removal efficiency of BHA and COD in the solution is 89.23% and 70.31%, respectively.

Discussion on the Chelation Precipitation and Flocculation Mechanisms

To further analyze the interaction between BHA and Fe(III) during the chelation reaction under the optimized chelating conditions, the chelated sediments were lyophilized and measured by FTIR. FTIR analysis of BHA and chelated BHA-Fe was shown in Fig. 3.

It can be seen from Fig. 3, the C-N peak at 1152 cm^{-1} and N–O peak at 1100 cm^{-1} in BHA-Fe flocs after chelation reaction have no deviation compared with the BHA solid particles. However, the stretching superposition vibration peaks of O–H moves from 3310 cm^{-1} to 3447 cm^{-1}. The stretching vibration peak of N–H in the complex shifts from 3062 cm^{-1} to 2810 cm^{-1} before chelation and the C=N peak of BHA was separated and moved to 1585 cm^{-1}. This indicates that the coordination between BHA and Cu ions is carried out on the oxygen or nitrogen atoms [9]. In addition, the stretching vibration peak of C-N shifts from 1162 cm^{-1} to 1150 cm^{-1}, and the benzene ring skeleton also moves to a higher wavenumber. And the vibration peak of Fe–O appears at 475 cm^{-1}. Based on the FTIR analysis, the equation of Fe(III) chelating reaction with BHA is described as follows:

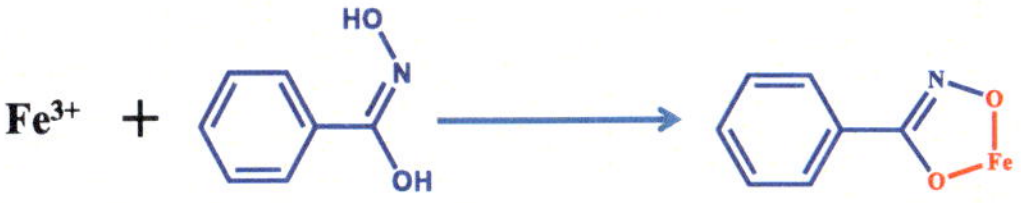

(3)

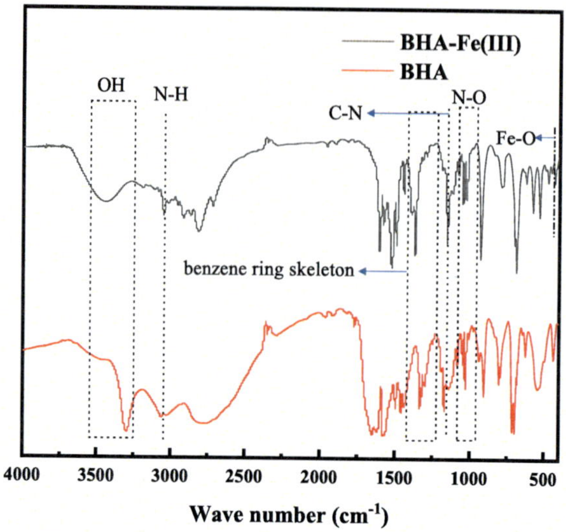

Fig. 3 FTIR analyses of BHA and BHA-Fe floc

Zeta Potential Analysis of Chelating Granules

In the flotation process, the surface potential of precipitated particles in solution has a great influence on the flotation efficiency [5]. Therefore, to explore the surface potential of precipitated particles produced by the chelation process, effect of surfactant on Zeta potential at the surface of flocs under a various chelating pH range was studied. The result is shown in Fig. 4.

According to Fig. 4a, the Zeta potential of the particles is negative at pH range of 7–10. With the increase of pH, the content of OH- in the solution increases and

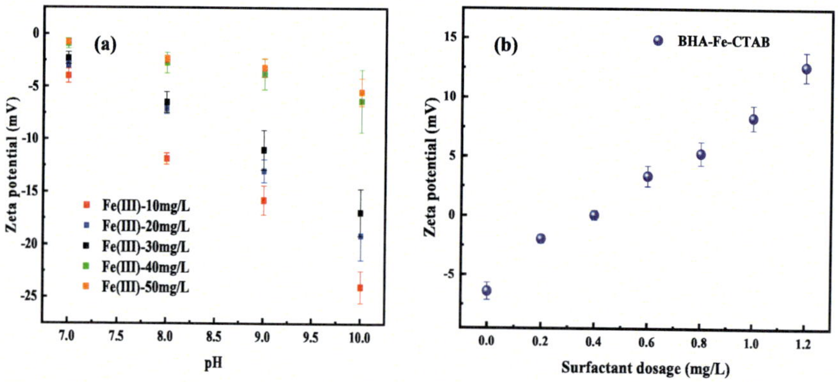

Fig. 4 **a** Zeta potential on the surface of flocs under different chelating conditions; **b** surface potential of the particles after CTAB addition under the optimal chelating conditions

the electronegativity increases. Under the optimal chelating conditions, the surface Zeta potential of the particles is −6.48 ± 0.99 eV. Figure 4b shows the surface Zeta potential of the particles after adding surfactant CTAB as the regulator under the optimal chelating conditions. As a cationic surfactant, CTAB has a positive charge and can neutralize the surface potential of the precipitated particles. Therefore, with the increase of CTAB addition, the surface potential of the particles changes from negative to positive. The particle surface potential is 0.075 eV as 0.4 mg/L CTAB is added, and this potential is very close to the "0 electric point." In such a case, the attraction force between particles is greater than the repulsive force, and the flocs have a better aggregate effect, which is conducive to the subsequent flotation separation.

Flotation Removal of the BHA-Fe Floccules

After adding CTAB, column flotation equipment was used to treat the chelated solution. Under the chelating conditions of pH 7, Fe(III) 30 mg/L, and CTAB 0.4 mg/L, the effects of NP-40 dosage, air flow rate, and flotation time on the removal of BHA were investigated, and the corresponding results are shown in Fig. 5.

Figure 5a shows the BHA removal rate and COD content of the solution after 40 min flotation with addition of foaming agent NP-40. With the increase of the concentration of NP-40 added, the flotation efficiency is gradually enhanced. When the concentration of NP-40 is 0.7 mg/L, the flotation efficiency reaches the highest and the removal efficiency of BHA in the solution is 89.83%, while the COD in the solution is 42.39 mg/L. After that, when the content of NP-40 is increased, the removal efficiency of BHA in the solution remains stable, while the COD begins to rise gradually since superfluous NP-40 is residual in the solution. Therefore, the optimal NP-40 dosage is 0.7 mg/L.

Figure 5b shows that the flotation efficiency gradually increases with the increase of the flow rate through the flotation. When the gas flow rate is 80 mL/min, the removal rate of BHA is 89.97% and the COD in the solution is 43.4 mg/L. After

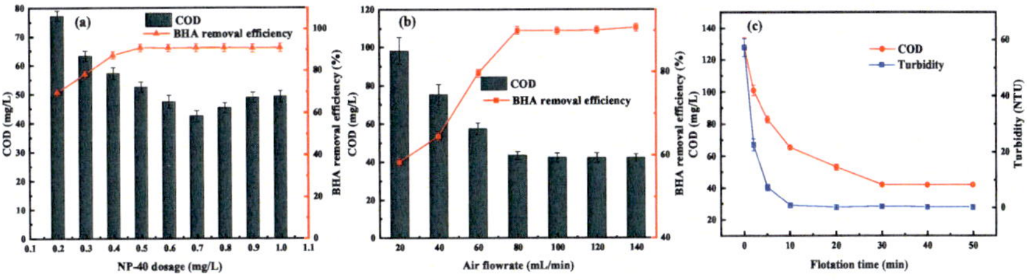

Fig. 5 **a** Effect of NP-40 addition on BHA removal rate and COD in flotation solution; **b** effect of flotation gas flow rate on BHA removal rate and COD in solution after flotation; **c** effect of flotation time on COD and turbidity of solution after flotation

that, the flotation efficiency remains stable when the flow rate continues to increase. Therefore, the optimal flow rate in the flotation process is 80 mL/min.

Under the optimal chelating conditions, the amount of NP-40 is 0.8 mg/L and the flow rate is 80 mL/min, the changes of turbidity and COD content of the solution with the flotation time are shown in Fig. 5c. With the increase of flotation time, the turbidity and COD of the solution decrease at first and then gradually flattening out. When the reaction time reaches 30 min, the turbidity of the solution is 0.47 NTU and the COD is 42.15 mg/L. After that, the turbidity and COD of the solution remain stable when the reaction time continues to increase. Therefore, the flotation time is selected as 30 min.

Conclusions

In this study, based on the easy chelation of soluble organic BHA and Fe(II), a facile and efficient removal method of BHA organic contaminants from the industrial beneficiation wastewaters via the chelation, flocculation, and precipitation flotation process was proposed. The microbubble flotation technology was successfully applied to separate the BHA-Fe sediments from the wastewaters using CTAB as surfactant and NP-40 as frother. Over 87% BHA has removed under the optimal chelation, flocculation and flotation conditions: chelation with Fe(III) concentration of 30 mg/L at pH of 7 for 30 min; flocculation with CTAB concentration of 0.4 mg/L for 20 min; followed by flotation with NP-40 of 0.7 mg /L and air flow rate of 80 mL/min for over 30 min. After flotation treatment, 70.32% of COD were removed from the wastewater, the COD content of the solution is reduced to 42.15 mg/L, and the turbidity is 0.47 NTU. The industrial BHA-bearing wastewaters after precipitation flotation treatment met the wastewater direct emission requirements. Eventually, precipitation flotation can effectively treat wastewater containing organic BHA.

Acknowledgements The authors wish to express their thanks to the National Natural Science Foundation of China (No. U2004215), the Innovation Talents Support Program in University of Henan Province (No. 20HASTIT012), the China Postdoctoral Science Foundation (No. 2019TQ0289), and the Key Scientific Research Project Plan of Henan Colleges and Universities (No. 20A450001).

References

1. Bertanza G, Collivignarelli C, Pedrazzani R (2001) The role of chemical oxidation in combined chemical-physical and biological processes: experiences of industrial wastewater treatment. Water Sci Technol 44:109–116
2. Buglyo P, Nagy EM, Farkas E, Sovago I, Sanna D, Micera G (2007) New insights into the metal ion-peptide hydroxamate interactions: metal complexes of primary hydroxamic acid, derivatives of common dipeptides in aqueous solution. Polyhedron 26:1625–1633

3. Feng QC, Zhao WJ, Wen SM, Cao QB (2017) Activation mechanism of lead ions in cassiterite flotation with salicylhydroxamic acid as collector. Sep Purif Technol 178:193–199
4. Griffith DM, Szocs B, Keogh T, Suponitsky KY, Farkas E, Buglyo P, Marmion CJ (2011) Suberoylanilide hydroxamic acid, a potent histone deacetylase inhibitor; its X-ray crystal structure and solid state and solution studies of its Zn(II), Ni(II), Cu(II) and Fe(III) complexes. J Inorg Biochem 105:763–769
5. Guo SH, Li G, Qu JH, Liu XL (2011) Improvement of acidification on dewaterability of oily sludge from flotation. Chem Eng J 168:746–751
6. Jiang YR, Pan YF, Chen D, Wang FP, Yan LL, Li GY, Xue YL (2012) A theoretical study of the effect of carboxyl hydroxamic acid on the flotation behavior of diaspore and aluminosilicate minerals. Clays Clay Miner 60:52–62
7. Meng XS, Khoso SA, Lyu F, Wu JQ, Kang JH, Liu H, Zhang QP, Han HS, Sun W, Hu YH (2019) Study on the influence and mechanism of sodium chlorate on COD reduction of minerals processing wastewater. Miner Eng 134:1–6
8. Peng WJ, Chang LP, Li PY, Han GH, Huang YK, Cao YJ (2019) An overview on the surfactants used in ion flotation. J Mol Liquids 286
9. Rio-Echevarria IM, White FJ, Brechin EK, Tasker PA, Harris SG (2008) Surface binding vs. sequestration; the uptake of benzohydroxamic acid at iron(III) oxide surfaces. Chem Commun 286:4570–4572
10. Rybarczyk P, Kawalec-Pietrenko B (2021) Simultaneous removal of Al, Cu and Zn ions from aqueous solutions using ion and precipitate flotation methods. Processes 9
11. Tian MJ, Gao ZY, Khoso SA, Sun W, Hu YH (2019) Understanding the activation mechanism of Pb2+ ion in benzohydroxamic acid flotation of spodumene: Experimental findings and DFT simulations. Miner Eng 143
12. Wang FP, Zhan GP, Jiang YR, Guo JN, Yin ZG, Feng R (2013) Theoretical evaluation of flotation performance of carboxyl hydroxamic acids with different number of polar groups on the surfaces of diaspore (010) and kaolinite (001). J Mol Model 19:3135–3142
13. Wu HY, Huang YF, Liu BB, Han GH, Su SP, Wang WJ, Yang SZ, Xue YB, Li SQ (2021) An efficient separation for metal ions from wastewater by ion precipitate flotation: probing formation and growth evolution of metal-reagent flocs. Chemosphere 263
14. Yan S, Zhang JQ, Diao WZ (2014) Flotation reagent progress and application overview. Mach Electron Control Eng 441:76–79
15. Zana R (2002) Dimeric and oligomeric surfactants. Behavior at interfaces and in aqueous solution: a review. Adv Coll Interface Sci 97:205–253

Recovery of Molybdenum from Metallurgical Wastewater by Fe(III) Coagulation and Precipitation Flotation Process

Bei Zhang, Bingbing Liu, Yuanfang Huang, Guihong Han, Yifan Du, and Shengpeng Su

Abstract Molybdenum (Mo) is an irreplaceable alloying element of steel, resulting in the vigorous development of Mo metallurgy industry. However, abundant Mo-bearing wastewaters with low concentrations are concomitantly generated, which is difficult to recycle and gives rise to the waste of resources. In this work, recovery of Mo from metallurgical wastewater by Fe(III) chelation and precipitation flotation process was investigated. The effects of pH value, Fe(III), HA dosage, and reaction time on the Mo recovery efficiency from wastewater were systematically studied. The results showed that the molybdate in the solution can be coagulated by Fe(III) to form precipitation. HA can effectively increase the sizes of precipitation particles and improve flotation separation of Mo. After optimization, over 99.6% Mo was removed under the optimal coagulation and flotation condition. This technology can realize the effective recovery of molybdate from wastewater.

Keywords Molybdenum · Metallurgical wastewater · Precipitation flotation · Recovery · Flocculation precipitation

Introduction

Molybdenum is one of the essential trace elements for the survival of plants and animals, and it is also an important strategic resource in the development of the national economy [1, 2]. However, in the mining and metallurgical processing of Mo minerals, there is a large amount of ultra-low concentration Mo-containing wastewater, which is difficult to recover, resulting in a serious waste of resources and Mo pollution [1–3]. Moreover, according to the pollutant discharge standard for the inorganic chemical industry issued in 2015, the total Mo content in the industrial

B. Zhang · B. Liu (✉) · Y. Huang · G. Han · Y. Du · S. Su
School of Chemical Engineering, Zhengzhou University, Zhengzhou 450001, Henan, China
e-mail: liubingbing@zzu.edu.cn

G. Han
e-mail: hanguihong@zzu.edu.cn

A. Lazou et al. (eds.), *REWAS 2022: Developing Tomorrow's Technical Cycles (Volume I)*, The Minerals, Metals & Materials Series,
https://doi.org/10.1007/978-3-030-92563-5_26

wastewater discharge process should not be higher than 0.5 mg/L. Therefore, it is imperative to study an economical and efficient method to recover molybdate from ultra-low concentration Mo-containing wastewater.

The most stable valence states of Mo in the natural environment are + 4 and +6, and it mainly exists in the form of MoO_4^{2-} in natural water and industrial wastewater [3]. At present, the methods of recovering Mo from Mo-containing wastewater are mainly chemical precipitation [4–6], ion exchange [7–10], solvent extraction [11–13], and adsorption [14–17]. However, the chemical precipitation method is only suitable for the treatment of high concentration solution, and the treated solution still contains a large amount of molybdate, which cannot meet the required discharge standards [4, 5]. The solvent extraction method is easy to form the second phase in the process of recovering Mo, resulting in secondary pollution [11, 12]. The ion exchange method and adsorption method are widely used in the treatment of ultra-low concentration Mo-containing wastewater, but the materials required are expensive and difficult to recover [7, 8, 14, 15].

Precipitation flotation is an effective method to remove ions from low concentration wastewater, in which the flocculant is added into the solution and reacts with MoO_4^{2-} to generate fine precipitation particles, and they are removed by flotation [18]. It has the advantages of simple process, low cost, and short reaction time [19, 20]. In this study, the precipitation flotation method was used to study the flocculation of precipitated particles by adding HA in the process of Fe^{3+} treatment of molybdenum wastewater. The effects of pH value, amount of HA, reaction time, and other parameters on the molybdenum recovery rate in wastewater were studied. The technology can realize the effective recovery of molybdate from wastewater and reduce its harm to the environment and humans body.

Materials and Methods

In this work, the MoO_4^{2-} solution is prepared from sodium molybdenum dihydrate ($Na_2MoO_4 \cdot 2H_2O$, Tianjin Yongda Chemical Reagent Co., Ltd.). The Fe^{3+} is used for in-situ preparation of metal hydroxyl coagulant adsorbents, and it is from iron chloride hexahydrate ($FeCl_3 \cdot 6H_2O$, www.rhawn.cn). Humicacid (HA, Tianjin Yongda Chemical Reagent Co., Ltd.) is used as a flocculating agent to increase particle sizes. Cetyl Trimethyl Ammonium Bromide (CTAB, www.rhawn.cn) is used as the collecting agent to recover the precipitation particles from the solution. Sodium hydroxide (NaOH, Tianjin Yongda Chemical Reagent Co., Ltd. China) and hydrochloric (HCl, Sinopharm Chemical Reagent Co., Ltd. China) are used to adjust the pH value of the solution. Notably, all the chemical reagents are analytical grade, and deionized water is used to prepare the solution.

Figure 1 shows the experimental flowsheet for removal of Mo from waste solutions by precipitation flotation. Firstly, MoO_4^{2-} reacts with Fe^{3+} to form fine precipitate under the condition of adding alkali; then, HA addition can increase the particle size

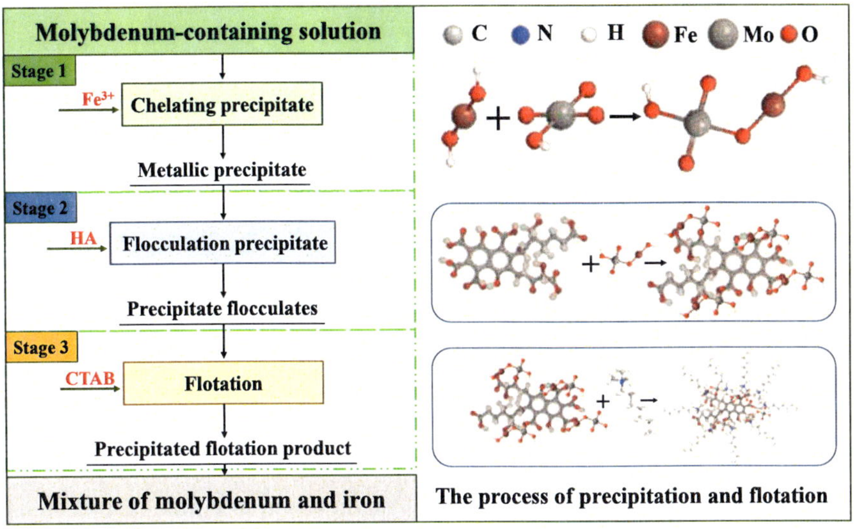

Fig. 1 Experimental flowsheet for removal of Mo from waste solutions by precipitation flotation

of precipitated particles. Finally, a collector is added to change the hydrophilicity of the particle surface and the precipitated particles are recovered by flotation.

Results and Discussion

Precipitation Transformation of Molybdate with Adding of Fe(III)

In this paper, Fe(III) is used as a precipitant to study the precipitation of molybdate under different pH values. The experimental conditions are as follows: initial Mo concentration of 10 mg/L, temperature of 25 °C, rotation speed of 150 r/min, n(Mo):n(Fe) of 1:5. After reaction for a while, the solution is filtrated through 0.22 um microporous water system filter membrane. The supernatant of the solution is taken for analysis of ion contents. The experimental results are shown in Fig. 2.

It can be seen that pH value has a great influence on the precipitation of molybdate in solutions. As the pH value is at a lower value of 3, there is no obvious solid formation in the solution, and the precipitation rate of molybdate is only about 25%. With the increase of pH value from 3 to 9, the molybdate precipitation rate firstly increases and then decreases with a maximum value at pH of 5, which the molybdate precipitation rate is 94.1%. When pH > 5, the precipitation rate of molybdate in the solution begins to decrease, which is mainly due to the Fe^{3+} added is in the form of $Fe(OH)_3$ under alkaline conditions. Moreover, when the pH is over 4, the content of

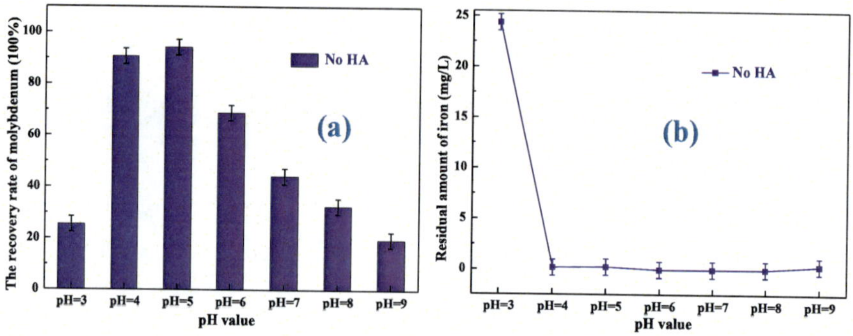

Fig. 2 Effect of pH value on precipitation and transformation of molybdate radical. **a** Recovery rate **b** Residual amount of iron (n(Mo):n(Fe) of 1:5; stirring speed of 150 r/min; temperature of 25 °C)

iron in the solution is always lower than 0.5 mg/L, which meets the iron emission standard required in the Three Wasting Discharge Standard. Therefore, the pH value of 5 is recommended for the following experiments.

Influence of HA Addition on the Precipitation

As can be seen from Fig. 2, the molybdenum precipitate rate reaches the maximum at pH of 5, while there is still 0.766 mg/L Mo residual in the solution, which could not meet the industrial discharge standard of molybdenum. Moreover, the precipitation particles in the solution are relatively small, and it is difficult to recover through filtration, sedimentation, centrifugal, and other methods. HA can chelate with iron to form a flocculent precipitate with larger particles [21]. Therefore, in this work, HA is added as a flocculant to increase the particle size of sediment for recovery. After adding HA, adjusting the solution again to pH of 5. The reaction conditions are as follows: initial Mo concentration of 10 mg/L, temperature of 25 °C, rotation speed of 150 r/min, n(Mo):n(Fe) of 1:5, and pH of 5. The experimental results are shown in Figs. 3 and 4.

Figure 3 shows the influence of HA addition on precipitation particles at pH of 5. As can be seen, the particle size in the solution is only 5.1 μm without HA addition, and it is difficult to recover the precipitation from the solution. After adding HA, the particle sizes of the precipitated particles in the solution increase dramatically. The particle sizes of the precipitated particles reach the maximum of 83.75 μm as HA content is 35 mg/L. However, the particle sizes of precipitated particles in the solution begin to decrease when the HA is over 35 mg/L.

In addition, HA addition can also increase the molybdate precipitation rate in the solution and reduce the content of iron ions, as shown in Fig. 4. With the addition of HA, the precipitation rate of molybdate increases gradually. When HA content is

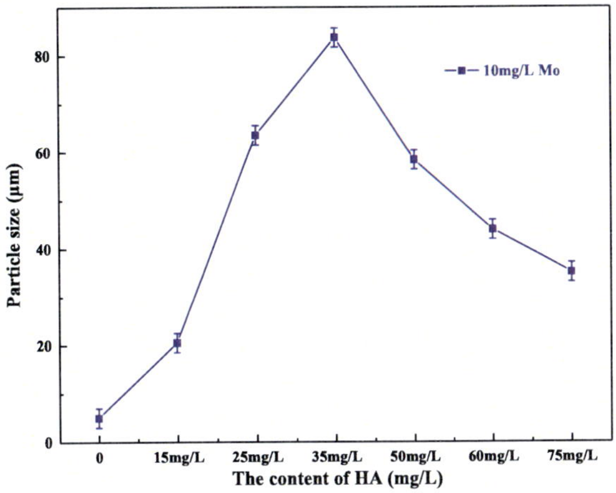

Fig. 3 Influence of HA addition on precipitation particles. ((n(Mo):n(Fe) of 1:5; stirring speed of 150 r/min; temperature of 25 °C; pH of 5)

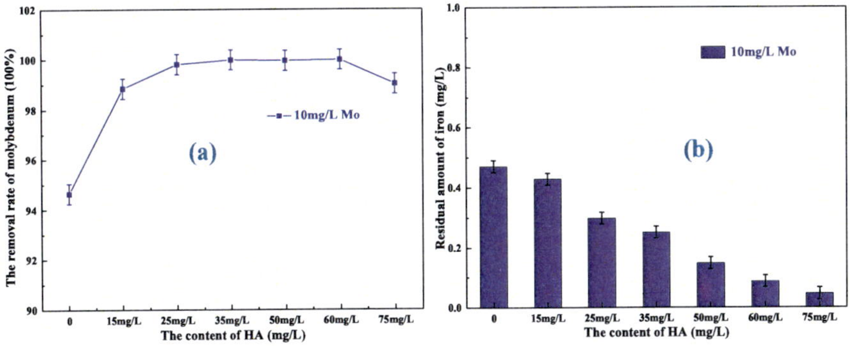

Fig. 4 Influence of HA addition on precipitation and transformation of molybdate. **a** Recovery rate **b** Residual amount of iron ((n(Mo):n(Fe) of 1:5; stirring speed of 150 r/min; temperature of 25 °C; pH of 5)

increased from 0 to 35 mg/L, the molybdate precipitation rate increases from 94.1% to 99.9%. The reason is that part of molybdate in the solution reacts with iron ions to form complex ions with charge. In addition, HA addition makes the complex iron react with HA and precipitate, resulting in the improvement of the precipitation rate of molybdenum. What is important is that the molybdenum content in the solution after flocculation with HA and Fe is only 0.023 mg/L, which is in line with the industrial emission standards. Figure 4(b) displays the residual content of iron after flocculation with HA and Fe. As observed, the residual iron content decreases continuously with the increase of HA dosage from 0 to 75 mg/L. All the iron contents in the purified solution after flocculation are below 0.5 mg/L.

Flotation Recovery of Precipitated Products

Surfactant is of great significance to precipitation flotation. It can change the hydrophobicity of precipitate and improve the hydrophobicity of particulate matter, to facilitate the combination of precipitate and bubble in the flotation process. Surfactants are divided into cationic, anionic, and nonionic types. Because HA is electronegative in solution, cationic surfactant CTAB is used in flotation research. Specific experimental conditions are as follows: initial Mo concentration of 10 mg/L, temperature of 25 °C, rotation speed of 150 r/min, precipitant dosage n(Mo):n(Fe) of 1:5, pH of 5, HA addition of 35 mg/L. The experimental results are shown in Figs. 5 and 6.

As can be seen from Fig. 5, the molybdate precipitation rate decreases slightly after the addition of surfactant. This is mainly because the cationic surfactant reacts with the carboxyl group of HA deprotonation. In the flocculation process, the interaction

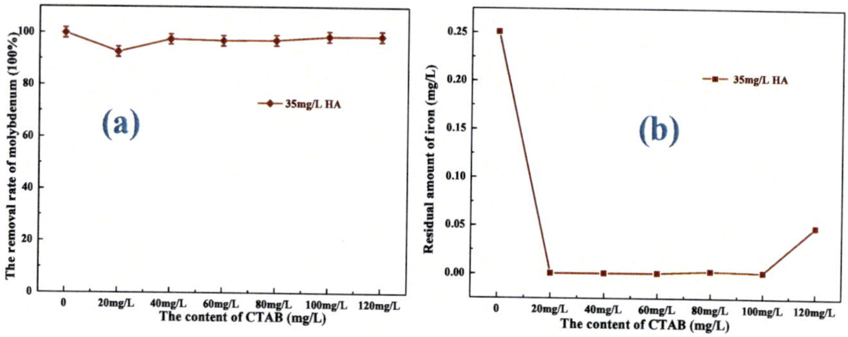

Fig. 5 Influence of CTAB addition on the flotation recovery of precipitation. **a** Recovery rate **b** Residual amount of iron ((n(Mo):n(Fe) of 1:5; stirring speed of 150 r/min; temperature of 25 °C; pH of 5)

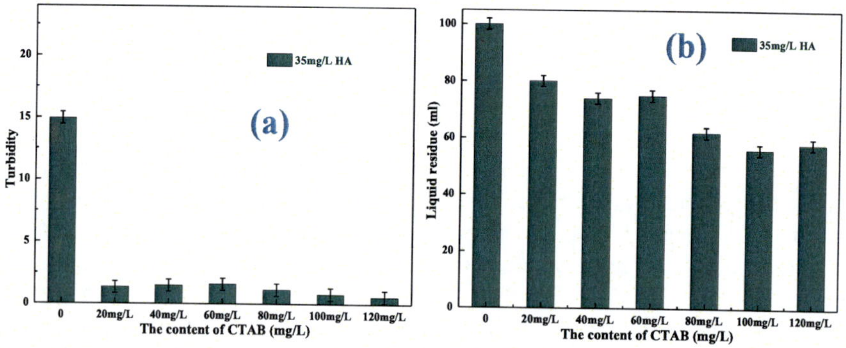

Fig. 6 Influence of CTAB addition on the turbidity and liquid residue of precipitation flotation. **a** Turbidity; **b** Liquid residue ((n(Mo):n(Fe) of 1:5; stirring speed of 150 r/min; temperature of 25 °C; pH of 5)

sites of iron ion and HA are the carboxyl functional group. Therefore, when the concentration of cationic surfactant in the solution increases, it preempts the binding sites in the free composite particles and HA, leading to some molybdate radicals being free in the solution, and thus, the precipitation rate of molybdate radicals is reduced. In addition, the content of iron ions in the solution is always lower than 0.5 mg /L.

Figure 6 shows the influence of CTAB addition on the turbidity and liquid residue after precipitation flotation. As observed, the precipitation flotation has a satisfactory effect at the addition amount of cationic surfactant of 20 mg/L. Under this condition, the turbidity of the solution is 1.325. However, there is still 80 ml liquid residue in the column after flotation separation. It's indicated that the solution loss is lower and the separated precipitate has a lower water content, which is conducive to further treatment.

Conclusions

(1) pH value has a great influence on the precipitation of molybdate in solution. The molybdate precipitation rate reaches the maximum at pH of 5, and the corresponding molybdate precipitation rate is 94.1%. Moreover, the content of iron in the solution is always lower than 0.5 mg/L, meeting the iron emission standard required in the Three Wasting Discharge Standard.

(2) HA addition can increase the particle size of the precipitated particles in the flocculation process. The particle size of precipitated particles reaches 83.75 μm when the addition of HA is 35 mg/L. HA addition can also increase the molybdate precipitation rate in the solution and reduce the content of iron ions. As HA content is 35 mg/L, the molybdate precipitation rate reaches 99.9%.

(3) Flotation experiment demonstrates that the precipitation flotation has a satisfactory effect when the addition amount of cationic surfactant is 20 mg/L, and the turbidity of the purified solution is only 1.325. It's indicated that the separated precipitate has a lower water content, which is conducive to further treatment.

Acknowledgements The authors wish to express their thanks to the National Natural Science Foundation of China (No. U2004215, 51774252, 51974280, 51674225), the Innovation Talents Support Program in University of Henan Province (No. 20HAST1T012), the China Postdoctoral Science Foundation (No. 2019TQ0289), and the Key Scientific Research Project Plan of Henan Colleges and Universities (No. 20A450001).

References

1. Huang F, Wang DH, Wang CH, Chen ZH, Yuan ZX, Liu XX (2014) The characteristics and metallogenic regularity of molybdenum ore resources in China. J Geol 88:2296–2314
2. Ji GY, Deng HJ, Wang CN, Jiang AL, Han J (2016) Status quo of China molybdenum ore resources and screening of physical geological data of state-level molybdenum deposits. China's Min 25:139–145
3. Fan Y, Zhou TF, Zhang DY, Yuan F, Fan Y, Ren Z (2014) Spatiotemporal distribution and metallogenic background of molybdenum deposits in China. J Geol 88:784–804
4. Ma Y, Wu W, Han HD (2011) Study on the removal of molybdenum from water by chemical precipitation. Water Supply Technol 5:26–27+31
5. Zhang JX, Feng X, Cao X (2019) Removal of molybdenum from aqueous solution by co-precipitation method. J Shenyang Univ Chem Technol 33:97–103
6. Huang YF, Zhang B, Liu BB, Su SP, Han GH, Wang WJ (2021) Clean and deep separation of molybdenum and rhenium from ultra-low concentration solutions via vapidly stepwise selective coagulation and flocculation precipitation. Sep Purif Technol 267:118632
7. Liu W, Ding L, Zhu P, Li J, Wei S (2019) Static separation of rhenium and molybdenum by Purolite A172 resin. Chin J Nonferrous Metals 29:1519–1527
8. Zhang J, Yao L, Lu G, Deng G (2010) dynamic separation of molybdenum and Rhenium by D-314 Resin. Chin J Rare Metals 34:85–89
9. Guo WJ, Shen YH (2016) Recovery of molybdenum and vanadium from acidic sulfate leach solution of blue sludge by ion exchange. Environ Prog Sustainable Energy 35:156–160
10. Fathi MB, Rezai B, Alamdari EK, Alorro RD (2018) Equilibrium modeling in adsorption of Re and Mo ions from single and binary aqueous solutions on Dowex 21K resin. Geosyst Eng 21:73–80
11. Gerhardt NI, Palant AA, Petrova VA, Tagirov RK (2001) Solvent extraction of molybdenum (VI), tungsten (VI) and rhenium (VII) by diisododecylamine from leach liquors. Hydrometallurgy 60:1–5
12. Thanh TH, Hong NT, Lee MS (2017) Separation of molybdenum(VI), rhenium(VII), tungsten(VI), and vanadium (V) by solvent extraction. Hydrometallurgy 171:298–305
13. Parhi PK, Park KH, Kim HI, Park JT (2011) Recovery of molybdenum from the sea nodule leach liquor by solvent extraction using Alamine 304–1. Hydrometallurgy 105:195–200
14. Shan W, Fang D, Zhao Z, Shuang Y, Ning L, Xing Z, Xiong Y (2012) Application of orange peel for adsorption separation of molybdenum(VI) from Re-containing industrial effluent. Biomass Bioenerg 37:289–297
15. Xiong Y, Chen C, Gu X, Biswas BK, Shan W, Lou Z, Fang D, Zang S (2011) Investigation on the removal of Mo(VI) from Mo-Re containing wastewater by chemically modified persimmon residua. Biores Technol 102:6857–6862
16. Lei Z, Xue Y, Li P (2009) Molybdenum(VI) and Rhenium(VII) separation on Nano-TiO_2. Chin J Appl Chem 26:1362–1366
17. Xiong Y, Lou ZN, Shan WJ, Ren FQ, Li YX, Zang SL (2013) Recovery of Mo(VI) from Re(VII)-containing wastewater by chemically modified orange waste: competitiveness and selectivity. Chin J Inorganic Chem 29:1717–1723
18. Huang YF, Han GH, Liu J, Chai WC, Wang WJ, Yang SZ, Su SP (2016) A stepwise recovery of metals from hybrid cathodes of spent Li-ion batteries with leaching-flotation-precipitation process. J Power Sources 325:555–564
19. Wu HY, Huang YF, Liu BB, Han GH, Su SP, Wang WJ, Yang SZ (2021) An efficient separation for metal-ions from wastewater by ion precipitate flotation: probing formation and growth evolution of metal-reagent flocs. Chemosphere, 263
20. Wu HY, Wang WJ, Huang YF, Han GH, S. Yang SZ, (2019) Comprehensive evaluation on a prospective precipitation-flotation process for metal-ions removal from wastewater simulants. J Hazard Mater 371:592–602
21. Huang YF, Han GH, Liu J, Wang WJ (2016) A facile disposal of Bayer red mud based on selective flocculation desliming with organic humics. J Hazard Mater 301:46–55

Recent Trend on the Studies of Recycling Technologies of Rare Earth Metals

Osamu Takeda, Xin Lu, and Hongmin Zhu

Abstract There is uncertainty in the supply of rare earth resources. Technical developments on the resource saving of rare earths and the alternate materials are important challenges, and the development of recycling technologies is also important for securing stable supply of rare earth resources. The waste of neodymium magnet that is the industrially-important product containing rare earth metals is currently recycled by a hydrometallurgical method. However, the hydrometallurgical method generates a large volume of waste solution and consumes a large amount of energy. In order to develop an environmentally-sound recycling process, various studies on pyrometallurgical methods have been conducted. One of the authors developed the molten metal extraction method and the flux remelting method. The pyrometallurgical methods are anticipated as the recycling processes with small waste generation and low energy consumption. Research and development on environmentally-sound recycling technologies for rare earths has to be advanced from environmental aspects as well as economic aspects.

Keywords Recycling · Rare earths · Pyrometallurgical methods

Introduction

The import prices of neodymium (Nd) and dysprosium (Dy) in Japan are shown in Fig. 1 [1, 2]. The prices of Nd and Dy in January 2005 were approximately 9 dollars·kg^{-1} and 50 dollars·kg^{-1}, respectively. They drastically increased in 2011 due to the shortage of resource supply and reached approximately 470 dollars·(kg-Nd)$^{-1}$ and 3700 dollars·(kg-Dy)$^{-1}$ at maximum. The import prices gradually decreased because of the diversification of resource supply and the advancement in the technological developments of resource savings and alternative materials.

O. Takeda (✉) · X. Lu · H. Zhu
Graduate School of Engineering, Tohoku University, 6-6-02 Aramaki-Aza-Aoba, Aoba-ku, Sendai 980-8579, Japan
e-mail: takeda@material.tohoku.ac.jp

A. Lazou et al. (eds.), *REWAS 2022: Developing Tomorrow's Technical Cycles (Volume I)*, The Minerals, Metals & Materials Series,
https://doi.org/10.1007/978-3-030-92563-5_27

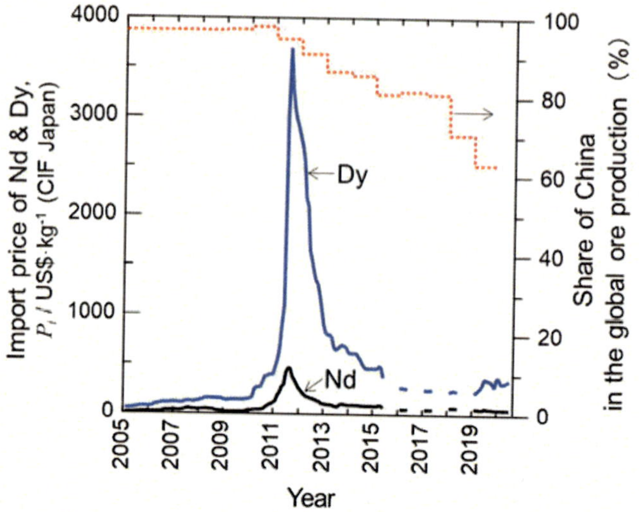

Fig. 1 Import prices of Nd and Dy in Japan. The share of China in the global ore production is also shown

The share of China in the production of rare earth ore and materials (oxides and metals) significantly increased in 1990s, and the share in the ore production reached approximately 98% in 2010. The detail of resource problem is described in a review [3]. The risk on the resource supply of rare earths was acknowledged by some specialists, but the preparation for the shortage of resource supply was insufficient. After the generation of the shortage, the resource supply was diversified in the world. The share of China in the global ore production gradually decreased and that was 63% in 2019. However, the resource supply of heavy rare earths such as Dy that are indispensable for high-performance magnets has not been diversified because the economically-competitive mines for heavy rare earths are concentrated in China.

Recently, the production volume of rare earth ore in Myanmar drastically increases. The share of Myanmar in the global ore production reached 12% in 2020 (Fig. 2 [4]). The rare earth ore mined in Myanmar is transported to China, and the rare earths are separated and refined there. Although China is the biggest supplier of rare earth ore in the world, China enlarges the import of rare earth ore because the domestic demand expands. The trend of import by China is now attracted attention.

As described above, there is uncertainty in the supply of rare earth resources. Technical developments on the resource savings of rare earths and the alternate materials are important challenges, and the development of recycling technologies is also important for securing stable supply of rare earth resources. The present article focuses on the recycling of the waste of Nd magnet that is the industrially-important product containing rare earth metals. Recent trend on research and development of recycling technologies of rare earth metals is briefly introduced. The

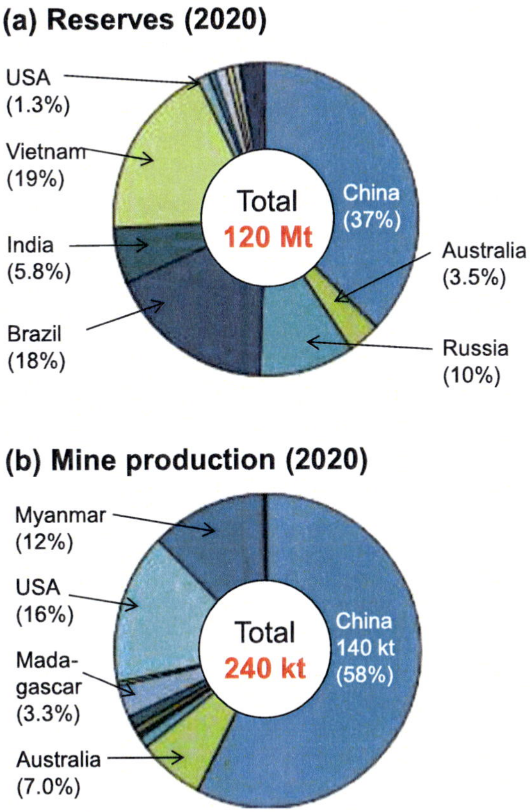

Fig. 2 **a** Reserves of rare earths and **b** annual mine production of rare earths in 2020 (USGS Mineral Commodity Summaries 2021)

research conducted by the authors, particularly the development of pyrometallurgical processes for recycling, is also introduced.

Current Status of Recycling of Rare Earths

In order to understand the recycling of rare earths, it is important to understand the metallurgy of rare earth metals. The detail is described in a book [5]. The reviews by Tanaka et al. [6] and Binnemans et al. [7] overview the entire status of recycling of rare earth-containing wastes. The detail of pyrometallurgical recycling methods including newly-developed processes is reviewed in a review [3].

Recycling technologies for rare earth metal wastes including Nd magnet wastes are classified as follows: (1) "materials recycling methods" in which scrap materials are charged into smelting processes as raw material, (2) "alloy recycling methods" in which the materials are regenerated into master alloys for magnet production, and

(3) "magnet recycling methods" in which magnet alloys are reused in their current form. The detail of classification is explained in a review [3].

The industrial recycling method practically adopted is a hydrometallurgical method (the materials recycling methods classified above) in which the waste is entirely dissolved by acids, and the solution is returned to a primary metallurgical process producing virgin materials [8]. The hydrometallurgical method is advantageous in the removal of highly-concentrated impurity and the mutual separation of rare earths. However, because not only rare earth elements but also iron as the major component is dissolved by acids, a large volume of waste solution is generated, and the environmental burden is high. Furthermore, the secondary raw materials such as an in-house magnet scarp and a sludge generated in fabrication process of magnet are mixed in primary raw materials, and they are reduced by molten salt electrolysis which consumes a large amount of energy. As a result, the recycling process is currently operated in some countries in which the environmental regulations are not strict and the energy cost is low.

The hydrometallurgical method abovementioned cannot be adopted in developed countries in which the environmental regulations is strict. Therefore, various studies on the development of pyrometallurgical methods [9–14] have been conducted. Recently, the processes using molten oxides as the extracting agent of rare earth oxides are actively studied [15–19].

Studies Involving the Authors

The flow of recycling processes for Nd magnets developed by one of the authors is shown in Fig. 3. Processes shown in Figs. 3a and 3b "molten metal extraction method" [20] classified as the materials recycling methods and "flux remelting method" [21] classified as the alloy recycling methods, respectively.

In the molten metal extraction method, Nd in the magnet scrap is directly extracted in metallic state without oxidation. The principle is schematically shown in Fig. 4. Rare earth metals are selectively extracted from the magnet scrap using molten collector metals such as magnesium (Mg) and silver (Ag). The scraps of Mg and Ag can be also used as the collector metals. Namely, a waste is recycled using another waste. This technology is termed as "scrap combination" and regarded as an environmental technology. As an example of results in the extraction experiment using molten Mg, Nd was extracted from Nd magnet alloy containing 31 mass% of Nd with 95% of extraction ratio [22]. Pure Nd was then obtained by removing Mg from Mg–Nd alloy in vacuum evaporation. Previously, Xu et al. also developed the molten metal extraction method with a different approach [23]. Even currently, similar studies are conducted by various researchers [24–26].

In the flux remelting method, rare earth oxides contained in Nd magnet are reacted with molten fluorides, in which the formed rare earth oxyfluoride is removed by

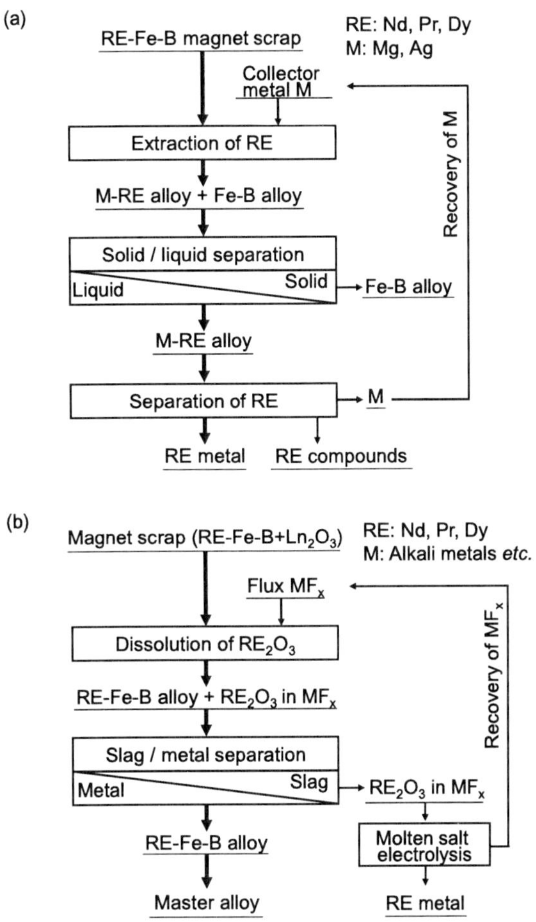

Fig. 3 Recycling processes for Nd magnets developed by one of the authors. **a** Molten metal extraction method and **b** flux remelting method

dissolving into the molten fluorides. The principle is schematically shown in Fig. 5. The refined alloy is used as the master alloy for the magnet fabrication. The separated rare earth oxide in the molten fluoride is recycled as rare earth metal by molten salt electrolysis. This method is simple, and most of component in magnet alloy is recycled without oxidation. The energy consumption is relatively low, and the industrial availability is high. As an example of results in the refining experiment, oxygen concentration in Nd magnet alloy was decreased from approximately 5000 to 200 mass ppm [21]. In this method, most of components in magnet alloy is not returned to an upstream step in the primary metallurgical process, but that is converted to a master alloy for the magnet fabrication, which enables a drastic energy saving in recycling. This method is anticipated as a recycling process with a short cyclic path.

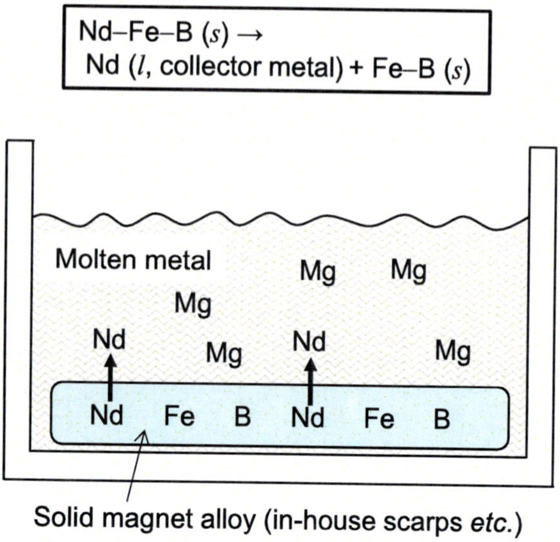

Fig. 4 Schematic illustration of principle in the selective extraction of rare earths by molten metal (Mg is shown as an example of the collector metal)

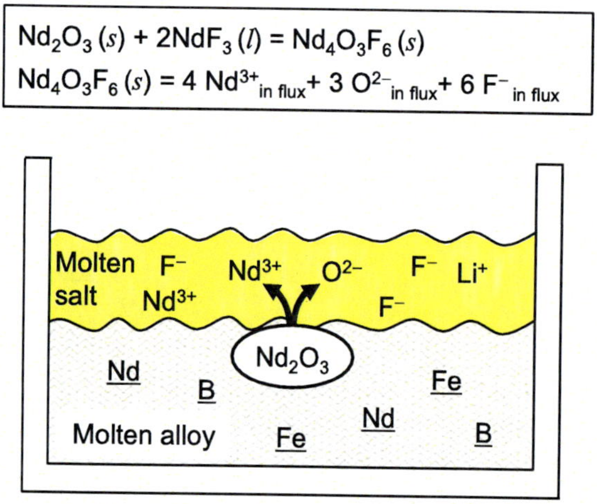

Fig. 5 Schematic illustration of principle in the dissolution removal of rare earth oxide by molten flux (LiF–NdF_3 is shown as an example of the flux)

Future Perspective

Nd magnet is indispensable for advanced mobilities such as hybrid vehicles and electric vehicles, wherein rare earth metals are critical elements. Namely, rare earth metals are indispensable for green technologies. On the other hands, mining and smelting of rare earth metals cause heavy environmental burden in the mining and smelting places [3]. Research and development on environmentally-sound recycling technologies for rare earths has to be advanced from environmental aspects as well as economic aspects.

References

1. Shirayama S, Okabe TH (2011) Current status of rare earth elements and their recycling processes. J. Soc. Automotive Engineers Jpn 65:87–94 (in Japanese)
2. Industrial rare metal, annual review, No. 129 (2013), p 40; No. 130 (2014), p 42; No. 131 (2015), p 38; No. 132 (2016), p 46; No. 133 (2017), p 37; No. 134 (2018), p 39; No. 135 (2019), p 46; No. 136 (2020), p 38. Arumu Publisher, Tokyo (in Japanese)
3. Takeda O, Okabe TH (2014) Current status on resource and recycling technology for rare earths. Metall Mater Trans E 1A:160–173
4. US. Geological Survey (2021) Mineral commodity summaries (USGS, 2021)
5. Takeda O, Uda T, Okabe TH (2014) Rare earth, titanium group metals, and reactive metals production. In: Seetharaman S (ed) Treatise on process metallurgy, vol 3, Elsevier Inc., Oxford, UK, pp 995–1069
6. Tanaka M, Oki T, Koyama K, Narita H, Oishi T (2013) Recycling of rare earths from scrap. Handb Phys Chem Rare Earths 43:159–210
7. Binnemans K, Jones PT, Blanpain B, Gerven TV, Yang Y, Walton A, Buchert M (2013) Recycling of rare earths: a critical review. J Clean Prod 51:1–22
8. Nakamura E (2006) Recycling of rare earths. In: Harada K (ed) Alternate materials and recycling of rare metals. CMC Publishing, Tokyo, pp 296–304 (in Japanese)
9. Adachi G, Murase K, Shinozaki K, Machida K (1992) Mutual separation characteristics for lanthanoid elements via gas phase complex with alkaline chlorides. Chem Lett 21:511–514
10. Uda T, Jacob KT, Hirasawa M (2000) Technique for enhanced rare earth separation. Science 289:2326–2329
11. Uda T (2002) Recovery of rare earths from magnet sludge by $FeCl_2$. Mater Trans 43:55–62
12. Asabe K, Saguchi A, Takahashi W, Suzuki RO, Ono K (2001) Recycling of rare earth magnet scraps: Part I, carbon removal by high temperature oxidation. Mater Trans 42:2487–2491
13. Shirayama S, Okabe TH (2018) Selective extraction and recovery of Nd and Dy from Nd–Fe–B magnet scrap by utilizing molten $MgCl_2$. Metall Mater Trans B 49B:1067–1078
14. Oishi T, Konishi H, Nohira T, Tanaka M, Usui T (2010) Separation and recovery of rare earth metals by molten salt electrolysis using alloy diaphragm. Kagaku Kogaku Ronbunshu 36:299–303 (in Japanese)
15. Saito T, Sato H, Ozawa S, Yu J, Motegi T (2003) The extraction of Nd from waste Nd–Fe–B alloys by the glass slag method. J Alloys Compd 353:189–193
16. Tokita Y, Shibata E, Iizuka J, Nakamura T (2013) Separation of rare earth elements and recovery of Fe–B alloy from neodymium magnet using molten flux. In: Abstracts of spring meeting of the mining and materials processing Institute of Japan, pp 99–100 (in Japanese)
17. Hoshi H, Miyamoto Y, Furusawa K (2014) Technique for separating rare earth elements from R–Fe–B magnets by carbothermal reduction method. J Jpn Inst Metals 78:258–266 (in Japanese)
18. Abrahami ST, Xiao Y, Yang Y (2015) Rare-earth elements recovery from post-consumer hard-disc drives. Trans Inst Min Metall C 124:106–115
19. Bian Y, Guo S, Jiang L, Liu J, Tang K, Ding W (2016) Recovery of rare earth elements from NdFeB magnet by VIM-HMS method. ACS Sustain Chem Eng 4:810–818
20. Okabe TH, Takeda O, Fukuda K, Umetsu Y (2003) Direct extraction and recovery of neodymium metal from magnet scrap. Mater Trans 44:798–801
21. Takeda O, Nakano K, Sato Y (2014) Recycling of rare earth magnet waste by removing rare earth oxide with molten fluoride. Mater Trans 55:334–341
22. Takeda O, Okabe TH, Umetsu Y (2006) Recovery of neodymium from a mixture of magnet scrap and other scrap. J Alloys Comp 408–412:387–390
23. Xu Y, Chumbley LS, Laabs FC (2000) Liquid metal extraction of Nd from NdFeB magnet scrap. J Mater Res 15:2296–2304

24. Chae HJ, Kim YD, Kim BS, Kim JG, Kim T (2014) Experimental investigation of diffusion behavior between molten Mg and Nd–Fe–B magnets. J Alloys Compd 586:S143–S149
25. Moore M, Gebert A, Stoica M, Uhlemann M, Löser W (2015) A route for recycling Nd from Nd–Fe–B magnets using Cu melts. J Alloys Compd 647:997–1006
26. Akahori T, Miyamoto Y, Saeki T, Okamoto M, Okabe TH (2017) Optimum conditions for extracting rare earth metals from waste magnets by using molten magnesium. J Alloys Compd 703:337–343

Leaching of Rare Earth Elements from Phosphogypsum Using Mineral Acids

Sicheng Li, Monu Malik, and Gisele Azimi

Abstract Rare earth elements (REEs) are critical metals for modern and emerging green technologies. Their increasing demand and limited supply have sparked the research on their recovery from secondary resources. The current study is focused on developing a hydrometallurgical process for the extraction of critical REEs from a waste byproduct, called phosphogypsum, and on elucidating the mechanism of the extraction process. Three types of mineral acids are used for the leaching, and a systematic study is utilized to assess the effect of operating parameters and to determine the optimum operating conditions. Thermodynamic modeling and solubility investigation shows the strong correlation between phosphogypsum solubility and leaching efficiency and the leaching process mechanism. Characterization results indicate that REEs can exist as isomorphous substitutions and/or separate phases inside phosphogypsum crystal. Based on these results, the destruction of phosphogypsum lattice is required to achieve improved extraction.

Keywords Circular economy · Hydrometallurgy · Leaching · Phosphogypsum · Process optimization · Rare earth elements

Introduction

Around the world, there is a growing push towards renewable energy generation and electrification of the transportation sector. With the increasing demand for wind turbines and electric vehicles, some rare earth elements (REEs) that are the building block of permanent magnets have been identified as critical elements [1]. The global market size of REEs was valued at $2.80 billion (USD) in 2018, and the demand

---

S. Li · M. Malik · G. Azimi (✉)
Department of Chemical Engineering and Applied Chemistry, Laboratory for Strategic Materials, 200 College Street, Toronto, ON M5S 35, Canada
e-mail: g.azimi@utoronto.ca

G. Azimi
Department of Materials Science and Engineering, University of Toronto, 184 College Street, Toronto, ON M5S 34, Canada

A. Lazou et al. (eds.), *REWAS 2022: Developing Tomorrow's Technical Cycles (Volume I)*, The Minerals, Metals & Materials Series,
https://doi.org/10.1007/978-3-030-92563-5_28

for REEs was predicted to increase by 10.4% annually between 2019 and 2025 [2]. Considering the increasing demand, some REEs such as dysprosium (Dy) and neodymium (Nd) were predicted to face supply shortages in the foreseeable future [1]. Concerns regarding the availability of these REEs are due to their limited geological resources, adverse effects of primary-resource mining, and extraction complexities [3]. Therefore, many countries around the world have initiated activities to extract these elements from alternative sources. Industrial waste streams, such as phosphogypsum (PG), are considered as one of the potential supply sources.

The PG is the byproduct of phosphoric acid production in the fertilizer industry by digesting a concentrated slurry of pulverized phosphate ores using sulfuric acid (reaction (1)) [4]. The PG mainly consists of calcium sulfate dihydrate or gypsum ($CaSO_4 \cdot 2H_2O$), calcium sulfate hemihydrate ($CaSO_4 \cdot 0.5H_2O$), and anhydrite ($CaSO_4$).

$$Ca_{10}(PO_4)_6F_2 + 10H_2SO_4 + 20H_2O \rightarrow 6H_3PO_4 + 10CaSO_4 \cdot 2H_2O + 2HF \quad (1)$$

The average concentration of REEs is around 0.01–0.4 wt% in PG [5]. Although the REE concentration per unit of PG is low, the total amount of REEs that can be extracted from this secondary resource is appreciable considering the vast supply of PG, with 100–280 million tonnes global annual generation [6]. Also, clean PG after trace metals removal (including REEs) could be used in the construction and fertilizer industries, thus offering added benefits. However, currently, only around 15% of PG is productively used in construction and agriculture industries, and the rest is landfilled or stored in stacks, losing the commercial benefits and causing potential environmental concerns [7].

The extraction process is one of the major steps in the REEs recovery, which conventionally relied on pyrometallurgy and/or hydrometallurgy, and the latter is preferred [8, 9]. Previous studies have the disadvantages of either low leaching efficiency, long leaching time, or high temperature [10, 11]. Besides low REE leaching efficiency, the occurrence of REEs in PG is another challenge. Although a few studies suggested some possibilities, the conclusions are still controversial [12]. Furthermore, because of the difficulty in determining the REEs occurrence, there is a lack of study on the mechanism behind the leaching process.

In the current study, a systematic investigation was performed to extract REEs (Y, Nd, Dy, and Ce) from a Canadian PG using three common mineral acids over a wide range of conditions. Solubility analysis, thermodynamic modeling, and kinetic analysis were used to explain the extraction trends and to identify the leaching process mechanism. The fundamental studies and novel findings from this study shed light on the physicochemical mechanisms behind REE leaching from PG, helping usher in an efficient extraction process that can be extended to any type of PG feeds with various REE concentrations.

Methods and Materials

Feeds and Materials

Phosphogypsum was obtained from Nutrien Ltd.'s fertilizer operations located in Redwater, Alberta, Canada. Concentrated hydrochloric acid (ACS Reagent Grade, 36.5–38.0 wt% Assay), nitric acid (ACS Reagent Grade, 68.0–70.0 wt% Assay), and sulfuric acid (ACS Reagent Grade, 95.0–98.0 wt% Assay) were purchased from VWR.

Experimental Procedure

The as-received PG sample contains moisture; thus, it was dried at 40 °C for 24 h, followed by grinding by mortar and pestle to a fine powder. The leaching experiments were conducted inside 250 mL Erlenmeyer flasks placed inside a temperature-controlled water bath (Fisher Scientific ISOTEMP 4100 H21P) with continuous stirring at 500 rpm. The pre-determined amounts of solid PG and the acid solution were added, based on the desired solid to liquiud (L/S) ratio. The flask was covered with a rubber stopper to prevent the loss of water by evaporation. Samples were taken at the end of leaching time, the leaching solution was diluted with 5 wt% HNO_3 using a Hamilton Microlab 600 diluter/dispenser system (Hamilton Company, Reno, NV, USA), and the concentration was measured using ICP-OES.

Characterization

The concentration of REEs in raw PG was determined through aqua-regia digestion (MARS6 Xpress microwave digestion system) followed by inductively coupled plasma optical emission spectroscopy (ICP-OES) measurements. As PG is sourced from mineral ores, it contains some primary radionuclide. To quantify the concentration of the radioactive elements (thorium: ^{232}Th and uranium: ^{238}U) in PG, the concentration of digested PG was measured by inductively coupled plasma mass spectrometry (ICP-MS, Thermo Scientific iCAP Q).

Design of Experiments and Empirical Model Building

For all three mineral acids, factorial design of the experiment (DOE) and empirical modeling were utilized to assess the effect of the operating parameters on the REE leaching efficiency. The studied factors were selected as: leaching temperature

Table 1 List of experimental factor levels for the leaching response surface trials

Experimental factor	Units	– 1 level	0 Level	+1 Level
X_1: Leaching temperature	°C	45	65	85
X_2: Acid concentration	mol L^{-1}	0.5	1.5	2.5
X_3: Liquid to solids ratio	mL g^{-1}	10	20	30

(X_1), acid concentration (X_2), and liquid to solid (L/S) ratio (X_3). The coded levels corresponding to each parameter are listed in Table 1.

For all three acid systems, a full factorial design was designed, and the experimental data from each trial was fitted with an empirical model (Eq. 2) with variance assessment at $\alpha = 0.05$ significance level. The model parameters were fitted by multiple linear least squares regression (Eq. 3).

$$\hat{y}_i = \hat{\beta}_0 + \hat{\beta}_1 X_1 + \hat{\beta}_2 X_2 + \hat{\beta}_3 X_3 + \hat{\beta}_{12} X_1 X_2 + \hat{\beta}_{13} X_1 X_3 + \hat{\beta}_{23} X_2 X_3 + \hat{\beta}_{11} X_1^2 + \hat{\beta}_{22} X_2^2 + \hat{\beta}_{33} X_3^2 \quad (2)$$

$$\hat{\beta} = \left(X^T X\right)^{-1} \left(X^T Y_i\right) \quad (3)$$

where $\hat{y}_i$ is the extraction efficiency, X_1 to X_3 is the experimental parameters, $\widehat{\beta}$ is model parameter vector including each of the model parameters ($\widehat{\beta}_0, \widehat{\beta}_1, \widehat{\beta}_2, \widehat{\beta}_3, \ldots$), X is the matrix, and Y_i is the response vector.

Results and Discussion

Characterization Results of Raw PG

As shown in Fig. 1a, the total content of REEs in raw PG is 295.21 mg/kg, and the concentrations of the radioactive elements (U and Th) are low, at 32.17 and 1.36 mg/kg, respectively. Considering the economic value of elements and the abundance in PG, Nd, Y, Dy, and Ce are mainly focused. Figure 1b indicates the concentrations of other elements in PG, and calcium (Ca) and sulfur (S) are the main components, accounting for 25.3 wt% and 19.9 wt%, respectively.

Acid Leaching Results

Leaching of the REEs of interest was very fast for all three mineral acids, as the maximum leaching efficiency was reached within 20 min. Empirical models (Eqs. 2

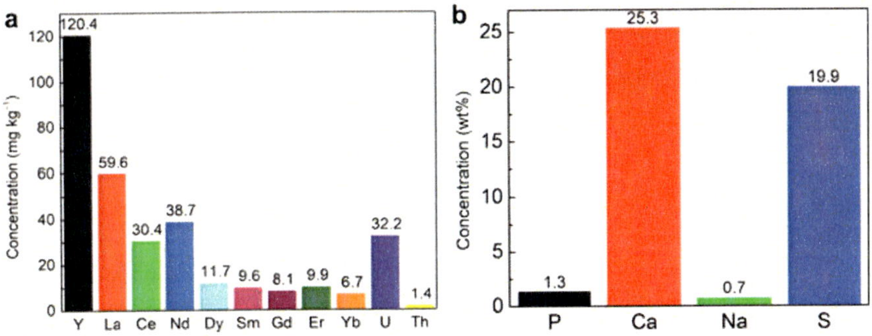

Fig. 1 Characterization of the raw PG sample: **a** content of REEs and radioactive elements (U, Th); **b** content of other elements in PG

and 3) were conducted to study the effects of operating parameters on the leaching process. It was found that acid concentration (X_2) and L/S ratio (X_3) have a positive effect on the extraction in the HCl system, whereas temperature (X_1) has a negligible effect. A similar trend was observed in the HNO_3 system, while temperature (X_1) has a positive effect in the H_2SO_4 system. Based on the overall leaching results, the optimized condition of the HCl system occurs at 45 °C, 2.5 M concentration, and 29.8 mL/g L/S ratio. For HNO_3 and H_2SO_4 system, the high leaching efficiency was achieved at much higher temperature (85 °C), and the optimal operating conditions of acid concentration and L/S ratio are 2.1 M, 1.3 M, and 27.7 mL/g, 30 mL/g, respectively. The best leaching efficiency of targeted REEs in each acid system (under optimized condition) is presented in Fig. 2, which indicates that HCl is the best leachant with the highest leaching efficiency (94.6%, 98.5%, 86.1%, 99.2% for Nd, Y, Dy, and Ce, respectively) under mild temperature condition (45 °C). Additionally, compared with HNO_3, HCl is less environmentally harmful and less expensive.

Fig. 2 Highest leaching efficiency was achieved for Nd, Y, Dy, and Ce in HCl, HNO_3, and H_2SO_4 system at 45 °C, 85 °C, and 85 °C, respectively

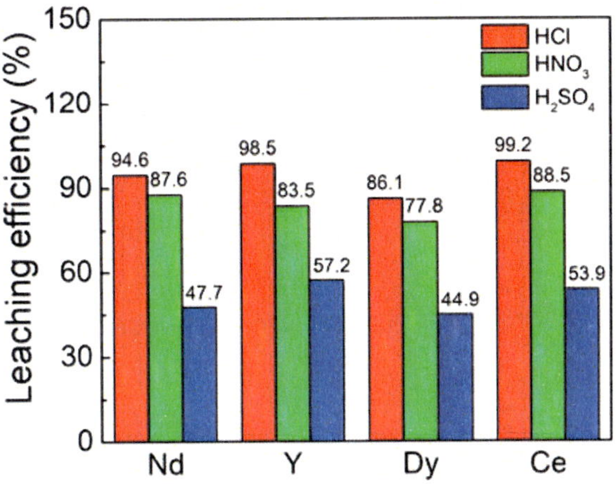

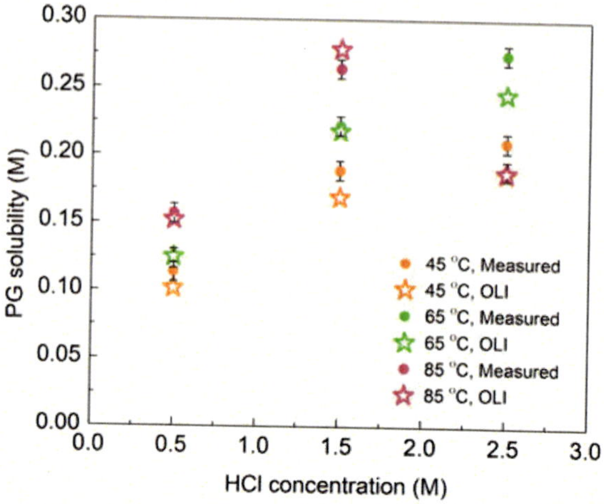

Fig. 3 Measured PG solubility and OLI simulation results in the HCl system as a function of temperature and acid concentration

Thermodynamic Modelling and Solubility Investigation

To further study the chemistry and mechanism behind the leaching process (focused on the best leachant–HCl system), the solubility of rare earth elements in HCl system was calculated from thermodynamic models, employing OLI software (Version 9.6) with modification on database developed by Azimi et al. [13]. As shown in Fig. 3, from the prediction via OLI, the solubility of gypsum increases with increasing HCl concentration; however, a significant drop is observed at 85 °C by increasing the acid concentration from 1.5 to 2.5 M. This decrease could be attributed to two reasons: the transformation of gypsum to anhydrite at high temperature and high acid concentration, and the solvation of more ions in the solution which leads to fewer water molecules available to participate in dissolution [13, 14]. The gypsum solubility trends explain the effect of operating parameters on REE extraction, which indicates the direct link between gypsum dissolution and REE leaching (Eq. 4 in general).

$$\mathrm{REE - CaSO_4 \cdot nH_2O_{(s)} + HCl_{(aq)} \rightarrow CaSO_{4(aq)} + CaCl_{2(aq)} + H_3O^+_{(aq)} + (REE)Cl_{3(aq)}} \qquad (4)$$

where n = 0, 0.5, and 2 correspond to anhydrite, hemihydrate, and gypsum, respectively. As a result, increasing the solubility of gypsum leads to matrix destruction, allowing more REEs to be extracted.

Mechanistic Investigation—Proposed Mechanism

On the basis of experimental leaching and thermodynamic modeling results, it is concluded that the destruction of PG lattice is required to achieve high leaching

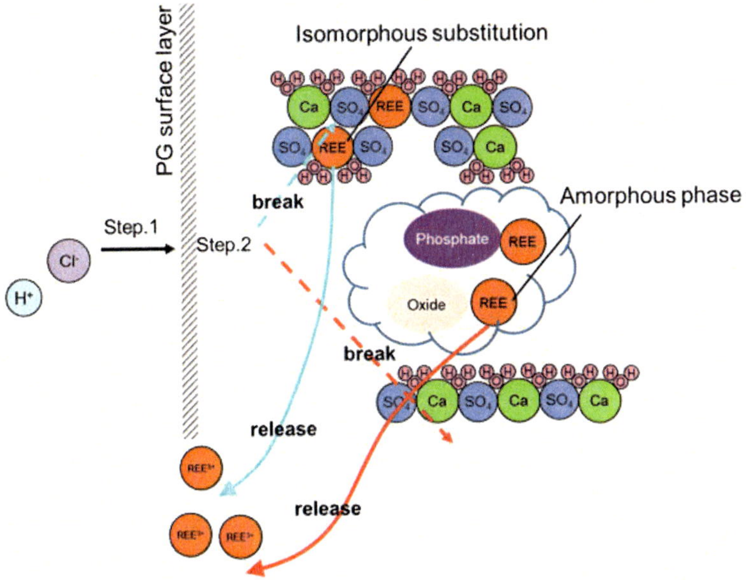

Fig. 4 Schematic diagram of the proposed leaching mechanism

efficiency. Although there are different possibilities of REEs occurrence, the leaching of REEs is highly correlated with the dissolution of gypsum regardless of whether they exist as their own phases or incorporate in PG lattice. The REEs that form separate phases are more easily leachable compared with the other case [12], which could cause the difference in the leaching efficiency of different REEs.

Overall, to improve the leaching process of REEs from PG, gypsum solubility is considered as an important factor, as its dissolution destructs the lattice, allowing the release of REEs substituting Ca^{2+} ions (isomorphous phase) and/or existing as separate phases (amorphous phase) within the PG lattice. The acid leaching mechanism is proposed as following steps (Fig. 4):

1. Acid diffuses and reaches the surface of PG particles.
2. Acid breaks the PG lattice (the bond between calcium and sulfate ion). Both REE phases inside PG and the isomorphous REEs substituting the Ca^{2+} ions are released into an aqueous solution due to the dissolution of the PG lattice.

Conclusions

In this study, the effect of various operating parameters on the leaching of REEs from PG was investigated for three mineral acid systems, and the optimum operating conditions were determined through systematic experimental and theoretical approaches based on the design of experiment. The HCl was identified as the best leachant, and the optimum operating conditions were determined to be 2.5 M HCl at 45 °C and 29.8 mL/g liquid to solid ratio which resulted in the highest extraction for all four targeted REEs. The developed HCl leaching process results in significantly higher leaching efficiency at a lower temperature compared with previous literature. Moreover, leaching results are explained via thermodynamic modeling and solubility investigation, which further elucidates the leaching mechanism.

References

1. Zhou B, Li Z, Chen C (2017) Global potential of rare earth resources and rare earth demand from clean technologies. Minerals 7. https://doi.org/10.3390/min7110203
2. Rare earth elements market size (2019). Share|Industry Growth Report, 2025
3. Ganguli R, Cook DR (2018) Rare earths: a review of the landscape. MRS Energy Sustain 5:1–16. https://doi.org/10.1557/mre.2018.7
4. Koopman C, Witkamp GJ (2000) Extraction of lanthanides from the phosphoric acid production process to gain a purified gypsum and a valuable lanthanide by-product. Hydrometallurgy 58:51–60. https://doi.org/10.1016/S0304-386X(00)00127-4
5. Habashi F (1985) Recovery of the lanthanides from phosphate rock. J Chem Technol Biotechnol Chem Technol 35(A):5–14. https://doi.org/10.1002/jctb.5040350103
6. Binnemans K, Jones PT, Blanpain B et al (2013) Recycling of rare earths: a critical review. J Clean Prod 51:1–22. https://doi.org/10.1016/j.jclepro.2012.12.037
7. Liang H, Zhang P, Jin Z, DePaoli D (2017) Rare earths recovery and gypsum upgrade from Florida phosphogypsum. Miner Metall Process 34:201–206. https://doi.org/10.19150/mmp.7860
8. Doyle FM (2005) Teaching and learning environmental. Hydrometallurgy 79:1–14. https://doi.org/10.1016/j.hydromet.2004.10.022
9. Kumar M, Kumari A, Panda R et al (2016) Hydrometallurgy Review on hydrometallurgical recovery of rare earth metals. Hydrometallurgy 165:2–26. https://doi.org/10.1016/j.hydromet.2016.01.035
10. Walawalkar M, Nichol CK, Azimi G (2016) Process investigation of the acid leaching of rare earth elements from phosphogypsum using HCl, HNO_3, and H_2SO_4. Hydrometallurgy 166:195–204. https://doi.org/10.1016/j.hydromet.2016.06.008
11. Rychkov VN, Kirillov EV, Kirillov SV et al (2018) Recovery of rare earth elements from phosphogypsum. J Clean Prod 196:674–681. https://doi.org/10.1016/j.jclepro.2018.06.114
12. Lambert A, Anawati J, Walawalkar M et al (2018) Innovative application of microwave treatment for recovering of rare earth elements from phosphogypsum. ACS Sustain Chem Eng 6:16471–16481. https://doi.org/10.1021/acssuschemeng.8b03588
13. Azimi G, Papangelakis VG, Dutrizac JE (2008) Development of an MSE-based chemical model for the solubility of calcium sulphate in mixed chloride-sulphate solutions. Fluid Phase Equilib 266:172–186
14. Azimi G, Papangelakis VG, Dutrizac JE (2007) Modelling of calcium sulphate solubility in concentrated multi-component sulphate solutions. Fluid Phase Equilib 260:300–315. https://doi.org/10.1016/j.fluid.2007.07.069

Recovery of Terbium, Europium, and Yttrium from Waste Fluorescent Lamp Using Supercritical Fluid Extraction

Jiakai Zhang and Gisele Azimi

Abstract Recycling of waste electrical and electronic equipment (WEEE) has been receiving significant attention around the world. Here, we develop an environmentally sustainable process that uses supercritical carbon dioxide as the solvent along with a small volume of tributyl-phosphate (TBP) nitric acid adduct as the chelating agent to recover rare earth elements (REEs) from fluorescent lamp waste. We show that mechanical activation using oscillation milling increases extraction efficiency. We elucidate the process mechanism by characterizing the solids before and after the process using transmission electron microscopy (TEM) and X-ray photoelectron spectroscopy (XPS). We show that Al^{3+} and Ca^{2+} cations from the Al_2O_3 and $Ca_5(PO_4)_3OH$ (hydroxyapatite) present in the fluorescent lamp waste result in competing reactions with REEs with TBP-HNO_3 adduct; thus, REE extractions from real fluorescent lamp waste are less than what has been reported from synthetic feeds. Management of fluorescent lamp waste leads to sustainability of biosphere and circular economy.

Keywords Supercritical fluid extraction · Fluorescent lamp phosphors · Rare earth elements · Oscillation milling · Transmission electron microscopy · X-ray photoelectron spectroscopy · UV visible spectroscopy

Introduction

Some rare earth elements (REEs) are considered critical elements because of their growing demand in various products including electric vehicles, wind turbines, and catalysts. For sustaining a healthy development of electrified transportation and renewable power generation, a stable supply of REEs is important. Currently, over 70% of the supply of REEs come from China, causing other countries dependent on them for their acquisition [1]. In addition, mining for REEs from primary resources

J. Zhang · G. Azimi (✉)
University of Toronto, 200 College Street, Toronto, ON M5S35, Canada
e-mail: g.azimi@utoronto.ca

A. Lazou et al. (eds.), *REWAS 2022: Developing Tomorrow's Technical Cycles (Volume I)*, The Minerals, Metals & Materials Series,
https://doi.org/10.1007/978-3-030-92563-5_29

poses a significant impact to the environment. To reduce supply risks and minimize environmental damage recycling of REEs from end-of-life products is highly desirable.

Waste fluorescent lamp (FL) is an ideal secondary resource that contains up to 30 wt% REEs [2]. The concentration of REEs in FL is more than ten times higher than the minimum industrial grade primary ores (1.0–2.0%) [3]. The FL dominated the lighting market in past few decades due to its energy consumption advantages (up to 75%) and longer life expectancy (up to 10 times) compared with incandescent lamps. As a result, there is a large quantity of waste FL that can be recycled. The FL phosphors are mainly made of three types of compounds: red phosphor (Y_2O_3: Eu^{3+}); blue phosphor ($BaMgAl_{10}O_{17}$: Eu^{3+}); and green phosphors ($LaPO_4$: Ce^{3+}, Tb^{3+}) [4].

Conventional extraction processes are based on hydrometallurgy and/or pyrometallurgy, which create significant environment burden and generate large volumes of hazardous waste due to heavy reliance on acids, organic solvents and high energy consumption [5, 6]. Moreover, most studies have focused on the recovery of yttrium and europium from red phosphor [2, 6–8]. Blue and green phosphors are discarded as they are resistant to acid attack. However, these phosphors contain considerable amount of terbium that is the most critical REE in FL waste.

Supercritical fluid extraction (SCFE) is an evolving green technology that can extract metals from solid and liquid phases efficiently. Previous studies have shown that SCFE can be used to recover REEs from post-consumer products, such as neodymium-iron-boron magnet, nickel metal hybrid battery, and synthetic fluorescent lamp waste [3, 9, 10]. However, there is lack of a comprehensive study on using SCFE for REEs recycling from real fluorescent lamp waste. Moreover, extraction of Tb has been low in previous studies (less than 10%) [11]. A fundamental investigation of the process mechanism and a through characterization are required to explain extraction difficulties, so a corresponding solution can be found to improve extraction. The reported high extraction for Y and Eu (over 90%) by SCFE from synthetic feed materials could not be achieved in real FL samples due to presence of other materials, such as aluminium oxide (Al_2O_3) and hydroxyapatite ($Ca_5(PO_4)_3OH$), since the cations can compete with REEs in complexation reactions.

To address the knowledge gaps mentioned above, this study investigates SCFE for REEs recovery from real fluorescent lamp waste using supercritical (sc)-CO_2 as a solvent with a small amount of tributyl-phosphate-nitric-acid (TBP-HNO_3) adduct as the chelating agent. Effect of a pre-treatment step using oscillation milling on extraction is investigated. In addition, a through characterization of solids before and after the pre-treatment process is conducted for elucidating the process mechanism using X-ray photoelectron spectroscopy (XPS) and X-ray diffraction (XRD).

Materials and Methods

Materials

Ontario Lamp Recyclers Inc. provided waste fluorescent lamp phosphors. The sample was sieved to remove glass, only particles with size less than 75 μm was used in the experiments. The following chemicals are used: tributyl phosphate (TBP $\geq$ 98%, VWR); concentrated nitric acid (15.7 M, 70 wt%, VWR); carbon dioxide (CO_2, grade 4.0, Linde Canada); calcium hydroxyphosphate (reagent grade, Sigma-Aldrich); aluminium oxide ($\geq$99% metal basis, Alfa Aesar); yttrium oxide europium doped (99% metal basis, Sigma-Aldrich); cerium phosphate (99%, Alfa Aesar); europium trinitrate hexahydrate (99.9% rare earth oxide basis, Thermo Fisher Scientific).

TBP-HNO_3 Preparation

TBP-HNO_3 adduct was made by mixing TBP and 70 wt% HNO_3. Then the solution was left for gravity separation for 5 min, where the top organic solution corresponding to the TBP-HNO_3 adduct was collected.

Mechanical Activation

Mechanical activation in this study refers to a high energy oscillation milling process, where the phosphor sample was introduced to a tungsten carbide grinding jar in an oscillation ball mill by Retsch MM400. The system oscillates at 30 Hz for 60 min.

SCFE Process

The SCFE process was conducted in a high-pressure reactor system manufactured by Supercritical Fluid Technology Inc. Feed materials including the FL samples and the adducts were added in to the reactor first, then liquid CO_2 was added until the system reaches designed temperature and pressure (a detailed drawing can be found in the previous work) [12]. After extraction, CO_2 carried organometallic compounds to a collection vessel while the system depressurized to 1 atm. The residue left in the reactor was collected for calculating extraction efficiency, as following:

$$E(\%) = \frac{\Delta C}{C_i} 100\% \quad (1)$$

where ΔC is the concentration difference between the sample of before and after extraction, and C_i is the concentration of unextracted sample.

Characterization

The morphology of FL samples that before and after mechanical activation was illustrated using scanning electron microscopy (FE-SEM, Hitachi SU5000) and high-resolution transmission electron microscopy/scanning transmission electron microscopy (HRTEM/STEM, FEI Titan 80–300 LB). X-ray diffraction (XRD, Rigaku MinFlex 600) was used for obtaining crystallography information of the FL sample before and after mechanical activation.

Results and Discussion

Characterization

The FL sample was first digested in aqua-regia using a microwave digestor (MARS Xpress System) at 200 °C, followed by Inductively Coupled Plasma Optical Emission Spectrometry (ICP-OES Perkin Elmer Optima 8000) analysis for determining elemental composition. The results are summarized in Table 1.

Mineralogical information of the FL samples was obtained using X-ray diffraction (XRD) (Fig. 1). The XRD results show that the sample contains yttrium europium oxide ($(Y_{0.95}Eu_{0.05})_2O_3$); lanthanum phosphate ($LaPO_4$); cerium phosphate ($CePO_4$); terbium phosphate ($TbPO_4$); aluminum cerium oxide ($CeAl_{11}O_{18}$); barium magnesium aluminum oxide ($BaMgAl_{10}O_{17}$); and hydroxylapatite ($Ca_5(PO_4)_3OH$).

Table 1 Elemental composition analysis of the FL sample

Element	Y	La	Ce	Eu	Tb
wt%	28.0	4.7	3.7	1.9	1.3
Element	Sr	Sb	Ca	Al	Ba
wt%	1.6	0.1	10.2	2.0	0.4

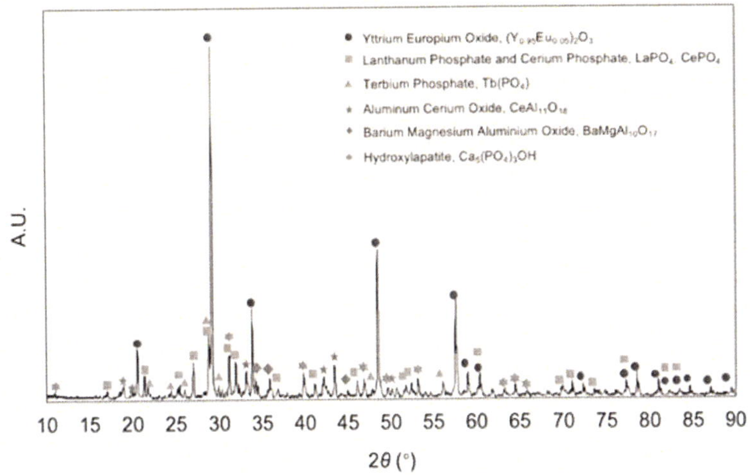

Fig. 1 XRD spectrum of the feed fluorescent lamp phosphors

Effect of Mechanical Activation on Extraction

In this work, first SCFE was directly used to extract Y, Eu, and Tb from the FL sample. However, the extraction results were low. The extraction conditions are shown in Table 2, and the highest extractions were 34%, 34%, and 16% for Y, Eu, and Tb, respectively. To increase the extraction, mechanical activation using high-energy oscillation milling was utilized. The extraction results for mechanically activated sample under various conditions are summarized in Table 2. As shown, the extraction results were improved to 73.2%, 71.2%, and 51.3% for Y, Eu, and Tb, respectively.

Table 2 Overview of the experimental design matrix with corresponding processing parameters and extraction efficiencies for Y, Eu, and Tb from mechanically activated samples

Run ID	Temperature X_1 (°C)	Pressure X_2 (MPa)	Time X_3 (h)	S: CA X_4 (g/mL)	Agitation X_5 (rpm)	Extraction efficiency (%)		
						Y	Eu	Tb
1	40	20.7	1	1.8	1500	40.6	37.0	7.4
2	60	20.7	1	0.2	750	70.8	66.8	50.9
3	40	31.0	1	0.2	1500	62.3	58.7	39.0
4	60	31.0	1	1.8	750	40.3	15.7	16.6
5	40	20.7	3	1.8	750	44.3	31.2	3.0
6	60	20.7	3	0.2	1500	64.0	53.0	32.2
7	40	31.0	3	0.2	750	73.2	71.2	51.3
8	60	31.0	3	1.8	1500	53.3	49.1	24.8

Effect of Mechanical Activation on Physical and Chemical Properties

To elucidate the mechanism behind higher extraction after mechanical activation, the effect of mechanical activation on physical properties including morphology and crystal structure was investigated. Figure 2a and b show SEM images of FL before and after mechanical activation. Mechanical activation resulted in a considerable reduction in particle size and an increase in surface roughness. The original FL sample consisted of similar sized smooth quasi-spherical particles (Fig. 2a). However, after mechanical activation for 60 min at 30 Hz, the friction, shear, impacts, and collisions effects of milling resulted in irregularly shaped particles with rough surfaces (Fig. 2b). In addition, crystal structure and crystallite size were illustrated using TEM. As can be seen, Fig. 2c shows that mechanical activation creates nano-sized crystallites, whereas the non-mechanically activated sample in a perfect lattice structure. The atoms are organized in a repeating array over the entire extent of the sample without nano-crystallites. Characterization of the samples before and after mechanical activation showed that mechanical activation results in decreased peak intensity and increased peak broadening. A previous study has also reported that mechanical activation can cause changes of crystallite size and introduction of strain within crystal lattice [5, 13]. According to Scherrer's equation, crystallite size (D) is inversely proportional to peak width:

$$\beta = \frac{K\lambda}{D\cos\theta} \tag{2}$$

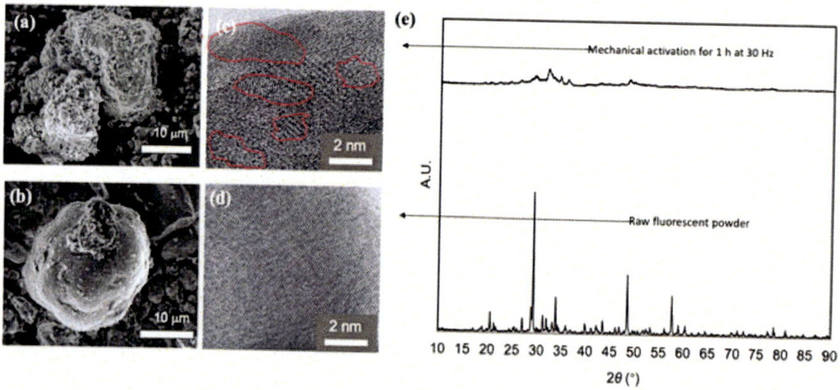

Fig. 2 Effect of mechanical activation on FL phosphor particle morphology, surface characteristics, crystallography: **a** the mechanically activated sample (60 min at 30 Hz); **b** the non-mechanically activated sample; **c** creation of nano-crystallites in the mechanically activated sample; **d** ordered atom orientation in the non-mechanically activated sample **e** XRD spectra of the FL sample before and after mechanical activation

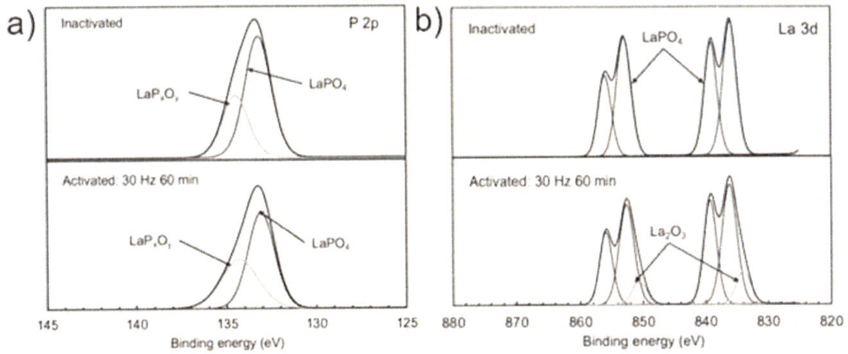

Fig. 3 XPS spectra of inactivated and activated FL samples: **a** P 2p; **b** La 3d

where K, λ, and θ are the constant of proportionality, wavelength, and diffraction angle, respectively. Hence, the peak broadening suggests a reduction of crystallite size after mechanical activation. In this study, the Halder-Wagner method was utilized for calculating crystallite size before and after mechanical activation because it can be used for peak profile analysis by convolution of Lorentzian function and the Gaussian function for acquiring more accurate modeling results [14]. Calculated results show that the average crystallite size reduced from 50 nm for non-mechanically activated sample to 8 nm for mechanically activated sample. Moreover, the intrinsic strain of mechanically activated sample increased by about four times in comparison with that of non-mechanically activated sample. Therefore, it is concluded that mechanical activation can reduce the crystallite size of the sample to nanoscale, generating more grain boundaries which provides more pathways for the diffusion of the lixivinat into the particle, resulting in increased extraction efficiency.

Mechanical activation also causes decomposition to the FL sample. Considering phosphorus (P) is mostly present in green phosphor ($LaPO_4$:Ce, Tb), and the P 2p spectra of $TbPO_4$, $CePO_4$, and $LaPO_4$ overlap to each other, $LaPO_4$ was used for identifying this effect using XPS. As shown in the XPS spectra in Fig. 3a, the proportion of $LaPO_4$ decreased from 74 to 66% after mechanical activation. This decrease is because of the decomposition of $LaPO_4$ into LaP_xO_y and La_2O_3, [15] according to the following reaction:

$$LaPO_4 \rightarrow \mu_1 LaPO_4 + \mu_2 LaP_xO_y + \mu_3 La_2O_3 \tag{3}$$

where μ_1, μ_2, and μ_3 are stoichiometric coefficients.

The La_2O_3 peak in Fig. 3b confirms the formation of REE oxide from the decomposition reaction, which is consistent to P 2p analysis. The generated La_2O_3 is more dissolvable in sc-CO_2 than its phosphate form. Therefore, mechanical activation improves extraction efficiency by destructing crystal structure and creating more dissolvable phases such as REE oxides.

Table 3 Composition of five different synthetic mixtures and REEs extraction efficiency for extraction mechanism exploration

Mixture label	$TbPO_4$, (mg)	$Y_{1.92}Eu_{0.08}O_3$, (mg)	Al_2O_3, (mg)	$Ca_3(PO_4)_2$, (mg)	SiO_2, (mg)	Extraction, (%)		
						Y	Eu	Tb
1	75.1	0.0	0.0	0.0	924.9	–	–	91
2	0.0	439.5	0.0	0.0	560.5	96	93	–
3	75.1	439.5	0.0	0.0	485.4	99	99	84
4	75.1	439.5	44.4	0.0	441.0	33	30	42
5	75.1	439.5	0.0	134.1	351.1	70	68	71

Extraction Interferences by Non-Phosphor Sample

Based on the XRD results in Fig. 1, the FL phosphor sample contains hydroxyapatite ($Ca_5(PO_4)_3OH$) and aluminum oxide (Al_2O_3). This was not considered in a previous study on the recovering of REEs from synthetic FL waste which reported high extraction for Y and Eu [10]. To identify this effect, five synthetic mixtures were used to investigate if hydroxyapatite and aluminum oxide could affect the extraction process, shown in Table 3. As shown, the extraction of REEs in the first three mixtures without the presence of hydroxyapatite and aluminum oxide was high, around 90%, which is similar to the previous study [10]. However, the presence of hydroxyapatite in mixture 4 severely decreased the extraction to about 30–40%, which is comparable with the results obtained from SCFE of the non-mechanically activated samples in our study. This is because of the formation of REE phosphates. A previous study reported that dissolved REEs and phosphate ions from the apatite can form insoluble REE phosphate in acid leaching [15]. In addition, introduction of calcium ion can cause competition reaction, where the calcium ions compete with REEs to bind with the chelating agent. In mixture 5, it was also observed that the presence of aluminum oxide decreases the extraction by 20–30%. This can also be because of competition reactions in which aluminum cations bind with the chelating agent; thus, there are less chelating agent molecules available for REEs. Further analysis can be made based on stoichiometry after determination of the coordination number of TBP to aluminum and calcium, which are currently not available.

Conclusions

A process using supercritical carbon dioxide as the solvent and a small amount (less than 10 v/v%) of TBP-HNO_3 adduct as the chelating agent for extracting Y, Eu, and Tb from real FL waste is developed. Mechanical activation through high-energy oscillation milling is utilized to improve the extraction results. Characterization based on SEM, TEM, XRD, and XPS shows that physical changes and decomposition are induced by mechanical activation. Using XRD and calculated crystallite size, it is shown that mechanical activation creates nano-sized crystallites with grain boundaries which provide more pathways for the diffusion of supercritical CO_2 and adduct into the FL particles thus improving extraction. Chemical changes were observed from XPS analysis which showed REEs phosphates are transformed into REE oxides that are more amenable to SCFE. Using the developed process, more than 70% Y and Eu and more than 40% Tb extraction was achieved which is considerably higher than the previously reported results in the literature [10].

References

1. Pavón S, Lapo B, Fortuny A, Sastre AM, Bertau M (2021) Recycling of rare earths from fluorescent lamp waste by the integration of solid-state chlorination, leaching and solvent extraction processes. Sep Purif Technol 272. https://doi.org/10.1016/j.seppur.2021.118879
2. Wu Y, Wang B, Zhang Q, Li R, Yu J (2014) A novel process for high efficiency recovery of rare earth metals from waste phosphors using a sodium peroxide system. RSC Adv 4(16):7927. https://doi.org/10.1039/c3ra46381h
3. Zhang J, Anawati J, Yao Y, Azimi G (2018) Aeriometallurgical extraction of rare earth elements from a NdFeB magnet utilizing supercritical fluids. ACS Sustain Chem Eng 6:16713–16725. https://doi.org/10.1021/acssuschemeng.8b03992
4. Gijsemans L, Forte F, Onghena B, Binnemans K (2018) Recovery of rare earths from the green lamp phosphor LaPO4:Ce3+, Tb3+ (LAP) by dissolution in concentrated methanesulphonic acid. RSC Adv 8(46):26349–26355. https://doi.org/10.1039/C8RA04532A
5. Van Loy S, Binnemans K, Van Gerven T (2017) Recycling of rare earths from lamp phosphor waste: enhanced dissolution of LaPO4:Ce3+, Tb3+ by mechanical activation. J Clean Prod 156:226–234. https://doi.org/10.1016/j.jclepro.2017.03.160
6. Tunsu C, Petranikova M, Ekberg C, Retegan T (2016) A hydrometallurgical process for the recovery of rare earth elements from fluorescent lamp waste fractions. Sep Purif Technol 161:172–186. https://doi.org/10.1016/j.seppur.2016.01.048
7. Yurramendi L, Gijsemans L, Forte F, Aldana JL, del Río C, Binnemans K (2019) Enhancing rare-earth recovery from lamp phosphor waste. Hydrometallurgy 187:38–44. https://doi.org/10.1016/j.hydromet.2019.04.030
8. Tunsu C, Petranikova M, Gergorić M, Ekberg C, Retegan T (2015) Reclaiming rare earth elements from end-of-life products: a review of the perspectives for urban mining using hydrometallurgical unit operations. Hydrometallurgy 156:239–258. https://doi.org/10.1016/j.hydromet.2015.06.007
9. Yao Y, Farac NF, Azimi G (2017) Supercritical fluid extraction of rare earth elements from nickel metal hydride battery. ACS Sustain Chem Eng. https://doi.org/10.1021/acssuschemeng.7b03803
10. Shimizu R, Sawada K, Enokida Y, Yamamoto I (2005) Supercritical fluid extraction of rare earth elements from luminescent material in waste fluorescent lamps. J Supercrit Fluids 33(3):235–241. https://doi.org/10.1016/j.supflu.2004.08.004
11. Kunanusont N, Zhang J, Watada K, Shimoyama Y, Azimi G (2021) Effect of organophosphorus ligands on supercritical extraction of neodymium from NdFeB magnet. J Supercrit Fluids 170(2020):105128. https://doi.org/10.1016/j.supflu.2020.105128
12. Van Loy S, Binnemans K, Van Gerven T (2018) Mechanochemical-assisted leaching of lamp phosphors: a green engineering approach for rare-earth recovery. Engineering 4(3):398–405. https://doi.org/10.1016/j.eng.2018.05.015

13. Nath D, Singh F, Das R (2020) X-Ray diffraction analysis by Williamson-Hall, Halder-Wagner and aize-strain plot methods of CdSe nanoparticles-a comparative atudy. Mater Chem Phys 239(2019):122021. https://doi.org/10.1016/j.matchemphys.2019.122021
14. Ivanova OP, Naumkin AV, Vasilyev LA (1996) An XPS study of compositional changes induced by argon ion bombardment of the $LaPO_4$ surface. Vacuum 47(1):67–71. https://doi.org/10.1016/0042-207X(95)00180-8
15. Kim R, Cho H, Han KN, Kim K, Mun M (2016) Optimization of acid leaching of rare-earth elements from Mongolian apatite-based ore. Minerals 6(3). https://doi.org/10.3390/min6030063.

Recycling Technologies Developed at KIGAM

Jin-Young Lee, Kyeong Woo Chung, Shin Shun Myung, Soo-Kyung Kim, Hong-In Kim, Ho-Seok Jeon, and Rajesh Kumar Jyothi

Abstract Korea Institute of Geoscience and Mineral Resources (KIGAM), which is a unique government-funded Geoscience research organization in Korea, has researched and devised many technologies on mineral processing and extractive metallurgy for more than 70 years. In addition, several recycling processes have much studied from lab scale to pilot plant scale for the last two decades, and many technologies for the recycling of secondary resources, such as lithium batteries, neodymium magnet, catalyst, and so on, have been developed based on pyrometallurgy, hydrometallurgy, and electrometallurgy. Especially, some of them were already transferred to industry and commercialized in Korea. In the presentation, details of the recycling technologies at KIGAM will be introduced and discussed.

Keywords Spent catalyst recycling · Spent battery recycling · Spent magnets recycling · Automobile scraps recycling

Introduction

Korea is a global leader in electronics and automobile industrial products supply. At the same time, it has very limited natural resources. Moreover, Korea's population densities are very high. The above said three major reasons are initiated the subject recycling-reuse at KIGAM. Many industries in Korea approach us to handle the manufacturing waste as well as scrap. KIGAM having the several research divisions, among all mineral resources research, is one of the division. Researchers categorized mainly three types: Mineral processing, extractive metallurgists, and industrial materials preparation teams.

The moto of the research is either natural resource (e.g. monazite) to industrial material (magnets) preparation or secondary resources to sustainable industrial material preparation via extractive metallurgical technologies. Numerous wastes (or)

J.-Y. Lee (✉) · K. W. Chung · S. S. Myung · S.-K. Kim · H.-I. Kim · H.-S. Jeon · R. K. Jyothi
Korea Institute of Geosciences and Mineral Resources (KIGAM), Daejeon 34132, Korea
e-mail: jinlee@kigam.re.kr

A. Lazou et al. (eds.), *REWAS 2022: Developing Tomorrow's Technical Cycles (Volume I)*, The Minerals, Metals & Materials Series,
https://doi.org/10.1007/978-3-030-92563-5_30

scraps are handle successfully at KIGAM; those are spent catalysts, waste permanent magnets, spent batteries (from e-waste and electric vehicles), etc. The technologies develop from bench scale to pilot scale; finally, technology transfers to small and medium scale industries in Korea.

In recycling subject, most favorable option always is prevention of as much as waste followed by minimization. In second option, reuse the minimized materials then go for recycling and energy recovery processings. The least favored option is disposal in a proper place and proper method. Once generated the waste it will gives another challenge is landfilling problem. Korea having much mountain area of total land, it will rises the landfilling problem and environmental damage by waste. To face these two major issues, one is less natural resources and another one is landfilling problem; the resources recycling tool can reach national demand of rare and precious by using environmentally friendly routes.

Spent Catalyst Recycling

Selective catalytic reduction of NO_x, commonly referred to 'SCR', has been developed and was proven in various industrial stationary applications including thermos-power plant using fossil fuels, incineration, and waste plants. In the recent decades, particular interests arose for NO_x emission from power plants and transportations due to the harmful effects including smog and particulate matter generation. Therefore, spent SCR catalysts contributed a significant amount of solid waste in accordance with an increasing world demand of SCR catalysts. Among the catalyst systems, V_2O_5-WO_3/TiO_2 catalyst is known to have excellent catalytic activity and durability. In this study, hydrometallurgical process for recovery of vanadium and tungsten from spent SCR catalyst was developed. Process sequence starts by air roasting for a deconstruction of SCR catalyst structure, decomposition of $CaWO_4$ (leaching of vanadium), leaching of tungstic acid using ammonia solution, and crystallization of ammonium paratungstate (APT, *aq*). From these processes, APT (*s*) was obtained as a final product. On the other hand, raffinate from decomposition process of roasted feedstock was used as feed solution for the recovery of vanadium by precipitation method. As a result, vanadium was recovered as V_2O_5 (*s*).

Currently, due to the threat to human's health and environmental hazard, nitrogen oxides pose a set of new; stricter regulations have been developed for industries and transportation air emissions. The rigorous policies in the air emissions from an industry have led to an increasing demand of catalysts to mitigate the effect of NOx. Currently, the most efficient SCR catalyst for stationary applications (power plants and big industrial complexes) contains around 0.5–1.5% vanadium oxide (V2O5) as the main catalytic agent, 7–10% tungstic oxide (WO3) to improve the thermal stability, and 70–80% rutile (TiO2) as the supporting oxide. The main advantages of this type of catalyst are its thermal resistance, long lifespan, regeneration capabilities, and resistance to poisons and sulfur acids that might be generated from the presence of humidity in the air. The usual SCR catalyst lifespan ranges from three years depending on its composition, but the lifespan can be almost doubled by chemical or heat regeneration. However, despite the long lifespan and the several methods to prevent poisoning, eventually spent catalyst cannot be regenerated anymore and need to be discarded. The present method to discard spent SCR catalyst is in landfills, which are of great burden for environment and human health. Vanadium and tungsten are considered of high toxicity; therefore, the policies for the discarding of this waste have become harsher; in addition, primary metal sources (mining) are an environmental taxing activity which makes the recovery of the title metals an urgent matter in the industrial, environmental, economical, and recycling fields.

The amount of spent SCR catalyst generated is around 11,000 tons per year in Korea, from that 9600 tons are landfilled. The spent SCR catalyst discarded has 7,800 ton of TiO2, 900 ton of WO3 and 100 ton of V2O5, which can be recovered and reused. Before our project development, all of the spent SCR catalyst was landfilled. KIGAM made a memorandum of understanding (MoU) between various power plant companies to supply the spent SCR catalyst free of cost; thus, in KIGAM, an environmentally friendly hydrometallurgical method to recover the valuable metals such as titanium, vanadium and tungsten was developed. The hydrometallurgical process development is described as flow-chart for three valuable metals extraction from spent SCR catalyst leach liquors (Fig. 1). Many reports were published in top tier research journals on this subject [1–5].

Spent NdFeB Magnets Recycling

The pilot plant was established and utilized two processes developed to recover rare earth elements (REEs) from NdFeB waste magnets and manufacturing scraps. The first recovery process starts with the oxidation of the powders and leaching in sulfuric acid solution. Then, REEs were selectively precipitated as double sulfate salts and re-dissolved in hydrochloric acid through REEs hydroxide. The obtained REEs chloride solution was treated by solvent extraction process [6–8], and finally, high-purity REEs compounds (99.9% purity) were synthesized from purified REEs chloride solutions (Fig. 2).

The other utilized methods consist in a caustic digestion followed by a thermal oxidation process. Hydrochloric acid leaching of the resultant powder gave rise to high-purity REEs chloride solution. Solvent extraction techniques are applied, and REEs compounds of 99.9% purity are synthesized (Fig. 3).

Spent-Batteries Recycling

This research aims to introduce the results of a commercialization process study on the stable dismantling and heat treatment of waste mixed batteries by mixing lithium-ion batteries, lithium primary batteries, manganese alkaline batteries, nickel

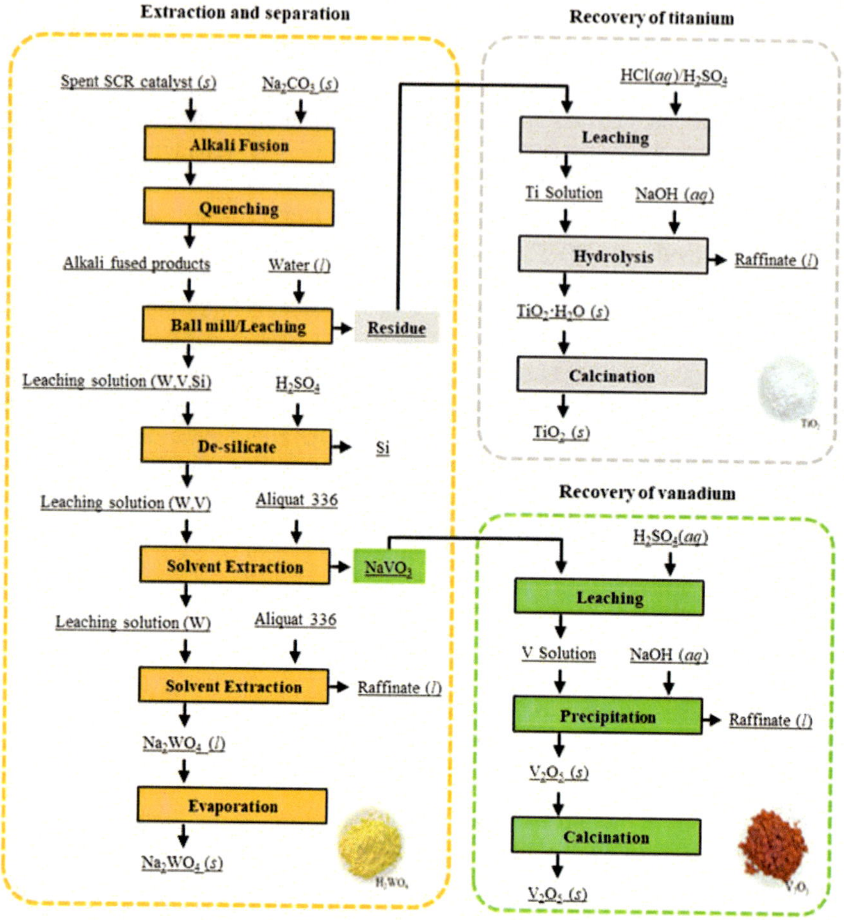

Fig. 1 Hydrometallurgical process development for Ti, V, and W extraction from spent SCR catalyst leach liquors

hydride batteries, etc. First, in the case of lithium-based mixed batteries, the ratio of lithium-ion batteries and lithium primary batteries is mixed at 4:1, a stable heat treatment furnace is used to explode at a temperature higher than 400°C, and then the crushing and grinding process is conducted. This heat treatment furnace allows disposal of 200 kg/batch waste cells. Also, through a stable heat treatment process, electrolytes in the battery that are harmful to the human body can be evaporated and removed, while separating separators and plastics that adversely affect crushing and grinding can be removed.

The obtained powder is injected into the leaching equipment system. In the leaching process, 1 mol/L H_2SO_4 and 10 vol % H_2O_2 are added to the sulfuric acid reduction leaching and wastewater reduction leaching process, and Al, Cu, Fe, etc. contained as impurities are controlled to be less than 10 mg/L. The solution with

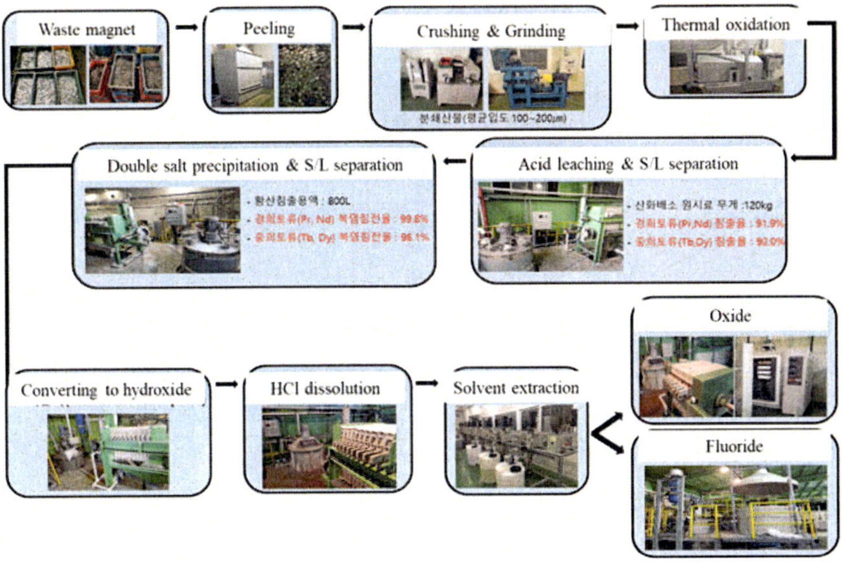

Fig. 2 Standard process for recycling process for NdFeB waste magnet

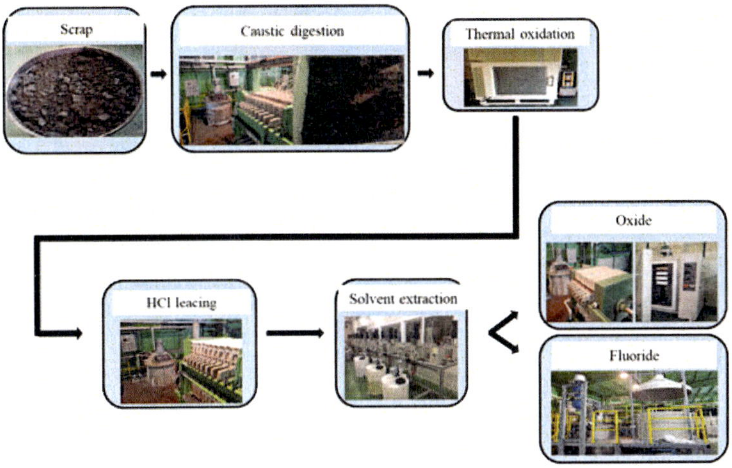

Fig. 3 Process for recycling process for NdFeB waste magnet

controlled impurities is applied to the solvent extraction system process with PC88A to selectively separate and extract cobalt and manganese nickel. After recovering cobalt and manganese, the nickel was recovered as $Ni(OH)_2$ by precipitation process or nickel was extracted by solvent extraction process at pH higher than 6.4. Finally, lithium left in the solution was concentrated to 18 g/L and then was precipitated as Li_2CO_3 with efficiency over 88%. The remaining 12% of lithium was precipitated

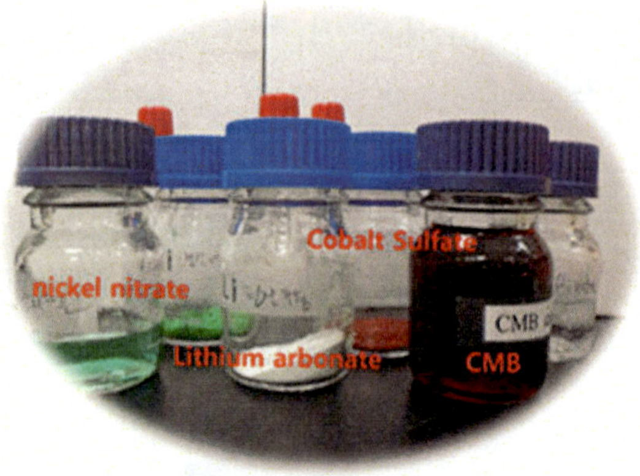

Fig. 4 Recovered valuable metals from spent LiBs

as Li_3PO_4 at room temperature with efficiency over 90%. Therefore, the recovery efficiency of lithium was more than 96%. For non-lithium-based batteries, a solution with controlled impurities is obtained through the wastewater reduction process and then subjected to a solvent extraction process. Zinc metal and manganese-zinc sulfate can be obtained by solvent extraction using D2EHPA as the extractant. In this new development process, approximately 90% of the valuable metals such as cobalt, manganese, nickel, lithium, and zinc was recovered (Fig. 4).

Recently, as the use of electric vehicles has increased, recycling of large-capacity batteries for automobiles has become an imperative research subject. Large-capacity lithium-ion batteries are becoming an energy source for electric vehicles such as HEVs, PHEVs, and EVs, and the weight of lithium-ion battery packs used in electric vehicles range from 50 to 400 kg, and the total energy of electric vehicles is about 4–60 Kwh per pack. For Tesla Model S, it is about 85 Kwh [9].

Unlike the recycling of small lithium-ion batteries, which are currently commercialized, large-capacity lithium secondary batteries are dangerous to process because they pose a high fire risk due to their high-energy capacity. Therefore, they require a lot of time and resources in dismantling, crushing, and sorting processes, and for the recovery of the valuable metals contained in the cells. It is difficult to secure economic feasibility because of the low content of metals and the presence of impurities in the cells. Therefore, a recycling study with its economic feasibility was conducted (Fig. 5).

Copper, aluminum, and plastic separated during crushing become reusable through a physical sorting process, and a mechanical sorting system is established to ensure the complete separation by material.

To separate the active material from the lithium-ion battery used in electric vehicles, crushing, pulverization, and sorting are employed. Then, a high-purity nickel and cobalt solution is produced through leaching, impurity removal, and electrolytic extraction. The recycled nickel and cobalt sulfate are used as raw materials to manufacture new lithium secondary batteries. Based on the hydrometallurgical process,

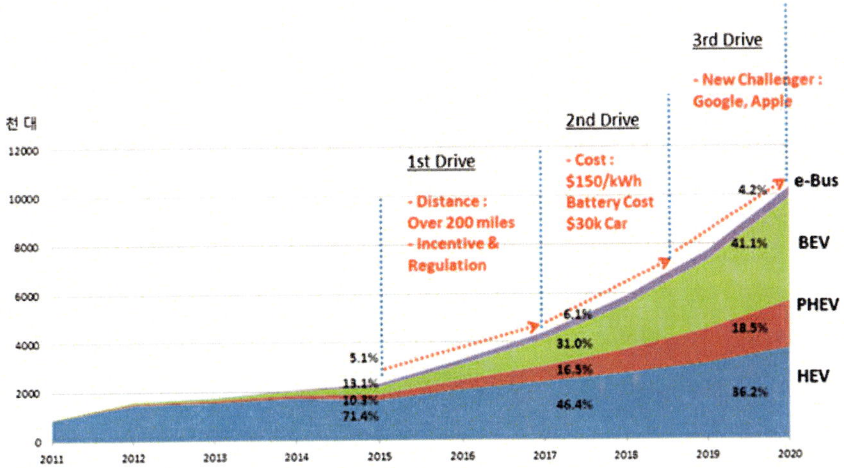

Fig. 5 Growth prediction electric vehicle market 2020 (Unit: Thousands)* (*Source* SNE Research)

we have obtained a complete recycling and commercialization technology for waste lithium secondary batteries from electric vehicles (Fig. 6).

By applying the discharging, dismantling, crushing, and separation processes of lithium secondary battery packs for electric vehicles, we built an industrial facility with a capacity of 50 units/day yielding a process that can dismantle 15,000 units/year. The technology can be applied for large-capacity industries treating waste lithium secondary battery packs from automobiles.

Electronics Industry Waste Recycling

Recently, eco-friendly metal recovery technology is an important and innovative technology that facilitates waste liquid treatment without requiring a great addition of chemical reagents. From an economic and environmental perspective, electrolysis has vast advantages over other purification techniques due to the use of simple electrical equipment and the low chemical reagents input. The purpose of this study is to establish and commercialize an eco-friendly economically sustainable gold recovery technology from secondary waste by cyclone electrolytic recovery technology (Fig. 7) [10–12].

In this study, 95% of gold was recovered with a purity of more than 95%. A continuous bench-scale cyclone electrolytic system with a processing capacity of 300 tons/year was built for the PCB waste (Fig. 8).

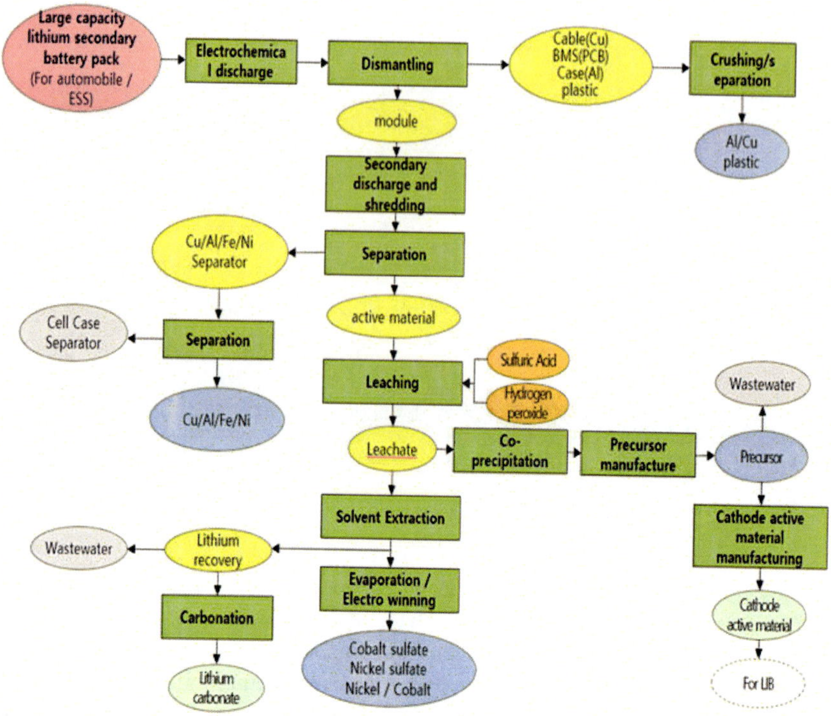

Fig. 6 Process diagram of lithium-ion battery pack for electric vehicles

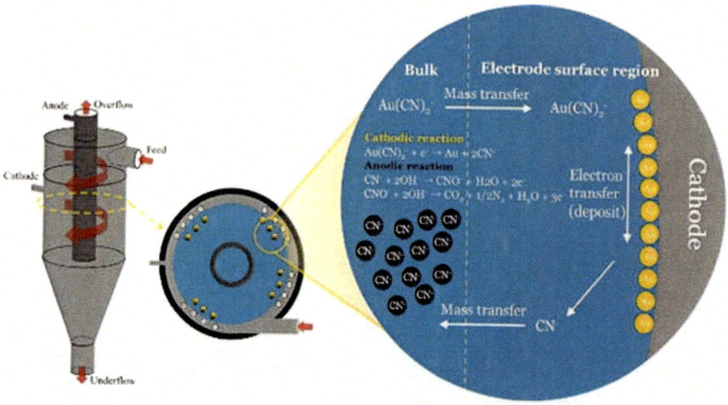

Fig. 7 Schematic of electrochemical recovery of low concentration of Au by hydrocyclone electrowinning

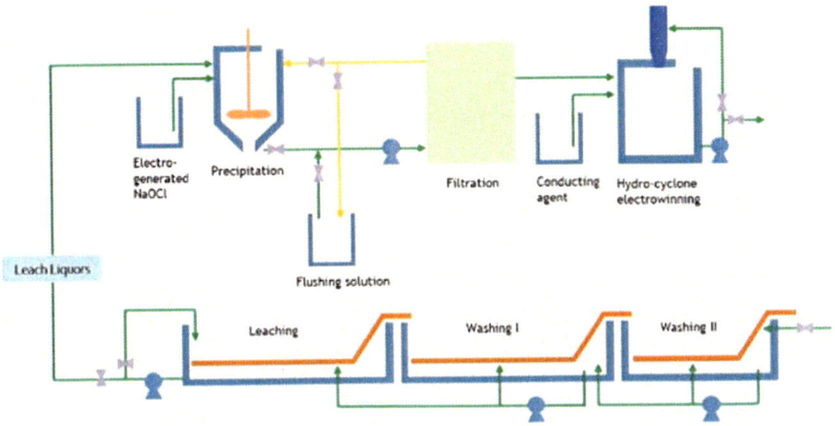

Fig. 8 Schematic of continuous bench-scale cyclone electrolytic system

Automobile Scraps and Shredder Residue Recycling

Development of physical separation system for Automobile Shredder Residues (ASR) recycling of ELVs: Recycling of ELVs (End-of-Life Vehicles) is becoming an imperative research subject due to the increase of the amount of the ELVs in the world. This has led to series research efforts on the recycling of the ELVs that have resulted in technologies being developed and transferred to industry. In general, the reusable parts are recovered in advance during the dismantling process of ELVs. The remains are shredded, and then the ferrous and non-ferrous metals are separated from the shredded scraps. Through this process, about 75% of ELV is recycled, while the remaining 25% is still landfilled or incinerated. The remaining 25% are called Automobile Shredder Residues (ASR). Many technologies for ASR recycling, such as compaction, solidification, pyrolysis, and gasification, have been researched and industrialized [13, 14]. However, the compression and solidification method causes a great burden to the environment due to air pollution. Pyrolysis and gasification incur not only the problem of air pollution but also they are costly processes. Therefore, physical separation techniques and processes show more advantages for ASR recycling. In this research, a physical separation system combining unit separation techniques has been developed for ASR recycling. The recycling rate was calculated by mass balance analysis of products recovered through this system, and the total recycling rate reached about to 86% (Fig. 9).

Fig. 9 Photograph of physical separation facility for ASR recycling of ELVs

Acknowledgements This study was supported by the R&D Center for Valuable Recycling (Global-Top R & D Program) of the Ministry of Environment (Project Number: 2019002230001), Korea.

References

1. Sola ABC, Parhi PK, Lee JY, Kang HN, Jyothi RK (2020) Environmentally friendly approach to recover vanadium and tungsten from spent SCR catalyst leach liquors by Aliquat 336. RSC Adv 10:19736–19746
2. Choi IH, Moon G, Lee JY, Jyothi RK (2019) Alkali fusion using sodium carbonate for extraction of vanadium and tungsten for the preparation of synthetic sodium titanate from spent SCR catalyst. Sci Rep 9:12316
3. Choi IH, Moon G, Lee JY, Kumar JR (2018) Extraction of tungsten and vanadium from spent SCR catalyst for stationary application by pressure leaching process. J Clean Prod 197:163–169
4. Choi IH, Moon G, Lee JY, Kumar JR (2018) Hydrometallurgical processing of spent SCR catalyst for recovery of tungsten. Hydrometallurgy 178:137–145
5. Choi IH, Moon G, Lee JY, Jyothi RK (2018) Spent V2O5-WO3/TiO2 catalyst processing for valuable metals by soda roasting-water leaching. Hydrometallurgy 175:292–299
6. Yoon HS, Kim CJ, Chung KW, Kim SD, Kumar JR (2015) Process development for recovery of dysprosium from permanent magnet scraps leach liquor by hydrometallurgical techniques. Can Metall Q 54:318–327
7. Yoon HS, Kim CJ, Chung KW, Kim SD, Kumar JR (2015) Recovery process development for the rare earths from permanent magnets scraps leach liquors. J Braz Chem Soc 26:1143–1151
8. Jeon JH, Yoon HS, Kim CJ, Chung KW, Jyothi RK (2021) Environmentally sound technology development for processing of rare earth elements from waste permanent magnets synthetic leach solutions: recovery and separation perspectives. Sep Purif Technol 275:119225
9. Ambrose H, Jimmy OD (2021) Electric vehicle batteries: addressing questions about critical materials and recycling. Cambridge, MA: Union of Concerned Scientists. https://www.ucsusa.org/resources/ev-battery-recycling

10. Bae MK, Lee JC, Lee HJ and Kim SK (2020) Recovery of nitric acid and gold from gold-bearing aqua regia by tributyl-phosphate. Sep Purif Technol 235:116154
11. Bae MK, Lee HJ, Yoo KK, Kim SK (2021) Copper(I) selective chemisorption on magnetite (Fe_3O_4) over gold(I) ions in chloride solution with cyanide. Hydrometallurgy 201:105560
12. Lee LHJ, Jung M, Bae MK, Lee EK, Jin H, Mishra B (2020) Recovery of oil from ferrous grinding swarf of automobile industry by aqueous washing process. Waste Manage 111:51–57
13. Kim BU, Han OH, Jeon HS, Baek SH, Park CH (2017) Trajectory analysis of copper and glass particles in electrostatic separation for the recycling of ASR. Metals 434:1–13
14. Jeon HS, Lee H, Lee KH, Baek SH (2015) Present condition of end-of-life vehicles recycling of Germany by EU-COM-table. J Korea Soc Waste Manag 32:115–122

Yellow Phosphorus Production from Phosphoric Acid by Carbothermic Reduction

Huafang Yu, Ryoko Yoshida, Yasushi Sasaki, and Tetsuya Nagasaka

Abstract Phosphorous is an essential element for agriculture and industry and is a non-renewable resource. Especially yellow phosphorus is a critical material for advanced industrial technology, but phosphorus resources were not produced in Japan, and all depend on imports. It has been suggested, however, that the remaining accessible reserves of phosphate ore will be depleted within 50 years. Therefore, alternative resources for phosphate ore must be found. In this research, we have developed a process that enables the production of high-purity yellow phosphorus from domestic unused phosphorus resources such as steelmaking slags. The process consists of two parts: (1) the production of crude phosphoric acid from wastes such as steelmaking slag; (2) producing high-purity yellow phosphorus by low-temperature carbothermic reduction of phosphoric acid (H_3PO_4). The details of the carbothermic reduction of phosphoric acid are presented in this paper. Yellow phosphorus is commercially produced by carbothermic reduction of phosphate ore in an electric arc furnace at more than 1673 K. In the newly developed system, gaseous P_4O_{10} evaporated from H_3PO_4 is successfully reduced to yellow phosphorus using a carbon-packed bed at less than 1273 K. To meet the depletion of phosphate ore, the proposed process in this study to produce yellow phosphorus by carbothermic reduction of H_3PO_4 that are extracted from dephosphorization slags will be one of the practical and economical solutions.

Keywords Carbothermic reduction · Dephosphorization slags · Phosphoric acid · Yellow phosphorus

---

H. Yu (✉) · R. Yoshida · Y. Sasaki · T. Nagasaka
Graduate School of Engineering, Tohoku University, Miyagi 980-8579, Japan
e-mail: huafang.yu.e8@tohoku.ac.jp

R. Yoshida
Dowa Holdings Co., LTD, Akita 017-0202, Japan

Y. Sasaki
Institute of Multidisciplinary Research for Advanced Materials, Tohoku University, Miyagi 980-8577, Japan

A. Lazou et al. (eds.), *REWAS 2022: Developing Tomorrow's Technical Cycles (Volume I)*, The Minerals, Metals & Materials Series,
https://doi.org/10.1007/978-3-030-92563-5_31

Introduction

Phosphorus is an essential element in food production [1]. Phosphorus is also used as an important material in a wide range of industrial fields such as an electronic component (semiconductor GaP, secondary battery cathode material, etc.), automobiles (steel plate surface treatment liquid, and secondary battery electrolyte), pharmaceuticals (osteoporosis treatment), and plastics. After World War II, humans have rapidly consumed phosphorus ore resources for fertilizer to produce food, so the depletion of phosphate ore with high quality and ease to mine is rapidly progressing [2].

As already mentioned, yellow phosphorus is critically important to produce sophisticated industrial products such as semiconductor electronic parts and pharmaceuticals. With existing technology, about 300 million kWh of electricity is required to produce about 20,000 tons of yellow phosphorus that Japan consumes annually. The huge electrical demand is simply due to the high-temperature reduction of phosphate ore at more than 1673 K. Due to this enormous energy consumption, current yellow phosphorus-producing countries are limited to China, the United States, Kazakhstan, and Vietnam. After the United States banned the export of yellow phosphorus as a strategic commodity from 1996, other yellow phosphorus-producing countries also limited the yellow phosphorus export. Thus, Japan's imports of phosphorus ore and yellow phosphorus are becoming increasingly severe year by year.

From the research of Matsubae et al. [3], the phosphorus of 110.5 kt as phosphorus ore is imported, and the comparable amount of phosphorus (93.07 kt) is accumulated in steel slag. Namely, the amount of phosphorus in phosphate ore is almost comparable to that in the imported iron ore. Thus, the exploitation of steel slags [3–13] and sewage sludge [14, 15] as a secondary phosphorus resource and are an emerging research project, and many studies have been carried out now. Phosphorus extraction from slags as crude H_3PO_4 has been extensively studied, and the details of the results are found elsewhere. As already mentioned, all of the yellow phosphorus is imported into Japan. Thus, if yellow phosphorus is also possibly produced from steelmaking slags, it is a great benefit. Therefore, the investigation to develop a new economical process to make yellow phosphorus by reducing crude H_3PO_4 extracted from slags has been carried out to replace the current production one. However, yellow phosphorus production from H_3PO_4 has not yet been studied. This study aims to confirm the feasibility of yellow phosphorus production by low-temperature carbothermic reduction of phosphoric acid (H_3PO_4) with reasonable reaction rates.

Experimental

Apparatus

The experimental system for the reduction comprises a reactor, a yellow phosphorus water bath trap, a silica gel tube for dehydration of exhaust gas, and a gas flow meter.

The vertical electric furnace is composed of two heating parts with a length of 30 cm that can control the temperatures independently. Each heating unit corresponds to (1) a heating unit for vaporizing phosphoric acid and (2) a heating unit to reduce vaporized phosphorus compounds. A quartz reaction tube (inner diameter: 32 mm, length: 1200 mm) was used as a reactor. The lower end of the tube was connected to a thin quart tube.

In the reactor quartz tube, activated carbon as a reducing agent was filled to a length of about 40 cm. The activated carbon was almost spherical, and the diameter is about 2–5 mm. The upper 30 cm of the packed bed was placed in heating zone B, and the lower 10 cm of the packed bed (zone C) was placed outside the heated zone B. The role of zone C is to enhance the collisions among P_4 gas molecules to produce liquid yellow phosphorus. Region A above the packed bed in zone B was filled with phosphoric acid-containing activated carbon with a height of about 15 cm. Phosphoric acid (H_3PO_4) is chemical agent grade, and it contains about 15% water. Phosphoric acid-containing activated carbon was prepared by immersing phosphoric acid in activated carbon.

To recover the produced gaseous yellow phosphorus as a liquid, a thin quartz tube at the lower end of the quartz tube was placed in a flask filled with hot water (343 K), and the gas mixture was bubbled into the flask to collect the gaseous yellow phosphorus as a liquid yellow phosphorous. The melting point of yellow phosphorus is 317.3 K. To avoid the escape of gaseous yellow phosphorus from the flask, the discharged gas was further bubbled into another flask filled with bromine water to trap and completely remove yellow phosphorus. The exhaust gas discharged from the second flask was passed through a dehydration tube filled with silica gel, and then the total flow rate of exhaust gas was measured every minute using a gas flow meter.

Procedure

Initially, the air in the system is replaced by introducing Ar with a flow rate of 20 cm^3/min, and then the temperature of the reduction reaction zone B filled with activated carbon was raised to 1273 K. When the temperature reached 1273 K, the temperature of region A in which the phosphoric acid-containing activated carbon was found to increase to about 573 K even without power input at that zone. In the temperature raising process, the outlet gas flow rate was continuously measured with the flow meter.

The preliminary experiments found that unreacted phosphorous compounds adhered to the reactor wall when the total flow rate exceeded more than 200 cm^3/min. Therefore, to avoid the unreacted phosphorous compounds, the temperature of upper heating zone (zone A) was carefully raised manually by keeping the exhaust gas flow rate below 200 cm^3/min up to 973 K. After that temperature, the unreacted phosphorous compounds were not observed even if the outlet gas flow rates exceeded more than 200 cm^3/min.

When the temperature of zone A reaches about 773 K, it was observed that droplets started to adhere to the wall surface of the lower part of the quartz tube, and they coalesced and trickled down along the wall surface of the quartz tube and captured in the flask.

After five hours, the power was turned off in zones A and B, and the experiment was terminated. After the furnace temperature reached room temperature, the samples collected in the flask were analyzed by Raman spectroscopy. In this study, gaseous phosphorus oxide existed in the packed bed, and yellow phosphorus is presented by P_4 and P_4O_{10}, respectively.

Results and Discussion

Typical results of the temperature changes of heating zone A and the exhaust gas flow rate with time during the reduction of H_3PO_4 are shown in Fig. 1. The total H_3PO_4 (15% H_2O) weight in the activated carbon (20 g) was about 40 g. The total amount of carbon packed in the reduction zone is about 70 g. The reduction temperature (heating zone B) was 1273 K. The Ar with a flow rate of 20 cm^3/min was always supplied during the experiment to transport the evaporated P_4O_{10} (g) and H_2O (g) from zone A to reduction zone B.

The exhaust gas flow rate is supposed to practically reflect the overall reaction rates. The exhaust gas flow rate fluctuated at less than 673 K. This is simply due to the difficulty of temperature control at this temperature range. As already mentioned, the unreacted phosphorous compounds, which suggested to be P_2O_5, [16], were

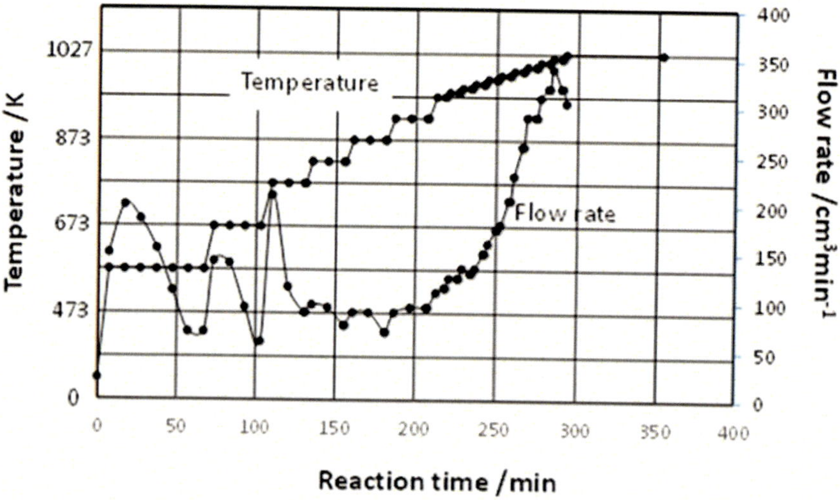

Fig. 1 Variation of temperature at upper heating zone and the total exhaust gas flow rate as a function of time

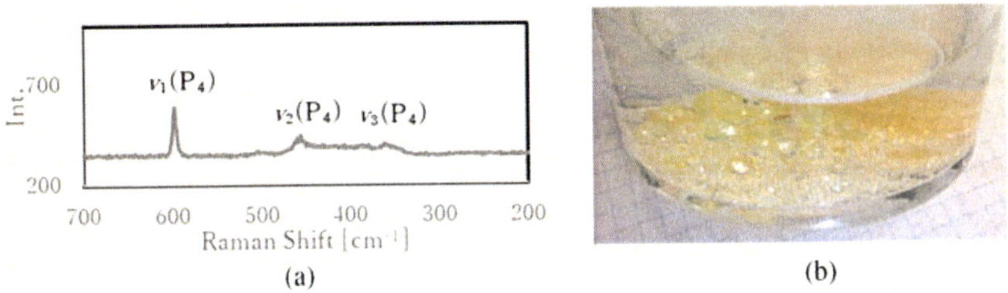

Fig. 2 **a** Raman spectroscopy of generated yellow phosphorus; **b** overview photo of the generated yellow phosphorus

adhered to the reactor wall when the total flow rate exceeded more than 200 cm^3/min. Therefore, the exhaust gas flow rate was carefully controlled by manually adjusting the temperature of zone A up to 973 K. This manually adjusting is the reason for the fluctuation at a lower temperature range. After reaching 973 K, the unreacted phosphorous compounds were not observed even the outlet gas flow rates excessed over more than 200 cm^3/min.

From Fig. 1, the exhaust gas flow rate started to increase rapidly after the zone A temperature reached around 1100 K. This increase is simply due to the increase of the P_4O_{10} supply rate since P_4O_{10} vaporizes rapidly at more than 1139 K. The gas flow rate gradually decreased since P_4O_{10} supply rate decreased after reaching to 973 K.

The collected sample in the flask was analyzed by Raman spectroscopy (Fig. 2a), and all the peaks well correspond to these standard yellow phosphorus [17]. Thus, the sample produced in the present study was confirmed to be yellow phosphorus, and the produced yellow phosphorus was shown in Fig. 2b.

Generally, the recovery rate of yellow phosphorus is around 50% evaluated from the weight difference between supplied H_3PO_4 and formed P_4. After the experiment, 12% of phosphorus was detected remaining in the activated carbon, which indicates the longer heating time is necessary in order to improve the recovery rate. Moreover, the smaller flow rate of carrier gas was taken into consideration to a conducive longer time for phosphoric acid/P_4O_{10} vapor reacting with activated carbon in order to achieve a higher recovery rate.

Conclusions

In the present work, the feasibility of the production of yellow phosphorus by carbothermic reduction of H_3PO_4 was confirmed. For the implementation of this process, however, further investigations, especially the quantitative kinetics approach, are certainly required.

Acknowledgements The authors are thankful to the Japan Science and Technology Agent with the No. JPMJMI17C2 for financial support.

References

1. Nakakubo T, Tokai A, Ohno K (2012) Comparative assessment of technological systems for recycling sludge and food waste aimed at greenhouse gas emissions reduction and phosphorus recovery. J Clean Prod 32:157–172
2. Cordell D, Drangert JO, White S (2009) The story of phosphorus: global food security and food for thought. Glob Environ Chang 19(2):15–64
3. Matsubae-Yokoyama K, Kubo H, Nakajima K, Nagasaka T (2009) A material flow analysis of phosphorus in Japan—the iron and steel industry as a major source. J Ind Ecol 13(5):687–705
4. Du CM, Gao X, Ueda S, Kitamura S (2018) Distribution of P_2O_5 and Na_2O between solid solution and liquid phase in the CaO-SiO_2-Fe_2O_3-P_2O_5-Na_2O slag system with high P_2O_5 content. Metall Mater Trans B 49(1):181–187
5. Du CM, Gao X, Ueda S, Kitamura S (2018) Recovery of phosphorus from modified steelmaking slag with high P_2O_5 content via leaching and precipitation. ISIJ Inter 58(5):833–841
6. Du CM, Gao X, Ueda S, Kitamura S (2018) Optimum conditions for phosphorus recovery from steelmaking slag with high P_2O_5 content by selective leaching. ISIJ Inter 58(5):860–868
7. Du CM, Gao X, Ueda S, Kitamura S (2018) Effect of Fe^{2+}/T.Fe ratio on the dissolution behavior of P from steelmaking slag with high P_2O_5 content. J Sustain Metall 4:434–454
8. Iwama T, Du CM, Gao X, Kim SJ, Ueda S, Kitamura S (2018) Extraction of phosphorus from steelmaking slag by selective leaching using citric acid. ISIJ Inter 58(7):1351–1360
9. Yokoyama K, Kubo H, Mori K, Okada H, Takeuchi S, Nagasaka T (2006) Separation and recovery of phosphorus from steelmaking slags with the aid of strong magnetic field. Tetsu-to-Hagane 92(11):683–689
10. Yu HF, Miki T, Sasaki Y, Nagasaka T (2020) Crystallography of the high-temperature Ca_2SiO_4-$Ca_3P_2O_8$ solid solutions. Metall Mater Trans B 51:3007–3015
11. Kubo H, Matsubae K, Nagasaka T (2009) Magnetic separation of phosphorus enriched phase from multiphase dephosphorization slag. Tetsu-to-Hagane 95(3):300–305
12. Matsubae K, Kajiyama J, Hiraki T, Nagasaka T (2011) Virtual phosphorus ore requirement of Japanese economy. Chemosphere 84(6):767–772
13. Yu HF, Lu X, Miki T, Sasaki Y, Nagasaka T (2021) Phosphorous recovery from Ca_2SiO_4–$Ca_3P_2O_8$ solid solution by carbothermic reduction. J Sustain Metall 7:459–469
14. Li R, Teng W, Li Y, Wang W, Cui R, Yang T (2017) Potential recovery of phosphorus during the fluidized bed incineration of sewage sludge. J. Clean Production 140:964–970
15. Egle L, Rechberger H, Zessner M (2015) Overview and description of technology for recovering phosphorus from municipal waste water. Resour Conserv Recy 105:325–346
16. Brown EH, White CD (1952) Vapor pressure of phosphoric acids. Ind Eng Chem 44(3):615–618
17. Edwards HGM (1993) Vibrational Raman spectroscopy and force constants of phosphorus P_4. J Mol Struct 295:95–100

Efficient Steel Mill Dust Recycling—Aiming for Zero Waste

Juergen Antrekowitsch and Gustav Hanke

Abstract Steel mill dust recycling today is an important business. However, most processes in operation hardly fulfil future requirements from the environmental point of view. In general, these treatment facilities show a certain CO_2-footprint due to carbothermal reduction and produce high amounts of residues which often have to be landfilled. The paper discusses possibilities to optimize state of the art processes regarding the realization of a zero waste strategy and the minimization of the CO_2-footprint. Furthermore, new developments are evaluated in respect of energy consumptions and environmental awareness. As an example, the "2-step-Dust-Recycling" process, an own development of the University of Leoben in Austria, is analyzed regarding its potential to meet requirements of present and future environmental legislation. Finally, the way how different options of steel dust treatment influence the overall zinc cycle, especially the role of zinc oxide from such sources as substitute for primary concentrates, is described.

Keywords Zero waste · By-products · Hydrogen · Reduction · Circular economy · Electric arc furnace dust · Climate change

Introduction

Steel mill dust recycling contributes essentially to the worldwide zinc production. As for every industrial sector, environmental related questions about CO_2-footprint, energy consumption, and waste generation are fundamental also for this sector. When trying to answer the questions facing the current state of the art, not many satisfying answers can be found. Due to the fact that the majority of processes currently in operation are based on carbothermal reduction and are focusing on the recovery of only one metal (zinc), future requirements for sustainable and environmental processing can hardly be fulfilled.

J. Antrekowitsch (✉) · G. Hanke
Chair of Nonferrous Metallurgy, University of Leoben, Leoben, Austria
e-mail: Juergen.antrekowitsch@unileoben.ac.at

A. Lazou et al. (eds.), *REWAS 2022: Developing Tomorrow's Technical Cycles (Volume I)*, The Minerals, Metals & Materials Series,
https://doi.org/10.1007/978-3-030-92563-5_32

The following chapters discuss state of the art processes, optimization potentials, and possible future alternatives for the efficient recycling of steel mill dust from electric arc furnace steel production. Furthermore, the importance of zinc oxide from dust recycling compared to concentrates out of ores is analyzed.

State of the Art Processes

The Waelz-Process

The Waelz-kiln is the by far dominating facility for the treatment of electric arc furnace dust (EAFD) out of steel making. In a rotary kiln, the dust is reduced by carbon at temperatures of about 1100 °C. Zinc is separated, re-oxidized in the off-gas stream, and collected in the filter house. The relatively simple one-step process, that does not require additional energy input beside the carbon necessary for reduction, is nowadays listed as best available technology.

However, beside zinc oxide, a lot of residue, the so called Waelz slag, is produced which hardly finds a utilization and has to be landfilled in many regions of the world. With this, also the iron content is lost and not recoverable anymore. The carbon amount required can be found in the range of 140–180 kg per ton of treated EAFD causing a specific CO_2-amount of roughly 500 kg. Because the recovered zinc is the only product, it carries the whole carbon footprint of the process. Beside this, a relatively bad product quality, due to high halogen contents, is typical for the Waelz concept [1].

Therefore, two of the three above mentioned future indicators for an environment-friendly process, the CO_2-footprint and the waste generation, have to be seen as critical.

Alternative Processes

Most of the alternatives that made it to industrial scale within the last three decades are based on carbothermal reduction as well. Examples are the Primus process (multiple hearth furnaces in combination with an electric furnace), the PIZO process (induction furnace), or the rotary hearth furnace.

However, they show a similar carbon consumption and the same bad product qualities. At least, the PIZO-concept and the Primus process recover iron and therefore minimize the remaining slag. The rotary hearth aims for the generation of a kind of direct reduced iron (DRI), which is of rather poor quality and can hardly be used in steel industry. The production of a second product, e.g., an iron alloy, splits the produced CO_2-amount and with this minimizes the carbon footprint for the generated main product zinc oxide. The reduction of iron oxide of course needs energy, which is for the first two mentioned processes covered by electricity. In this case, the question is, if the electricity is produced carbon neutral or has a CO_2-footprint [1].

However, the two targets "low energy consumption" and "zero waste" can hardly be reached together because iron reduction is essential for the minimization of residues but requires energy, especially when talking about pyrometallurgical technologies.

Strategies to Minimize the CO_2-Footprint

In the following section, concepts that allow a minimization of the CO_2-footprint in EAFD-treatment are discussed.

Hydrogen as Reducing Agent

The most popular idea is the use of hydrogen which in best case should substitute 100% of the carbon and brings the CO_2-footprint to almost zero.

Unfortunately, this idea shows several difficulties. The Waelz-kiln is the dominating technology, and an implementation of a gaseous reducing agent into such a facility is nearly impossible. The process in general mixes EAFD with solid carbon carriers which react with the oxides when reaching the reaction zone, where the temperature allows the reduction of zinc oxide. To implement hydrogen, it must be introduced via a very long lance or by nozzles in the furnace jacket. A lance with an appropriate length of more than 15 m is hard to realize because of material stability and the movement of the solid charge where the hydrogen must be blown in somehow. Also, the installation of nozzles which have to be positioned in the refractory lining and fed with hydrogen with special mechanisms because of the rotating furnace is problematic. Even if problems related to engineering could be solved, the efficiency of the introduced hydrogen would be very poor.

In case of alternative processes where a liquid iron/slag bath is present, an introduction of hydrogen by lances or, e.g., porous plugs would by possible from the technical point of view. How efficient hydrogen bubbles act in a liquid bath must be questioned.

The best option would be to make use of a retort and realize a reduction of pelletized EAFD. However, also in this case, it is important to evaluate how efficient hydrogen is and which yield could be realized. Nevertheless, for such a concept, most experience out of direct reduction of iron ore is available.

Finally, the price of hydrogen is the main issue which makes hydrogen hardly competitive for the moment and in near future. Comparing the price for the required hydrogen with the fossil carbon carrier including CO_2-certificates, hydrogen is still about eight times more expensive than usual carbon carriers in a Waelz-kiln. Even taking very positive forecasts into account where the costs for hydrogen (production) drop down to 2 USD per kg in 2050 [2], utilization of hydrogen will be about two times more cost intensive than standard carbon carriers including CO_2-certificate costs (based on current prices of about 70 USD/t) [3]. Only if CO_2-certificate prices climb to 150 USD per ton and hydrogen can be produced for 2 USD per ton, the costs of hydrogen would be equal to those for currently used fossil carbon carriers for the treatment of one ton of EAFD.

All these calculations do not take into account additional costs coming up with the utilization of hydrogen. These are:

- Safety measures when applying hydrogen in a process and facing a hydrogen containing off-gas
- Using hydrogen does not allow the release of energy as it is done when CO is generated and can be burned to CO_2. Therefore, additional energy supply is required.
- Due to difficulties with hydrogen storage, continuous supply for processes are only possible when applying on site hydrogen production.

Summarizing, hydrogen is of course one strategy to minimize the CO_2-footprint, but depending on the type of process, it will take many decades before hydrogen is competitive from the economic perspective.

Charcoal Out of Biomass

Charcoal out of biomass is per definition CO_2-neutral and with this would be an alternative for carbon carriers with fossil origin. Also, the general physical form is the similar and therefore does not change much, regarding process concept.

However, two important characteristics that make charcoal different from, e.g., metallurgical coke are:

- the lower physical strengths and
- the higher reactivity due to a higher surface and porosity.

This might not be a problem for processes based on liquid metal bath/slag, where a higher reactivity could be even positive. For shaft furnace processes, the low strength combined with high reactivity is an exclusion criterion for a possible substitution [4].

For Waelz-kilns, the low physical strength does not disturb but high reactivity and a higher amount of volatiles lead to losses during heating up. This fraction combusts in the off-gas stream and causes higher off-gas temperatures and with this a loss of energy. Nevertheless, a partial substitution is realistic and can at least minimize the CO_2-footprint [4].

Regarding economics, charcoal shows higher costs in production but when taking costs for CO_2-credits into account, utilization of charcoal is not much more expensive. This means, compared to hydrogen, charcoal would already be competitive now and not only in 20 years.

However, an important question is the availability of biomass and if charcoal can be produced locally, avoiding long distance transports.

Hydrometallurgical Approaches

For many decades, hydrometallurgy has been one strategy when developing alternatives for EAFD-treatment. The idea to recover zinc by leaching and avoid high temperatures always supported the hope for a less energy consuming process.

Unfortunately, the presence of a certain percentage of zinc in form of zinc ferrite, which is rather difficult to leach, minimizes the chance for a simple hydrometallurgical concept. If the ferrite is leached too, a complex iron-separation must be implemented. Some concepts try to leach most of the material with the aim to recover different metals. However, the process is getting more complex, and with this, the costs increase. Especially the treatment of wastewater, either by chemicals or evaporation, causes additional costs and consumes energy. Some of the developments require an amount of energy for drying and evaporation which is in the same range as the one of pyrometallurgical concepts [5].

Even though there have been many activities within the last 30 years, no concept proofed success and made it to industrial scale.

CO_2-footprint and energy consumption today are well known concerns. The environmental hazard that might be caused by complex hydrometallurgical processes also has to be taken into account, when deciding if such a process could be an environmentally responsible and sustainable alternative.

The Importance of Product Quality

In general, product quality is of course important for every industrial process; in case of steel mill dust recycling, this quality plays a very important role.

Unfortunately, the quality of the produced zinc oxide for the above described processes is low, and with this, it can be only used as substitute for concentrates out of primary ores. EAFD always carries a certain number of halogens in various compounds. These impurities follow, due to their low vaporization temperature, the zinc oxide product and limit its use in primary zinc industry. This is only possible after a washing step that removes a major number of halogens but often still leaves too high amounts of fluorine behind. These oxides must be fed into the roaster for further halogen removal. Because of this, requirement and remaining amounts of fluorine and chlorine only allow a substitution of 15–20% of the input of a primary smelter.

This limitation is not the only critical aspect when utilizing such secondary zinc oxides in primary zinc industry. The necessity to have a roasting, a complex leaching procedure followed by purification, and an energy intensive winning electrolysis leaves the question open, if there would not be a more straight forward way to generate a final product out of EAFD-treatment.

Anyway, the key to a higher sophisticated product is the efficient separation of halogens. Nevertheless, it must be underlined that for primary zinc industry, such secondary concentrates are still a highly welcome input material due to their low iron content. Even though the halogens cause certain problems, these materials often dilute the high iron values out of primary concentrates and minimize related problems like required dumps for iron precipitation residues.

The Two Step Dust Recycling (2sDR) Process

Based on many of the above discussed future goals but mainly with the aim to create a zero waste concept, the 2sDR-process has been developed at the University of Leoben together with the companies ARP and Primetals Austria.

As the name defines, the process includes two pyrometallurgical steps. The flowchart below describes the concept in general.

After agglomeration to avoid carry over in the first step, the material is charged into a short drum furnace, where it is treated at 1100 °C under oxidizing conditions to get rid of lead and halogens. The clinkered material is charged in hot stage onto a metal bath in an electric furnace to reduce zinc and iron. Zinc evaporates and re-oxidizes above the furnace while iron and possible alloying elements accumulate in the iron bath. The slag can be modified to meet requirements of either cement or construction industry [6].

The zinc oxide which is produced with a high quality can enter primary zinc industry. Due to the higher purity compared to other zinc oxides, there is no necessity to treat the concentrate in the roaster and direct leaching can be applied. The quality

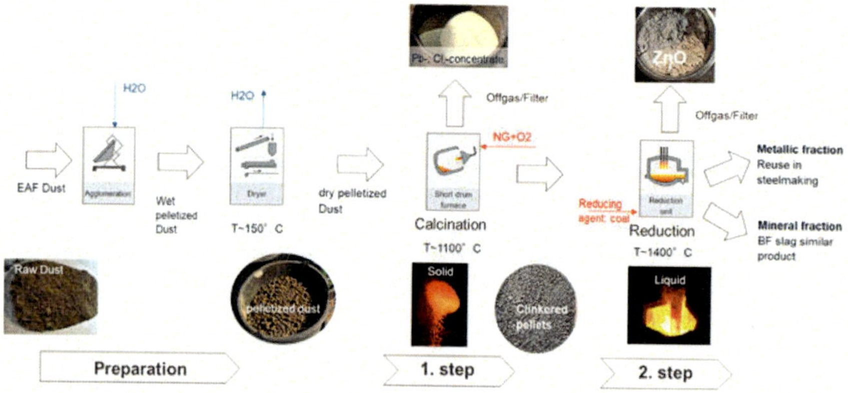

Fig. 1 Basic principle of the 2sDR-process

Table 1 Analytics of main products generated in the pilot scale trials

Zinc oxide

	ZnO	Fe_2O_3	Cl	F	Pb	K	rest
%	>93	2–4	<0.1	<0.03	<0.3	<0.3	3–4

Iron alloy

	Fe	C	S	Cu	P	Mn	Si
%	>95	2–3	<0.8	<0.6	<0.1	<0.3	<0.1

of iron is high enough to go back to steel industry as scrap substitute. Slag which is generated in a comparable small amount is of a quality that allows a further utilization. Lead chloride from the first step represents a product of low quality and value but can be treated in specialized areas of lead industry [6].

The process has been verified in pilot scale treating batches of 1000 kg, essential process steps are patented. Table 1 shows the quality of the produced zinc oxide as well as the one of the iron product.

Referring to the questions discussed in the introduction, the following can be stated:

- Zero waste can be realized; at least, the amount of possible residues is very low.
- The amount of carbon used as reducing agent is similar to the one of Waelz-kilns but in case of 2sDR. an iron product is formed aside the zinc oxide leading to a lower specific CO_2-footprint
- Also. the utilization of biochar has been successfully tested. The higher reactivity as well as the lower physical strength does not have a negative impact
- Required energy is provided mainly by electricity. In best case (like, e.g., possible in Austria), it is 100% out of "green" sources and does not have a relevant CO_2-footprint
- Overall energy consumption is of course higher compared to the state of the art Waelz process but is related to more products and with this at the end realizes a more sustainable energy utilization.
- Investigations are ongoing to implement hydrogen into the concept to allow an alternative CO_2-neutral solution for the future when hydrogen prices are in a more rational range.

A very important additional advantage is the higher quality of the zinc product. This quality does not only offer to skip the often limiting roasting process when utilizing it at primary zinc smelters but also could allow a further amelioration of the quality and the production of zinc compounds which may allow other, more profitable, markets.

Summary

Zero waste has been an important aim in metallurgy for many years. For a few years, now the CO_2-footprint topic occupies the scientific community in combination with possible utilization of hydrogen. State of the art processes for steel mill dust recycling do not really fulfil any of these two requirements what might cause difficulties in future when facing high landfilling costs and rising CO_2-certificate prices.

The 2sDR-process tries to realize a zero waste concept including a low CO_2-footprint. Energy is used efficiently, and the high zinc oxide quality offers more sophisticated utilization either in primary zinc industry or maybe alternative markets.

In any case, current and future requirements regarding environmental responsibility and climate protection will force the metallurgical business to optimize or even change processes. To make use of by-products as an important metal source for the future, such as steel mill dust, it will need highly sophisticated concepts to be competitive and economic feasible.

References

1. Antrekowitsch J, Rösler G (2015) Steel mill dust recycling in the 21st century. In: Proceedings of XXIX. microCAD international multidisciplinary scientific conference. Miskolc, Hungary
2. Hydrogen insights 2021 report, the Hydrogen Council, 2021. https://hydrogencouncil.com/wp-content/uploads/2021/02/Hydrogen-Insights-2021-Report.pdf
3. Daily carbon prices, EMBER, Internet. https://ember-climate.org/data/carbon-price-viewer/
4. Griessacher T, Antrekowitsch J, Steinlechner S (2012) Charcoal from agricultural residues as alternative reducing agent in metal recycling. Biomass Bioenerg 39:139–146
5. Southwick L (2008) From dust to iron: new directions in EAF dust processing. In: Proceedings of scrap substitutes and alternative ironmaking V. Baltimore
6. Auer MJ (2019) Antrekowitsch und G. Hanke: "2sDR" an innovative mini mill concept for EAF-dust recycling. WASTES—solutions, treatments, opportunities

Characterization and Thermal Treatment of Electric Arc Furnace Dusts Generated During Steel Production in Peruvian Industries

Mery C. Gómez-Marroquín, José Carlos D´Abreu, Roberto de Avillez, Sonia Letichevsky, Abraham J. Terrones-Ramires, and Kim J. Phatti-Satto

Abstract Self-reduction is a pyrometallurgical treating process that aims to valuable metal recovery from mining-metallurgical industry wastes, mainly from steelmaking industries. Electric Arc Furnace Dusts (EAFD) are still the most attractive materials to be tested in using this technique, due to their high magnetite and franklinite/zinc ferrite contents. This research will address the reuse of these co-products in steel plants, providing added value to this material that until now is constituted as an environmental liability of considerable economic importance in steelmaking industries. Chemical and microstructural analysis has determined high contents of iron and zinc from magnetite and franklinite/zinc ferrite. Iron was present in the non-stoichiometric form of "hapkeite" ($Fe_{1.34}Si_{0.06}$) in both EAFD 1 and EAFD 2. A rare appearance of Moissanite CSi –2H was also found in EAFD 1. Thermogravimetric evaluations allowed elimination of almost 15% of volatile matter at 1000 °C in EAFD 1. EAFDs were partially reduced and showed a high porosity, which would make it possible for the recovery of its main metal content by carbothermic self-reduction. Proximate analysis and carbon dioxide reactivity of two reductants were tested for evaluating the behavior of selected reductants in carbothermic self-reduction of EAFDs using a procedure given by the Steelmaking and Ironmaking Group of DEQM PUC/RJ. This mixture included 85% (EAFD + coal), 6% CPV ARI, and 9% water. Operational Diagram of Phase Predominance (ODPP) from the Zn–Fe–C–O system was used to calculate the required carbon and to guarantee the occurrence of the global chemical reactions of carbothermic reduction either in franklinite/zinc ferrite as in magnetite by 100% CO and temperatures between 1000 and 1100 °C. In these conditions, self-reducing briquettes of EADF 2 lost more weight so reacted faster than EAFD 1. Finally, reactions rates of carbothermic self-reducing briquettes EAFDs were very fast during the first 5 min and retarded from 5 to 40 min.

M. C. Gómez-Marroquín (✉) · A. J. Terrones-Ramires · K. J. Phatti-Satto
National University of Engineering, 210, Túpac Amaru Ave. Rímac, LIMA 25 Lima, Perú
e-mail: mgomezm@uni.edu.pe

J. C. D´Abreu · R. de Avillez · S. Letichevsky
Pontifical Catholic University of Rio de Janeiro, 225, Marquês de São Vicente Street, Gávea, RJ 22451-900, Brazil

A. Lazou et al. (eds.), *REWAS 2022: Developing Tomorrow's Technical Cycles (Volume I)*, The Minerals, Metals & Materials Series,
https://doi.org/10.1007/978-3-030-92563-5_33

Keywords Self-reduction · Carbothermic reduction · Electric arc furnace dusts-EAFD · Steelmaking residues · Processing and extraction · Environment effects · Recycling and secondary recovery

Introduction

Self-reduction is a pyrometallurgical treating process that aims to recover valuable metal recovery from mining-metallurgical industry wastes, mainly from steelmaking industries. Electric Arc Furnace Dusts (EAFD) are still the most attractive materials to be tested in using this technique, due to their high magnetite and franklinite/zinc ferrite contents. EAFD is the main solid waste of EAF steelmaking, 10 to 20 kg EAFD for each one-ton of steel produced by EAF is produced [1]. EAFD recycling involves finding the optimal briquette composition, in the form of an agglomerate made at high pressure (self-reducing briquette), for recovering some elements such as Fe, Zn, Pb, Cr, V, and W [2, 3]. As it is known, carbothermic self-reduction evidences higher reduction efficiency compared with conventional carbothermic reduction. While, the applicability of bio-reductants to smelting is promising, challenges remain, and the pre-conditioning of the main reductants for recovering the unrecoverable metals from steelmaking wastes. According to previous information, mineral metallurgical coke has exhibited lower reduction efficiency than charcoal, bio charcoal, or wood charcoal [4]; for that, they can act as bio-reductants in carbothermic self-reduction eco-sustainable processes via microwave energy (heating) or thermogravimetric analysis [5]. The aim of this paper was to characterize EAFDs using different techniques such as chemical analysis, X-ray fluorescence spectrometry (XRF), X-ray diffraction (XRD), scanning electron microscopy (SEM) including X-rays energy dispersive spectrometry (EDS), thermogravimetric analysis (TGA, DTG, and DSC), and physical determinations (specific gravity and average particle size), and to treat EAFDs by carbothermic self-reduction using a previous proximate analysis and carbon dioxide reactivity of coals, a mixture of 85% (EAFD + coal), 6% CPV ARI, and 9% water to make self-reducing briquettes, ODPP from the Zn--Fe–C–O system with optimal conditions:100%CO and temperatures between 1000 and 1100 °C. This research also will address the reuse of these co-products in steel plants, providing added value to this material that until now is constituted as an environmental liability of considerable economic importance in steelmaking industries.

Experimental Procedures

The experiments that were part of this research involved a mechanical preparation and complete characterization of two types of samples of Peruvian Electric Arc Furnace Dusts–EAFDs (EAFD 1 and EAFD 2) and coals (wood charcoal-WCh and mineral metallurgical coke-MMC). In addition, thermal treatments involving carbothermal self-reduction techniques using a mixture of EAFD, coal, CPV ARI, and water were conducted to evaluate the behavior of the agglomerates or self-reducing briquettes.

Chemical Compositions

Chemical analyses were developed using an Optical Emission Spectrometer Jarrell Ash USA model Ebert 3.4 m for determining the elemental composition and an Atomic Absorption Spectrometer Shimadzu (FAAS) model AA-7000 for estimating of the main oxides of EAFDs.

X-Rays Diffraction Analysis

XRD analyses of EAFDs were performed using a D8 Discover diffractometer, Bruker, with CuK-α radiation ($\lambda = 1.5418$ Å), Ni filter, and Lynxeye detector, operating at 40 mA and 40 kV of XRD patterns were collected in the 2θ range of 10–90° with 0.02° increment and time of 3 s/step. An energy discriminator was employed to reduce the iron fluorescence. The identification of the crystalline phases was carried out by comparison with standard patterns with the EVA program, Bruker. The crystal structure of the most probable compounds was taken from the Inorganic Crystal Structure Database (ICSD). The experimental diffraction patterns were fitted by the Rietveld method with fundamental parameters using the TOPAS program, Bruker.

SEM–EDS Observations

Scanning Electron Microscope (SEM), FEI Quanta 650 high vacuum with 80 amps of current and 20 kV voltage were used to analyze the main morphologies present in EAFD. The SEM was equipped with, SEM-EDAXEDS detector and the results were analyzed using the quantification software TEAM.

Thermogravimetric Evaluations

Thermal characterization of EAFD 1 was performed using Differential Calorimetric Scanning Analysis—DSC, Thermal Differential Gravimetric Analysis—DTG and Thermal Gravimetric Analysis—TG from the room temperature to 1000 °C. This equipment: NETZSCH STA 449 F3 was composed of combined techniques of DSC, TG, and DTG, which are certified with ASTM E1131-08 Standard. The tests were carried under nitrogen atmospheres with a flow of 50 mL/min and a heating rate of 10 K/min.

Physical Determinations

Some physical properties of EAFDs were estimated, i.e., specific gravity and average particle size. The specific gravity was calculated using the pycnometer method. For the case of the granulometric analysis for estimation of the average particle size, a series of ASTM meshes were used, m14, m20, m30, m50, m70, m100, m140, m200, m270, m400, and background of -m400.

Proximate Analysis and Carbon Dioxide Reactivity Tests of Coals

Proximate analysis of coals was performed in order to determine moisture, volatile matter, ash and fixed carbon percentages according to ASTM standards: 7582, D2013, and D346, which consisted of heating 1 g of wood charcoal-WCh and mineral metallurgical coke-MMC in triplicate (three samples of coals in their respective crucibles) and to heat them at 900° C for 5 h into the electric furnace and thus eliminate not only all the humidity also volatile matter present in the sample of coals [6]. On the other hand, based on the ASTM D 5341 standard, carbon dioxide reactivity tests of dried samples of wood charcoal-WCh and mineral metallurgical coke-MMC were analyzed. 5 g were taken from these two samples of reductants, in the form of a cylindrical agglomerated briquette. These cylindrical agglomerated briquettes were prepared using a hydraulic press PAVITEST-brand with a maximum load of 24,000 kg-f, at a briquetting pressure of 6000 kg-f. These agglomerates were taken over a crucible and carried into a COMBUSTOL tubular electric furnace brand previously heated to a temperature of 1020 °C. The carbon dioxide reactivity test took place for 120 minutes under a constant 90 mL/min flow of carbon dioxide (CO_2) without. These test was carried out for 15 min a 50 mL/min flow of nitrogen (N2), as before as after this experience [6]. To determine the reactivity of these coals, formula (1) was used.

$$R = \left(\frac{A - B}{A}\right) \times 100 \tag{1}$$

where: A is the initial mass and B is the final mass of the coal sample before the CO_2 reactivity test, and R is the CO_2 reactivity.

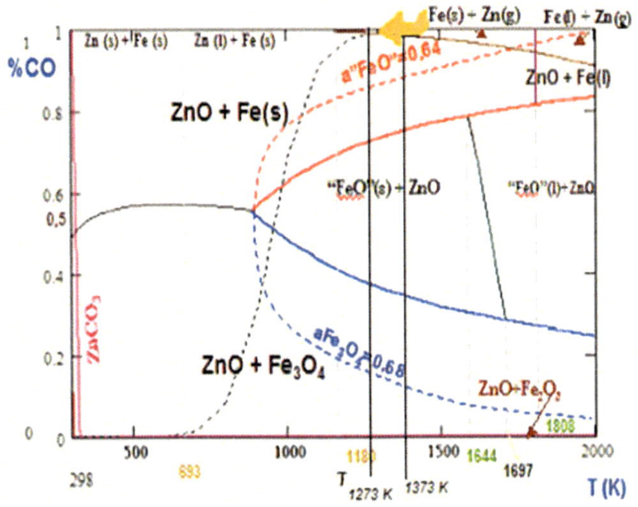

Fig. 1 ODPP: Zn–Fe–C–O, indicating the optimal thermodynamics conditions for the carbothermic self-reduction of EAFDs: %CO = 100 or PCO = 1 between 1000 °C (1273 K) and 1100 °C (1373 K) [7]

Thermodynamics Approach of Carbothermic Self-Reduction

In Fig. 1, the ODPP of the Zn–Fe–C-O system is shown, resulting from the superposition of two other ODPPs (Zn–C–O and Fe–C–O), where the equimolar composition of the ferrite of zinc and the values of the activities achieved by other reducible species in CO and CO_2 atmospheres such as wustite and magnetite is shown. In addition, the tracing of two isotherms, one at 1000 °C (1273 K) and the other at 1100 °C (1373 K), was used to locate the zone of the predominance of Zn (g) and Fe (s) phases for % CO = 100 or PCO = 1, in order to guarantee the complete metallization of the self-reducing briquettes to be formed, which are also observable in ODPP [7].

Routine for Making Self-reducing Briquettes with a Mixture of EAFD, Coal, CPV ARI, and Water

Self-reducing briquettes were made according a to procedure given by the Steelmaking and Ironmaking Group of DEQM PUC/RJ Brazil. It consists in preparing a mixture of 85% (EAFD + coal) (w/w), 6% CPV ARI (w/w), and 9% water (w/w) [8]. This routine is shown in Fig. 2.

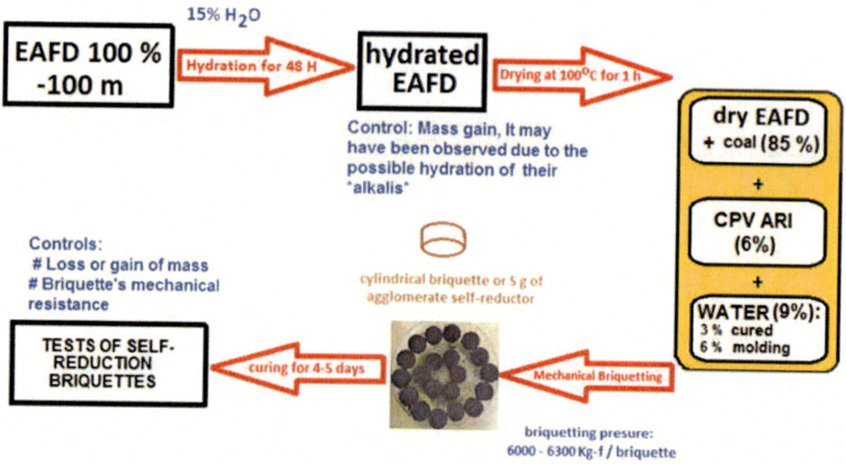

Fig. 2 Routine for making self-reducing briquettes from the mixture of EAFD, Coal, CPV ARI and water

Carbothermic Self-Reduction Tests

Carbothermic self-reduction tests using two types of EAFDs (EAFD 1 and EAFD 2) and only one coal (wood charcoal-WCh) were conducted in a COMBUSTOL tubular electric furnace nitrogen gas flow between 50 and 70 mL/min., continuously to avoid oxidation of the sample with the outside air. The working temperatures were 1000 °C and 1100 °C, and the times were 5, 10, 20, 30, 40 min, with about 30 min of preheating. The conversions or carbothermic self-reduction percentages of EAFDs were evaluated through their weight loss expressed as a percentage, as indicated in formula (2).

$$\alpha = 100\left(\frac{\Delta W_t}{W_i}\right) = \left(\frac{W_i - W_t}{W_i}\right) \tag{2}$$

where: Wi is the initial weight of the sample, Wt is the final weight of the reduced sample at time t, ΔWt is the total loss weight at time t, and α is the carbothermic self-reduction percentage.

Results and Discussion

Chemical Compositions

In Table 1, chemical compositions of the samples were determined for some elements: Zn, Fe, Cu, Cd, Pb, Ni, As, Cr, and Hg, and oxides: SiO_2, Al_2O_3, CaO, MgO, MnO, Na_2O, and K_2O.

X-Ray Diffraction Analysis

Figure 3 shows experimental and Rietveld refinement X-ray diffraction patterns (Y calculated) of EAFD 1 and EAFD 2, and their corresponding difference.
In Tables 1 and 2, chemical and microstructural characterizations have determined high contents of iron and zinc in both materials EAFD 1 and EAFD 2. The presence of iron and zinc is observed in their main oxides: franklinite/zinc ferrite, magnetite, and hematite. Also, iron and silicon were found in the form of a non-stoichiometric phase called "hapkeite" ($Fe_{1.34}Si_{0.06}$) in EAFD 1 and EAFD 2. A rare appearance of Moissanite CSi–2H was also confirmed in the XRD of EAFD 1. The most abundant phase present in the EAFDs was magnetite which is between 57.74 and 63.12%. The other two more significant present phases are franklinite/zinc ferrite-$ZnFe_2O_4$ and zinc oxide/zincite-ZnO. However, it is not possible to distinguish magnetite and franklinite/zinc ferrite by X-ray diffraction, since both oxides (magnetite and franklinite/zinc ferrite) have the same space group and further identical crystal structure and lattices (Fd-3mZ). On the other hand, carbon is still detected as carbonates and may be also present as an amorphous phase which was not observed in the X-ray diffraction patterns. EAFD 1 has more carbon content than EAFD 2. The Goodness of Fit (GOF) of the Rietveld method refinement showed a good adjustment because it was between 1 and 1.5.

SEM–EDS Observations

Figure 4 show the elemental mapping of SEM images of the EAFD 1 sample: (a) selected SEM image, (b) mapping by various chemical elements, and (c) EDS of the selected SEM image for identified elements in certain analyzed spots or areas in EAFD 1 sample. Several circular grains probably of metallic iron Fe (ccc) are observed, because they show some typical spherical morphologies of metallic iron partially oxidized to the form of magnetite and franklinite/zinc ferrite. SEM images also show clustered branches with large internal poly-directional voids within other acicular aggregates of possible zincite or zinc oxide. As magnetite is the most abundant mineralogical phase contained in EAFDs, probably it has been formed at the expense of metallic iron, forming spherical concentric layers and others in the form

Table 1 Chemical compositions of elements and oxides of EAFD 1 and EAFD 2

Samples	% Zn	% Fe	% Cu	% Cd	% Pb	% Ni	% As	% Cr	% C	% SiO_2	% Al_2O_3	% CaO	% MgO	% MnO	% Na_2O	% K_2O	Hg (ppm)
EAFD1	5.45	21.95	0.19	0.08	1.15	0.02	0.01	0.42	0.04	5.50	0.75	2.80	7.24	3.21	2.06	0.71	1.75
EAFD2	16.03	31.20	0.20	0.04	1.09	0.02	0.01	0.81	0.01	2.78	0.02	4.81	2.47	2.10	1.75	0.80	0.72

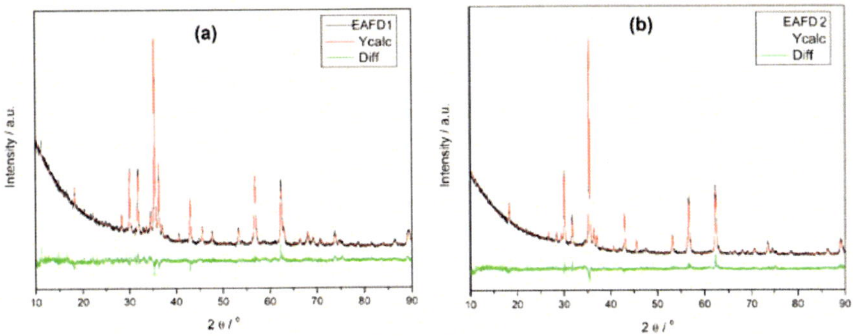

Fig. 3 XRD patterns black of EAFD 1 (**a**) and EAFD 2 (**b**) samples including the fitting (red) and difference (green) obtained by the Rietveld refining method (Color figure online)

Table 2 Summarizes detailed information on the crystal phases obtained by Rietveld method refinement from EAFD 1 and EAFD 2 samples

Samples	Phase	Chemical formula	Phase %wt	Space group	LVOLIB/nm	GOF	R_{wp}	Rexp
EAFD1	Franklinite	$ZnFe_2O_4$	16.94	Fd-3mZ	69.57	1.37	5.79	4.23
	Fe-ccc	Fe	0.52	Im-3 m	16.84			
	Zincite	ZnO	10.04	P63mc	33.081			
	Pyrolusite	MnO_2	5.67	P42/mnm	5.21			
	Hapkeite	$Fe_{1.34}\ Si_{0.66}$	2.53	Pm-3 m	59.07			
	Moissanite	CSi-2H	2.54	P63mc	12.73			
	Magnesite	$MgCO_3$	4.01	R-3cH	59.77			
	Magnetite	Fe_3O_4	57.74	C1c1	37.90			
EAFD2	Franklinite	$ZnFe_2O_4$	22.56	Fd-3mZ	90.12	1.29	5.18	4.02
	Zincite	ZnO	2.87	P63mc	41.74			
	Pyrolusite	MnO_2	3.81	P42/mnm	94.50			
	Hapkeite	$Fe_{1.34}\ Si_{0.66}$	1.68	Pm-3 m	102.95			
	Calcite	$CaCO_3$	3.88	R-3cH	40.97			
	Magnetite	Fe_3O_4	63.12	C1c1	41.55			
	Quartz	SiO_2	1.47	P3221	284.25			
	Hematite	Fe_2O_3	0.60	R-3cH	–			

of whiskers due to the presence of the gas bubble blow or burst reducer in the solidification of steels [9]. While, franklinite/zinc ferrite cluster type micrometric and homogeneous agglomerate can be also observed [7]. Franklinite/zinc ferrite may appear during the cooling of the gas excretion system of the EAF from a solid–solid reaction between zincite and hematite at temperatures between 600 and 700 °C, which is why franklinite/zinc ferrite is always associated with these two compounds

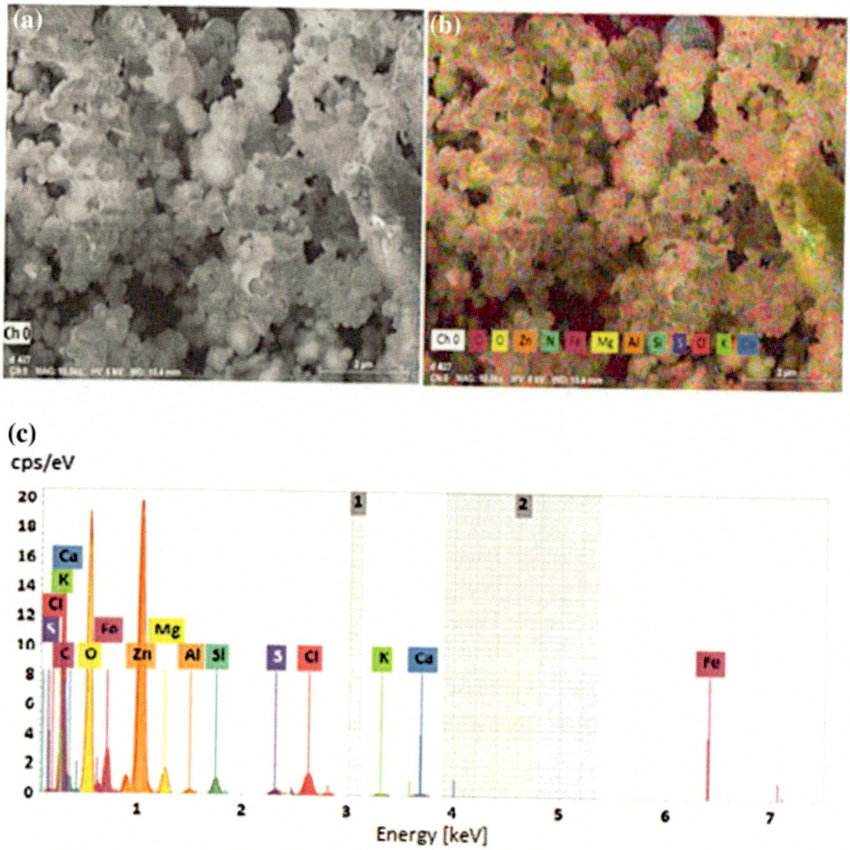

Fig. 4 Selected SEM image (**a**), mapping multi-elemental (**b**), and quantification of elements identified by EDS of selected SEM image of EAFD 1 sample (**c**)

(zincite and hematite) also showing cluster or cluster-like granular morphologies [10]. The elemental mapping of the selected SEM image show a significant presence of zinc because the EDS was taken from an area of the sample richer in zinc than iron. SEM–EDS analysis in the EAFD1 sample showed the absence of Cu, Cd, Pb, Ni, As, and Cr in chemical analysis.

Table 3 summarizes EDS results of elemental mapping and quantification of elements identified in Fig. 4c.

Table 3 EDS of elemental mapping and quantification of elements identified in Fig. 4c

Element	At. no	Net	Mass, %	Mass Norm., %	Atom, %	Abs. error [%] (1 sigma)	Rel. error [%] (1 sigma)
C	6	65,833	12.66	12.95	27.34	1.49	11.76
O	8	106,175	27.54	28.18	44.66	3.12	11.32
Zn	30	162,541	33.92	34.71	13.46	2.00	5.91
Fe	26	242,541	12.93	13.23	6.01	1.65	12.74
Mg	12	12,197	1.34	1.37	1.43	0.09	7.03
Al	13	3162	0.36	0.36	0.34	0.04	12.26
Si	14	8984	1.76	1.80	1.62	0.11	5.98
S	16	3652	0.64	0.66	0.52	0.06	8.69
Cl	17	17,138	4.07	4.16	2.98	0.21	5.05
K	19	1798	0.82	0.84	0.55	0.07	9.00
Ca	20	2061	1.70	1.74	1.10	0.13	7.90

Table 4 Physical determinations of EAFD 1 and EAFD 2

Samples	Specific gravity (10^3 kg/m^3)	Average particle size
EAFD 1	3.71	95%−m100
EAFD 2	3.43	26.74%+m10

Physical Determinations

In Table 4, the results of physical determinations of EAFD 1 and EAFD 2 are shown.

Proximate Analysis and Carbon Dioxide Reactivity Tests

Table 5 exhibits the results of proximate analysis and reactivity to carbon dioxide (CO_2) tests of coals used as reductants in carbothermic self-reduction of EAFDs.

As expected, wood charcoal-WCh showed greater reactivity and a significant amount of volatile matter and fixed carbon, characteristics that will significantly favor the reduction of iron and zinc oxides present in EAFDs in atmospheres of CO and CO_2, in this case, carbothermic self-reduction magnetite and franklinite/zinc ferrite.

Table 5 Proximate analysis of coals and carbon dioxide reactivity test of two types of reductants samples

Samples	% Moisture	% Volatile matter	% Ash	% Fixed carbon	% Reactivity
MMC	4.91	5.16	80.01	9.92	91.40
WCh	7.38	39.68	41.41	11.53	95.62

Table 6 Mass loss values registered during heating of EAFD 1

Sample	Initial mass sample, mg	Lost mass (%)			Residual mass, %
		245 °C	785 °C	1000 °C	
EAFD 1	11.90	1.10	3.98	15.24	84.76

Thermogravimetric Evaluations

Thermogravimetric evaluations showed three significant exothermic peaks that are related to decomposition temperatures and probably reduction of iron and zinc oxides. These temperatures were 189, 720, and 942 °C. The initial 1.10 % mass loss of EAFD 1 was due to the presence of water, as well as some volatile matter, followed by 3.98% of mass loss due to volatilization of some metals, CO, and CO_2, and 15.24% of mass loss for the liberation of gases Zn, CO, and CO_2. Therefore, it is observed that exothermic reactions occurred when heating the EAFD 1 sample from room temperature to 1000 °C, and it is known that the iron oxide and zinc oxide carbothermic reduction follows the Boudouard reaction [11]. Two chemical reactions could have occurred in this process: (a) the instantaneous decomposition of franklinite/zinc ferrite, present in EAFD 1, to their constituent oxides, iron oxide (Fe_2O_3-hematite) and zinc oxide (ZnO), and (b) carbothermic reduction of some of its oxides: franklinite/zinc ferrite (16.94 % $ZnFe_2O_4$), zinc oxide (10.04 % ZnO) and magnetite (57.74 % Fe_3O_4), taking advantage its low content of C and iron, 0.04 % and 21.95 %, respectively. Table 6 shows the weight losses of the EAFD1 sample at different temperatures.

Carbothermic Self-reduction Tests

From this previous quantification by chemical and structural analysis, we have: (a) EAFD 1, contains metallic iron (0.52% Fe-ccc), 21.95% total Fe and 84.72% of reducible oxides, and (b) EAFD 2, containing Fe_2O_3 (0.60%), 31.20% total Fe and 89.16% of reducible oxides. Figure 5 shows carbothermic self-reduction curves of EAFD samples using wood charcoal (WCh) in two temperatures and five times of thermal treatment which were carried out using optimal thermodynamics conditions determined and shown in Fig. 1. Carbon used in self-reduction of EAFD 1 and EAFD 2 was wood charcoal because it holds a major reactivity in front of CO_2 compared with mineral metallurgical coke. As it can be observed, EADF 2 samples lost more weight and so reacted faster than the EAFD 1 sample. Also, reactions rates of carbothermic self-reduction of EAFDs were very fast during the first 5 min and retarded from 5 to 40 min. It was notable that the reaction rates increased with the increment of self-reduction times and temperatures. These conversions or carbothermic self-reduction percentage, α, turned out to be very low, and at the same time, they did not improve significantly as the carbothermal self-reduction time and temperature

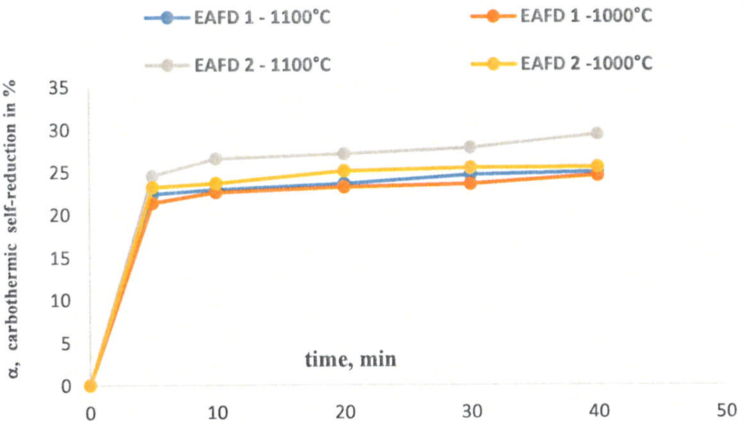

Fig. 5 Carbothermic self-reduction curves by CO of EAFDs samples

were increased, probably due to the insufficient amount of carbon in the mixture to guarantee the complete metallization of the self-reducing briquettes. Despite this, great release of volatile materials from the briquettes was observed, including gases such as Zn, CO, and CO_2 that emanated mainly from the central part, both from the top and bottom of the briquette. Due to the high reaction kinetics that these geometries theoretically observe (flat circular surfaces of cylindrical briquettes), most of the briquettes presented deep and long fractures. A poly-directional growth of acicular crystals of whitish to yellowish color of probabe zincite or redeposited zinc present in both EAFDs was observed on all the flat surfaces of the briquette and even the base of the crucible. This yellow color varied in intensity depending on the time and temperature of the self-reduction, for example, the self-reduced briquettes of EAFD 2 light yellow and those of EAFD 1 more intense yellow color. An alternative treatment route for these EAFDs should include a prior magnetic concentration to separate magnetite of franklinite/zinc ferrite then to perform individual carbothermal self-reduction of magnetite and franklinite/zinc ferrite.

Conclusions

Chemical and structural characterizations have determined high contents of iron and zinc in both materials EAFD 1 and EAFD 2. The presence of iron and zinc is observed in their main oxides: franklinite/zinc ferrite, magnetite, and hematite. Also, iron and silicon were found in the form of a non-stoichiometric phase called "hapkeite" ($Fe_{1.34}Si_{0.06}$) in EAFD 1 and EAFD 2. Both samples of EAFD 1 and EAFD 2 are partially reduced as magnetite and franklinite/zinc ferrite phases and have a high porosity, which would make carbothermal self-reduction possible. Thermodynamic considerations included ODPP from the Zn–Fe–C–O system to calculate required carbon for carbothermic self-reduction of magnetite and franklinite/zinc ferrite. These optimal thermodynamics conditions for the process were 100% CO or PCO = 1 atm and a range of temperature between 1000 °C (1273 K) and 1100 °C (1373 K). Atomic Absorption and DRX analyses had determined high contents of iron and zinc. Iron was present in the non-stoichiometric form of "hapkeite" ($Fe_{1.34}$ $Si_{0.06}$). in both EAFD 1 and EAFD 2. A rare appearance of Moissanite CSi –2H was also found in EAFD 1. Thermogravimetric evaluations allowed the elimination of almost 15% of volatile matter at 1000 °C in EAFD 1. From this previous quantification by chemical and structural analysis, we have: (a) EAFD 1, contains metallic iron (0.52% Fe-ccc) of 21.95% total Fe, for that 84.72% of reducible oxides while, (b) EAFD 2, contains to Fe_2O_3 (0.60%) of 31.20% total Fe, in consequence, 89.16% of reducible oxides, by 100%CO at 1000 and 1100 °C of temperature. As expected, wood charcoal-WCh showed greater reactivity and a significant amount of volatile matter and fixed carbon, characteristics that will favor the self-reduction of magnetite and franklinite/zinc ferrite present in EAFDs. Carbon used in self-reduction of EAFD 1 and EAFD 2 was WCh because it holds a major reactivity in front of CO_2 compared with MMC. EADF 2 sample lost more weight so reacted faster than the EAFD 1 sample in this range of temperature. Reaction rates of carbothermic self-reduction of EAFDs were very fast during the first 5 min and retarded from 5 to 40 min. An alternative treatment route for these EAFDs should include a prior magnetic concentration to separate magnetite of franklinite/zinc ferrite then to perform individual carbothermal self-reduction of magnetite and franklinite/zinc ferrite.

Acknowledgements Authors thank the Peruvian Agency CONCYTEC/FONDECYT for the financial assistance granted for the training of the main researcher in techniques of self-reduction and X-ray Diffraction Laboratory, Lab XRD at Chemical Engineering and Materials Science Department (DEQM) of PONTIFICAL UNIVERSITY OF RIO DE JANEIRO (PUC/RJ), Brazil, without which these programmed experiences could not have been carried out and two Peruvian Steelmaking Companies for the background and the samples provided to this research.

References

1. Wang J, Zhang Y, Cui K, Fu T, Gao J, Hussain S, AlGarni TS (2021) Pyrometallurgical recovery of zinc and valuable metals from electric arc furnace dust–a review. J Clean Prod 298:126788. https://doi.org/10.1016/j.jclepro.2021.126788
2. Coradini DSR, de Lima ARR, Deike R, Telles VB, de Oliveira JR (2020) Comparative study of Cr, V and W recovery from a self-reduction briquette and a steel-briquette system. J Market Res 9(4):7508–7517
3. Ri SC, Chu MS, Chen SY, Liu ZG, Hong H (2016) Self-reduction mechanism of coal composite stainless steel dust hot briquette. J Iron Steel Res Int 23(4):314–321
4. Ye Q, Li G, Peng Z, Lee J, Lin X, Rao M, Jiang T (2019) Microwave-assisted self-reduction of composite briquettes of zinc ferrite and carbonaceous materials. Powder Technol 342:224–232
5. de Medeiros, GA, da Silva LM, Fernandes MT, dos Santos DGDS, de Castro JA (2021) Analysis of the carbothermic reduction of iron ore-coke composite mixtures by microwave heating. Mater Res 24(6):e20210284

6. Alvarado QJE (2012) Reatividade ao CO_2 de Carvoes Minerais de Biomasas e Coques. Dissertação de Mestrado do Departamento de Engenharia de Materiais. Programa de Pós-gradução em Engenharia de Materiais e Processos Químicos e Metalúrgicos—DEMa da Pontificia Universidade Católica de Rio de Janeiro, PUC-Rio. 2012. Orientador: Prof. D'Abreu José Carlos
7. Gómez-Marroquín MC (2008) Characterization and zinc ferrite contained into steelmaking dusts reduction, by CO-CO_2 gás mixtures. Doctoral thesis—department of science of the materials and metallurgy, Pontifical Catholic University of Rio de Janeiro. DCMM—PUC-Rio, 2008, 230 p. Adviser: Prof. D'Abreu José Carlos
8. Alvarado QJE (2017) Reciclagem dos pós do despoeiramento a sêco de aciaria LD e de alto forno na refrigeração de aço líquido e nos processos de autorredução. Tese de Doutorado do Departamento de Engenharia de Materiais. Programa de Pós-gradução em Engenharia de Materiais e Processos Químicos e Metalúrgicos - DEMa da Pontificia Universidade Católica de Rio de Janeiro, PUC-Rio. 2017, 195 p. Orientador: Prof. D'Abreu José Carlos
9. Buboltz Ferreira F, Deves Flores B, Osorio E, Faria Vilela AC (2018) Carbothermic reduction of electric arc furnace dust via thermogravimetry. Metallurgy and materials. REM Int Eng J Ouro Preto 71(3):411–418
10. Gómez-Marroquín MC (2004) Contribution to the study of the formation and reduction of zinc ferrite. Thesis dissertation—department of science of the materials and metallurgy, Pontifical Catholic University of Rio de Janeiro. DCMM—PUC-Rio, 2004, 179 p. Adviser: Prof. D'Abreu José Carlos
11. Omran M, Fabritius T, Heikkinen EP (2019) Selective zinc removal from electric arc furnace (EAF) dust by using microwave heating. J Sustain Metall 5(3):331–340

Investigation of Hydrometallurgical Recycling Parameters of WC–Co Cutting Tool Scraps

Hakan Kuşdemir, Kagan Benzesik, Ahmet Turan, and Onuralp Yücel

Abstract The aim of the study is recovery of WC–Co cutting tool scraps through hydrometallurgical methods. Scraps contain 83.67 wt.% W, 5.56 wt.% C, and 8.76 wt.% Co. The study consists of two progressive stages. The first stage is based on selective leaching of cobalt and separation from tungsten carbide. The second stage is the precipitation of cobalt from leach solution. Before experiments, scraps were grinded and sieved to lower than 250 μm particle size. In the first stage, WC–Co powders were leached in HNO_3 solution. In order to achieve maximum dissolution of cobalt, acid concentration, temperature, stirring rate, and raw material particle size parameters were investigated. In the second stage of the study, dissolved cobalt ions from leach solutions were precipitated as $Co(OH)_2$ compound by using NaOH solution. Solution pH and temperature parameters were examined to reach maximum cobalt recovery from the solution. According to results, the optimum leaching conditions were found to be 2M HNO_3, 25 °C, 800 rpm, 2 h, and 1/10 solid/liquid ratio. Cobalt dissolution efficiency was detected as 94.87% under these conditions. Besides, dissolved cobalt in the leach solution was precipitated with an efficiency of more than 99.99% in the form of cobalt hydroxide by the addition of NaOH reagent at pH = 11 and 50 °C.

Keywords Recycling · Hydrometallurgy · Sustainability

Introduction

WC–Co based materials, also called cemented carbides, are generally used in metal cutting and forming industries because of their outstanding mechanical properties. Moreover, they are also commonly used in petroleum and gas exploration works

H. Kuşdemir (✉) · K. Benzesik · O. Yücel
Metallurgical and Materials Engineering Department, Faculty of Chemical and Metallurgical Engineering, Istanbul Technical University, 34469 Istanbul, Maslak, Turkey

A. Turan
Materials Science and Nanotechnology Engineering Department, Faculty of Engineering, Yeditepe University, 34755 Istanbul, Atasehir, Turkey

A. Lazou et al. (eds.), *REWAS 2022: Developing Tomorrow's Technical Cycles (Volume I)*, The Minerals, Metals & Materials Series,
https://doi.org/10.1007/978-3-030-92563-5_34

and in mining industries [1]. WC–Co based materials are produced by the liquid phase sintering process [2]. Cobalt is added as a binder metal which increases the facture toughness of the WC main matrix. Generally, 5–20 wt.% cobalt is added to cemented carbides. While increasing cobalt ratio reduces the hardness and abrasion resistance of the material, it increases the fracture toughness of the material [3]. Thus, mechanical properties can be adjusted by cobalt content of WC–Co based materials depending on the application area of the final product.

Both tungsten and cobalt are defined as strategic metals because of their critical application areas in industry [4]. Tungsten is widely used in cemented carbide industry, steel alloys, and mining industry. Cobalt is also critical metal for the rechargeable battery industry. Moreover, magnet and stainless-steel industries are other important cobalt usage areas. It is estimated that cobalt abundance in the earth's crust is approximately 20 ppm [5]. In addition, tungsten abundance is less than 1.5 ppm which is lower than that of cobalt [3]. Hence, worldwide reserves of tungsten and cobalt are limited. Also, even high concentration tungsten and cobalt ores contain much lower metal content than a typical cemented carbide scrap. Tungsten and cobalt extraction involves complex process steps and excessive chemical usage which affects the environment negatively. Therefore, the recovery of these critical metals is essential in terms of economic and environmental concerns [6]. Due to its commercial importance, cemented tungsten carbide scrap recycling has been investigated for many years. There are many different types of conventional recycling methods in the literature. Each of them has specific metallurgical processes: pyro, hydro, and electrometallurgical processes or combination of them.

Cemented carbide recycling methods are generally divided into three categories which are direct, semi-direct, and indirect. In the direct recycling method, recycled products and scrap have the same chemical composition. Thus, the purity of the product is very high. Physical, chemical, and thermal applications or combination of them are used for that method. In the semi-direct method, the binder is separated from the main matrix by selective dissolution in a solution. When WC is suitable for re-use, the recovered binder material cannot be used directly. Indirect recycling processes are used to obtain pure tungsten products. By using chemical modification techniques, tungsten in the scrap transforms into ammonium paratungstate intermediate compound. After that, tungsten production methods from ores are applied to reach pure tungsten powder. Selection of WC–Co recycling method is applied in consideration of scrap characteristics, cost, and environmental concerns [7].

Selective cobalt leaching from cemented carbide scrap tenders an environmentally friendly process in view of its lower energy consumption and environmental impact due to the reduced number of process steps [8]. The aim of this study is recovery of tungsten and cobalt through hydrometallurgical route. The study consists of two subsequent stages. The first stage is based on the selective leaching of cobalt and separation from tungsten carbide. The second stage is the precipitation of cobalt from leach solution. To achieve maximum recycling efficiency, leaching and precipitation process parameters were investigated.

Experimental Procedure

Raw materials were supplied from Körfez Döküm Metal Machining Department. The scraps were used for turn lathes to cut different types of steel casting and white cast iron materials. Before experiments, scraps were grinded in a WC-based vibratory cup mill for 15 min. After grinding operations, scraps were screened by a Retzch brand sieve analyzer. Lower than 250 μm particle size powders were selected in order to use in further experiments. The objective of grinding process was to enhance leaching efficiency by increasing specific surface area. Table 1 shows the chemical analysis of scraps (AAS), and the XRD pattern of the raw material was given in Fig. 1. Table 2 shows the sieve analysis of the raw powders which were used in leaching experiments. Moreover, a typical bulk cutting tool scrap was investigated from outer surface to inner surface by using scanning electron microscopy. Figure 2 shows SEM micrograph and EDS analysis of a typical cutting tool scrap. From the micrograph, a typical scrap had approximately 4–6 μm thick multilayer TiN, Al_2O_3, and TiC coatings.

Table 1 Chemical composition of cutting tool scraps

Element	W	C	Co	Ti	Fe	Cr	Ni	Ta	Al
Wt.%	83.67	5.56	8.76	0.82	0.42	0.36	0.17	0.14	0.10

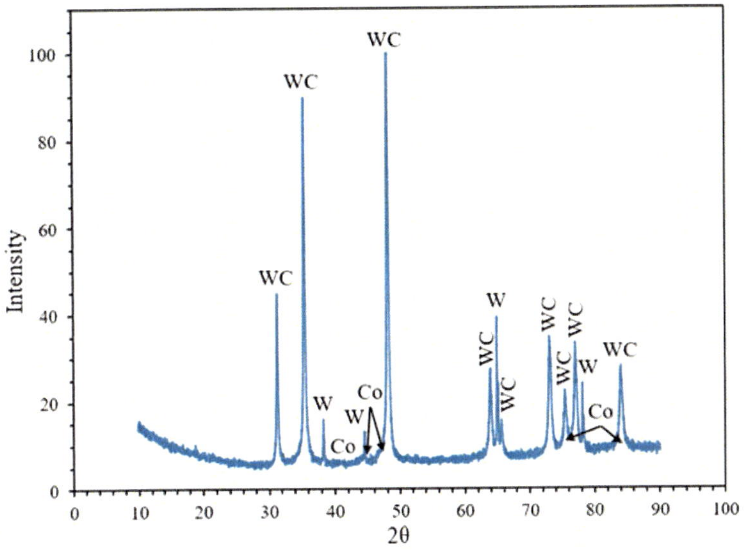

Fig. 1 XRD pattern of cutting tool scraps

Table 2 Sieve analysis of the raw material

Particle size, μm	Σ percentage retaining, %	Σ percentage passing, %
−250 + 106	67.04	100.00
−106 + 45	16.92	32.96
−45	16.04	16.04
Total	100.00	

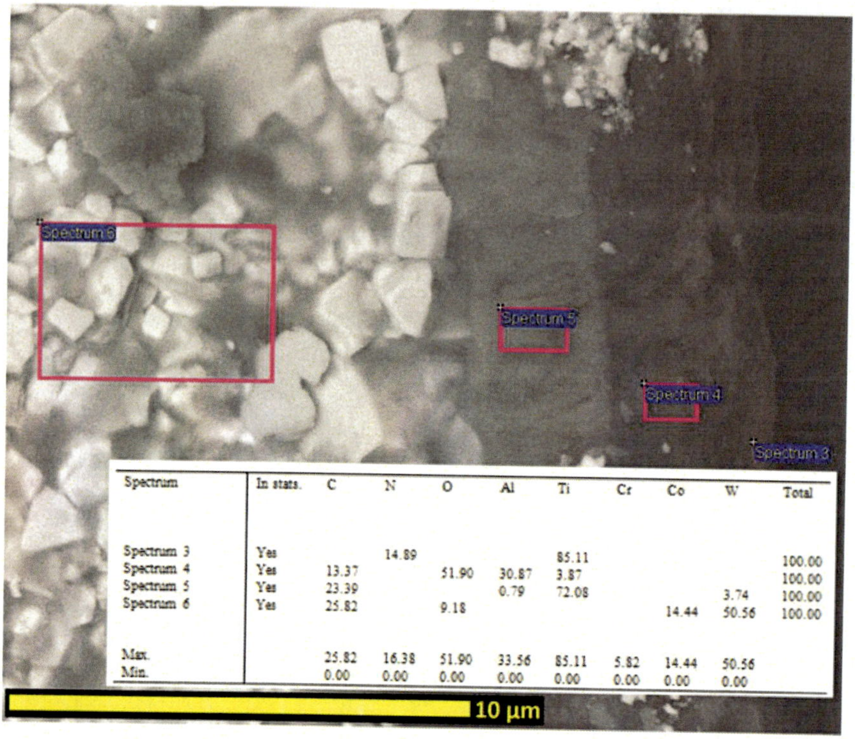

Spectrum	In stats.	C	N	O	Al	Ti	Cr	Co	W	Total
Spectrum 3	Yes		14.89			85.11				100.00
Spectrum 4	Yes	13.37		51.90	30.87	3.87				100.00
Spectrum 5	Yes	23.39			0.79	72.08			3.74	100.00
Spectrum 6	Yes	25.82		9.18				14.44	50.56	100.00
Max.		25.82	16.38	51.90	33.56	85.11	5.82	14.44	50.56	
Min.		0.00	0.00	0.00	0.00	0.00	0.00	0.00	0.00	

Fig. 2 SEM micrograph and EDS analysis of a typical cutting tool scrap

In the first stage of experiments, grinded waste cutting tools were leached with a HNO_3 solution. Leaching experiments were carried out on a Daihan brand heater-magnetic stirrer equipped with a K-type stainless-steel contact thermometer. The solution was stirred from the top with a teflon-plated propeller with adjustable speed and stirring direction. Acid concentration, leaching temperature, and stirring rate parameters were studied separately. Moreover, solid particle size on cobalt solubility efficiency was investigated.

In the second stage of the experiments, dissolved cobalt ions were precipitated as $Co(OH)_2$ compound. 2M NaOH solution was slowly added to HNO_3 acid concentration until reaching previously determined pH values. The aim of this process is to obtain maximum cobalt precipitation efficiency from cobalt pregnant leach solution. Temperature and pH parameters were investigated in precipitation experiments.

After experiments, all samples were filtered off and analyzed by Perkin Elmer Analyzed 800 brand atomic absorption spectrometer (AAS). Moreover, PANanalytical PW3040/60 brand X-ray diffractometer was used for phase identification of raw materials and products.

Results and Discussion

In the first stage, WC–Co powders leached in HNO_3 solution. The solid/liquid ratio was 1:10 g/ml in all leaching experiments. The purpose of the leaching process is the dissolution of cobalt entirely in acid solution. Thus, the remaining undissolved solid (filter cake) would consist of pure WC. To achieve that, the leaching process parameters were investigated. First, different acid concentrations of 0.5, 1 and 2M were tried to evaluate the effect of acid concentration in cobalt solution. Figure 3 shows that 2M HNO_3 solution had better cobalt solubility than 0.5 and 1M concentrations.

In order to investigate the effect of temperature change on cobalt dissolution, experiments were done between 25–60 °C (average particle size <250 μm, 2 h, 800 rpm). Figure 4 shows the effect of temperature on the dissolution of cobalt in the 2M HNO_3 solution. As seen from the figure, cobalt solubility decreases from 86.98 to 60.31% between the temperatures 25–60 °C. This can be attributed to the formation of a passive layer due to the chemical behavior of cobalt at relatively high temperatures, and this passive layer prevents the dissolution of cobalt by forming a protective layer between the liquid and the solid [8].

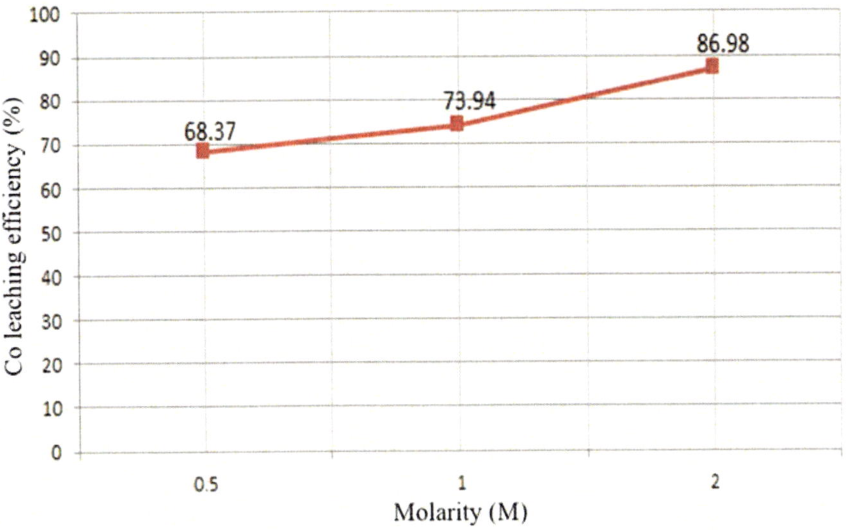

Fig. 3 Correlation between acid concentration and cobalt leaching efficiency

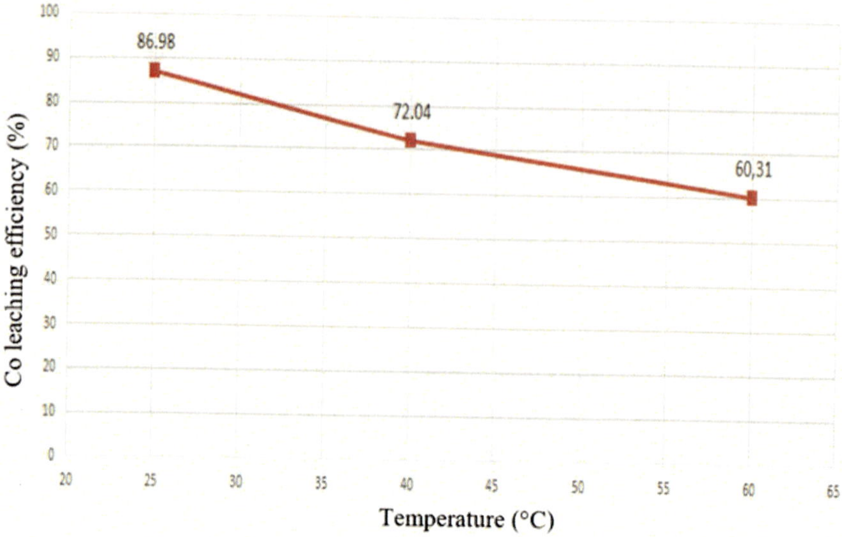

Fig. 4 Correlation between temperature and cobalt leaching efficiency

In the next stage of leaching experiments, the effect of stirring rate on cobalt solubility was investigated (Fig. 5). Different stirring rates were applied from 600 to 1200 rpm (average particle size <250 μm, 2 h, 800 rpm). 800 rpm stirring rate provided better solubility than 600 rpm and 1200 rpm stirring rates. It was observed

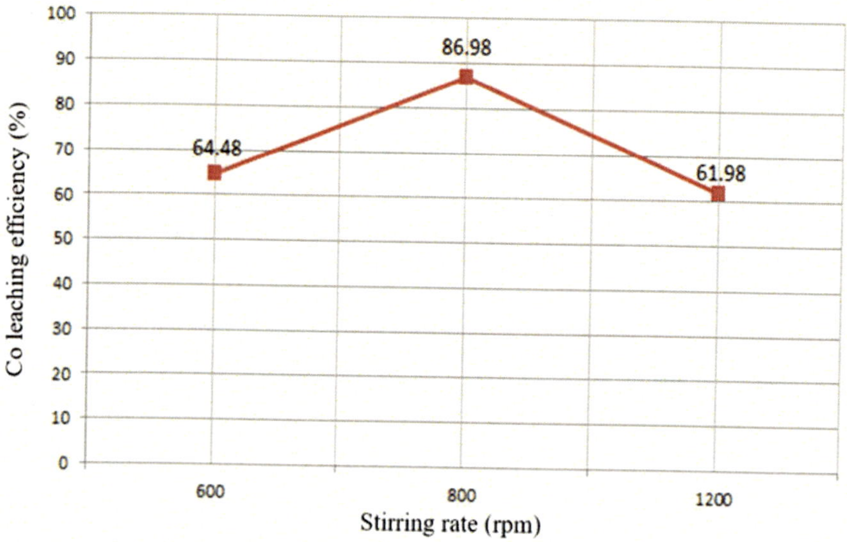

Fig. 5 Correlation between stirring rate and cobalt leaching efficiency

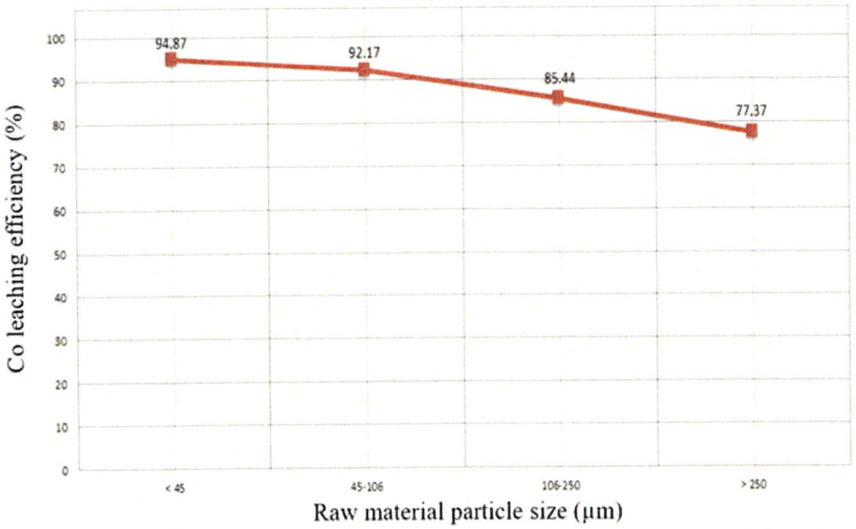

Fig. 6 Correlation between raw material particle size and cobalt leaching efficiency

that mixing effect was not enough at 600 rpm, and there were some settled down particles at the bottom of the beaker. Furthermore, at 1200 rpm, the formation of vortex prevented an effective mixing.

In the last series of the leaching experiments, effect of solid particle size on cobalt leaching was investigated. Experiments were applied to different particle size WC–Co powders under the standard conditions of 2M HNO_3 solution, 25 °C temperature, 800 rpm stirring rate, and for 2 h duration. As it seen Fig. 6, decreasing solid particle size affected cobalt dissolution positively. As the reduction of solid particle size increases the contact area between solid–liquid, an increase in dissolution efficiency was observed. Therefore, the best cobalt solubility efficiency was observed with powders having less than 45 μm particle size (94.87%).

The XRD analysis of the leaching residue of the experiment with the most efficient dissolution (raw material particle size less than 45 μm) was shown in Fig. 7. According to the results, no cobalt peak was found in the filter cake. Therefore, it also shows that cobalt was solved in the solution at a very high rate.

In the precipitation experiments, cobalt pregnant leaching solutions were used. The aim of the process was to precipitate all cobalt ions in the solution as $Co(OH)_2$ compound. For this purpose, 2M NaOH solution was added to initial leach solutions up to certain pH values to investigate. First, pH-cobalt precipitation efficiency correlation was investigated. Precipitation processes were applied at 7.5, 9 and 11 pH values. Results displayed that increasing pH value had a positive impact on cobalt precipitation from the solution. Therefore, 11 pH was determined as optimum value (Table 3). In the next stage, temperature effect was examined. Experiments were carried out at 25 and 50 °C temperatures at pH = 11 value. Increasing temperature positively affected the precipitation efficiency (Table 4). Moreover, XRD analysis

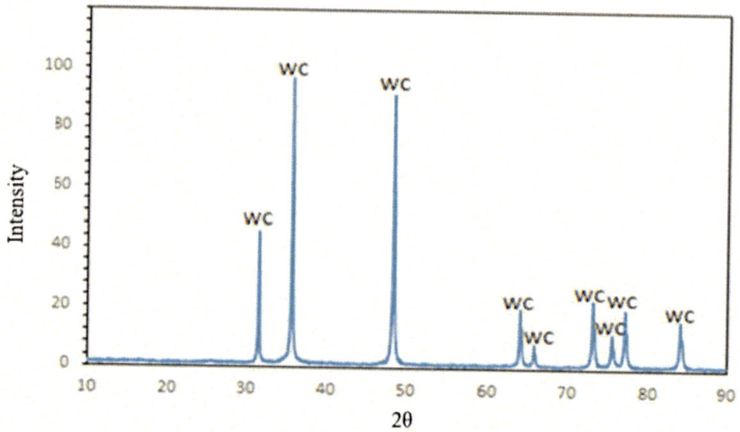

Fig. 7 XRD pattern of the filter cake (raw material particle size, <45 μm)

Table 3 Correlation between pH and Co precipitation efficiency

pH	Temperature (°C)	Co content in the solution (ppm)	Co precipitation efficiency (wt.%)
Initial leach solution		8.380	–
7,5	25	851.2	89.84
9	25	109.7	98.69
11	25	0.2216	99.99

Table 4 Correlation between temperature and Co precipitation efficiency

pH	Temperature (°C)	Co content in the solution (ppm)
11	25	0.2216
11	50	0.0822

results of precipitates showed that while at pH = 11 and 25 °C, precipitated product had amorphous structure, at 50 °C, precipitate turned into crystalline structure (Fig. 8). As a result, the highest cobalt precipitation efficiency from leach solution was determined as greater than 99.99% at pH = 11 and at 50 °C process temperature.

Thermogravimetric analysis was applied to $Co(OH)_2$ products which were precipitated at 25 and 50 °C. The $Co(OH)_2$ powders were heated from the room temperature up to 1000 °C. The aim of that analysis was to observe the thermal treatment of precipitated products. TGA results were given in Fig. 9. According to the analysis, both products displayed similar thermal properties. With the beginning of heating (1st region), the weight of both $Co(OH)_2$ precipitates reduced because of separation of some absorbed water in the structure. When temperature reached 600–650 °C (2nd region), it was observed that $Co(OH)_2$ turns into Co_2O_3 as a result of thermal decomposition reaction which causes removing of some amount of water and oxygen from

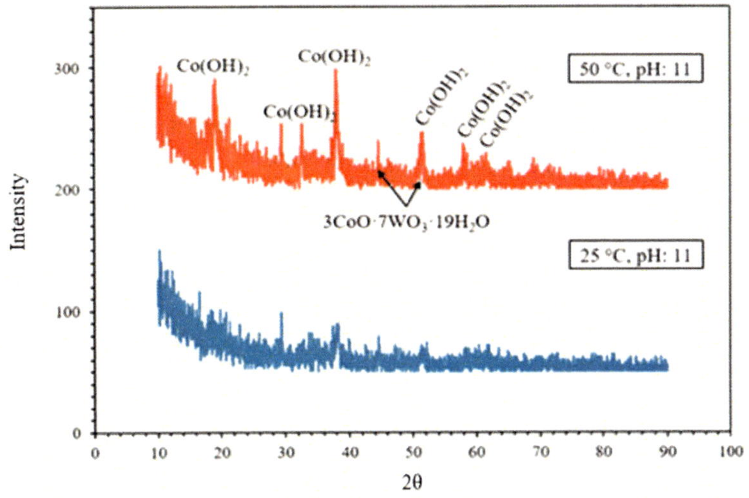

Fig. 8 XRD pattern of precipitated compounds

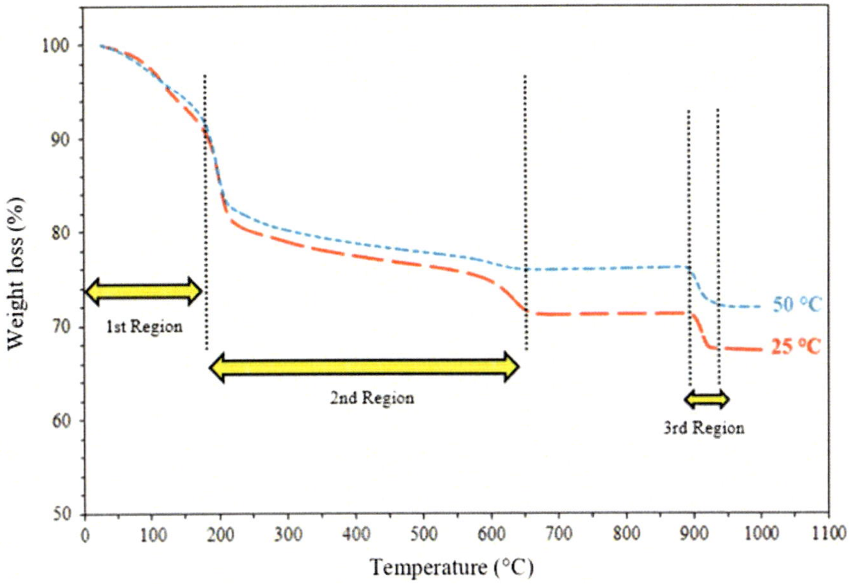

Fig. 9 TGA results of precipitated $Co(OH)_2$ compounds

$Co(OH)_2$. Also, additional weight losses in both samples at 900–930 °C happened. It was predicted that the reason of weight loss at 900–930 °C (3rd region) was the oxygen loss due to the transformation of Co_2O_3 into CoO.

Conclusion

In the leaching experiments, acid concentration, temperature, stirring rate, and raw material particle size parameters were investigated in order to determine optimum leaching parameters for selective cobalt solubility from grinded WC–Co cutting tool scraps. As a result of acid concentration experiments, 2M HNO_3 solution showed better cobalt solubility efficiency than 0.5 and 1M solutions. Therefore, optimum acid concentration was determined as 2M. After that, leaching processes were conducted at 25, 40, and 60 °C temperatures. Results showed that increasing leaching temperature affected negatively cobalt solubility. Thus, 25 °C was determined as optimum process temperature. In the next step, in order to reach optimum stirring rate parameter for selective leaching of cobalt, 600, 800, and 1200 rpm stirring rates were applied. According to the results, 800 rpm stirring rate provided better solubility efficiency. Consequently, 800 rpm stirring rate was selected as optimum stirring rate. In the final step of the leaching experiments, raw material particle size and cobalt dissolution efficiency relationship was investigated. In the experiments, −45, 45–106, 106–250, and +250 µm WC–Co powders were used. According to the results, decreasing particle size had a positive impact on cobalt solubility efficiency. Therefore, −45 µm were determined as optimum raw material particle size.

In the leaching experiments, the highest cobalt solubility efficiency was observed as 94.87% in 2M HNO_3 solution, at 25 °C, at 800 rpm stirring rate for 120 min duration, and for 1/10 solid–liquid ratio for the average particle size of less than 45 µm.

In the precipitation experiments, cobalt pregnant leaching solutions were used. In the first series of the precipitation experiments, pH-cobalt precipitation efficiency correlation was investigated. Precipitation processes were applied at 7.5, 9 and 11 pH values. Results showed that increasing pH value had a positive impact on cobalt precipitation from the solution. Therefore, 11 pH was determined as optimum value. In the next series, temperature effect was examined. Experiments were carried out at 25 and 50 °C temperatures at pH = 11 value. According to the results, increasing temperature positively affected the cobalt precipitation. Moreover, XRD analysis of solids showed that whereas the precipitate (pH = 11 and at 25 °C) had an amorphous structure, precipitate turned into crystalline structure at pH = 11 and at 50 °C. In those conditions, the highest cobalt precipitation efficiency from leaching solution was determined as greater than 99.99% (pH = 11 and 50 °C). Thus, these process parameters were selected as optimum cobalt precipitation parameters.

With this study, WC cutting tools, which are considered as waste, were re-evaluated, and Co was recovered. Cobalt recovery from cutting tool scraps will provide an alternative route to cobalt production from ore, which is found in limited quantities in the earth's crust. In addition, as a result of this process, the high energy requirement and the wastes generated in the enrichment processes of cobalt from ore may be eliminated.

References

1. Katiyar PK, Randhawa NS (2020) A comprehensive review on recycling methods for cemented tungsten carbide scraps highlighting the electrochemical techniques. Int J Refract Metal Hard Mater 90:105251
2. García J, Ciprés VC, Blomqvist A, Kaplan B (2019) Cemented carbide microstructures: a review. Int J Refract Metal Hard Mater 80:40–68
3. Lassner E, Schubert WD (1999) Tungsten properties, chemistry, technology of the element, alloys, and chemical compounds, 351–353. Kluwer Academic/Plenum Publishers, New York
4. Demirtaş M, Turan A, Acar E, Yücel O (2017) Kritik Hammaddeler. Metalurji 183:28–33
5. Hazen RM, Hystat G, Golden J, Hummer DR (2017) Cobalt mineral ecology. Am Miner 102:108–116
6. Freemantle, C.S., Sacks, N. (2015). Recycling of cemented tungsten carbide mining tool scrap. J South Afr Inst Min Metall 115:1207–1213.
7. Shemi A, Magumise A, Ndlovu S, Sacks N (2018) Recycling of tungsten carbide scrap metal: a review of recycling methods and future prospects. Miner Eng 122:195–205
8. Gürmen S, Friedrich B (2004) Recovery of cobalt powder and tungsten carbide from cemented carbide scrap—part I: kinetics of cobalt acid leaching. World Metall-ERZMETALL 57(6):113–117

Part IV
REWAS 2022: Sustainable Production and Development Perspectives

The UK Transforming the Foundation Industries Research and Innovation Hub (TransFIRe)

Mark Jolly, Anne P. M. Velenturf, Konstantinos Salonitis, and Sanjooram Paddea

Abstract Transforming Foundation Industries Research and Innovation hub (TransFIRe) was developed in response to the UK Government Industrial Strategy Challenge Fund call to transform the Foundation Industries: Chemicals, Cement, Ceramics, Glass, Metals, and Paper. These industries produce 75% of all materials in the UK economy and are vital for the UK's manufacturing and construction industries. Together, the Foundation Industries are worth £52 Bn to the UK economy and produce 28 Mt of materials per year, accounting for about 10% of the UK's total CO_2 emissions. TransFIRe is a consortium of 20 investigators from 12 institutions, more than 50 companies, and 14 NGO, and government organisations related to the sectors, with expertise across the FIs as well as material flows and energy mapping, life cycle sustainability, circular economy, industrial symbiosis, computer science, AI and digital manufacturing, management, social science, and technology transfer. This paper will introduce the Foundation Industries, present the three work streams through which transformative change will be enabled, and initial plans for including a diversity of stakeholders.

Keywords Foundation industries · Sustainable transition · Circular economy · Industrial symbiosis

---

M. Jolly (✉) · K. Salonitis · S. Paddea
Cranfield University, Cranfield MK43 0AL, UK
e-mail: m.r.jolly@cranfield.ac.uk

K. Salonitis
e-mail: k.salonitis@cranfield.ac.uk

S. Paddea
e-mail: sanjooram.paddea@cranfield.ac.uk

A. P. M. Velenturf
University of Leeds, Leeds LS2 9JT, UK
e-mail: a.velenturf@leeds.ac.uk

A. Lazou et al. (eds.), *REWAS 2022: Developing Tomorrow's Technical Cycles (Volume I)*, The Minerals, Metals & Materials Series,
https://doi.org/10.1007/978-3-030-92563-5_35

Background

A £166.5 million funding package transforming the Foundation Industries (FIs) was announced by the UK Government on the 24th May 2021 via UK Research and Innovation (UKRI) and will support development of technologies in carbon capture, greenhouse gas removal, and hydrogen, while also helping find solutions to decarbonise the UK's polluting sectors including manufacturing, metals, energy, and waste. The FIs; Chemicals, Cement, Ceramics, Glass, Metals, and Paper produce 75% of all materials in the UK economy and are valued at £52 Bn p.a. However, they produce at least 10% of the UK CO_2 emissions but are vital for the UK's manufacturing and construction industries.

This investment will help put the UK at the forefront of the green technologies of the future, while supporting British industries to lower costs, remain competitive, and protect jobs as they improve their energy efficiency and transition to a green economy. The investment, awarded to innovators, businesses, academics, and heavy industry right across the UK, will accelerate the delivery of the critical game-changing technologies needed to further drive Britain's climate change ambitions, with a projected 60,000 new jobs across the UK. This was linked to the publication of the UK Prime Minister's Ten Point Plan for a Green Industrial Revolution in November 2020 [1].

UKRI has allocated £4.7 M with a further £4.7 M matched funding for a research and innovation hub for three years to work with universities, research organisations and industry, to assist with: technology development and transfer, new business developments, and new opportunities in materials and technologies to help achieve the Net Zero 2050 target. **Trans**forming **F**oundation **I**ndustries **Re**search and Innovation hub (TransFIRe) was developed in response to the UKRI call [1].

TransFIRe is a consortium of 20 researchers from 12 institutions, more than 50 companies, and 14 non-governmental organisations related to the sectors, with expertise across the Foundation Industries as well as material flow and energy mapping, life cycle sustainability, circular economy, industrial symbiosis, computer science, AI and digital manufacturing, management, social science and technology transfer. Led by Cranfield University, the consortium includes investigators from Bangor, Cambridge, Cardiff, Durham, Edinburgh, Exeter, Leeds, Northumbria, Sheffield Hallam, and York universities, and the British Geological Survey.

The research programme will develop a self-sustaining hub of expertise to support the Foundation Industries' transformation into non-polluting, resource efficient modern competitive manufactories working in harmony with the communities in which they are situated, providing attractive places to be employed with unparalleled Equality, Diversity, and Inclusion (ED&I) performance.

The foundation materials industries are the root of all supply chains providing fundamental products into the industrial sector, often in a vertically-integrated fashion. They are all water, resource and energy-intensive, often requiring high temperature processing; they share processes such as grinding, heating, and cooling; they produce voluminous and often pernicious waste streams and heat; and they have low profit margins, making them vulnerable to foreign competition.

However, there is little transfer of knowledge horizontally; why would a paper mill researcher talk to a steel maker? There have been some traditional uses of waste materials/by-products as raw materials in other sectors, but little cross-fertilisation of technologies. In the nineteenth century, blast furnace slag was used in road building for kerbs, and more recently, Ground Granulated Blast Furnace Slag (GGBFS) has been used to improve concrete, originally by improving its resistance to sulphates and chlorides. More recently, still it is used as a low-carbon partial replacement for cement and as a value-added raw material in glass manufacture—whilst the paper industry pays the agricultural industry to spread its waste sludge on fields as a soil improver. But could this material be used as a raw material for other products with further processing? In the past, flaring of oil refinery gas was seen as part of the process. Now, it is avoided to reduce carbon emissions. How might this waste heat, from oil refineries or from cooling molten metals, glass, fired ceramics, or cement clinker, be put to good use in communities of businesses?

Partners

A list of co-investigators is given in Table 1. The Universities of Bangor, Cardiff, Exeter, Leeds, Northumbria, and Sheffield Hallam are key in the fundamental materials understanding of the specific FI sectors, and six of the PDRAs funded by the programme will have specific materials expertise. These PDRAs will spend extended periods of time on secondment at partner universities and companies, applying their expertise to support the site-based materials expert teams as they work on the analysis of a particular process unit operation. The PDRAs will also spend time working with their peers in the other sectors specifically working on cross-sector case studies. This will help maximise knowledge transfer between the industrial sectors, as well as being an efficient way of providing the multidisciplinary capability needed to ensure success in attaining project targets in each processing sector.

The other collaborating research institutions, Cranfield University, British Geological Survey, and the Universities of Cambridge, Durham, Edinburgh, Leeds, and York, will focus on the cross-sector challenges of circular economy, materials and energy mapping, AI and data, life cycle analysis, management and governance, and environment and social aspects. The remaining PDRAs will develop or have existing expertise in the supporting topics of the hub and will spend significant periods of time across both the hub universities and commercial partners, as appropriate. This approach will be comparable to the EU Marie Sklodowska-Curie academic mobility programmes. Each PDRA will be allocated a mentor from industry as well as a line manager from one of the university partners. Developing a cohort approach to the development of the PDRAs is key to the success of the hub and to the knowledge transfer to the FIs.

Table 1 List of co-investigators

Role/expertise	Name	Organisation
Director; process simulation and sustainable manufacture	Prof. Mark Jolly	Cranfield University
Deputy director; life cycle analysis	Prof. Kostas Salonitis	Cranfield University
Technical programme manager; industrial symbiosis and circular economy	Dr. Anne Velenturf	University of Leeds
WS1 lead; energy and materials mapping	Dr. Jonathan Cullen	University of Cambridge
WS2 lead; glass	Prof. Paul Bingham	Sheffield Hallam University
WS3 lead; management and operations	Prof. Peter Ball	University of York
ED&I lead; computer science and AI	Prof. Sue Black	University of Durham
Cement	Prof. Phil Purnell	University of Leeds
Ceramics	Prof. Shaowei Zhang	University of Exeter
Chemicals	Prof. Justin Perry	Northumbria University
Chemicals	Dr. Matt Unthank	Northumbria University
Metals	Prof. Rossi Setchi	Cardiff University
Metals	Prof. Sam Evans	Cardiff University
Paper	Dr. Graham Ormondroyd	University of Bangor
Materials resources	Mr. Clive Mitchell	NERC British Geological Survey
Materials resources	Dr. Evi Petavratzi	NERC British Geological Survey
Materials resources	Mr. Tom Bide	NERC British Geological Survey
Environment and society	Dr. Steve Cinderby	University of York
Environment and society	Dr. Gary Haq	University of York
Environment and society	Prof. Steve Yearley	University of Edinburgh

Research Plan

TransFIRe will focus on three major work streams (Fig. 1), the direction of which will remain flexible as projects and case studies start to produce results with the effort being targeted in the direction with the consortium Stakeholder Steering Group sees as bringing the best return for the government investment.

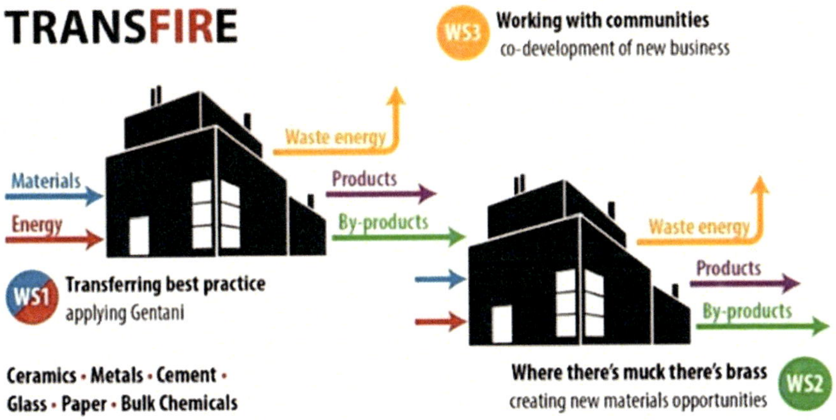

Fig. 1 Schematic showing the three work streams of TransFIRe

WS1: Transferring Best Practice—Applying "Gentani"

The Foundation Industries share many common processes (i.e. comminution, granulation, drying, cooling, heat exchange, and materials handling) and equipment (i.e. steam systems, furnaces, mechanical drives). Most of these are energy, material, and water intensive. Using the Gentani philosophy, "the minimum resource needed to carry out a process", this research will explore best practice for common processes and equipment and identify the theoretical and practical limits for resource use (energy, materials, and water) and minimum environmental impacts (dust, emissions, and waste) across sectors. This unique approach will identify cross-sector opportunities for improving plant performance. The research will be underpinned by the mapping material streams along the UK Foundation Industry supply chains, including raw materials inputs, products, by-products, waste streams, and trade flows. The British Geological Survey Centre for Sustainable Mineral Development, which is the global leader in the compilation, provision, and analysis of mineral statistics, will provide spatial and statistical minerals information on the UK raw materials used by the FIs. This will be used to identify common cross-sector processes and equipment and explore resource efficiency options and environmental impact reduction. Evidence suggests that most industrial processes use energy at levels well beyond the theoretical minimum process energy requirements. For example, the production of steel from iron requires a minimum energy of 6.7 GJ/t, with an additional 1.2 GJ/t [2] for melting, yet average global energy use is 22.6 GJ/t [3]. Cement requires only 1.8 GJ/t but typically uses 4.6 GJ/t, while flat glass requires a minimum 3 GJ/t [4], which is well below the current energy use of 7 GJ/t. Studies exploring the reduction of energy are typically focused only in one sector and often overlook material efficiency strategies such as improving process yields, reducing waste material, and utilising by-products. The analysis in this work stream will take a different approach, exploring material and energy efficiency opportunities in processes and equipment which are common across Foundation Industries. This will facilitate knowledge transfer and allow the learning of best practice resource use and waste minimisation to be shared across the sectors. In the first year, WS1 will hold workshop sessions across the hub to identify modelling approaches (e.g. material flow analysis (MFA), value stream mapping (VSM), life cycle assessment (LCA), exergy analysis, agent-based modelling and discrete event simulation) and focuses (e.g. energy, material, reaction energy, phase changes, utilities, water use and emissions) used to analyse processes and equipment in the Foundation Industries. Technical Working Groups (TWGs) will be set up for each material, comprised of researchers with specific materials, energy and processing expertise, industry experts, and representatives from trade associations, policymakers and non-profits, to identify the hot spots of resource usage within each sector and to locate "Gentani" to drive improvements. Finally, the material streams for UK Foundation Industries will be mapped, from raw materials to final products and including trade flows.

WS2: Where There's Muck There's Brass—Creating New Materials and Process Opportunities [5]

Fundamental to the transformation of our FIs will be the development of new and novel materials and processes that enable cheaper, lower-energy, lower-carbon, and at the same or even exceeding quality products. For example, in the ceramic and glass fields, chemistries to enable lower processing temperatures and advanced materials for enhanced processing (e.g. new glass refining materials, ceramic sintering aids, and enhanced particle morphologies including consolidation, to provide improved heat transfer and longer furnace lifetimes, with lower-energy demand and CO_2 emissions) will be investigated. Another example will be drawing together experience from the chemicals and paper sectors to share knowledge to develop biopolymer alternatives across both FIs exemplifying the use of sustainable sourcing and atom economical chemistries. Comparable research approaches to this, for each of the represented FIs, will be taken. The research will also investigate relationships between the new materials chemistries and their underlying physics (thermodynamics) via lab-based experiments to establish the practical viability of each approach by comparison versus benchmarks representing current best practice with consideration of new technologies in the context of industrial capability requirements. The use of hydrogen for thermally-driven processes across the Foundation Industries will also be explored, collaborating initially with research entities that have investigated hydrogen power in their sector (e.g. Project Partners Glass Futures and Pilkington-NSG for glass [6], the SUSTAIN Manufacturing Hub [7] for steel and the Low Carbon Hydrogen (HyPER) project led by Cranfield [8]) and transferring this knowledge across all FI sectors—working in partnership with these and other UK entities.

Wastes arising from the FIs and their associated supply chains, from mineral waste from primary extraction which is one of the largest UK waste flows (80 Mt pa), to back-end wastes generated by FI manufacturing operations and post-user wastes, represent a vast, and only partially-tapped UK resource. The challenges associated with finding applications for these wastes are the lack of data on their properties (physical, mineralogical and chemical), the absence of production statistics, inconsistent quality, and its occurrence in multiple locations, sometimes distant from the potential consuming FIs. Data on resource stocks and flows obtained in WS1 and WS2.1 will be combined with e.g. spatial information about waste sources and locations, mineral workings (active, inactive), availability, impacts, and supply constraints to identify opportunities minimisation and valorisation of these wastes. Much of the UK supply is vertically integrated and co-located with the FIs, for example, clay pits supplying brick works, limestone quarries supplying lime, glass, and cement factories providing a useful starting point. In WS2 through a combination of fundamental research and focussed technology development, the hub will directly address these needs. Raw material waste streams will be mapped and those that could potentially be used as raw materials for other sectors in the industrial landscape with little or no further processing (i.e. industrial symbiosis) will be identified. Streams will be identified where there is potential to add value by "upcycling" waste through further processes (e.g. by application of separation technologies), and multiple approaches from simple sizing to magnetic, wind sifting, and advanced separation technologies available through multiple project partners will be applied to new wastes and in new FI sectors to develop new materials, new potential applications/end-users and alternative by-products from innovative processing technologies with less environmental impact. This will require novel industrial symbioses and relationships, new processing technologies, sustainable and circular business models and governance arrangements. Our holistic approach (i) is essential for success and (ii) has never previously been undertaken from a UK Foundation Industry perspective before, emphasising the breadth, depth, and quality of our proposed research.

WS3: Working With Communities—Co-Development of New Business

FIs have great potential for community socio-economic benefit. Large volumes of warm air and water are produced across the sectors providing opportunities for low-grade energy capture. The 46 UK paper mills produce over 12 Mt of product per annum. Specific energy per tonne is 12.5 GJ, leading to the potential recoverable energy up to 150 PJ. For every tonne of metal melted, 1 GJ of energy is released when it cools to room temperature. In the UK, about 8 Mt of metal is melted p.a. [9, 10] giving about 8 PJ of energy into the atmosphere. A Volvo cast iron foundry in Skövde, Sweden, has developed a new process and innovative building in which excess heat from all processes is captured and used for office heating or shared with the local community. Energy is used from renewable electrical resources and is half that of equivalent factories across Europe: 2.1 GJ/t (590 kWh/t) compared to upwards of 3.6 GJ/t (1000 kWh/t) [11, 12]. There are also no carbon emissions. Potential synergies with waste hydrogen from future hydrogen-fuelled FIs will also be investigated, linking with work in WS2 and with other UK hydrogen studies.

In WS3, the focus will be on four key areas: (1) the promotion and dissemination of innovative co-development business practices, (2) community co-design approaches, (3) place-based new business opportunities, and (4) holistic impact-assessment metrics.

The research in **(1) Business practices** drives later activities as well as draws from them. At a transactional level, good practice is shared. At a research level, it is asked "how do we foster a culture of sharing and collaboration", akin to how small breweries readily share knowledge, resources, and capital. We want to learn "how" to share as well as influence "who" shares. We want to know how EDI advance impacts this. We will work with regional agencies (e.g. Local Enterprise Partnerships (LEP)), initiatives (e.g. Circular Yorkshire [12]) and support organisations (e.g. Make UK [13]) for routes to wider business to investigate local opportunities. We will examine new ways to disseminate innovative practice across sectors and not only through firms and researchers, but through communities and employees' organisations, leading to community-to-community learning.

For **(2) Working with communities and companies**, we will identify the potential for co-located businesses, social enterprises and community benefits in eight specific locales selected in the first 6 months. We ask "how does stakeholder co-design impact on value creation" by adapting co-design and co-investigation methods used in AHRC "connected communities" research, engaging communities in devising ideas and investigating proposals. Examples range from the use of waste heat (for community buildings and food production) to other resources (materials, water) to education (sharing STEM knowledge), noting the scope for positive ED&I community impacts.

Progressing from this, we engage in **(3) Business opportunities** by working across the FI sector, wider industry, and solution providers from concept to experimentation. We ask "How can 'place' foster value creation". Using participatory mapping the potential co-located businesses, social enterprises and community benefits can be studied spatially, and considered from societal, environmental, gender-impact, technical, business, and governance perspectives. Place opportunities could include resources (e.g. waste heat and CO_2 for horticulture and the move to biomass as a starting point for new industrial scale processing), valorising wastes (akin to AB Sugar fostering new resource exploitation partnerships) and knowledge (akin to Sony, Wales co-locating innovative businesses). We work beyond the incumbent supply chain to seek opportunities for new actors, apply value-creation tools, devise circular business models, and identify funding sources for prototypes and demonstrators. Demonstrators will also be online to promote concepts widely.

Finally, in **(4) Impact assessment,** we examine the range of impact-assessment tools currently available (including assessing equality, gender, wellbeing). Our focus is on "performance measurement and lifecycle theories"; we will validate metrics through engagement with community groups. We ask "which tools are best suited to provide a holistic set of metrics that capture social (including ED&I), environmental, technical, and economic performance" across the FIs; these metrics will allow us to devise and measure our responsible research and innovation (RRI) strategies.

Equality, Diversity, and Inclusion (ED&I)

Equality, Diversity, and Inclusion is at the heart of the hub and is critical in establishing and maintaining a good working environment within it. We have an award-winning, experienced lead in this area in Professor Sue Black at the University of Durham. Building ED&I experience and capability from the inside of the project, we will also work with external organisations that have key expertise in specific areas to supplement and deepen our knowledge of ED&I issues within Science, Technology, Engineering, and Maths in general and the Foundation Industries in particular. A number of the partner companies (e.g. Encirc360 [14]) already have proactive strategies for gender diversity in place from which we will take best practice and disseminate across the sectors.

It is now becoming increasingly apparent and widely accepted that ED&I are of paramount importance in the success of any organisation or project. The Foundation Industries in general are behind the curve in this respect. As part of this project, we will design and implement a cutting-edge ED&I plan with the aim not only of informing the participating organisations but also to inform best practice within UK Foundation Industries and beyond. We will review the current best practice, both within and without the Foundation Industries in the UK and overseas, using these findings to inform our strategy. As this project brings together highly skilled individuals from the many areas of expertise within the Foundation Industries, it provides a unique opportunity to formulate a comprehensive plan meeting the needs of all in the sector.

This project aims to make a distinct and tangible difference in the Foundation Industries and their perception from an ED&I perspective. Alongside working hard within the project as a whole to champion and support ED&I, our successes will be shared through social and traditional media. Highlighting key role models identified within the lifetime of the project with the aim of changing the public perception of the suitability of a career in the Foundation Industries for women, BAME, and all minority groups and people of diverse socio-economic backgrounds.

Stakeholder Engagement

Transformative change in the Foundation Industries requires the adoption of far-reaching changes in practices from a variety of stakeholders across the UK. Developing an effective strategy for stakeholder engagement and impact delivery, integrated with the research programme, is hence of vital importance. TransFIRe's engagement and impact strategy will be based on participation process management, building on best practice in previously funded UKRI research programmes [15].

Stakeholders can be engaged at increasing participation levels from informing to listening, consulting, co-producing, co-deciding, and complete self-determination. There is an ethical motivation for stakeholder participation, given that transformative change will have a major impact on companies and their communities. Participation processes can offer stakeholders equal opportunity to contribute to the changes that will affect them. Moreover, higher participation levels have been associated with greater commitment to solutions, and they can also hold essential ingredients for change such as information, experience, access to facilities, contacts, finance, networks, and influence.

In a participation process, stakeholders can be brought together to explore, share, and negotiate perceptions with regard to, in the case of TransFIRe, the transformation of Foundation Industries. During this process, stakeholder views start to overlap, allowing for the identification of shared solution spaces within which transformative changes can take shape. Participation processes are geared towards increasing the quality, legitimacy, and speed of transformative change.

TransFIRe aims to engage the full stakeholder community involved in transforming the Foundation Industries in interdisciplinary research for solutions that can be rapidly deployed within these industries. The TransFIRe proposal was co-produced with stakeholders who signed up as project partners. After securing the funding, these project partners were involved in the development of a detailed engagement and impact strategy, while new stakeholders were also identified and invited to join TransFIRe. The following steps were taken to inform the strategy (based on [15, 16]):

- Clarified the exact challenges that were to be solved, targets to be achieved, and for whom and why. This happened through the Technical Working Groups, ED&I Advisory Committee, Stakeholder Steering Group, International Advisory Committee, Foundation Industries Forum, and open events.
- Evaluated resource availability. While initial funds and person-time were secured via UK Research and Innovation and project partner contributions, the required expertise and contacts evolved with the clarification of the challenges.
- Analysed the context using the "PESTEL" approach covering the political-economic system, markets, societal trends (e.g. public narrative, community perceptions, others aiming for change), technological potential for change, environment (e.g. climate change, biodiversity), and legislation.
- Revisited the stakeholder analysis and built up relations to get to know them and gain insights into their perspectives, motivations, interests, concerns, wishes, relations, and learning and innovation behaviour with the aim to understand how TransFIRe could help them.
- Developed a tacit understanding of the "space for change" and where there might be areas of overlapping perceptions that could function as a springboard for change.

- Developed a public identity that is being communicated consistently.
- Involved the TransFIRe team in the coproduction of plans for engagement and impact, and identify how the team needs to be supported in the delivery of the stakeholder engagement and activities that are likely to make a real difference in practice.

Based on an initial understanding of stakeholder perceptions, a targeted programme for research and innovation was detailed using a Participatory Situational Analysis (PSA) [17]. PSA is a structured approach to bring a diverse group of stakeholders with widely differing and sometimes conflicting perspectives into a controlled environment to co-produce a shared vision for transformative change, identify drivers and barriers for realising the vision, identify stakeholders who are in control of making the most of drivers and/or helping to overcome barriers, agree upon potential actions for the stakeholders who are in control of unlocking transformative change, and identify any gaps in expertise that may constrain the actions from being taken.

During the delivery of the engagement and impact strategy and technical programme, stakeholders continue to be involved via the Technical Working Groups and the delivery of case studies, i.e. tranches of research projects, as well as the governance bodies of TransFIRe. Knowledge transfer is enhanced by secondments of researchers to project partners in industry, the co-mentoring of researchers by industry and academic advisors, and supporting the career progression of researchers into industry.

Progress towards delivery of the vision and objectives of TransFIRe is monitored and regularly evaluated based upon a set of Key Performance Indicators, and the Engagement and Impact Strategy and Technical Programme are adapted when needed.

The hub will act as a springboard for Early-Career Researchers (ECR). The key investigators have extensive track records of supporting ECRs into fellowships and permanent posts, and mentoring support will be given to help develop fellowship proposals that are consistent with our strategy. The alignment of the hub with the strategic plans of the parent universities and our project partners will ensure that our researchers are in prime place to apply for the posts in both academia and industry that are certain to emerge as the CE agenda gathers pace.

Summary

The Transforming Foundation Industries Research and Innovation hub was developed in response to the UK Government Industrial Strategy Challenge Fund call to transform the Foundation Industries: Chemicals, Cement, Ceramics, Glass, Metals, and Paper. TransFIRe is a consortium of 20 investigators from 12 institutions, 49 companies, and 14 NGO and government organisations related to the sectors, with expertise across the FIs as well as energy mapping, life cycle and sustainability, industrial symbiosis, computer science, AI and digital manufacturing, management, social science, and technology transfer.

Our vision is to build a proactive, multidisciplinary research, and practice driven research and innovation hub that optimises the flows of all resources within and between the FIs. The hub will work with communities where the industries are located to assist the UK in achieving its Net Zero 2050 targets and transform these industries into modern manufactories which are non-polluting, resource efficient and attractive places to be employed. TransFIRe will initially focus on three major challenges: (1) Transferring best practice—using the philosophy Gentani (minimum resource needed to carry out a process). (2) Creating new materials and process opportunities that enable cheaper, lower-energy and lower-carbon products. (3) Working with communities—co-development of new business and social enterprises. This research will highlight issues of equality, diversity and inclusiveness, investigating the potential from societal, environmental, technical, business, and governance perspectives.

At the end of the funding, we will expect to have created a "core" cross-sector team of researchers and industrial stakeholders, a body of fundamental research, and a set of governance and mechanisms enabling us to work together across the UK Foundation Industries for the next 15 to 20 years.

Acknowledgements The Transforming the Foundation Industries Research and Innovation hub (TransFIRe (EP/V054627/1)) is funded by UKRI (https://gow.epsrc.ukri.org/NGBOViewGrant.aspx?GrantRef=EP/V054627/1).

References

1. HMG (18 November 2020) The ten point plan for a green industrial revolution. https://assets.publishing.service.gov.uk/government/uploads/system/uploads/attachment_data/file/936567/10_POINT_PLAN_BOOKLET.pdf
2. Fruehan RJ, Fortini O, Paxton HW, Brindle R (2020) Theoretical minimum energies to produce steel for selected conditions. Carnegie Mellon University, Pittsburgh, PA
3. Hernandez AG, Paoli L, Cullen JM (2018) How resource-efficient is the global steel industry? Resour Conserv Recycl 133:132–145
4. Rue DM, Servaites J, Wolf W (2007) Inbdustrial glass bandwidth analysis. Gas Technology Institute, Des Plaines, IL
5. Lombardi RD, Laybourn P (2012) Redefining industrial symbiosis: crossing academic-practitioner boundaries. J Ind Ecol 16(1):28–37
6. Glass Futures (2021) https://www.glass-futures.org/. Accessed 6 September 2021
7. SUSTAIN (2021) Sustain, Future steel manufacturing hub. https://www.sustainsteel.ac.uk/. Accessed 6 September 2021
8. HyPER (2021) The HyPER project leading the development of a low carbon hydrogen production process in the UK, https://hyperh2.co.uk/. Accessed 6 September 2021
9. American Foundry Society (December 2019) Modern casting, vol 109, no 12. https://www.qgdigitalpublishing.com/publication/?m=55001&i=637245&p=6&ver=html5. Accessed 10 November 2020

10. Statista Research Development (17 June 2020) Steel production in the UK-Statistics & Facts. https://www.statista.com/topics/4939/steel-production-in-the-uk/. Accessed November 2020
11. Jolly M, Salonitis K (April 2017) Primary manufacturing, engine production and on-the-road CO_2: how can the automotive industry best contribute to environmental sustainability? In: 38th international Vienna motor symposium, Vienna, Austria
12. York & North Yorkshire Local Enterprise Partnership (2021) Circular Yorkshire. https://www.businessinspiredgrowth.com/circular-yorkshire/. Accessed 6 September 2021
13. Make UK (2021) Make UK, The Manufacturer's Organisation. https://www.makeuk.org/. Accessed 6 September 2021
14. Women with Bottle, Women in Manufacturing Forum, Encirc360
15. Velenturf APM, Jopson J, Marshall R (June 2019) Delivering knowledge exchange and impact: a case study of the Resource Recovery from Waste programme-invited presentation at UK Research and Innovation. UKRI, Swindon, UK
16. Breman B, Pleijte M, Ouboter S, Buijs A (2008) Participatie in Waterbeheer: Een vak Apart (Participation in water management: a specialist subject). In: Alterra Wageningen UR, Wageningen, The Netherlands
17. Velenturf APM, Marshall R, Suarez-Suarez A, Christensen H, Yu E, Falagan C, Sapsford DH, Gomes HI, Mayes W (2018) Participatory situational analysis: how can policy and regulation support resource recovery? Synthesis report. Resource Recovery from Waste, University of Leeds, Leeds

A Review of Comprehensive Utilization of Copper Slag of China

Jun Hao, Zhi'he Dou, Ting'an Zhang, and Kun Wang

Abstract With the development of the economy and the increasing annual output of copper, the treatment of copper slag has become a more and more critical issue. Copper slag contains valuable metals such as iron, copper, gold, silver, lead, and zinc, which have extremely high recycling value. Physical, hydrometallurgical and pyrometallurgical methods for valuable metals recovery from copper slag are particularly reviewed in this paper. Besides, this paper then introduces a novel technique proposed by Northeastern University of China which can achieve high-value and slag-free treatment of molten copper slag. It is much preferred over the current methods. This method significantly increases the added value of the product and greatly reduces energy consumption. This paper reviews the comprehensive utilization of copper slag and provides some recommendations for the large-scale and green utilization of copper slag in the future.

Keywords Copper slag · Energy saving · Utilization · Copper recovery technologies

Introduction

Copper smelting methods mainly include pyrometallurgical process and hydrometallurgical process. Most copper mines in China are sulphide ores which are not suitable for the hydrometallurgical process. Therefore, more than 95% of copper is produced by the pyrometallurgical process. Every 1 ton of copper produced in the smelting process is accompanied by an average of 2–3 tons of copper slag [1]. Copper sulfide concentrate undergoes complex physical and chemical reactions in a high-temperature oxidizing atmosphere at 1150–1250 °C. Copper is enriched in matte, and associated iron oxides are enriched with gangue to form copper slag, which belongs

J. Hao · Z. Dou (✉) · T. Zhang · K. Wang
Key Laboratory of Ecological Metallurgy of Multi-metal Intergrown Ores of Ministry of Education, Special Metallurgy and Process Engineering Institute, Northeastern University, Shenyang 110819, Liaoning, China
e-mail: douzh@smm.neu.edu.cn

A. Lazou et al. (eds.), *REWAS 2022: Developing Tomorrow's Technical Cycles (Volume I)*, The Minerals, Metals & Materials Series,
https://doi.org/10.1007/978-3-030-92563-5_36

Fig. 1 Copper production and copper slag emissions of China in recent ten years

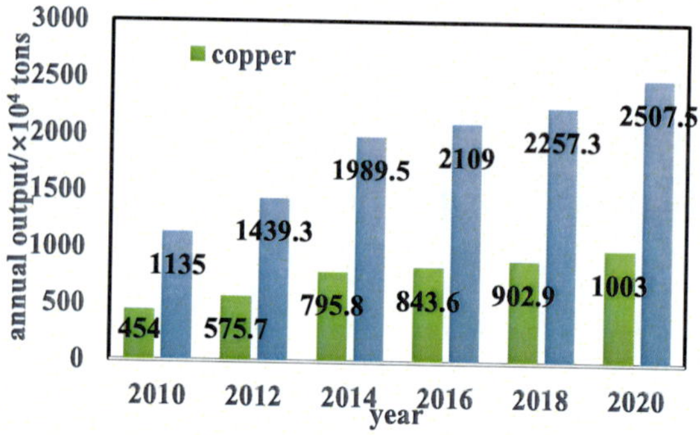

to FeO-CaO-SiO_2 series minerals [2]. Copper slag emissions in China are increasing year by year, as shown in Fig. 1. According to statistics, cumulative stockpiles of copper slag in China are more than 300 million tons. Copper slag not only contains a large amount of Fe but is also rich in various valuable metal elements such as Cu, Pb, Zn, and Co, which is a valuable secondary resource [3]. The effective recovery of copper and iron resources in copper slag not only improves the economic benefits of the copper industry but also relieves the iron ore resource pressure faced by the continuous development of the steel industry. More importantly, it is conducive to resource conservation and environmental protection [4].

So far, many researchers have conducted research on the comprehensive utilization of copper slag and proposed a variety of methods for treating copper slag. In this paper, the properties, research status, and valuable metal recovery of copper slag in China are reviewed, and the recovery status of copper and iron from copper slag is reviewed in detail. Furthermore, the relevant research methods are briefly described, whose advantages and disadvantages are analyzed, and some reasonable suggestions are given. Finally, a resource-based, high-value, and slag-free copper slag treatment technology is proposed in this paper, which can provide a new idea for the industrial application of copper slag in the future.

Properties of Copper Slag

The copper slag is black or brown, with metallic luster on the surface, and the internal structure is hard and brittle, fragile but difficult to grind. Its typical physical and mechanical properties are shown in Table 1 [5]. The microscopic morphology of the copper slag under the electron microscope is shown in Fig. 2 [6]. The phases of copper slag are mainly iron olivine phase ($2FeO \cdot SiO_2$) and magnetite phase (Fe_3O_4) [7]. Copper in copper slag mainly exists in the form of chalcocite (Cu_2S), accounting for 80–90%, and the content of metallic copper and copper oxide is generally very small. Iron in copper slag mainly exists in iron olivine, magnetite, zinc-iron spinel

Table 1 Typical physical and mechanical properties of copper slag [5]

Appearance	Black and glassy	Hardness	6–7 Moh
Unit weight	2800–3800 (kg/m^3)	Moisture	<5%
Absorption, %	0.13	Water soluble chloride	<50 ppm
Bulk density	144–162 (lb/ft^3)	Abrasion loss, %	24.1
Conductivity	500 µs/cm	Sodium sulphate soundness loss, %	0.90
Sp. gravity	2.8–3.8	Angle of internal friction	40–53

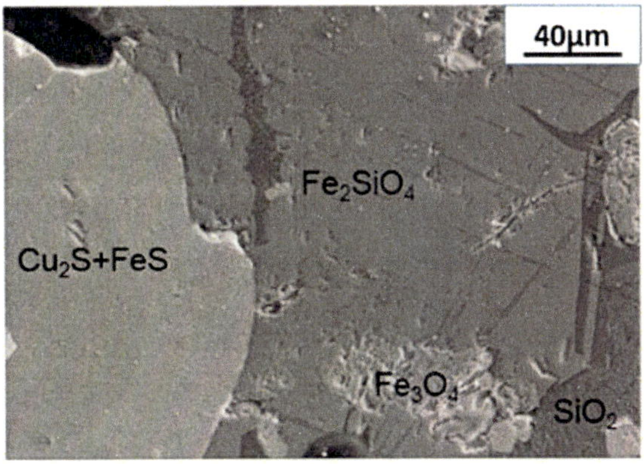

Fig. 2 Microscopic morphology of the copper slag [6]

($ZnFe_2O_4$), and a very small amount of iron exists in the form of hematite and metallic iron. Zinc in copper slag occurs in zinc-iron spinel with content of 1–3% [8], besides, lead exists in copper slag in the form of lead alum ($PbSO_4$) [9].

The copper content of copper slag is 0.5–4.5% or even higher, which is much higher than the 0.2–0.3% mining grade of copper mines in China. The total iron content in copper slag is about 40%, which is significantly higher than the mineable grade of iron ore in China (TFe > 27%) [10].

In the smelting process, in addition to the properties of copper concentrate affecting the properties of copper slag, smelting method additives and cooling system have a great impact on the mineralogy of copper slag. The different smelting methods have a great influence on the composition of copper slag, but it basically belongs to acid slag, in which the copper content is 0.42–4.60 wt.% and the iron content is 29–44.4 wt.% [11].

Overall Utilization of Copper Slag

In cement industry, copper slag can be used as raw material for cement production and mineralizer for cement clinker production. Copper slag can mix and ground with gypsum and cement clinker to produce cement. This process uses a large

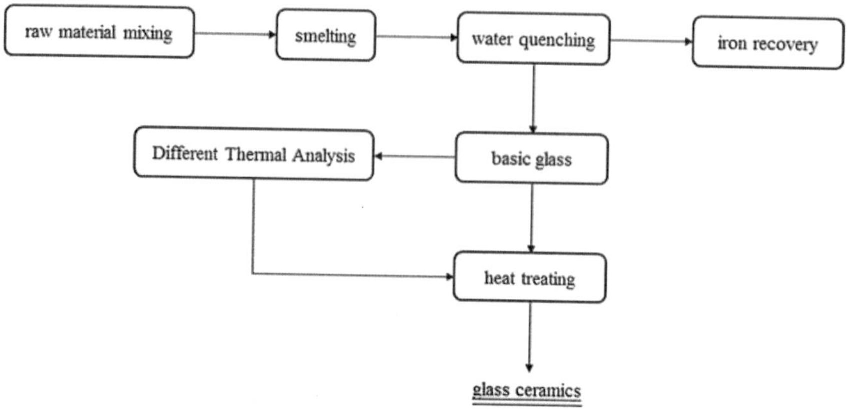

Fig. 3 Process flow chart of preparing glass–ceramics from copper slag [14]

amount of copper slag (about 60–70% of the total cement raw materials), and energy consumption can be reduced by 50%. The process has low investment cost and can be used to produce cement products with high later strength, low heat of hydration, low shrinkage, abrasion resistance, and conform to national standards [12]. In the process of cement clinker production, iron powder is generally used as a mineralizer, and copper slag contains a large amount of divalent iron, which can be used as a mineralizer for cement clinker production [13].

In the construction industry, the main applications of copper slag include the production of bricks and insulation panels, the production of cast stone, the preparation of glass–ceramics, the preparation of concrete and mortar, and the use of roadbeds and roadbeds. Lin et al. [14] took a certain copper slag as raw material to recover iron by smelting reduction method and then heat-treated the remaining slag to prepare glass–ceramics. The process flow chart is shown in Fig. 3.

Copper slag is widely used in cement and construction industry. Although it can temporarily relieve the storage pressure, heavy metal ions in building materials may cause secondary pollution to the environment. In addition, valuable elements in copper slag cannot be enriched and separated, which is a waste of resources in a sense.

The Depletion of Copper Slag

The depletion of copper slag refers to the continuous recovery of copper from the copper slag to reduce the copper content in the slag. The depletion methods of copper slag mainly include beneficiation method, leaching method, and fire method [15].

Beneficiation Method

Beneficiation is the most widely used dilution method in industry at present. Beneficiation dilution refers to the separation, enrichment, and recovery of valuable elements by physical methods such as gravity separation, magnetic separation, or flotation according to the hydrophilic magnetic properties of the surface of the valuable metal phases and the difference of the density of each phase in the slag. Copper can be recovered by flotation and iron can be recovered by magnetic separation in the beneficiation method.

The hot copper slag needs to be cooled before beneficiation. There are two ways of cooling copper slag: slow cooling and water quenching. Different cooling speed will determine the particle size and mineral composition of copper minerals in copper slag. Studies have shown that the slow cooling of copper slag above the phase transition temperature of 1080 is beneficial to the aggregation and growth of sulfide or metallic copper particles in the slag [16]. Yan [17] proposed that a copper concentrate with a copper grade of 23–25% and a copper recovery rate of 88–90% can be obtained. The method was to cool slowly at 1–3 °C/min between 1250 and 1000 °C and rapidly cool below 1000 °C. Then through grinding and flotation, the copper content in the copper slag was reduced to 0.2–0.3%.

The iron in the copper slag mainly exists in the iron olivine phase. Due to the high content of SiO_2 in the slag, it cannot be directly used in the traditional blast furnace process, and because of the close interlocking of the minerals, it is difficult to separate gangue by traditional beneficiation method to produce iron concentrate. Therefore, many researchers added grinding or selective reduction processes before starting beneficiation. Yang and Yu [18] used a multi-stage grinding, subsequent roasting, and flotation method to recover iron from copper slag. The process flow chart is shown in Fig. 4. When the copper slag grinding fineness was 0.074 mm, the roasting temperature was 1000 °C, the roasting time was 1 h, and the final iron ore concentrate was obtained. The final iron grade and recovery rate were 63.16% and 60.39%, respectively.

Leaching Method

The leaching method is direct leaching by adding acid–base leaching agents or indirect leaching using sulfation and chlorination methods, and oxidizing leaching with leaching microorganisms. Compared with the fire method, the leaching method has lower energy consumption, less air pollution, better separation selectivity, and is more suitable for processing low-grade copper slag. However, the disadvantages of the leaching method are that the amount of acid and alkali reagents used is large, and the equipment has high requirements for acid and corrosion resistance. Besides, the leaching speed and cycle are slow and long, respectively, and the most important is environmental protection issues such as wastewater treatment.

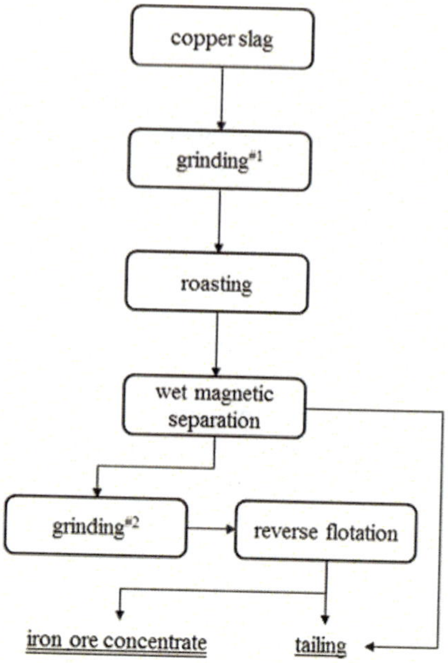

Fig. 4 Multi-stage grinding-magnetic-flotation process [18]

Banza et al. [19] proposed to use oxidation-leaching-extraction process to treat copper slag, oxidizing and leaching the slag with a mixed solution of H_2O_2 and H_2SO_4 under normal pressure, and then using extractant to recover valuable metals step by step. The recovery rate of copper can reach 80%. Some researchers used oxidation-leaching-solution extraction technology to treat copper slag, selected oxidants (commonly used H_2O_2 and Cl_2) according to the recovered elements, then used H_2SO_4 and H_2O_2 mixed solution to oxidize the slag under normal pressure [20], and then extracted the leachate step by step with an extractant to obtain valuable metals; finally, the recovery rates of Cu, Co, and Zn were 80%, 90%, and 90%, respectively. Herreros et al. [21] studied reverberatory slag and flash slag, using chlorine leaching method, and the leaching rate of copper reached 80–90%.

Leaching method is more effective in recovering valuable metal elements from copper slag; however, it will bring a series of problems such as waste of water resource and pollution. Compared with traditional leaching method, bacterial leaching method has the characteristics of lower cost and being eco-friendly; however, the reaction speed of bacterial leaching is slow and leaching cycle is long. At present, there are relatively few studies on the bacterial leaching of copper slag.

Fire Method

Fire method is achieved by returning copper slag to remelting and reducing-matting smelting. For example, electric furnace dilution technology was first used, and slag vacuum dilution technology was developed on this basis [22]. Both processes can reduce the copper content in the copper slag to less than 0.5% and discard the slag directly. In order to promote the rapid enrichment of molten copper droplets more effectively, researchers considered the effect of electric field and studied the electric field enrichment method; the highest copper enrichment rate can reach more than 80% [23].

The electric furnace method, the vacuum depletion technology, and the electric field enrichment method are all physical ways to separate the copper in the copper slag; they have an effect on the metal droplets only but are not applicable to the copper oxide and copper sulfide in the copper slag. Reddy et al. [24] adopted the reduction method to recover the metallic copper, reduced the copper oxide, and limited the reduction of iron(II) oxide as much as possible at the same time. The recovery rate of metallic copper has reached more than 85%, but the problem of recovery of copper sulfide has not been solved. Chen et al. [25] studied the new process of reducing-sulfurizing-stirring-heating fire method to strengthen the depleted copper slag. The process flow is shown in Fig. 5. At 1300 °C, by adding pyrite and crushed coal, blast mixing and clarification, the copper content of the copper slag was depleted from 1.27 to 0.47%, which met the requirements of discarding slag.

Some scholars have conducted research on smelting reduction ironmaking of copper slag. Li et al. [1] proposed to control the basicity of molten slag to 1.6 (the mass ratio of CaF_2/CaO is 0.1) for 30 min under nitrogen protection at 1575 °C, the iron recovery rate in the copper slag is 89.28%, the sulfur content in the molten iron is 0.039%, and the phosphorus content is 0.087%, which effectively solves the problem of high sulfur content in the molten iron from copper slag smelting reduction.

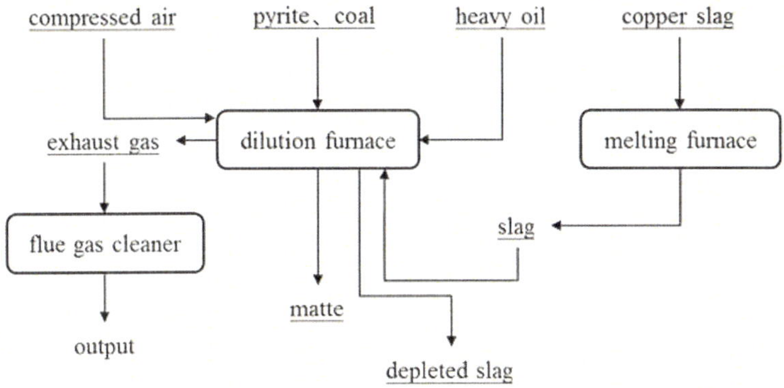

Fig. 5 Intensified dilution process flow chart [25]

A Novel Process for Copper Slag Treatment

A common problem exists in the above copper slag depletion processes, which is that each process only focuses on recycling a specific component. For example, processes such as flotation and leaching focus on the recovery of copper, while magnetic separation and direct smelting reduction processes only focus on the recovery of iron. In order to realize the comprehensive utilization of various valuable components such as copper and iron in the copper slag at the same time, a novel process named "moderately deplete-vortex reduction to prepare copper-containing antibacterial stainless steel" was proposed by the team of professor Zhang from Northeastern University. The technical flow chart is shown in Fig. 6.

This technology is characterized by high-value and slag-free recovery of copper slag. Specifically, a depleting agent which is mainly a mixture of pyrite, lime, and fluorite is added to the high-temperature molten copper slag for moderate depletion, and the copper matte enters the converter for blowing, and Pb and Zn are concentrated

Fig. 6 Process flow chart of preparing antibacterial stainless steel containing copper

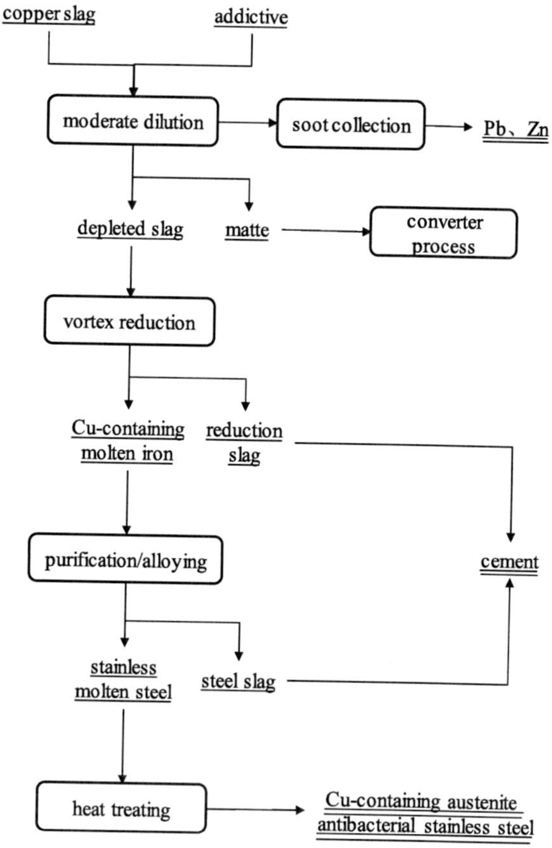

in the smoke and dust in the form of lead oxide and zinc oxide, respectively. The high-temperature liquid depleted slag is directly transferred into the reduction furnace, and then slag-forming agent and reducing agent such as carbon and natural gas are added for smelting reduction to obtain copper-containing molten iron. According to different copper slag depletion process requirements, the total addition ratio of depleting agent and reducing agent accounts for 3–8 wt.% of the copper slag. After refining and heat treatment, antibacterial stainless steel is obtained. The reduced tailings fully meet the requirements of the cement industry, and this technology truly realizes the resource utilization and high-value utilization of copper slag.

The advantages of this technology are low raw material costs, significant energy savings, and significant emission reduction effects. The technology has completed the pilot scale experiment of treating copper slag of ten thousand tons per year, and more than 50% of the reduced water quenched slag is used as cement clinker, and its performance is better than the activity of the 95-level slag. The prepared 30Cr13Cu3 grade copper-containing antibacterial stainless steel is shown in Fig. 7, which meets the requirements of national standards, as shown in Table 2. The copper content in the precipitate phase reaches 91.2%, and the antibacterial property is better than that of the commercially available antibacterial stainless steel.

The market for copper-containing antibacterial stainless steel is huge. Current annual output of stainless steel is about 21 million tons, and the market price of copper-containing antibacterial stainless steel is 28000–30,000 yuan per ton. According to the conservative estimate that antibacterial stainless steel occupies 10% of stainless steel, the current market size of the former can reach more than 20–30 billion yuan. If the stainless steel of catering appliances, bathroom equipment, and daily hardware are all made of antibacterial stainless steel in the future, the market value will be more than 200 billion yuan, and the social and economic benefits will be huge.

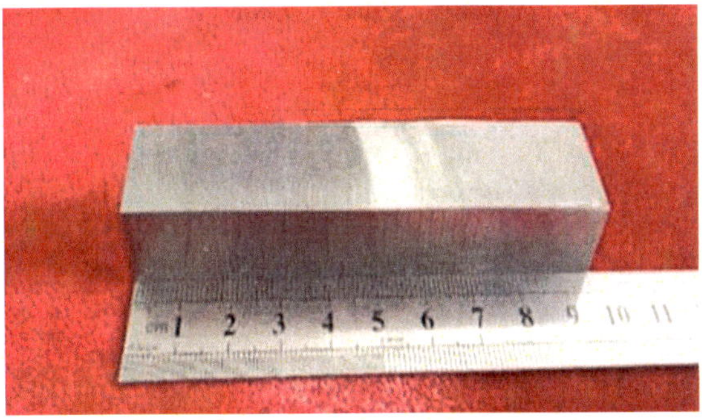

Fig. 7 Experimentally prepared copper-containing antibacterial stainless steel

Table 2 Comparison of the composition of antibacterial stainless steel with the national standard

Elements	C	Si	Mn	P	S	Cu	Cr
Sample	0.31	0.027	0.085	0.031	0.026	3.66	13.04
National standard	0.26–0.35	≤1.00	≤1.00	≤0.035	≤0.030	2.5–4.0	12.0–14.0

Conclusions

The long-term and large-scale storage of copper slag will not only occupy a large amount of land but also cause adverse effects on the environment. The use of copper slag in the construction industry can greatly alleviate the pressure of stockpiling, but this is a low-value-added utilization method and may cause secondary pollution to the environment. At present, the process such as beneficiation method, reduction-magnetic separation method, and oxidative leaching method, which are most used in industry, has the characteristics of low energy consumption, low air pollution, and good separation selectivity, and it is more suitable for processing low-grade copper slag. However, this kind of process requires high acid and alkali resistance of the equipment, and at the same time, it will cause different degrees of water pollution. In addition, most of the process flow is complicated, and the treatment cycle is long. Therefore, the current depletion methods are only aimed at the recovery of a particular valuable component of copper or iron and cannot realize the comprehensive utilization of the valuable component of copper slag.

"moderately deplete-vortex reduction to prepare copper-containing antibacterial stainless steel" is a new process to treat copper slag. Compared with the existing copper slag depleted method, this technology can realize comprehensive utilization of copper, iron, lead, zinc, and other valuable components in the copper slag and transform it into high-value stainless steel products. This technology utilizes the waste heat of copper slag and can contribute to energy saving and emission reduction. In addition, this technology combines the recovery of valuable metals with reduced slag in the production of cement clinker. Besides, it has a strong ability to process copper slag and is easy to realize industrialization. Therefore, it is a promising new choice for the future use of copper slag as a resource and high-value utilization.

Acknowledgements This work was financially supported by the National Natural Science Foundation of China (U1702253, 51774078 and U1908225) and the Fundamental Research Funds for Central Universities (N182515007 and N2025004).

References

1. Li L, Hu JH, Wang H (2011) Study on smelting reduction ironmaking of copper slag. Chin J Process Eng 11(1):65–71
2. Peng RQ (2004) Copper metallurgy. Central South University Press, Changsha, China
3. Zhao K, Gong XR, Li J (2016) Thermodynamics of recovering iron, copper, zinc in copper slag by direct reduction method. Chin J Environ Eng 10(5):2638–2646
4. Jiang PG, Wu PF, Hu XJ (2016) Copper slag comprehensive utilization development and new technology is put forward. Chi Min Mag 25(2):76–79
5. Gorai B, Jana R K, Premchand (2003) Characteristics and utilisation of copper slag—a review. Resour Conserv Recycl 39(4):299–313
6. Zhang BJ (2017) Alternative reduction of copper matte in reduction process of copper slag. ISIJ Int 57(5):775–781

7. Wang HY, Zhang XX, Zhang WT (2021) Formation mechanism of silica solid solution during carbothermal reduction of copper slag. Chin J Nonferrous Met 31(6):1591–1600
8. Han B, Dong X, Zhang GH (2015) Process mineralogy study on a copper slag. Cons Util Miner Res 1:63–68
9. Zhang HW, Fan XX, Wu HL (2013) Experimental research on pressure leaching of copper using sulfuric acid. Hydrometallurgy 32(5):305–308
10. Guo XJ, Ni XM, Ma D (2017) Copper slag treatment and comprehensive utilization. Nonferrous Met Eng Res 38(2):23–26
11. Wu L, Hao YD (2015) The investigation of utilization status of copper slag resources and efficient utilization. China Nonferrous Metall 2:61–64
12. Zhang RL (2001) A study on the comprehensive utilization of flotation tailings in flash smelting furnace. Nonferrous Met Sci Eng 1:31–34
13. Lai XS, Huang HJ (2017) Current status of the comprehensive utilization technology of copper slag. Met Mine 11:205–208
14. Lin Q, Yang ZH, Xie HJ (2012) Research on preparation of glass ceramics with copper slag. Bull Chin Silic Soc 5:1204–1207
15. Cao HY, Zhang L, Fu NX (2009) Review of copper slag impoverishment. J Mater Metall 1:33–39
16. Wang GH (2014) Copper smelting slag slow cooling technology research and practice. Copper Eng 04:27–30
17. Yan M (2011) On the design of automatic spraying system for copper slag slow cooling. Min Metall 32:95–96
18. Yang C, Yu H (2014) Experimental study of recovery of iron from copper smelting slag. Multipurp Util Miner Resour 5:55–58
19. Banza AN, Gock E, Kongolo K (2002) Base metals recovery from copper smelter slag by oxidizing leaching and solvent extraction. Hydrometallurgy 1:63–69
20. Shen HT, Forssberg E (2003) An overview of recovery of metals from slags. J Waste Manage 23(10):933–949
21. Herreros O, Quiroz R, Manzano E (1998) Copper extraction from reverberatory and flash furnace slags by chlorine leaching. Hydrometallurgy 49(1–2):87–101
22. Du QZ, Duan YX, Huang ZJ (1995) New method and mechanism of copper smelting slag depletion. Nonferrous Met 3:17–19
23. Fang LW, Hong X, Li CG (2006) A study on setting behavior of copper drops in copper slag under electric field. Shanghai Met. 6:28–31
24. Reddy RG, Prabhu VL, Mantha D (2006) Recovery of copper from copper blast furnace slag. Miner Metall Process 23(2):97–103
25. Chen HQ, Li PX, Liu SG (2006) Study on the strengthening depletion of copper from copper smelter slag by pyro-process. Hunan Nonferrous Met 3:16–18

Economics-Informed Material System Modeling of the Copper Supply Chain

John Ryter, Xinkai Fu, Karan Bhuwalka, Richard Roth, and Elsa Olivetti

Abstract Material production drives an increasingly large fraction of CO_2-equivalent emissions. Material efficiency strategies such as recycling serve to reduce these emissions. However, prior analyses of such strategies do not include economically induced rebound effects, overestimating the associated environmental benefits. We present a dynamic supply chain simulation model for copper through 2040 incorporating inventory-driven price evolution, dynamic material flow analysis, and life cycle assessment alongside mine-level economic evaluation of opening, closing, and capacity utilization decisions. We show that increases in recycling suppress raw material prices, driving increases in demand that limit primary production reduction and offset ~45% of the potential environmental benefits. Sufficiently small recycling increases and policy reversals were found capable of increasing mining and CO_2-equivalent emissions. This model was expanded to accommodate regional variations and assess the impacts of China's solid waste import ban and the COVID-19 pandemic, demonstrating the need for further investment in secondary markets.

Keywords Industrial ecology · Material efficiency · Life cycle assessment · Recycling economics · Inventory-driven price formation · Material system modeling

---

J. Ryter · X. Fu · E. Olivetti (✉)
Department of Materials Science and Engineering, Massachusetts Institute of Technology, 77 Massachusetts Ave. 8-403, Cambridge, MA 02139, USA
e-mail: elsao@mit.edu

J. Ryter
e-mail: ryterj@mit.edu

K. Bhuwalka · R. Roth
Materials Systems Laboratory, Materials Research Laboratory, Massachusetts Institute of Technology, 77 Massachusetts Ave. 8-403, Cambridge, MA 02139, USA
e-mail: bhuwalka@mit.edu

R. Roth
e-mail: rroth@mit.edu

A. Lazou et al. (eds.), *REWAS 2022: Developing Tomorrow's Technical Cycles (Volume I)*, The Minerals, Metals & Materials Series,
https://doi.org/10.1007/978-3-030-92563-5_37

Introduction

Reducing the impacts of climate change necessitates reduction of CO_2 emissions [1], but continued economic development is closely tied to increasing materials demand [2]. Material production drives an increasingly large fraction of CO_2-equivalent emissions, accounting for 15% of global emissions in 1995 and reaching 23% in 2015 [3]. Compounding these issues, the environmental impacts of primary materials extraction are projected to continue increasing as ore grades decline, particularly as mining, smelting, and refining processes are further relegated to regions with less social, economic, and political power to enforce environmental protections [4, 5].

Material efficiency strategies such as recycling serve to reduce these emissions, while also maintaining supply chains for renewable energy materials [2]. However, prior analyses of such strategies do not include economically induced rebound effects, where increased recycling provides excess raw material to a supply chain, lowering prices and incentivizing increased consumption. This increase in consumption, combined with slow mine response to changing economic conditions, inhibits the complete displacement of primary material. Because the primary environmental impact reduction mechanism for recycling is the displacement of primary production, excluding the economic rebound effect overestimates recycling's environmental benefits [6]. Standard practice in life cycle assessment assumes recycled material displaces primary material on a one-to-one basis using the avoided burden method [7]. To quantify the efficiency of a given scrap supply change in displacing primary production, displacement, d, has been defined as the cumulative reduction in primary production divided by the cumulative scrap supply increase. The resulting net environmental impact per tonne of increased scrap supply is then $E_{\text{net}} = E_{\text{secondary}} - d * E_{\text{primary}}$, where E_{primary} is the environmental impact per tonne of primary production.

Policies governing recycling have grown increasingly common in recent years. Circular economy-oriented efforts to limit resource consumption and reduce emissions in China (Circular Economy Promotion Law, 2008) and the EU (Circular Economy Action Plan, 2015) work to increase material use intensity, reuse, and recycling [8–11]. Countries including China, Malaysia, and Thailand have introduced bans on solid waste imports in efforts to limit air, soil, and water pollution and toxicity [12–14]. Alongside these shifts, environmental, social, and governance conflicts have become the primary near-term sources of metal supply risk [15], with climate change increasing water and infrastructure risk at mines as well [16]. Reducing supply chain environmental impacts necessitates decreasing reliance on primary material extraction, and recycling may do so by decreasing mine production, delaying mine opening, and hastening mine closure [17].

This model has previously been used to show that increases in recycling suppress raw material prices, driving increases in demand that limit primary production reduction and offsetting on average ~45% of the potential environmental benefits from 2018 to 2040 [18]. With declining ore grades and increasing demand, best-case scrap scenarios still produced increasing environmental impacts relative to 2018, requiring simultaneous global implementation of best available techniques to produce annual emissions reductions. In addition, this model was expanded to accommodate regional variations and assess the impacts of China's solid waste import ban and the COVID-19 pandemic [19]. The solid waste import ban induced a redistribution of scrap availability and consequently primary and secondary copper refining, leading to increasing emissions of CO_2 and other pollutants both within China and globally, which were demonstrated reversible by limiting China's copper concentrate imports. Assessing various supply chain disruptions showed that mines were more resilient to supply chain disruptions, leading to increasing CO_2 emissions per mass copper in final products. In this work, we present additional results from the two model formulations above, providing additional detail around small-scale scrap supply increases and describing the impacts of increases in scrap demand. These scenarios enable assessment of potential recycling policies at the regional, local, or corporate levels.

Methods

Determining the extent to which recycling can facilitate these changes requires accounting for market effects, discrepancies in supply chain response rates, and economic rebound. We developed a dynamic economics-informed model of the copper material system, incorporating inventory-driven price evolution, dynamic material flow analysis (dMFA), and life cycle assessment (LCA) alongside mine-level economic evaluation of opening, closing, and capacity utilization decisions. We divide the material system into five modules: price formation, mine production, refinery production, scrap supply, and semi-fabricated goods manufacturer (semi-fabricator) production. Each module relies on and produces supply–demand (SD) or price information for other modules, advancing annually from 2018 to 2040. All price parameters are reported in 2017 constant US dollars (USD). This model was tested, validated, and improved by industry experts through more than 20 interviews including miners, smelters, refineries, semi-fabricators, scrap yards and dealers, traders, market analysts, and other scientists. Scrap supply is calculated using standard dMFA methods based on Glöser et al. [20], and LCA was performed using EcoInvent V3 and TRACI 2.0 with mine impacts scaling with ore grade according to Northey et al. [21–23]. For more model details, including data sources and statistical validation, see Ryter et al. [19].

Price Formation

The price formation module is comprised of cathode (refined copper) price, treatment charges and refining charges (TCRC), and scrap spreads (cathode price minus scrap prices). Cathode price evolution is a function of the difference between cathode supply (refinery production plus solvent extraction-electrowinning (SX-EW) mine production) and cathode demand (semi-fabricator cathode consumption). We developed an autoregressive distributed lag (ARDL) model, using oil price to control for non-SD effects, and the resulting equation used within the model is shown in Eq. (1).

$$P_{\text{cathode},t} = P_{\text{cathode},t-1} - 0.461\left(S_{\text{cathode},t-1} - D_{\text{cathode},t-1}\right) \quad (1)$$

where $P_{\text{cathode},t-1}$, $S_{\text{cathode},t-1}$, and $D_{\text{cathode},t-1}$ represent London Metal Exchange (LME) cathode price in USD per metric ton (USD/t), supply, and demand in the preceding year.

TCRC is a charge levied by refineries on mines, and serves as an indicator of copper concentrate SD imbalance. Concentrate supply is determined by concentrate mine production, while concentrate demand is determined by refinery concentrate consumption. An ordinary least squares (OLS) regression model was used to evaluate the TCRC-SD relationship, with the result shown in Eq. (2).

$$\text{TCRC}_t = \text{TCRC}_{t-1} + 0.164(S_{\text{Conc},t-1} - D_{\text{Conc},t-1}) \quad (2)$$

where TCRC_t is the annual TCRC of year t (in USD/t payable copper), $S_{\text{Conc},t-1}$ is the world concentrate supply in year $t-1$, and $D_{\text{Conc},t-1}$ is concentrate demand in year $t-1$.

With high correlation between cathode price and scrap prices and expecting scrap SD imbalance to affect scrap price only relative to cathode price, we treat scrap economics using scrap spread, the difference between cathode price and scrap price. With poor data availability on historical scrap supply and high correlation between cathode price and scrap spread, we estimated the scrap SD effect on scrap spread to be smaller than the cathode price effect. We estimated this value as the product of cathode price elasticity to cathode SD and scrap spread elasticity to cathode price. Among the many scrap grades informing decision making throughout the supply chain, we found No.2 copper scrap (94–99 wt.% Cu) to have the most statistical significance, and its evolution is described by Eq. (3).

$$\Delta\text{Spread}_{\text{No.2},t} = 0.184\Delta P_{\text{Cat},t} + 0.0845(S_{\text{Scrap},t-1} - D_{\text{Scrap},t-1}) \quad (3)$$

where $\text{Spread}_{\text{No.2},t}$ is the difference between cathode price and No.2 price in year t, $P_{\text{Cat},t}$ is cathode price in year t, and $S_{\text{Scrap},t-1}$ and $D_{\text{Scrap},t-1}$ represent supply and demand for all scrap grades in the preceding year.

Mine Production

We model opening, closing, and capacity utilization (CU) evolution of individual mines, with currently-operating mines from S&P Capital IQ Pro and an incentive pool of opening mines created by resampling those opening 2015–2018 with ±10% perturbation [24]. We calculated each mine's total cash margin (TCM) as cathode price minus total cash cost, which determines mine profitability [25]. Ore production evolves in the near term as mines change CU, which is modeled according to Eq. (4) following a generalized method of moments (GMM) dynamic panel regression model.

$$\text{Mine Capacity Utilization} = \begin{cases} 0.4, \text{ if in ramp up or ramp down} \\ 0.75, \text{ if not in ramp and TCM} < 0 \\ \text{CU}_0 * \left(\frac{\text{TCM}}{\text{TCM}_0}\right)^{0.024}, \text{ else} \end{cases} \tag{4}$$

where mine CU is assumed equal to 0.4 in ramp up and ramp down periods, which last three years after the opening decision and one year after the closing decision respectively. CU_0 is the 2018 average CU, while TCM_0 is assumed to be the median TCM of all operating mines in 2018. Mine closure occurs when the net present value (NPV) of entering ramp down in the current year exceeds the anticipated NPV of entering ramp down in the following year. The projected cash flow is calculated using the maximum cathode price of the preceding five years, with NPV calculated as in Eq. (5).

$$\text{NPV} = \sum_{t=0}^{T} \frac{C_t}{(1+\delta)^t} \tag{5}$$

where C_t is the total cashflow expected in year t with t an integer from 0 to T, δ is the discount rate set at 10% [25, 26], T is 1 for ramp down beginning in year 0 and 2 for ramp down beginning in year 1, where C_t is equal to the reclamation cost in year T. In all other years, C_t is a function of cathode price, TCRC, CU, ore grade, and several mine-level exogenous variables. Mine opening occurs when the internal rate of return (IRR) of a mine selected from the incentive pool exceeds 15%, as calculated by solving for r in Eq. (6).

$$\text{NPV}_r = \sum_{t=0}^{T} \frac{C_t}{(1+r)^t} = 0 \tag{6}$$

where C_t includes three years of development capital expenditures, estimated cashflows, and reclamation costs, while T is the year of mine closure. Cashflows are estimated by simulating the lifetime of the mine using the trailing three-year average of cathode price and TCRC. The number of mines selected from the incentive pool for opening evaluation was tuned such that annual mining production 2018–2040 grew at the 2001–2011 linear growth rate, with cathode price and TCRC held constant. This subsample size time series was held constant in all other scenarios.

Refinery Production

Refinery production is separated into primary and secondary refineries, where primary refineries consume only copper concentrate and secondary refineries consume concentrate and scrap, with the scrap fraction termed the secondary ratio (SR). Separate GMM panel models were developed for primary CU, secondary CU, and SR, with Eqs. (7)–(9) used for model evolution.

$$\mathrm{PCU}_t = \mathrm{PCU}_{t-1} \cdot \left(\frac{\mathrm{TCRC}_t}{\mathrm{TCRC}_{t-1}}\right)^{0.057} \tag{7}$$

$$\mathrm{SCU}_t = \mathrm{SCU}_{t-1} \cdot \left(\frac{\mathrm{TCRC}_t}{\mathrm{TCRC}_{t-1}}\right)^{0.153} \tag{8}$$

$$\mathrm{SR}_t = \mathrm{SR}_{t-1} \cdot \left(\frac{\mathrm{TCRC}_t}{\mathrm{TCRC}_{t-1}}\right)^{-0.197} \cdot \left(\frac{\mathrm{Spread}_t}{\mathrm{Spread}_{t-1}}\right)^{0.316} \tag{9}$$

where PCU_t is the CU for the primary-only refinery at time t, SCU_t is secondary refinery CU at time t, SR_t is the SR for the secondary refinery at time t, and Spread_t is the difference between cathode price and No.2 scrap price at time t.

Demand

The demand module forecasts future world copper demand as a function of cathode price, using the product of an intensity component and a volume component. Volumes (e.g. number of vehicles sold per year) are considered independent of copper price, being primarily driven by macroeconomic indicators such as gross domestic product (GDP) and population, while intensities (e.g. kg Cu per vehicle) are function of cathode price. We model demand and intensities according to Eqs. (10) and (11), with intensity following a Bayesian regression method.

$$D_{s_i,r_j,t} = I_{s_i,r_j,t} \cdot V_{s_i,r_j,t} \tag{10}$$

$$\Delta\log\left(I_{s_i,r_j,t}\right) = \beta_{0,s_i} + \beta_{s_i}\Delta\log\left(P_t^{'}\right) + \beta_{r_j}\Delta\log\left(P_t^{'}\right) + \beta_{\mathrm{GDP}}\Delta\log\left(\mathrm{GDP}_{r_j,t}\right) \tag{11}$$

where subscript s_i is the sector index for sector i, r_j the region index for region j and year$t$. D_t, the total copper content (demand) in final products for year$t$, is broken down into five sectors (construction, electrical, industrial, transportation, consumer and others) and five regions (China, EU, Japan, North America, and rest of world (RoW)).

$I_{s_i,r_j,t}$ and $V_{s_i,r_j,t}$ are the corresponding intensity and volume values. β_{0,s_i} represents sector-specific intensity reduction or dematerialization due to technological growth, β_{s_i} and β_{r_j} are sector- and region-specific price responses, with $P_t' = \frac{P_{t-1}+P_{t-1}}{2}$ the first lag of the trailing two-year average cathode price. β_{GDP} is the intensity response to regional per capita GDP,$GDP_{r_j,t}$. All β values except β_{GDP} were expected to be negative, with β_{GDP} expected positive.

Scenario Development

We have conducted two sets of scenarios using this framework. The first set comprises changes in scrap collection rates and sorting efficiencies such that scrap collection and sorting increase from baseline for some duration before plateauing (plateau scenarios). Collection rates and sorting efficiencies range from current values up to maximum values estimated based on technological and logistical limitations of these systems, termed opportunity. With 0% of recycling opportunity captured, collection and sorting remain at current values, while 100% of opportunity corresponds with ~2.8 million metric tons (Mt) additional scrap supply in 2019, or 24% of estimated 2018 scrap consumption. Here, we include only the smallest 50 of these scenarios, with annual scrap supply increases ranging from ~5 to 35 kt. These scenarios represent policy outcomes at smaller scales than previously reported, enabling estimation of corporate- or local-scale recycling policy impacts. To measure the efficiency of scrap supply changes in reducing mine production, we define displacement, d, as the cumulative decrease in mining divided by the cumulative increase in the scrap supply, as shown in Eq. (12).

$$d = \frac{-\sum_{t\in(2018\ldots2040)}\left(P_{\text{Mining,scenario},t} - P_{\text{Mining,baseline},t}\right)}{\sum_{t\in(2018\ldots2040)}\left(S_{\text{Scrap,scenario},t} - S_{\text{Scrap,baseline},t}\right)} \tag{12}$$

where P_{Mining} represents mine production for either the scenario or baseline, S_{Scrap} represents the scrap supply for either the scenario or baseline, and t represents time, an integer from 2018 to 2040.

In the second set of scenarios, we introduce minimum recycled content mandates at 0, 5, 10, 15, or 20% of global copper and brass alloy semi-fabricated goods production, and introduce and vary collection rate elasticities to scrap spread. These mandates included only directly remelted copper scrap, as secondary refined copper is chemically and economically indistinguishable from primary. Collection rate evolution occurred in each year according to Eq. (13).

$$\text{CR}_t = \text{CR}_{t-1} * \frac{\text{Spread}_t}{\text{Spread}_{t-1}}^{\text{elas}_{\text{CR}}} \tag{13}$$

where CR_t is the collection rate for any sector in year t, $Spread_t$ is the No.2 scrap spread at year t, and $elas_{CR}$ is the collection rate elasticity to scrap spread, which took the value of zero, −0.5, or −1.1. Given limited data availability on collection rate response to scrap spread, these values were used as sensitivity parameters to explore possible market responses. Scrap and refined copper demand are determined by the blending optimization model described in previous work [19], with constraints set on minimum recycled content.

Results

Small-Scale Scrap Supply Increases

To permit estimation of corporate- or locality-level increases in recycling, we present small-scale annual scrap supply increase results in Fig. 1. These annual increases last 5–20 years, meaning that longer durations lead to larger total scrap supply increases. It is evident from the color differences in Fig. 1a that these larger total scrap supply increases produce larger maximum annual CO_2-equivalent emissions reductions, as mines respond slowly and reduce production more rapidly when the scrap supply change lasts long enough to be coincident with mine decision making. Figure 1b shows that larger annual increases in scrap supply correlate with higher displacement, with only the shortest durations appearing as low-displacement outliers. This behavior is further verified in Fig. 1c, where more efficient CO_2-equivalent emissions reductions may be accomplished with larger and longer-duration scrap supply increases.

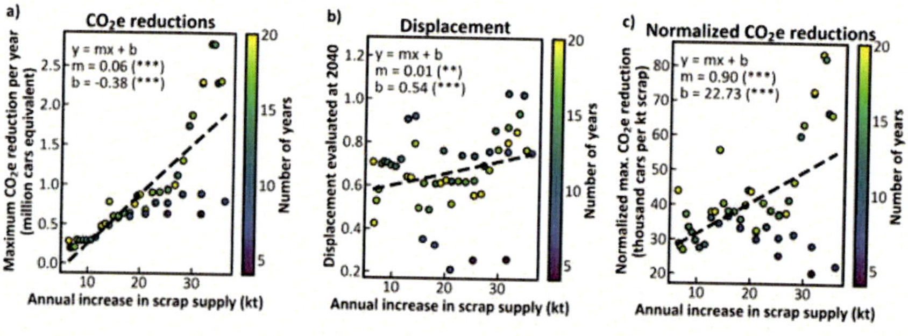

Fig. 1 **a** Maximum annual CO_2-equivalent emissions reductions from 2018 to 2040, using 4.6 metric tons CO_2/year tailpipe emissions for the average passenger vehicle [27]. **b** Displacement values evaluated at 2040. **c** The maximum annual CO_2-equivalent emissions reduction from 2018 to 2040 normalized by the annual increase in scrap supply. Color bars represent the duration of the scrap supply increase before it plateaus. Regression line parameter p-values are given as: $*p < 0.1$; $**p < 0.05$; $***p < 0.01$. kt = thousand metric tons, CO_2e = CO_2-equivalent emissions

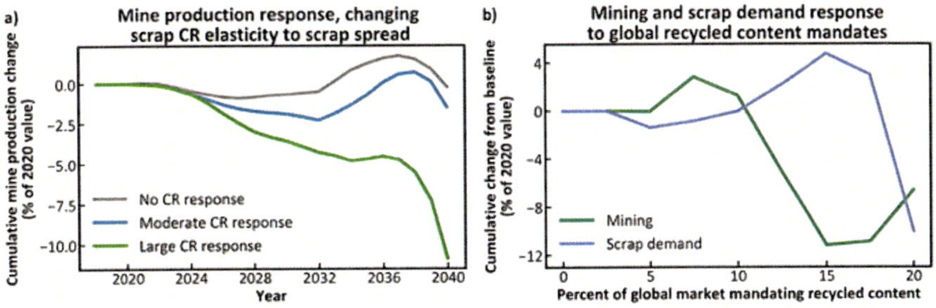

Fig. 2 **a** Mine production response for changes in collection rate elasticity to No.2 scrap spread for a 17.5% recycled content mandate. **b** The cumulative change in mine production and scrap demand from 2018 to 2040 relative to the baseline scenario, as a function of the percent of the global market mandating recycled content using the large collection rate elasticity to No.2 spread. These scenarios use CR = collection rate

Changes in Scrap Demand

Because many copper applications require refined copper, directly remelted copper scrap is relatively limited in its applications. As a result, direct melt scrap consumption is largely redistributed among different copper alloys. The mine production response with no collection rate response to scrap spread in Fig. 2a reflects this redistribution, with very little impact on mine production. However, the introduction of collection rate response to No.2 spread enables substantial impacts on the market. With high collection rate response, it is evident that larger recycled content mandates produce larger mine production reductions, where economic rebound keeps these changes small for smaller mandates. In the largest recycled content mandate scenario, we hypothesize that the collection rate elasticity to No.2 scrap spread is insufficient to keep up with rising demand, inducing sufficient rebound to reduce scrap demand and inhibit mine production reduction.

References

1. Hoegh-Guldberg O et al (2018) Impacts of 1.5 °C global warming on natural and human systems. Intergovernmental Panel on Climate Change
2. Elshkaki A et al (2016) Copper demand, supply, and associated energy use to 2050. Glob Environ Chang 39:305–315
3. Hertwich E (2019) The carbon footprint of material production rises to 23% of global greenhouse gas emissions
4. Northey S et al (2014) Modelling future copper ore grade decline based on a detailed assessment of copper resources and mining. Resour Conserv Recycl 83:190–201
5. Azapagic A (2004) Developing a framework for sustainable development indicators for the mining and minerals industry. J Clean Prod 12(6):639–662
6. Geyer R et al (2016) Common misconceptions about recycling. J Ind Ecol 20(5):1010–1017
7. Earles JM, Halog A (2011) Consequential life cycle assessment: a review. Int J Life Cycle Assess 16(5):445–453

8. Standing Committee of the National People's Congress (2008) Circular economy promotion law of the People's Republic of China. Congressional-executive commission on China: People's Republic of China
9. European Commission (2015) First circular economy action plan, E. Commission. European Union
10. Geng Y, Sarkis J, Ulgiati S (2016) Sustainability, well-being, and the circular economy in China and worldwide. Science 6278(Supplement):73–76
11. Lieder M, Rashid A (2016) Towards circular economy implementation: a comprehensive review in context of manufacturing industry. J Clean Prod 115:36–51
12. Ren Y et al (2020) Life-cycle environmental implications of China's ban on post-consumer plastics import. Resour Conserv Recycl 156:104699
13. Xia Y (2018) China's environmental campaign: how China's war on pollution is transforming the international trade in waste. NYU J Int Law Politics 51:1101
14. Gregson N, Crang M (2019) Made in China and the new world of secondary resource recovery. Environ Plan A Econ Space 51(4):1031–1040
15. Jowitt SM, Mudd GM, Thompson JF (2020) Future availability of non-renewable metal resources and the influence of environmental, social, and governance conflicts on metal production. Commun Earth Environ 1(1):1–8
16. Northey SA et al (2017) The exposure of global base metal resources to water criticality, scarcity and climate change. Glob Environ Chang 44:109–124
17. Zink T, Geyer R, Startz R (2018) Toward estimating displaced primary production from recycling: a case study of U.S. aluminum. J Ind Ecol 22(2):314–326
18. Ryter J et al, Assessing recycling, displacement, and environmental impacts using a copper econometric material system model. J Ind Ecol. Under Review
19. Ryter J et al (2021) Emission impacts of China's solid waste import ban and COVID-19 in the copper supply chain. Nat Commun 12(1):1–13
20. Glöser S, Soulier M, Tercero Espinoza LA (2013) Dynamic analysis of global copper flows. Global stocks, postconsumer material flows, recycling indicators, and uncertainty evaluation. Environ Sci Technol 47(12):6564–6572
21. Northey S, Haque N, Mudd G (2013) Using sustainability reporting to assess the environmental footprint of copper mining. J Clean Prod 40:118–128
22. Wernet G et al (2016) The ecoinvent database version 3 (part I): overview and methodology. Int J Life Cycle Assess 21(9):1218–1230
23. Bare J (2011) TRACI 2.0: the tool for the reduction and assessment of chemical and other environmental impacts 2.0. Clean Technol Environ Policy 13(5):687–696
24. S&P Global Market Intelligence (2019) Metals & mining data, subscription. S&P Capital IQ Pro
25. S&P Global Market Intelligence (2019) Mine economics methodology, market intelligence metals & mining database. S&P Capital IQ Pro 2021
26. Runge IC (1998) Mining economics and strategy. SME
27. (2018) Greenhouse gas emissions from a typical passenger vehicle. https://www.epa.gov/greenvehicles/greenhouse-gas-emissions-typical-passenger-vehicle. Accessed August 18, 2020

Sustainability-Based Selection of Materials for Refractory High Entropy Alloys

Xinyi Wang, Annalise Kramer, Christopher Glaubensklee, Haoyang He, and Julie M. Schoenung

Abstract Refractory high entropy alloys (RHEAs) are considered candidate materials for high-temperature applications beyond nickel superalloys. In addition to good mechanical behavior, sustainability-related materials selection and design should be considered at an early stage of development, which will allow the application of RHEAs to expand. In this work, we present an alloy design framework for RHEAs from three different perspectives. We evaluate resource availability based on geographic concentration for refractory elements and generate a supply risk index for each element. We then evaluate material prices and price volatility for each element. We also estimated an aggregate price of select alloys and compared them to their yield strength at room and high temperatures. A proper balance of improved performance and economics is considered. Finally, we evaluated environmental, physical, and human health hazards for refractory elements and used the Green Screen® for Safer Chemicals methodology to assign benchmark scores to each element.

Keywords Refractory high entropy alloys · Sustainability · Materials selection

Introduction

Materials with superior mechanical properties at elevated temperature are in high demand for many applications, such as aerospace and nuclear industry. Recently, a new type of material called high entropy alloys (HEA) has been developed due to their extraordinary mechanical properties [1, 2]. Unlike traditional alloys, these

X. Wang
Materials and Manufacturing Technology, University of California, Irvine, CA 92697, USA

A. Kramer
School of Physical Sciences, University of California, Irvine, CA 92697, USA

C. Glaubensklee · H. He · J. M. Schoenung (✉)
Department of Materials Science and Engineering, University of California, Irvine, CA 92697, USA
e-mail: Julie.Schoenung@UCI.edu

A. Lazou et al. (eds.), *REWAS 2022: Developing Tomorrow's Technical Cycles (Volume I)*, The Minerals, Metals & Materials Series,
https://doi.org/10.1007/978-3-030-92563-5_38

HEAs contain five or more elements in equal or near-equal atomic percentage (at%). HEAs usually have a simple face-centered cubic (FCC) or body-centered cubic (BCC) crystal structure due to high configurational entropy of mixing [1–4]. Among currently reported HEAs, HEAs based on refractory elements are considered candidate materials for high-temperature structural applications beyond Ni superalloys. The first two refractory high entropy alloys (RHEAs), MoNbTaW and MoNbTaWV, were invented by Senkov et al. [5]. Later researchers have broadened the scope of RHEAs. These RHEA alloys now generally comprise nine refractory elements (Cr, Hf, Mo, Nb, Re, Ta, V, W, and Zr) and can also contain other elements (such as Al, Ti, Si, Co Ni, C, and N) [6, 7]. In addition to better mechanical behaviors, developments should consider other aspects of design at an early stage, thus allowing RHEAs to expand the role metal alloys play in a variety of applications. Fu et al. proposed a selection framework for HEA materials based on price or supply availability [8]. Previous studies have shown that it is a struggle to find non-toxic elements with a low enough density and cost to be usable in the design of HEAs [9]. Therefore, elements must be evaluated to not only assess their mechanical behavior but also their potential to make the alloy more sustainable (e.g., by using fewer chemical hazards).

In the current study, availability, prices, and chemical hazards of refractory elements (Mo, Nb, Ta, W, Zr, V, Re, Ti, Hf) are analyzed. This study is divided into three main sections. The resource availability section describes geographic concentration for refractory elements. The material price section performs elemental screening of prices and evaluates price volatilities for refractory elements and common RHEA compositions. A proper balance of improved properties and economic effects is established. The chemical hazard assessments section evaluates hazard traits that range from physical hazards and environmental hazards to human health issues.

Analysis

Resource Availability

Analysis of resource availability is of great importance in exploring material design. There are several factors that can impact the resource availability of materials: geographic distribution of sources, ore grade, recycling rate, etc. [10]. In the current study, data on domestic mine production and import and export resources are obtained from the U.S. Geological Survey (USGS) [11–14]. Figure 1 shows the cumulative fraction of global mine production volume for refractory elements as a function of location. Poor geological distribution means the material resource is more affected by local politics, monopsony market conditions, and physical constraints in the supply chain. According to guidelines from the U.S. Department of Justice, moderate levels of concern exist when individual suppliers reach market shares of approximately 30%, and high levels of concern exist when market shares approach or exceed 40% [15, 16]. By this standard, V, W, Nb, Ti, and Re appear to be particularly vulnerable to

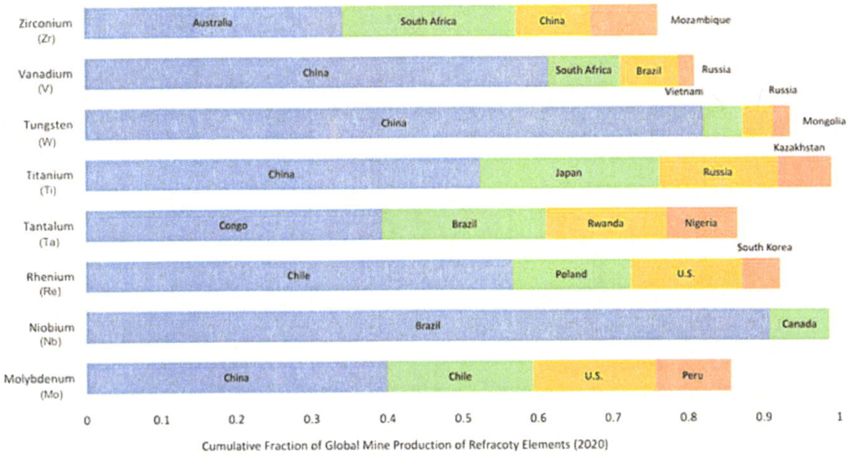

Fig. 1 Geological distribution of global mine production of refractory elements

availability risks. Over 50% of mine production for V, W, and Ti takes place in China, while Brazil controls over 90% of the mine production of Nb and Chile controls over 50% of the mine production of Re.

Table 1 shows the supply risk index for eight refractory elements, as determined by the British Geological Survey [17]. The risk index provides a simple indication of the relative risk in the year 2015 for the supply of eight refractory elements. Supply risk index values range from 1 (green—very low risk) to 10 (red—very high risk). The score for each element is determined by several factors that might affect availability, for example, location of current production and reserves, the political stability of locations, recycling rate, etc. [17]. The supply risk indexes for V, W, and Mo are relatively high compared to other elements. Thus, special consideration should be taken when designing a new RHEA with any of these elements.

Refractory metal elements carry additional supply risk if they are co-mined with another metal because this leads to a dependency on the market for the parent metal system [8, 18]. For example, Hf is the by-product of producing Zr; Zr and Hf are typically contained in zircon at a ratio of 36:1. A detailed analysis of co-mining is

Table 1 Relative supply risk index for refractory elements [16]. Low values are preferred

Element	V	W	Mo	Re
Relative Supply Risk Index	8.6	8.1	8.1	7.1
Element	Ta	Nb	Zr	Ti
Relative Supply Risk Index	7.1	6.7	6.4	4.8

beyond the scope of this study but should be considered when selecting elements for use in RHEAs.

Material Price

Materials price is an important metric for RHEA design feasibility. Annual price data are obtained from the USGS historical statistics for mineral and material commodities [19]. Values averaged over the 10-year period of 2007–2017 are from lowest to highest (in units of $/mole): Al (0.06), Cr (0.13), Ti (0.53), V (1.32), Nb (2.63), Mo (3.05), Zr (3.46), W (6.96), Ta (39.38), Hf (98.89), and Re (780.56). Since prices of metallic elements may change significantly over time, assessment of price volatility is also presented. Price volatility can be calculated using the following two steps: a. Calculate logarithmic price changes—by taking the natural log of the quotient of the period-2 price divided by period-1 price, as shown in Eq. 1; b. calculate the standard deviation for price change, as shown in Eq. 2. Figure 2 displays the price volatility for refractory metal elements for the 10-year period of 2007–2017. A higher value of price volatility indicates that the element price has a larger degree of change during the past 10 years.

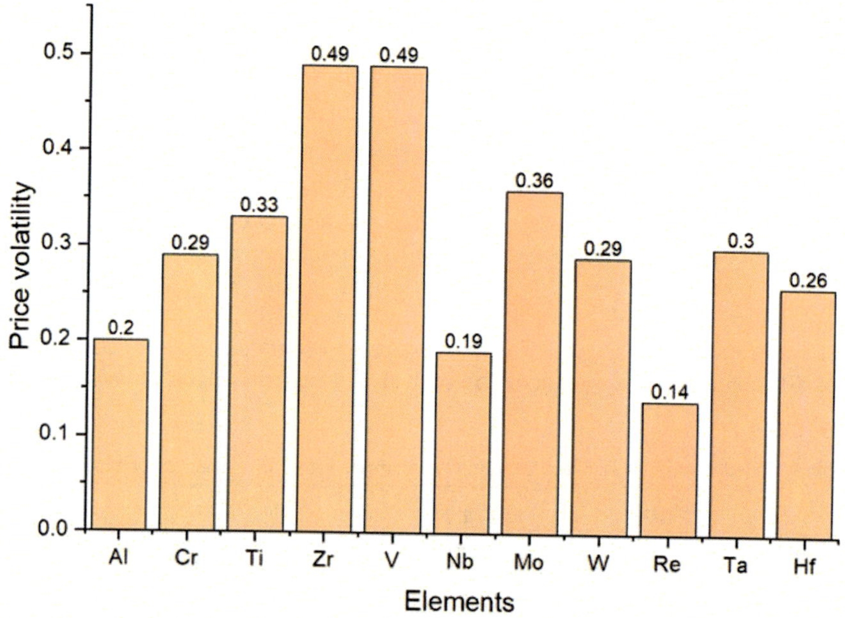

Fig. 2 Elemental price volatility assessment for the 10-year period 2007–2017 (based on data from [18]). Low values are preferred

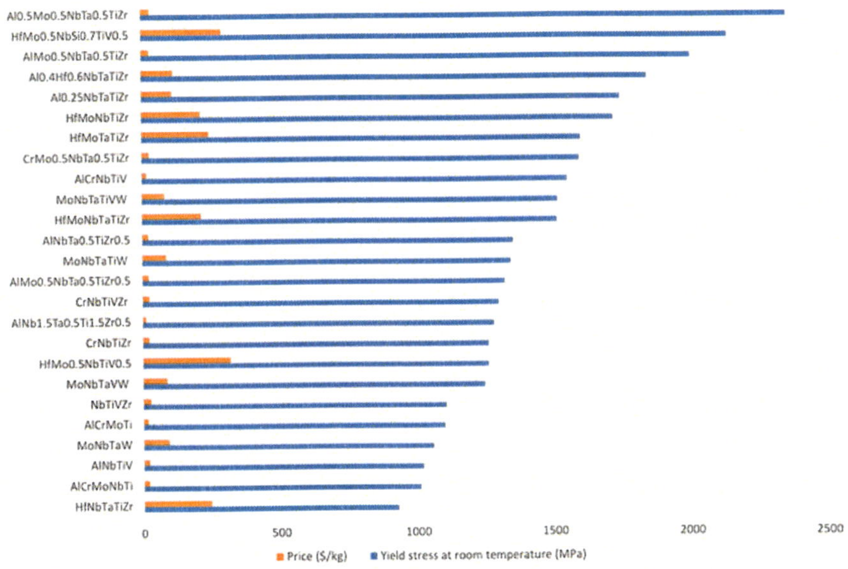

Fig. 3 Estimated alloy price (based on data from [19]) and yield stress values at room temperature [6] for refractory high entropy alloys (RHEAs)

$$\text{Logarithmic Price Change} = \text{In}(P1/P2) \tag{1}$$

$$\text{Standard Deviation} = \sqrt{\frac{\Sigma(x-\overline{x})^2}{n}} \tag{2}$$

To consider the domain of RHEAs, both economic feasibility and mechanical properties are considered. An estimated alloy price is calculated for 25 common compositions for RHEAs by using a weighted sum of the elemental prices [19]. Mechanical properties for the RHEAs are derived from the literature [6]. Figures 3 and 4 show the estimated price values and room temperature and high-temperature strength for each composition, respectively. For room temperature applications, Al-containing RHEAs, such as $AlMo_xNbTa_yTiZr_z$, can achieve both high strength and low price. Materials containing Hf are relatively expensive and are not necessary for room temperature applications. At high temperature, $AlMo_xNbTa_yTiZr_z$ still has a better combination of low price and high strength. At high temperature, the addition of Hf and Ta can significantly increase material's strength. Designers might avoid elements such as V or Zr since they cannot provide a high strength at high temperature. If an appropriate balance between economic feasibility and improved mechanical properties can be achieved, RHEAs may have wider applications.

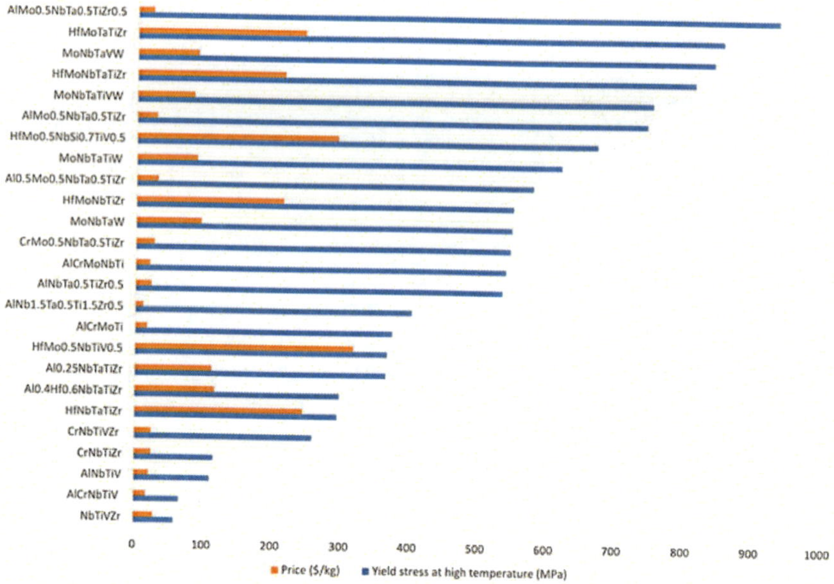

Fig. 4 Estimated alloy price (based on data from [19]) and yield stress values at high temperature (1000 °C) [6] for refractory high entropy alloys (RHEAs)

Chemical Hazard Assessment

In this study, data on environmental hazards, physical hazards, and human health hazards for all ten elements (we include Al, Cr, and Ti as they are commonly added to RHEAs to achieve better mechanical properties, and we exclude Re due to its extremely high price) are collected from different sources: Globally Harmonized System of Classification and Labelling of Chemicals (GHS)—Japan [20], European Chemicals Agency (ECHA) [21], U.S. Environmental Protection Agency (US-EPA) [22], and Canadian Environmental Protection Act (CEPA)—Domestic Substances List (DSL) [23]. One hazard assessment tool is used to score and rank the metallic elements in this study: GreenScreen® for Safer Chemicals (GS). Based on standards established by the United Nations' Globally Harmonized System of Classification and Labeling of Chemicals [24] and the U.S. Environmental Protection Agency's Design for the Environment Program [25], 17 chemical hazard traits associated with human health and the environment from the GS framework were evaluated in this study as shown in Table 2 [26]. The 17 traits are grouped into five hazard groups: priority health effects (PE), human health (HH), environmental fate (EF), ecological (Eco), and physical (Phy), where the hazard of each specific trait is integrated into an overall hazard rating in each group. The hazard trait of "Bioaccumulation" is neglected in this study because all substances of interest are metals. According to the GS methodology, hazard thresholds of low, medium, high, or very high are assigned to each hazard trait. A "not detectable" (ND) label is assigned to any hazard trait

Table 2 Chemical hazard traits used for GreenScreen®-based assessment

Group I human	Group II and II* human	Ecotox	Fast	Physical
Carcinogenicity (c)	Acute mammalian toxicity (AT)	Acute aquatic toxicity (AA)	Persistence (P)	Reactivity (RX)
Mutagenicity and genotoxicity (M)	Systemic/organ toxicity (ST)	Chronic aquatic toxicity (CA)	Bioaccumulation (B)	Flammability (F)
Reproductive toxicity (R)	Neurotoxicity (N)			
Developmental toxicity (D)	Skin sensitization (SnS)			
Endocrine activity (E)	Respiratory sensitization (SnR)			
	Skin Irritation (IrS)			
	Eye Irritation (IrE)			

where there is insufficient information to assess the hazard for the element. The GS decision logic is then applied to the hazard data, and a "GS-based" benchmark score is determined [26]. We refer to our results as "GS based" because our analysis has not been independently reviewed by a toxicologist. Despite the existence of some data gaps and data inconsistencies, GS-based benchmark scores were determined for each element. The chemical hazard data and BM scores for the ten refractory metal elements are provided in Table 3 (note that "ND" means "no data"). All ten refractory metal elements are classified as either BM-1 "Avoid, Chemical of High Concern" (Al, Cr, and V) or BM-2 "Use but Search for Safer Substances" (Hf, Mo, Nb, Ta, Ti, W, and Zr). The hazard traits that lead to the BM-1 benchmark scores include "high" carcinogenicity for Cr and V and "very high" chronic aquatic toxicity for Al. Additional hazard traits of concern include high acute toxicity for Mo; high systemic toxicity for Al, Cr, and Hf; high skin sensitization or respiratory sensitization for Al,

Table 3 Hazard data and GS-based benchmark scores for refractory elements

	Group I Human					Group II and II* Human									Ecotoxicity and Fate				Physical Hazards	
							ST		N											
Elements/CAS	C	M	R	D	E	AT	Single	Repeated	Single	Repeated	SnS*	SnR*	IrS	IrE	AA	CA	P	B	Rx	F
W 7440-33-7	ND	ND	ND	ND	ND	ND	M	ND	ND	ND	ND	ND	M	M	vL	ND	vH	ND	M	H
BM 2							GHS Japan						ECHA	GHS Japan	German FEA		EC-CEPA DSL		H252 H228 GHS Japan	
Hf 7440-58-6	ND	ND	ND	ND	ND	M	M	H	ND	ND	ND	ND	H	H	ND	L	ND	ND	H	H
BM 2						H302 H312 ECHA	H335 ECHA	H373 ECHA					GHS Japan			German FEA			H251 ECHA	H228 GHS Japan
Ta 13981-95-8	M	ND	ND	ND	ND	ND	M	ND	ND	ND	ND	ND	ND	ND	vL	ND	vH	ND	vH	H
BM 2	MAK						GHS Japan								German FEA		EC-CEPA DSL			
Zr 7440-67-7	ND	ND	ND	ND	ND	ND	M	ND	ND	ND	H	ND	ND	ND	vL	ND	vH	ND	vH	H
BM 2							GHS Japan				MAK				German FEA		EC-CEPA DSL		EU GHS	
Mo 7439-98-7	ND	ND	M	ND	ND	H	M	ND	ND	ND	ND	ND	H	H	ND	ND	vH	ND	M	H
BM 2			GHS New Zealand			US EPA PPT	GHS Japan						GHS Japan				EC-CEPA DSL		ECHA	H228 GHS Japan
V 7440-62-2	H	H-vH	ND	ND	ND	ND	ND	ND	ND	ND	ND	H	H	H	ND	ND	vH	ND	ND	M
BM 1		MAK										DHI	ECHA				EC-CEPA DSL			ECHA
Nb 7440-03-1	ND	ND	ND	ND	ND	ND	ND	ND	ND	ND	ND	ND	H	H	ND	ND	vH	ND	ND	H
BM 2													GHS Japan				EC-CEPA DSL			GHS Japan
Al 7429-90-5	L	M	L	L	M	L	H	H	ND	M	L	H	L	L	vH	vH	vH	L	H	H
BM 1		MAK			TEDX	US EPA	H370 H372 GHS Japan				AOEC Asthmagens				US EPA	US EPA	EC-CEPA DSL		H261 GHS Japan	H250 ECHA
Cr 7440-47-3	H	M	M	ND	M	ND	H	ND	ND	ND	H	H	H	M	M	M	vH	ND	H	H
BM 1	ECHA	GHS Japan	GHS New Zealand		TEDX		H371 GHS Japan				GHS Japan				H400 H413 EU		EC-CEPA DSL		H272 H228 ECHA	
Ti 7440-32-6	ND	ND	ND	ND	ND	ND	M	ND	ND	ND	ND	ND	ND	ND	vL	ND	vH	ND	vH	H
BM 2							GHS Japan								German FEA		EC-CEPA DSL		H252 H250 H228 ECHA	

Cr, V, and Zr; high skin irritation or eye irritation for Cr, Hf, Mo, Nb, and V; very high acute aquatic toxicity for Al; high or very high reactivity for Al, Cr, Hf, Ta, Ti, and Zr; and high flammability for all the elements except V.

Conclusions

In this work, we present availability, price, and chemical hazard analysis for common elements used in RHEAs. We provide resource, economic, and environmental alloying screening principles that needs to be considered by the metallurgical community. A proper balance between mechanical behavior and sustainable viability is necessary for the expansion of RHEAs application. Methods employed in this study can also be applied to materials selection for other alloy systems. Given the high potential for carcinogenicity, as categorized in Group I Human Health Hazards, Cr and V are identified as BM-1 elements, and therefore, their use should be avoided, if possible. On the other hand, alloys using Ta and Ti are considered better candidates since they do not pose any human health hazards, instead they present only physical hazards which can be minimized if processed with proper safety precautions. Designers should avoid Re due to its extremely high price. To help mediate the scarcity and cost concerns of elements in RHEAs, such as Hf and V, future research should focus on improving the recycling infrastructure and reuse capabilities of these elements. At the moment, scrap collection is one of the largest hindering factors due to geographic dispersity [9].

Acknowledgements This work was supported by the Lincoln Dynamic Foundation, World Institute for Sustainable Development of Materials (WISDOM) at the University of California, Irvine. The authors thank Dr. Amir Saeidi for helpful discussions.

References

1. Cantor B, Chang ITH, Knight P, Vincent AJB (2004) Microstructural development in equiatomic multicomponent alloys. Mat Sci Eng A 375–377:213
2. Yeh JW, Lin SJ, Chin TS, Gan JY, Chen SK, Shun TT, Tsau CH, Chou SY (2004) Formation of simple crystal structures in Cu-Co-Ni-Cr-Al-Fe-Ti-V alloys with multiprincipal metallic elements. Metall Mater Trans A 35:2533–2536
3. Miracle DB, Senkov ON (2017) A critical review of high entropy alloys and related concepts. Acta Mater 122:448–511
4. George EP, Raabe D, Ritchie RO (2019) High entropy alloys. Nat Rev Mat 4:515–534
5. Senkov ON, Wilks GB, Scott JM, Miracle DB (2011) Mechanical properties of $Nb_{25}Mo_{25}Ta_{25}W_{25}$ and $V_{20}Nb_{20}Mo_{20}Ta_{20}W_{20}$ refractory high entropy alloys. Intermetallics 19:698–706
6. Senkov ON, Miracle DB, Chaput KJ, Couzinie JP (2018) J Mater Res 33(19)
7. Wu J, Yang Z, Xian J, Gao X, Lin D, Song H (2016) Thermodynamic properties of refractory high entropy alloys. J Alloys Compd 682:773

8. Fu X, Schuh CA, Olivetti EA (2017) Materials selection considerations for high entropy alloys. Scr Mater 138:145–150
9. Cann JL et al (2021) Sustainability through alloy design: Challenges and opportunities. Prog Mater Sci 117:100722. https://doi.org/10.1016/j.pmatsci.2020.100722
10. Alonso E, Gregory J, Field F, Kirchain R (2007) Material availability and the supply chain: risks, effects, and responses. Amer Chem Soc 41(19):6649–6656
11. U.S. Geological Survey (2013) Mineral commodity summaries 2013: U.S. Geological Survey, 198 p 121
12. U.S. Geological Survey (2018) Mineral commodity summaries 2018: U.S. Geological Survey, 200 p 125. https://doi.org/10.3133/70194932
13. U.S. Geological Survey (2020) Mineral commodity summaries 2020: U.S. Geological Survey, 200 p 121
14. U.S. Geological Survey (2021) Mineral commodity summaries 2021: U.S. Geological Survey, 200 p 1–200. https://doi.org/10.3133/mcs2021
15. Hirschman AO (1945) National power and the structure of foreign trade. University of California Press, Berkeley, CA
16. Federal Trade Commission (1992) Horizontal merger guidelines. Department of Justice, Washington, DC
17. British Geological Survey (2015) Risk list 2015: British Geological Survey, 200 p 1–8
18. Nassar N, Brainard J, Gulley A, Manley R, Matos G, Lederer G, Bird L, Pineault D, Alonso E, Gambogi J, Fortier SM (2015) Sci Adv 1(3): e1400180
19. "National Minerals Information Center." Historical Statistics for Mineral and Material Commodities in the United States, www.usgs.gov/centers/nmic/historical-statistics-mineral-and-material-commodities-united-states
20. "Tungsten." NITE, 2016, www.nite.go.jp/en/chem/chrip/chrip_search/cmpInfDsp?cid=C004-783-80A&slScNm=CI_07_001&bcPtn=3
21. "Niobium." C&L Inventory, www.echa.europa.eu/information-on-chemicals/cl-inventory-database/-/discli/notification-details/130894/862080
22. EPA, Environmental Protection Agency, www.epa.gov/
23. Canada, Environment and Climate Change. "Government of Canada." Domestic Substances List-Canada.ca,/Gouvernement Du Canada, 11 Aug 2020, www.canada.ca/en/environment-climate-change/services/canadian-environmental-protection-act-registry/substances-list/domestic.html
24. United Nations (2011) Globally Harmonized System of Classification and Labelling of Chemicals (GHS), Fourth rev. New York and Geneva
25. "History of Safer Choice and Design for the Environment." EPA, Environmental Protection Agency, www.epa.gov/saferchoice/history-safer-choice-and-design-environment
26. "GreenScreen® Method." GreenScreen® For Safer Chemicals, Jan 2018, www.greenscreenchemicals.org/learn/full-greenscreen-method

Life Cycle Sustainability Assessment of Repair Through Wire and Arc Additive Manufacturing

Emanuele Pagone, Joachim Antonissen, and Filomeno Martina

Abstract Extending the useful life of a product through repair can significantly reduce the environmental impact associated with its production, and it can be less resource intensive than other environmentally virtuous practices like recycling. Wire and arc additive manufacturing (WAAM) appears to be a promising approach in this context, being characterized by high-resource efficiency, flexibility to perform repairs, and having recently gained industrial maturity. In this work, a methodology to assess the life cycle environmental sustainability of repaired products through WAAM will be presented with a real-world, industrial case study.

Keywords Sustainability · Additive manufacturing · Modeling and simulation

Introduction

Wire and arc additive manufacturing (WAAM) is a fusion and wire-based additive manufacturing (AM) technology that uses a robotic arm to build, layer upon layer, a desired shape [1]. It is characterised by relatively high material deposition rates that makes it well-suited to produce medium-to-large, custom-made components. WAAM is a relatively novel process that has gained considerable industrial attention thanks to its potential to reduce cost and environmental impact in comparison with traditional subtractive approaches. In particular, many of the mentioned advantages stem from lower material utilisation, making the process significantly more efficient [2, 3]. Furthermore, other important business drivers for the adoption of WAAM are customisation and faster time to market. In particular, WAAM creates

E. Pagone (✉)
Sustainable Manufacturing Systems Centre, Cranfield University, Cranfield MK43 0AL, UK
e-mail: e.pagone@cranfield.ac.uk

J. Antonissen
Guaranteed, J.F. Kennedylaan 3, Zelzate B-9060, Belgium

F. Martina
WAAM3D, 5 Thornton Chase, Milton Keynes MK14 6FD, UK

A. Lazou et al. (eds.), *REWAS 2022: Developing Tomorrow's Technical Cycles (Volume I)*, The Minerals, Metals & Materials Series,
https://doi.org/10.1007/978-3-030-92563-5_39

the opportunity to simplify the manufacture of complex assemblies substituting many subcomponents (that would otherwise be traditionally required) as well as produce objects not constrained by a pre-defined bounding volume [4]. The industrial deployment of WAAM is being considered for several important sectors including aerospace, marine, construction, oil and gas, railway, and heavy industry [5].

Considering the novel nature of WAAM, its environmental impact has been studied only to a limited extent in the scientific literature. For example, it has been estimated that energy consumption for thin-walled, steel parts is 34% lower in comparison with conventional milling when including materials primary production but excluding recycling [6]. Other studies have tried to define frameworks to assess energy consumption for additive processes in general, starting from first principles (e.g. mass and energy conservation), but, although useful, they are not specific enough for industrial applications [7].

WAAM products life cycle assessment (LCA) has been sporadically investigated in the scientific literature. For example, a comparative "cradle-to-gate" (i.e. excluding life cycle phases beyond to shipment to the customer) analysis has been performed proposing a modelling framework able to estimate cumulative energy consumption and carbon dioxide equivalent emissions of steel parts [8]. Earlier works included the LCA through the Eco-Indicator 99 methodology of direct additive laser manufacturing for titanium alloys suitable for small and complex parts that showed an environmental impact reduction as high as 70%, but excluded any quantitative considerations about repair routes [9]. Product LCA of less-productive, powder-based, additive processes has been studied as well: electron beam melting (EBM) has been compared to conventional processes using the cumulative exergy demand and "CML 2 Baseline 2000" methods to assess the environmental impact in combination with dimensionless criteria to support decision making [10]. Another work focussed on novel materials aimed at improving the environmental performance, in a life cycle perspective, considering generically additive manufacturing technologies [11].

As shown, although the LCA of WAAM products has been carried out in the scientific literature, no quantitative works can be found that have comparatively assessed WAAM routes to repair as an alternative to purchasing new parts conventionally manufactured. Such analysis is particularly interesting because it combines two rather peculiar characteristics of WAAM: (a) enable repair of products that otherwise need to be scrapped and (b) perform such repair in a resource-efficient manner. On the other hand, the scientific question arises about the fact that such repair through WAAM might be offset by the cumulative environmental impact required for the transportation of the damaged products. The current work aims at addressing this question with an industrial case study that paves the way to a future, more comprehensive assessment framework.

Methods

The use of parts of industrial interest may generate various defects (e.g. holes, cracks, segregation, inclusions, surface marks, and notches). This study considers morphological damage repaired using WAAM. Such defects are normally superficial, and their repair comprises three main steps: removal of defects by machining, deposit material to fill the groove by WAAM, and finishing by machining.

The initial part of this study performs a comparative LCA to manufacture the product from the extraction of resource until it reaches the customer (i.e. including the environmental burden of shipment but excluding use). Such analysis allows to better contextualise the results of a second phase that compares the cumulative environmental impact of repair by WAAM with the conventional manufacture of a new item (including transportation in both cases). The LCA is carried out in four stages as specified by ISO standard 14,044 [12].

1. Goal and scope definition: aim, scope, and level of detail of the study are defined;
2. Life cycle inventory analysis: it defines the inventory of input and output metrics to assess the study previously defined;
3. Life cycle impact assessment: opportunely combining the mentioned inventory, the impact can be finally quantitatively assessed;
4. Life cycle interpretation: an analysis of the results obtained to enable decision making.

The environmental impact parameters considered are provided by Ansys Granta EduPack [13] specifying a range of typical minimum and maximum values. The analysis is performed considering first mean values of such ranges and then combining the "best" and "worst" case scenario for each manufacturing route that are used to construct error bars. One exception to the above is the environmental impact metrics of the material deposition step of the WAAM process that have been estimated from industrial laboratory measurements. Furthermore, for the assessment of the transportation phase, the intermediate value of environmental impact is not exactly the arithmetic mean of the "best" and "worst" case scenarios but simply the value of a plausible shipment route that falls within the two more extreme cases.

Environmental indicators provided by the database are normalised by mass of material, thus the calculation process performs first material flow calculations to estimate the characteristic mass at each LCA step and then computes appropriately the relevant impacts. To this end, an application written in modern Fortran stores the LCA routes in the form of a mathematical direct acyclic graph that is first topologically sorted and then determines the material flows applying simple mass conservation principles in conjunction with a scalar root finder (if necessary by the arrangement of steps).

Case Study

The case study is a steel disk that drives a crank using a protruding pin. A simplified 3D drawing of the part is shown in Fig. 1, whereas similar types of steel are used to manufacture the part (Table 1) according to the conventional and WAAM routes (Fig. 2). The pin is deposited by the WAAM process starting from a wire, and all other geometrical features are obtained by machining. For the shipping phase, the final customer is located in Norway, and the manufacturing plants are considered in Spain for the conventional case and in Belgium for the WAAM one. An estimate of the distances considered for each mode of transportation is summarised in Table 2 detailing different scenarios (see Methods in Sect. 2). The chosen environmental indicators are the cumulative energy consumption and the amount of equivalent carbon dioxide emissions.

The results (Fig. 3) show clearly how the material efficiency of WAAM has a significant impact on the overall environmental impact of manufacturing. In fact, steel primary production (i.e. more rigorously, a collection of steps combining a number of raw materials) dominates clearly the total result for both metrics. It is notable also that, even in the unlikely combination of the worst case scenario for WAAM, it still outperforms conventional manufacturing in terms of cumulative energy consumption, whereas equivalent carbon dioxide emissions show a slight overlap of error bars.

However, it is the repair scenario that shows the significant extent of the environmental benefit of enabling repair through WAAM (Fig. 4). The relevant route includes the shipment of the damaged disk back to Belgium, the repair (removing and re-depositioning the pin), and finally the return of the component to Norway

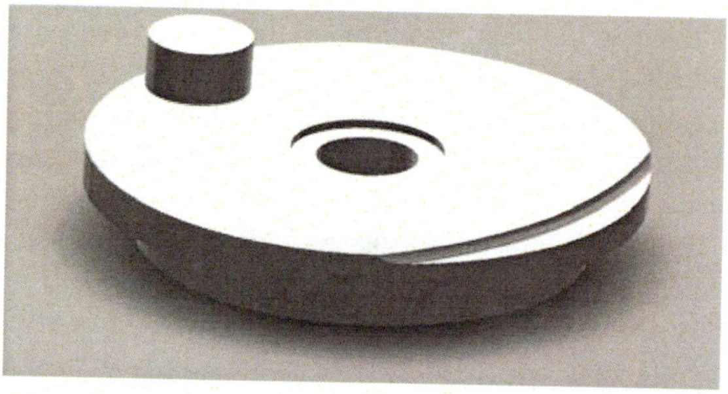

Fig. 1 Simplified CAD model of the driver disk considered as a case study

Table 1 Materials of the driver disk used for the case study. The WAAM wire is used to deposit the pin on a cylindrical substrate

Part	Material
Conventional disk	34CrNiMo6
WAAM wire	ER70
WAAM substrate	YS600

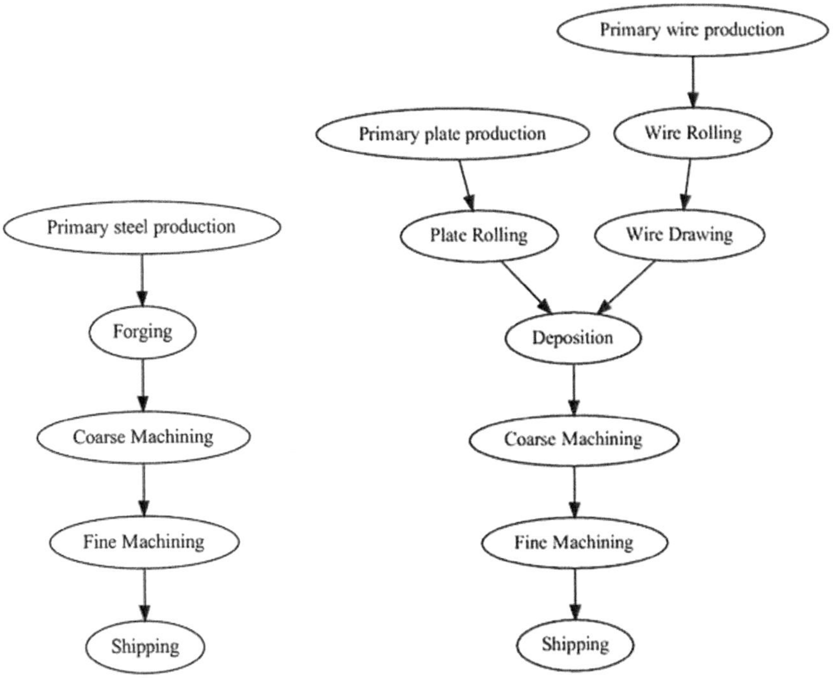

Fig. 2 Conventional (left) and WAAM (right) manufacturing routes to assess the environmental impact of the driver disk used for the case study

Table 2 Assumptions about the shipping modes and relevant distances for the shipment from the manufacturer to the customer of the driver disk considered as a case study

Option	Route	Mode	Distance (km)
Mostly air	Conv	Air freight	2150
		2-axle truck	50
	WAAM	Air freight	1065
		2-axle truck	50
Mostly sea	Conv	Coastal freight	4410
		2-axle truck	30
	WAAM	Coastal freight	970
		4-axle truck	210
		2-axle truck	30
Mostly road	Conv	6-axle truck	2413
		Coastal freight	165
	WAAM	6-axle truck	1250
		Coastal freight	165

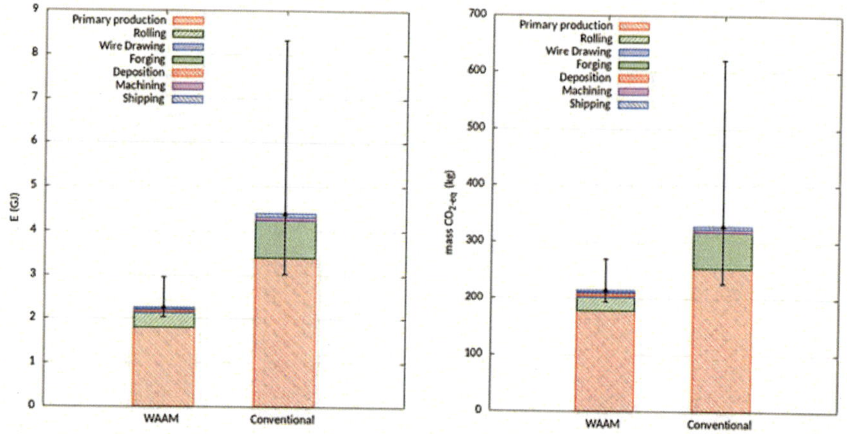

Fig. 3 Cumulative energy consumption (left) and cumulative carbon dioxide equivalent emissions (right) associated with the conventional and the WAAM routes to manufacture the driver disk used for the case study

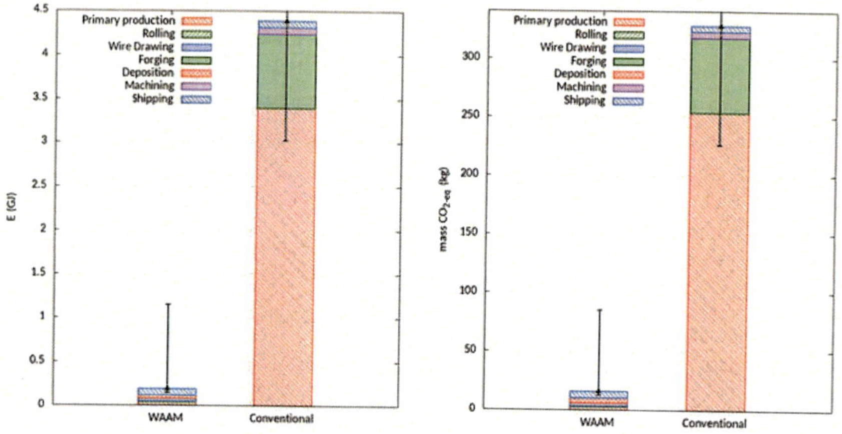

Fig. 4 Cumulative energy consumption (left) and cumulative carbon dioxide equivalent emissions (right) associated with the repair using WAAM of the driver disk used for the case study and production of a new one by conventional manufacturing

(see Fig. 5) according to the same modes of transportation and distances of Table 2. However, it would be also possible to install WAAM robots on site reducing even further the environmental impact.

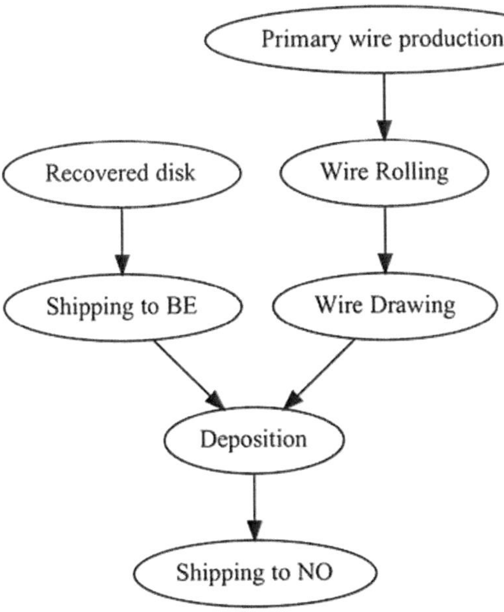

Fig. 5 Repair through WAAM route to assess the environmental impact of the driver disk used for the case study. BE: Belgium; NO: Norway

Conclusion and Further Work

This work has presented an approach to quantify the environmental benefit of repair opportunities of industrial components enabled by wire and arc additive manufacturing (WAAM) in a life cycle perspective. A provisional calculation framework has been outlined and, then, illustrated considering a case study based on real data of a steel disk driver. Product life cycle phases from use and beyond were excluded by the scope of the study, but the impact of transportation between the manufacturer and the customer was included. It could be observed that the material efficiency of WAAM in comparison with conventional, subtractive manufacturing routes determines a clear advantage on the associated cumulative energy consumption and cumulative equivalent carbon dioxide emissions. When the repair of the disk (re-depositioning its pin) is enabled by WAAM, the avoided manufacture of an entire new component reduces significantly (i.e. by orders of magnitude) the environmental impact of the component.

To further extend this study, the following future development aspects are envisioned.

- Consider parts characterised by different geometries and materials;
- Follow a more realistic and rigorous treatment of uncertainty avoiding to conservatively combine "best" and "worst" case scenarios of the input parameters. To this goal, both an analytical and numerical (i.e. Monte Carlo) approach can be considered;
- Include reuse and recycling of products in the life cycle phases;

- Consider more comprehensive methods for the life cycle impact assessment (e.g. ReCiPe2016 [14]);
- Include and combine other metrics of industrial interest to assess also financial, productivity, responsivity, and quality aspects.

Acknowledgements The foundational work carried out by Mr. Hrushikesh Sawant as part of his MSc thesis project is gratefully acknowledged.

References

1. Wu B, Pan Z, Ding D, Cuiuri D, Li H, Xu J, Norrish J (2018) A review of the wire arc additive manufacturing of metals: properties, defects and quality improvement. J Manuf Process 35:127–139
2. Priarone PC, Pagone E, Martina F, Catalano AR, Settineri L (2020) Multi-criteria environmental and economic impact assessment of wire arc additive manufacturing. CIRP Ann 69(1):37–40
3. Bekker ACM, Verlinden JC (2018) Life cycle assessment of wire + arc additive manufacturing compared to green sand casting and CNC milling in stainless steel. J Clean Prod 177:438–447
4. Frazier WE (2014) Metal additive manufacturing: a review. J Mater Eng Perform 23:1971–1928
5. Busachi A, Erkoyuncu J, Colegrove P, Martina F, Ding J (2015) Designing a WAAM based manufacturing system for defence applications. Procedia CIRP 37:48–53
6. Campatelli G, Montevecchi F, Venturini G, Ingarao G, Priarone P (2020) Integrated WAAM-subtractive versus pure subtractive manufacturing approaches: an energy efficiency comparison. Int J Precis Eng Manuf Green Technol 7(1):1–11
7. Peng T (2016) Analysis of energy utilization in 3D printing processes. Procedia CIRP 40:62–67
8. Priarone PC, Campatelli G, Montevecchi F, Venturini G, Settineri L (2019) A modelling framework for comparing the environmental and economic performance of WAAM-based integrated manufacturing and machining. CIRP Ann 68(1):37–40
9. Serres N, Tidu D, Sankare S, Hlawka F (2011) Environmental comparison of MESO-CLAD process and conventional machining implementing life cycle assessment. J Clean Prod 19(9–10):1117–1124
10. Paris H, Mokhtarian H, Coatanéa E, Museau M, Ituarte IF (2016) Comparative environmental impacts of additive and subtractive manufacturing technologies. CIRP Ann 65(1):29–32
11. Faludi J, Van Sice CM, Shi Y, Bower J, Brooks OMK (2019) Novel materials can radically improve whole-system environmental impacts of additive manufacturing. J Clean Prod 212:1580–1590
12. International Organization for Standardization (2018) BSI Standards Publication Environmental Management—Life cycle assessment—Requirements and Guidelines. ISO 14044:2018
13. Ansys Granta (2021) Ansys Granta EduPack. https://www.ansys.com/en-gb/products/materials/granta-edupack
14. Huijbregts MAJ, Steinmann ZJN, Elshout PMF, Stam G, Verones F, Vieira MDM, Hollander A, Van Zelm R (2016) ReCiPe2016: a harmonized life cycle impact assessment method at midpoint and endpoint level. RIVM Report 2016–0104. Bilthoven, The Netherlands

The REMADE Institute: R&D to Accelerate the Transition to a Circular Economy

Edward J. Daniels

Abstract The REMADE Institute invests in research and development to develop technology solutions to enable the increased remanufacturing and recycling of metals, polymers, fibers, and e-waste. Increasing the recycling and remanufacturing of these materials can significantly contribute to an increase in energy efficiency, an increase in materials use efficiency, and a reduction of GHG emissions in the domestic manufacturing sector. The current R&D portfolio of REMADE is structured across five activities including systems analysis and integration, manufacturing materials optimization, design for Re-X, remanufacturing and end-of-life reuse, and recovery and recycling. This paper outlines the mission of the Institute, provides an overview of the structure and approach in developing the R&D portfolio, and describes a few of the projects and their beneficial energy, materials, and environmental impacts from our broad R&D portfolio.

Keywords Circular economy · Recycling · Remanufacturing · Metals · Plastics · Polymers · Fibers · E-waste · R&D · Education and workforce development

Introduction

The Reducing EMbodied-energy And Decreasing Emissions (REMADE) Institute is a Department of Energy (DOE) sponsored Manufacturing USA Institute [1]. Today, manufacturing accounts for about 25% of US energy consumption. With improvements in materials production, processing, and end-of-life (EoL) disposition, the USA could significantly increase manufacturing energy efficiency, which could also yield substantial economic savings. To help realize these opportunities, the REMADE Institute was launched in January 2017.

REMADE brings together industry innovators, academic researchers, and national laboratories to enhance the nation's industrial competitiveness and lead the transition to a circular economy in the USA.

E. J. Daniels (✉)
The REMADE Institute, 150 Lucius Gordon Drive, West Henrietta, NY 14586, USA
e-mail: edaniels@remadeinstitute.org

A. Lazou et al. (eds.), *REWAS 2022: Developing Tomorrow's Technical Cycles (Volume I)*, The Minerals, Metals & Materials Series,
https://doi.org/10.1007/978-3-030-92563-5_40

REMADE Mission

REMADE's mission is to support the early stage applied research and development of key industrial platform technologies that could dramatically reduce the embodied energy and carbon emissions associated with industrial-scale materials production, processing, and end-of-life (EOL) disposition.

REMADE is a growing membership organization. To achieve its mission, the Institute and its 130 industry, academia, trade association, and national laboratory members focus on increasing the recovery, reuse, remanufacturing, and recycling of metals, fibers, polymers, and electronic waste (e-waste).

The four primary goals of the Institute are as follows:

1. Develop technologies capable of reducing energy and emissions through a reduction in primary material consumption and an increase in secondary feedstock use in energy-intensive industries
2. Develop technologies capable of achieving "better than cost and energy parity" for key secondary materials
3. Promote widespread application of new enabling technologies across multiple industries, and
4. Educate, train, and develop the incumbent and future workforce to support deployment of REMADE technologies.

Technology Research and Development

To achieve the first three of these goals, the technology research agenda for the Institute is structured across the five (5) REMADE nodes.

The *systems analysis and integration node* identifies strategic opportunities to reduce the embodied energy and emissions associated with materials production and processing and evaluates the economic impact of new technologies or changing demand patterns at a project, company, sector, or national level.

The *design for Re-X node* develops application domain-specific frameworks (such as design for remanufacturability) and creates tools that enable design engineers to understand how their design choices will impact the ability to reuse, remanufacture, recover, or recycle products, components, and materials.

The *manufacturing materials optimization node* develops processes, sensing technologies, and simulation tools that enable manufacturers to increase their use of secondary and cross-industry feedstocks without loss of performance or properties, reuse scrap generated during manufacturing, and reduce in-process losses.

The *remanufacturing and end-of-life reuse node* develops and improves technologies for characterizing the condition of products and components, identifies (the most) cost-effective approaches for core and component processing, and develops repair technologies to restore component to like-new condition.

The *recycling and recovery node* develops and matures technologies to increase the availability of secondary feedstocks by developing tools and technologies to economically collect, recover, sort, separate, purify, and reprocess metals, polymers, fibers, and e-waste.

Education and Workforce Development (EWD)

Achievement of the fourth goal is enabled through the REMADE EWD program. REMADE has developed a roadmap to guide the development and execution of the EWD program [2].

To support the transition to a more circular economy through design, recycling, reuse, remanufacturing, and systems analysis improvements, industry needs a workforce educated and trained in the technologies, processes, and analysis methods critical to design, decision making, and implementation; however, current workforce programs and training materials are geared toward traditional manufacturing. Through its EWD training offerings, REMADE prepares the incumbent and future workforce to deploy and use REMADE-relevant technologies that currently exist or will result from REMADE technology development projects.

The EWD program offers technology webinars, short courses, and a Tiered Certificate Pathway. EWD offerings are available on demand and can also be presented in person.

Selected webinars that have been presented include

- Design for Reman—Real World Challenges and Opportunities
- New Technology Recovers Aluminum from Industrial Scrap Metal
- Quantitative Non-Destructive Evaluation of Fatigue Damage Based on Multi-Sensor Fusion
- Pushing the State of the Art Through Innovation in Scrap Sorting and Impurity Removal
- Data-Driven Design Support for Re-X of High Value Components in Industrial and Agricultural Equipment.

Short courses that are available include:

- Crash Course in Plastics Recycling: Overview of existing recycling technologies capable of converting plastic waste into high-value products
- End-of-Life and the Circular Economy: Discover options for recovering end-of-life plastics and areas where there is opportunity to improve
- Introduction to Remanufacturing: Introduction to reman and EOL reuse, related technologies, and environmental impacts

- Condition Assessment for Reman: Overview of inspection technologies, methods, capabilities, and limitations
- Design Capture and Design for Reman: Understanding design capture, integration, and the remanufacturing design framework.

Tiered Certificate Pathways are developed by grouping similar short courses together to provide a deeper level of understanding by competency level of a broad technology subject, such as mechanical recycling or remanufacturing. To achieve a formal certificate from the REMADE Institute, students need to complete each of the short courses for the associated Tiered Certificate Pathways. The Institute created Tiered Certificate Pathways to address the lack of established certification programs for employers to assess workers Re-X knowledge and skills. Tiered Certificate Pathways that are presently available include

- Fundamentals of remanufacturing
- Fundamentals of mechanical recycling of plastics.
- Emerging trends in plastics recycling.

Additional Tiered Certifcate Pathways are being developed.

Technology R&D Portfolio

REMADE has issued five Requests for Proposals (RFP) to date. The proposals that were submitted in response to RFP 5 are under review as of this writing and are not discussed in this paper.

Topics that are solicited in RFPs are based on the REMADE technology roadmap [3]. The roadmap is updated periodically and identifies the high-priority research activities, which are a subset of research activities deemed to be the most impactful/important activities that REMADE should pursue.

Including all projects through RFP 4, the REMADE technology R&D portfolio includes 60 projects, with a total funding (including partner cost-share) of $52.2 million.

The projects and the participating REMADE members that comprise the R&D portfolio, to date, are highlighted for each of the five REMADE nodes.

Systems Analysis and Integration

Mapping the Materials Base for REMADE—Yale University, Unilever, Institute of Scrap Recycling Industries, Sunnking, Massachusetts Institute of Technology.

- This project will develop a materials flow baseline for REMADE materials (metals, fibers, polymers, and e-waste) to support measurement of the impact of future technology improvements through REMADE projects. A harmonized and validated set of data for metals, fibers, polymers, and e-waste will be developed within a consistent framework that allows comparisons of material efficiencies across REMADE materials at all life cycle stages.

Other systems analysis and integration projects include

- *Dynamic Systems Analysis of PET and Olefin Polymers in a Circular Economy—Michigan Technological University, Idaho National Laboratory, Resource Recycling Systems, Yale University, Chemstations, Honeywell UOP*
- *A Dynamic Techno-economic Systems Modeling Framework for U.S. Fiber Recycling—Northwestern University, Yale University, Institute of Scrap Recycling Industries*
- *Identifying Strategies to Maximize Benefit of Fiber Recovery through Systems Quantification—Massachusetts Institute of Technology, The American Forest and Paper Association, WestRock, Graphic Packaging*

Design for Re-X

Design for Remanufacturing—Rochester Institute of Technology, Caterpillar Inc., Remanufacturing Industries Council.

- This project is focused on working directly with remanufacturing industry leaders to create a set of pragmatic "design for remanufacturing" rules that would allow design engineers to integrate remanufacturing considerations in their component and part designs and pave the way for integration of these design rules across various engineering tools and CAD platforms currently in use to enable improvement in component and part manufacturability. These design rules will be verified on existing parts and CAD file(s) provided by the industrial partner to identify potential changes to improve the part manufacturability.

Material and Vehicle Design for High-Value Recycling of Aluminum and Steel Automotive Sheet—University of Michigan, Ford Motor Company, Novelis, Argonne National Laboratory, The Institute of Scrap Recycling Industries, The Aluminum Association, Light Metal Consultants.

- This project seeks to increase automotive sheet metal EOL (post-consumer) recycled content, thus reducing vehicle embodied energy and primary feedstock consumption. The objectives are to produce a new analytical *design for recycling* tool tailored for automotive metal sheet and to generate new knowledge on how EOL sheet recycling is affected by vehicle design (e.g., alloy specification), recycling system infrastructure (e.g., deployment of emerging separation processes), and sheet manufacturing process decisions (e.g., temperature profiles informed by new Integrated Computational Materials Engineering (ICME) tools).

Other design for Re-X projects include

- *Analysis and Design for Sustainable Circularity of Barrier Film in Sheet Molding Composites*—The Ohio State University, Kohler Co., National Renewable Energy Laboratory, Arizona State University
- *Development of an Industrially Relevant RE-SOLAR Design Framework*—University of Pittsburgh, University of California-Irvine, National Renewable Energy Laboratory, First Solar
- *Quantification of Financial and Environmental Benefits Tradeoffs in Multi-Generational Product Family Development Considering Re-X Performances*—University of Illinois at Urbana-Champaign, Iowa State University, Deere and Company, Green Electronics Council
- *Design Iteration Tool to Sustain Remanufacturability*—Iowa State University, Danfoss.

Manufacturing and Materials Optimization

Supramolecular Interfacial Reinforcement for Manufacture Utilizing Mixed Secondary Plastic Feedstock—The University of Akron, Braskem.

- This project seeks to develop a compatibilizer for MPO (mixed polymeric olefins, i.e., PE and PP) which are difficult to separate from each other and are incompatible. The compatibilized MPO will compete with virgin PE and virgin PP.

Achieving 100% Recycling Aluminum in Die Casting Applications—The Ohio State University, Alcoa USA Corp., North American Die Casting Association, CompuTherm LLC.

- The goal of this project is to achieve 100% use of recycled aluminum in die casting applications by realizing the following two objectives: 1) substitute 100% secondary materials for primary alloys in structural die castings with no degradation in properties and 2) improve the mechanical properties of the current secondary alloys for non-structural applications.

Other manufacturing materials optimization projects include

- *Upcycling Polyethylene Waste into Value-Added Nylons for Sustainable Automotive Manufacturing*—Michigan State University, American Chemistry Council, BASF
- *Chemical Conversion and Process Control for Increased used of Polyethylene and Polypropylene Secondary Feedstocks*—University of Massachusetts-Lowell, Massachusetts Institute of Technology, SER North America LLC, iMFLUX Inc.
- *Development of a Castable High Strength Secondary Aluminum Alloy from Recycled Wrought Aluminum Scrap*—University of Illinois at Urbana-Champaign, Eck Industries Inc.

- *Increasing Melt Efficiency and Secondary Alloy Usage in Aluminum Die Casting*—The Ohio State University, Alcoa USA Corp., North American Die Casting Association
- *Biological and Bio-Mechanical Technologies for Recycled Fibers to Regain Fiber Quality and Increase Secondary Feedstock in High Value-Added Paper Grades*—Western Michigan University, Idaho National Laboratory, Graphic Packaging International, WestRock Company

Remanufacturing and End-Of-Life Reuse

Rapid Damage Identification to Reduce Remanufacturing Costs—Iowa State University, John Deere and Company.

- The objective of this project is to develop and validate a remanufacturability assessment method that will support decision making about the viability of remanufacturing a component. The proposed method is based on development of machine learning (ML) techniques for recognizing different types of component damage, embedding developed ML algorithms in low-cost, damage identification hardware for use in-process at the remanufacturing factory floor and using this in-process technique to develop a real-time estimate of remanufacturing costs for a component. Although most high-value, metal-alloy components can be remanufactured, sufficiently accurate and rapid decision-making support tools are needed to significantly reduce remanufacturing costs and increase the throughput and volume of remanufactured components.

Epoxy/Silicon Potting Material Removal for Greater Recovery of Circuit Boards—Rochester Institute of Technology, Caterpillar Inc., CoreCentric Solutions.

- More cost-effective technologies are needed to remove coating or potting materials from circuit boards to enable repair and reuse. Two alternative technologies, laser ablation and micro-media blasting, will be tested and evaluated to quantify cost effectiveness relative to industry specified cost targets.

Other remanufacturing and end-of-life reuse include

- *Development of Additive Manufacturing Material and Process Technologies to Improve the Re-Manufacturing Efficiency of Commercial Vehicle Tires*—Virginia Polytechnic Institute, Arizona State University, Michelin North America, Nike Inc., Sealed Air Corporation
- *Remanufacturing of Surface-Hardened Steel Components by Ultrasonic Surface Modification*—Rochester Institute of Technology, Caterpillar, Inc., University of Pittsburgh
- *In-situ Nondestructive Evaluation of In-flight Particle Dynamics and Intrinsic Properties for Thermal Spray Repairs*—Iowa State University, John Deere

- *Remaining Life Determination*—Rochester Institute of Technology, University of Illinois at Urbana-Champaign, Caterpillar Inc.
- *Condition Assessment of Used Electronics*—Rochester Institute of Technology, Caterpillar Inc., CoreCentric Solutions.

Recycling and Recovery

Scalable High Shear Catalyzed Depolymerization of Multilayer Plastic Packaging—University of Massachusetts-Lowell, Michigan State, Unilever, American Chemistry Council, National Renewable Energy Laboratory.

- Industry is increasingly combining layers of different polymer materials to construct highly functional, lightweight packaging (e.g., to extend food life). These multilayer films are unfortunately less recyclable than single layer films. This project will investigate catalytic depolymerization as a cost-effective approach to process these films into higher value products suitable for use in a variety of applications.

Pushing the State of the Art in Steel Recycling through Innovation in Scrap Sorting and Impurity Removal—Colorado School of Mines.

- Increasing the utility of steel scrap through innovation in sorting and impurity removal will increase the use of secondary feedstock and achieve cost parity for secondary materials for steel products. This study will investigate (1) physical methods such as optical sorting to upgrade scrap steel and (2) chemical or metallurgical treatment methods to remove or neutralize the effect of impurities in molten steel.

Other recycling and recovery projects include

- *Selective Recovery of Elements from Molten Aluminum Alloys*—Phinix, LLC, Worcester Polytechnic Institute, Kingston Process Metallurgy, Smelter Service Corporation, Certified Flux Solutions, LLC
- *Rapid Sorting of Scrap Metals with Solid State Device*—University of Utah
- *Development of New Cost-Effective Methods for Removing Trace Contaminants in Recycled Metals*—The Ohio State University, Alcoa, Computherm
- *Smart Additive Manufacturing Towards Use of Recycled Paper Fibers for Producing High-quality Fiber-Reinforced Plastic (FRP) Composites*—University of Iowa, Impossible Objects, Inc.
- *New Approaches to Improve Deinking Flotation to Increase the Availability of High-Quality, Low-Cost Recycle Paper Fibers*–Virginia Polytechnic Institute
- *Determining Material, Environmental and Economic Efficiency of Sorting and Recycling Mixed Flexible Packaging and Plastic Wrap*—American Chemistry Council, Resource Recycling Systems, Idaho National Laboratory

- *Chemical Recycling of Mixed PET/Polyolefin Streams Through Sequential Pyrolysis and Catalytic Upgrading*—The Pennsylvania State University, Northwestern University, Shaw Group Industries, Inc., Process Systems Enterprise, Inc.
- *Recycling of PET in Sustainable Food Packaging Systems*—MuCell Extrusion LLC, Plastilene SAS a Plastilene Group Company, Wingate Packaging, Sugar Creek Packaging Co., Center for Innovative Food Technology, The Ohio State University
- *Chemical Recycling of Mixed Plastics and Valuable Metals in the Electronic Waste Using Solvent-Based Processing*—University of Massachusetts-Lowell, Sunnking, Inc., Institute of Scrap Recycling Industries.

REMADE Technical Performance Metrics

The REMADE Institute has established four target technical performance metrics (TPM) for the technology R&D portfolio [4]. The target values of the metrics are as follows:

1. Reduction in embodied energy use: 1018 PJ/y
2. Reduction in (GHG) emissions: 51.2 MMT/y of CO_2 equivalent
3. Increase in secondary materials use: 41.2 MMT/y
4. Decrease in primary materials use: 41.2 MMT/y.

Following the selection of projects from each RFP, REMADE evaluates the estimated potential impact of each project, and then the overall portfolio impacts on the four TPMs of embodied energy savings, decreases in primary materials use, and increases in secondary materials use and reduction in GHG emissions. The current technology R&D portfolio is estimated to enable a reduction in embodied energy of about 920 PJ/y, a reduction of GHG of about 50 MMT/y, a reduction in primary materials use, and an increase in secondary materials use of about 27 MMT/y.

Conclusions

Today, REMADE is the only Institute in the USA dedicated to leading the transition from an unsustainable linear economy to a sustainable circular economy. REMADE will continue to lead this transition and will continue to invest in research and workforce development to enable the development of the technology and workforce that is needed to increase the remanufacturing and recycling of materials in the US economy and transition to a circular economy.

Acknowledgements This material is based upon work supported by the US Department of Energy's Office of Energy Efficiency and Renewable Energy (EERE) under the Advanced Manufacturing Office Award Number DE-EE0007897 awarded to the REMADE Institute, a division of Sustainable Manufacturing Innovation Alliance Corp.

Disclaimer This report was prepared as an account of work sponsored by an agency of the US Government. Neither the US Government nor any agency thereof, nor any of their employees, makes any warranty, express or implied, or assumes any legal liability or responsibility for the accuracy, completeness, or usefulness of any information, apparatus, product, or process disclosed, or represents that its use would not infringe privately owned rights. Reference herein to any specific commercial product, process, or service by trade name, trademark, manufacturer, or otherwise does not necessarily constitute or imply its endorsement, recommendation, or favoring by the US Government or any agency thereof. The views and opinions of authors expressed herein do not necessarily state or reflect those of the US Government or any agency thereof.

References

1. The REMADE Institute
2. EWD_Roadmap_2020.pdf (squarespace.com)
3. Technology Roadmap—The REMADE Institute
4. Impact Report—The REMADE Institute

Assessing the Future Resource and Environmental Impacts of China's Aluminum Industry: Implications of Import and Export Transition

Shupeng Li and Tingan Zhang

Abstract Aluminum is widely used in buildings, transportation, and home appliances. However, primary aluminum production is a resource-, energy-, and emission-intensive industrial process. At present, China is the world's largest producer of aluminum. Under China's new national development pattern of the "internal–external dual cycle", China's aluminum industry (AID) future development may also need to be adjusted. This study combines material flow analysis, life cycle assessment and scenario analysis to investigate the potential of resource conservation, energy saving, and emission reduction for China's AID till 2030 under the transition of import and export trade. The results show that nearly 40% of China's annual aluminum production entered the inventory in use in other parts of the world through trade between 2010 and 2017. In the business as usual (BAU) scenario, the bauxite consumption of China's AID will increase from 170 Tg in 2017 to 291 Tg in 2030, with an annual growth rate of 4.2%. Compared with the BAU scenario, the demand for bauxite in the two scenarios of reducing exports of aluminum products will be reduced by 27% (Scenarios A) and 47% (Scenarios B) in 2030, respectively. In addition, there are obvious benefits in terms of water saving, energy saving, and emission reduction under Scenarios A and Scenarios B. Therefore, promoting the transformation of imports and exports can effectively decrease the external dependence on bauxite of China's AID and is also an important means to achieve carbon peaking by 2030.

Keywords Material flow analysis · Life cycle assessment · Scenario analysis · Aluminum · China

---

S. Li · T. Zhang (✉)
School of Metallurgy, Northeastern University, Shenyang 110819, China
e-mail: zta2000@163.net

S. Li
e-mail: neu_lisp@163.com

Key Laboratory of Ecological Metallurgy of Multi-Metal Intergrown Ores of Ministry of Education, Shenyang 110819, China

A. Lazou et al. (eds.), *REWAS 2022: Developing Tomorrow's Technical Cycles (Volume I)*, The Minerals, Metals & Materials Series,
https://doi.org/10.1007/978-3-030-92563-5_41

Introduction

Aluminum is the third most abundant element in the earth's crust, after oxygen and silicon. Due to its excellent physical and chemical properties, metallic aluminum is regarded as a critical primary material for social development and economical construction. Since 1990, China has ushered in an explosion of primary aluminum production. At present, China's primary aluminum output accounts for more than 50% of the world's total output, reaching about 40 Tg. However, the production of primary aluminum is one of the most energy- and emission-intensive industrial processes. Therefore, the subject of investigating the future resource and environmental impact of China's AID has aroused intense interest from national and global stakeholders.

Some studies have been developed to investigate the flow and stock of aluminum [1–4]. Among them, the issue of aluminum flow on the global level has been widely concerned, including the mapping from liquid to goods [1], the centennial evolution of stocks [3], and the secondary aluminum reserves [4]. There are also some studies that have conducted national-level material flow analysis, such as the USA [5, 6], Austria [7, 8], and Italy [9]. In particular, the material flow analysis of aluminum in China has also been deeply reported [10–16]. In addition, previous studies focusing on the environmental issues of aluminum production can be divided into two categories. One is to analyze various environmental impacts of aluminum production based on life cycle assessment [17–20]. The other is to explore the energy consumption and GHG emission reduction potential of the aluminum industry [21–25].

Nevertheless, most of these studies do not quantitatively analyze the resource and environmental impact of the transformation for imports and exports, especially the reducing in aluminum exports, in the China's AID. The reasons for the prosperity of China's primary aluminum production since 1990 can be attributed to two, robust domestic consumption and export demand [14]. At present, China's aluminum exports may be weakened due to the global pandemic of COVID-19. Therefore, the new national development model of the "internal–external dual cycle" is proposed by the Chinese government. China's AID also needs to make corresponding adjustments, that is, to moderately reduce aluminum exports to give priority to domestic demand. As a result, the feasible benefits of reduction of aluminum exports to address the challenges of energy conservation and emissions reduction should be presented clearly to the stakeholders. To fill this gap, this paper has analyzed reducing aluminum exports on energy and resource conservation and emissions reduction from 2017 to 2030 under three scenarios: the business as usual (BAU), the moderate reduction (Scenario A), and the enhanced reduction (Scenario B).

Methodology

Modeling Domestic Demand and Scrap Generation

Similar to previous studies [26], the regression analysis is used to predict China's domestic aluminum demand from 2017 to 2030, using per capita GDP and urbanization rate as explanatory variables. The form of the regression equation is as follows:

$$D(t) = \alpha_0 + \sum\nolimits_{i=1}^{n} \alpha_i X_i(t) + \varepsilon(t) \tag{1}$$

where $D(t)$ is the domestic demand of aluminum at time t; n is the number of explanatory variables; $X_i(t)$ are the explanatory variables at time t; α_i are the regression model parameters; and ε_i is the residuals of the regression model.

When the life of aluminum-containing products ends, aluminum will flow out of in-use stock to produce aluminum scrap. The amount of aluminum scrap produced can be calculated by multiplying the total consumption of aluminum-containing products by their lifetime distribution function. Detailed descriptions of the estimates of future aluminum demand and scrap generation can be found in our published research [27].

$$S_{i,t} = \sum\nolimits_{\tau=a}^{b-1} D_{i,t-\tau} * P(\tau) \tag{2}$$

$$OS_t = \sum\nolimits_{i=1}^{7} S_{i,t} \tag{3}$$

where $S_{i,t}$ is the amount of aluminum scrap generation for the i sector in the year t; $D_{i,t-\tau}$ is the amount of aluminum demand for the i sector in the year $t-\tau$; $P(\tau)$ is the lifetime distribution of the different aluminum-containing product; a and b represent the upper and lower limits of the product life value, respectively; and OS_t is the total amount of old aluminum scrap generation in the year t. The historical data of domestic aluminum consumption were obtained from the International Aluminum Institute (IAI) [28]. The information of GDP and urbanization rate was obtained from the China Statistical Yearbook [29].

Scenario Setting

This study introduces R to quantitatively describe the relationship between domestic consumption and the production of aluminum. R is the ratio of domestic aluminum consumption to total output in a specific year, and its expression is as follows:

$$R_t = \frac{D_{t,\,\text{consumption}}}{O_{t,\,\text{total}}} \tag{4}$$

where R_t is the ratio of the trade structure in the year t;$D_{t,\,\text{consumption}}$ is the amount of aluminum flowing into the domestic in-use stock in the year t; $O_{t,\,\text{total}}$ is the total output of aluminum including primary aluminum and recycled aluminum in the year t.

The R-values of China, North America, and Europe from 1980 to 2017 be shown in Table 1. The R-value in North America and Europe gradually increased from 0.6 in 1980 to 1.4 in 2017. Correspondingly, these two regions have also changed from a net exporter of aluminum to a net importer of aluminum. China has always been a net exporter of aluminum ($R < 1$). Especially since 2010, the R-value has hovered around 0.6, indicating that about 40% of the aluminum produced in China is exported to foreign countries through trade. The experience of North America and Europe shows that the export volume of aluminum produced in China is likely to gradually decrease in the future with the upgrading of the industry. In addition, due to the global pandemic of COVID-19, China has proposed a new model of national development of the "internal–external dual cycle". China's AID also needs to make corresponding adjustments, that is, to moderately reduce aluminum exports to give priority to domestic demand. To analyze the impact of reduced exports, we established three scenarios: the business as usual (BAU), the moderate reduction (Scenario A), and the enhanced reduction (Scenario B). The BAU scenario assumed that the R-value remains unchanged at 0.62 until 2030; this means that China maintains the current high export situation of aluminum products unchanged. Scenario A assumed that the R-value will increase from 0.62 in 2017 to 0.80 in 2030; this means that China is still a net exporter of aluminum, but the net export volume of aluminum is greater than the Scenario B. Scenario B assumed that the R-value will increase from 0.62 in 2017 to 1.0 in 2030; this means that China's aluminum net export volume will decrease to 0 by 2030.

(1) The total output of aluminum. The total amount of aluminum produced under various scenarios in the future can be calculated.

$$O_{t,\,\text{total}} = D_{t,\,\text{consumption}} * R_t \tag{5}$$

Table 1 R-value of China, North America, and Europe

Year	1980	1985	1990	1995	2000	2005	2010
China	0.77	0.91	0.86	0.94	0.77	0.74	0.66
North America	0.62	0.86	0.73	1.10	0.96	1.31	1.17
Europe	0.60	0.72	0.86	1.01	1.00	1.12	1.26
Year	2011	2012	2013	2014	2015	2016	2017
China	0.63	0.62	0.61	0.61	0.55	0.61	0.62
North America	1.18	1.29	1.28	1.36	1.41	1.42	1.43
Europe	1.25	1.23	1.20	1.28	1.24	1.31	1.36

(2) Recycled aluminum output. The mathematical formula of recycled aluminum production is as follows.

$$O_{t,\text{recycled}} = \left(OS_{t,\text{domestic scrap}} + IS_{\tau,\text{import scrap}}\right) \times 0.75 \tag{6}$$

where $O_{t,\text{recycled}}$ represents the output of recycled aluminum in the year t; $OS_{t,\text{domestic scrap}}$ and $IS_{\tau,\text{import scrap}}$ represent the amount of recycled aluminum produced from domestic scrap and imported scrap in the year t, respectively; and *0.75* is the conversion rate of recycled aluminum produced from aluminum scrap.

(3) Primary aluminum output. According to the law of conservation of quality, the calculation formula of primary aluminum output is as follows.

$$O_{t,\text{primary}} = O_{t,\text{total}} - O_{t,\text{recycled}} \tag{7}$$

where $O_{t,\text{primary}}$ represents the output of the primary aluminum in the year t.

(4) Resource consumption and environmental emission. The value of resource consumption, energy consumption, and waste emission in the aluminum production can be obtained by the following three equations:

$$RS_t = O_{t,\text{primary}} \times rsp + O_{t,\text{recycled}} \times rsr \tag{8}$$

$$ES_t = O_{t,\text{primary}} \times esp + O_{t,\text{recycled}} \times esr \tag{9}$$

$$WE_t = O_{t,\text{primary}} \times wep + O_{t,\text{recycled}} \times wer \tag{10}$$

where RS_t, ES_t, and WE_t are resource consumption, energy consumption, and waste emission of aluminum industry in China in the year t; *rsp*, *esp*, and *wep* are resource consumption, energy consumption, and waste emission of per ton primary aluminum production, respectively; *rsr*, *esr*, and *wer* are resource consumption, energy consumption, and waste emission of per ton recycled aluminum production, respectively.

Results and Discussion

Domestic Demand and Scrap Generation

The domestic demand, scrap generation, and primary aluminum production in China during 2017–2030 are shown in Fig. 1. In the future, China's domestic aluminum demand and scrap generation will continue to increase. Aluminum demand will increase from 29 Tg in 2017 to 54 Tg in 2030, with an annual growth rate of 4.9%.

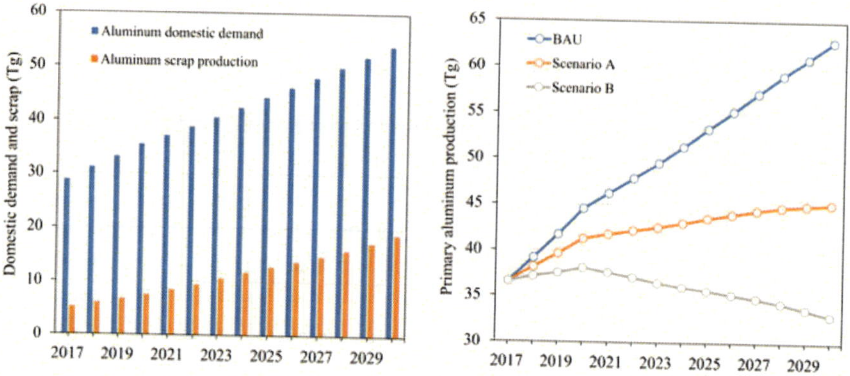

Fig. 1 Domestic demand, scrap generation, and primary aluminum production during 2017–2030

The scrap generated will increase from 5.2 Tg in 2017 to 18.9 Tg in 2030, and its average annual growth rate is twice the domestic demand, at 10.5%. In addition, the measures of reducing China's aluminum exports can effectively decrease primary aluminum output. Under the BAU scenario, China's primary aluminum production will climb from 37 Tg in 2017 to 63 Tg in 2030. Under the moderate reduction scenario (Scenario A), China's primary aluminum output increased slower and eventually remained around 45 Tg. Under the enhanced reduction scenario (Scenario B), China's primary aluminum production will peak at 39 Tg in 2020 and then will be reduced to 33 Tg by 2030. For sustainable development, the primary aluminum production capacity in China has now begun to be restricted and set its upper limit to be about 45 Tg. Therefore, it is a feasible measure to moderately reduce the export volume of aluminum products under China's policy of expanding domestic demand. In addition, compared with other parts of the world, China's primary aluminum production has a higher greenhouse gas (GHG) emission factor. Reducing China's aluminum exports can promote the transfer of this part of the production capacity to regions rich in renewable energy (such as Norway, Iceland, and Australia), thereby reducing the greenhouse gas emission burden of aluminum production from the perspective of the global industrial chain.

Resource Consumption and Environmental Emission

Due to increasing aluminum production, China is already heavily dependent on imported bauxite. The main reasons for this are as follows: firstly, China's alumina refining enterprises abandoned domestic bauxite due to its relatively low grade; secondly, the rapid growth of aluminum demand puts pressure on bauxite resources. China therefore relies on bauxite imports to meet demand. Currently, more than 50% of the bauxite required for the production of alumina in China needs to be imported. However, the overuse of bauxite has tremendous implications for the resource utilization and environmental management. We assumed that the production of 1 t of aluminum required 4.6 t of bauxite during 2017–2030 [19]. The results of this paper indicate that the bauxite consumption of China's AID will reach 291 Tg by 2030 under the BAU scenario. Compared with the baseline scenario, the demand for bauxite under Scenario A and Scenario B will be reduced by 27 and 47% in 2030, respectively. Also, Scenario A saves 526 Tg bauxite between 2017 and 2030, and Scenario B saves 938 Tg. On the premise of giving priority to meeting China's domestic demand, a moderate reduction in aluminum exports is an important measure to alleviate the shortage of bauxite resources.

In this paper, the water consumption, energy consumption, GHG emissions, SO_2 emissions, and NO_X emissions of the production of 1 t primary aluminum required 10.06 m^3, 144,612 MJ, 14,773 kg CO_2-eq, 53.5 kg, and 40.4 kg, respectively [19]. The water consumption, energy consumption, GHG emissions, SO_2 emissions, and NO_X emissions of the production of 1 t recycled aluminum required 0.88 m^3,9211.78 MJ, 665 kg CO_2-eq, 1.2 kg, and 1.79 kg, respectively [19]. The conservation of water resources has been highlighted as a critical pillar to shape the aluminum industry towards sustainable development. This paper assesses the water resources saving benefits brought by reducing aluminum exports for China's AID. In 2030, the fresh-water consumption is 6.5×10^8, 4.7×10^8, and 3.5×10^8 m^3 under the BAU scenario, Scenario A, and Scenario B, respectively (Fig. 2). Under Scenario A, the energy consumption, and the emissions of GHG, SO_2, and NO_X are 6.67 EJ, 667.88 Tg, 2.43 Tg, and 1.85 Tg. Under Scenario B, the energy consumption, and the emissions of GHG, SO_2, and NO_X are 4.91 EJ, 497.11 Tg, 1.78 Tg, and 1.36 Tg.

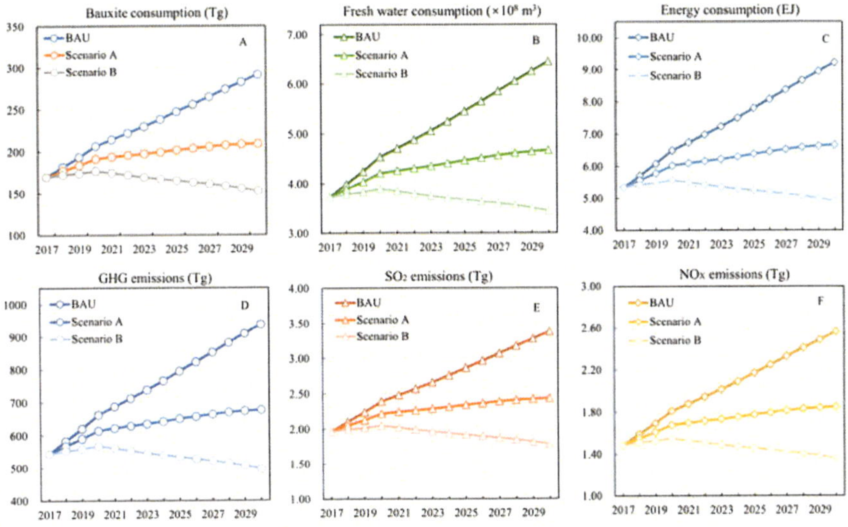

Fig. 2 Resource input, energy consumption, and environmental emissions under different scenarios

Discussion and Suggestions

Compared with other significant aluminum producing countries, the proportion of China's primary aluminum production for export currently is relatively high, about 40%. Under the constraints of sustainable development, a moderate reduction in the proportion of China's primary aluminum exports is particularly important for energy conservation and emission reduction. Additionally, during the vital process of structural adjustment and developing a circular economy, a shift of aluminum production from bauxite based to scrap based is essential. The development strategy of prioritizing domestic demand and moderately reducing foreign exports can increase domestic aluminum in-use stocks, thereby increasing future aluminum scrap production. Figure 3 shows the production process of primary aluminum and secondary aluminum. Compared with bauxite based, the scrap-based method to produce aluminum has apparent advantages in terms of resource input, energy consumption, and environmental emissions. Under three scenarios, China will be confronted with a substantial increase in domestic scrap during 2020–2030 due to the vast consumption of aluminum products since 1990. However, there is no well-developed policy to manage and process such vast aluminum scrap. Firstly, the establishment of a cycle is constrained by insufficient domestic aluminum scrap. Domestic scrap aluminum was produced 20–30 years ago when aluminum output was minimal. Secondly, the aluminum scrap industry is poorly managed and uses backward techniques. For example, some scrapped vehicles are privately refitted to extend their lives. This unsafe practice may cause some loss of aluminum scrap. In addition, some aluminum scrap is processed by illegal small-scale factories in poor

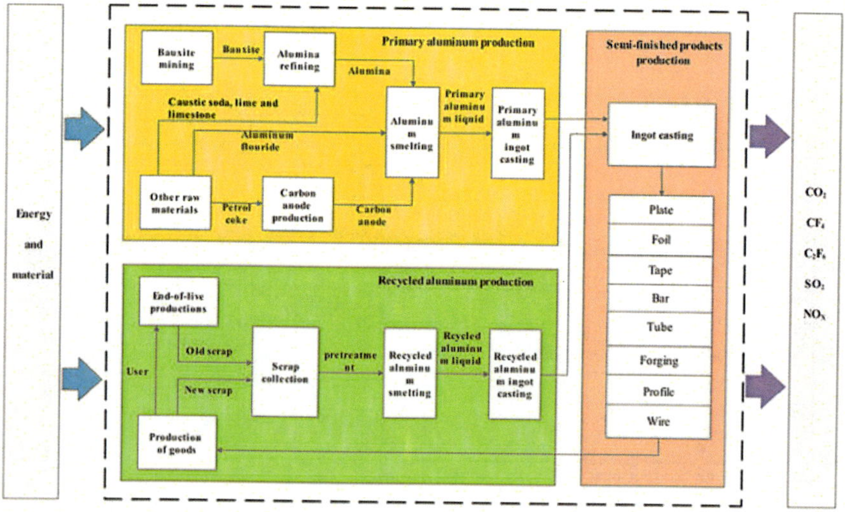

Fig. 3 Production flow chart of primary aluminum and secondary aluminum

technological conditions, which reduces its recycling efficiency and quality. Moreover, compared with other countries, the current aluminum production in China mainly relies on coal power. Increasing the use of clean energy is essential for the decarbonization of China's AID and the reduction of pollutants.

Based on the above analysis, we have put forward some policy recommendations to promote the sustainable development of the aluminum industry. The first is to formulate policies to restrict the export of low-end aluminum materials to ensure the scientific development of the aluminum industry. For economic and environmental reasons (Fig. 2), it is necessary for China to restrict the export of aluminum products. In the baseline scenario, China's primary aluminum production will rise to 57 Tg in 2030. Second, strict energy consumption and pollution regulations should be established to promote the sustainable development of China's AID. The energy densities of refining and smelting in China were 12,054 MJ/t-Al_2O_3 and 13,579 kWh/t-Al in 2018, respectively; the current world's advanced levels were 8938 MJ/t-Al_2O_3 and 13,200 kWh/t-Al, respectively [27]; thus, the application of new advanced technologies can also effectively reduce the environmental load of China's AID. According to our predictions, the total energy consumption and exhaust gas emissions in BAU Scenario are as high as 9.22 EJ and 943.82 Tg, respectively. If the export volume of aluminum scrap is reduced (Scenario A), energy resources and environmental pressure will be significantly reduced (energy consumption 6.67 EJ, exhaust gas emissions 681.16 Tg). Third, advanced technology, well-trained workers, and strict industry standards are indispensable to ensure the quality and supply of aluminum products. Reducing exports may have a negative impact on the economic benefits of China's AID. China's AID needs to complete the transformation from the export of low-end raw materials to the manufacturing of high-end aluminum products. At present, the added value of aluminum products exported by our country is relatively low. By improving the quality, it is possible to reduce the export volume while taking into account economic benefits. For example, replace the exported products from raw materials (aluminum ingots, aluminum foil, etc.) to aluminum-containing products such as automobiles and airplanes.

Conclusions

In this study, the use of scenario analysis was used to discuss the impact of reducing exports on the resource and energy consumption and pollution emissions of China's AID. If there are no restrictions, China's aluminum demand may exceed 60 Tg by 2030. Although the output of aluminum scrap will gradually increase in the future, due to huge demand, primary aluminum production will continue to rise to 58 Tg by 2030. The moderate reduction in exports is critical to the sustainable development of China's AID. The upper limit of primary aluminum production capacity set by the Chinese government is 45 Tg. In Scenario A, China's primary aluminum production is basically stable at 40–45 Tg during 2020–2030. Therefore, it is a reasonable choice to control the ratio of aluminum net exports to total production at about 20%. In the future, the government should issue relevant policies to promote the development of China's aluminum processing industry to enhance the technological content and added value of aluminum products. At the same time, the promotion of energy-saving and emission reduction technologies and the recycling of aluminum waste are essential measures to reduce the environmental burden of China's AID.

Acknowledgements This work was supported by the National Natural Science Foundation of China (Grant Nos. 51874078 and 51874094).

References

1. Cullen JM, Allwood JM (2013) Mapping the global flow of aluminum: from liquid aluminum to end-use goods. Environ Sci Technol 47(7):3057–3064
2. Liu G, Bangs CE, Müller DB (2013) Stock dynamics and emission pathways of the global aluminium cycle. Nat Clim Chang 3(4):338–342
3. Liu G, Müller DB (2013) Centennial evolution of aluminum in-use stocks on our aluminized planet. Environ Sci Technol 47(9):4882–4888
4. Maung KN, Yoshida T, Liu G et al (2017) Hashimoto S. Assessment of secondary aluminum reserves of nations. Resour Conserv Recycl 126:34–41
5. Liu G, Bangs C, Müller DB (2011) Unearthing potentials for decarbonizing the U.S. aluminum cycle. Environ Sci Technol 45(22):9515–22
6. Chen WQ (2018) Dynamic product-level analysis of in-use aluminum stocks in the United States. J Ind Ecol 22(6):1425–1435
7. Buchner H, Laner D, Rechberger H et al (2015) Dynamic material flow modeling: an effort to calibrate and validate aluminum stocks and flows in Austria. Environ Sci Technol 49(9):5546–5554
8. Buchner H, Laner D, Rechberger H et al (2014) In-depth analysis of aluminum flows in Austria as a basis to increase resource efficiency. Resour Conserv Recycl 93:112–123
9. Ciacci L, Chen W, Passarini F et al (2013) Historical evolution of anthropogenic aluminum stocks and flows in Italy. Resour Conserv Recycl 72:1–8
10. Dai M, Wang P, Chen W et al (2019) Scenario analysis of China's aluminum cycle reveals the coming scrap age and the end of primary aluminum boom. J Clean Prod 226:793–804
11. Ding N, Yang J, Liu J (2016) Substance flow analysis of aluminum industry in mainland China. J Clean Prod 133:1167–1180
12. Yue Q, Wang H, Lu Z et al (2014) Analysis of anthropogenic aluminum cycle in China. Trans Nonferrous Met Soc China 24(4):1134–1144
13. Yue Q, Wang H, Lu Z (2012) Quantitative estimation of social stock for metals Al and Cu in China. Trans Nonferrous Met Soc China 22(7):1744–1752
14. Chen W, Shi L (2012) Analysis of aluminum stocks and flows in mainland China from 1950 to 2009: exploring the dynamics driving the rapid increase in China's aluminum production. Resour Conserv Recycl 65:18–28
15. Wang J, Graedel T (2010) Aluminum in-use stocks in China: a bottom-up study. J Mater Cycles Waste Manag 12(1):66–82
16. Chen W, Shi L, Qian Y (2010) Substance flow analysis of aluminum in mainland China for 2001, 2004 and 2007: exploring its initial sources, eventual sinks and the pathways linking them. Resour Conserv Recycl 54(9):557–570
17. Yang Y, Guo Y, Zhu W et al (2019) Environmental impact assessment of China's primary aluminum based on life cycle assessment. Trans Nonferrous Met Soc China 29(8):1784–1792

18. Farjana SH, Huda N, Mahmud M (2019) Impacts of aluminum production: A cradle to gate investigation using life-cycle assessment. Sci Total Environ 663:958–970
19. Zhang Y, Sun M, Hong J et al (2016) Environmental footprint of aluminum production in China. J Clean Prod 133:1242–1251
20. Nunez P, Jones S (2016) Cradle to gate: life cycle impact of primary aluminum production. Int J Life Cycle Assess 21(11):1594–1604
21. Guo Y, Yu Y, Ren H et al (2020) Scenario-based DEA assessment of energy-saving technological combinations in aluminum industry. J Clean Prod 260:121010
22. Li Q, Zhang W, Li H et al (2017) CO_2 emission trends of China's primary aluminum industry: a scenario analysis using system dynamics model. Energy Policy 105:225–235
23. Liu Z, Geng Y, Adams M et al (2016) Uncovering driving forces on greenhouse gas emissions in China' aluminum industry from the perspective of life cycle analysis. Appl Energy 166:253–263
24. Hao H, Geng Y, Hang W (2016) GHG emissions from primary aluminum production in China: Regional disparity and policy implications. Appl Energy 166:264–272
25. Zhang W, Li H, Chen B et al (2015) CO_2 emission and mitigation potential estimations of China's primary aluminum industry. J Clean Prod 103:863–872
26. Elshkaki A, Lei S, Chen W (2020) Material-energy-water nexus: modelling the long term implications of aluminum demand and supply on global climate change up to 2050. Environ Res 181:108964
27. Li S, Zhang T, Niu L et al (2021) Analysis of the development scenarios and greenhouse gas (GHG) emissions in China's aluminum industry till 2030. J Clean Prod 290:125859
28. International Aluminum Institute (2021) Global material flow model. http://www.world-aluminium.org/publications/. Accessed 4 May 2021
29. National Bureau of Statistics of China (2021) China statistical yearbook. http://www.stats.gov.cn/tjsj./ndsj/. Accessed 13 May 2021

Potential Health Impact Assessment of Large-Scale Production of Batteries for the Electric Grid

Haoyang He, Shan Tian, Chris Glaubensklee, Brian Tarroja, Scott Samuelsen, Oladele A. Ogunseitan, and Julie M. Schoenung

Abstract Battery storage technologies such as redox flow batteries (RFBs) and lithium-ion batteries (LIBs) are appealing candidates for large-scale energy storage requirements to support the integration of renewable energy into electric grids. To ensure that their environmental benefits outweigh the environmental costs of producing battery storage systems, it is vital to assess the potential health impacts of battery materials and waste emissions during production. Here, we present a case study based on life cycle impact assessment (LCIA) to characterize the toxicity hazard associated with the production of six types of battery storage technologies including three RFBs [vanadium redox flow battery (VRFB), zinc-bromine flow battery (ZBFB), and the all-iron flow battery (IFB)], and three LIBs [lithium iron phosphate (LFP), lithium nickel cobalt manganese hydroxide (NCM), and lithium manganese oxide (LMO)]. USETox® v2.0 (USETox®) was used for LCIA and we found higher impacts found higher impacts on human health outcomes for the production of LIBs than for RFBs, noting that uncertainties associated with the characterization factors demand caution in interpreting the results. Overall, the study provides (1) a comprehensive evaluation of life cycle impacts for materials, components, and systems associated with the production of burgeoning six battery energy storage technologies and (2) an important foundation for the identification of battery

H. He · C. Glaubensklee · J. M. Schoenung (✉)
Department of Materials Science and Engineering, University of California, Irvine, CA, USA
e-mail: Julie.Schoenung@UCI.edu

S. Tian · S. Samuelsen
Department of Mechanical and Aerospace Engineering, University of California, Irvine, CA, USA

S. Tian · B. Tarroja · S. Samuelsen
Advanced Power and Energy Program, University of California, Irvine, CA, USA

B. Tarroja
Department of Civil and Environmental Engineering, University of California, Irvine, CA, USA

O. A. Ogunseitan
Department of Population Health and Disease Prevention, University of California, Irvine, CA, USA

A. Lazou et al. (eds.), *REWAS 2022: Developing Tomorrow's Technical Cycles (Volume I)*, The Minerals, Metals & Materials Series,
https://doi.org/10.1007/978-3-030-92563-5_43

technologies with lower potential negative impacts associated with integrating energy storage in strategies for upscaling renewable energy sources.

Keywords Battery · Energy storage · Energy grid · Renewable · Life cycle impact assessment · Human health impact

Introduction

In order to mitigate the adverse environmental impacts of conventional non-renewable energy resources, it is necessary to deploy energy storage systems to manage variable renewable resources. For example, California has implemented policies such as the Renewables Portfolio Standard (RPS) and Senate Bill 100 to promote the use of renewable resources such as wind and solar for its long-term carbon mitigation goal [1, 2]. Energy storage systems, which are designed to temporarily store the electrical energy from renewable resources in a form that can be converted back for electricity generation, can enable a high renewable penetration level on the electric grid. Among the various energy storage systems developed, lithium-ion batteries are the most popular candidates given their high energy density and lower cost [3]. Flow batteries are also potential candidates due to their separation of energy capacity and power capacity, which can be easily scaled up for the electric grid [4]. In general, the implementation of renewable energy systems could reduce environmental impacts by substituting for non-renewable sources. However, the manufacturing of batteries, including the extraction of raw materials, processing, and assembly, may have a negative environmental impact. In recent years, research projects focusing on sustainability assessments of batteries are emerging under the pressure to promote renewable energy. For example, methods including life cycle assessment, materials flow analysis, and risk assessment have been applied to address issues on environmental impact, supply chain, and chemical exposure, respectively [5–9]. Nevertheless, systematic assessment designed to evaluate the human health effects across the whole battery production process is still lacking. Therefore, there is a need to perform these assessments to better support materials selection decisions regarding the production of grid-connected energy storage systems.

The goal of this study is to investigate the potential human health impact in terms of the life cycle of battery systems, components, and materials during the battery production chain. In total, we selected six types of batteries for energy storage for evaluation, which include three redox flow batteries (RFBs).

- Vanadium redox (VRFB),
- Zinc-bromine (ZBFB), and
- All-iron (IFB);

and three lithium-ion batteries (LIBs):

- Lithium iron phosphate (LFP),
- Lithium nickel cobalt manganese hydroxide (NCM), and

- Lithium manganese oxide (LMO).

We considered opportunities to minimize the use of toxic chemicals through life cycle impact assessment (LCIA) methods to assess the associated compositional materials used to manufacture flow batteries and lithium-ion batteries.

Materials and Methods

Battery Products for Energy Storage

Information on battery components and materials used for RFBs and LIBs are provided in Tables 1 and 2, respectively. The data for those batteries are collected from recently published literature with the information on battery energy density and materials mass percentage highlighted [10–12]. In this study, only the materials and components that are essential for battery functions are considered, while peripheral components such as battery management system, powder conditioning system, and

Table 1 Major components and primary materials used for the three flow batteries—vanadium redox flow battery (VRFB), zinc-bromine flow battery (ZBFB), and all-iron flow battery (IFB)

Battery type	VRFB [10]	ZBFB [10]	IFB [10]
Energy density	18.68 kg/kWh	14.40 kg/kWh	32.52 kg/kWh
Cell stack component			
Bipolar plate	Graphite	Titanium	Graphite
	Polyethylene	Polyethylene	Bisphenol-A epoxy-based vinyl ester resin
Electrode	Carbon fiber felt	/[a]	Carbon fiber felt
Membrane	Nafion®	/[a]	Polyethylene
Cell frame	E-glass fiber	Polyethylene	E-glass fiber
	Polypropylene		Polyester resin
Current collector	Copper	Titanium	Aluminum
Electrolyte storage component			
Electrolytes	Vanadium pentoxide	Zinc bromide	Iron (II) chloride
	Hydrochloric acid		Potassium chloride
	Sulfuric acid	Bromine	Manganese chloride
			Hydrochloric acid
Tank	Polyethylene	Polyethylene	Isophthalic polyester
Pipe	Polyethylene	Polyethylene	Polyvinylchloride

/[a] No extra electrode and membrane materials required for ZBFB

Table 2 Major components and primary materials used for the three lithium-ion batteries—lithium iron phosphate (LFP), lithium nickel cobalt manganese hydroxide (NCM), and lithium manganese oxide (LMO)

Battery type	LFP [11]	NCM [11]	LMO [12]
Energy density	7.86 kg/kWh	6.15 kg/kWh)	6.80 kg/kWh
Cell stack component			
Cathode	Lithium iron phosphate	Lithium nickel cobalt manganese hydroxide	Lithium manganese oxide
	Carbon black	Carbon black	Carbon black
	Polytetrafluoroethylene	Polytetrafluoroethylenc	Latex
	N-methyl-2-pyrrolidone	*N*-methyl-2-pyrrolidone	
Cathode substrate	Aluminum	Aluminum	Aluminum
Anode	Graphite	Graphite	Graphite
	Polytetrafluoroethylene	Polytetrafluoroethylene	Carbon black
	N-methyl-2-pyrrolidone	*N*-methyl-2-pyrrolidone	Latex
Anode substrate	Copper	Copper	Copper
Electrolytes	Lithium Hexafluorophosphate	Lithium Hexafluorophosphate	Lithium Hexafluorophosphate
	Ethylene carbonate	Ethylene carbonate	Ethylene carbonate
Separator	Polyethylene, low density	Polyethylene, low density	Silica sand
			Phthalic anhydride
			Acetone
	Polypropylene	Polypropylene	Polyethylene, low density
			Hexafluoroethane
			Polyvinyl fluoride

battery shell and packages are not included in the assessment. To ensure consistency of the comparative assessment, all the results presented are normalized to per kilowatt-hour (kWh) energy capacity of a given battery system.

Life Cycle Impact Assessment

We conducted life cycle impact assessments (LCIA) to evaluate the human health impacts associated with potentially hazardous emissions from the production of the six battery technologies. To construct the life cycle inventory (LCI), data compiled from the Ecoinvent database were used as input for the SimaPro LCA software to generate outcomes [13, 14]. USETox® version 2.0 (USETox®) model (endorsed by the Society for Environmental Toxicology and Chemistry (SETAC) and the United Nations Environment Program) was chosen as the LCIA methodology for this study

[15–17]. The outputs of USETox® are based on sets of chemical characterization factors (CFs) consisting of fate, exposure, and effect parameters. The mid-point indicator, which uses the number of disease cases as a comparative toxic unit, was adopted to assess cancer and non-cancer effects. Two sets of chemical CFs, including *'recommended'* and *'interim'*, were both considered. Unlike the recommended CFs on which a scientific consensus has been reached, the interim CFs still have high uncertainties related to the parameters describing fate and exposure due to insufficient data or flawed models [18]. Additionally, we conducted a detailed analysis showing the impact sources attributed to various battery components and materials to highlight the main drivers on human health impact.

Results and Discussion

Potential Impacts of Energy Storage Battery Production

In this section, we present the results of the human health impact assessment based on the fate, exposure, and effect factors associated with the chemical emissions during battery production. The assessment focused on cancer and non-cancer effects determined by using *recommended* and *interim* CFs; two scenarios are considered: (1) *Recommended + Interim* CFs and (2) *Recommended-Only* CFs. The interim factors include chemicals with CFs with no universal consensus, and, for simplification, metals are the major materials characterized using interim CFs due to their high use in the six battery technologies.

In Fig. 1, we present the normalized LCIA results by disease case per kWh. The results show that LIBs generally have impact scores higher than RFBs though the energy density of LIBs is much higher. The results based on *Recommended + Interim* CFs show that the impact of the non-cancer effect is approximately one magnitude higher than cancer effects with LMO showing the highest impact score and IFB showing the lowest. Note that low scores are preferred. The relative ranking of the six battery technologies is the same for both cancer and non-cancer effects. The results with the *Recommended-Only* CFs show that cancer and non-cancer effects are the same order of magnitude, with the LFP and NCM showing higher impact scores. Interestingly, LMO is now ranked to be the lowest among the six battery technologies in contrast to the previous results using *Recommended + Interim* CFs.

The uncertainties associated with the results based on different CFs warrant explanation. Since various CFs are directly correlated with their corresponding materials, we investigated the detailed contributions attributed to different battery components and each specific material. The results are presented in Fig. 2. These flow charts show the percentage contribution of each battery component with the accompanying materials, which the components and materials with a contribution less than 0.5% are merged into the categories of other components and other materials in order to highlight the major contributors. We first focus on the findings for LMO due

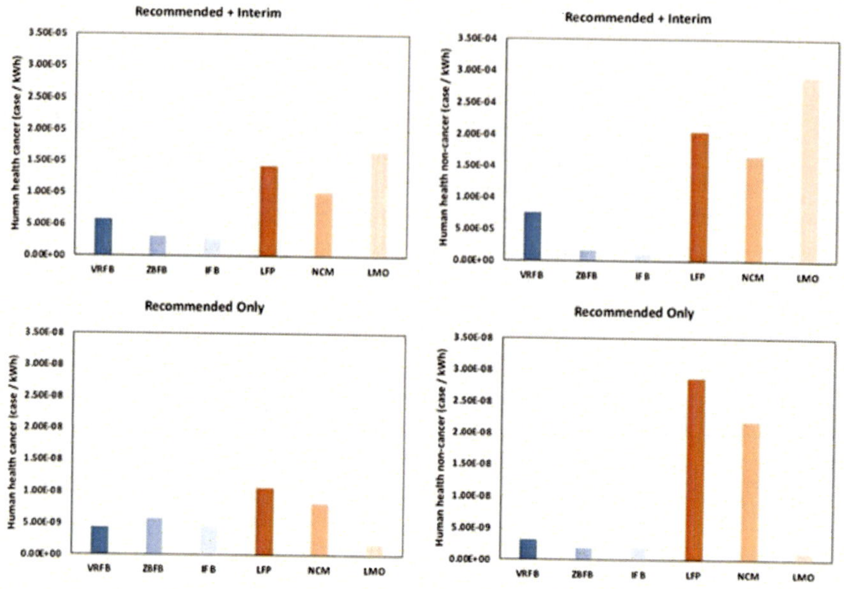

Fig. 1 USETox® results showing cancer and non-cancer effects for the six battery energy storage systems under the two scenarios: *Recommended + Interim* characterization factors (CFs) and *Recommended-Only* CFs

to the large variations observed in the previous LCIA results. When considering the *Recommended + Interim* CFs, copper is the highest contributor dominating the health impact, especially for the non-cancer effect. Since the CFs for copper are categorized as *interim* CFs, this explains the large impact reduction of LMO when using the *Recommended-Only* CFs. For VRFB, another important contributor to human health impact (especially for cancer effect) is vanadium pentoxide, which is the active material used in the VRFB system. When considering the results based on *Recommended-Only* CFs, the impact contributed by Nafion® membrane increases. The major contributing materials for the ZBFB are titanium, zinc bromide, and bromine regardless of what types of CFs were applied. For the IFB, the impact results are more uniformly distributed to each component and material, while the impact contribution of several polymer resins that are not relevant to the battery functional performance is slightly higher. For LFP and NCM, the contributing components and materials are similar since the original data for these batteries were extracted from the same study. Interestingly, the major contributors for LFP and NCM, no matter what types of CFs considered, are not the cathode active materials. When considering the scenario using *Recommended + Interim* CFs, copper is among the highest contributors. However, when considering only the recommended CFs, polytetrafluoroethylene, which is used as the electrode binder, tends to be the highest contributor for both of these two batteries.

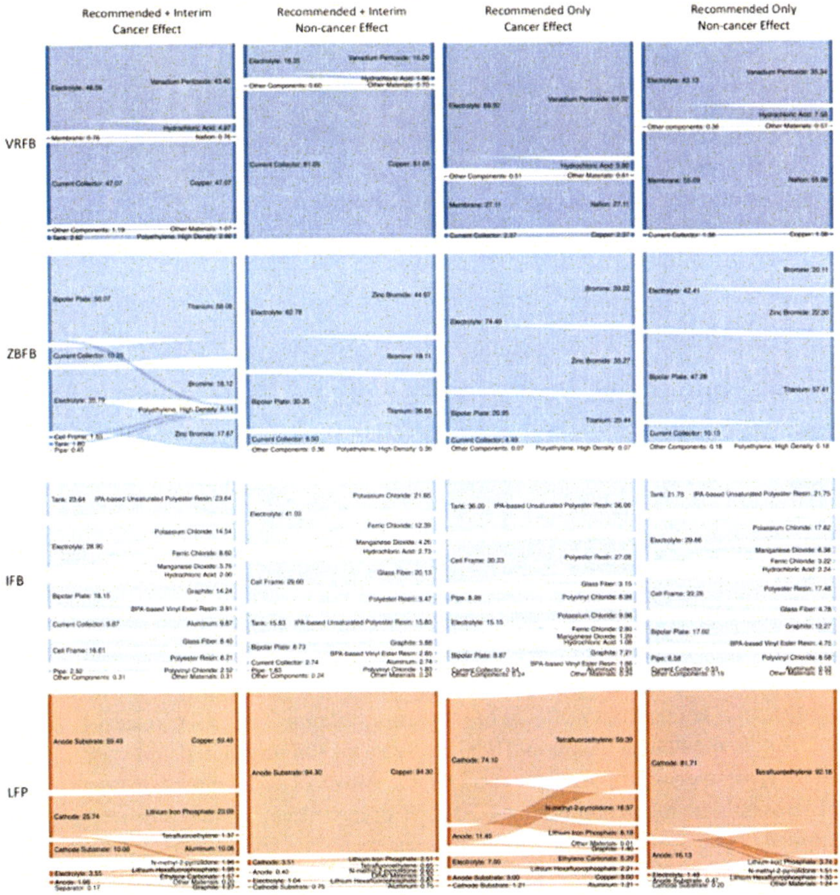

Fig. 2 Numbers represent percentage contributions of each component and material to the total impact of their associated batteries as generated by USETox® results for cancer and non-cancer effects considering the recommended and interim characterization factors

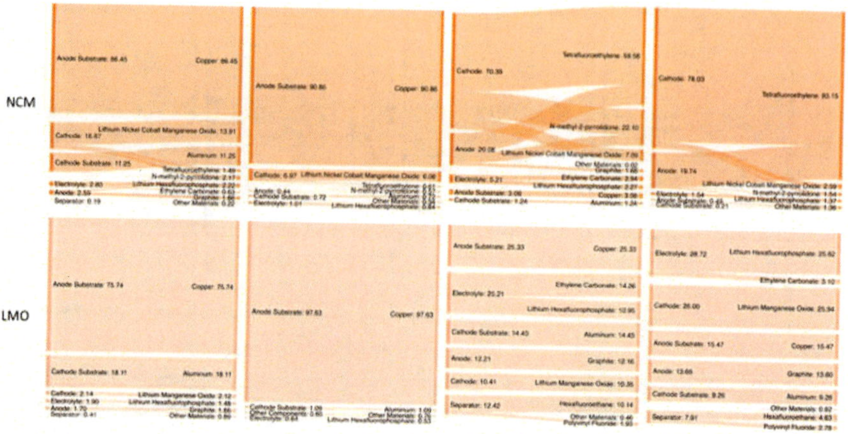

Fig. 2 (continued)

Conclusions

This assessment of potential health impacts of battery production materials successfully highlighted the major contributors of impact for each battery system, so that safer alternatives could be identified to decrease the potential impact of batteries installed in a scaled-up renewable energy grid systems. The LCIA results based on USETox® indicate that the human health effects can be substantial and vary for each battery type given the different material choices. In most cases, the total impact results of LIBs could be much higher than RFBs. With the expansion to consider the contribution of impacts attributed to components and materials, the use of metals, especially copper, would be the major contributors among several metal-based batteries when using *Recommended + Interim CFs*. If solely looking at the *Recommended-Only CFs*, complex synthetic polymers would instead be the major contributors.

Acknowledgements The authors would like to acknowledge the funding provided by the California Energy Commission under Agreement # EPC-16-039. The authors would also like to acknowledge the funding from the Lincoln Dynamic Foundation World Institute for Sustainable Development of Materials (WISDOM) at the University of California, Irvine.

References

1. California Senate Bill No. 32, 2015–2016. California Global Warming Solutions Act of 2006: emissions limit
2. California Senate Bill No. 100, 2017–2018. California Renewables Portfolio Standard Program: emissions of greenhouse gases. 2017–2018
3. Wood DL III, Li J, Daniel C (2015) Prospects for reducing the processing cost of lithium-ion batteries. J Power Sources 275:234–242

4. Tian S, He H, Kendall A, Davis SJ, Ogunseitan OA, Schoenung JM, Samuelsen S, Tarroja B (2021) Environmental benefit-detriment thresholds for flow battery energy storage systems: a case study in California. Appl Energy 300:117354
5. Liang Y, Su J, Xi B, Yu Y, Ji D, Sun Y, Cui C, Zhu J (2017) Life cycle assessment of lithium-ion batteries for greenhouse gas emissions. Resour Conserv Recycl 117:285–293
6. Weber S, Peters JF, Baumann M, Weil M (2018) Life cycle assessment of a vanadium redox flow battery. Environ Sci Technol 52(18):10864–10873
7. Olivetti EA, Ceder G, Gaustad GG, Fu X (2017) Lithium-ion battery supply chain considerations: analysis of potential bottlenecks in critical metals. Joule 1(2):229–243
8. Whitehead AH, Rabbow TJ, Trampert M, Pokorny P (2017) Critical safety features of the vanadium redox flow battery. J Power Sources 351:1–7
9. Park YJ, Kim MK, Kim HS, Lee BM (2018) Risk assessment of lithium-ion battery explosion: chemical leakages. J Toxicol Environ Health, Part B 21(6–8):370–381
10. He H, Tian S, Tarroja B, Ogunseitan OA, Samuelsen S, Schoenung JM (2020) Flow battery production: materials selection and environmental impact. J Clean Prod 269:121740
11. Majeau-Bettez G, Hawkins TR, Strømman AH (2011) Life cycle environmental assessment of lithium-ion and nickel metal hydride batteries for plug-in hybrid and battery electric vehicles. Environ Sci Technol 45(10):4548–4554
12. Notter DA, Gauch M, Widmer R, Wager P, Stamp A, Zah R, Althaus HJ (2010) Contribution of Li-ion batteries to the environmental impact of electric vehicles. Envrion Sci Technol 44(17):6550–6556
13. Ecoinvent, version 3.4. https://www.ecoinvent.org
14. SimaPro. https://simapro.com
15. Hauschild MZ, Huijbregts M, Jolliet O, MacLeod M, Margni M, van de Meent D, Rosenbaum RK, McKone TE (2008) Building a model based on scientific consensus for life cycle impact assessment of chemicals: the search for harmony and parsimony. Environ Sci Technol 42(19):7032–7037
16. Henderson AD, Hauschild MZ, van de Meent D, Huijbregts MA, Larsen HF, Margni M, McKone TE, Payet J, Rosenbaum RK, Jolliet O (2011) USEtox fate and ecotoxicity factors for comparative assessment of toxic emissions in life cycle analysis: sensitivity to key chemical properties. Int J Life Cycle Assess 16(8):701–709
17. Rosenbaum RK, Bachmann TM, Gold LS, Huijbregts MA, Jolliet O, Juraske R, Koehler A, Larsen HF, MacLeod M, Margni M, McKone TE (2008) USEtox—the UNEP-SETAC toxicity model: recommended characterisation factors for human toxicity and freshwater ecotoxicity in life cycle impact assessment. Int J Life Cycle Assess 13(7):532–546
18. USETox® Frequently Asked Questions. https://usetox.org/faq#t23n119

Validation of an Innovative On-Line *Legionella* Detection Technology in Water-Cooling Systems

Marie-Christine Simard and Geneviève Doyer

Abstract Bacteria of the *Legionella* family are known to grow in water-cooling systems, where conditions are favourable. Because these bacteria can cause Legionnaires' disease (*Legionellosis*), the owners of such systems are bound by legal obligations. Currently, the only accepted method for the legal requirements for quantifying *Legionella* in all type of water-cooling systems—growth on a culture medium—can take up to 14 days to obtain results, which does not allow proactiveness. The technology tested in this project, the BioAlert Lp15™ by BioAlert Solutions, is a fully automated, on-site equipment that samples, quantifies, and sends a *Legionella* result within four hours. The technology showcase project described herein has demonstrated the applicability of the equipment in heavy industry environments as it was possible to establish a control and action chart based on the daily results obtained with the BioAlert Lp15™. This instrument provides numerous benefits, such as a lower risk of *Legionella* exposure, better understanding of the *Legionella* metabolism and its influencing factors, improved operational practices and, finally, reduced impacts on production.

Keywords Legionella pneumophila · BioAlert Lp15™ · Cooling systems

Introduction

Legionella bacteria are at the root of Legionnaires' disease, a severe, acute lung infection with pneumonia-like symptoms which can kill immunodeficient individuals [1]. The *Legionella* genus (*Legionellaceae* family) numbers 59 species and 70 serotypes, of which nearly half have been linked to human diseases [2]. They

M.-C. Simard (✉)
Rio Tinto, Arvida Research and Development Centre, 1955 Mellon Boulevard, Building 110, Jonquiere, QC G7S 4K8, Canada
e-mail: marie-christine.simard@riotinto.com

G. Doyer
Rio Tinto, Casting Operational Excellence, 1955 Mellon Boulevard, Building 100, Jonquiere, QC G7S 4K8, Canada

A. Lazou et al. (eds.), *REWAS 2022: Developing Tomorrow's Technical Cycles (Volume I)*, The Minerals, Metals & Materials Series,
https://doi.org/10.1007/978-3-030-92563-5_44

develop in humid environments (water-cooling systems) at temperatures between 25 and 45 °C. The *Legionella pneumophila* genus is associated with over 85% of all infections. They belong to 15 different serotypes, and serotype 1 represents more than 61% of reported cases [3]. In the USA, reported *Legionella* cases have increased ninefold since the year 2000 (Fig. 1) [4]. The owners of water-cooling facilities are bound by legal requirements for monitoring and managing *Legionella* bacteria levels in their systems, in accordance with existing public health standards. Table 1 shows the mandatory legal compliance in Quebec [5].

The currently only legal accepted method for quantifying *legionella* in water-cooling systems is by colony growth on culture medium. However, this method, which is carried out in accredited laboratories, requires a delay of 7 to 14 days before the analysis results are available. This prevents effective monitoring of the system

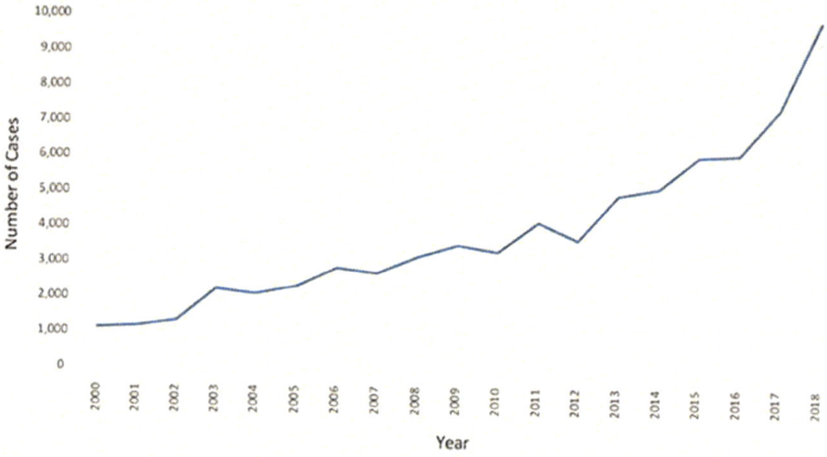

Fig. 1 Increase in cases of Legionnaires' disease in the USA, 2000–2018

Table 1 Legal compliance for the control of *Legionella* contamination in water-cooling systems in Quebec

Legionella pneumophila contamination level	Risks	System condition	Specific action
<10,000 CFU/L	Negligible	Under control	Normal operation
10,000–100,000 CFU/L	Low	Under monitoring	Process optimisation
100,000–1,000,000 CFU/L	High	Early contamination	Preventive actions: identify causes, apply corrective actions, effectiveness check
>1,000,000 CFU/L	Critical	High contamination	Immediate corrective actions required: shut down airflow (fans), apply emergency decontamination (12–72 h), effectiveness check

and endangers people in the vicinity of the installations. In addition, poor monitoring of *Legionella* and poor process treatment control can have significant consequences:

- social repercussions: reputation, employee health, citizen health;
- economic repercussions: production slowdowns and losses, increased processing costs, early equipment and system wear, legal action;
- environmental repercussions: overuse and disposal of chemical products, significantly higher water consumption, lower energy efficiency in towers.

In order to reduce *Legionella* contamination hazards and their negative repercussions, the installation of an on-site, on-line detection, and monitoring system can lead to improved risk control. Within the framework of this technology showcase project, three monitoring instruments were installed in three separate aluminium smelters in the Saguenay—Lac-Saint-Jean region of Quebec, with the aim of assessing the benefits and limits of the technology in heavy industry environments. This paper briefly illustrates the technology and compares it with the accredited method, describes the assessment project, and follows up with a presentation and discussion of some results.

Legionella Pneumophila: Quantification and the Bioalert Lp15^{tm} Technology

The standard method, which is also the reference method herein, is the culture method (AFNOR NF T90-431) [6]. Water-cooling system samples, with or without dilution, are deposited on specific culture media to foster the development of *Legionella spp.*, the full range of *Legionella* species. The colonies are counted, and the results are expressed in Colony Forming Units (CFU)/L. Subsequent steps are needed to confirm the presence of *Legionella pneumophila*. Bacteria development requires several days and, consequently, results are usually only available in 5–10 days' time. In theory, only viable *Legionella* cells can develop in a culture environment, but shipping times, the presence of other microorganisms and chemicals can create interferences, which greatly limit the reproducibility of the tests.

Genetic material quantification (qPCR) in genomic units (GU/L) has gained in popularity to evaluate *Legionella* level in industrial water-cooling systems as it yields faster results. However, the method quantifies the full range of genetic material in the samples: living bacteria, dead bacteria, and cell-free DNA [7]. Whiley and Taylor (2016) show that 26/28 qPCR studies detected higher counts of *Legionella* than culture media [8]. Therefore, the method tends to overestimate *Legionella* counts [9], which can lead to inappropriate corrective actions having repercussions on processes and production. Various research works have focused on developing new *Legionella* detection and counting methods in recent years: fluorescence in situ (FISH), solid-phase cytometry, optical spectroscopy, DNA electrochemical probe, immunology

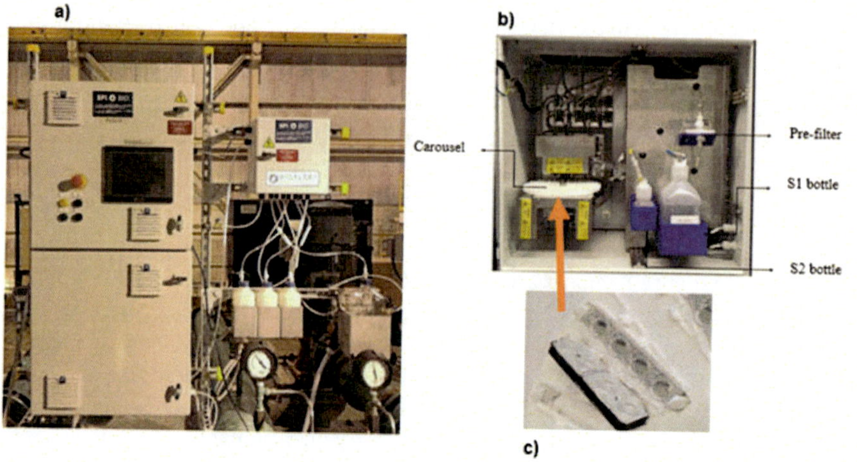

Fig. 2 Presentation of the BioAlert Lp15™: **a** Outside view, including the power supply housing (top), sampling and analysis housing (bottom), and external module (right); **b** inside view, including the pre-filter, reagent bottles, and carousel **c** test-cartridge

methods (ELISA, etc.), etc. [8]. The latter are difficult to automate, making them unsuitable for industrial environments.

BioAlert Lp15™ is an on-line *Legionella* detection and measurement solution for cooling systems. This unique, patented technology measures quantities of *Legionella pneumophila* bacteria from serotypes 1–15, living or dead and whose form is intact, by capturing them in a filter and quantifying the genetic material (DNA).

The device (Fig. 2a) is equipped with a sampling system that can take water samples at a given point in the water system either by initiating an analysis upon request or according to a defined schedule. The water samples are then analysed using cartridges housed within the instrument (Fig. 2c). The cartridges are inserted in a carousel that can accommodate up to 16 cartridges (Fig. 2b). Following each test, the carousel changes position, making a new cartridge available for the next test. Once all the cartridges have been consumed, the equipment requires a refill of consumables (carousel test-cartridges, reagent bottles and a pre-filter). [10]

The device also has an external module (Fig. 2a) allowing to test other sampling points. It is also used to clean the equipment.

Technological Showcase

The devices were installed in three Saguenay—Lac-Saint-Jean smelters, in the casting water treatment centre. The casting of various aluminium products requires processed and cooled water, to which casting lubrication oil is added to lubricate the casting moulds. The water that has been used in casting becomes heated and oil charged. It is routed to a treatment centre where the oil is removed using a clarification and flotation system. The water then recirculates in the system and flows through cooling towers before it can be reused in casting. The physico-chemical matrix of the process water favours the proliferation of *Legionella* pathogens due to the system design, flowrate variations according to production, nutrient intake, biofilm formation, and the possibility of dead arms.

Prior to installing the device, a compatibility study was carried out by BioAlert Solutions to ensure their equipment was compatible with our process water matrix at various sampling points. The results showed that the process water did not create interference with the quantification method.

To ensure the strength and validity of the technology, the BioAlert Lp15™ results were compared weekly with the results of culture tests (AFNOR NF T90-431) [6] by an ISO/CEI 17,025: 2005 accredited laboratory (Genochemia, Saguenay). Comparative study samples were taken at the same time as the BioAlert Lp15™ samples and at the same sampling points.

BioAlert Lp15™ can detect when there is an upward trend in the *Legionella* counts (Fig. 3) and therefore allows for rapid planning and implementation of corrective actions. During summer (May to September), an extra water-cooling system is required for efficiency. Figure 3 shows that after start up of the additional system, *Legionella* counts increased significantly. This finding made it possible to improve operational practices.

The comparative results with the culture tests (Fig. 3) gave us the tools needed to improve our internal decision-making process—the risk-control chart based on legal requirements (Table 1). The control chart uses a colour code which guides our decision-making process in terms of corrective actions and risk management according to *Legionella* counts in the systems. Required actions are identified using a risk scale. Adding the BioAlert Lp15™ results to the monitoring tool represents a

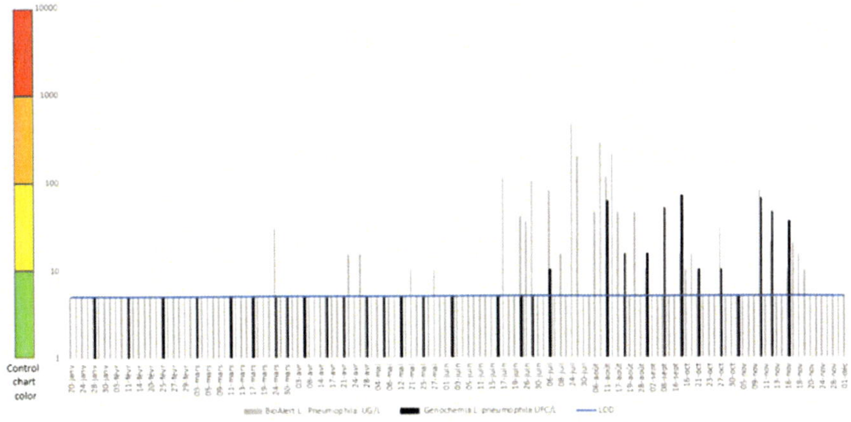

Fig. 3 Chronological evolution of BioAlert Lp15™ *Legionella pneumophila* results compared with culture results at Grande-Baie casting water-cooling system

complementary parameter for process control, thus accelerating the decision-making process. The risk-control chart must be adapted to each process since several site-specific factors can affect detection levels.

The results also showed that the device is highly efficient in detecting malfunctions and/or instabilities in a process and/or water quality change, which can influence *Legionella* proliferation.

As an example, Fig. 4 shows the effects of stopping biocide injections, as well as the effectiveness of shock treatments on *Legionella* level.

Despite a first shock treatment, a second such treatment was required a few weeks later in order to bring conditions under control. This observation was also made at the other sites equipped with BioAlert Lp15™ devices. BioAlert Lp15™ allows taking proactive actions if the process becomes unstable, thus avoiding potential major outbreaks.

There is no perfect equivalence between the BioAlert Lp15™ and culture results. Based on Table 2, it can be said that correspondence is 74% (42% + 32%). This gap can be due to several factors, such as fundamental differences between the two

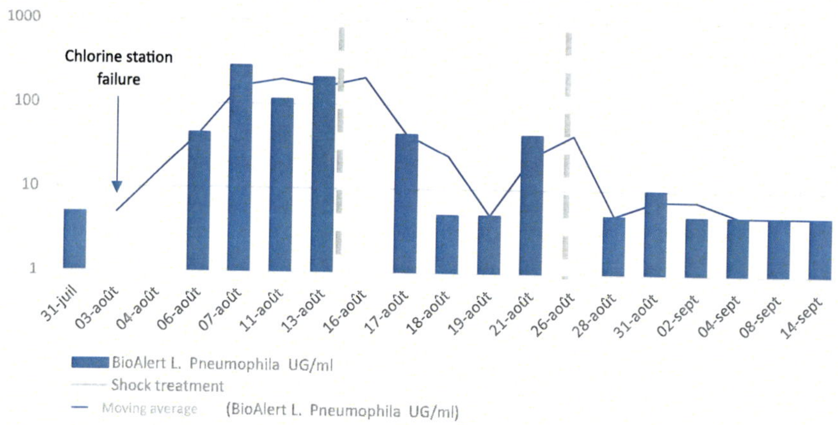

Fig. 4 *Legionella* count variations in process water treated at the Grande-Baie smelter due to the failure of the chlorine pump, corrective actions, and shock treatment

Table 2 Equivalence/comparative matrix of BioAlert Lp15™ and culture results

		Culture	
		-	+
BioAlert	-	42%	13%
	+	13%	32%

- : < 5 UG ou UFC/mL

+ : > 5 UG ou UFC/mL

methods, which are expressed in different units: CFU/L for culture, and GU/L for BioAlert Lp15™. The culture method assesses the development potential of bacteria while qPCR quantifies genetic material. According to Whiley and Taylor (2016), the two main factors influencing bacteria development in culture media are the growth of undesirable microorganisms and the presence of viable but non-culturable bacteria (VBNC). When under stress (chlorine, temperature, etc.), bacteria can fall into a dormancy state, which can slow their development in a culture medium, and thus lead to underestimated counts [8]. In addition to the differences between the two methods, sampling and processing times greatly affect culture tests, as opposed to BioAlert Lp15™, whose process is repeatable. All these factors together are the cause of variations between tests. Genetic material measurements, such as BioAlert Lp15™, yield results that are more stable, with less variation over time. The results have shown that BioAlert Lp15™ is an effective complementary tool for the management of *Legionella* proliferation in water-cooling systems.

Summary

The BioAlert Lp15™ technology allows monitoring *Legionella* counts in water-cooling systems, with direct, on-site sampling. This technology showcase project has shown that the device is compatible, reliable, and durable for use in heavy industry environments. The instrument yields fast *Legionella* count variation results and can detect process instabilities and malfunctions, thus allowing to better understand the conditions that influence the metabolisms involved in *Legionella* proliferation.

Acknowledgements The authors thank BioAlert Solutions for their participation in this showcase project and the Ministère de l'Économie et de l'Innovation du Québec for the financial contribution. Also, a special thanks to all the people who believed and participated in this project: technicians from the research center and casting water treatment plant, industrial hygienists, process supervisors, and site managers.

References

1. World Health Organization (2007) LEGIONELLA and the prevention of legionellosis. In: World Health Organization. https://www.who.int/water_sanitation_health/emerging/legionella.pdf
2. Springston JP, L Yocavitch (2017) Existence and control of *Legionella* bacteria in building water systems: a review. J Occupat Environ Hygiene 14:124–134
3. Reimer AR, Au S, Schindle S, Bernard KA (2010) *Legionella pneumophila* monoclonal antibody subgroups and DNA sequence types isolated in Canada between 1981 and 2009: laboratory component of national surveillance. Eur J Clin Microbiol Infect Dis Off Public Eur Soc Clin Microbiol 29:191–205

4. Centers for Disease Control and Prevention (CDC) (2021) History, burden and trends. In: *Legionella* (Legionnaires' Disease and Pontiac Fever). https://www.cdc.gov/legionella/about/history.html
5. Régie du bâtiment du Québec (2014) Règlement modifiant le Code de sécurité intégrant des dispositions relatives à l'entretien d'une installation de tour de refroidissement à l'eau. In : Gazette officielle du Québec. http://www2.publicationsduquebec.gouv.qc.ca/dynamicSearch/telecharge.php?type=1&file=61543.pdf
6. Association Française de Normalisation (AFNOR) (2003) Recherche et dénombrement de *Legionella spp.* et de *Legionella pneumophila*, NF T90–431
7. International organization for standardization (ISO) (2019) Water quality—detection and quantification of Legionella spp. and/or Legionella pneumophila by concentration and genic amplification by quantitative polymerase chain reaction (qPCR), ISO/TS 12869:2019
8. Whiley H, Taylor M (2016) *Legionella* detection by culture and qPCR: comparing apples and oranges. Crit Rev Microb 42:65–74
9. Toplitsch D, Platzer S, Pfeifer B, Hautz J, Mascher F, Kittinger C (2018) *Legionella* detection in environmental samples as an example for successful implementation of qPCR. Water 10:1–11
10. BioAlert Solutions (2021) Notre produit. In: BioAlert Solutions. https://bioalert.ca/notre-produit/?lang=fr

Environmental Benefits of Closing the Solar Manufacturing and Recycling Loop: Preparation of Solar Manufacturing Inventories

Robert Flores, Haoyang He, Parikhit Sinha, Garvin Heath, Paul W. Leu, and Julie M. Schoenung

Abstract The cumulative global solar panel waste stream is projected to reach between 60 and 78 million tonnes by 2050. Steps towards developing, demonstrating, and implementing processes that recover glass, metals, and semiconductor materials from end-of-life solar panels have already been taken. However, these processes result in the downcycling of most secondary solar materials. Critically, the costs and benefits of capturing these secondary materials for use in new solar panels are unknown. To evaluate the environmental benefit associated with solar material recycling and reuse in next generation panels, prior inventories must be updated and prepared for integration with recycling processes to examine the benefit of closing the material loop between solar panel end-of-life and new panel manufacturing. This current work describes steps taken to upgrade existing inventories that detail the manufacturing of cadmium-telluride (CdTe) panels from the Ecoinvent -v2.2 life cycle inventory database to Ecoinvent -v3. During this update, material inventories were modified to capture different realistic material supply chains within the constraints of Ecoinvent -v3. Materials discussed in detail in this work include primary flat glass, aluminum, steel, copper, and CdTe. This work demonstrated that environmental indicators such as embodied carbon, acidification, and terrestrial eutrophication associated with solar

R. Flores
Department of Mechanical and Aerospace Engineering, University of California, Irvine, CA 92697, USA

H. He · J. M. Schoenung (✉)
Department of Materials Science and Engineering, University of California, Irvine, CA 92697, USA
e-mail: Julie.Schoenung@UCI.edu

P. Sinha
First Solar, Tempe, AZ 85281, USA

G. Heath
National Renewable Energy Laboratory, Golden, CO 80401, USA

P. W. Leu
Department of Industrial Engineering, University of Pittsburgh, Pittsburgh, PA 15260, USA

A. Lazou et al. (eds.), *REWAS 2022: Developing Tomorrow's Technical Cycles (Volume I)*, The Minerals, Metals & Materials Series,
https://doi.org/10.1007/978-3-030-92563-5_45

panel production can be reduced by 25% to over 40% through improved primary material sourcing.

Keywords CdTe solar panels · Life cycle analysis · Solar panel materials · Recycling · Life cycle inventory

Introduction

The cumulative global solar panel waste is projected to reach between 60 and 78 million tonnes by 2050 [1]. Assuming that solar cell architecture and bulk material composition remain stable over the next two decades, 80 to 95% of solar panel waste will consist of glass, aluminum, copper, and other metals [2]. These materials can potentially be captured such that material quality is maintained [3–7], allowing for reuse in next generation solar panels. Panel recycling processes have been demonstrated at scale for cadmium-telluride (CdTe) panels [8] and are under development for most other semiconductor types, including silicon [9]. Aside from recovery of semiconductor material from CdTe panels [7], it is unclear if the material recovered through these processes is of sufficient quality to supplant primary material used in new solar panels.

Processes required to maintain material grade after recycling must clearly result in isolated secondary material outputs. Steps towards accomplishing this task are more readily accomplished for metals used in panel frames, mounts, and wiring where material can be recovered through non- or minimally destructive processes. Analogous steps for the capture of clean glass cullet suitable for flat glass used in solar panels are not as apparent due to (1) the intimate layering of glass, semiconductor material, and other panel layers and (2) the strict glass cullet purity requirements for use in float glass processes [10]. The ultimate application of these processes will depend on the technical, economic, and environmental attributes compared against all other competing options.

Techno-economic analyses and life cycle assessments of solar panel recycling have been documented extensively [11–13] and have focused on capturing the benefits associated with the capture of secondary materials [14–16]. Understandably, the focus has been on establishing economic and environmental value for solar panel recycling processes, not in recommending how secondary material should be used. Additionally, others articles have argued that panel recycling techniques should focus on the recovery of rare minerals and metals for reuse [17], allowing for the downcycling of bulk materials such as glass and aluminum. Regardless, the economic and environmental benefits and costs associated with recovery processes that maintain material quality are not known and have not been predicted.

This current effort is focused on predicting the value of recovering secondary material from end-of-life (EoL) panels, followed by the recycling and processing of the panel material for reuse in next generation solar panels. The focus of this work is on establishing the environmental impacts through life cycle analysis (LCA) of a solar panel from procurement of materials for manufacturing through EoL panel recovery, recycling, and disposal. The primary motivation for the consideration of LCA prior to techno-economic feasibility is to establish if and how much of an environmental benefit can be generated from closing the loop between panel material recovery and new panel manufacturing. The focuses of this work are the Series 4 and Series 6 (S4 and S6, respectively) [18, 19] panels manufactured by First Solar (FS). These panels are a prime choice for analysis due to the extensive documentation of panel manufacturing, installation [11, 12] and recycling processes [20, 21]. This recycling process includes capture of bulk panel material (glass, aluminum, copper, etc.), as well as the recovery of cadmium and tellurium semiconductor materials.

In order to predict the environmental benefit associated with secondary material recovery and reuse in new solar panels, the following tasks must be accomplished:

1. Existing solar panel manufacturing life cycle inventories must be analyzed and updated in preparation for integration with secondary material recycling and production.
2. Life cycle inventories for recycling and processing of secondary material for use in new solar panels must be predicted and developed.
3. Secondary material use scenarios in new solar panels must be developed.
4. The life cycle analysis of replacing primary material in solar panels with secondary material derived from end-of-life solar panel recycling must be evaluated.

This current work is focused on Task #1, while Tasks #2–4 are still ongoing and will be reported on in a future publication. The purpose of analyzing and updating existing solar panel manufacturing life cycle inventories is to consider primary material inventories used to form the product inventory. In particular, the focus is on inventories associated with material that can be recycled and potentially reused in future panels. A successful update of the FS manufacturing inventory will generate a set of inventories that account for differences in how and from where primary material is sourced. The goal of this current work is to identify and select different primary material sources that can reasonably provide material input for S4 and S6 panels.

The FS panel inventories adapted in this work were originally presented in Sinha et. al. [12]. The remaining paper is organized as follows: Sect. "Current CdTe Panel Life Cycle Inventories" discusses current life cycle inventories that are adopted and modified in this work, Sect. "Updates to CdTe Panel Life Cycle Inventories" describes how these inventories are updated in preparation for upcoming work, and Sect. "Life Cycle Analysis Results" presents results on how changes in the manufacturing inventory affect the environmental burden associated with solar panel manufacturing.

Current CdTe Panel Life Cycle Inventories

The two panels considered in this work are the FS S4 and S6 panels. The S4 and S6 panels have an area of 0.72 and 2.47 m^2, a panel mass density of 16.67 and 14.17 kg per m^2, and a peak power density of 155 and 175 watts per m^2, respectively [12]. These cells consist of a solar photovoltaic laminate that can be mounted to a racking or tracking system. Three key differences between the S4 and S6 panels are as follows: (1) S6 panels are mounted on an aluminum frame while S4 panels are not, (2) S6 front and back glass is thinner than S4, and (3) the cadmium sulfide layer present in S4 panels has been replaced with a cadmium selenide layer in S6 panels. The first and second changes are captured in life cycle inventories. The third change to the semiconductor mixture is not captured in the literature, and an updated inventory is not available for this work. We assume that this change will not significantly affect LCA results for the following two reasons: the total mass fraction of semiconductor material within the solar laminate (the cell prior to frame attachment or mounting) is small (0.001 and 0.002 for the S4 and S6 laminates, respectively), and both the cadmium sulfide and cadmium selenide layers make up less than 10% of total semiconductor material.

Prior efforts by the International Energy Agency (IEA) Photovoltaic Power Systems (PVPS) have produced life cycle inventories for manufacturing the S4 panels based on actual FS manufacturing data [2], mounting and wiring processes, and recycling based on actual FS recycling data [16]. This information was combined for the S4 panel in [12], which presents LCI information for both manufacturing and recycling. Using this reference in combination with [11] provides the entire LCI for the S4 panel.

Complete data from S6 manufacturing were not available from FS at the time of this study due to continuing developments and refinements to the S6 manufacturing process. An estimation of the manufacturing and recycling inventory for the S6 panel is provided in [12]. Additional estimates include that the addition of the aluminum frame reduced mounting material requirements by 33%, and that the wiring inventory used for S4 panels can be scaled by area to the S6 panel.

Both the S4 and S6 inventories account for location through region-specific inventories associated with energy use and infrastructure-related inventories based on panel production capacities of 86.2% in Malaysia and the remainder in the USA [12]. Inventories associated with materials and chemicals used during panel manufacturing use definitions associated with production and sourcing in Europe and Switzerland. All prior inventories used definitions from the Ecoinvent v2.2 database and determined environmental impacts using the ILCD 2011 Midpoint V 1.09 impact method with long-term emissions excluded [2, 12, 16]. The functional unit used in this prior work is kWh electricity production assuming a 30-year lifetime and electricity production of 975 kWh per kW solar, or an 11% capacity factor based on average European deployment [12]. In total, the S4 and S6 panel inventories include the solar photovoltaic laminate, mounting equipment used in a rooftop installation, and back panel wiring. The S6 inventory also includes the aluminum frame that the solar PV laminate is attached to.

Updates to CdTe Panel Life Cycle Inventories

The goal of updating the CdTe manufacturing inventory is to identify and select potential primary material production pathways that can be used to source primary material input for S4 and S6 panels. This step is taken in preparation for quantifying the impact of recovering and reusing secondary material from end-of-life solar panels. The inventory update is accomplished by transitioning from the inventory set used in prior work, Ecoinvent-v2.2, to Ecoinvent-v3 [22]. A major difference between these two inventory databases is the addition of new region and country-specific material production inventories in Ecoinvent -v3. Any change in country or region-specific code was made to capture the possibility of sourcing primary material from a different location. The process for selecting geographically unique inventories was made by (a) using U.S. Geological Survey material production information or comparable information to determine major material production hubs, followed by (b) selection of primary material production pathways that produce sufficiently different environmental impact results or can be readily integrated with secondary material production processes. While all primary material and chemical inputs captured in prior inventories [12] were examined, only materials associated with glass, aluminum, copper, steel, cadmium, and tellurium are discussed in this work. Beyond these six materials, little to no difference in environmental impact was observed when examining all remaining materials and chemicals used during panel manufacturing. Analysis and results used to select updated material inventories were done using results from the ILCD 2011 Midpoint V 1.09 impact method with long-term emissions excluded, or the same method as used under the Ecoinvent -v2.2 inventory. Note that these prior inventories [12], which were designed to capture the impacts of actual solar panel manufacturing, used a mix of primary and secondary material input based on supplier data, while this work assumes only primary material input.

Flat Glass

Flat glass material used in the front and back glass sheets accounts for over 70% of laminate mass for both S4 and S6 panels. Within the Ecoinvent -v3 database, there are only two definitions for flat glass production: global and Europe. Additionally, there are negligible differences between the current -v3 and prior -v2.2 life cycle impact results under the ILCD method. Due to the limited options, the global inventory was selected for use in the updated inventory. Note that inventory selection does not match well versus global flat glass production. As of 2015, 50% of flat glass was produced in China, 12% in Europe, 8% in North America, and 4% in South East Asia [23]. The major sources of air pollution impacts are due to combustion-related processes that occur during flat glass production. Additionally, total pollutant emissions per unit glass depend heavily on post-combustion and emission release controls [3], which can vary significantly between regions. This mismatch between potential sources and suppliers of flat glass and Ecoinvent inventories could result in underestimated environmental burdens associated with flat glass production. However, further exploration of flat glass production in Central and Southeast Asia falls outside the scope of the current work.

Aluminum

Aluminum is the primary material used in solar panel mounts, contributing approximately 2.8 and 1.9 kg per m^2 of installed S4 and S6 panels, respectively, and 11% of S6 panel mass is due to the aluminum frame. According to the 2020 Minerals Yearbook from the U.S. Geological Survey (USGS) [24], 57% of all finished aluminum production occurred in China, 4.8% in Canada, and 1.5% in the USA. These countries ranked first, third, and ninth in total aluminum production. Furthermore, according to the 2018 Minerals Yearbook, approximately half of all 2018 aluminum imports into the USA originated from Canada [25].

The Ecoinvent -v3 database includes inventories for China (CN), North America excluding Quebec (NA), and Quebec (QB). For comparison against Ecoinvent -v2.2 aluminum production, aluminum production in Europe (EU) was also considered. The embodied carbon emissions and mineral and fossil resource depletion totals for these four aluminum sources are shown in Fig. 1. These inventories include the mining and processing of bauxite into alumina (shown as “primary material processing” in purple and “primary materials” in yellow in Fig. 1), followed by the conversion

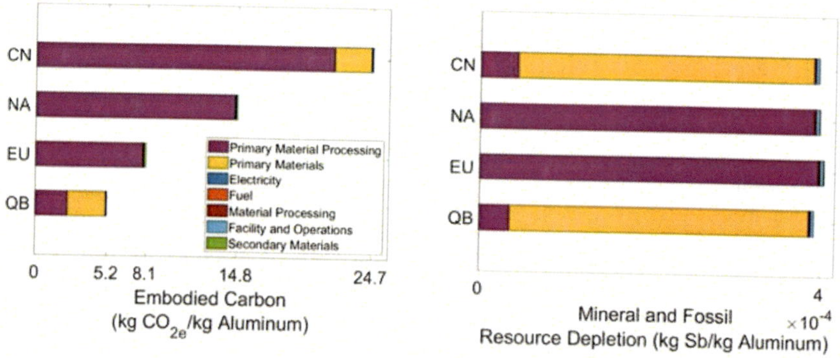

Fig. 1 Embodied carbon and mineral and fossil resource depletion for primary aluminum production from China, North America, Europe, and Quebec, Canada, from the Ecoinvent -v3 database [22]

of alumina to aluminum (all other processes shown in Fig. 1). Depending on detail and quality of data, allocation of environmental burdens associated with alumina production between resource extraction processes and the actual resource varies drastically between regions. Regardless, both carbon emission and mineral depletion indicators show that the primary burden associated with aluminum production is caused by the production of alumina.

These results also show significant differences in carbon emissions between regions with aluminum produced in China resulting in nearly five times more carbon emissions than aluminum produced in Quebec. These results are indicative of all other environmental indicators captured in the ILCD method (e.g., eutrophication, acidification, ozone formation, etc.) except for mineral and fossil resource depletion. Closer examination of these indicators shows that fuel inputs for electricity combustion processes are key drivers in environmental impact aside from mineral and fossil resource depletion. Energy input in China uses a combination of coal and heavy oil, while Quebec receives electricity from hydroelectric power plants. Both North America and European aluminum use energy mixtures that fall between these two extremes. These differences in primary energy sources are responsible for major differences in environmental impact between these regions.

Mineral and fossil energy depletion shows similar levels of depletion regardless of region. Under the Ecoinvent database, the primary driver of mineral and fossil depletion is the mining of bauxite for primary aluminum. Conversion efficiencies of bauxite into aluminum are similar across all regions, resulting in a similar level of mineral resource depletion between the regions. For this current work, CN and QB primary aluminum production were adopted as two potential material sources. FS data do not include sufficient detail to resolve origination of aluminum used in panel racks and mounts. However, Fig. 1 shows that these two locations span the environmental impacts associated with primary aluminum use.

Steel

Steel components are used to assemble frame and mounting assemblies. The steel mass fraction in both S4 and S6 panels is approximately 0.05. Region-specific steel production inventories are provided for steel production in India, Europe, and across the globe. According to the USGS, nearly 69 and 56% of pig iron and raw steel were produced in China. Like glass, the lack of detailed regional or country-specific steel inventories creates another potential source of error. Regardless, steel production can be updated to reflect potential shifts in steel production. If all steel components used in FS solar panels are 18|8 steel (18% chromium, 8% nickel, and remainder as iron), two production pathways for steel production are documented in Ecoinvent -v3: oxygen converter and electric arc furnace pathways. Both inventories are taken for "global" production. Since results are taken from a global production perspective, ILCD indicator values are similar across all environmental impact categories. An example of this is shown in Fig. 2, which shows embodied carbon for both production

Fig. 2 Embodied carbon associated with the production of 18|8 steel through a converter and arc furnace pathway, from the Ecoinvent -v3 database [22]

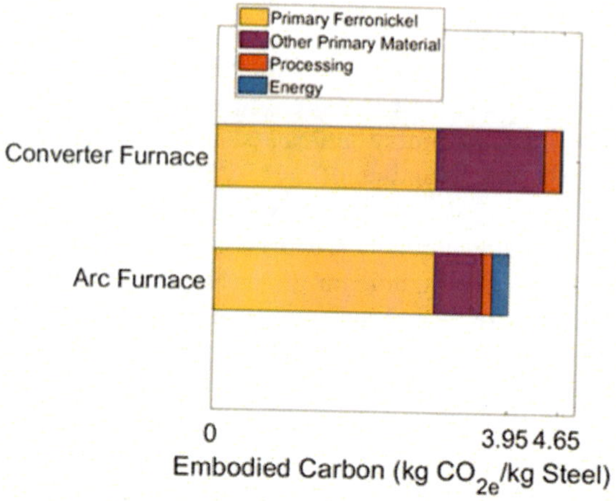

pathways. This figure is indicative only of both pathways yielding similar environmental impacts and does not signify that the converter furnace pathway produces steel with a lower environmental impact.

Ultimately, differences between these two production pathways are small when considering all material, chemical, and energy inputs required for S4 and S6 manufacturing. Due to the negligible difference, total system results presented in Sect. "Life Cycle Analysis Results" only consider steel components produced using a converter furnace. However, the arc furnace is included in preparation for addressing the recovery and recycling of steel components, which are typically processed using an electric arc furnace.

Copper

Copper is primarily used in S4 and S6 wiring and is also used in small quantities in panel interconnections. Copper contributes between 0.03 and 0.04 of FS panel mass fractions. According to the USGS, Chile and Peru produce 40% of global copper [25]. The next closest country, China, produces 8% of the global copper supply. Furthermore, Ecoinvent copper production definitions span a wide range of mining and extraction operations where copper is a primary driver for mineral extraction and some where it is a by-product of other mining operations. The result is a dilute set of copper production pathways with no clear dominant copper source. Without further information on copper sourcing for FS panels, two copper production pathways are adopted: primary copper from South America and global copper production. Total carbon emissions for these two copper sources are shown in Fig. 3. In general, these results are indicative of other ILCD indicator results. Similar to aluminum production in Quebec, the South American copper inventory receives hydroelectric

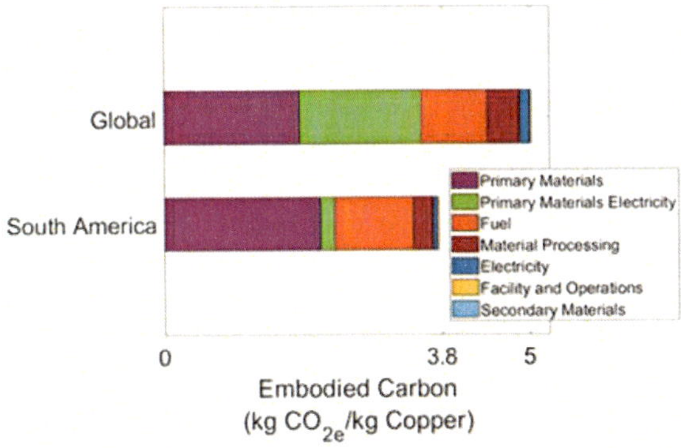

Fig. 3 Embodied carbon associated with the global and South American production of copper, from the Ecoinvent -v3 database [23]

energy, generally resulting in a lower overall impact than comparable copper production pathways. Exceptions to this are mineral and fossil resource depletion, marine eutrophication, freshwater eutrophication, and terrestrial eutrophication, which are all driven by copper ore mining processes.

Similar to steel, copper makes up a small fraction of total mass contributed towards both S4 and S6 panels, and the difference between considered copper production pathways is small. As a result, only global copper is considered in subsequent results.

Cadmium Telluride

Cadmium telluride makes up only a tiny fraction of overall solar panel mass fraction. According to reports on solar panel life cycle development, CdTe life cycle inventories were developed to capture the environmental impacts of CdTe solar panel production [26]. Two CdTe production pathways are available in Ecoinvent -v3, which match the v2.2 inventory. The global CdTe production inventory was adopted for use in this work.

Functional Units

The final change to the life cycle inventory is a shift in functional unit from per kWh produced to per kW nameplate capacity. This shift serves two purposes. First, it eliminates latitudinal, installation orientation, and local weather effects that affect solar panel output and life cycle analysis results. Second, a kW functional unit captures panel improvements that result in increased panel capacity, allowing for the analysis to capture structural and chemical design changes.

Life Cycle Analysis Results

S4 and S6 life cycle inventory modifications described in Sect. "Updates to CdTe Panel Life Cycle Inventories" were applied to the inventory described in [12] and evaluated using the ILCD 2011 Midpoint V 1.09 impact method. Results were generated for two scenarios: one where aluminum was sourced from China (scenario name "Global Materials") and one where aluminum was sourced from Quebec (scenario name "QB Aluminum"). Both scenarios assume all new material that steel components are made using a converter furnace and that copper components are sourced through the "global" pathway. Results are shown for embodied carbon, acidification, terrestrial eutrophication, and mineral and fossil resource repletion, all versus the nameplate capacity of the applicable solar panel. Results only indicate the environmental burden due to panel manufacturing and installation and do not include any burden associated with panel end-of-life.Results for the S4 panel are shown in Fig. 4, which shows total impacts per kW separated by installed panel component. These components include the laminated solar cell, mounting material, wiring, and installation. These results indicate a drop in overall impact due to sourcing aluminum from low impact production centers, predicting a 36.7%, 24.7%, and 30.6% reduction in embodied carbon, acidification, and terrestrial eutrophication, respectively. Shifting aluminum sourcing, however, does not reduce mineral and fossil resource depletion, which remains virtually unchanged. Note that a shift in steel sourcing from a converter furnace to arc furnace would yield negligible changes in life cycle analysis results across more indicators. Changes in copper sourcing, however, can have a measurable impact due to the significant contributions to environmental burdens such as acidification and eutrophication.

Results for the S6 panel are shown in Fig. 5. The effect of sourcing aluminum from a low impact production center results in larger decreases than observed for the S4 panel, reducing embodied carbon, acidification, and terrestrial eutrophication by 43.5%, 29%, and 36.5% respectively. These larger reductions are due to design differences in the S6 panel versus S4. Specifically, the reduction in glass thickness and addition of an aluminum frame makes the S6 panel more sensitive to changes to aluminum.

These results also demonstrate the improved environmental impact due to a shift from the S4 to S6 panel, replicating results produced by Sinha and Wade in [12]. Close comparison of these prior and current results shows differences in total reduction and, in certain instances, disagreement in results. These differences stem from the scope of the current and prior work. Sinha and Wade considered panel manufacturing, installation, operation, and end-of-life including panel recycling. Modeling of panel recycling included attribution of benefits derived from material recycling, reducing the overall burden associated with S4 and S6 operation. The current work only considers panel manufacturing. As a result, agreement can be found when comparing the updated panel inventory impacts versus the impacts generated through the manufacturing process described in [12]. Recycling benefits will be addressed in subsequent work that includes potential modifications to improve panel recyclability.

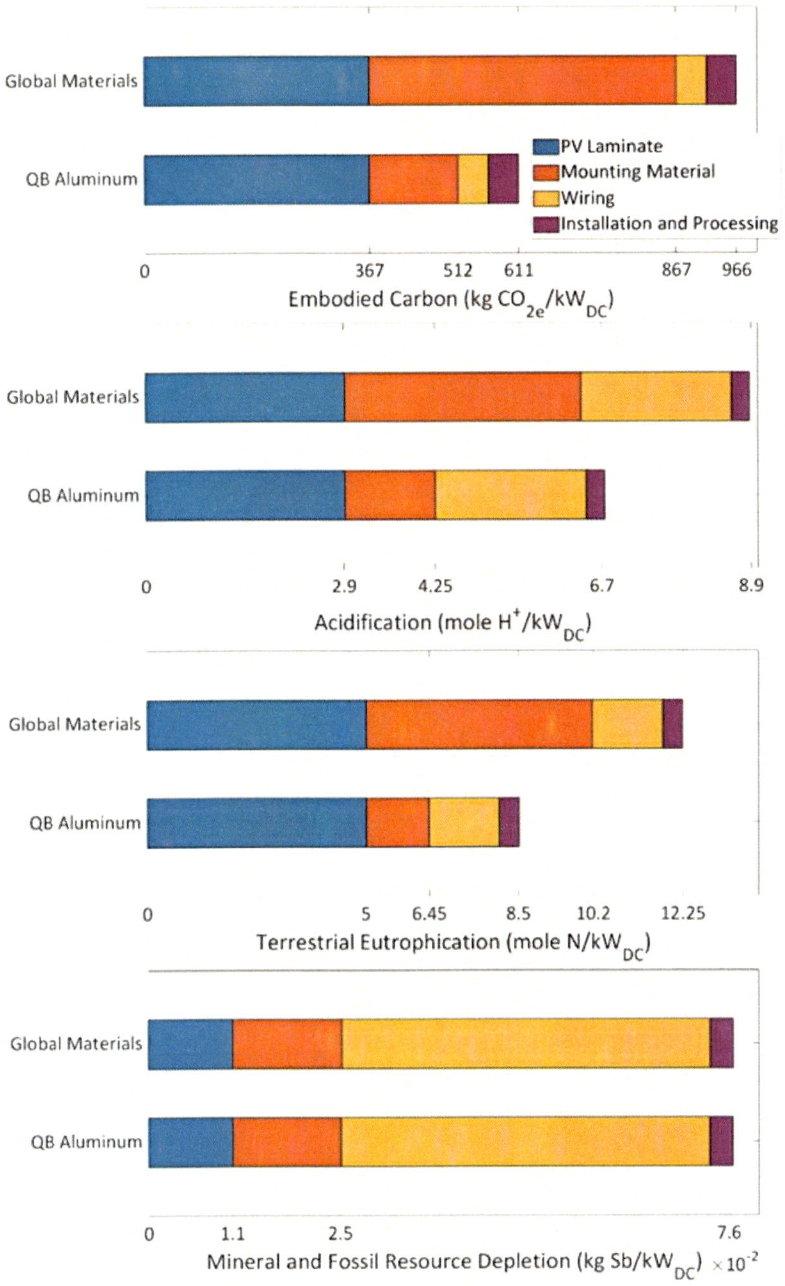

Fig. 4 Updated LCA results for the S4 panel, showing embodied carbon, acidification, terrestrial eutrophication, and mineral and fossil resource depletion per kW capacity

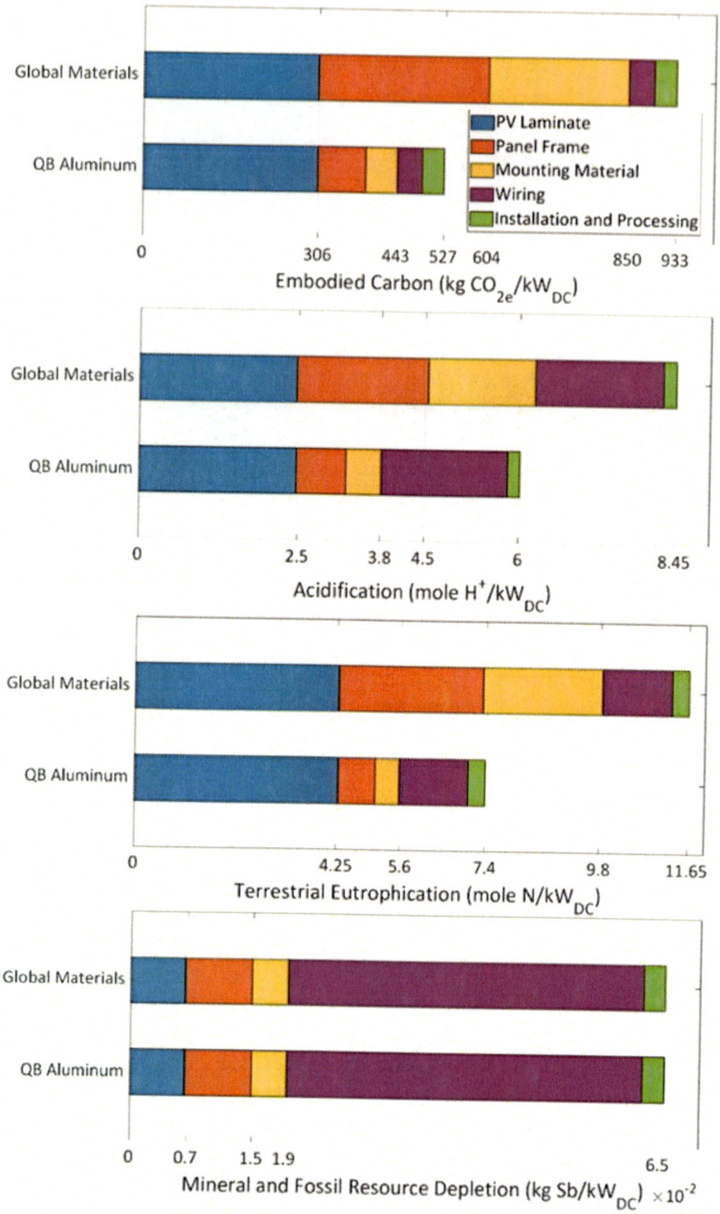

Fig. 5 Updated LCA results for the S6 panel, showing embodied carbon, acidification, terrestrial eutrophication, and mineral and fossil resource depletion per kW capacity

Summary and Conclusion

There is a potential opportunity to further reduce the environmental impact of solar energy by recycling end-of-life panel material for use in next generation solar cells. The environmental benefit associated with closing the material loop between old and new solar panels has not been estimated. This work is the first step towards developing this estimation by updating and preparing existing solar panel manufacturing inventories for integration with secondary material inventories associated with solar panel recycling. This work focused on the Series 4 and Series 6 panels manufactured by First Solar, both of which can be recycled to recover nearly all glass, aluminum, steel, copper, and semiconductor material. The update focused on selecting realistic material production pathways that create a range of environmental impacts associated with panel production. It was assumed during inventory development that only primary material is used during manufacturing to establish an upper environmental impact boundary (actual aluminum, steel, and copper material tend to include a mix of primary and secondary material input). During this process, it was found that selection of the primary aluminum production pathway was critical to establishing manufacturing scenarios with a high and low environmental impact. Testing the results showed that informed sourcing of primary materials can reduce environmental impact by 25% to over 40%. The updated inventories also produced similar environmental impacts as preceding inventories for the S4 and S6 panels, showing agreement between the current and previously vetted inventories. Future work will use these updated inventories presented in this work to quantify the impacts of secondary material recovery and the reuse of recovered solar materials in next generation solar panels.

Acknowledgements The authors gratefully acknowledge the financial support for this work, which was provided by the REMADE Institute, project number 18-01-DE-07, and the University of California, Irvine. Valuable technical discussions with our project manager Ed Daniels are also acknowledged.

References

1. Weckend S, Wade A, Heath G (2016) End of life management: solar photovoltaic panels. (No. NREL/TP-6A20-73852). National Renewable Energy Lab.(NREL), Golden, CO (United States)
2. Stolz P, Frischknecht R, Wyss F, de Wild-Scholten M (2016) PEF screening report of electricity from photovoltaic panels in the context of the EU product environmental footprint category rules (PEFCR) Pilots v. 2.0. Treeze Ltd. SmartGreen Scans, Uster
3. Schorcht F, Kourti I, Scalet BM, Roudier S, Sancho LD (2013) Best available techniques (BAT) reference document for the production of cement, lime and magnesium oxide. European Commission Joint Research Centre Institute for Prospective Technological Studies, Luxembourg
4. Aluminumn Association (2013) The environmental footprint of semi-finished aluminum products in North America. A life cycle assessment report

5. Broadbent C (2016) Steel's recyclability: demonstrating the benefits of recycling steel to achieve a circular economy. Int J Life Cycle Assess 21:1658–1665. https://doi.org/10.1007/s11367-016-1081-1
6. Villalba G, Segarra M, Fernández AI, Chimenos JM, Espiell F (2002) A proposal for quantifying the recyclability of materials. Resour Conserv Recycl 37:39–53. https://doi.org/10.1016/S0921-3449(02)00056-3
7. Sinha P, Cossette M, Ménard JF (2012) End-of-life CdTe PV recycling with semiconductor refining. In: 27th European photovoltaic solar energy conference and exhibition, pp 4653–6
8. Ravikumar D, Sinha P, Seager TP, Fraser MP (2016) An anticipatory approach to quantify energetics of recycling CdTe photovoltaic systems. Prog Photovoltaics Res Appl 24:735–746. https://doi.org/10.1002/pip
9. Chowdhury MS, Rahman KS, Chowdhury T, Nuthammachot N, Techato K, Akhtaruzzaman M et al (2020) An overview of solar photovoltaic panels' end-of-life material recycling. Energy Strateg Rev 27:100431. https://doi.org/10.1016/j.esr.2019.100431
10. Rodriguez Vieitez E, Eder P, Villanueva A, Saveyen H (2011) End-of-waste criteria for waste paper: technical proposals. JRC Sci Tech Rep:1–99. https://doi.org/10.2791/7150
11. Fthenakis V, Kim HC, Frischknecht R, Raugei M, Sinha P, Matthias S (2011) Life cycle inventories and life cycle assessment of photovoltaic systems. https://doi.org/10.1111/j.1365-2117.2010.00499.x
12. Sinha P, Wade A (2017) Addressing hotspots in the product environmental footprint of CdTe photovoltaics. In: 2017 IEEE 44th photovoltaic specialist conference (PVSC) (IEEE, 2017), pp 2005–2010
13. Ravikumar D, Sinha P, Seager TP, Fraser MP (2015) An anticipatory approach to quantify energetics of recycling CdTe photovoltaic systems. https://doi.org/10.1002/pip
14. Held M (2009) Life cycle assessment of CdTe module recycling. In: 24th European photovoltaic solar energy conference. Hamburg, Germany, pp 2370–5
15. Held M, Ilg R (2011) Update of environmental indicators and energy payback time of CdTe PV systems in Europe. Prog Photovoltaics Res Appl 19:614–626
16. Stolz P, Frischknecht R, Wambach K, Sinha P, Heath GA (2018) Life cycle inventory of current photovoltaic module recycling processes in Europe (No. NREL/TP-6A20-73845). National Renewable Energy Lab.(NREL), Golden, CO (United States)
17. Brenner W, Bednar N, Biermayr P, Adamovic N (2018) Standardization and life cycle cost assessment approach in circular economy for photovoltaic waste. In: 2018 3rd International conference smart sustain technology split 2018, pp 1–6
18. First Solar (2018) Series-4V3-Module-Datasheet.ashx. http://www.firstsolar.com/-/media/First-Solar/Technical-Documents/Series-4-Datasheets/Series-4V3-Module-Datasheet.ashx. Accessed 25 Sept 2020
19. First Solar (n.d.) Series-6-Datasheet.ashx. http://www.firstsolar.com/-/media/First-Solar/Technical-Documents/Series-6-Datasheets/Series-6-Datasheet.ashx. Accessed 25 Sept 2020
20. Stolz PP, Frischknecht R, Wambach K, Sinha P, Heath G (2018) Life cycle assessment of current photovoltaic module recycling. Int Energy Agency Power Syst Program Rep (IEA-PVPS T12):13
21. Stolz P, Frischknecht R, Wambach K, Sinha P, Heath G (2016) Life cycle assessment of photovoltaic module recycling. Uster, CH Treeze Ltd
22. Wernet G, Bauer C, Steubing B, Reinhard J, Moreno-Ruiz E, Weidema B (2016) The ecoinvent database version 3 (part I): overview and methodology. Int J Life Cycle Assess 21:1218–1230. https://doi.org/10.1007/s11367-016-1087-8
23. Butler JH, Hooper PD (2019) Glass waste, 2nd edn. Elsevier Inc. https://doi.org/10.1016/b978-0-12-815060-3.00015-3
24. U.S. Geological Survey & Orienteering, Survey (2020) Mineral commodity summaries, 2020. Government Printing Office
25. U.S. Geological Survey (2020) 2018 Minerals yearbook
26. Jungbluth N, Stucki M, Frischknecht R, Tuchschmid M, Doka G, Vollmer M (2009) Life-cycle inventories-Photovoltaics. Swiss Cent Life Cycle Invent Dübendorf, CH

Techno-Economic Analysis of Material Costs for Emerging Flow Batteries

Haoyang He, Shan Tian, Brian Tarroja, Branden Schwaebe, Scott Samuelsen, Oladele A. Ogunseitan, and Julie M. Schoenung

Abstract To ensure that the production cost of battery energy storage systems for the electric grid does not compromise the environmental benefits gained from the substitution of traditional fossil fuels, it is important to evaluate and manage the cost feasibility of the feedstock materials used in battery production. In this study, we present a techno-economic analysis to evaluate the cost of materials in three emerging redox flow battery products: vanadium pentoxide redox flow batteries (VRFB), zinc-bromine flow batteries (ZBFB), and all-iron flow batteries (IFB), with a focus on primary materials used in functional components. Furthermore, we performed sensitivity analysis for selected materials to explore the uncertainty due to dynamic variation in market prices. The normalized results indicate that the major cost contributors for each battery type vary significantly over time, and the historical variations in material prices could largely affect the battery system production cost. Thus, material costs should be considered as a key attribute in material selection and product design for installing flow battery technologies in the electric grid.

Keywords Battery · Energy storage · Energy grid · Renewable · Techno-economic analysis · Material prices

---

H. He · B. Schwaebe · J. M. Schoenung (✉)
Department of Materials Science and Engineering, University of California, Irvine, CA 92697, USA
e-mail: Julie.Schoenung@UCI.edu

S. Tian · S. Samuelsen
Department of Mechanical and Aerospace Engineering, University of California, Irvine, CA 92697, USA

S. Tian · B. Tarroja · S. Samuelsen
Advanced Power and Energy Program, University of California, Irvine, CA 92697, USA

B. Tarroja
Department of Civil and Environmental Engineering, University of California, Irvine, CA 92697, USA

O. A. Ogunseitan
Department of Population Health and Disease Prevention, University of California, Irvine, CA 92697, USA

A. Lazou et al. (eds.), *REWAS 2022: Developing Tomorrow's Technical Cycles (Volume I)*, The Minerals, Metals & Materials Series,
https://doi.org/10.1007/978-3-030-92563-5_46

Introduction

Among the current battery technologies on the market, redox flow batteries (RFBs) are positioned to serve as a key component for renewable energy grid evolution, namely high energy capacity due to a physical design with separate charging/discharging and storage subsystems, and are attractive for renewable energy systems. Additional advantages of RFBs include a large depth of discharge, minimal degradation, and a long lifespan [1]. The cost and revenue associated with the RFB systems are critical to their widespread application. While previous research has focused on analyzing the potential for cost reduction in RFBs and identifying possible practices for enhancing the revenue of RFB operation. However, most of the published studies have focused on the relatively mature vanadium pentoxide redox flow batteries (VRFB), whereas RFBs with different chemical combinations such as zinc-bromine flow batteries (ZBFB) and all-iron flow batteries (IFB) are yet to be explored. In this study, we assess the material costs associated with flow battery production of not only VRFB, but also zinc-bromine flow batteries (ZBFB) and all-iron flow batteries (IFB). Based on material inventories provided by the manufacturers, we applied a techno-economic analysis (TEA) approach. We also provide a component cost distribution for each flow battery type, including sensitivity analysis to account for temporal variations in the price of raw materials.

Methodology

Techno-Economic Analysis Model

To perform cost sensitivity analyses, we relied on methods of techno-economic analysis (TEA) [2–4]. The goal is to investigate and understand the cost contributors for flow battery systems because these technologies are relatively early in their commercial deployment. Therefore, this section focuses on material costs because these costs are associated with each flow battery type, while the other costs (e.g., utilities, labor, and fixed costs) vary from manufacturer to manufacturer and business strategy and are not able to be assessed with currently available data. The cost assessment is performed for the three flow batteries based on manufacturer-provided materials inventories, and the unit materials costs are derived from various published sources. The cost distribution by battery component is determined to highlight the major cost drivers in battery systems. Lastly, uncertainty due to price variability is evaluated.

Material Cost Data and Uncertainty

For the TEA model, data on the prices of key materials used in the flow battery systems are required. Gathering material cost information that complies with data quality and reliability standards, however, can be difficult. The cost of materials is subject to the dynamics of global markets and trade, causing these values to vary over time. The sources for price information in our case can be classified into four types: (1) international market prices, (2) United States (U.S.) import prices, (3) prices published in the literature, and (4) retail prices.

(1) The "international market price" is suitable for materials that are traded as bulk commodities, where their prices are continuously monitored and updated in international trade, such as metals like copper and aluminum.
(2) The "US import price" is collected based on the price of goods imported to the USA. These are well documented by several US governmental institutes and databases such as the United States Geological Survey (USGS) and Statista.
(3) The "literature price" is based on price values found in the published literature for the material costs of flow battery production. The advantage of using literature data is that the cost information is complete even for materials that are difficult to track to a market, and these data are peer reviewed. Due to the lack of original studies and primary data in these studies, however, much of the cost information in literature studies are predicted values, while some are cited from previous publications. These values, therefore, may not capture the dynamic price variations to reflect the current situation.
(4) The "retail price" is the material price collected from the vendors who purchase those materials from upstream supply chains and sell them directly to the commercial end-users.

Collectively, we refer to these four types as "market prices". Due to the variations in the market prices for materials, we conducted sensitivity analysis to explore the uncertainty in market price variability for selected materials, specifically: vanadium pentoxide, titanium, bromine, and carbon fiber felt. A three-point estimation is applied based on a pessimistic price (worst case), most likely price (current value) and optimistic price (best case). The three-point estimation creates an approximate probability distribution to predict the outcomes of future events, e.g., materials price, when only limited information is available. In this study, a double-triangular distribution is used, as shown in Fig. 1. The notation "a" is the optimistic price which represents the best case, "b" is the pessimistic price indicating the worst case, and "m" is the most likely price, which indicates the current price [5]. With the three value points determined, a weighted average (E) as expected price and a standard deviation (SD) can be calculated as follows:

$$\mathrm{E} = (\mathrm{a} + 4\mathrm{m} + \mathrm{b})/6 \tag{1}$$

$$\mathrm{SD} = (\mathrm{b} - \mathrm{a})/6 \tag{2}$$

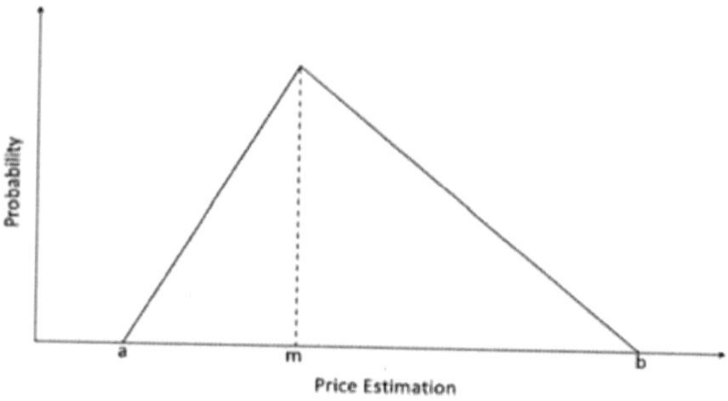

Fig. 1 Example of a probability distribution used for three-point estimation

Results and Discussion

Baseline Cost Analysis

Vanadium Pentoxide Flow Battery

The material costs and the associated distribution by component for the VRFB system are provided in Table 1 and Fig. 2. Due to the high cost of vanadium pentoxide and its use as the major species in the electrolyte, the cost of electrolyte accounts for 80% of the total material cost. Surprisingly, the material cost for the whole cell stack, which is related to the power capacity, only contributes 4% to the total material cost. When only considering the power capacity component, the Nafion® membrane is the highest contributor and accounts for 55% of the material cost for the power capacity subsystem.

Zinc-Bromide Flow Battery

The material costs and the associated distribution by component for the ZBFB system are provided in Table 2 and Fig. 3. The power capacity components comprise the largest share of total material costs as the cell stack accounts for 33% of the total material cost. The electrolyte and the bipolar plate are identified as material cost drivers, as they account for 29% and 22% of the total material cost, respectively.

All-Iron Flow Battery

The material costs and the associated distribution by component for the IFB system are provided in Table 3 and Fig. 4. Contrary to the VRFB and ZBFB, the battery management system in IFB contributes to the largest share of the material cost at 49% of the total material cost. The cell stack accounts for 25% of the total material cost, while the electrolyte only accounts for 5%.

Table 1 Material price information for materials used in the VRFB system

Battery technology	VRFB			
Component	Price	Unit	Data type	Data source
Bipolar plate				
Graphite	1.58	$/kg	2015 Market average	USGS [6]
Polyethylene, low density	1.22	$/kg	2017 Import average	STATISTA [7]
Cell frame				
Polypropylene	1.84	$/kg	2017 Import average	STATISTA [8]
Glass fiber	2.00	$/kg	Estimated value	Amirhossein et al. [9]
Electrode				
Carbon felt paper	237.60	$/kg	Literature value	Minke et al. [10]
Membrane				
Nafion®	937.53	$/kg	Literature value	Minke et al. [10]
Cell stack accessories				
Steel, low alloyed	0.69	$/kg	2019 Market average	Worldsteelprice [11]
Copper	6.61	$/kg	2018 Vendor value	USGS [12]
Polyvinylchloride	0.97	$/kg	2019 Market average	Investing.com [13]
Electrolyte				
Vanadium pentoxide	35.75	$/kg	2019 Market average	USGS [14]
Hydrochloric acid	0.13	$/kg	2018 Market instant	ICIS [15]
Sulfuric acid	0.06	$/kg	Literature value	Minke et al. [16]
Water	0.00241	$/kg	Government value	[17]
Tank				
Polyethylene, high density	1.26	$/kg	2017 Import average	STATISTA [18]
Pipes				
Polyethylene, high density	1.26	$/kg	2017 Import average	STATISTA [18]
Pump	13.46	$/kW	Literature value	Minke et al. [19]
Inverter	112.13	$/kW	Literature value	Minke et al. [16]
Battery management system				
Aluminum	2.54	$/kg	2018 Import average	USGS [20]
Titanium	30.00	$/kg	2019 Market instant	TRICORMETALS [21]
Power control system	150.00	$/kW	Literature value	Minke et al. [19]
Balance of plant accessories				
Steel, low alloyed	0.69	$/kg	2019 Market average	Worldsteelprice [11]

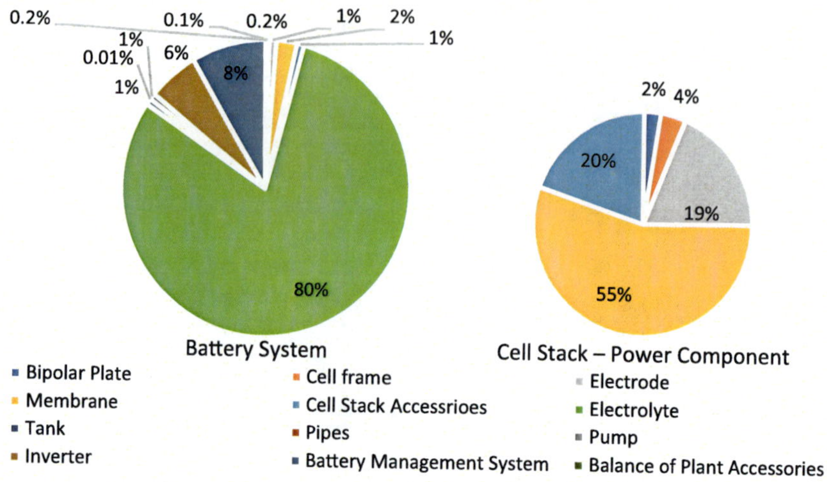

Fig. 2 Materials cost distributed by component in the VRFB system

Sensitivity Analysis Due to Material Price Variations

The cost analyses presented in this study are driven by material prices; however, the price of raw materials used in the three flow battery systems is subject to historical variations due to fluctuations in the global markets. Therefore, in this section, we describe sensitivity analysis of variations in raw material prices based on historical fluctuations. The three-point estimation values (described in Sect. "Techno-Economic Analysis Model") are summarized in Table 4 for the four materials assessed. With the three price points determined, a weighted average (E) as the expected market price is calculated with a standard deviation available to reflect the variations. The larger the price gap between the pessimistic and optimistic price, the larger the standard deviation identified.

The results of the sensitivity analysis are presented in Fig. 5. The total material cost of the VRFB system due to the variations in vanadium pentoxide price (Fig. 5a) ranges between \$186 and \$648 per kWh when using different price points. The expected price value is estimated to be \$466/kWh with a standard deviation value of \$77/kWh. Thus, changes in the price of vanadium pentoxide will greatly affect the VRFB system cost. The baseline value, \$491/kWh, is near the upper value in the range, reflecting that recent prices for this material have been higher than historical values. The influence of titanium and bromine prices on the ZBFB cost is shown in Fig. 5b, c, respectively. The expected ZBFB cost (\$158/kWh) is higher than the baseline estimated value (\$153/kWh), relative to titanium sensitivity, whereas the expected cost for the bromine scenario (\$149/kWh) is lower than the baseline estimation, but these differences are relatively small. The material cost deviations due to variations on titanium price are larger than the case for bromine, and both are relatively small compared to the vanadium pentoxide for VRFB. For the IFB, the

Table 2 Material price information for materials used in the ZBFB system

Battery technology	ZBFB			
Component	Price	Unit	Data type	Data source
Bipolar plate				
Titanium	30.00	$/kg	2019 Market instant	TRICORMETALS [21]
Polyethylene, high density	1.26	$/kg	2017 Import average	STATISTA [18]
Cell frame				
Polyethylene, high density	1.26	$/kg	2017 Import average	STATISTA [18]
Cell stack accessories				
Steel, low alloyed	0.69	$/kg	2019 Market average	Worldsteelprice [11]
Titanium	30.00	$/kg	2019 Market instant	TRICORMETALS [21]
Polyethylene, high density	1.26	$/kg	2017 Import average	STATISTA [18]
Electrolyte				
Bromine	4.90	$/kg	2017 Import average	USGS [22]
Zinc	3.20	$/kg	2018 Import average	USGS [21]
Water	0.00246	$/kg	Government value	[17]
Tank				
Polyethylene, high density	1.26	$/kg	2017 Import average	STATISTA [18]
Pipes				
Polyethylene, high density	1.26	$/kg	2017 Import average	STATISTA [18]
Pump	13.46	$/kW	Literature value	Minke et al. [19]
Inverter	112.13	$/kW	Literature value	Minke et al. [16]
Battery management system				
Steel, low alloyed	0.69	$/kg	2019 Market average	Worldsteelprice [11]
Aluminum	2.54	$/kg	2018 Import average	USGS [20]
Copper	6.61	$/kg	2018 Vendor value	USGS [12]
Power control system	150.00	$/kW	Literature value	Minke et al. [19]
Balance of plant accessories				
Steel, low alloyed	0.69	$/kg	2019 Market average	Worldsteelprice [11]
Aluminum	2.54	$/kg	2018 Import average	USGS [20]
Titanium	30.00	$/kg	2019 Market instant	TRICORMETALS [21]

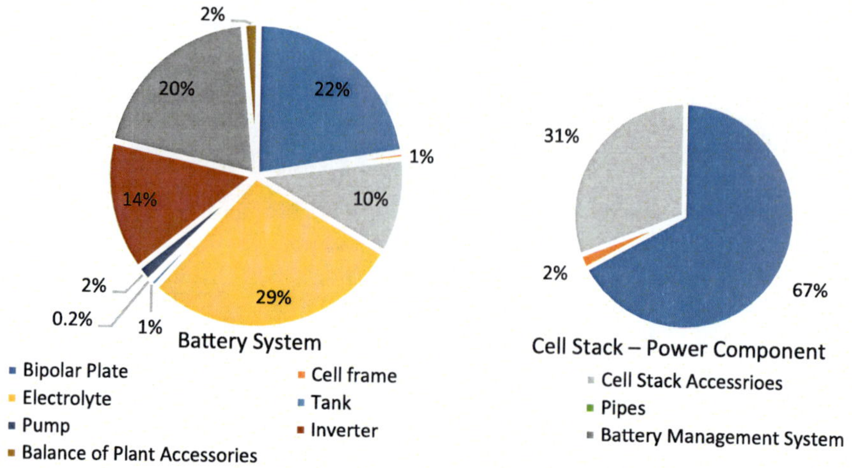

Fig. 3 Materials cost distributed by component in the ZBFB system

variations in material cost due to price changes for carbon fiber felt are provided in Fig. 5d. The expected material cost ($191/kWh) is slightly lower than the baseline estimated value ($196/kWh). The decrease in material cost using the optimistic price is larger than the increase in material cost when using a pessimistic value, which indicates the future price is likely to further lower the cost.

Conclusions

The cost of materials to produce RFBs was analyzed in this study with a focus on materials used in different functional components. Among the three flow batteries selected for assessment, the VRFB shows the highest cost for materials due to the high unit price and quantity of vanadium pentoxide used as battery electrolytes. For ZBFB and IFB, total costs of material are comparable, but the major cost contributors are different. Electrolytes and the bipolar plate are among the largest cost contributors for ZBFB while; in contrast, the battery management system is the highest cost contributor in IFB. The uncertainty associated with the dynamic variations in market prices also affects the total material cost of given battery products, and the price intervals for total material cost are largely dependent on the upper and lower bounds based on observed price ranges.

Table 3 Material price information for materials used in the IFB system

Battery technology	IFB			
Component	Price	Unit	Data type	Data source
Bipolar plate				
Graphite	1.58	$/kg	2015 Market average	USGS [6]
Polypropylene	1.84	$/kg	2017 Import average	STATISTA [8]
Cell frame				
Polyester resin	3.36	$/kg	2017 Import average	STATISTA [23]
Glass fiber	2.00	$/kg	Estimated value	Amirhossein et al. [9]
Electrode				
Carbon felt paper	237.6	$/kg	Literature value	Minke et al. [10]
Membrane				
UHMW polyethylene	595.88	$/kg	2019 Vendor value	Sigma Aldrich [24]
Cell stack accessories				
Steel, low alloyed	0.69	$/kg	2019 Market average	Worldsteelprice [11]
Aluminum	2.54	$/kg	2018 Import average	USGS [20]
EPDM Gasket	2.50	$/kg	Literature value	Viswanathan et al. [25]
Electrolyte				
Iron chloride	0.35	$/kg	2018 Market instant	ICIS [26]
Potassium chloride	0.27	$/kg	2019 Market average	Indexmundi [27]
Manganese dioxide	2.21	$/kg	2015 Import average	USGS [28]
Hydrochloric acid	0.13	$/kg	2018 Market instant	ICIS [15]
Water	0.00186	$/kg	Government value	[17]
Tank				
Polyester resin	3.36	$/kg	2017 Import average	STATISTA [23]
Pipes				
Polyvinylchloride	0.97	$/kg	2019 Market average	Investing.com [13]
Pump	13.46	$/kW	Literature value	Minke et al. [19]
Inverter	112.13	$/kW	Literature value	Minke et al. [16]
Battery management system				
Carbon felt paper	237.6	$/kg	Literature value	Minke et al. [12]
Power control unit	150.00	$/kW	Literature value	Minke et al. [19]
Balance of plant accessories	None			

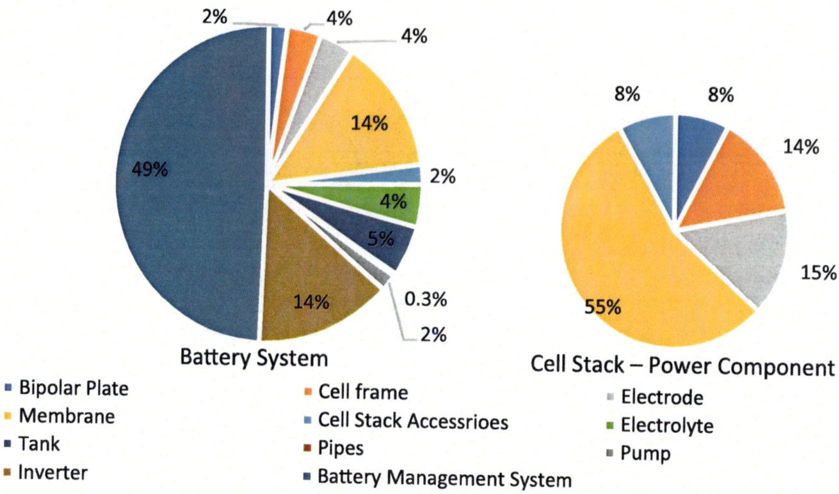

Fig. 4 Materials cost distributed by component in the IFB system

Table 4 Price estimation values ($/kg) for selected materials used in flow batteries

Value	Vanadium pentoxide [14]	Titanium [21]	Bromine [22]	Carbon fiber felt [10]
Range of years for available price data	1991–2019	1971–2019	1991–2017	2004–2017
Current—price data for most recent year	35.75	30.00	4.90	237.60
Pessimistic—historical maximum price observed	50.00	62.65	6.00	280.00
Optimistic—historical minimum price observed	8.00	17.65	1.00	80.00
Weighted average (E)	32.33	33.38	4.43	218.40
Standard deviation (SD)	7.00	7.50	0.83	33.33

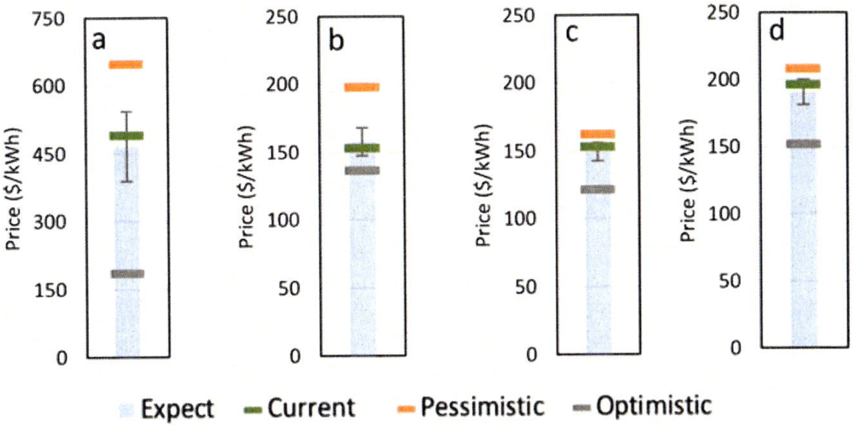

Fig. 5. Sensitivity of flow battery material cost due to variations in the material price: **a** vanadium pentoxide for VRFB, **b** titanium, **c** bromine for ZBFB, and **d** carbon fiber felt for IFB

Acknowledgements The authors would like to acknowledge funding provided by the California Energy Commission under Agreement # EPC-16-039. The authors would also like to acknowledge funding from the Lincoln Dynamic Foundation World Institute for Sustainable Development of Materials (WISDOM) at the University of California, Irvine.

References

1. He H, Tian S, Tarroja B, Ogunseitan OA, Samuelsen S, Schoenung JM (2020) Flow battery production: materials selection and environmental impact. J Clean Prod 269:121740
2. Kang H-Y, Schoenung JM (2006) Economic analysis of electronic waste recycling: modeling the cost and revenue of a materials recovery facility in California. Environ Sci Technol 40(5):1672–1680
3. Ye J, Schoenung JM (2004) Technical cost modeling for the mechanical milling at cryogenic temperature (cryomilling). Adv Eng Mater 6(8):656–664
4. Isaacs JA, Tanwani A, Healy ML, Dahlben LJ (2010) Economic assessment of single-walled carbon nanotube processes. J Nanoparticle Res 12(2):551–562
5. ProjectManagement.com (2020) 3-Points estimating. https://www.projectmanagement.com/wikis/368763/3-Points-Estimating. Accessed 06 Jan 2020
6. Olson DW (2016) USGS 2016 minerals yearbook-Graphite
7. LLDPE (Linear Low-Density Polyethylene): production, price and its properties. https://www.plasticsinsight.com/resin-intelligence/resin-prices/lldpe/. Accessed 06 Jan 2020
8. Polypropylene Production Capacity, Market and Price. https://www.plasticsinsight.com/resin-intelligence/resin-prices/polypropylene/. Accessed 06 Jan 2020
9. Glass fiber-Wikipedia. https://en.wikipedia.org/wiki/Glass_fiber. Accessed 06 Jan 2020
10. Minke C, Turek T (2018) Materials, system designs and modelling approaches in techno-economic assessment of all-vanadium redox flow batteries-a review. J Power Source 376:66–81. https://doi.org/10.1016/j.jpowsour.2017.11.058
11. Latest Free Steel Prices. https://worldsteelprices.com/. Accessed 06 Jan 2020
12. USGS 2019 Mineral Commodity Summaries–Copper. https://doi.org/10.3133/fs20143004. Accessed 14 Jan 2020

13. Polyvinyl Chloride Futures Historical Prices-Investing.com. https://www.investing.com/commodities/pvc-com-futures-historical-data. Accessed 06 Jan 2020
14. Vanadium Statistics and Information. https://www.usgs.gov/centers/nmic/vanadium-statistics-and-information. Accessed 06 Jan 2020
15. Price and market trends: US hydrochloric acid supply poised to tighten | ICIS. https://www.icis.com/explore/resources/news/2018/02/01/10189415/price-and-market-trends-us-hydrochloric-acid-supply-poised-to-tighten/. Accessed 06 Jan 2020
16. Minke C, Kunz U, Turek T (2017) Techno-economic assessment of novel vanadium redox flow batteries with large-area cells. J Power Sources 361(1):105–114
17. Commercial Water Rates-Utilities|seattle.gov. https://www.seattle.gov/utilities/businesses-and-key-accounts/water/rates/commercial-water-rates. Accessed 06 Jan 2020
18. HDPE Production Capacity, Price and Market. https://www.plasticsinsight.com/resin-intelligence/resin-prices/hdpe/. Accessed 06 Jan 2020
19. Minke C, Dorantes Ledesma MA (2019) Impact of cell design and maintenance strategy on life cycle costs of vanadium redox flow batteries. J Energy Storage 21:571–580
20. USGS 2018 Mineral Commodity Summaries–Aluminium (2013). https://www.usitc.gov/publications/332/pub4703.pdf. Accessed 14 Jan 2020
21. Titanium Cost Comparison-Tricor Metals. https://www.tricormetals.com/cost-comparison.html. Accessed 06 Jan 2020
22. USGS 2018 Mineral Commodity Summaries-Bromine. https://s3-us-west-2.amazonaws.com/prd-wret/assets/palladium/production/mineral-pubs/bromine/mcs-2018-bromi.pdf. Accessed 06 Jan 2020
23. Polyester Production, Price and Market Forecast. https://www.plasticsinsight.com/resin-intelligence/resin-prices/polyester/. Accessed 06 Jan 2020
24. Polyethylene-U.H.M.W. (UHMW PE), sheet, thickness 2 mm, size 500 × 500 mm | Sigma-Aldrich. https://www.sigmaaldrich.com/catalog/product/aldrich/gf36466623?lang=en&region=US. Accessed 06 Jan 2020
25. Parr G (2005) Annual Forum 2005 trade and uneven development: opportunities and challenges import parity pricing: a competitive constraint or a source of market power? In: Tips forum 2005 import parity pricing: a competitive constraint or a source of market power?
26. US Ferric Chloride Producers Raise Prices in Healthy Market|ICIS. https://www.icis.com/explore/resources/news/2001/08/27/145944/us-ferric-chloride-producers-raise-prices-in-healthy-market/. Accessed 06 Jan 2020
27. Potassium Chloride-Monthly Price-Commodity Prices-Price Charts, Data, and News-Index Mundi. https://www.indexmundi.com/commodities/?commodity=potassium-chloride&months=12. Accessed 06 Jan 2020
28. USGS 2015 Minerals Yearbook–Manganese. https://prd-wret.s3-us-west-2.amazonaws.com/assets/palladium/production/atoms/files/myb1-2015-manga.pdf. Accessed 06 Jan 2020

Part V
REWAS 2022: Automation and Digitalization for Advanced Manufacturing

Digitalizing the Circular Economy (CE): From Reactor Simulation to System Models of the CE

Markus A. Reuter and Neill Bartie

Extended Abstract

Embracing the circular economy, this paper will discuss recent work that scales reactor technology to system models with relevant digital twins. The aspects that are discussed are best explained by the figure below.

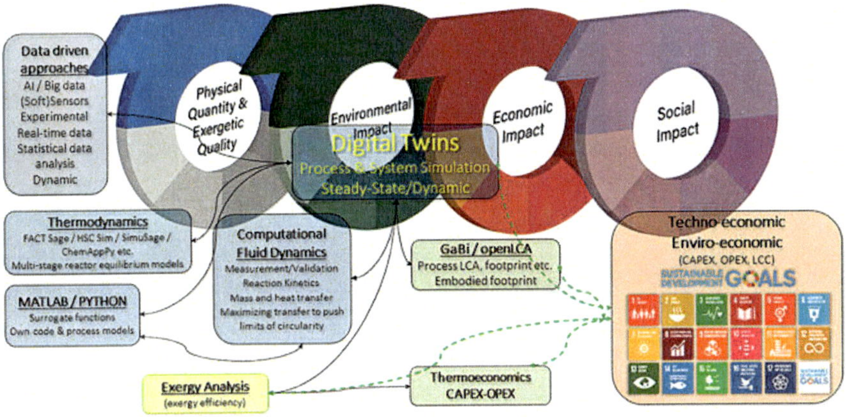

By the combination of different tools and methods (from AI, CFD, mass- and heat transfer, kinetics, industrial experience, etc.) and integrating these into suitable digital platforms, this paper will analyze different circular economy systems also in

M. A. Reuter (✉) · N. Bartie
SMS Group, Eduard-Schloemann-Str. 4, 40237 Düsseldorf, Germany
e-mail: markusandreasreuter@icloud.com

Technical University of Munich, Am Essigberg 3/II, 94315 Straubing, Germany

A. Lazou et al. (eds.), *REWAS 2022: Developing Tomorrow's Technical Cycles (Volume I)*, The Minerals, Metals & Materials Series,
https://doi.org/10.1007/978-3-030-92563-5_47

terms of complete supply chains. With a focus on for example thermoeconomics, exergy dissipation of the systems will be quantified by suitable digital twins for PV cell manufacture, battery technology, etc. Integration with impact assessment approaches will show how to minimize the impact of complete supply chains and show which systems produce the lowest footprint products. In addition, the link of the digital twins to the sustainability development goals (SDGs) of the United Nations will be elaborated on.

In the quest of digitalizing our metallurgical and CE systems better, this paper will further discuss among others [1–3]:

- improving the slag solutions models for more complex slag,
- measurement of kinetic data to understand the effect of slag composition and processing flow conditions on the exchange of materials between different phases in order to push systems to the thermodynamic limits,
- integration of fundamental and industrial know-how into multi-compartment dynamic equilibrium-based simulation models of our reactors to better direct materials into the correct phases and thence metallurgical infrastructures,
- integration of the above into large system models so that one can understand the footprint of metals supply chains better and supply the footprint data to populate environmental databases with specific and well-defined data,
- understand the interaction of complex molten slag solutions with refractories and therefore optimize reactor integrity,
- understand the effect of injection technology interacting with furnace slags,
- provide the detail to help populate environmental and economic physics and technology-based data (where required), to realize the noble goals of the SDGs as well as the circular economy, and
- streamline the real-time data (and sensors) and data types and their qualities that can interact directly with the digital twins via big data approaches.

The above points for example will help to ensure the CE system can be driven to its thermodynamic/exergetic, technological, and therefore economic limit, at the same time revealing our industry's key enabling role towards a greener circular economy society.

References

1. Reuter MA, van Schaik A, Gutzmer J, Bartie N, Abadías Llamas A (2019) Challenges of the circular economy—a material, metallurgical and product design perspective. Ann Rev Mater Res 49:253–274
2. Bartie N, Abadías Llamas A, Heibeck M, Fröhling M, Volkova O, Reuter MA (2020) The simulation-based analysis of the resource efficiency of the circular economy–the enabling role of metallurgical infrastructure. Miner Process Extr Metall (TIMM C) 129(2):229–249
3. Bartie NJ, Cobos-Becerra YL, Fröhling M, Schlatmann R, Reuter MA (2021) The resources, exergetic and environmental footprint of the silicon photovoltaic circular economy: assessment and opportunities. Resour Cons Recycl 169:105516

Factors to Consider When Designing Aluminium Alloys for Increased Scrap Usage

Luca Montanelli, Eric R. Homer, and Elsa Olivetti

Abstract For a significant shift in alloy design to happen, the aluminium industry needs to explore a broader range of compositional and processing dimensions. This work provides the background to guide the design of new alloys, especially where opportunities are present to improve recyclability. To achieve this, a blending model with an integrated material flow analysis will be optimised over compositional space to inform alloy compositions that enable higher quantities of scrap use. Blending models inform the scrap usage of a candidate alloy when it is set in a predetermined market landscape. Due to their computational cost, machine learning optimisation methods such as Bayesian optimisation will be employed to focus the design process. The optimisation will be subject to constraints that are based on compositions, phase combinations, and relevant properties to ensure that the alloys not only maximise the amount of scrap use but also meet technical requirements. This manuscript describes the background content that will inform this work.

Keywords Aluminium · Alloy design · Bayesian optimisation · Recycling

Introduction

Design of aluminium alloys has historically focused on performance improvements without direct incorporation of sustainability considerations in the process. However, based on decarbonisation and recycling goals set by the aluminium industry, a significant shift must happen in the design process [1]. In addition, as scrap streams become

L. Montanelli (✉) · E. Olivetti
Department of Materials Science and Engineering, Massachusetts Institute of Technology, Cambridge, MA, USA
e-mail: montanel@mit.edu

E. Olivetti
e-mail: elsao@mit.edu

E. R. Homer
Department of Mechanical Engineering, Brigham Young University, Provo, UT, USA
e-mail: eric.homer@byu.edu

A. Lazou et al. (eds.), *REWAS 2022: Developing Tomorrow's Technical Cycles (Volume I)*, The Minerals, Metals & Materials Series,
https://doi.org/10.1007/978-3-030-92563-5_48

increasingly sought after, we must offer opportunities for a broader range of scrap compositions to be used by remelters. Recycled streams are accompanied by impurities including other metals or glass from mixed collection and incomplete component separation or organic materials and minerals from coatings used on specific products. When there is repeated recycling of the material (a goal in more closed systems), there may be accumulation of these impurities over time if not properly controlled or accounted for in the scrap beneficiation processes. Impurities are adjusted through dilution with primary or higher-quality materials, but this presents a challenge in the desire to use increased scrap quantities. So, there is an important trade-off at play in the desire to decarbonize materials production: as we approach saturation of materials, stock emissions reductions are further enabled by the use of old scrap, but as we have the opportunity to increase circularity (more material coming out of use than is demanded), we must track the potential for impurity accumulation more closely. Together with ever more powerful computational tools, this need calls for the exploration of a broader range of compositional and processing dimensions, especially where opportunities are present to improve recyclability.

We hypothesize that a blending optimization framework over a broad compositional space can inform alloy compositions that enable higher quantities of scrap use. Blending models are optimisation algorithms that are used in alloy production to calculate the lowest cost to produce alloys considering primary and secondary material flows. This approach incorporates a set of material flows (with their compositions, quantities, and prices) as well as a set of demanded alloys (with their compositional specifications and quantities). The output is the mapping from the former to the latter that leads to the lowest cost. Typically, scrap materials are less expensive than their primary counterparts, so by minimising the cost to produce an alloy, in most cases, scrap use is maximised. Blending models can also offer insight into how agents within a system may respond to different scrap availability or demand scenarios by varying the material flows and recycled alloy demands. This is especially useful to consider the design for different market regions or different years [2]. As a result, blending models allow for a unique insight into how new alloy compositions can be produced when set into a market landscape which, when using real-world data, simulates real-world performance.

However, due to blending models' high number of inputs and outputs, computational cost, and lack of continuity or derivability, it would be infeasible to perform regression and apply the common property prediction paradigm. Instead, machine learning optimisation methods such as Bayesian optimisation (BO) will have to be employed. BO is a black-box optimisation method that does not assume any functional form and uses a sequential design strategy where samples are selected and evaluated at each iteration until a stopping criterion is met [3]. It is an efficient algorithm that aims at minimising the number of calls to the expensive function it optimises. Furthermore, it has the property of being very flexible to the types of functions that are optimised as the goal is only to find the maximum (or minimum) and not perform regression. For applications in alloy development, the optimisation will be subject to constraints that are based on compositions, phase combinations, and relevant properties to ensure that the alloys not only maximise the amount of scrap use but also meet technical requirements.

We hypothesize that by combining efficient blending optimisation with recyclability and property considerations, one can design alloys that use more scrap while producing desirable alloys. This proceedings contribution provides relevant background content to support this hypothesis.

Critical Features for Recycling-Friendly Alloy Design

The set of work that forms the basis for the modelling framework proposed includes approaches that have optimised computational design, assessment of scrap tolerance, broad studies on accumulation of elements in secondary materials streams, as well as experimental work to assess links between properties, processing, and composition.

Material Flow Analysis

Given the broad interest in circular economy and stated goals by companies to achieve high levels of recycled content in their products, recycling may be ultimately limited by accumulation of undesirable materials constituents or materials characteristics inherently caused by reprocessing. With respect to aluminium, elements that have been referenced as problematic include Si, Mg, Ni, Zn, Pb, Cr, Fe, Cu, V, and Mn. Plant operators must specifically identify approaches in production to avoid accumulation of these elements. These elements may form intermetallic compounds that can negatively impact processing downstream, particularly in the case of wrought alloys. Even minor additions of iron, for example, increase constituent particles because of the reduced solid solubility of iron [4]. Current approaches for alloy design or modelling scrap utilisation do not accurately reflect the dynamics of accumulation because these models do not explicitly consider the actions of recyclers in the system or link alloy design to overall expected material flows. There are two benefits in inherently linking material flow analysis (MFA) to alloy design: first, models that ignore the actions of agents through blending may misestimate expected accumulation and therefore undervalue upgrading technologies that might mitigate accumulation; second, through explicit accounting of the expected future trends, alloys can be designed with a view into the composition of the available scrap. Previous work has raised concerns that recycling of a resource degrades that resource and modelled the implication of these accumulation mechanisms through MFA. This has been done either at aggregate product levels or by tracking individual elements, but all through dynamic MFA strategies [5, 6]. Buchner et al. used dynamic MFA to quantify aluminium-based circularity within a specific geography aggregating at the level of wrought versus cast Al scrap [7]. They point to the need for separation between scrap streams for these two products to avoid excess of mixed scrap streams, particularly when export of lower-quality scrap is constrained. Hatayama and co-authors found that effective scrap sorting on end-of-life vehicles, in particular, addresses the challenges of decreased demand for cast alloys [8]. Other work has calculated the accumulation of elements over time through composition thresholds and tracking alloys and products through their lifetimes for aluminium [9], but also other metals such as steel [10, 11]. Designing for recycling-friendly alloys in the context of elemental accumulation has been limited in the published literature, but the modelling approach explored in by Gaustad et al. provides insight into which elements lend themselves most to incorporation of scrap [12]. However, there is a gap in this literature that links these concepts. Methodologically, we choose to do that linking through the use of efficient blending optimisation linked to overall material flows.

Experimental Alloy Design

Meeting sustainability goals and managing the accumulation of elements in the recycling stream are two issues that have been present in the research on the experimental design of aluminium alloys. For example, a key strategy in achieving energy goals is to design aluminium alloys for better operability at higher temperatures enabling higher efficiencies, in the case of electrical cables and heat exchangers, or the lightweighting of cars and planes by replacing currently used steel alloys [13]. However, current commercial aluminium alloys are limited under load-bearing conditions to temperatures of around 220 °C. Beyond this, the creep resistance of those alloys begins to suffer due to phase transformation and precipitate coarsening, amongst other effects. Extensive research has been done in this field, and better creep resistance can now be achieved through the promotion of $L1_2$ or α-Al(Mn,Mo)Si precipitates with microadditions of Sc, Zr, Er, or Mo, respectively. Using such methodology, operable temperatures have reached the 400–450 °C range [14]. Alternatively, the effects of impurities have been investigated, most notably in Al-Si cast alloys that are commonly used in the automotive industry and are particularly sensitive to the accumulation of iron in the scrap supply [15, 16]. This is because of the presence of the β-Al9Fe2Si2 intermetallic phase affecting negatively the castability and tensile properties of those alloys. Amongst the few strategies to deal with the β phase is the addition of Ni, Cu, or Mn to destabilise the detrimental intermetallic phase by changing its composition or promoting other phases [17–20]. However, experimental research is inherently limited in scope and time. The classic trial-and-error approach is too time consuming and is limited to problems that are already present, with little consideration of or means to predict future problems such as accumulation.

Computational Alloy Design

The optimal computational design of an alloy, for example to optimise a certain property, is usually done via regression where models are created to relate known information about the alloys with the targeted properties. Those models are then used to perform property prediction to find better alloys. This methodology, however, only works in cases where sufficient data is available for models to be good estimators of properties. In our case, the input space of a blending model coupled with an MFA is very large, as it contains the compositions, quantities, and prices of scrap and demand material in addition to the common MFA parameters. Therefore, where large amounts of data are not available, or, equivalently, the feature space is too large, formulating the design process as an optimisation problem becomes more efficient than property prediction. Such methodology has been applied in the field of high entropy alloys where the common trial-and-error design methodology that has dominated metallurgy for centuries is inefficient because of the large size of the design space, which includes many elements with significant variability [21–23]. It has also been applied to various types of alloys to, for example, increase hardness, yield strength, or multiple mechanical properties at once [24–26]. However, little research has considered sustainability or recyclability in the design process. Barnett et al. introduced the concept of scrap tolerance via a compositional flexibility metric in their discussion of high entropy alloys, but the optimisation process was still exclusively focused on the maximisation of precipitation hardening [23].

Integrated Design

To meet the decarbonisation and recycling goals set by the industry, it is imperative to incorporate recyclability considerations in the design of new aluminium alloys, with explicit inclusion of dynamics associated with elements that may accumulate in scrap reprocessing. The common experimental design approach that has been employed extensively until now has managed to give answers for certain specific alloys and cases but lacks the outlook and integration needed to properly manage future flows problems such as accumulation. Instead, the shift in alloy design towards computational optimisation can provide a unique opportunity to introduce sustainable design of alloys either by considering relevant metrics as constraints or, even better, objectives. We must employ techniques looking at a broader range of compositions such as blending models and MFA to properly assess the extent to which a new alloy can be produced with more scrap and mitigate recycling problems of the future. Drawing on the methodologies of previous research in alloy design by optimisation, we propose to optimise an integrated blending model with MFA to consider the next generation of design challenges that the aluminium industry will face to reach a fully circular economy.

Framework for Alloy Design Through Optimization

The modelling framework presented here contains three areas not previously explored across one line of inquiry. Those are as follows: (1) the high cost of the objective function, (2) the extensive use of constraints, and (3) the high dimensionality of the design space. Genetic algorithms are usually the algorithm of choice in material science problems, but they often require a lot of iterations to converge and do not work well with constraints, especially if the design space is non-convex [21, 22, 24, 25]. In contrast, this work will use BO, a more efficient algorithm that can implement constraints easily even when considering mixed continuous and discrete variables such as compositions and processing parameters [27]. Finally, the design space needed to evaluate blending models on is of high dimensionality. This is because, to properly quantify the uncertainties in scrap supply and alloy production, the compositional ranges (lower and upper compositions) of each element present in an alloy must be considered. The design space therefore lies in 20 dimensions when considering only the nine most common alloying elements as well as aluminium, and this is without adding the dimensions for processing parameters. However, for all of its advantages, BO is usually said to reach its limit at around 20 dimensions. At higher dimensions, BO proves to be an intrinsically hard problem that requires the use of different methodologies such as special acquisition function or lower-dimensional embeddings to function properly [28–30].

We summarise these gaps as an optimisation problem with the following form:

$$\max_{x \in \mathbb{R}^n} \quad B(x) \tag{1}$$

$$s.t. \quad l_i \leq c_i(x) \leq h_i \quad \forall i \tag{2}$$

where the vector x is the n dimensional descriptor of an alloy, $B(x) : \mathbb{R}^n \rightarrow \mathbb{R}$ is a function of a blending model and MFA that outputs the scrap usage of x based on a predetermined scenario, and $c_i(x) : \mathbb{R}^n \rightarrow \mathbb{R}$ are the constraints imposed on the optimisation, bounded by the values l_i and h_i. The modelling framework will be implemented in the following stages:

1. Integration of the blending model and the MFA together into a function $B(x)$ along with the necessary demand scenarios and representation of the current material flow landscape (quantities, prices, compositions of primary and secondary material).
2. Development of the optimisation algorithm, which includes the selection of the best alloy descriptor, best acquisition function for BO, and any methodologies used to manage the high dimensionality, cost, and discontinuity of $B(x)$.
3. Selection of the constraints $c_i(x)$ to define the design space. As stated previously, those will fall into one of three categories: compositions, phase combinations, and properties. Compositional constraints ensure that the vector descriptor is a valid alloy, but they can also be used to limit the quantities of elements or

limit the design space to certain alloy series. Phase constraints will implement CALPHAD calculations to ensure that x is a solid solution alloy with a controlled amount of unwanted phases. Finally, property constraints direct the optimisation towards alloys with desired specifications.

The scope of the project is ultimately to provide with a complete design methodology to enable aluminium alloys with higher scrap usage. By specifying a range for any property constraints for which a model is implemented, the output of the optimisation can be tailored to fit a wide range of specifications. Furthermore, to reflect the real-world scrap usage of an alloy, the integrated blending model and MFA would need to represent the current material flow landscape as best as possible. This includes the possibility of modelling different demand scenarios such as the rise in electrical vehicle consumption or increased incentives to recycling. This requires data on quantities, prices, and compositions of post-consumer and post-industrial scrap for different sectors, regions, and years. This data has already been collected as part of other previous projects and can therefore be used in this analysis too. The more granular the data, the more possibilities there is to design alloys for specific regions or demand scenarios. Additionally, the availability of data for multiple years enables the model to assess the scrap usage of an alloy throughout its lifetime therefore incorporating the effect of future material flows, such as accumulation, in the analysis.

References

1. The Aluminum Association (2011) Aluminum: the element of sustainability
2. Gaustad G, Olivetti E, Kirchain R (2011) Toward sustainable material usage: evaluating the importance of market motivated agency in modeling material flows. Environ Sci Technol 45:4110–4117. https://doi.org/10.1021/es103508u
3. Shahriari B, Swersky K, Wang Z et al (2016) Taking the human out of the loop: a review of Bayesian optimization. Proc IEEE 104:148–175. https://doi.org/10.1109/JPROC.2015.2494218
4. Wagstaff SR, Wagstaff RB, Allanore A (2017) Tramp element accumulation and its effects on secondary phase particles. In: Ratvik AP (ed) Light metals 2017. Springer International Publishing, Cham, pp 1097–1103
5. Chen W-Q, Graedel TE (2012) Dynamic analysis of aluminum stocks and flows in the United States: 1900–2009. Ecol Econ 81:92–102. https://doi.org/10.1016/j.ecolecon.2012.06.008
6. Modaresi R, Müller DB (2012) The role of automobiles for the future of aluminum recycling. Environ Sci Technol 46:8587–8594. https://doi.org/10.1021/es300648w
7. Buchner H, Laner D, Rechberger H, Fellner J (2017) Potential recycling constraints due to future supply and demand of wrought and cast Al scrap—a closed system perspective on Austria. Resour Conserv Recycl 122:135–142. https://doi.org/10.1016/j.resconrec.2017.01.014
8. Hatayama H, Daigo I, Matsuno Y, Adachi Y (2012) Evolution of aluminum recycling initiated by the introduction of next-generation vehicles and scrap sorting technology. Resour Conserv Recycl 66:8–14. https://doi.org/10.1016/j.resconrec.2012.06.006

9. Løvik AN, Müller DB (2016) A material flow model for impurity accumulation in beverage can recycling systems. In: Grandfield J (ed) Light metals 2014. Springer International Publishing, Cham, pp 907–911
10. Daehn KE, Cabrera Serrenho A, Allwood JM (2017) How will copper contamination constrain future global steel recycling? Environ Sci Technol 51:6599–6606. https://doi.org/10.1021/acs.est.7b00997
11. Nakamura S, Kondo Y, Nakajima K et al (2017) Quantifying recycling and losses of Cr and Ni in steel throughout multiple life cycles using MaTrace-alloy. Environ Sci Technol 51:9469–9476. https://doi.org/10.1021/acs.est.7b01683
12. Gaustad G, Olivetti E, Kirchain R (2010) Design for recycling. J Ind Ecol 14:286–308. https://doi.org/10.1111/j.1530-9290.2010.00229.x
13. Raabe D, Tasan CC, Olivetti EA (2019) Strategies for improving the sustainability of structural metals. Nature 575:64–74. https://doi.org/10.1038/s41586-019-1702-5
14. Cann JL, De Luca A, Dunand DC et al (2021) Sustainability through alloy design: challenges and opportunities. Prog Mater Sci 117:100722. https://doi.org/10.1016/j.pmatsci.2020.100722
15. Fabrizi A, Ferraro S, Timelli G (2013) The influence of Sr, Mg and Cu addition on the microstructural properties of a secondary AlSi9Cu3(Fe) die casting alloy. Mater Charact 85:13–25. https://doi.org/10.1016/j.matchar.2013.08.012
16. Yang H, Ji S, Fan Z (2015) Effect of heat treatment and Fe content on the microstructure and mechanical properties of die-cast Al–Si–Cu alloys. Mater Des 85:823–832. https://doi.org/10.1016/j.matdes.2015.07.074
17. Basak CB, Babu NH (2017) Influence of Cu on modifying the beta phase and enhancing the mechanical properties of recycled Al-Si-Fe cast alloys. Sci Rep 7:5779. https://doi.org/10.1038/s41598-017-05937-2
18. Basak CB, Hari Babu N (2017) Improved recyclability of cast Al-alloys by engineering β-Al9Fe2Si2 Phase. In: Ratvik AP (ed) Light metals 2017. Springer International Publishing, Cham, pp 1139–1147
19. Basak CB, Meduri A, Hari Babu N (2019) Influence of Ni in high Fe containing recyclable Al-Si cast alloys. Mater Des 182:108017. https://doi.org/10.1016/j.matdes.2019.108017
20. Shabestari SG (2004) The effect of iron and manganese on the formation of intermetallic compounds in aluminum–silicon alloys. Mater Sci Eng A 383:289–298. https://doi.org/10.1016/j.msea.2004.06.022
21. Menou E, Toda-Caraballo I, Rivera-Díaz-del-Castillo PEJ et al (2018) Evolutionary design of strong and stable high entropy alloys using multi-objective optimisation based on physical models, statistics and thermodynamics. Mater Des 143:185–195. https://doi.org/10.1016/j.matdes.2018.01.045
22. Rickman JM, Chan HM, Harmer MP et al (2019) Materials informatics for the screening of multi-principal elements and high-entropy alloys. Nat Commun 10:2618. https://doi.org/10.1038/s41467-019-10533-1
23. Barnett MR, Senadeera M, Fabijanic D et al (2020) A scrap-tolerant alloying concept based on high entropy alloys. Acta Mater 200:735–744. https://doi.org/10.1016/j.actamat.2020.09.027
24. Menou E, Ramstein G, Bertrand E, Tancret F (2016) Multi-objective constrained design of nickel-base superalloys using data mining- and thermodynamics-driven genetic algorithms. Modelling Simul Mater Sci Eng 24:055001. https://doi.org/10.1088/0965-0393/24/5/055001
25. Li S, Kattner UR, Campbell CE (2017) A computational framework for material design. Integr Mater Manuf Innov 6:229–248. https://doi.org/10.1007/s40192-017-0101-8
26. Li J, Zhang Y, Cao X et al (2020) Accelerated discovery of high-strength aluminum alloys by machine learning. Commun Mater 1:1–10. https://doi.org/10.1038/s43246-020-00074-2
27. Zhang Y, Apley DW, Chen W (2020) Bayesian optimization for materials design with mixed quantitative and qualitative variables. Sci Rep 10:4924. https://doi.org/10.1038/s41598-020-60652-9
28. Li C, Gupta S, Rana S et al (2018) High dimensional bayesian optimization using dropout. arXiv:180205400 [stat]

29. Ling J, Hutchinson M, Antono E et al (2017) High-dimensional materials and process optimization using data-driven experimental design with well-calibrated uncertainty estimates. Integr Mater Manuf Innov 6:207–217. https://doi.org/10.1007/s40192-017-0098-z
30. Moriconi R, Deisenroth MP, Kumar KSS (2020) High-dimensional Bayesian optimization using low-dimensional feature spaces. arXiv:190210675 [cs, stat]

An Automated Recycling Process of End-of-Life Lithium-Ion Batteries Enhanced by Online Sensing and Machine Learning Techniques

Liurui Li, Maede Maftouni, Zhenyu James Kong, and Zheng Li

Abstract This paper attempts to address key challenges to automate unit operations (e.g., disassembling and sorting battery components at the cell level) of the lithium-ion battery direct recycling process. In our previous publications, we introduced the design and prototype of an automated disassembly system that can separate cell cases, metal tab, cathode, anode, and separators of a LIB pouch cell with minimum human intervention. In this paper, we take one step further to integrate industrial vision cameras and sensors into the prototyped system which allows real-time process defect detection and corrective action. Specifically, this paper focuses on online sensor integration in the electrode sorting step for separating cathode and anode electrode sheets. The electrode sorting and separation method developed here simplifies the subsequent materials extraction and purification operations of the LIB direct recycling.

Keywords Lithium-ion battery · Smart manufacturing · Battery recycling · Machine learning · Online sensing

Introduction

The estimated annual demand for lithium-ion batteries (LIBs) in 2025 will reach 408 GWh due to the market expansion of electric vehicles (EVs) [1]. The lifespan of EV battery packs ranges from 4.5 to 14.5 years depending on their operating conditions [2, 3]. Therefore, a tremendous amount of end-of-life (EOL) LIBs will be generated in the foreseeable future which creates severe concerns on environmental sustainability and the security of the critical materials supply chain [4]. Meanwhile, the EOL LIBs are a potential resource of valuable metals (e.g., Ni, Mn, Li, or Co) [5]. Therefore,

L. Li · Z. Li (✉)
Department of Mechanical Engineering, Virginia Tech, Blacksburg, VA 24061, USA
e-mail: zhengli@vt.edu

M. Maftouni · Z. J. Kong
Department of Industrial and Systems Engineering, Virginia Tech, Blacksburg, VA 24061, USA

A. Lazou et al. (eds.), *REWAS 2022: Developing Tomorrow's Technical Cycles (Volume I)*, The Minerals, Metals & Materials Series,
https://doi.org/10.1007/978-3-030-92563-5_49

it is important to develop an effective recycling process that is ecologically friendly and economically feasible for EOL LIBs.

The incumbent EOL LIBs recycling methods combine mechanical pretreatments and metallurgical processes [6]. Mechanical pretreatments comprised steps of discharging battery packs, dismantling packs into cells, and separating materials of single cells. Metallurgical processes that consist of pyro-, hydro-, and bio-metallurgy are generally downstream procedures of mechanical pretreatments [4]. Despite that the metallurgical processes are widely practiced in industry, hazardous gas emissions, acid waste, and high-energy consumption issues have always been barriers toward truly sustainable closed-loop recycling [7]. In recent years, a direct recycling approach that regenerates cathode materials proved to be feasible [8–10]. While the coated electrode materials recovery rate for metallurgical processes is only around 75% [11], the coated material recovery rate through the direct recycling process can reach as high as 97.4% if electrode sheets can be separated and extracted with their integrity well preserved [12]. Although direct recycling is more eco-friendly and energy conserving with high materials recovery rate, the development of direct recycling does face some challenges including complex unit operations, high recycling production cost, and requirement for battery presorting when scaling up the operation. Specifically, the existence of anode powders and mixed current collector metals in extracted EOL cathode powder during the direct recycling process increases the complexity of downstream processing and negatively impacts the electrochemical performance of the final recycling product [13].

To simplify the material extraction and purification in direct recycling, an automatic disassembly system has been designed and prototyped for dismantling and separating cathode sheets, anode sheets, separators, and Al laminated film housing from lithium-ion pouch cells in our previous report [14]. Compared to the destructive pretreatment widely adopted in the metallurgical process, this proposed system has a great potential to achieve a higher coating material recovery rate as well as yield purer cathode powder. In this paper, we take one step further to integrate industrial vision cameras and sensors into the prototyped system which allows real-time process defect detection and corrective action. Specifically, this paper focuses on online sensor integration in the electrode sorting step for separating cathode and anode electrode sheets. The electrode sorting and separation method developed here simplifies the subsequent materials extraction and purification operations of the LIB direct recycling.

Disassembly Sequence and System Overview

Figure 1 shows the continuous process for direct regeneration of cathode materials recycled from E LIBs in our laboratory. As part of this direct regeneration strategy, our proposed single battery disassembly system has great potential to ensure the automatic separation of Al laminated films, separators, cathode sheets, and anode sheets with their well-preserved integrity. Before being fed into the disassembly

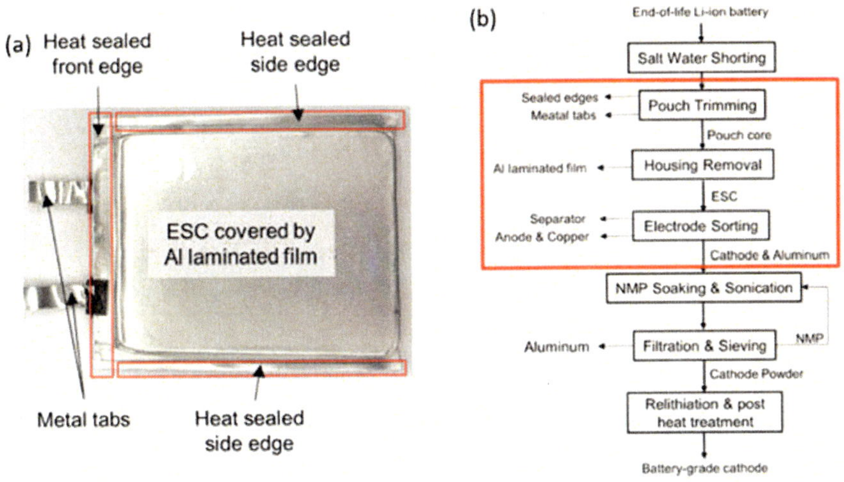

Fig. 1 **a** Configuration of a pouch LIB and **b** continuous process for recovery of cathode coating from EOL LIBs [14]

system, EOL LIBs need to be fully discharged in saltwater to avoid any explosions or fire hazards. Then, three sealed edges along with metal tabs are cut off from the core area sequentially. The remaining folded Al laminated housing film needs to be stretched from both sides by external forces to extract the electrode separator compound (ESC). Separators need to be unfolded and continuously fed forward. For EOL LIBs, electrode sheets tend to attach to the separator due to the surface tension of the electrolyte or the aging of the electrodes. Thus, specialized skiving tools are needed to scrape cathode and anode sheets off from opposite sides of the separator.

The system is physically designed and built into three modules as shown in Fig. 2 pouch trimming module, housing removal module, and electrode sorting module. These modules can achieve the automated disassembly of LIB and finish three key corresponding steps encircled in Fig. 1. All mechanisms or toolsets involved in these three modules can be divided into three categories: end effectors, fixtures, and transporters. Each module contains several customized key apparatuses to achieve the designed connection removing plan. These apparatuses are usually called end effectors. Special fixtures are also needed to hold the position of targeted assemblies or subassemblies while the end effectors operate. Between each module, the remaining subassemblies such as Al film housing covered ESC and the ESC itself need to be transported between fixtures located in adjacent modules by transporters. The cooperating of end effectors, fixtures, and transporters assures the success of connection removing as well as material separation in our automated disassembly line. Detailed design of these key apparatuses is introduced in [14].

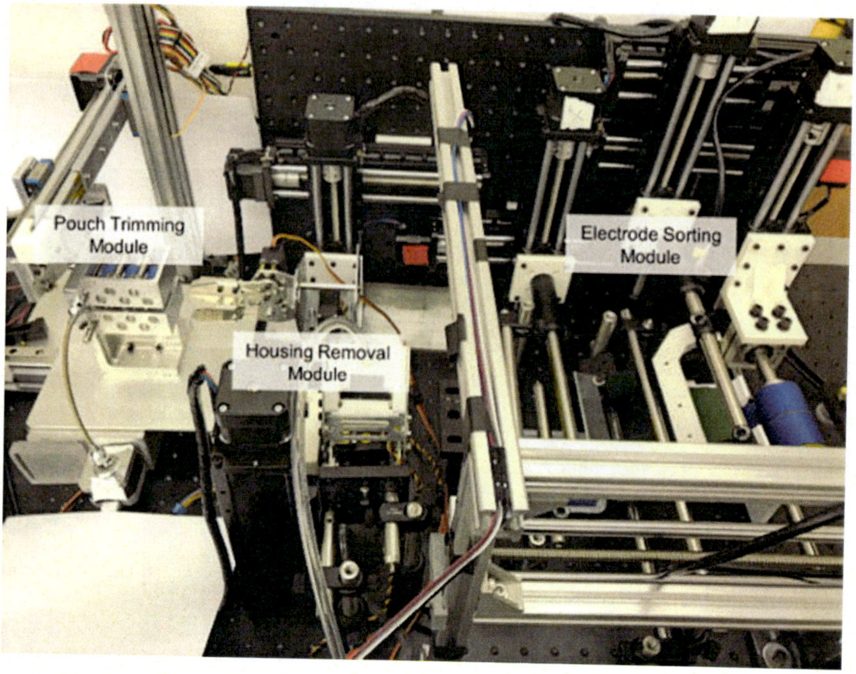

Fig. 2 Prototyped LIB pouch cell disassembly system overview [14]

Electrode Sorting Module

The separation of cathodes, anodes, and separators is a critical process for any LIBs recycling process. It directly influences the purity and recovery rate of the recycled materials. Our proposed electrode sorting strategy extracts cathode sheets and anode sheets, respectively, without applying destructive forces. By automatically stretching and feeding the Z-folded separator, cathode sheets and anode sheets attached on opposite sides of the separator are scraped off by specialized toolsets as the schematic shown in Fig. 3a. Since commonly used PVDF binder can be removed by either dissolving in organic solvents or decomposing at temperatures above 400 °C, multiple combinations of chemical, thermal, and mechanical treatments will then be used to break the adhesion between Al foil and cathode coating. The prototype of this module is shown in Fig. 3b. The functionality of this module was verified in the dummy cell test [14].

The sensor-integrated sorting module was assembled in a fume hood so that the electrode sorting process of a real ESC from LIB pouch cell could be recorded and analyzed accordingly. Three FLIR S USB 3 mono industrial cameras with 1.6 MP resolution and 226 FPS as shown in Fig. 4a are integrated. These cameras integrate the industry's most advanced sensors within a 29 mm * 29 mm * 30 mm cube. Both

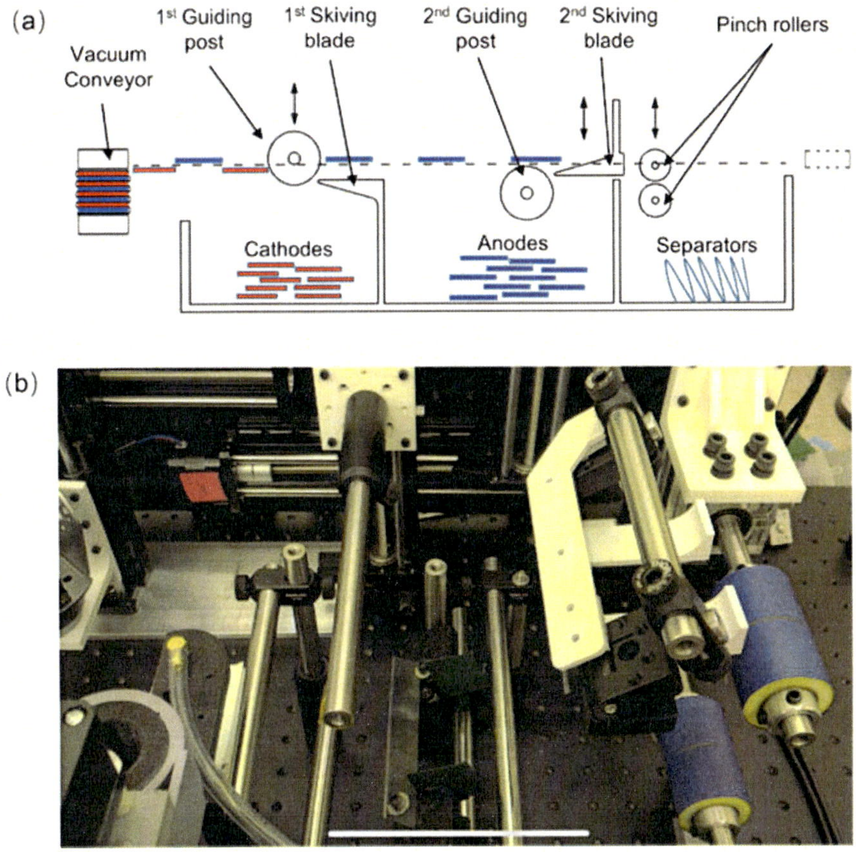

Fig. 3 **a** Schematic and **b** prototype of the electrode sorting module

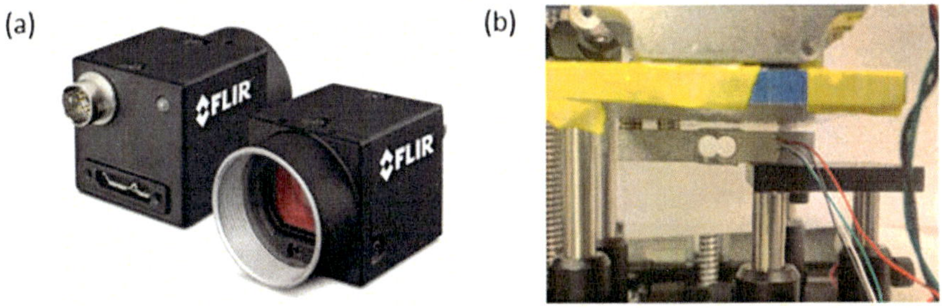

Fig. 4 Vision sensor network components: **a** FLIR S USB3 mono industrial cameras with 1.6 MP resolution and 226 FPS and **b** tension sensor modified from strain gauge-based load cell

automatic and precise manual control modes over image capture and on-camera pre-processing are available, thus enabling us to record videos and capture images with trigger signals from LABVIEW. A load cell that measures force and outputs the force signal as an electrical signal is also integrated into the sorting module. The selected load cell as shown in Fig. 4b uses a strain gauge to detect load changes, and hydraulic or pneumatic load cells are also favorable choices. The strain gauge-type load cell usually integrates four strain gauges in a Wheatstone bridge. Such bridge circuits originally have two balanced legs. When external load deforms the strain gauge, the electrical signal changes can be captured by the HX711 24 bit precision ADC module that is connected to LabVIEW via Arduino Mega 2560 R3 board.

Figure 5 shows the schematic of the sensor-integrated electrode sorting module with minimum changes on the separator feeding mechanism. The original pinch roller toolset is replaced by an initiative roller for the convenience of recording the load change on the roller which equals the resistance force applied to the separator from guiding posts and skiving blades. The overview of the upgraded sorting module is shown in Fig. 6. The first industrial camera focuses on the ESC unfolding area as indicated in Fig. 6d. The second industrial camera monitors the first guiding post and the first skiving blade from the above so that the separation process of the electrodes positioned on top of the separator can be monitored as shown in Fig. 6b. Ideally, the third industrial camera should focus on the separation process of the electrodes attached to the bottom of the separator. However, due to the space restriction, the camera is set to 45° tilted from the horizontal direction as shown in Fig. 6c. In this sensor-integrated system, any electrodes separation defect (e.g., failed to be skived off from the separator) can be detected through the semi-transparent wet separator. The output of the load cell is shown in Fig. 6e, and the negative value between 5 and 10 s at the very beginning indicates a calibration step by applying a standard 100 g weight to the load cell.

The control UI constructed within LabVIEW is shown in Fig. 7. The image acquisition section as encircled in Fig. 7a allows us to integrate as many industrial cameras as needed into the control software so that future upgrades on the entire prototyped disassembly system are convenient. Figure 7b displays the real-time data collection

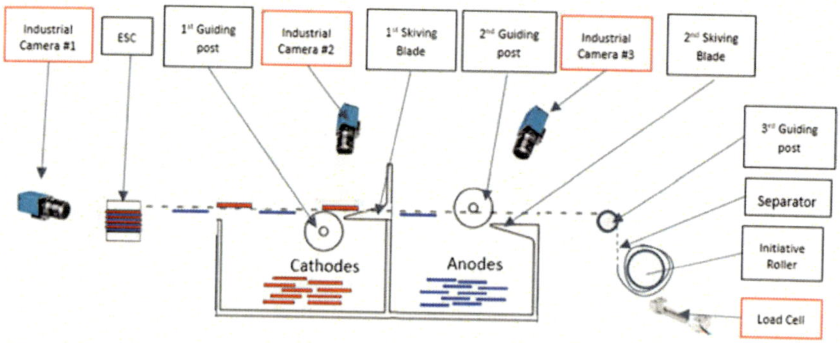

Fig. 5 Schematic of the sensor-integrated electrode sorting module

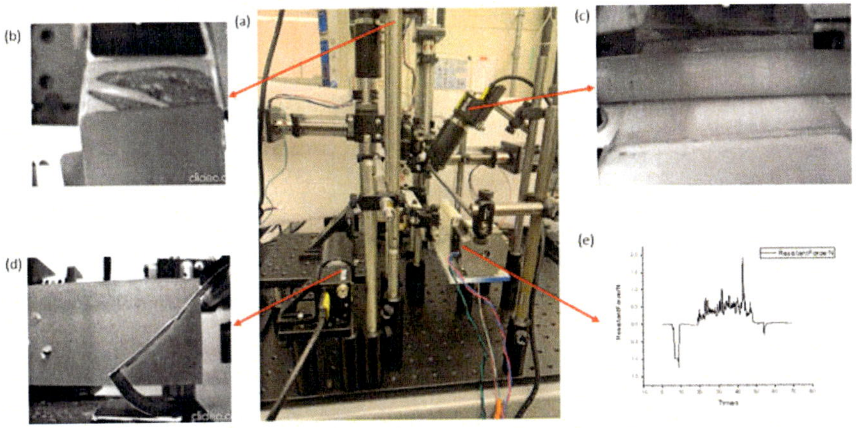

Fig. 6 Sensor-integrated system operating with EoL H605060 LIB: **a** system overview, **b** the operating frame of industrial camera #2, c the operating frame of industrial camera #3, **d** the operating frame of industrial camera #1, and **e** the recorded resistant force from the load cell

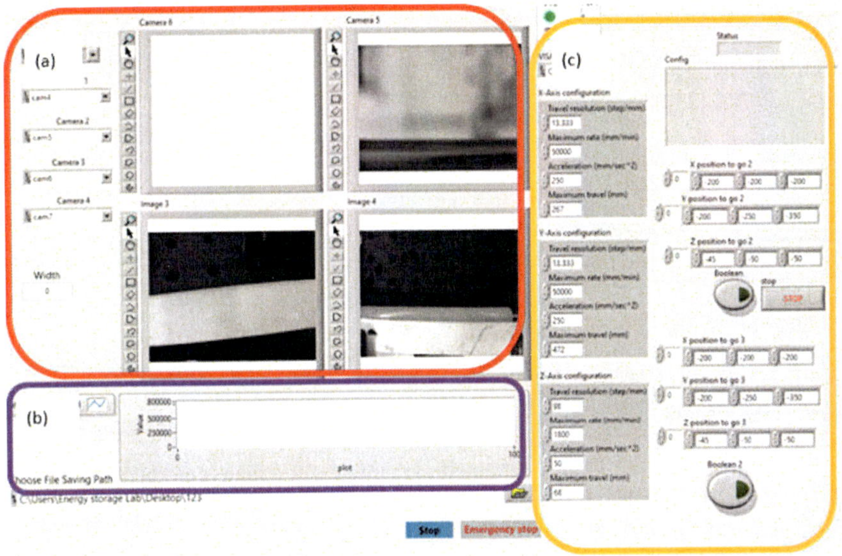

Fig. 7 LabVIEW control unit front user interface: **a** image acquisition section, **b** load cell reading section, and **c** linear motion control section

from the load cell with the autoscale adjusting function. Figure 7c is the linear motion control section integrated with functions of the position restoration, precise linear positioning, and arbitrary speed adjustment.

The Closed-Loop Controller

The closed-loop controller is implemented in the EOL LIB disassembly system to execute the corrective action in real time based on the sensor data. The condition of the EOL batteries can vary dramatically due to their different operating parameters. Thus, a closed-loop controller that can adjust parameters to correct the anomaly or defect is needed to improve the electrode separation quality. Figure 8 shows the overall architecture of the closed-loop controller. The physical part can be directly transferred from the prototyped disassembly system introduced in the previous sections with an integrated sensor network. The cyber part established within MATLAB or Python environments will include two steps: defects diagnosis and defects mitigation actions. For the electrode sorting module, the separator from an EOL LIB can be fragile, and small cracks shown in Fig. 9 can lead to the breakage of the separator and eventually compromise the sorting process by having a discontinuous separator. If such cracks

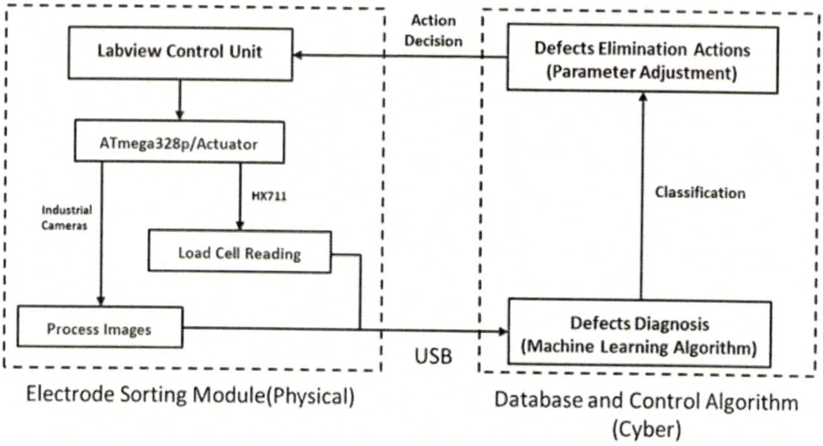

Fig. 8 Control architecture of the closed-loop process monitor for the electrode sorting module

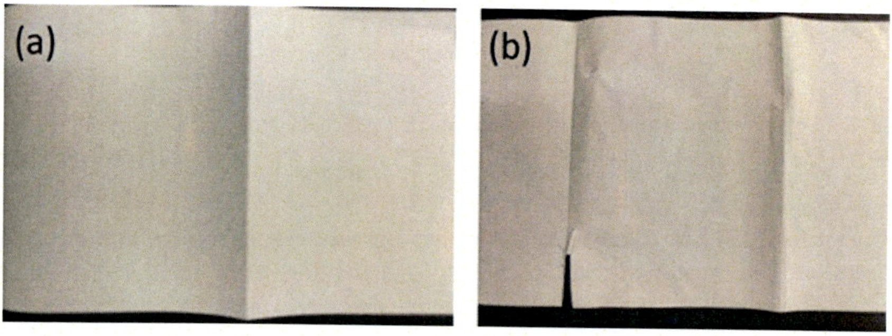

Fig. 9 Input images of **a** intact separator and **b** cracked separator

can be detected by the industrial cameras at the nascent stage, machine learning algorithms will be able to identify them by classifying the captured images. Then the corresponding system operating parameters can be adjusted to either decrease the pulling force applied to the separator or simply shut down the operation and call for human intervention. Thus, the implementation of this closed-loop controller will enable the machine to cognize the possible uncertainty and apply necessary parameter adjustment correspondingly in real time. An example of detecting defects as shown in Fig. 9 and adjusting machine parameters according to the physical model of the apparatus is demonstrated here.

The convolutional neural network (CNN) is used to identify the defects by continuously classifying images captured by integrated industrial cameras during the electrode sorting process which shows promising results in image classification by automatically discovering the separator interconnections [15]. Two image labels for the cracking defect are defined as "new" and "crack" in the preliminary CNN classification model development. Around 750 images for each label are used as the training data set, and 350 images for each label are used as the validation set. Images are downsized from the resolution of 4032 * 3024 to 200 * 1150 to minimize the model training time. Hyperparameters in a CNN model, such as learning speed, depth of the neuron network, and batch size, determine the neuron network structure and setting. These parameters are tuned to the best possible condition based on the classification accuracy of the validation set before initiating the model training utilizing the training set. Here the Bayesian hyperparameter tuning method, which can smartly explore the space of potential choices of hyperparameters in CNN model by deciding which combination to explore next based on previous observations, is adopted. The trained CNN model will label each image with the highest probability value as the classification result shown in Fig. 10.

The trained CNN model is then tested by an 1120 mixed image pool containing images from both categories. A confusion matrix is utilized to demonstrate the effectiveness of trained CNN models. The matrix is a summary of prediction results on the classification to provide an insight into the errors being made by the CNN model. Figure 11a gives the basic structure of a 2 * 2 confusion matrix which is also the matrix size for cracking defect detection since only two labels are involved. P in predicted class and actual class represents that the observation is positive and N represents the negative observation. Here in the cracking defect detection task, P represents the "crack" separator and N represents the "new" separator. True positive (TP) represents a positive observation and a positive prediction. True negative (TN) represents a negative observation and a negative prediction. Both TP and TN are favorable results for the observation that agrees with the prediction. False negative (FN) represents a positive observation but is predicted as negative, and false positive (FP) represents a negative observation but predicted as a positive. Thus, FN and FP are major sources of inaccuracy and should be avoided as much as possible. With TP, TN, FN, and FP available, the accuracy of the CNN classification model can be calculated by Eq. (1). With the confusion matrix of the preliminary algorithm indicated in Fig. 11b, the accuracy of the trained CNN model proved to be 98.12%, which is a satisfying result.

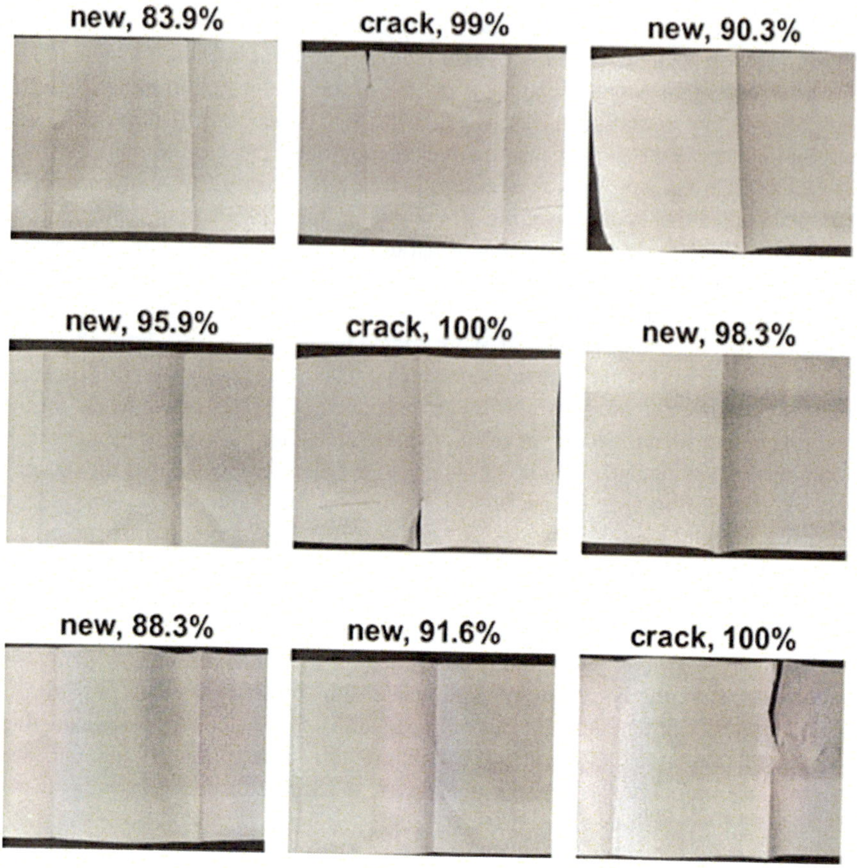

Fig. 10 Examples of image classification result and probability value

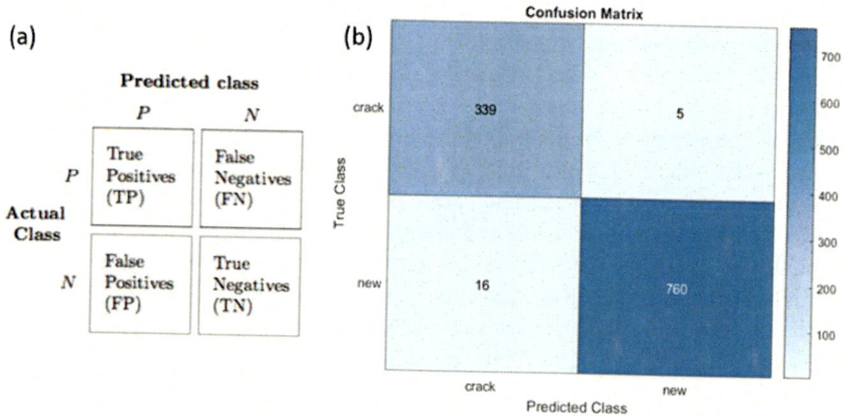

Fig. 11 **a** Structure of the confusion matrix and **b** the confusion matrix of the classification model

$$Accuracy = (TP + TN)/(TP + TN + FN + FP) = (TP + TN)/(P + N) \quad (1)$$

The last step to fulfill the closed control loop is to enable the real-time machine parameter adjustment based on the physical model of the mechanism so that detected defects can be mitigated or even eliminated. After detecting the "crack" defect, a series of actions will be taken to reduce the tension applied to the separator along its feeding direction. These actions may include decreasing the separator tilting angle, reducing the feeding speed of the separator, and increasing the blade tilting angle, which will be further verified by the physical model of the mechanism and a series of design of experiments. The commands for machine parameter adjustment will be sent from MATLAB or Python to LabVIEW, thus the continuous feeding of the separator can be assured by this control loop. The closed control loop for the entire LIB disassembly system will be completed in the future development of this project.

Conclusion

An automated disassembly system for EOL Z-folded pouch LIBs has been introduced in which cathode sheets, anode sheets, separators, and Al laminated film housing are automatically separated. The preliminary effort to integrate sensor networks and implement machine learning and closed-loop control is investigated and reported to improve the robustness of the module. Compared to the destructive pretreatment, the integrity of cathode sheets can be well preserved which will greatly benefit downstream direct recycling processes. Further research activities will focus on implementation of closed-loop control for the entire LIB electrode sorting system.

Acknowledgements This work was funded by the Department of Mechanical Engineering at Virginia Tech and Alfred P. Sloan Foundation under award G-2020-12651.

References

1. Curry C (2017) Lithium-ion battery costs and market. Bloomberg New Energy Finance 5
2. Peterson SB, Apt J, Whitacre J (2010) Lithium-ion battery cell degradation resulting from realistic vehicle and vehicle-to-grid utilization. J Power Sources 195:2385–2392
3. Ellingsen LAW et al (2014) Life cycle assessment of a lithium-ion battery vehicle pack. J Ind Ecol 18:113–124
4. Lv W et al (2018) A critical review and analysis on the recycling of spent lithium-ion batteries. ACS Sustain Chem Eng 6:1504–1521. https://doi.org/10.1021/acssuschemeng.7b03811
5. Sun X, Hao H, Zhao F, Liu Z (2017) Tracing global lithium flow: a trade-linked material flow analysis. Resour Conserv Recycl 124:50–61
6. Ferreira DA, Prados LMZ, Majuste D, Mansur MB (2009) Hydrometallurgical separation of aluminium, cobalt, copper and lithium from spent Li-ion batteries. J Power Sources 187:238–246

7. Zhang X et al (2018) Toward sustainable and systematic recycling of spent rechargeable batteries. Chem Soc Rev 47:7239–7302
8. Li X, Zhang J, Song D, Song J, Zhang L (2017) Direct regeneration of recycled cathode material mixture from scrapped $LiFePO_4$ batteries. J Power Sources 345:78–84
9. Song X et al (2017) Direct regeneration of cathode materials from spent lithium iron phosphate batteries using a solid phase sintering method. RSC Adv 7:4783–4790
10. Shi Y, Chen G, Chen Z (2018) Effective regeneration of $LiCoO_2$ from spent lithium-ion batteries: a direct approach towards high-performance active particles. Green Chem 20:851–862
11. Diekmann J et al (2017) Ecological recycling of lithium-ion batteries from electric vehicles with focus on mechanical processes. J Electrochem Soc 164:A6184–A6191
12. Hanisch C et al (2015) Recycling of lithium-ion batteries: a novel method to separate coating and foil of electrodes. J Clean Prod 108:301–311
13. Kim D-S et al (2004) Simultaneous separation and renovation of lithium cobalt oxide from the cathode of spent lithium ion rechargeable batteries. J Power Sources 132:145–149
14. Li L, Zheng P, Yang T et al (2019) Disassembly automation for recycling end-of-life lithium-ion pouch cells. JOM 71:4457–4464
15. LeCun Y, Bengio Y, Hinton G (2015) Deep learning. Nature 521:436–444

Steel Production Efficiency Improvements by Digitalization

Markus Schulte

Extended Abstract

Steel is one of the most important engineering and construction materials and plays a critical role in the overall economic development. However, iron and steel industry is one of the most energy-intensive industries which accounts for >20% of global industrial energy use and about a quarter of industrial CO_2 emissions in the world. Energy efficiency improvements are one driver for production optimizations, but efficiency improvements in steel production can be reached in several ways like equipment availability improvements as well as yield increases or optimized production schedules.

Using digitalization for these improvements in steel production requires reliable data pipelines from the sensors and automation systems to the applications. An additional data collection and storage layer between the IT and OT systems helps to process these data and inserts them in a compressed form in a data warehouse. The applications can access the data in that additional layer through a generic API. This data warehouse, which unifies data from different data sources, is the key enabler for the successful and quick deployment of value adding applications.

Our integrated energy platform has the capabilities to manage and optimize energy consumption as well as to track the carbon footprint on a product level. It helps to reduce overall consume of resources by process efficiency modeling, simulation, and forecasting.

Process data are used to train AI/ML models which can be utilized in the area of asset optimization. Especially in the current steel market situation, the availability of the equipment is critical for steel producers. Forecasting of the time-to-failure of critical components helps to reduce the unplanned downtimes, to extend the lifetime

M. Schulte (✉)
Brussels, Belgium
e-mail: markus.schulte@sms-group.com

A. Lazou et al. (eds.), *REWAS 2022: Developing Tomorrow's Technical Cycles (Volume I)*, The Minerals, Metals & Materials Series,
https://doi.org/10.1007/978-3-030-92563-5_50

of the assets, and to avoid catastrophic failures. Pattern in the data of hundreds of signals indicates upcoming equipment failures in early stages and helps to trace down the root cause of the issue. This is has successfully been developed for several use cases for electric arc furnaces, e.g. for refractory relines, taphole failures, and panel failures.

Despite the fact that digitalization drives and enables these efficiency improvements and data are the foundation for the value of these applications, the understanding of the physics and processes in metallurgy as well as of automation are required for a successful application in the field. Especially when the data basis is small or not even existing, the physical models and measurement capabilities play an inherent role to close that gap. In this way, we could develop smart applications, which, e.g. optimize mechanical properties, reduce the amount of virgin material consumption by increased consume of recycled material and the optimization of heat treatment times and temperatures for reduced energy waste.

These types of applications get more and more accepted in the steel industry as they have shown their value for the producers. The use of these applications starts usually for small and isolated processes, but the future is heading towards a more holistic optimization for production areas, entire plants, and even enterprises which will increase the impact of AI/ML for this industry for the next years significantly.

Refractory Lifetime Prediction in Industrial Processes with Artificial Intelligence

Nikolaus Voller, Christoph Pichler, Christine Wenzl, and Gregor Lammer

Abstract This paper deals with refractory lifetime prediction by using artificial intelligence (AI). Through the effective use of process parameters, obtained from various operational processes within the Industrial (Cement/Lime, Non-Ferrous Metals, Process Industries, Foundry) and Steel sector, an AI model is generated. With the assistance of modern surveying technology, a correlation can be identified between process parameters and refractory wear. Using this method, suitable prediction of the refractory lifetime, as well as the wear mechanism, is possible. In addition, maintenance cycles can be adjusted, and the optimal maintenance intensity of the operated furnaces can be ensured. With our intelligent Automated Process Optimization (APO) solution, the prediction of refractory lifetime and wear mechanism can be done in real time. Thus, we can provide value-added service to our customers.

Keywords APO · Refractory wear · Lifetime prognosis · Artificial intelligence · Process optimization

Introduction

Through the ongoing digitalization process in different industries and the transition to Industry 4.0 applications based on artificial intelligence are becoming increasingly important. The popular term "Industry 4.0" is often used in different contexts as well

N. Voller (✉) · C. Wenzl · G. Lammer
RHI Magnesita, Vienna, Austria
e-mail: nikolaus.voller@rhimagnesita.com

C. Wenzl
e-mail: christine.wenzl@rhimagnesita.com

G. Lammer
e-mail: gregor.lammer@rhimagnesita.com

C. Pichler
RHI Magnesita, Leoben, Austria
e-mail: christoph.pichler@rhimagnesita.com

A. Lazou et al. (eds.), *REWAS 2022: Developing Tomorrow's Technical Cycles (Volume I)*, The Minerals, Metals & Materials Series,
https://doi.org/10.1007/978-3-030-92563-5_51

as with different meanings. In the following, Industry 4.0 has its focus on intelligent production, processes, and products [1]. Modern production mills aim to be smart factories and enables networking between people, machines, and products [2, 3]. As the amount of recorded data from various processes is constantly increasing, decisions based on automated, self-optimized systems will increase rapidly in the future [4]. The aim of such systems, especially the patented Automated Process Optimization (APO), is to increase safety, optimize production, and lower production losses [5]. In addition, the lifetime of the refractory material itself is always extremely difficult to assess. The right time must be chosen in order both to avoid wasting refractory material that is still usable and to prevent the furnace from a break-through that would cause extensive damage. Decisions for maintenance actions or process modifications are still mostly made by humans based on their experience. Therefore, the correlation between process parameters and refractory wear based on this AI system is extremely valuable, and the demand for using APO is raising too [6].

In the following, the applicability and requirements to set up such a system, the data processing, as well as a practical example of a self-optimized system are highlighted.

Applicability

Originally, the APO system was developed for the steel industry, and first successful applications are in operation. The important aspect is that APO is also applicable for all different types of furnaces in industrial processes where continuous process data is generated. The key point is to fulfill the minimum requirement as shown in Fig. 1.

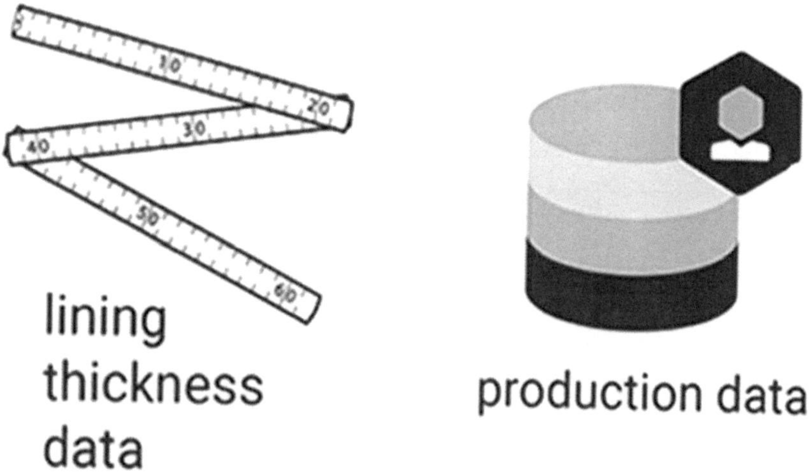

Fig. 1 Minimum requirements for APO

On the one hand, a measuring device must be used to generate the necessary lining data. In the best case, a modern technology, like laser scans, is used. To save as much time as possible and reduce the thermal shock for the refractory, a laser scan can be used under hot conditions, as the furnace does not need to be cooled down. However, since this variant needs a special laser system, that withstands the high temperature, a scan in cold conditions can also be applied. On the other hand, relevant production data, as temperature, process time, and/or charging mixture, is needed to run the application properly. This could be historical data at the beginning of the implementation and real-time data at a later stage. The APO system is learning based on this input data. Therefore, the more valuable the input data is, the better the results will be in the prediction of refractory wear and lifetime.

Data Processing

The APO system is built on artificial intelligence combined with machine learning [1, 7] and is calculating a digital twin of the refractories in the furnace [5]. Figure 2 shows the data processing pipeline for an APO system.

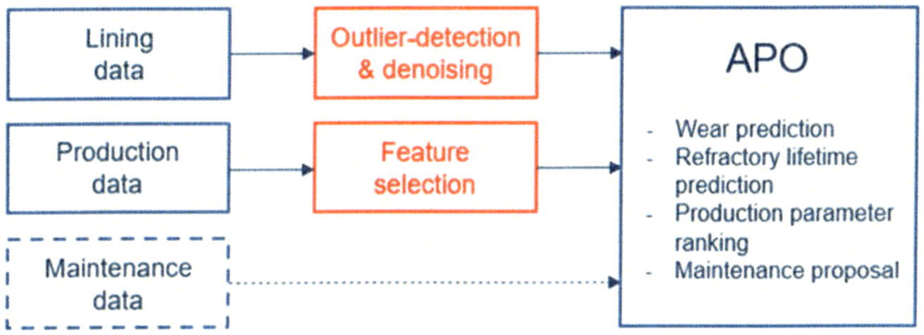

Fig. 2 APO data processing pipeline [1]

The minimum requirements to establish an APO system are lining and production data. By combining these data sources, the AI model (digital twin) is generated. Modern machine learning and artificial intelligence methods are used to process this data [1]. The system evolves in an agile manner, and additional maintenance data enhances this self-learning process. Important is the selection of the most useful production parameters (feature selection). For many data analysis methods, selecting the appropriate features has become increasingly valuable [8–10]. The aim is to exploit only influential production features to achieve the best possible refractory wear prediction [7]. Furthermore, detection of outliers in the lining data packages takes place in order to denoise them. This ensures higher accuracy in the prediction process. Especially with small data packages, outliers cause disadvantages for the training and the self-development of the APO system [7]. A more detailed explanation of the data processing steps to obtain a suitable APO model is provided in [1, 7, 11].

In addition to selecting the most significant parameters for wear, it is also possible to manually force parameters that affect the production process itself. With this capability, not only refractory wear can be related to the data, but it also allows different production problems or scenarios to be linked.

Practical Example/Implementation

To best demonstrate the functions and capabilities of the APO system, a more hands-on approach is useful. Therefore, a practical example based on a short rotary kiln is chosen. This kiln can be used in different industrial production lines, like aluminum, lead, copper, and many others. After the relining of the furnace, the first lining data is generated through a laser scan. In the meantime, historical production data is transferred by the APO data processing pipeline. During normal production, new process data is transferred continuously to further enhance the self-learning system. If maintenance work takes place as part of the production campaign, a new laser scan can be used to obtain up-to-date data on the refractory material inside the kiln. At the end of the campaign, the used lining is measured again.

All gained information and predictions will be displayed in the APO dashboard that the end-user can adapt according to specific needs. Besides important parameters, details about remaining heat cycles, and critical refractory areas, also the refractory wear prediction are visible on the dashboard. An example of such an APO dashboard with lifetime prognosis is shown in Fig. 3.

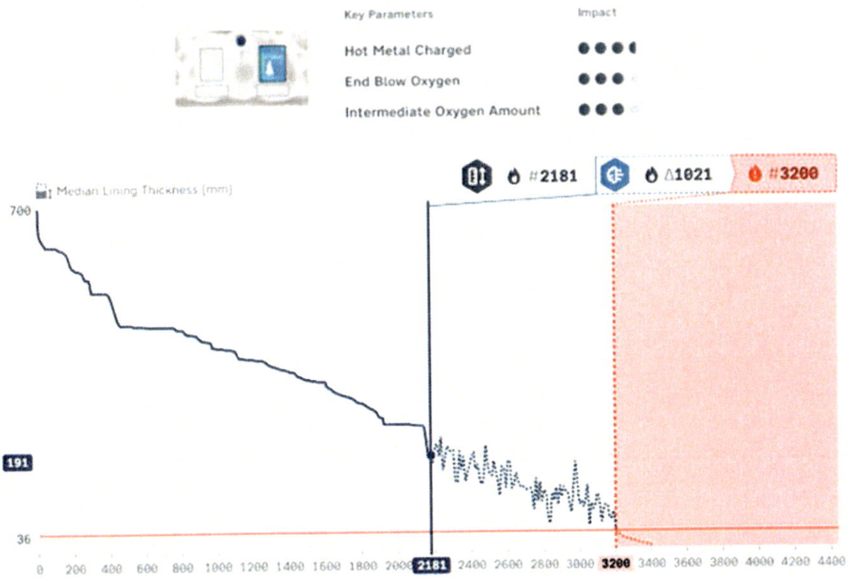

Fig. 3 Refractory wear and lifetime prognosis dashboard

The median lining thickness is plotted on the ordinate and the number of process cycles on the abscissa. The thin vertical lines represent additional lining scans. Based on the correct process parameters and the lining data, the APO system can predict the number of process cycles remaining before relining is necessary. Moreover, the predicted refractory wear, as well as the predicted remaining process cycles (#1021), is visible between the already finished process cycles (#2181) and the maximum possible process cycles predicted (#3200). Furthermore, the most influencing key parameters, as well as the respective area analyzed, are displayed on this APO dashboard.

All required information about refractory wear and important process parameters are displayed on this dashboard and can be accessed by the operator anywhere via tablet or mobile device. Thus, optimal furnace and refractory performance can be achieved.

Results and Achievements

The APO system can improve our customers' production process and save resources, time, and money too, as already proven by successful installations at several of RHIM's steel customers. The implementation of APO generates benefits from various improvements: On the one hand, process cycles times can be improved and tap-to-tap times reduced through process optimization. Moreover, these improvements in process control result in cost savings through more efficient use of the production process. On the other hand, optimal usage of the existing refractory material can be obtained. As a result, maintenance tasks can be better planned, and any sudden furnace breakouts can be prevented. Both aspects save customer's valuable time and resources. In general, a total benefit of approximately 15% to 20% can be achieved. This primarily relates to the increased furnace availability.

In addition, the APO system provides a visualization of the furnace wear profile, as shown in Fig. 4. This enables a better understanding of the differently influenced areas in the furnace and promotes quick operator intervention when critical situations occur.

Fig. 4 Virtual profile of the furnace's refractory wear

Conclusion

The ability to predict refractory wear is a significant advantage of our APO technology. Furthermore, the understanding of the process can be improved and optimized by the appropriate selection of parameters (feature selection). The use of APO thus improves not only the monitoring of the refractory wear mechanisms but furthermore also the process control. Subsequently, it allows for optimization of production, as well as safety standards in the operational business.

The decisive factor is the use of suitable production and lining data to create the APO model. The customer dashboard can be customized to meet specific needs, allowing an easy and quick overview of key indicators and the visualized profile of the furnace's refractory wear.

Due to the wide range of applications in countless different industrial furnaces, numerous customers can benefit from the implementation of the APO technology.

References

1. Lammer G, Lanzenberger R, Rom A, Hanna A, Forrer M, Feuerstein M, Pernkopf F, Mutsam N (2017) Advanced data mining for process optimizations and use of A.I. to predict refractory wear and to analyze refractory behavior. In: AISTech 2017, conference proceedings, pp 1195–1207
2. Brecher C et al (2011) Integrative Produktionstechnik für Hochlohnländer. Springer, Berlin, Heidelberg, Heidelberg, pp 17–81
3. Womack JP, Jones DT, Roos D (1990) The machine that changed the world: The story of lean production. Simon and Schuster, New York, pp 1–325
4. Yan HS, Xue CG (2007) Decision-making in self-reconfiguration of a knowledgeable manufacturing system. Int J Prod Res 45(12):2735–2758
5. Magnesita RHI APO: wear lifetime prediction with the power of AI. https://www.rhimagnesita.com/apo-wear-lifetime-prediction-with-the-power-of-ai/
6. Steiner R, Lammer G, Spiel C, Jandl C (2017) Refractories 4.0, BHM 2017, vol 162, no 11, pp 514–520
7. Viertauer A, Mutsam N, Pernkopf F, Gantner A, Grimm G, Winkler W, Lammer G, Ratz A, Persson M (2019) Refractory lifetime prognosis for RH degassers. In: UNITECR 2019, conference proceedings, pp 154–157
8. Jain AK, Zongker D (1997) Feature selection: evaluation, application, and small sample performance. IEEE Trans Pattern Anal Mach Intell 19(2):153–158
9. Kohavi R, John GH (1997) Wrappers for feature subset selection. Artif intell 97(1–2):273–324
10. GuyonI, Elisseeff A (2003) An introduction to variable and feature selection. J Mach Learn Res: 1157–1182
11. Pudil P, Novovicová J, Kittler J (1994) Floating search methods in feature selection. Pattern Recogn Lett 15(11):1119–1125

Digitalization for Advanced Manufacturing Through Simulation, Visualization, and Machine Learning

Chenn Zhou, John Moreland, Armin Silaen, Tyamo Okosun, Nick Walla, and Kyle Toth

Abstract Computer simulation, visualization, and machine learning are increasingly playing key roles in the digitalization of advanced manufacturing processes. These technologies can be used to create cutting-edge physics-based and data-driven tools for real-time decision making to address critical issues related to energy efficiency, carbon footprint, and other pollutant emissions, productivity, quality, operation efficiency, maintenance, and more. They can also provide fundamental understanding and practical guidance for process design, troubleshooting, and optimization, new process development and scale up, as well as workforce development. The Center for Innovation through Visualization and Simulation (CIVS) at Purdue University Northwest, established in 2009, has used these technologies to develop and implement digitalization for Advanced Manufacturing in partnerships with steel and other industries.

Keywords Computer simulation · Visualization · Machine learning

Introduction

Computer simulation, visualization, and machine learning are increasingly playing key roles in the digitalization of advanced manufacturing processes. These technologies can be used to create cutting-edge physics-based and data-driven tools for real-time decision making to address critical issues related to energy efficiency,

C. Zhou (✉) · J. Moreland · A. Silaen · T. Okosun · N. Walla · K. Toth
Center for Innovation Through Visualization and Simulation, Purdue University Northwest, 2200 169th Street, Hammond, IN 46323, USA
e-mail: czhou@pnw.edu

A. Lazou et al. (eds.), *REWAS 2022: Developing Tomorrow's Technical Cycles (Volume I)*, The Minerals, Metals & Materials Series,
https://doi.org/10.1007/978-3-030-92563-5_52

carbon footprint, and other pollutant emissions, productivity, quality, operation efficiency, maintenance, and more. They can also provide fundamental understanding and practical guidance for process design, troubleshooting, and optimization, new process development and scale up, as well as workforce development. The Center for Innovation through Visualization and Simulation (CIVS) at Purdue University Northwest, established in 2009, has used these technologies to develop and implement digitalization for Advanced Manufacturing in partnerships with steel and other industries.

Methodology

CIVS has developed an integrated and application-driven methodology to apply digital technologies to solve real-world industrial problems. This methodology integrates numerical simulation techniques such as computational fluid dynamics (CFD) with virtual reality (VR) and augmented reality (AR) visualization to create 3-D interactive virtual processes, which can be further integrated with sensor data, machine learning, and reduced-order modeling (ROM) to create 3-D digital twins as illustrated in Fig. 1.

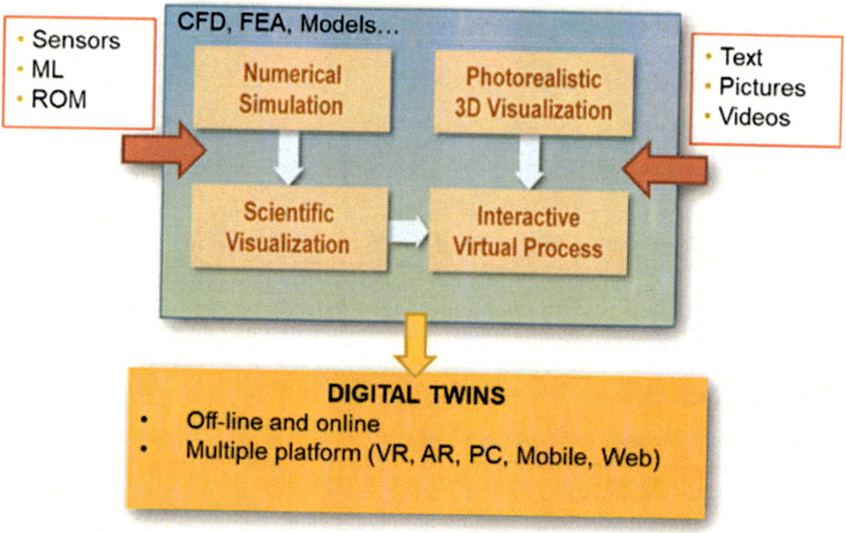

Fig. 1 CIVS integrated methodology

Examples

Integrated Virtual Blast Furnace

A key component in integrated steel mills: the blast furnace is an extremely complex energy-intensive process with complex dynamic and physical–chemical phenomena, operating at more than 1500 °C. The operability and observability of the process are also rather limited. Due to the complexity of the process and difficulties in attaining sensor measurements, modeling and simulation are promising tools for further improvements. Over the past 20 years, blast furnace CFD models have been developed and integrated with advanced visualization technology to create virtual blast furnaces (VBF) of real blast furnaces at Purdue University Northwest in collaboration with the steel industry. These VBFs have been used for troubleshooting, optimization, design, and training, resulting in significant reductions in cost and BF downtime.

Recently, a large-scale collaborative effort, funded by the Department of Energy Advanced Manufacturing Office, has been initiated to develop and implement a novel, next-generation, physics-based and data-driven integrated virtual blast furnace (IVBF) using advanced modeling and simulation, high-performance computing (HPC), state-of-the-art artificial intelligence (AI) methods, 3-D visualization, and novel sensor technology (see Fig. 2). The IVBF will allow operators and engineers to conduct real-time visualization and analytics for monitoring BF performance and internal states, rapid off-line modeling to quickly adapt to changing conditions and investigate "what-if" scenarios and high-fidelity analysis for evaluating new technologies. Once completed, the IVBF will provide the U.S. steel industry with a

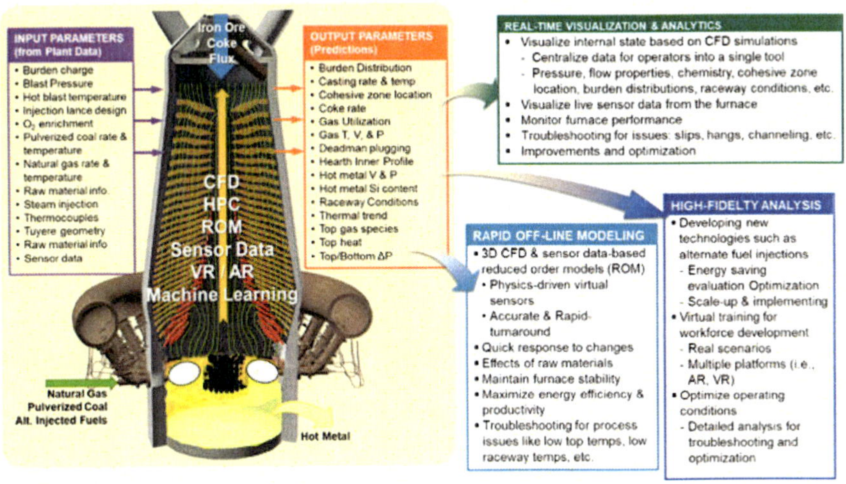

Fig. 2 Integrated virtual blast furnace (IVBF)

unique and powerful tool for improving BF operation, stability, energy efficiency, and carbon footprint.

Smart Ladle

Refining of liquid steel in a ladle is a key process in steelmaking primarily due to its overall impact on operational efficiency and performance of finished products. The ladle is also used for transportation of liquid steel from primary steelmaking (EAF or BOF) sites to the ladle metallurgy facility (LMF) or the ladle treatment station (LTS), where the steel chemistry is refined and its temperature is maintained before it is transported to the caster. Some facilities also feature a degassing stage between the LMF/LTS and casting. Variations from the optimum casting temperature can affect the casting process. Steel temperatures that are too high increase the likelihood of breakouts, and casting would need to be slowed down to allow proper solidification. Too-low temperatures can result in nozzle clogging due to premature solidification. Thus, temperature variations can impact casting quality, equipment integrity, and safety, potentially leading to significant production losses and/or safety hazards. Therefore, it is crucial to have an accurate steel temperature control.

In order to achieve consistent casting quality and maximize productivity, a smart ladle has been created using a deep learning network to develop quantifiable relationships between the casting temperature and various factors during the ladle refining process to enable predictions of casting temperature and precise adjustments to steel temperature prior to the ladle reaching the casting stage of the production process. The smart ladle is developed based on the facility at SDI Butler Division. It can provide information on the heat loss in the ladle between the LMF and the caster as well as information on the tundish temperature behavior for that future heat. The predictions from the smart ladle tool have a mean absolute error and RMSE close to the accuracy of the thermocouples used to take temperature measurements, discounting extreme outliers. Further model robustness is being developed to include additional data as well as enable the model to handle datasets from other production facilities (Fig. 3).

Caster Digital Twin

Digital twins are becoming increasingly viable across many industries and are set to play a very impactful role in Industry 4.0. Some examples have focused on data-driven models, where data is shown primarily in tablature or graph format, and in others, CAD modeling is used to show a virtual version of a piece or an array of equipment to show a whole process visually. Under a grant from the Clean Energy and Smart Manufacturing Innovation Institute (CESMII), a prototype of a 3-D digital twin has been developed for usage as part of operations at a caster in the Cleveland Cliffs Burns Harbor Plant. The caster digital twin incorporates different information

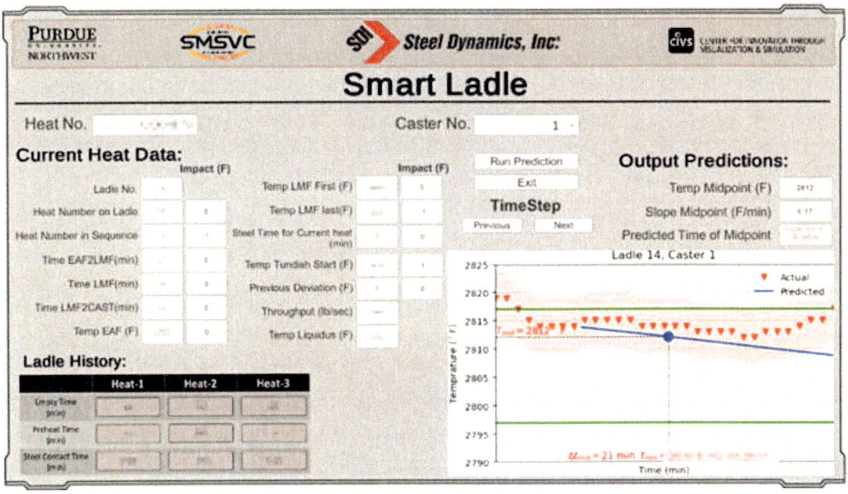

Fig. 3 Smart ladle

Fig. 4 Caster digital twin

and features from across many different casting HMIs, combining them into one concise interface that can be modular. The resulting digital twin solution can be modified for many other processes outside of casting as well (Fig. 4).

Summary

Advanced digital technologies such as simulation, visualization, and machine learning have demonstrated great promise for smart manufacturing. The integration of these technologies can provide physics-based and data-driven tools for the following applications:

- Real-time visualization and analytics for monitoring and optimizing process performance.

- Rapid off-line modeling for fast and accurate predictions, allowing operators to quickly adapt to changing conditions and investigate "what-if" scenarios.
- High-fidelity analysis for developing new technologies and virtual training.

Computational Methodology to Simulate Pyrometallurgical Processes in a Secondary Lead Furnace

Vivek M. Rao, Vineet Kumar, Alexandra Anderson, Joseph Grogan, and Prashant K. Jain

Abstract Pyrometallurgical recovery of nonferrous metals involves a combination of thermally intensive transformations during exothermic gas-phase reactions, endothermic decomposition of solid charge, and melting of simpler solids. In the recovery of secondary lead, simultaneous thermal effects in a reverberatory-style furnace cause a melt pool to accumulate at the bottom, with lighter solids (slag) floating above and gaseous products from decomposition of the charge diffusing through the gas–slag interface. Species from oxy-fuel combustion of natural gas, species profiles from smelting reactions, and the formation of a melt pool consisting primarily of lead are simulated via a time-averaged formulation. Predictions of outflow are compared with preset inflow profiles to ensure conservation of mass. Thermal profiles for solid, liquid, and gas phases are presented by species. A novel method is implemented to model the latent heat of fusion using a heterogeneous chemical reaction. The simulation is conducted in Simcenter STAR-CCM+ v. 16.02.009-R8.

Keywords CFD · Smelting · Lead · Melting · Oxy-fuel combustion · DEM

This manuscript has been authored by UT-Battelle, LLC, under contract DE-AC05-00OR22725 with the US Department of Energy (DOE). The US government retains and the publisher, by accepting the article for publication, acknowledges that the US government retains a nonexclusive, paid-up, irrevocable, worldwide license to publish or reproduce the published form of this manuscript, or allow others to do so, for US government purposes. DOE will provide public access to these results of federally sponsored research in accordance with the DOE Public Access Plan (http://energy.gov/downloads/doe-public-access-plan).

V. M. Rao (✉) · V. Kumar · P. K. Jain
Oak Ridge National Laboratory, 1 Bethel Valley Road, Oak Ridge, TN 37830, USA
e-mail: raovm@ornl.gov

A. Anderson · J. Grogan
Gopher Resource LLC, 6505 Jewel Avenue, Tampa, FL 33619, USA

A. Lazou et al. (eds.), *REWAS 2022: Developing Tomorrow's Technical Cycles (Volume I)*, The Minerals, Metals & Materials Series,
https://doi.org/10.1007/978-3-030-92563-5_53

Introduction

Computational simulations of chemical recovery processes require the implementation of suitable validated models to account for the thermally driven physical and chemical transformations. Commercial software packages implement finite-element and finite-volume formulations of conservation laws to model and simulate the transport of species, energy, and momentum encountered during pyrometallurgical processes. These models are developed for generic applications and do not necessitate compatibility with other models developed to resolve isolated effects, as inferred by the authors in an exhaustive review of relevant literature. For instance, the recovery of nonferrous metals involves heterogeneous chemical reactions, gas-phase reactions, and melting of certain solid products. However, the model for melting–solidification was developed in the Eulerian multiphase (EMP) mixture framework and is not inherently compatible with particle reaction models that are used to simulate heterogenous chemical reactions between particulate solid (Lagrangian multiphase [LMP]) and gas (EMP) phases [1]. As such, the record of published simulation work in pyrometallurgical recovery is largely restricted to gas–solid interactions. The effect of residual heat on consequent melting of simpler solids has conventionally been simulated as an isolated effect.

Work published on computational fluid dynamics (CFD) models applied to the pyrometallurgical industry has given little attention to the dynamics of the gas-burden interface and metal-phase transformation. Much emphasis has been devoted to the recovery of Al and Cu, with a sparsity of studies on Pb. Current modeling and simulation efforts of thermal effects in secondary Pb furnaces have been restricted to emulating heat from the pyrochemical reactions by defining artificial volumetric heat sinks above the gas-burden interface in pseudo-transient formulations [2, 3]. Although this approach allows the multiphysics CFD model to inform the design or operational changes based on the distributions within the gas region, it limits the model's ability to predict the effect these changes might have on furnace efficiency or productivity because the simulation does not provide any spatial or temporal resolution of the consumption of heat by charge to the furnace. Therefore, the multiphysics model cannot predict the actual production of the metallurgical species of interest without using additional empirical models or correlations. Furthermore, this treatment of the gas-burden interface could lead to inaccuracies caused by not fully accounting for factors such as melting, heat loss, and phase change of the charge material. Development of a robust, high-fidelity method for resolving the interaction between the gas space and furnace burden regions will allow CFD simulations to produce information on improving the efficiency and productivity of reverberatory furnaces as a function of design changes.

Background on the Recovery of Secondary Pb

Reverberatory furnaces are commonly utilized for the pyrometallurgical processing of secondary Pb. A reverberatory furnace is powered by the exothermic oxy-fuel combustion (OFC) of a gas such as natural gas, and heat from combustion is absorbed by a solid charge, which causes endothermic decomposition of larger molecules (sulfates, carbonates, hydroxides) to simpler solids. Of these simpler solids, some are elemental (Pb, As, Cd, Sn, Sb), and some are molecular (oxides, sulfides). OFC of natural gas is commonly initiated by ignition of the air–fuel mixture at lower operating temperatures between 300 and 473 K, whereas decomposition of charge material has been experimentally indicated to occur primarily between 800 and 1473 K [4]. Over this large range of temperatures, several significant effects contribute to the thermal balance of the reverberatory furnace. Thermophysical properties of various solid compounds in the charge material vary with temperature. This variation is greater for elemental components than for molecular compounds. The presence of moisture in the feed (up to 5%) is a common result of unit operations upstream of the reverberatory furnace. This moisture evaporates before larger ionic solids decompose, and water vapor contributes to the species profile of flue gas emanating from OFC of natural gas. In this case, water vapor can further participate in the validated four-step mechanism for OFC of hydrocarbons in natural gas [5, 6]. Although the presence of moisture is quantitatively insignificant, it can delay the transfer of heat to the burden (i.e., the accumulated mass of charge inside the reverberatory furnace). The presence of moisture within the burden can create localized dry spots if the burden is not uniformly heated by burners.

Chemical reduction of metals is typically accomplished in the presence of carbon from coke, gaseous carbon monoxide (CO[g]), and gaseous carbon dioxide (CO_2[g]) for heterogeneous transformations. Petrochemical coke is routinely added to chemically reduce larger molecules in the solid charge to simpler molecular or elemental species. Petrochemical coke has various constituents: carbon and moisture are the most plentiful; therefore, they are of highest thermal relevance to the recovery processes. The moisture from coke must evaporate, which requires latent heat of vaporization to be absorbed from OFC of natural gas and adds to the species profile of flue gas. Carbon in the coke could be further oxidized by free oxygen in the product of combustion (POC) if that free oxygen is not fully utilized by natural gas or nitrogen to form nitrous oxide and gaseous nitric oxide ($NO_{x(g)}$). Sulfates in the solid charge are reduced by carbon in the coke to release CO_2(g), and simpler sulfides and oxides form elemental Pb and gaseous sulfur dioxide ($SO_{2(g)}$).

The recovery of secondary Pb requires a measurable amount of Pb product in the slag layer [7, 8]. Various chemical methods have been explored to isolate Pb product from the slag; however, the underlying reasons for entrainment of product Pb in the slag are not well understood. Therefore, the nonuniformities in the following processes must be understood: (i) the distribution of temperature in the burden; (ii) the rate of decomposition of Pb-bearing compounds relative to their temperature inside the burden; (iii) the entrainment created by bubbling burden gases and drag forces from the flue gas, which flocculate relatively heavier product Pb into the slag layer; and (iv) entrainment of unreacted burden downstream of the converging flames from burners. Thermal decomposition reactions of Pb-bearing compounds with available experimental thermochemistry involve lead sulfate ($PbSO_4$) [4], lead carbonate ($PbCO_3$) [9], and lead sulfide (PbS) [10, 11]. According to the Arrhenius model, the chemical kinetics data for these reactions range from 1×10^7 to 2×10^8 J/kmol for activation energies and 1×10^5 to 1×10^{12} J/kmol for pre-exponential factors. Despite this available literature, existing measurements are insufficient for the thermochemical response of Pb-bearing compounds over the range of temperatures encountered in the recovery of secondary Pb. Therefore, deterministic approaches from statistical thermodynamics are being used to inform the thermochemical responses of a given molecule without relying solely on the Arrhenius theory for kinetic data [12, 13].

Without full insight into the sequence of effects in recovery processes, formation of chemical complexes [14] and entrainment of product in the slag [8] have been shown to reduce productivity. Therefore, strategies have been developed to recover more desired product from slags [7] or improve thermal distribution conversion of desired products and process efficiency. These strategies have been demonstrated using CFD simulations for aluminum reverberatory furnaces via intermittent directed heat from transient heat burners [15]. Ineffective transfer of heat from burners can lead to the formation of soot, an undesired POC from any flammable fuel source. A methodology to simulate chemical reactions in the gas phase (OFC of natural gas, soot formation), solid phases (heterogeneous decomposition), and melting of solid products in the recovery of nonferrous metals is demonstrated using secondary Pb processing as an example. Simcenter STAR-CCM + v. 16.02.008-R8 was used for simulation on high-performance computing resources at Oak Ridge National Laboratory.

Methodology

Computational Domain

The combustor volume (1 × 3 m), shown in Fig. 1, is supplied with natural gas by isolated streams of natural gas, enriched air, and technical oxygen and is fitted with a 0.25 m outflow pipe. Particle decomposition is simulated in two steps: injection and settling. The particles are first continuously injected using a fictitious conical volume, assigned to type "Random Injector" in Simcenter STAR-CCM+ [1]. The injection stops when the number of injected particles approaches 100,000, allowing the charge particles to settle into a burden. The injection cone is not meshed and is artificially mapped to the physical domain for the purpose of particle injection.

A finite-volume mesh was generated for a symmetric model of the 3D geometry using volumetric controls to refine the shape and location of the expected flame near the burners and the burden. A representation of the finite-volume mesh is shown in Fig. 2. The generated mesh comprised 811,514 polyhedral cells. The cell count was

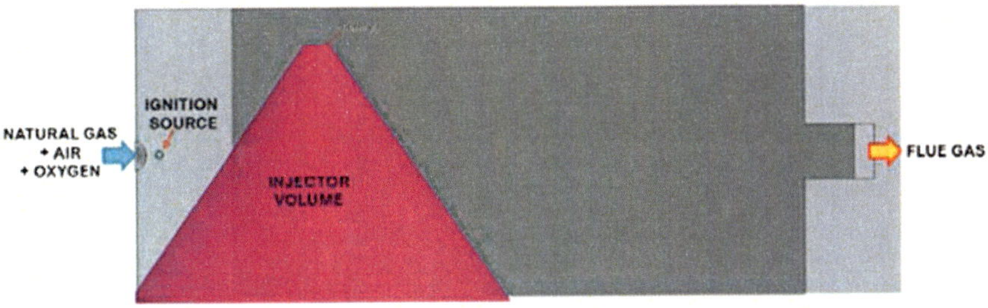

Fig. 1 Computational representation of the reverberatory furnace

Fig. 2 Finite-volume representation of the simulated domain (2D)

reduced to a 2D representation of 97,213 cells to reduce the time and CPU count of computation.

Solver Setup

In this study, the reverberatory furnace is powered by OFC of natural gas. Natural gas is assumed to comprise methane, ethane, propane, carbon dioxide, and nitrogen. Chemical reactions for OFC are implemented based on the benchmark study for hydrocarbons using the modified four-step Jones–Lindstedt scheme. OFC of an alkane, C_nH_{2n+2}, follows the steps listed in Eqs. (1)–(5). Each step is described by a rate equation that was verified and validated for the International Flame Research Foundation 0.8 MW furnace as a benchmark [16] and supported by prior studies [6, 17]. Equations (1)–(5) summarize the values used to simulate the kinetic scheme for methane, ethane, and propane (i.e., $n = 1, 2, 3$) in Simcenter STAR-CCM+ . The eddy break-up model was used for gas-phase reactions in a hybrid formulation, wherein the mean reaction rate for each species was limited by the mixing length of eddies and the net reaction rate from chemical kinetics. The eddy break-up ignitor was used to ignite the fuel mixture at the throat of the burner.

$$C_nH_{2n+2} + \frac{n}{2}O_2 \rightarrow nCO + (n+1)H_2 \quad \frac{d[C_nH_{2n+2}]}{dt} = AT^be^{-\frac{E}{RT}} \cdot [C_nH_{2n+2}]^{0.5}[O_2]^{1.25} \tag{1}$$

$$C_nH_{2n+2} + \frac{n}{2}H_2O \rightarrow nCO + (2n+1)H_2 \quad \frac{d[C_nH_{2n+2}]}{dt} = AT^be^{-\frac{E}{RT}} \cdot [C_nH_{2n+2}][H_2O] \tag{2}$$

$$H_2 + \frac{1}{2}O_2 \rightarrow H_2O \quad \frac{d[H_2]}{dt} = AT^be^{-\frac{E}{RT}} \cdot [H_2]^{0.25}[O_2]^{1.5} \tag{3}$$

$$H_2O \rightarrow H_2 + \frac{1}{2}O_2 \quad \frac{d[H_2O]}{dt} = AT^be^{-\frac{E}{RT}} \cdot [H_2]^{-0.75}[O_2][H_2O] \tag{4}$$

$$CO + H_2O \rightarrow H_2 + CO_2 \quad \frac{d[CO]}{dt} = AT^be^{-\frac{E}{RT}} \cdot [CO][H_2O] \tag{5}$$

where in the system of units given by (J, kmol, K, s), A is the Arrhenius pre-exponent factor, T is the temperature, b is the temperature exponent, R is the universal gas constant, E is the activation energy of the hydrocarbon molecule, and t is time. The gas phase was represented by a multi-component gas model in Simcenter STAR-CCM+ and was developed in the Eulerian multiphase (EMP) framework. A single transport equation for the mixture of gaseous reactants and products was solved for each of the transport of species, energy, and flow using the coupled flow solver. Additionally, closure models for turbulence were enabled to solve for turbulent kinetic energy and specific dissipation rate, ω, to obtain the turbulent viscosity ratio using the SST k-omega (Menter) turbulence model. A wall treatment model was used to account for near-wall flows. A constant Courant-Friedrichs-Lewy condition number of 0.5 at a time step of 0.005 s was used to numerically stabilize the progress of the simulation. Radiative heat transfer was solved using the weighted sum of gray gases model in accordance with published recommendations [18].

The solid charge material was represented by a multi-component particle model in the Lagrangian multiphase (LMP) framework. This formulation allowed individual particles to be tracked throughout their life in the furnace. Thus, each charge particle contains a fixed proportion of larger Pb-bearing compounds such as lead sulfate ($PbSO_4$), lead carbonate ($PbCO_3$), lead hydroxide ($Pb(OH)_2$), and recycled Pb, as well as non-Pb-bearing compounds such as sodium sulfate (Na_2SO_4), sodium carbonate (Na_2CO_3), plastic, wood, As, Cd, Sn, Sb, and other trace elements. For this study, only the Pb-bearing compounds were considered in the charge because the primary goal was to study the formation of Pb, Pb(l), and the slag layer of simpler compounds from thermal decomposition of the charge. Carbon content in the coke,

Table 1 Summary of reactions in the burden

Reaction	Activation energy, *E* (J/kmol)	Temperature exponent, *n*	Pre-exponent, *A*	Rate of molar flux (kmol/m^2-s)	Diffusion coefficient (m^2/s)
$PbCO_3 \rightarrow PbO + CO_2(g)$	1.8×10^8	0	1×10^{12}	–	4.5×10^{-5}
$PbSO_4 + 2C \rightarrow PbS + 2CO_2(g)$	–	–	–	1×10^{-9}	–
$Pb(OH)_2 \rightarrow PbO + H_2O(g)$	1.36×10^7	0	1×10^{12}	–	4.5×10^{-5}
$2PbO + PbS \rightarrow 3Pb + SO_2(g)$	–	–	–	1×10^{-9}	–
$2PbO + C \rightarrow 2Pb + CO_2(g)$	–	–	–	1×10^{-9}	–
$Pb \rightarrow Pb(l)$	0	0	1	–	4.5×10^{-5}

moisture in the coke, and moisture in the charge material were also included within the multi-component particle. This setup allowed particle reactions (Table 1) to be simulated as a hybrid of the LMP–EMP framework. The discrete element method model is a subset of the LMP framework and was used to simulate accurate collisions and contacts between particle pairs and between particles and walls [1]. For heterogeneous chemical reactions involving more than one component from the LMP, Simcenter STAR-CCM+ does not allow the use of Arrhenius coefficients.

Results and Discussion

The simulation for OFC of natural gas was run until the outflow temperature stabilizes within ±10 °C. Figure 3 shows profiles of temperature, velocity, and mole fraction across the central axial plane of the furnace. Figure 4 illustrates how the presence of loaded charge material imposes a physical barrier which affects the trajectory and momentum of the flame.

Two-way coupling between Eulerian and Lagrangian phases is relayed by an equivalent Eulerian volume fraction of the Lagrangian phase (charge particles) in the furnace, as seen in Fig. 4. The charge material is denser (~7,500 kg/m^3) than the gas mixture (~1.5 kg/m^3); therefore, it deflects the impinging flame and initiates the breakup of larger eddies, thereby creating a warm front on the burden's exposed perimeter. The modified geometry is not an idealistic representation of true furnace operations, as evidenced by the cold zone created near the mouth of the burner owing to compressibility effects. Some direct heat from the flame is also observed to penetrate voids in the burden and create heating from within the depth. Heat from the flame is transferred to the surface of each multi-component particle and recorded as a surface value. Due to conduction, heat is transferred to particles at the bottom of

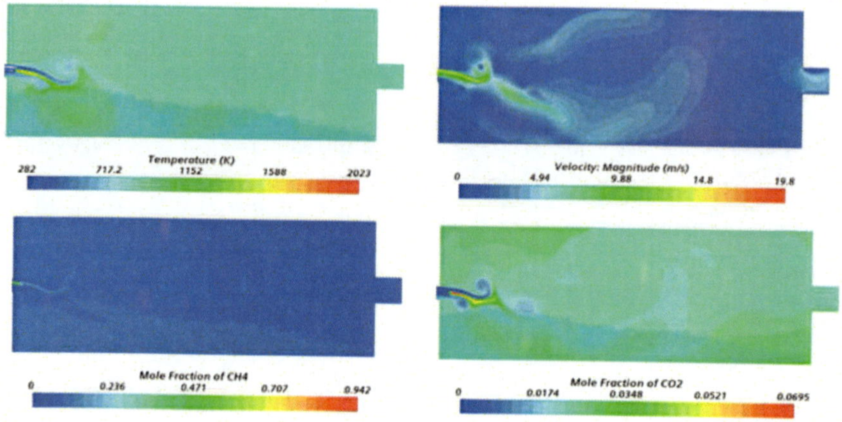

Fig. 3 Profiles of key species from OFC of natural gas

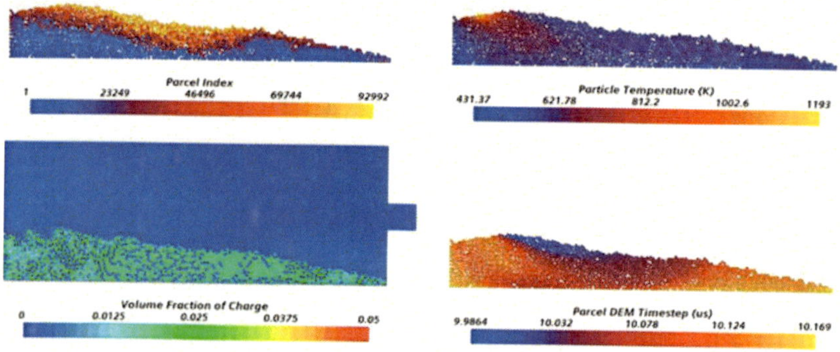

Fig. 4 Profiles of charge particles in the furnace

the burden, where the enthalpy of charge particles increases leading to endothermic chemical reactions and the release of gases, as well as melting of Pb. Much of the heat released by OFC is consumed by the formation of CO_2(g) and H_2O(g). The deflected flame creates recirculation zones above and below the burden that stabilize the flame. The trajectory of the flame continues along the exposed perimeter of the burden and is traced by product gases seen in Fig. 5 and the location of solid products in Fig. 6.

The qualitative distribution of Pb followed that of other solids in the charge. However, the latent heat of fusion was quickly absorbed, resulting in a phase change. The properties of Pb(l) were computed using temperature-dependent functions [19]. Figure 7 shows that Pb begins to melt at the top of the burden, and Pb(l) begins to trickle through voids in the burden, accumulating downstream at the floor of the

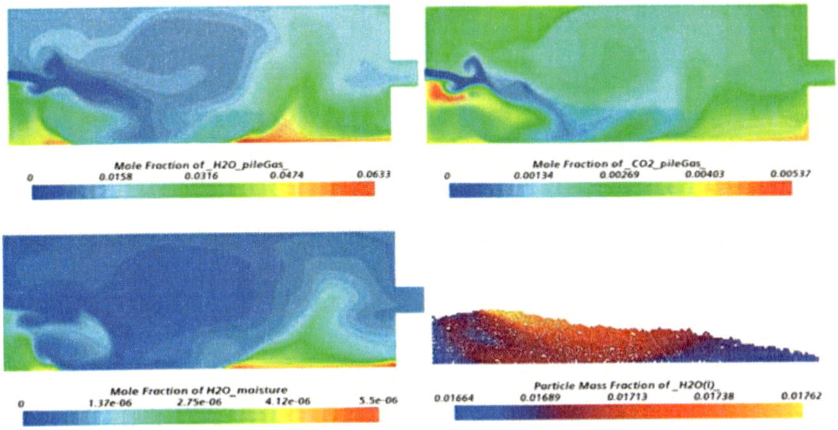

Fig. 5 Profiles of product gases and evaporation of moisture from particles

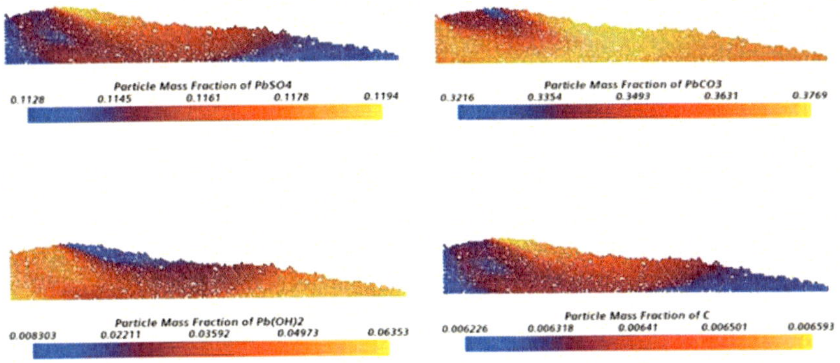

Fig. 6 Profiles of solid species in the furnace

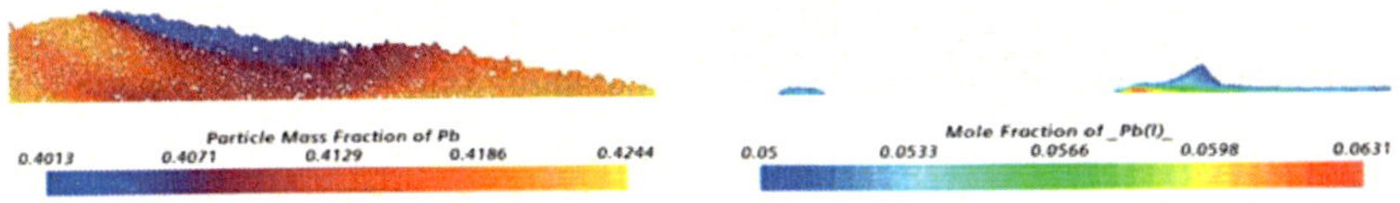

Fig. 7 Profiles of Pb and Pb(l) in the furnace

furnace, eventually forming a continuous layer. Note that, the reduction in mass fraction of Pb in the particle phase also occurs because of simultaneous thermal decomposition of other solids in the charge, which decreases in mass as they are consumed. Eventually, the mole fraction of Pb(l) on the floor of the furnace exceeds 0.9.

Conclusions

The setup, approach, and execution of simulated recovery processes for secondary Pb in a reverberatory furnace are summarized. The furnace was powered by OFC of natural gas, and the heat of reaction was used to smelt Pb-containing solids in the charge. Pb(l) forms by melting Pb in the charge and by melting Pb that was formed as a product of thermal decomposition of larger Pb-containing solids such as $PbSO_4$, $PbCO_3$, $Pb(OH)_2$, PbO, and PbS. Lead sulfate and lead oxide are chemically reduced by carbon in coke added to the charge material, whereas the other solids decompose spontaneously upon absorption of heat equivalent to the activation energy. These simultaneous changes were simulated in Simcenter STAR-CCM+ , and a method to model phase change of a single component from a multi-component particle mixture was implemented. Future work includes acquiring accurate data to characterize the kinetics of thermal decomposition and enhance the level of compatibility between EMP change models and heterogeneous chemical reactions.

Acknowledgements This work was sponsored by the US Department of Energy, Office of Energy Efficiency and Renewable Energy (EERE), Advanced Manufacturing Office, High Performance Computing for Energy Innovation Program (HPC4EI), under contract DE-AC05-00OR22725 with UT-Battelle, LLC. The HPC4EI Program is managed by Lawrence Livermore National Laboratory for the US Department of Energy.

References

1. SIEMENS STAR-CCM+ (2021) STAR-CCM+ Simcenter Documentation 2021.1
2. Anderson A, Grogan J, Bogin G, Taylor P (2018) Computational modeling of a secondary lead reverberatory furnace: effect of burden geometry. Extraction 2018:881–890
3. Anderson A (2017) Computational fluid dynamic modeling of a secondary lead reverberatory furnace. PhD thesis, Colorado School of Mines
4. Malecki S (2014) Thermogravimetric study of the reduction of basic lead sulphate. J Therm Anal Calorium 117:1091–1095
5. Jones WP, Lindstedt RP (1988) Global reaction schemes for hydrocarbon combustion. Combust Flame 73:233–249
6. Leung KM, Lindstedt RP, Jones WP (1991) A simplified reaction mechanism for soot formation in nonpremixed flames. Combust Flame 87:289–305
7. Nicol S, Grogan J, Davis B, Lebel T (2020) Slag reduction kinetics of a lead slag from a secondary lead smelter. In: Siegmund A et al (eds) PbZn 2020: 9th international symposium of lead and zinc processing. The minerals, metals & materials society, Pittsburgh. Springer, New York, p 41–50
8. Schriner D. Taylor P, Grogan J (2016) A review of slag chemistry in lead recycling. In: Advances in molten slags, fluxes, and salts: proceedings of the 10th international conference on molten slags, fluxes and salts (MOLTEN16)

9. Marshall AL, Bruzs B (1925) Heat of formation of lead carbonate. J Phys Chem 29(9):1184–1186
10. Sleeman PJ (1976) Oxidation of lead sulphide. PhD thesis, Plymouth Polytechnic
11. Merrick SJ (1957) A study of the "Roast-Reaction" of lead sulfide. MS thesis, The University of British Columbia
12. Piskulich ZA, Mesele OO, Thompson WH (2019) activation energies and beyond. J Phys Chem A 123:7185–7194
13. Xu Z, Sun X, Khaleel MA (2012) A generalized kinetic model for heterogeneous gas-solid reactions. J Chem Phys 137:074702
14. Jahanshahi S, Wright S (2017) Kinetics of reduction of CaO-FeOx-MgO-PbO-SiO2 slags by CO-CO2 gas mixtures. Metall Mater Trans B 48B:2017–2057
15. Sane A, Gangoli S, He X, Lawrence M, Hewertson R (2018) Effective use of oxy-fuel combustion in aluminum reverberatory furnaces. Light Metal Age 36
16. Yin, C (2011) Advanced Modeling of Oxy-Fuel Combustion of Natural Gas. Aalborg University, ForskEL 2009–1–0256
17. Andersen J, Rasmussen CL, Giselsson T, Glarborg P (2009) Global combustion mechanisms for use in CFD modeling under oxy-fuel conditions. Energy Fuels 23:1379–1389
18. Yin C, Rosendahl LA, Kær SK (2011) Chemistry and radiation in oxy-fuel combustion: a computational fluid dynamics modeling study. Fuel 90:2519–2529
19. Sobolev V (2011) Database of thermophysical properties of liquid metal coolants for GEN-IV. SCK CEN, Belgium. Scientific Report SCK CEN-BLG-1069

Development of Virtual Die Casting Simulator for Workforce Development

John Moreland, Kyle Toth, John Estrada, Junyi Chen, Na Zhu, and Chenn Zhou

Abstract High-pressure die casting is a complex manufacturing process that requires a highly developed work force. A virtual die casting machine has been developed for operators to provide a better understanding of how the machine works and how to deal with a variety of practical situations and issues that arise on the shop floor. Computational fluid dynamics (CFD) simulations have also been developed and integrated into the simulator to help die casters understand how parameters such as shot speed can affect the resulting quality of castings being produced. A virtual melter furnace is also being developed to learn and practice maintenance and safety procedures. The simulator was developed for virtual reality (VR) headsets and controllers, but is also usable on standard PC with mouse and keyboard. Development methodology and overview of simulator functionality will be discussed.

Keywords Virtual reality · Simulator · Training · Die casting

Introduction

High-pressure die casting is a complex and high-stakes process that relies on well-trained operators to produce high-quality parts. Much of the process is hidden from view, and new operators often have difficulty understanding what is happening inside the machine and do not intuitively realize the effect that operating condition changes can have on the resulting castings. To address this issue, a virtual reality simulator is being developed by the Center for Innovation through Visualization and Simulation (CIVS) at Purdue University Northwest in collaboration with the North American Die Casting Association (NADCA).

The die casting machine injects molten metal into a cavity, where it cools, solidifies, and is subsequently ejected and removed by robotic arm. Much of the process is automated, but the operators and maintenance technicians setup, initiate, and monitor

---

J. Moreland (✉) · K. Toth · J. Estrada · J. Chen · N. Zhu · C. Zhou
Purdue University Northwest, 2200 169th Street, Hammond, IN 46323, USA
e-mail: morelanj@pnw.edu

A. Lazou et al. (eds.), *REWAS 2022: Developing Tomorrow's Technical Cycles (Volume I)*, The Minerals, Metals & Materials Series,
https://doi.org/10.1007/978-3-030-92563-5_54

production, troubleshoot issues, and make adjustments as necessary to ensure that the castings produced by the machine are high quality and free of defects.

Virtual simulators enable operators and maintenance personnel to learn, interact, and practice setup, operating, and troubleshooting [1–4]. This virtual simulator enables die casting personnel to improve understanding of the die casting machine, how it operates, and the visualizing the results of simulated castings and changing operating conditions using computational fluid dynamics (CFD) and 3D visualization that are integrated into the simulator.

Methodology

Interactive Simulator

The interactive simulator uses a combination of 3D models, CFD visualizations, and interactive controls to enable users to learn, explore, and practice machine operation. CAD models of the die casting machine were optimized to reduce the complexity of geometry to allow for sufficient frame rate in virtual reality (Fig. 1). The Unity 3D game engine was used to develop the interaction system, placing the user in the die casting environment where they can explore and interact from a first-person point of view.

Controls were originally developed for virtual reality, enabling two-handed interaction with the environment while wearing a VR headset. The controllers are tracked in 3D space, allowing users to reach and touch virtual objects in the environment. Buttons on the controllers allow users to grab objects, turn switches, and operate controls on the die casting control panel and other interfaces within the die casting environment (Fig. 2). The current system is specifically tailored for the Oculus Quest 2 VR system (Fig. 3), but has also been modified to allow users to interact with mouse/keyboard and view on a standard monitor as well.

Fig. 1 Virtual die casting machine, and surrounding environment was modeled based on configurations used in industry, including die casting cell layout, safety guards, and support equipment

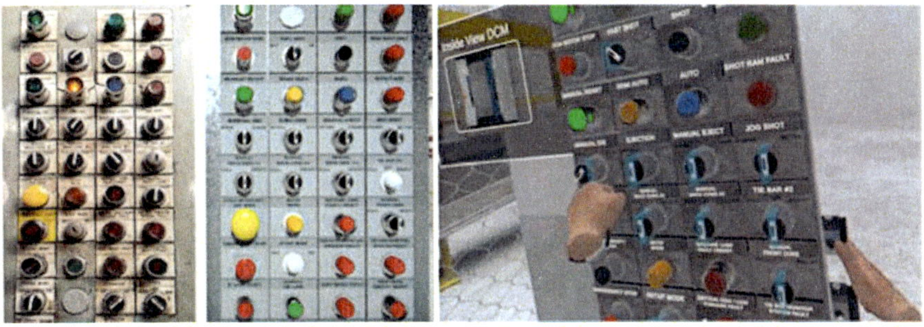

Fig. 2 Die casting control panel (left) was modeled in 3D (center) to control the virtual die casting machine. Users see their virtual hands in VR (right) and use controller buttons to trigger pointing and grabbing gestures to interact with buttons, levers, and toggle switches on the machine

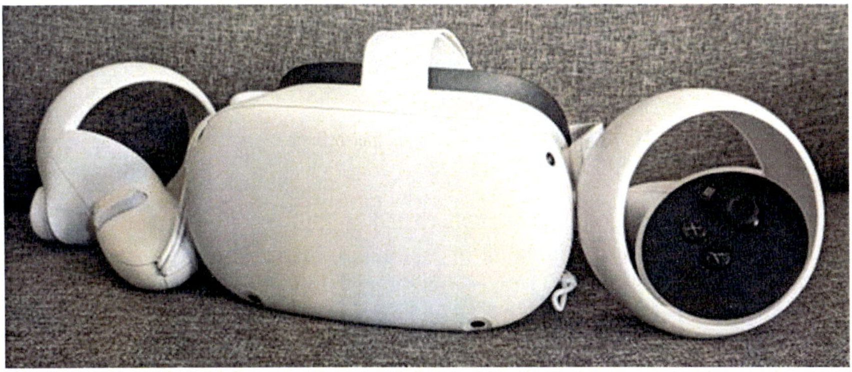

Fig. 3 Oculus Quest 2 VR system enables users to view the virtual environment on a tracked 3D headset and interact with the environment using two tracked 6-DOF controllers

CFD Simulation

Integrating CFD simulations into interactive simulations has been done in a variety of contexts and has been shown to improve understanding of complex processes [5–8]. A computational fluid dynamics (CFD) model of the high-pressure die casting process was developed in the CFD software STAR-CCM+. The volume of fluid (VOF) multiphase approach was adopted to model the transient two-phase (aluminum-air) flow. The computational domain includes the shot sleeve, the runners, and the casting cavity (Fig. 1a). Morphing mesh technique was used to model the movement of the plunger through the shot sleeve. The plunger velocity through the entire process was prescribed to match the three velocity cycles used in the real process, i.e. (a) slow cycle, (b) critical slow shot velocity (VCSS) cycle, and (c) fast cycle. The suitable velocity shot profile was established following the guideline published by NADCA [9]. The simulation describes the flow of aluminum and air during the entire die

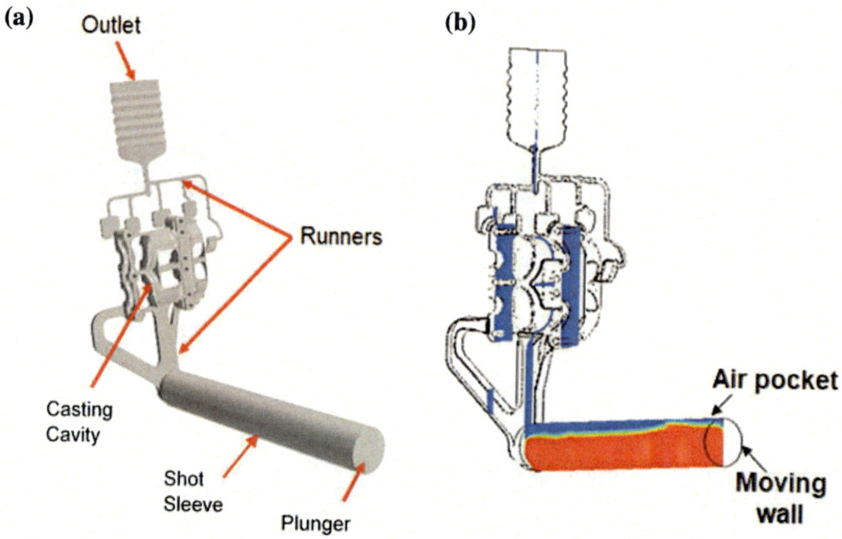

Fig. 4 **a** Computational domain and **b** Aluminum profile in the early stage of die casting process

casting process. The quality of the parts casted in the die is very dependent on the shot parameters, such as the amount of liquid aluminum charge in the shot sleeve and the velocity shot profile. The CFD model developed was used to simulate die casting with good and bad shot parameters. An example in Fig. 4 shows air pocket forms near the plunger due to improper velocity shot profile. This air pocket will later result in air entrapments in the casting, which leads to poor part quality.

Scenarios of different combinations of velocity shot profiles and different amounts of liquid aluminum charged into the shot sleeve to simulate casting with good and bad shot parameter were established. Animations of the liquid aluminum flow generated from these simulations were incorporated training simulator as shown in Fig. 5. The simulator allows users to make the machine transparent to see the internal operation and uses embedded questions to verify that users understand the impact of shot speed and amount of liquid aluminum on the resulting casting.

Results

The virtual die casting simulator includes a variety of functionality, educational, and practice scenarios, and supplemental information which can be launched and reviewed from the main menu (Fig. 6). New users begin with an embedded introduction video and interactive tutorial to teach them how to use the VR controllers to move around in the virtual environment and interact with menus, objects, and controls.

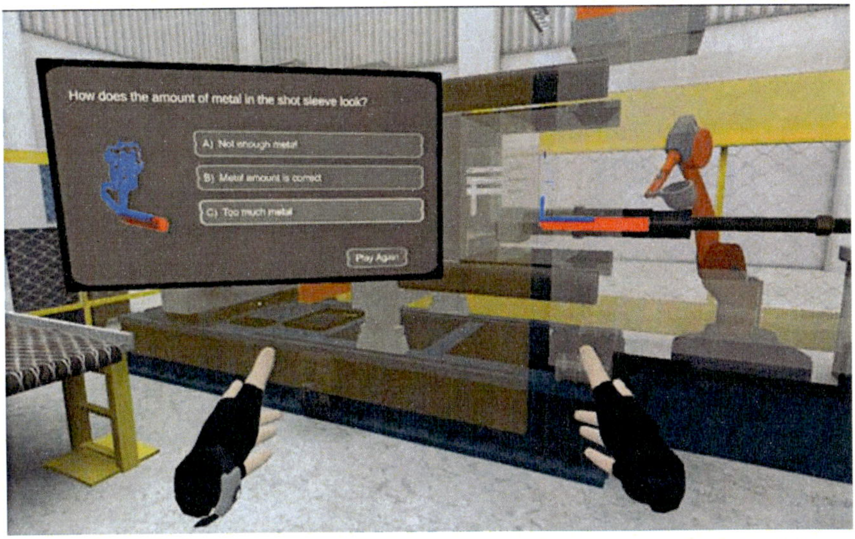

Fig. 5 Animation from CFD simulation embedded in the training simulator

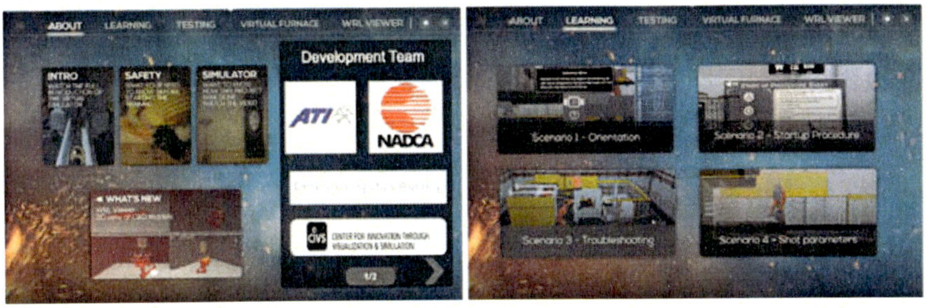

Fig. 6 Die casting simulator provides a variety of interactive scenarios for operators and maintenance personnel to learn about and practice in a virtual die casting environment

Once users have mastered the controls, they are introduced to the die casting machine and melting furnaces through an interactive orientation. During the orientation, users walk around and explore the die casting machine, furnaces, and surrounding equipment. Icons hover in front of important parts of the machine, prompting users to interact with them to learn about their function through pop-up information. Each item is automatically checked off on an orientation checklist, which the user must complete before moving on (Fig. 7).

After the user is familiar with main parts of the machine, learning scenarios teach them step-by-step procedures and operation of controls that startup the die casting machine and begin producing castings (Fig. 7). Troubleshooting scenarios task the user with going through the startup procedure, but introduce simulated issues that they must investigate and remedy in order to produce good parts. Instruction is embedded

Fig. 7 Interactive checklist guides users through orientation where they learn about the die casting machine (left). A startup procedure sheet guides users through the step-by-step process of safety and status checks and control sequences to turn on the die casting machine and begin producing parts (right)

within learning mode to teach users how to deal with issues. The interactive testing mode produces the same issues, but without guidance, verifying whether they can correct the issues and produce quality castings on their own.

Additional scenarios provide users with capabilities not possible in the real world, such as making parts of the machine invisible to see the inner workings, viewing simulated CFD visualizations animating inside the machine, and viewing and interacting with castings and other 3D models in an exploratory fashion that allows the user to get up close and in some cases put their head through solid objects to see the inside of castings.

Conclusion

The virtual die casting simulator provides operators and maintenance personnel with a resource for learning how to operate and troubleshoot the die casting machine, furnaces, and other support equipment. VR enables users to practice simulated scenarios in the safety of a virtual environment and gain experience dealing with issues before interacting with real equipment on the shop floor. The simulator has potential to improve workforce preparation and understanding of how the die casting machine operates and the effect that changing operating conditions can have on the resulting casting quality. Potential impacts include safety, improved part quality, reduced downtime, and improved tool life through better operation of the machine.

Acknowledgements The authors wish to thank DLA, ATI, NADCA, BuhlerPrince Inc., Mercury Marine, Schaefer and RCM, and all other collaborators for their contributions and feedback to the simulator development. This American Metalcasting Consortium (AMC) project is sponsored by the Defense Logistics Agency Troop Support, Philadelphia, PA and the Defense Logistics Agency Information Operations, J62LB, Research & Development (R&D), Ft. Belvoir, VA. The publication of this material does not constitute approval by the government of the findings or conclusion herein. Wide distribution or announcement of this material shall not be made without specific approval by the sponsoring government activity.

References

1. Doolani S et al (2020) A review of extended reality (xr) technologies for manufacturing training. Technologies 8(4):77
2. Choi M, Ahn S, Seo J (2018) VR-based investigation of forklift operator situation awareness for preventing collision accidents. Accid Anal Prev 136(2020):105404
3. Moreland J, Toth K, Zhou C (2018) Troubleshooting and safety simulator for wind turbine technician education. Advances in wind energy: exploring advanced materials & new repair techniques for the wind industry, Boulder, CO, June 5–6, 2018
4. Vergnano A, Berselli G, Pellicciari M (2017) Interactive simulation-based-training tools for manufacturing systems operators: an industrial case study. Int J Interact Des Manuf (IJIDeM) 11(4):785–797
5. Moreland J, Estrada J, Mosquera E, Toth K, Silaen A, Zhou C (2020) Integrating fluid simulation with virtual die casting machine for industry 4.0 and operator training. In: Light metals 2020. Springer, Cham, pp 1026–1031
6. Wang J, Moreland J, Tang G, Okosun T, Silaen AK, Zhou CQ (2017) Using numerical simulation and virtual reality visualization for steelmaking applications. In: Proceedings of 3rd ESTAD 2017 conference. Vienna, Austria
7. Zhou C, Wang J, Tang G, Moreland J, Fu D, Wu B (2016) Integration of advanced simulation and visualization for manufacturing process optimization. JOM 68(5):1363–1369
8. Zhou CQ (2015) Advanced computer simulation and visualization for industry training. In: 11th international conference on modern industrial training. Beijing, China, Oct 23–28, 2015
9. Bakemeyer H, Operating the die casting machine. NADCA Publication E-902. Arlington Heights, IL

Evolution of Process Models to Digital Twins

Alex Holtzapple, Sander Arnout, and Els Nagels

Extended Abstract

Process simulation in the mining and metals industry was introduced in the early 1960s, with the invention of the computer, and has been rapidly evolving since that time. Programming languages and calculation methodologies have been constantly expanding to encompass innovative technology and equipment into metallurgical software for use by engineers and operators. Moreover, with the introduction of graphical interfaces to process simulators, engineers and metallurgists are able to be directly involved in the creation of process flow diagrams that include detailed calculations for mass and energy balances. The next stage of evolution for process simulation has been dubbed "digital twins", though a clear definition of such has yet to be determined.

Simulation itself is a broad term and yields several forms and results, but one key rule is followed: there is a give and take between restricting the model architect to pre-built algorithms and ensuring trustworthy results. Microsoft Excel models, for example, include zero restrictions in their capabilities, but also provide zero assistance in model construction and technical affirmation of the results; the model architect is responsible in full for the content. Other types of simulators available include the aforementioned flowsheet models, as well as statistical-based programs along with any number of independently written functions in a variety of programming languages. For a given mineral processing plant, most value from these efforts is achieved when all models and simulations are in communication, whether directly or indirectly.

A. Holtzapple (✉)
Metsim International, LLC, Churubusco, IN 46723, USA
e-mail: alex@metsim.com

S. Arnout · E. Nagels
ProOpt International, Leuven, Belgium

A. Lazou et al. (eds.), *REWAS 2022: Developing Tomorrow's Technical Cycles (Volume I)*, The Minerals, Metals & Materials Series,
https://doi.org/10.1007/978-3-030-92563-5_55

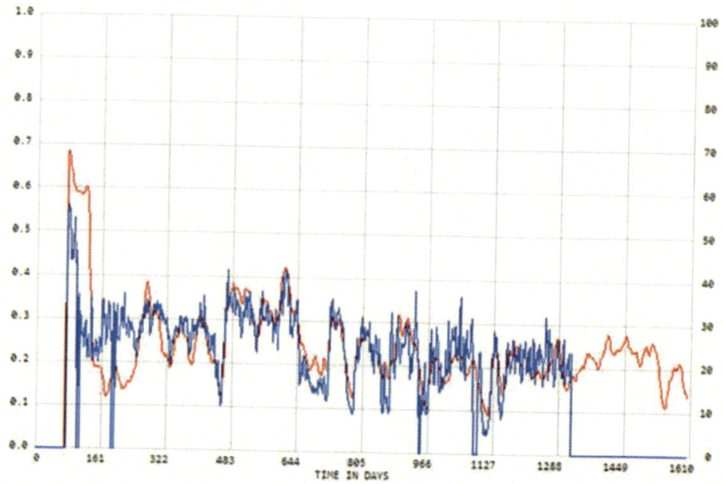

Fig. 1 Dynamic simulation results (red) plotted against actual operational data (blue) over a time period of ~4.5 years. As the model development progressed and further calibrations were included, the error between simulator and operations is decreased (Color figure online)

However, before a model or series of models can be implemented into a digital twin, some key requirements are necessary. Firstly, all models must be properly calibrated against available laboratory, pilot plant, or operational data to ensure that technically sound results are produced. As shown in Fig. 1, operational data should be plotted against simulation results and any areas of major difference addressed. Once calibrated and proven technically sound, the model should then be configured to provide a robust calculation routine; i.e. the model must run the calculations every time when called upon by an external program and provide the results in a timely manner.

The next requirement is an adequate data management system for: (i) collection and storage; (ii) data cleansing; (iii) interfile communication; and (iv) final reporting. One system in place for each of these four categories of data management ensures that no information is lost from one simulator to the next or from one discipline to another. Finally, the human–machine interface (HMI) must be selected for personnel access and display, in line with their responsibilities at the operation. For example, plant operators should have an HMI that is identical to that which is in the control room with limited abilities to adjust key inputs to the models, while the engineering team's HMI permits in-depth changes to the models along with the ability to forecast future operational conditions.

Once these pieces are in place, a master list of "tags" can be defined which represent the data moving through the entire system. Tags generated by one program or instrument may serve as the input tags for another program within the digital twin environment. One master list will ensure that all programs are communicating with the same data and in the same units, eliminating loss of information. Now, completely functional, training of all personnel on the access, use, and benefit of

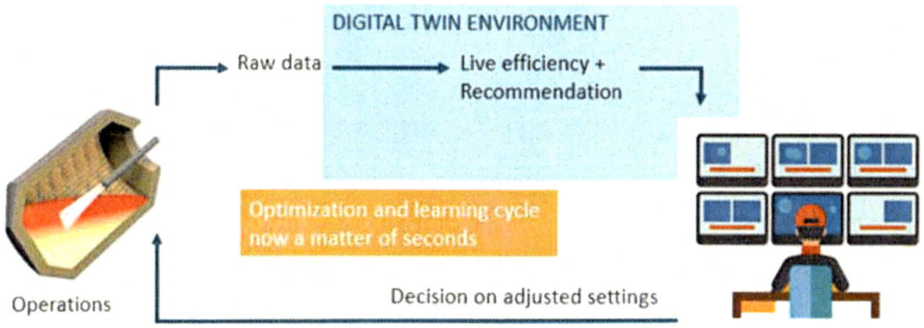

Fig. 2 Data flowchart from operations through the digital twin environment and back

the digital twin is essential, as is the continued maintenance of all systems utilized within its environment.

Figure 2 shows a simple schematic of data flow from the operating equipment through the digital twin environment for analysis and calculation, where results are then sent to the custom HMIs in place where decisions are made and setpoints are adjusted. Digital twins operate live time, whenever the plant is running, allowing for quicker optimization and decision making.

Key impacts on the operation when utilizing a digital twin for faster real-time adjustments and decision making include: better energy efficiency with improved control of flux and reagents; higher production yields due to less raw material waste; operation of major equipment more inline with specifications; decreased environmental incidents by maintaining impurity levels in waste streams and inventory; and finally, safer operations for all personnel when the plant is operating more consistently.

Continued evolution of process simulation models and how they are applied to daily operations is sure to change how mineral processing plants operate worldwide, with several potential advancements. Firstly, off-site control rooms for multiple plants may be constructed in more urban areas which are attractive for talented and experienced operators. Autonomous plant control may even be achieved where digital twin recommendations are proven and safe. Finally, the digital portion of the twin, including all data management, HMI, and simulation components may be constructed in parallel to plant construction, providing a platform for personnel training and likely decreasing the overall commissioning time.

Regardless of the specific programs utilized, constructing a multi-disciplined metallurgical, engineering, and operational team working together through a digital twin ensures reliable results and recommendations in an operation.

Audio Signal Processing for Quantitative Moulding Material Regeneration

Philine Kerst and Sebastian Tewes

Abstract As a natural product with limited resources, sand is of existential importance for various industries. Foundry technology in particular requires considerable quantities worldwide for sand moulds and cores. Depending on the process, the sand is recycled and thus reused. However, especially in the case of inorganically bound moulding sand, it is still frequently not recycled. Existing regeneration methods include mechanical, pneumatic, or combined processes. These have been developed, but have reached their analytical optimisation limits, as the processes are not transparent and make in situ analyses of moulding materials impossible. Within the scope of this research work, a methodology for the computer-aided processing of sound and image data with the help of convolutional neural networks (CNNs) was developed, which is to evaluate the non-measurable changes of the moulding material in the running process in real time via process acoustics. The aim is to optimise process control in terms of time, cost, and energy efficiency.

Keywords Machine learning · Convolutional neural network (CNN) · Audio signal processing · Spectrogram · Process technology · Friction regeneration · Moulding material

Introduction

As a natural resource, silica sand is a finite natural product. In the foundry industry, silica sand is essential for creating sand cores and casting moulds. In order to process this sand for new sand moulds after casting again, several processing steps must be carried out, such as the removal of metal residues and grain separation. Regeneration processes are already established procedures for the further preparation of the

P. Kerst (✉) · S. Tewes
Faculty of Engineering, Institute for Technologies of Metals (Itm), Mathematics for Engineers, University of Duisburg-Essen, Ruhr, Germany
e-mail: philine.kerst@uni-due.de

S. Tewes
e-mail: sebstian.tewes@uni-due.de

A. Lazou et al. (eds.), *REWAS 2022: Developing Tomorrow's Technical Cycles (Volume I)*, The Minerals, Metals & Materials Series,
https://doi.org/10.1007/978-3-030-92563-5_56

sand. Each regeneration process must be individually adapted to the moulding sand system used in foundries, because different moulding sand systems contain different pollutants that must be removed from the system [1]. A distinction is made between three regeneration processes, namely wet regeneration, thermal regeneration, and mechanical regeneration. These processes can also occur in combination to achieve higher quality results specific to the application [1].

The given problem is that the quality of the inorganically bound moulding material in the mechanical regeneration process under consideration can only be determined before or after regeneration using the usual quality criteria of moulding material analysis. These criteria include particle size distribution, bulk density, and strength values. A sample quantity that can be taken during operation through a small opening in the regeneration device is not sufficient to be able to carry out significant quality analyses. In the future, it should be possible to determine this non-measurable parameter of the moulding sand quality during the process to be able to control the process more efficiently. The feasibility of quantitative moulding sand determination during the process offers the possibility of considering environmentally relevant goals such as reduced energy requirements, reduction of CO_2 emissions, conservation of resources.

The focus for the quantitative determination of the moulding material quality lies on process acoustics. It is investigated whether the changes in the moulding material during the process can be detected acoustically. Audio signal processing offers a variety of analysis methods for feature recognition. Foregrounded is the analysis of spectrograms, which visualise the temporal behaviour of the frequency spectrum. With the help of deep learning, i.e., convolutional neural networks (CNNs), which are already successfully used in image recognition, spectrograms are to be analysed. The changes during the process are to be recognised and learned.

In Section "Sustainability", the focus lies on the environmental background, before the concept of research work and the applied methods of signal analysis as well as CNN are presented in the third section. Section "Results" deals with the data basis and analysis results. Section "Future Work" concludes with an overview of future research aspects and a summary in Section "Conclusion".

Sustainability

For silica, silica sand, and silica gravel, there is often no possibility of becoming part of a circular economy. While high-quality aggregates are used in the production of building materials, glass, and glass-fibre-reinforced plastics, among other things, the grains themselves are not reusable or recyclable in their original shape and morphology [2]. In foundry industry, silica sand is used to produce sand cores and moulds and remains available as a raw material during the process; it is, therefore, not consumed. This makes it even more important to focus on this industrial sector where the circular economy is possible. The number of foundries has not changed significantly over the last 15 years. Table 1 lists the number of foundries of selected countries from 2005 and 2019. This means that the raw material demand of finite

Table 1 Year-on-year comparison of the number of foundries in selected countries [3]

Country	2005	2019	Country	2005	2019
China	26,000	26,000	Korea	847	882
India	4500	4500	Germany	630	513
Japan	1708	1738	United Kingdom	422	420
USA	2380	1672	Canada	150	175
Russia	1900	1140	France	505	120
Italy	1077	1036	Switzerland	51	46
Brazil	1264	1017	Austria	45	38

sand resource is constant or even increasing. Added to this, the increasingly global climate targets are important.

A large part of the casting processes is carried out with sand moulds [4]. Silica sand has unique properties such as very high thermal stability, low thermal expansion, sufficient thermal shock resistance, and favourable chemical behaviour [1]. In combination with an application-specific binder system, silica sand can be pressed into sand moulds or sand cores that can withstand the high stresses of a casting process.

To produce moulds or cores, depending on the structure of the moulding material system, between 85 and 98% mould base materials (e.g., silica sands) and 2–10% moulding material binder are used along with possible additives [1]. The moulding material is exposed to high thermal and chemical stresses during a casting cycle. After the casting is removed from the mould, the moulding material remains. Binder residues are still present on the individual sand grains. This remaining moulding material could be mixed with new untreated sand and used again for mould or core production. However, this process is limited, and the moulding material becomes unusable after several cycles [4]. The contamination of the sands can carry significant landfill costs when disposed of, as at least the German waste management industry is run under strict environmental regulations. Possibilities to use foundry sands in other industries or productions are for example [5, 6]:

- As underfill material in road construction,
- In asphalt production,
- For filling slopes,
- As flowable fillings,
- For reinforcement of soils,
- In cement production.

Further information and details on the reuse of waste foundry sand are given in [6, 7]. In addition to unprocessed reuse, the regeneration of moulding sand plays an important role in circular economy. The regeneration of foundry sands is understood as the conversion into a mould base material with properties similar to new sand by removing the remaining binder residues through the regeneration process [1]. According to Bührig-Polaczek et al., it is imperative for ecological and economic reasons to regenerate old moulding materials in order to ensure their reuse for the production of cores or moulds [1]. As early as 1993, Flemming and Tilch also pointed out the necessity of regenerating moulding materials with the following arguments [1]:

- The availability of suitable silica sands is decreasing; in addition, the effort for extraction and preparation is increasing,
- Rising costs for the removal and storage of used sands and bulk materials in landfills,
- Emerging problems due to increasing waste volumes and legal requirements for landfill storage,
- Due to the low expansion rate of silica and the saving of binders, the use of reclaim can lead to technological advantages,
- Reuse of foundry sands is an effective protection of the environment and raw material resources.

In addition to moulding material preparation itself, optimisation of regeneration process in terms of energy efficiency and pollutant emission can also contribute significantly to sustainability. Worldwide, greenhouse gas emissions have increased from 30.74 billion tonnes (1990) to 46.77 billion tonnes of CO_2 equivalent (2017) [8]. In terms of the German industrial sector, greenhouse gas emissions peak for the iron and steel industry at 31.4 million tonnes of CO_2 equivalent in 2020 [9].

Concept and Methods

Setup

The process under consideration is a mechanical regeneration process with a so-called fluidised bed cleaner for inorganically bound moulding material. The equipment consists of a round body and a motor-driven shaft inside. Symmetrically attached to this shaft are friction bars. The mechanical regeneration of the moulding material takes place through the rotation of the friction tools and an additional vertical air flow, which flows through the system from bottom to top. A graphic representation of the system is shown in Fig. 1 (left). Frictional energy is generated between the sand grains of the moulding material, the moulding material, and the friction tools as well as between the moulding material and the shell of the plant during operation. The resulting thermal energy embrittles the binder shells, which are knocked off during operation and partly already removed via the exhaust air.

For comparability of the results, the friction generator is operated with a constant batch weight of 30 kg and always with constant parameter settings. Figure 1 (right) shows the process parameters that can currently be called up and saved. These include four key figures of the frequency converter (triangle), a total of nine temperature sensors (circle), as well as one vibration sensor (rhombus) and one flow sensor each

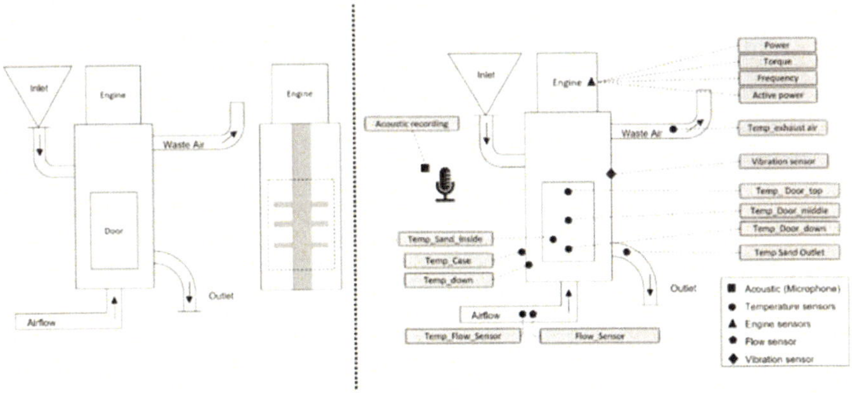

Fig. 1 Graphical representation of the friction generator (left) sensor system of the friction regenerator (right)

for measuring the incoming air volume (pentagon). In addition, each regeneration process is recorded with a microphone [Anua MIC- 900, USB capacitor microphone] (square).

New sand-like properties are the goal of any regeneration process. The central research question is: How long must the moulding material be regenerated to meet this goal? Typically, design of experiments (DoE) is set up for such problems, which involve many experiments to approximate an optimal regeneration time. However, the focus of this paper is on the acoustic signal. The process acoustics are investigated to determine whether the regeneration progress can be indirectly determined acoustically. Thus, the research question posed above can be reformulated as follows: When does moulding sand sounds like new sand again? To answer this question, each regeneration is recorded as a 16-bit stereo signal with a sampling rate of 44,100 Hz. The temporal subdivision of the recording belongs to the post-processing or preparation for the analysis with methods of audio signal processing.

Audio Signal Processing

General audio processing is still limited to rather basic signal models due to diverse and broad nature of audio signals. Audio classification is part of the larger problem of processing audio–visual data with important applications in digital libraries, professional media production, education, entertainment, and surveillance. Speech and speaker recognition can be considered a classic problem in audio retrieval and have attracted research attention for decades [10–12].

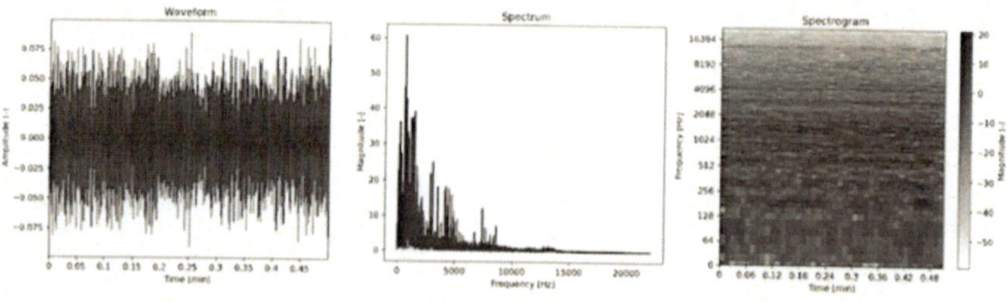

Fig. 2 Waveform (left). Spectrum (middle). Spectrogram (right)

The acoustic properties of sound events can be visualised in a time/frequency diagram of acoustic signals. In order to switch from classical waveform (amplitude/time) to a frequency/magnitude representation first, the fast Fourier transform (FFT) is performed. This generates the so-called magnitude spectrum. With the short-time Fourier transformation (STFT), time, frequency, and amplitude (due to colour) of the signal are represented graphically; this is called a spectrogram. Further information and the mathematical background to the transformations can be found in Chap. 2 of the following source [13].

Figure 2 shows the three different ways of representing an acoustic signal. A 0.5 s audio signal of the regeneration process is used. The signal is processed with the Python package *librosa* [14] with a frame length of 1024 and a hop length of 512.

Convolutional Neural Networks (CNNs)

CNNs form the cornerstone of deep learning (DL) methods and thus belong as a subfield to machine learning [15]. Like shallow neural networks, a CNN has several layers. A layer typically comprises three phases. The first phase performs multiple convolutions simultaneously to produce a series of linear activations [16]. In this process, the inputs of the layers are filtered to find useful features in these inputs. This filtering process is called convolution, hence the name of this type of neural network [11]. In the second phase, a nonlinear activation function is applied to each linear activation, such as the ReLU function (rectified linear activation function). This phase is also called the detector stage. In the last stage, a so-called pooling function is used to further modify the output of the layer [16]. The input data of a CNN can be, for example, (x, y, z) images, where x and y are the width and height of an image in pixels and z is the number of colour channels of each pixel. The output of a CNN corresponds to a vector whose length is equal to the number of categories to be classified [12].

Convolutional Layer

The goal of a convolutional layer is to extract features from the input and transfer them into feature maps, which are then passed on to the next layer, regardless of where the feature is located on the input image [12]. In image processing, two-dimensional convolutions are used to implement image filters. These are important, for example, to find a particular spot on an image or to find a particular feature in an image. The convolution layers tile an input sensor with a small window called a kernel. A kernel defines exactly the things a convolution operation should filter for and produces a strong response when it finds what it was looking for. The parameters learned in a convolutional layer correspond to the weight of the kernel of a layer. As CNN trains, the values of these filters are automatically adjusted to extract the most useful information for the task at hand [11].

Pooling Layer

The task of a pooling layer is to represent and reduce the spatial dimension of the activation tensor, but not the volume depth, in order to reduce the number of parameters and computations in the network. This is done in a non-parametric way, i.e., the pooling layer has no weights [11]. Specifically, one cell is created for each tile. This strategy allows, on the one hand, to keep control over the overfitting and, on the other hand, to adjust the computations in the memories. One of the most common pooling layers is max-pooling, which uses the max-operation, i.e., the maximum value for all nodes in their convolutional window. For image processing, the most common form of max-pooling is a pooling layer with filters of size 2×2. In this case, each max-pooling operation generates a maximum over 4 numbers (a 2×2 region). The depth dimension remains unchanged [15]. This procedure leads to the fact that the resolution becomes coarser with increasing depth of the CNN, but no information of the input data is lost in the CNN [12].

Fully Connected Layer

The fully connected dense layer serves as a classifier. Operations such as convolutional layer, pooling layer, and activation function layer map the raw data to the hidden feature space. The fully connected layer is then used to map the learned features to the sample labelling space [17]. The CNN is modelled in Python using Keras and TensorFlow, respectively, as the interface for recasting and executing machine learning algorithms [18].

The combination of audio signal processing with machine learning methods is already known. For example, audio signal processing methods have been used in medical research to distinguish wet and dry coughs based on the audio profile [19].

Image recognition with convolutional neural networks also supports areas in radiology [20]. The combination of audio signal analysis as well as deep learning has been used to detect snoring [21]. In addition to datasets containing environmental sounds or musical instruments, the first dataset to publish acoustic recordings of four different industrial machine sounds appeared in 2019. The data is available for research in the field of audio event recognition, especially for machine noises [22].

Results

For our experimental setup, three separate batches of new sand were loaded with an inorganic binder system. After the bonded grains had been separated, the moulding material was regenerated. The regeneration time was varied between the three batches: The first batch was regenerated for 5 min, the second one for 10 min, and the third one for 15 min. For each of these three test series, the mixing process (silica sand with binder) and the regeneration process were repeated ten times. The first and the last two minutes of the acoustic recordings from this procedure were then cut into 1-s partial recordings, resulting in a dataset of 1200 1-s records per phase (beginning phase or end phase of the process) per test series, out of which all records with interfering noise were removed. Each of the remaining records was then transformed into its spectrogram. Before any of these spectrograms were passed on to the CNN model, 10% of the data was randomly separated as test datasets. The remaining learning data was divided into 70% training and 30% validation data.

The used sequential CNN model starts with four layers, each with one convolutional layer and one pooling layer. The filters and step size remain constant for the layers. Zero-padding ensures that the image size also remains constant after each step. The output of these convolutional layers is then passed to the fully linked layers as a one-dimensional array. The softmax activation function for calculating probabilities for classification in the final stage of prediction is used in the last layer. Dropout layers are used only in part due to the relatively small size of the datasets. Keras offers eight optimisers (SGD, RMSprop, Adam, Adadelta, Adagrad, Adamax, Nadam, and Ftrl) that change the weights of the neurons after each epoch to minimise losses. For all three test lines, Agagrad delivered the most accurate results (cf. Table 2). The described model was trained in combination with different hyperparameters, and the prediction results were compared. Table 2 shows the specifications of the respective models that delivered the best result. Table 3 contains the corresponding confusion matrices, with Class 0 representing the beginning and Class 1 as the end phase of the process. The rows of the confusion matrix represent the actual labels, whereas the columns represent the predicted labels [11]. In particular, for the 5-min test series, 42.5% of the beginning phase records were correctly classified as such, while 2.5% of the end phase records were incorrectly predicted to stem from the beginning phase.

The accuracy of these predictions suggests that the beginning and the end of the regeneration process can indeed be distinguished using only the acoustic data from

Table 2 CNN models for 5-, 10-, and 15-min regeneration

Time of regeneration (min)	Optimizer	Dropout (%)	Kernal regularization	Training/validation	# of epochs	Batch size	Accuracy (%)	Stratification
5	Agagrad	No	Yes	70/30	50	20	0.90	Yes
10	Agagrad	0.5	Yes	70/30	50	20	0.82	Yes
15	Agagrad	0.5	Yes	70/30	20	50	0.88	Yes

Table 3 Confusion metrices of the prediction of 5, 10, and 15 min

	5 min		10 min		15 min	
	0	1	0	1	0	1
0	42.5%	7.5%	32.5%	17.5%	37.5%	12.5%
1	2.5%	47.5%	0.0%	50.0%	0.0%	50.0%

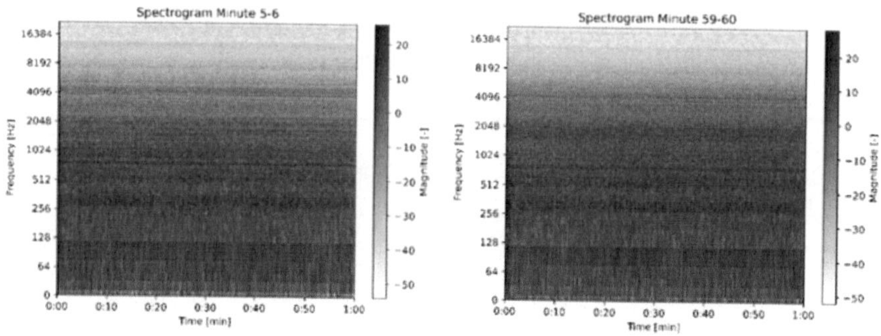

Fig. 3 Spectrograms of the first and last phases of one-hour regeneration

the process, even for relatively short regeneration times. The exact influence of the regeneration time on the quality of moulding material will need to be investigated in further tests. The following question needs then to be answered: When does moulding material sound like new sand again? For this purpose, new sand is regenerated and recorded for a total of one hour. Figure 3 shows the spectrograms of one minute from the beginning (left) and from the end phase (right) of the regeneration process. In the frequency range around 2048 Hz, the right spectrogram shows a recognisable change. In addition to the abrasion of the binder, the regeneration process always involves a certain amount of quartz abrasion, so that a change of the actual grain form or grain size takes place. The extent to which these changes are significant and must be considered is to be investigated in the further work of the research.

Future Work

Based on the fact that the grain morphology changes can be detected in the long-term process, it should be investigated whether a CNN can learn these differences. If this proves to be feasible, it should then be analysed at which points in time these changes emerge significantly from the data. In the case of temporal undercutting, time classes can be delimited via the regeneration of new sands, and a higher information content can be generated. Otherwise, the data in their entirety form builds the basis for the acoustic comparison of bonded and unbonded silica sand.

Another main aspect of future research will be the optimisation of individual (hyper-)parameters. This involves both, data processing in audio signal processing and fine-tuning of existing, as well as newly created models of the CNN.

The acoustic recordings have been transformed as 1-s-long snippets and passed to the CNN model. It is to be investigated whether there is a more optimal length of the audio recordings. Furthermore, the window size, the size of the individual frames, and their overlapping play a role. The representations of spectrograms for visualising audio content have a high dimensionality. More ideal would be low-dimensional features extracted from the spectrograms or even directly from the acoustic signal that contains only the important distinguishing features of the intended audio classes [10]. These features have to be examined, exactly those which can extract the essential differences of the audio signals. Which are the "best" features always depends on the application area as well as the way the signals are processed [23]. For the classification of audio signals, the following research papers and the features they contain will be used as suggestions [13, 24–26]. Selected features can be transferred to so-called feature vectors, which are also machine learning methods, and the features can be learned.

In addition to the hyperparameter optimisation of existing models or models to be created from scratch, it is planned to investigate whether equally good or even better results can be achieved with transfer learning. Some papers prove the usefulness of already trained networks, such as VGG16, VGG19 [27–30], GoogleNet, MobileNet, in which the basic structure with the learned weights is used, and only the last layers are adapted.

Conclusion

In this paper, a concept was presented to record process parameters that cannot be measured directly by means of acoustic signals. Designed for the regeneration process of inorganically bound moulding materials, the focus is on determining the moulding material quality during the process. The process recordings are prepared with methods of audio signal processing for the use of machine learning methods. Spectrograms of 1-s-long partial recordings of the process are used for transfer to a specially created CNN model. The main result shows that the initial and final phases of a regeneration process of moulding material can be captured and learned by the CNN model. In future experiments, new sand will be regenerated, recorded, and examined in order to answer the question posed at the beginning: when does moulding sand sound like new sand? Preliminary results show that changes in the acoustic signal during the process can be detected in the spectrograms. The aim is to significantly classify and learn these changes. With this comparative data, the regeneration process can be stopped exactly when the moulding material sounds like the original material again.

References

1. Flemming E, Tilch W (1993) Formstoffe und Formverfahren: Mit 118 Tabellen, Leipzig, Dt. Verl. für Grundstoffindustrie
2. Bührig-Polaczek A, Michaeli W, Spur G (2014) Handbuch Urformen, 2nd edn. Hanser, München
3. BGR—Bundesanstalt für Geowissenschaften und Rohstoffe (2020): Deutschland—Rohstoffsituation 2019, https://www.bgr.bund.de/DE/Themen/Min_rohstoffe/Downloads/rohsit-2019.pdf?__blob=publicationFile&v=. Accessed 24 Aug 2021
4. content.digitalpub.blue-soho.com, Anzahl der Gießereien nach ausgewählten Ländern weltweit im Jahresvergleich 2005 und 2019. https://de.statista.com/statistik/daten/studie/421564/umfrage/anzahl-der-giessereien-nach-ausgewaehlten-laendern-weltweit/. Accessed 11 Aug 2021
5. Doğan-Sağlamtimur N (2018) Waste foundry sand usage for building material production: a first geopolymer record in material reuse. Adv Civil Eng 1–10. https://doi.org/10.1155/2018/1927135
6. Bakis R, Koyuncu H, Demirbas A (2006) An investigation of waste foundry sand in asphalt concrete mixtures. Waste Manage Res: The J Int Solid Wastes Public Cleansing Assoc, ISWA 24(3):269–274.https://doi.org/10.1177/0734242X06064822
7. Siddique R (2008) Waste materials and by-products in concrete: with 174 tables. Springer, Berlin
8. Meeran Rawther Salim P, Siva Rama Krishna Prasad B (2020) A review on the usage of recycled sand in the construction industry, in sandy materials in civil engineering—usage and management, S. Nemati und F. Tahmoorian, Hg., IntechOpen. https://doi.org/10.5772/intechopen.92790
9. WIR, Höhe der weltweiten Treibhausgasemissionen in den Jahren 1990 bis 2017 (in Milliarden Tonnen CO_2-Äquivalent). https://de.statista.com/statistik/daten/studie/311924/umfrage/treibhausgasemissionen-weltweit/. Accessed 11 Aug 2021
10. Umweltbundesamt und Deutsche Emissionshandelsstelle, Treibhausgasemissionen des Industriesektors in Deutschland nach Branchen im Jahr 2020 (in Millionen Tonnen CO_2-Äquivalent). https://de.statista.com/statistik/daten/studie/1078829/umfrage/treibhausgasemissionen-der-deutschen-industrie-nach-branchen/. Accessed 11 Aug 2021
11. Prasad B (2008) Speech, audio, image and biomedical signal processing using neural networks. Springer, Berlin
12. Zafar I, Tzanidou G, Burton R, Patel N, Araujo L (2018) Hands-on convolutional neural networks with TensorFlow: solve computer vision problems with modeling in TensorFlow and Python. Birmingham: Packt Publishing Ltd. https://ebookcentral.proquest.com/lib/gbv/detail.action?docID=5504396
13. Liu Y et al (2016) Application of deep convolutional neural networks for detecting extreme weather in climate datasets. http://arxiv.org/pdf/1605.01156v1
14. Peeters G (2021) A large set of audio features for sound description (similarity and classification) in the CUIDADO project. http://recherche.ircam.fr/anasyn/peeters/ARTICLES/Peeters_2003_cuidadoaudiofeatures.pdf. Accessed 20 Aug 2021
15. McFee B, Metsai A, McVicar M, Balke S, Thomé C, Raffel C et al (2021) librosa/librosa: 0.8.1rc2, Zenodo. Accessed 22 Aug 2021. https://doi.org/10.5281/zenodo.4792298
16. Brunton SL, Kutz JN (2019) Data-driven science and engineering. Cambridge University Press
17. Goodfellow I, Bengio Y, Courville A (2018) Deep Learning: Das umfassende Handbuch: Grundlagen, aktuelle Verfahren und Algorithmen, neue Forschungsansätze, Frechen
18. Yang J, Li J (2017) Application of deep convolution neural network. In: 2017 14th international computer conference on wavelet active media technology and information processing (ICCWAMTIP). Chengdu, p 229–232. https://doi.org/10.1109/ICCWAMTIP.2017.8301485
19. TensorFlow Developers (2021) TensorFlow: Zenodo. Accessed 22 Aug 2021. https://doi.org/10.5281/zenodo.5645375

20. Chatrzarrin H, Arcelus A, Goubran R, Knoefel F (2011) Feature extraction for the differentiation of dry and wet cough sounds. In: 2011 IEEE international symposium on medical measurements and applications (MeMeA). Bari, Italy, p 162–166. https://doi.org/10.1109/MeMeA.2011.5966670
21. Yamashita R, Nishio M, Do RKG, Togashi K (2018) Convolutional neural networks: an overview and application in radiology, Insights into imaging, Jg. 9. Nb 4:611–629. https://doi.org/10.1007/s13244-018-0639-9
22. Xie J et al (2021) Audio-based snore detection using deep neural networks. In: Computer methods and programs in biomedicine, Jg. 200, p 105917. https://doi.org/10.1016/j.cmpb.2020.105917
23. Purohit H et al (2019) MIMII dataset: sound dataset for malfunctioning industrial machine investigation and inspection. http://arxiv.org/pdf/1909.09347v1
24. Wyse L (2017) Audio spectrogram representations for processing with convolutional neural networks. http://arxiv.org/pdf/1706.09559v1
25. Tzanetakis G, Cook P (2002) Musical genre classification of audio signals, IEEE Trans. Speech Audio Process., Jg. 10. Nb 5:293–302. https://doi.org/10.1109/TSA.2002.800560
26. Breebaart J, McKinney MF (2004) Features for audio classification in philips research, algorithms in ambient intelligence. In: Toolenaar F, Verhaegh WFJ, Aarts E, Korst J, Hg., Springer Netherlands, Dordrecht, pp 113–129. https://doi.org/10.1007/978-94-017-0703-9_6
27. Bhalke DG, Rama Rao CB, Bormane D (2016) Hybridisation of Mel frequency cepstral coefficient and higher order spectral features for musical instruments classification, archives of acoustics, Jg. 41. Nb 3:427–436. https://doi.org/10.1515/aoa-2016-0042
28. Simonyan K, Zisserman A (2014) Very deep convolutional networks for large-scale image recognition. http://arxiv.org/pdf/1409.1556v6
29. Choi K, Fazekas G, Sandler M, Cho K (2017) Transfer learning for music classification and regression tasks. http://arxiv.org/pdf/1703.09179v4
30. Tammina S (2019) Transfer learning using VGG-16 with deep convolutional neural network for classifying images. IJSRP, Jg. 9, Nb. 10, p 9420. https://doi.org/10.29322/IJSRP.9.10.2019.p9420

Determining the Bubble Dynamics of a Top Submerged Lance Smelter

Avinash Kandalam, Michael Stelter, Markus Reinmöller, Markus A. Reuter, and Alexandros Charitos

Abstract Top submerged lance (TSL) smelters are widely used in non-ferrous metallurgy to extract various primary and secondary materials. The technology has found wide application with regard to copper, lead and zinc, nickel, tin, while applications concerning iron and municipal solid waste treatment have been reported. As of 2019, there are about 66 known operating TSL plants globally. By controlling the air/fuel ratio (i.e. by lambda value), the TSL can be operated under oxidizing/reducing/inert conditions. The bubble dynamics play a crucial role in determining the reaction kinetics, splashing, turbulence, sloshing as well as the refractory and lance wear. This paper shows the efforts to determine the bubble dynamics in a cold TSL model using acoustic measurements and high-speed photography. The results show a correlation between bubble dynamics with respect to varying lance submersion depths, bath properties (varying viscosities and densities, i.e. glycerol/water mixtures), and lance flow rates (i.e. airflow).

Keywords Top submerged lance (TSL) · Extractive metallurgy · Acoustic measurement · Bubble dynamics · High-speed photography

Introduction

J. M. Floyd invented the top submerged lance (TSL) technology in the 1970s at the Commonwealth Scientific and Industrial Research Organization (CSIRO), Australia. Floyd aimed to create a flexible smelter with faster reaction kinetics and to recover metals from the slags. The initial TSL operations were used to recover tin from tin-containing slags from the reverberatory furnaces at Associated Tin Smelters (ATS),

A. Kandalam · M. Reinmöller
Institute of Energy Process Engineering and Chemical Engineering, TU Bergakademie Freiberg, Fuchsmühlenweg 9 D, 09599 Freiberg, Saxony, Germany

A. Kandalam · M. Stelter · M. A. Reuter · A. Charitos (✉)
Institute of Nonferrous Metallurgy and Purest Materials, TU Bergakademie Freiberg, Leipziger Strasse 34, 09599 Freiberg, Saxony, Germany
e-mail: alexandros.charitos@inemet.tu-freiberg.de

A. Lazou et al. (eds.), *REWAS 2022: Developing Tomorrow's Technical Cycles (Volume I)*, The Minerals, Metals & Materials Series,
https://doi.org/10.1007/978-3-030-92563-5_57

Australia. Later, due to its flexibility and efficiency, the TSL was used to recover various base metals such as copper, lead, zinc, nickel, iron, and aluminum. Due to enormous developments in TSL design over the years, equipment, and slag chemistry, today, TSL is also used to process municipal waste incinerator ash (MWIA) and recover PGM's, minor elements from slags and secondary resources (e.g., e-waste). The mechanism of the TSL technology is based on a lance carrying fuel and air/oxygen that is submerged into molten slag, where the combustion gases react with the bath/feed. The fuel can be of different forms, such as heavy fuel oil (HFO), natural gas, or milled coal. The air can be enriched with oxygen up to 80% to increase the combustion potential and oxygen activity/affinity/partial pressure. Apart from its fuel flexibility and faster reaction kinetics, the SO_2 and dust emissions from the TSL are lower than the traditional smelters, making it a safe operation for a clean environment. Further aspects of the TSL plant flowsheets, slag chemistry, and design can be read in [1–6] and are out of the scope for this article [7, 8].

This article focuses on understanding and predicting one of the most critical and challenging aspects of the TSL smelter—"bubble dynamics". The authors have published [9] preliminary results on acoustics and lance motion system in a water bath model at the European Metallurgical Conference in 2021. In general, TSL lance is considered as the heart of the technology through which the smelting of feed takes place. When the lance is vertically submerged into the molten bath, the combustible gases (CG) react with different interfaces such as CG-liquid, CG-solid, and CG-gas. These interactions will lead to oxidation or reduction reactions, energy transfer, and intense agitation of the slag layer (turbulence, splashing, and sloshing). Whenever the CG-liquid interacts (i.e. gas bubble surrounded with molten bath), bubbling occurs, and the gas surface area reacts with the molten bath. Therefore, the reaction kinetics are higher when more bubbles (i.e. large surface area) are generated (excluding the influence of external forces like turbulence). A schematic representation of this phenomenon is shown in Fig. 1. The left-side image shows the pilot-scale TSL smelter at the Institute of Nonferrous Metallurgy and Purest Materials (INEMET), TU Bergakademie Freiberg, Germany, and the right-side image projects the simplistic representation of bubble dynamics.

Although there have been significant improvements in the TSL smelter over the last five decades, the bubble dynamics is still a black box. The complexity in investigating the bubble dynamics in a real TSL plant is due to its harsh environment (i.e., off-gas dust, high temperatures, and splashing). These factors eliminate the usage/flexibility of traditional measurement techniques such as photography, laser/X-ray scanning, or sonar. In this article, a method of measuring the bubble dynamics using acoustics is demonstrated using a cold model. The data is further validated using high-speed photography.

On the contrary to experimental investigation, there has been a significant amount of work in computational fluid dynamics (CFD) modeling. These models give a better and deeper understanding of the bubble dynamics, but they are not validated with the industrial TSL plants. For example, within this project framework under virtual high-temperature conversion (Virtuhcon) at TU Bergakademie Freiberg, a significant amount of CFD work is carried out by D. Obiso [10–14], which is focused on TSL

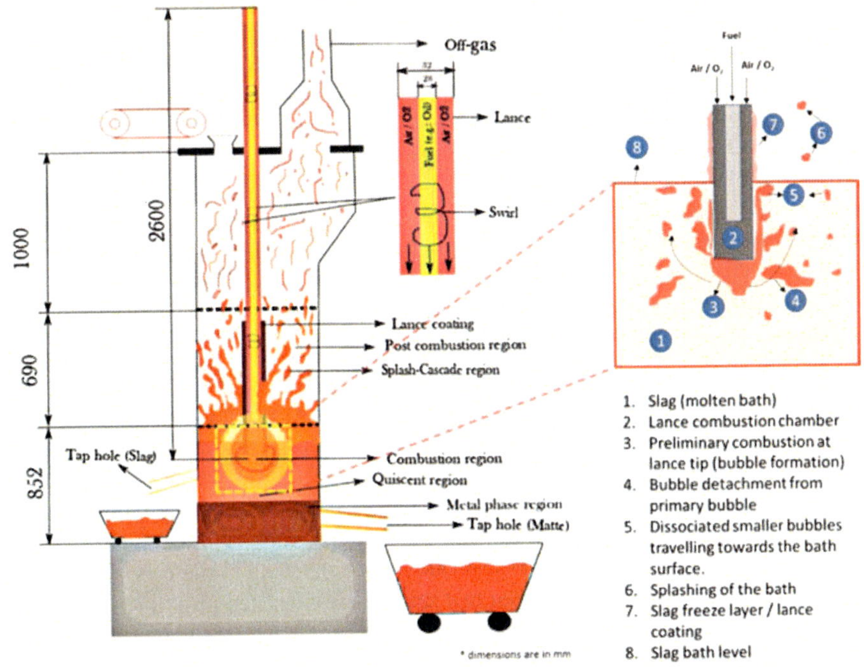

Fig. 1 Left: Schematic representation of the pilot-scale TSL smelter at Institute of Nonferrous Metallurgy and Purest Materials (INEMET), TU Bergakademie Freiberg, Germany. Right: Zoom-in illustration of bubble dynamics at the lance tip and bath surface

bubble dynamics and its influencing parameters. One of the latest developments [15] in the experimental investigation is visualizing the bubble dynamics in an opaque bath (i.e. Galinstan, GaInSn, a ternary alloy) using X-ray radiography. This methodology opens the possibility of investigating the bubble dynamics of different opaque viscous liquids and validating CFD models for better accuracy. Nevertheless, this technique is not feasible in a real TSL plant due to its high costs and safety concerns.

In this article, a method of determining the bubble dynamics is investigated on a cold model using acoustic measurements and high-speed photography. This methodology seems to overcome the aforementioned challenges and can also be adapted to real TSL plants because of contact-free measurement. The patented Metso Outotec LanceMotion (results not shown/discussed in this article) and Smelter Acoustics system used was investigated by Metso Outotec, Australia, over many years, starting at a small scale, resulting in adoption by some commercial users of TSL over the past few years. Their research and commercial plant work are undisclosed at this time. The authors thank Metso Outotec for providing the LanceMotion and Smelter Acoustics system to INEMET together with the Metso Outotec advanced control technologies (ACT) platform and sharing their expertise.

Experimental Set-Up

The results shown in this article are conducted on a cold model. The dimension of the cold model (i.e. cylinder and lance diameter) is similar to the geometry of the small-scale TSL smelter at INEMET (see Fig. 2, left). The inner diameter of the TSL cylinder (cold model) is 22 cm and is made of polymethyl methacrylate (PMMA). The inner diameter of the lance is 2 cm, and it is made of chrome steel. Air flows through the lance via a manual rotameter which can measure a range between 0.8–50 L/minute. The inlet air pressure is set around 1 bar.

The sensors set-up (LanceMotion and Smelter Acoustics system) and the ACT software are provided by Metso Outotec as a package within the CIC-Virtuhcon's Phase II strategic investment project framework. A LanceMotion sensor is attached to the lance to register any movement in the X, Y, and Z axes. The sampling rate of the motion sensor is fixed to 600 Hz. The motion sensor is coupled to Metso Outotec ACT software to generate JavaScript Object Notation (JSON) files, which store the lance movement in all directions with respect to time (results not discussed in this article). The high-speed camera "Imager MX" is from LaVision company, and the images are recorded at 600 fps (i.e., 600 Hz). The camera is connected to the LaVision image processing software, from which the raw images are extracted [16]. The condenser microphone "ECM8000" is from Behringer company [17]. The frequency response of the microphone is calibrated between 15 Hz and 20 kHz (with a sensitivity of −60 dB); however, the microphone can also record infrasound frequencies (i.e., <20 Hz) with less sensitivity. The microphone is connected to the analog to digital converter (ADC) manufactured by Focusrite, and the model is "Scarlett 18i20" [18].

Fig. 2 Left: Laboratory/small-scale TSL smelter at the Institute of Nonferrous Metallurgy and Purest Materials (INEMET), TU Bergakademie Freiberg. Right: Geometrically similar TSL cold model set-up

The data is processed through ACT software from the ADC, and the raw waveform audio file (WAV) files are stored in the computer.

Results

The bubble dynamics of the cold model are investigated under the following types of liquids:

a. 100% glycerin
b. Glycerin: distilled water = 80: 20 (vol. -%)
c. Glycerin: distilled water = 60: 40 (vol.-%)
d. Glycerin: distilled water = 40: 60 (vol.-%)
e. 100% distilled water

The post-processing and analysis are carried as follows:

i. *High-speed camera:* The stored images (600 fps) can be analyzed using LaVision software [16] to determine the bubble size, velocity, area, and trajectory. In this article, the results are visually determined together with the measured data from the sensors.
ii. *Microphone:* The recorded WAV files are processed using MATLAB [19] to filter frequencies and analyze signals. Traditionally, frequency is defined as the number of events repeating per unit of time. The microphone signal, i.e. the acoustic/wave frequency is categorized with respect to the wavelength of the sound wave (i.e., more oscillations = more frequency). In other words, the smaller bubble collapses faster, and the wavelength/wave period is shorter; therefore, the smaller bubbles tend to have higher wave frequency, and the opposite is true for the larger bubbles.

For each liquid type, the airflow (such as 60, 80, 100, 200, 300, and 600 L/hour) and the lance submersion depth from the bath surface (such as 2, 4, 6, 8 cm) are varied to investigate the effect on the bubble dynamics. This manuscript aims to project how the acoustics system can predict a TSL smelter's bubble dynamics and be integrated into the dynamic flowsheet models to control the process.

Bath Properties

The correlation between the volume fraction to the density and the dynamic viscosity is shown in Fig. 3 [20]. On x-axis, "1" represents 100% glycerin bath. It can be observed that the viscosity is significantly decreased with 20 vol.% of water addition to glycerin. On the other hand, the density steadily decreases with increasing the volume percentage of water. In comparison with the industrial TSL, a typical copper

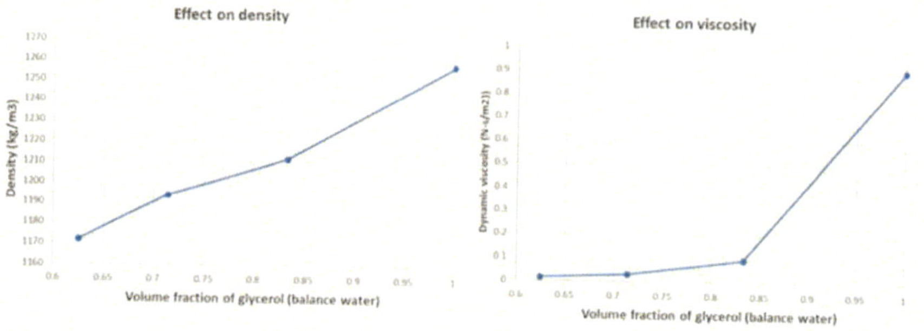

Fig. 3 Influence on density and viscosity with respect to the volume fraction of glycerin

slag at 1200 °C has a density of ~3485 kg/m^3 and a viscosity of ~0.175 N-s/m^2 (own measurements).

High-Speed Photography

In all experiments, the camera was set to capture 600 fps for 2 s (i.e., 1200 images, which consumes ~4 GB space). For example, Fig. 4 shows the bubble dynamics (i.e., bubble formation, detachment from the lance tip, trajectory, and collapse at the surface) for two extreme conditions (i.e., 100% distilled water and 100% glycerin) at a 60 l/h airflow rate through the lance and 2 cm lance submersion depth. For more viscous liquid (i.e. glycerin), the whole process time (i.e. bubble formation to bubble collapse) is 3.5 times higher than water. In addition, during the bubble collapse for lower viscous liquid (i.e. water), there is a tendency to form additional smaller bubbles at the bath surface.

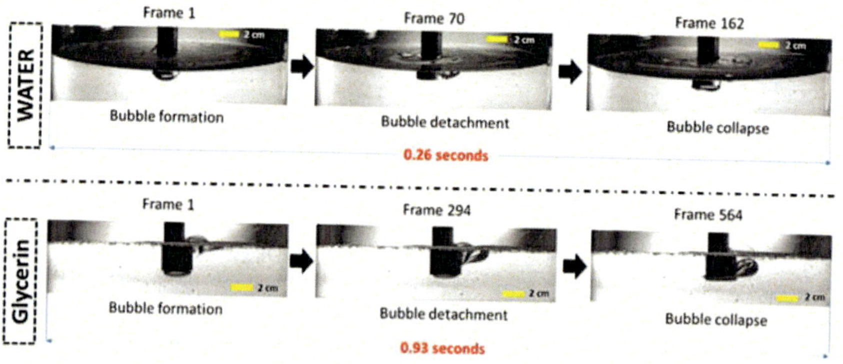

Fig. 4 Bubble dynamics of 100% distilled water (top) and 100% glycerin (bottom) under the conditions—60 L/hour airflow rate and 2 cm lance submersion

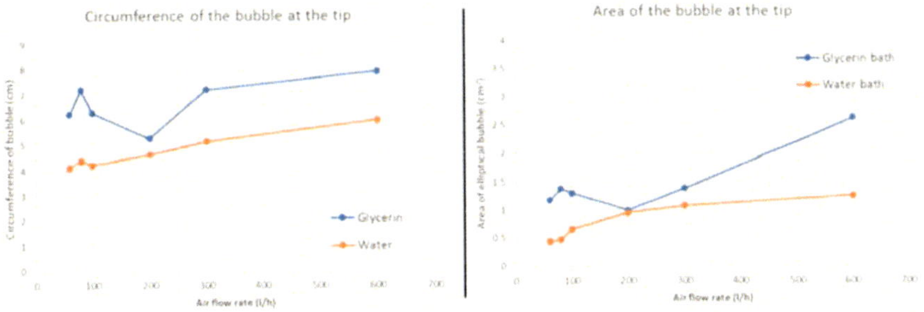

Fig. 5 Influence of airflow rate on the bubble formation at the lance tip for glycerin (blue) and water (orange) bath when the lance is submerged 2 cm (Color figure online)

Due to the buoyancy forces and stresses acting towards the bubble from the bath at lower flow rates, the bubble shape is assumed to be elliptical at the lance tip. Considering this assumption, the surface area/circumference of the bubble is calculated [21], as shown in Fig. 5. It should be noted that only "half-side of the ellipse" is in contact with the bath at the lance tip. Therefore, when calculating the surface area of the bubble, these values should be taken in half. Except for the 600 l/h airflow rate, the bubbles were intact during its path (i.e., from the lance tip until the bath surface). Hence, after the bubble is detached from the lance tip, both the sides of the ellipse will be in contact with the bath (i.e., the values in Fig. 5). In case of higher viscosities and lower bath submersion depths, the bubble might be attached to the lance during its trajectory towards to surface (see Fig. 4).

Acoustic Measurement

The raw acoustic signals can be processed in MATLAB to plot each experiment's amplitude versus time graph. As the surrounding noise is significantly less than the bubble collapse sound, filtering the signal is unnecessary. As shown in Fig. 6, 60 L/hour of air is fed to the lance, which is submerged 2 cm in various liquids. It can be observed that there is a pattern of peak formation, which represents the series of events during the bubble collapse. For example, the events (i.e. red circle in Fig. 6) include the sound wave moving through the air during the bubble rise, bubble expansion, bubble collapse, and bubble dissipation at the bath surface. Of course, the loudest event (i.e. highest peak/amplitude) occurs during the bubble collapse. This can be validated by comparing it with high-speed photography. For example, in a water bath model, the results from high-speed photography (see Fig. 4) show that the bubble collapse occurs every ~0.3 s, which is similar to the acoustic data as shown in Fig. 6 (see green rectangles). For 100% glycerin bath, the camera shows that the bubble collapse occurs ~1 s, but the acoustic signal does not show any change

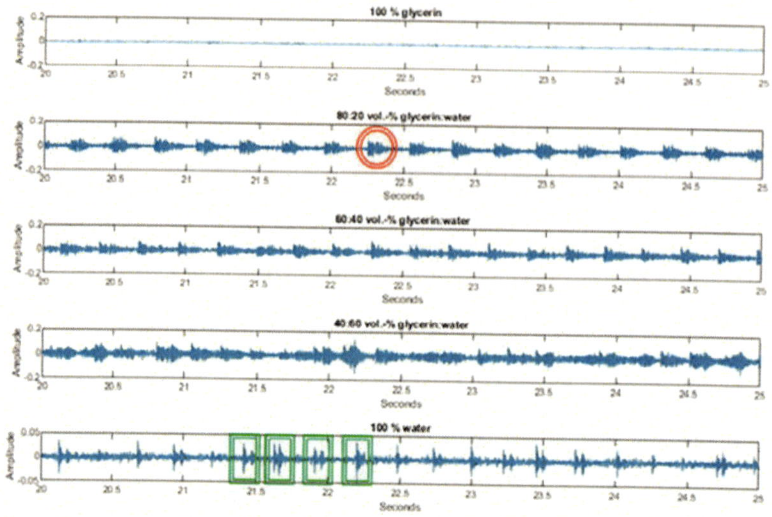

Fig. 6 Amplitude versus time plot for 60 L/hour airflow rate and 2 cm lance submersion under different bath properties

because the bubble collapse is unregistered by the microphone. This effect is due to the high surface tension and high viscosity, leading to a weak bubble collapse noise.

The acoustic signals are processed in MATLAB, and fast Fourier transformation (FFT) is carried out in order to differentiate various acoustic frequencies. The time duration for recording 1 WAV file is 60 s. The background noise is considered and subtracted from the real measurement. Apart from the general background noise from the environment, the noise/vibration of the air flowing through the lance is also deducted from the real signal in order to measure the noise of bubble collapse at the surface. For clarity of acoustic signal, the microphones are placed in close proximity to the bath surface (see Fig. 2). Figure 7 shows the influence of airflow rates through the lance in different viscous baths. Firstly, if the signal's amplitude is higher, the event occurs frequently within the time frame, i.e. the bubble collapsing at 200 Hz is formed more often at 600 l/h than 60 l/h. Additionally, in more viscous liquids, the bubble collapses at the surface happen less than in less viscous liquids because of higher surface tension. In this scenario, the bubble deformation occurs after it is detached from the lance tip, and at the surface, it dissipates into the bath in the form of ripples without proper mixture of the bath.

Figure 8 shows the influence of lance submersion depth at different airflow rates. Only the water bath model example is illustrated in Fig. 8. At lower flow rates, the bubble collapses can be distinguished throughout the smaller frequency range, while at higher flow rates, the dominant frequency is about 350 Hz.

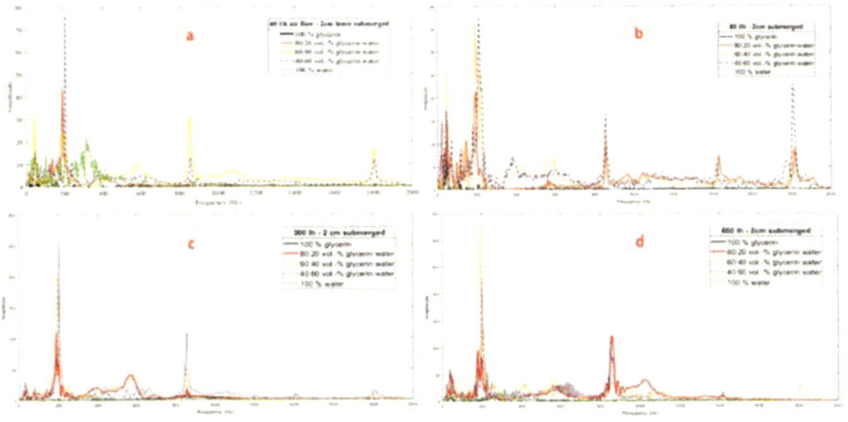

Fig. 7 Influence of the airflow rate (a = 60 l/h, b = 80 l/h, c = 300 l/h, and d = 600 l/h) on the bubble dynamics of different viscous bath

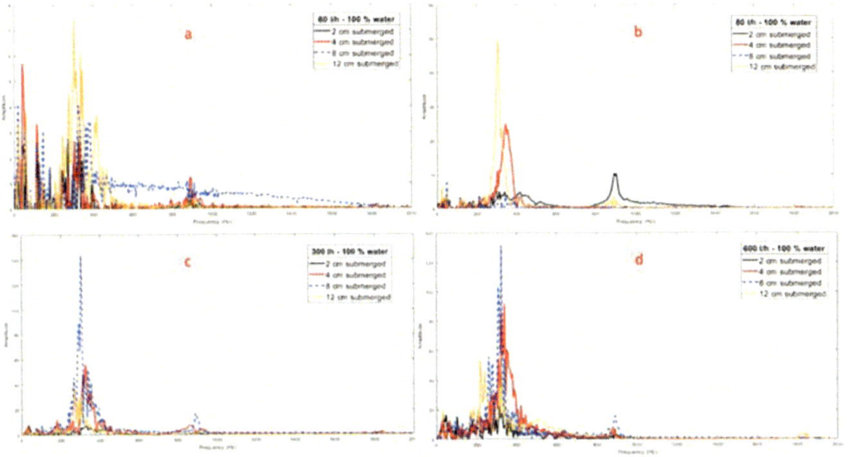

Fig. 8 Effect of the lance submersion depth of the bubble dynamics at different airflow rates (a = 60 l/h, b = 80 l/h, c = 300 l/h, and d = 600 l/h)

Discussions and Conclusions

This article illustrates a novel method of determining the bubble dynamics in a cold model, which may also be feasible under industrial TSL operations. It is observed that if the lance submersion depth increases, the bubble sphericity decreases, i.e., the bubble elongates and surface area increases. If the bath is less viscous, the bubble tends to break into secondary or ternary bubbles during the bubble trajectory towards the surface. The acoustic signal shows that 80 l/h (i.e. 0.08 Nm^3/h) airflow generates smaller bubbles than higher flow rates with the current lance configuration. At 600 l/h (i.e. 0.6 Nm^3/h), many bubbles are generated around 200 Hz, even at different viscous baths. This could be the primary bubble, which is collapsing near the lance surface. The effect on lance submersion depth is also evident in the acoustic signal. Whenever the lance is submerged deeper in a water bath, most bubbles are generated around 350 Hz and tend to form the lower end of the frequency spectrum. The lance motion system and high-speed photography support acoustic measurements (i.e. amplitude vs. time). Each photograph frame can be compared with the acoustic signal and lance motion data. However, compared to the industrial TSL, one could normalize the results by considering the flow rates, which range between 30,000–50,000 Nm^3/h with a lance inner diameter of ~0.5 m.

Acknowledgements The authors would like to thank The German Federal Ministry of Education and Research (BMBF) for funding this project within the framework of Center for Innovation Competence Virtuhcon (CIC-Virtuhcon Phase II, Project numbers: 03Z22FN12 and 03Z22FI1). In addition, the authors would like to acknowledge Metso Outotec (Mr. R. Matusewicz and Dr. J. Hoang from Australia) for their support in sharing their experience and installing the acoustics and lance motion system at INEMET, TU Bergakademie Freiberg.

References

1. Floyd JM (2005) Converting an idea into a worldwide business commercializing smelting technology. Metall Mater Trans B 36(B):557–575
2. Matusewicz RW, Reuter MA, Hughes SP (2010) Large scale copper smelting using Ausmelt TSL technology at the Tongling Jinchang Smelter. In: GDMB—international copper conference. Hamburg, Germany
3. Voigt P, Burrows A, Somerville M, Chen C (2017) Direct-to-Blister copper smelting with the ISASMELT process. In: TMS—8th international symposium on high-temperature metallurgical processing, USA
4. Alvear GR, Hunt SP, Zhang B (2006) Copper ISASMELT—dealing with impurities. In: TMS—Sohn International Symposium, USA
5. Bakker ML, Nikolic S, Mackey PJ (2011) ISASMELT TSL—applications for nickel. Miner Eng 24(7):610–619
6. Hughes S (2000) Applying Ausmelt technology to recover Cu, Ni and Co from slags. J Mater (JOM) 52(8):30–33
7. Glencore (2019) ISASMELT gives you flexible, clean smelting for the real world. https://www.isasmelt.com/en/Pages/home.aspx. Accessed 24 Aug 2021
8. Metso:Outotec (2021) Ausmelt TSL Process. https://www.mogroup.com/portfolio/ausmelt-tsl-process/?r=2. Accessed 24 Aug 2021
9. Kandalam A, Stelter M, Reinmöller M, Reuter MA, Charitos A (2021) Novel methods to determine the bubble dynamics of a top submerged lance (TSL) smelter. In: European metallurgical conference—GDMB, Clausthal-Zellerfeld, Germany
10. Obiso D, Kriebitzsch S, Reuter MA, Meyer B (2019) The importance of viscous and interfacial forces in the hydrodynamics of the Top Submerged Lance furnace. Metall Mater Trans B 50:2403–2420. https://doi.org/10.1007/s11663-019-01630-z
11. Obiso D, Akashi M, Kriebitzsch S, Meyer B, Reuter MA, Eckert S, Richter A (2020) CFD modeling and experimental validation of top submerged lance gas injection in liquid metal 1509–1525. Accessed 08 June 2020

12. Obiso D, Schwitalla DH, Korobeinikov I, Meyer B, Reuter MA, Richter A (2020) Dynamics of rising bubble in a quiescent slag bath with varying thermo-physical properties. Metall Mater Trans B 51:2843–2861. https://doi.org/10.1007/s11663-020-01947-0
13. Obiso D, Reuter MA, Richter A (2021) CFD investigation of rotational sloshing waves in a top submerged lance metal bath. Metall Mater Trans B 52:2386–2394. https://doi.org/10.1007/s11663-021-02182-x
14. Obiso D, Reuter MA, Richter A (2021) CFD investigations of bath dynamics in a pilot scale TSL furnace. Metall Mater Trans B 19:07
15. Akashi M, Keplinger O, Shevchenko N, Anders S, Reuter MA, Eckert S (2020) X-ray radioscopic visualization of bubbly flows injected through a top submerged lance into a liquid metal. Metall Mater Trans B 51(B):124–139
16. LaVision (2021) DaVis—software solution for intelligent imaging. https://www.lavision.de/en/products/davis-software/. Accessed 07 Sep 2021
17. Behringer (2021) Condenser Microphones ECM8000. https://www.behringer.com/product.html?modelCode=P0118. Accessed 13 Sep 2021
18. Focusrite (2021) Scarlett 18i20. https://focusrite.com/en/usb-audio-interface/scarlett/scarlett-18i20. Accessed 13 Sep 2021
19. MathWorks (2021) Matlab—R2020a at a glance. https://www.mathworks.com/products/new_products/release2020a.html. Accessed 07 Sep 2021
20. Cheng N-S (2008) Formula for the viscosity of a glycerol-water mixture. Indus Eng Chem Res 47(9):3285–3288. https://doi.org/10.1021/ie071349z
21. Richter W-D (2011) Ellipses numbers and geometric measure representations. J Appl Phys 17(2):165–179. https://doi.org/10.1515/JAA.2011.011

Part VI
REWAS 2022: Decarbonizing the Materials Industry

Pilot Scale Test of Flue Gas Recirculation for the Silicon Process

Vegar Andersen, Ingeborg Solheim, Heiko Gaertner, Bendik Sægrov-Sorte, Kristian Etienne Einarsrud, and Gabriella Tranell

Abstract Flue gas recirculation (FGR) for the silicon process may facilitate increasing the CO_2 concentration in the off-gas, which will be beneficial for potential future carbon capture. Lower oxygen concentration in the combustion gas will also reduce NO_X emissions. An existing 400 kVA submerged arc furnace (SAF) pilot setup was modified to be able to recirculate flue gas and equipped with gas analysis to monitor both the flue gas and the mixed combustion gas entering the furnace. Over a running period of 80 h, including 32 h of startup, twelve different combinations of FGR ratios and flow rates were tested using typical industrial raw materials. Increased CO_2 flue gas concentrations were successfully demonstrated with concentrations over 20 vol. % CO_2. Emissions of NO_X were shown to be reduced when isolating stable comparable periods within each tapping cycle.

Keywords Silicon · Flue gas recirculation · Silicon production off-gas

Introduction

Silicon is an essential element that is used in many of the products of modern society such as electronics, silicones, and for alloying other metals. So far, the only industrial way of producing silicon is through carbothermal reduction, where quartz is reduced to silicon with carbonaceous reductants at high temperatures in a submerged arc furnace. The silicon alloy is tapped from the bottom of the furnace, and the process also produces the combustible gases silicon monoxide (SiO), carbon monoxide (CO), and hydrogen (H_2) which combust in the furnace hood and are the source of the byproduct silica fume, as well as CO_2. A detailed description of production of high-silicon alloys can be found in the work of Schei et al. [1].

V. Andersen (✉) · K. E. Einarsrud · G. Tranell
Norwegian University of Science and Technology, Trondheim, Norway
e-mail: vegar.andersen@ntnu.no

I. Solheim · H. Gaertner · B. Sægrov-Sorte
SINTEF Industry, Trondheim, Norway

A. Lazou et al. (eds.), *REWAS 2022: Developing Tomorrow's Technical Cycles (Volume I)*, The Minerals, Metals & Materials Series,
https://doi.org/10.1007/978-3-030-92563-5_58

Carbon capture and storage (CCS) may reduce the environmental impact of producing silicon by preventing carbon dioxide from entering the atmosphere. Mathisen et al. [2] assessed the opportunity for carbon capture for the Norwegian silicon industries through a techno-economic investigation. Low concentration of CO_2 in the off-gas was one of the challenges identified, resulting in a high capture cost. Of the two plants investigated, one operation had an off-gas containing 1% CO_2, while the other operation had a concentration of 4%. Mathisen et al. suggested that flue gas recycling (FGR) could be a way of increasing CO_2 concentration, thereby decreasing capture cost.

Flue gas recycling is the concept of recirculating some of the combusted gases back to the process. Flue gases are mixed with combustion air and introducing into the furnace. This is a well-established emission control strategy for reducing NO_X emissions formed during combustion [3, 4]. NO_x formation in the silicon process has been described by Kamfjord [5] and Panjwani et al. [6], who also calculated potential NO_X reduction using FGR [7]. FGR can also be used in oxy-fuel combustion processes to avoid excessively high temperatures in the combustion zone, such as optimizing coal power plants for carbon capture [8].

Conventional silicon submerged arc furnaces have a semi-closed hood to capture the flue gases. The fume hood is operated at pressures lower than atmospheric pressure. Due to the pressure gradient, an excessive amount of surrounding air is drawn into the furnace hood, thereby ensuring that the combustion process is contained to the hood and off-gas system. Dilution with fresh air results in off-gas rich in oxygen and nitrogen, but dilute CO_2 concentrations. Reducing the amount of gas extracted from the furnace and thereby reducing dilution with air is the most effective way of increasing CO_2 concentrations but will be limited by either high off-gas temperature or loss of process containment.

With decreasing off-gas extraction, smoke and process gases would at some point start to leak out of openings in the hood. Closing the furnace has been investigated by both Elkem in the 1980s [9] and by Dow Chemical in the 90s [10] with mixed results. Still, almost all industrial silicon is produced in conventional furnaces with semi-closed hoods. Flue gas recirculation could be a potential solution for bridging the gap between a semi-closed and a fully closed furnace. The maximum off-gas temperature is primarily set by the downstream processes. Baghouse filtration, commonly used in silicon production, is usually limited to an off-gas temperature of around 250 °C. Installing a heat exchanger/boiler between the furnace and the filter is a solution which allows higher off-gas temperatures. Maximum off-gas temperature would be limited by fouling issues in the off-gas system between the furnace and boiler.

Experimental Setup

The pilot furnace, located at the NTNU/SINTEF laboratories in Trondheim, Norway, was lined with a monolithic high alumina refractory side lining and a silicon carbide bottom lining, as illustrated in Fig. 1. Tapping was done through a carbon block

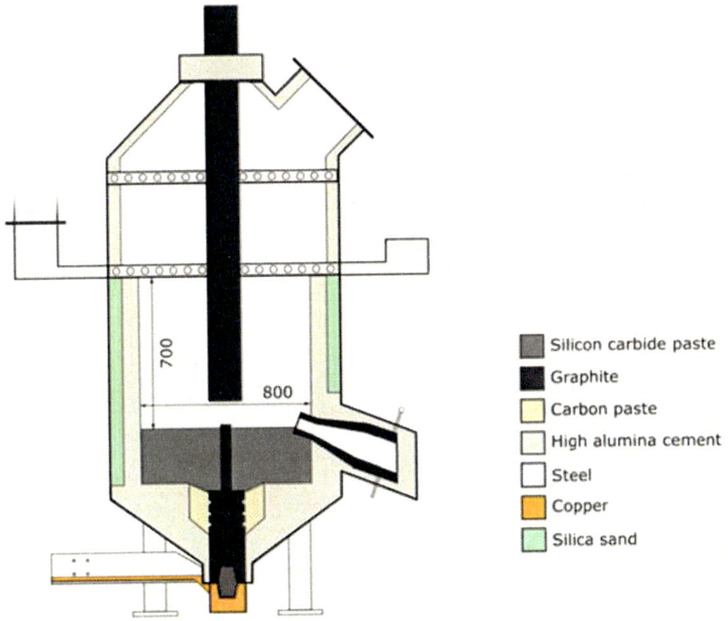

Fig. 1 Schematic illustration of furnace body and hood

tapping portal, using either iron rods in combination with sledgehammers or heavy-duty drill to open the taphole.

Electrical power was supplied through a 400 kVA electrical supply through a 6″ graphite electrode of EG90 quality, delivered by Tanso. The target electrical load was 160 kW, based on experiences with previous experiments [11], but in several cases had to be reduced to limit the temperature of the off-gas. Variations in electrical resistance and latency in electrode regulation also contributed to deviations in load. In practice, a setpoint for current was entered into the power supply, and the corresponding resistance, to achieve the right load, was adjusted by manually moving the electrode to the right height.

To investigate the effects of FGR for silicon production, an experimental matrix with two different flow rates and six different FGR rates for each flow was planned and shown in Table 1. The FGR rate was here defined as the ratio of flue gas to the total gas going into the furnace hood. A broad matrix with several FGR rates was selected to get an overview of which settings could be applicable for the industry and to test what was possible to achieve with the equipment in use. The experiment was conducted in a one-phase SAF as a continuous 3-day run with discontinuous tapping, where each tapping cycle of roughly 1.5 h was one separate experimental step. When comparing different steps, only the stable part of the tapping cycle was included. Temperatures within operational range, normal silica fume generation, and minimal concentration transients characterized the stable part of operation. The period right after charging was a transient period where gas composition stabilized, while the end

Table 1 Experimental matrix

Exp ID	Flow rate [Nm^3/h]	FGR rate [%]	Parallels completed
1	1000	0	3
2	1000	50	2
3	1000	70	2
4	1000	80	1
5	1000	88	2
6	500	0	3
7	500	45	3
8	500	60	3
9	500	70	1
10	500	78	3
11	1000	100	2
12	500	100	2

of the cycle was usually dominated by a collapsing charge and excessive blowing. These periods are omitted in the comparison since the process variations influence the results stronger than the different FGR rates of the experimental matrix.

A mix of industrially used raw materials was supplied by Elkem Thamshavn and Elkem Rana. Details of each material are shown in Table 2. The woodchips were dried in a heating cabinet at 100 °C for 24 h before use, while the other materials were used as received. Raw materials were premixed and added after each tapping from the top openings of the furnace. Before adding new raw materials, the charge was stoked with pneumatic chisels to break up crust formations. Carbon coverage was 100% after the initial filling of the furnace.

Figure 2 shows the off-gas system for the experiment. The off-gas system was modified to extract combusted process gas from the furnace hood, remove particulate matter through a baghouse filter, and distribute the gas either back to the furnace or to the smokestack, depending on the desired FGR rate. Pneumatically controlled butterfly valves were used for directional control, and fan F2 was installed to overcome the pressure drop in the recirculation line. A 4-m-long double-pipe

Table 2 Raw materials used for the experiment

		Coal	Coke	Charcoal	Woodchips	Quartz
Moisture	wt %	10.8	11.7	4.7	4.8	N/A
Fix C (DB)	wt %	57.7	92.0	85.2	25.0	N/A
Ash (DB)	wt %	1.5	2.0	2.1		1–2
Volatiles (DB)	wt %	40.8	6.0	12.7		N/A
Size fraction	mm	1–7	5–15	5–25	20–50	5–25
Share of C-mix	% Fix C	40	15	30	15	N/A

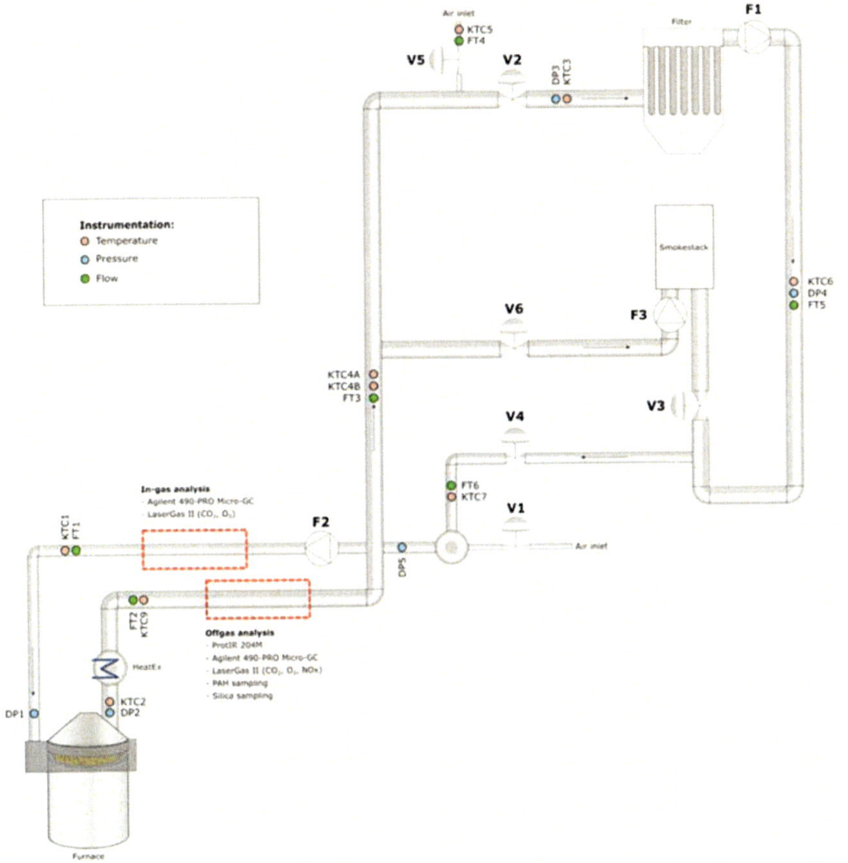

Fig. 2 Illustration of the off-gas system. Valves are denoted V#, and fans are denoted F#. Instrumentation placement is also shown, where pressure sensors are marked DP#, temperature measurements KTC#, and flow measurements FT#

heat exchanger cooled the off-gas before the analysis station. After gas analysis, a baghouse filter removed particulate matter before the gas was directed either to the smokestack or back to the furnace. The baghouse filter was operated with a maximum off-gas temperature of 180 °C, which turned out to be a limiting factor for the length of some experimental steps. Emergency ventilation to the smokestack through Valve 6 was implemented as a safety feature and used when temperatures exceeded the operational limit. Enclosure of the furnace hood was done by installing a distribution manifold over the already semi-closed hood. The furnace, combustion hood, and parts of the off-gas system are illustrated in Fig. 3.

Permanent openings in the furnace hood were closed, and with the manifold mounted to the furnace hood, the system was in principle closed. Baghouse cleaning with pressurized air pulsing was only performed in periods of charging and stoking, to avoid adding more air than necessary. Tapping gas was directed into the same

Fig. 3 Picture of the furnace and parts of the off-gas system during tapping

off-gas system but was only open in the short periods of tapping. Air was added to the system through the air inlet close to V1. Adding air here was done to make sure that air and flue gas were well mixed before entering the combustion hood.

Although efforts were made to make the system as sealed as possible, air leakage into the system was observed. Purging of measuring instrument added roughly 0.6 Nm^3/h N_2 gas and 2.4 Nm^3/h of air. Leakages of fresh air into the filter and through valves were considerably larger and were estimated to be in the range of 50–100 Nm^3/h at the highest FGR rates, calculated from the dilution of CO_2 at the off-gas analysis point to the inlet gas analysis point. The furnace hood was lined with refractory material and had to have large hatches to be able to access the furnace for charging and stoking, which proved hard to get airtight. To limit the leakage into the hood, pressure sensors DP1 and DP2 were placed close to the hood so that fans F1 and F2 could be operated at loads giving as close to ambient pressure as possible in the hood itself. The same principle was applied to limit air leakage in the baghouse filter.

Both the off-gas and the in-gas were analysed during the experiment. Off-gas was analysed with an Agilent 490-PRO Micro-GC, ProtIR 204 M combustion gas analyser, and LaserGas II instruments for measuring CO_2, O_2, and NO_x. Dust in the off-gas was measured with a LaserDust instrument. Inlet gas was analysed with another Agilent 490-PRO Micro-GC and two LaserGas II instruments, measuring CO_2 and O_2. Both LaserGas and LaserDust instruments were from NEO monitors. In addition, isokinetic sample extraction of silica samples and off-gas for PAH analysis was performed on the off-gas line.

The Agilent 490-PRO Micro-GCs used the Pro-station and Soprane II software to quantify $H_2, O_2, N_2, CO_2, CH_4$, and CO ($C_2H_4, C_2H_6$, COS). The GCs were equipped with two columns to separate the gas constituents and thermal conductivity detectors (TCD) to quantify the gas species. GC Channel 1 was equipped with 10 m Molsieve 5 Å column, operated at 110 °C with argon as carrier gas. The injector temp. was set to 90 °C and injection time of 50 ms. Channel 1 was back-flushed after 7 s.

The ProtIR 204 M FTIR analyser runs Protea Analyser Software (PAS) using the Protea combustion model with pre-defined analysis routines for 20 common combustion gases. The FTIR was used to quantify among others, CO, CO_2, and NO_x. An on-board zirconia sensor provides oxygen concentration results.

Results and Discussion

Over the course of the almost 80 h long experiment, 27 experimental steps were completed. A total of 7725 kWh electrical energy was used, where 725 kWh was used for baking the lining and 1075 kWh for heating during filling the furnace with raw materials and starting the process, leaving 5925 kWh distributed over 48 h for the actual production process. This gives an average load of 123 kW, well below the set point of 160 kW, which partly is explained by the down-time during tapping and charging. However, due to high temperatures in the off-gas system, the load was often reduced to reduce heat load. Load variations were also caused by varying electrode resistance and regulation latency. Average load in the periods of stable operation was 136 kW.

A total of 176.3 kg of silicon was tapped, and 65.41 kg of silica fume was collected in the baghouse filter. The total quartz consumption ended up at 800 kg for the full experiment, where 200 kg was used to fill the furnace. With residual raw materials left in the furnace during shut down, the actual consumption was estimated to 600 kg. The average power consumption was 33.6 MWh/tonne silicon and 9.9 MWh/tonne quartz. The silicon yield, calculated from the amount tapped Si divided by the amount of Si charged to the furnace as quartz, was 62% after the quartz used for filling the furnace was deducted. Note that, some silicon was left in the furnace during shut down. Since a significant amount of silica fume was released directly through the smokestack during periods of stoking, charging, and periods of too high off-gas temperatures, calculating silicon yield with regards to silica fume would give an erroneously high estimate.

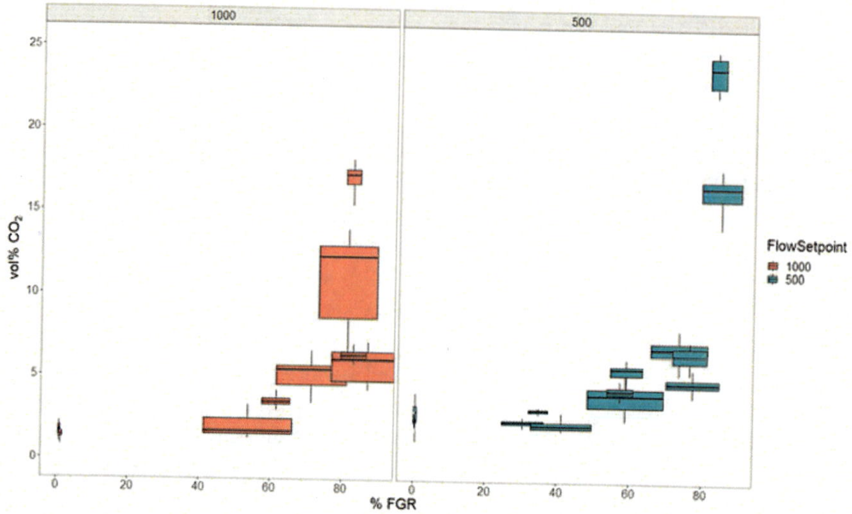

Fig. 4 Median boxplot of CO_2 concentration in off-gas, measured with the NEO LaserGas instruments, with increasing FGR rate where the FGR rate is based on CO_2 dilution (FGR4). Each column represents an experimental step

Figure 4 shows the CO_2 concentration in the off-gas with increasing FGR rate. CO_2 concentration was successfully increased with both flow rates in the experiment. The FGR ratio used in this figure is based on the FGR rate from the dilution of CO_2 from the off-gas to the in-gas. Correspondingly, O_2 levels in the combustion gas going into the furnace hood are shown in Fig. 5. In both boxplots in Figs. 8 and 9, each bar represents one experimental step, where the width of the bar shows the span of FGR rate within each step. The highest off-gas CO_2 concentration was achieved in the runs with the lower flowrate, which allowed for lower pressure differentials in the system, a reduced leakage of fresh air.

A decreasing generation of NO_x was observed with increasing FGR. The NO_X concentration also appeared to be correlated with the silica fume concentration in the off-gas. The periods with a high concentration of silica fume in the off-gas can be explained by direct blowing of SiO from the crater. As SiO combusts in the furnace hood, the high local temperatures would give a high rate of thermal NO_X production. The same effect was observed by Kamfjord [5] for industrial furnaces and has been modelled by Panjwani [7].

For future experiments, effort should be put into closing potential leakage points. This will allow better control of and higher FGR rates. An evaluation of the energy content in a future charge composition against the capacity of the heat exchanger in the system should also be done. For this experiment, the energy content in the charge mix was too high. The charcoal had a significant content of fines, and the woodchips were dried before use, reducing the moisture content from roughly 50% down to 4.75%. This resulted in too-high off-gas temperatures right after charging where the gas was too hot to handle for the filter and delayed start of the experimental

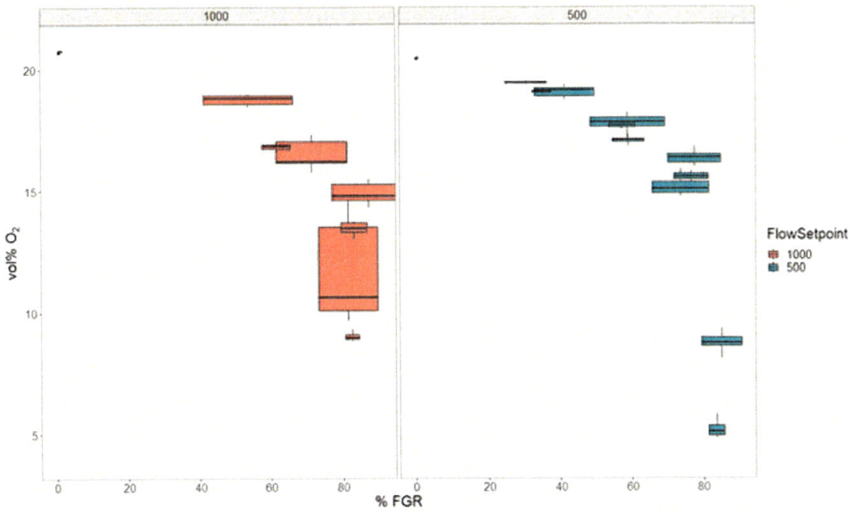

Fig. 5 Median boxplot of O2 concentration in off-gas, measured with the NEO LaserGas instruments, with increasing FGR rate where the FGR rate is based on CO2 dilution (FGR4). Each column represents an experimental step

step. Increasing the sampling time for each step could be done by either reducing the energy content in the charge or increasing the capacity of the heat exchanger in the system.

Conclusions

Flue gas recirculation has been successfully used to increase the CO_2 concentration in the off-gas from a pilot scale silicon process using typical industrial raw materials. Measured NO_X emissions were lower with decreasing oxygen content in the combustion gas. These findings are promising in terms of industrial application in larger scale and suggest that further research is warranted.

Acknowledgements This work has been funded by the research centers FME HighEFF (257632) and SFI Metal Production (237738) and the projects SiNoCO2—silicon production with no CO2 emissions (269431), PAHssion (295744), as well as the Climit supported project Elkem CCS (620106). The authors gratefully acknowledge the financial support from the Research Council of Norway and the user partners in the involved centers and projects.

References

1. Schei A, Tuset JK, Tveit H (1998) Production of high silicon alloys. TAPIR, Trondheim
2. Mathisen A, Normann F, Biermann M, Skagestad R, Haug AT (2019) CO_2 Capture opportunities in the Norwegian silicon industry. Paper presented at the 10th trondheim conference on CO_2 capture, transport and storage, Trondheim, Norway, June 2019
3. Turns S, Haworth DC (2020) An introduction to combustion: concepts and applications. McGraw-Hill, New York
4. Clean Air Technology Center and U.S. Environmental Protection Agency (1999) Nitrogen oxides (NOx), why and how they are controlled. U.S. Environmental Protection Agency, Research Triangle Park, North Carolina 27711, Technical Bulletin EPA 456/F-99–006R, Nov 1999
5. Kamfjord NE (2012) Mass and energy balances of the silicon process. PhD thesis, NTNU, Trondheim, Norway
6. Panjwani B, Olsen JE (2013) Combustion and mechanism of NOx formation in Ferrosilicon. In: Paper presented at the European Combustion Meeting, June 2013
7. Panjwani B, Pettersen T, Wittgens B (2020) Controlling flue gas temperature from ferro silicon submerged arc furnaces (SAF) using flue gas recirculation (FGR). In: Paper presented at the 14th international conference on CFD in oil and gas, metallurgical and process industries, Dec 2020
8. Fujimori T, Yamada T (2013) Realization of oxyfuel combustion for near zero emission power generation. Proc Combust Inst 34(2):2111–2130. https://doi.org/10.1016/j.proci.2012.10.004
9. Arnesen AG, Bjørdal J (1981) Split body closed FeSi75 furnace. In: Paper presented at the electric furnace conf, vol 38
10. Dosaj VD, Brumels MD, Haines CM, May JB (1992) Silicon smelting in a closed furnace
11. Solheim I, Jensen R, Kamfjord NE (2012) Equipment for pilot scale experiments of NOx emissions from the silicon process. In: Proceedings of 3rd international symposium on high-temperature metallurgical processing, Wiley, 2012, pp 33–40. https://doi.org/10.1002/9781118364987.ch5

Supercritical Carbonation of Steelmaking Slag for the CO_2 Sequestration

Jihye Kim and Gisele Azimi

Abstract Mineral carbonation of industrial byproducts is a promising carbon capture and storage technique to abate global warming. Steelmaking slag is the main byproduct of the steelmaking industry, and it is a potential source of alkaline oxides which can be transformed into carbonates. The carbonation of the steelmaking slag has proven to be a great countermeasure to sequester significant amounts of CO_2 emitted from the steelmaking process at the point sources while offering the environmental benefits of waste reduction. In this study, a supercritical carbonation process is developed to sequester CO_2 using steelmaking slag. Compared with conventional aqueous carbonation, this process has several advantages including higher reactivity, less waste generation, and better economic feasibility. A response surface methodology is utilized to assess the effect of operating parameters on carbonation efficiency and to optimize the process. Under the optimum conditions, the maximum CO_2 uptake of 213 gCO_2/kg_{Slag} is achieved. We believe that the findings of this study would help enable efficient CO_2 mitigation utilizing an efficient and environmentally sustainable process and thereby contribute to carbon neutrality and waste reduction.

Keywords Supercritical carbonation · Steelmaking slag · Mineral carbonation · Process optimization · Response surface methodology

Introduction

Iron and steel are essential materials that enable a modern way of life filled with buildings, automobiles, and bridges. Steelmaking slag is the main byproduct of the steel industry and is a potential source of alkaline oxides which can be transformed into carbonates [1, 2]. In 2019, world iron slag production was estimated to be

J. Kim · G. Azimi (✉)
University of Toronto, 200 College Street, Toronto, ON M5S35, Canada
e-mail: g.azimi@utoronto.ca

J. Kim
e-mail: jhye.kim@mail.utoronto.ca

A. Lazou et al. (eds.), *REWAS 2022: Developing Tomorrow's Technical Cycles (Volume I)*, The Minerals, Metals & Materials Series,
https://doi.org/10.1007/978-3-030-92563-5_59

between 320 and 384 million tons, with steel slag production estimated to be between 190 to 280 million tons [3]. Of all steel slags produced globally, about 14% are landfilled [4]. The main reasons behind the partial utilization of steel slags are their volume instability and their leaching behavior [5].

The carbonation of steelmaking slag can offer the benefits of improving the volume stability of the slag and producing enhanced construction materials with better inertness and corrosion resistance. Furthermore, considerable amounts of CO_2 emitted from the steelmaking industry can be sequestered at the point sources via steelmaking slag carbonation, and in turn, abating global warming. The steelmaking industry generated 1.85 tons of CO_2 for one ton of steel produced in 2018, and it accounts for 7–9% of the direct CO_2 emissions from global fossil fuel use [6]. To this end, the carbonation of slag has come to the forc as a breakthrough technique to mitigate the CO_2 gas emissions resulting from the steelmaking industry.

In this study, a supercritical carbonation process of steelmaking slag was developed for CO_2 sequestration. A response surface methodology was used to quantify the effect of four operating parameters, i.e., slag particle size, CO_2 pressure, reaction temperature, and water-to-slag ratio, on the carbonation efficiency and to optimize the process. The main contributor to enhance the carbonation efficiency was revealed, and the maximum CO_2 uptake was obtained under the optimum conditions.

Experimental

Materials

The steelmaking slag was obtained from an undisclosed source. Glass vials were supplied from VWR (USA). The liquid CO_2 tank was purchased from Messer Canada Inc. (Canada). Hydrochloric acid (36.5–38.0 vol.% assays) and nitric acid (68.0–70.0 wt% assays) were purchased from VWR. Milli-Q Integral water purification system (MilliporeSigma, Germany) was used to produce deionized water (0.055 μS/cm). A certified standard stock solution of iron, calcium, magnesium, silicon, manganese, aluminum, and chromium (1000 mg/L) was purchased from Inorganic Ventures, Inc. (USA) and used for calibration of the analytical instrument.

Experimental Procedure

For the supercritical carbonation of EAF slag, the EAF slag samples with different particle sizes were prepared using a jaw crusher (BB 200 Jaw Crusher, Retsch GmbH, Germany), ball mill (roll lab ball mill, The Galigher Co., USA), and ro-tap (RX-29 ro-tap, W.S. Tyler, USA). Afterward, the dried slag sample with a certain particle size was thoroughly mixed with water at different water-to-slag ratios. The glass vials containing the mixture of the slag sample and water were placed in an autoclave reactor (Parr 4560 mini-benchtop reactor-600 mL, Parr Instrument Company, USA). Next, the reactor was sealed and heated to a set temperature (ramp-up time of 30 min). In the meantime, the pressure reached a setpoint utilizing a supercritical CO_2 pump (GB40-CO2, Dongguan Usun Fluid Control Equipment Co., Ltd., China) and air compressor. The reaction time was maintained constant at 48 h, and no agitation was made. After the test was completed, the carbonated slag sample was dried at 80 °C for > 24 h and analyzed by different characterization techniques.

Factorial Design of Experiments and Empirical Model Building

A systematic study was carried out to investigate the effect of operating parameters, namely slag particle size (X_1), CO_2 pressure (X_2), reaction temperature (X_3), and water-to-slag ratio (X_4), on the CO_2 uptake. The primary and quadratic effects of one factor, alongside the second-order interactions between two factors, were quantitatively analyzed. Each parameter was coded to five levels (−2, −1, 0, +1, +2) to directly compare the effect of different parameters. Upper and lower limits of factor levels were determined on the basis of the preliminary experimental results and the operating ranges of comparable processes in the literature [7–10]. Table 1 shows the detailed information of operating parameters.

A response surface methodology was used to design the experiments and to build an empirical model. The experimental data were then fitted to the empirical model Eq. (1) using multiple linear least squares regression (mLLSR, Eq. (2)).

$$\begin{aligned}\hat{y} =& \hat{\beta}_0 + \hat{\beta}_1 X_1 + \hat{\beta}_2 X_2 + \hat{\beta}_3 X_3 + \hat{\beta}_4 X_4 + \hat{\beta}_{12} X_1 X_2 \\ &+ \hat{\beta}_{13} X_1 X_3 + \hat{\beta}_{14} X_1 X_4 + \hat{\beta}_{23} X_2 X_3 + \hat{\beta}_{24} X_2 X_4 \\ &+ \hat{\beta}_{34} X_3 X_4 + \hat{\beta}_{11} X_1^2 + \hat{\beta}_{22} X_2^2 + \hat{\beta}_{33} X_3^2 + \hat{\beta}_{44} X_4^2 \end{aligned} \quad (1)$$

Table 1 Factor levels for operating parameters for supercritical carbonation

Factor	X_1	X_2	X_3	X_4
Factor description	Slag particle size	CO_2 pressure	Reaction temperature	Water-to-slag ratio
Units	μm	bar	°C	mL/g
−2 level	−420 + 250	80	40	0.1
−1 level	−250 + 177	90	50	0.2
0 level	−177 + 105	100	60	0.3
+1 level	−105 + 74	110	70	0.4
+2 level	−74 + 44	120	80	0.5

where $\widehat{\beta}$ is the vector including each of the model parameters, β_0 corresponds to the baseline bias, β_1 corresponds to X_1 (slag particle size), β_2 corresponds to X_2 (CO_2 pressure), β_3 corresponds to X_3 (reaction temperature), β_4 corresponds to X_4 (water-to-slag ratio), β_{ij} corresponds to the second-order interaction of the two different factors, and $\widehat{\beta}_{ii}$ corresponds to the quadratic term.

$$\widehat{\beta} = \left(X^T X\right)^{-1}\left(X^T Y\right) \quad (2)$$

where X is the experimental design matrix and Y is the response matrix including the actual response (measured CO_2 uptake).

Morphological, Mineralogical, and Compositional Characterization

The inorganic carbon analysis (IC, SSM-5000A, Shimadzu, Japan) was carried out to calculate the CO_2 uptake of the solid sample. The compositional characterization was performed using alkali fusion digestion (Claisse LeNeo, Malvern Panalytical, UK; $Li_2B_4O_7$ 49.75 wt%, $LiBO_2$ 49.75 wt%, LiBr 0.50 wt%). The concentration of elements of interest (Fe, Ca, Mg, Si, Mn, Al, and Cr) in the solution was measured using inductively coupled plasma optical emission spectroscopy (ICP-OES, Optima 8000, PerkinElmer, USA).

Results and Discussion

Characterization of Steelmaking Slag

The elemental composition of the steelmaking slag was analyzed using alkali fusion followed by ICP-OES. As shown in Table 2, the steelmaking slag consists of 23.0 wt% iron, 15.3 wt% calcium, 6.4 wt% magnesium, 5.6 wt% silicon, 5.4 wt% manganese, 2.5 wt% aluminum, and 1.9 wt% chromium. The mineralogical characteristics of the slag were investigated with XRD in our previous work, and it was found that the

Table 2 Alkali fusion followed by ICP-OES results indicating the elemental composition of the steelmaking slag

Element	Fe (wt%)	Ca (wt%)	Mg (wt%)	Si (wt%)	Mn (wt%)	Al (wt%)	Cr (wt%)
Average	23.0	15.3	6.40	5.59	5.35	2.51	1.91
Standard deviation	0.78	0.44	0.25	0.17	0.17	0.11	0.07

slag is mainly composed of larnite (Ca_2SiO_4), brownmillerite ($Ca_2((Fe_{1.63}Al_{0.37})O_5)$, magnesioferrite ($Fe_2MgO_4$), wustite ($Fe_{0.944}O$), and hematite ($Fe_2O_3$) [11].

Effect of Operating Parameters and Empirical Model Building

A series of supercritical carbonation experiments was conducted based on the experimental matrix designed using a response surface methodology. The effect of operating parameters on the extrinsic CO_2 uptake was investigated, and the empirical model was built using multiple linear least squares regression (mLLSR). Figure 1 presents the ordered factor effect coefficients with enough significance ($\alpha = 0.05$) on the extrinsic CO_2 uptake of the steelmaking slag.

As can be seen in Fig. 1, the slag particle size (X_1) showed the most significant positive impact on the CO_2 uptake. This is understandable because the decrease of slag particle size increases the total surface area and carbonated volume of the slag particles, and in turn, the carbonation efficiency.

The CO_2 uptake was also positively affected by the CO_2 pressure (X_2). To explain this phenomenon, the thermodynamic modeling of the $CaCO_3$-$MgCO_3$-CO_2-H_2O system was conducted. The detailed explanations of the modeling procedure and results are presented in the authors' previous work [11]. The results showed that the concentration of dissolved CO_2 dramatically increases with increasing the partial pressure of CO_2. Because the equilibria of the overall carbonation reaction are governed by the concentration of dissolved CO_2 (Reaction (3)), this increase results in the increase of the carbonation efficiency.

The positive effect of the water-to-slag ratio (X_4) was captured because water participates in the CO_2 dissolution and dissociation reactions and is significantly consumed for producing silicic acid, which is a product of the carbonation reaction

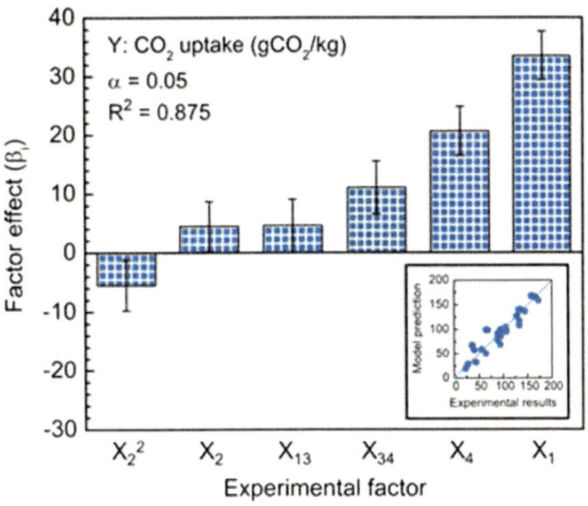

Fig. 1 Ordered chart of factor effect coefficients for the empirical model for the CO_2 uptake. The inset graph shows the correlation between the model predicted results and the experimental results

(Reaction (3)).

$$Ca_2SiO_{4(s)} + 2CO_{2(aq)} + 2H_2O_{(aq)} = 2CaCO_{3(s)} + H_4SiO_{4(aq)} \quad (3)$$

Optimization of the Supercritical Carbonation Process

Based on the factor effect coefficients (Fig. 1), the empirical model for the CO_2 uptake was built as presented in Eq. (4).

$$y = 99.4 + 33.5X_1 + 4.5X_2 + 20.7X_4 + 4.6X_1X_3 + 11.1X_3X_4 - 5.5X_2^2 \quad (4)$$

The supercritical carbonation process was then optimized using the ExcelSolver software with the objective of maximizing the CO_2 uptake in 48 h processing time. The conditions that satisfy this scenario were slag particle size of -105 + 74 μm, CO_2 pressure of 101.4 bar, temperature of 69.5 °C, and water-to-slag ratio of 0.444 mL/g_{slag}. Under these conditions, 189.0 ± 19.8 g_{CO2}/kg of the extrinsic CO_2 uptake was experimentally achieved, validating the high model accuracy (absolute relative deviation = 3.40%). Considering the intrinsic CO_2 uptake of the steelmaking slag, this slag proves to sequester a maximum of 213.4 g_{CO2}/kg via supercritical carbonation.

Conclusions

The current study is focused on developing a supercritical carbonation process of steelmaking slag to sequester CO_2. A systematic investigation was performed using a response surface methodology to study the effect of operating parameters, i.e., slag particle size, CO_2 pressure, reaction temperature, and water-to-slag ratio, on the carbonation efficiency of the steelmaking slag. On the basis of the results, the slag particle size was found to have the most significant positive impact on the CO_2 uptake. After the process optimization, the optimum operating conditions were determined to be slag particle size of -105 + 74 μm, CO_2 pressure of 101.4 bar, temperature of 69.5 °C, and water-to-slag ratio of 0.444 mL/g_{slag}, which resulted in 189.0 ± 19.8 g_{CO2}/kg extrinsic carbonation efficiency in 48 h processing time. The results of this study ascertain that the developed process offers several advantages including high carbonation efficiency, no use of chemicals, and minimal volumes of waste. It is expected that the findings of this study help enable the carbon sequestration using industrial wastes, in particular, steelmaking slag, thereby contributing to carbon neutrality and sustainable recycling.

References

1. Ghosh A, Chatterjee A (2008) Ironmaking and steelmaking: Theory and practice, 1st ed. PHI Learning Private Limited, New Delhi
2. Zhang J, Matsuura H, Tsukihashi F (2014) Processes for Recycling. Elsevier Ltd, Amsterdam
3. U.S. Geological Survey (2020) Mineral commodity summaries 2020: U.S. geological survey, 200p
4. Branca TA, Colla V, Algermissen D et al (2020) Reuse and recycling of by-products in the steel sector: recent achievements paving the way to circular economy and industrial symbiosis in Europe. Metals (Basel) 10:345. https://doi.org/10.3390/met10030345
5. Fisher LV, Barron AR (2019) The recycling and reuse of steelmaking slags—a review. Resour Conserv Recycl 146:244–255. https://doi.org/10.1016/j.resconrec.2019.03.010
6. World Steel Association (2020) Steel's contribution to a low carbon future and climate resilient societies, 6p
7. Fernández-Carrasco L, Rius J, Miravitlles C (2008) Supercritical carbonation of calcium aluminate cement. Cem Concr Res 38:1033–1037. https://doi.org/10.1016/j.cemconres.2008.02.013
8. Min Y, Jun YS (2018) Wollastonite carbonation in water-bearing supercritical CO_2: Effects of water saturation conditions, temperature, and pressure. Chem Geol 483:239–246. https://doi.org/10.1016/j.chemgeo.2018.01.012
9. Urbonas L, Leno V, Heinz D (2016) Effect of carbonation in supercritical CO_2 on the properties of hardened cement paste of different alkalinity. Constr Build Mater 123:704–711. https://doi.org/10.1016/j.conbuildmat.2016.07.040
10. Sugama T, Ecker L, Butcher T (2010) Carbonation of rock minerals by supercritical carbon dioxide at 250 °C, 23p
11. Kim J, Azimi G (2021) The CO_2 sequestration by supercritical carbonation of electric arc furnace slag. J CO2 Util 52:1–11. https://doi.org/10.1016/j.jcou.2021.101667

Carbon Footprint Reduction Opportunities in the Manganese Alloys Industry

Camille Fleuriault and Kåre Bjarte Bjelland

Abstract Production of manganese alloys is an energy intensive process associated with high carbon consumption. In the wake of the Paris agreement, manganese alloys producers have implemented new strategies to minimize the environmental footprint of smelting and refining activities. At Eramet Norway, process decarbonization is well underway with the goal to reduce CO_2 emissions by 43% within 2030. Initiatives cover the full manganese alloys process range, starting from the assessment of carbon neutral reagents and pre-processing of the manganese ore to downstream reuse of furnace offgas as energy vector. In collaboration with a rich grid of academic and industrial partners, Eramet Norway is especially investigating solutions for carbon capture and storage from manganese alloys furnace gases. Technical challenges are associated with the presence of deleterious components in the flue gas as well as process discontinuities, while economic feasibility is relying on maximizing upscale efficiencies, developing downstream networks, and securing governmental support.

Keyword Carbon capture · Manganese · Smelting · Sustainability

Introduction

Eramet Norway is a world leader in production of refined manganese alloys, with three smelters in Porsgrunn, Sauda, and Kvinesdal. Eramet Norway delivers approximately 530,000 tonnes of manganese alloys via smelting and refining of manganese ores. Ferromanganese and silicomanganese alloys are used for the production of steel mainly for the construction, machinery, automotive, and transportation sectors. Manganese is specifically used as an alloying element, for deoxidation of steel and immobilization of sulfur. At Eramet Norway Sauda, two closed 42 MW submerged arc furnaces produce approximately 270,000 tonnes of HC FeMn per year. Subsequent oxygen refining yields refined metal (MC FeMn and LC FeMn).

C. Fleuriault (✉) · K. B. Bjelland
Eramet Norway AS, Rolighetsvegen 11, 3933 Porsgrunn, Norway
e-mail: Camille.Fleuriault@eramet.com

A. Lazou et al. (eds.), *REWAS 2022: Developing Tomorrow's Technical Cycles (Volume I)*, The Minerals, Metals & Materials Series,
https://doi.org/10.1007/978-3-030-92563-5_60

As part of its commitment to reduce emissions by at least 43% by 2030 and 80% by 2050 (compared to 2005), Eramet Norway has adopted an ambitious climate roadmap covering all aspects of ferroalloys production, from raw material supply to process by-product utilization (1). At the Sauda plant, plans for carbon capture and storage (CCS), specifically, are underway.

The Environmental Footprint of the Manganese Alloys Industry

The main direct sources for greenhouse gases in the manganese ferroalloy production field are reducing agents used for the carbothermic reduction of manganese ore treatment process, consumption of electrodes, and calcination of fluxes such as dolomite or calcite. As most pyrometallurgical processes, sources such as electricity generation for the furnace, feed material, and finished product handling are also considerable. The energy consumed by all three smelters in 2020 for powering furnaces and auxiliary equipment reached nearly 2 TWh. Combined with other emissions, it represented approximately 1,5 tonnes of greenhouse gas per tonne of product, aligning with the average steel production emission factor (2). At the Eramet Sauda plant, approximately 320,000 tonnes of CO_2 are generated every year.

Strategies to Decrease the Environmental Footprint in the Manganese Alloys Industry

To achieve its 2030 and 2050 goals, Eramet Norway has identified several key initiatives and developed a climate action roadmap involving the following levers:

- Raw materials: optimization of feed blend composition by reducing the carbonate content in fluxes
- Biocarbon: replace fossil fuels by tailor made, high quality bioreagent for the carbothermic process
- NewERA: recovery and usage of waste energy carried in the furnace gas, as well as optimize ore-pretreatment via drying and fines recovery
- Pre-reduction: increase ore reduction inside the furnace and upstream of the smelting process
- CCS: capture and permanently store CO_2 emissions from the smelters
- CCU: capture and valorize unburnt furnace offgas as an energy carrier

Among these initiatives, carbon capture has the potential to halve CO_2 emissions. At the Norwegian level, the industry sector also has established a roadmap to reach carbon neutrality by 2050. Of this goal totaling 16.35 million tonnes reduction in CO_2 emissions, CCS is credited with a reduction potential of 34% (3).

The Carbon Capture and Storage Network in Norway

As of 2021, there are 135 commercial CCS facilities in the world, for a capture capacity of 149,3 Mtpa (4). Early on, Norway positioned itself as one of the precursor countries for CCS with the development and operation of the Sleipner (1996) and Snøhvit (2008) capture facilities. While most established CCS operations worldwide have been developed for the oil and gas industry, recent years have seen a diversification of processes and expansion of CCS plans to industries such as power generation, waste incineration, cement and iron and steel production. Norway, especially, beneficiates from a vast gas handling, transport and storage knowledge, as well as established infrastructures inherited from the oil and gas industry (Langship initiative and Northern Lights project for the enablement of cost-effective CCS in Northern Europe).

This context fosters the multiplication of hubs and clusters demonstrating the shift from the single lined emission to storage model, to the network model, combining sources and taking advantages of economies of scale (for compression and transport). Among them, Eramet Norway is a member of the Eyde cluster, a group of companies establishing commercial viability of CCS within the process industry with plans for pilot scale plants at two different sites.

Carbon Capture and Storage at Eramet Norway Sauda

After completing a scoping study for the selection of applicable technologies, Eramet Norway Sauda has entered a partnership with a technology provider for the installation of a pilot scale facility for carbon capture. Currently at the feasibility stage, the implementation of carbon capture in the ferromanganese industry must overcome two main challenges: achieve acceptable recovery rates and establish a cost-effective downstream network. The key parameters influencing the technical viability of CO_2 extraction from a ferromanganese furnace flue gas are:

- Feed gas characteristics and stability
  - High partial pressure of CO_2 in the gas
  - Feed gas conditioning to meet stringent capture technology requirements
  - Furnace stability, downtime, and feedstock variation management
- Capture reagent selectivity and ageing
  - Typical furnace gas impurities are detrimental to reagent performance (SO_x, NO_x, H_2...)
  - Reagent selectivity for FeMn furnace offgas blends has not been proven

The estimated cost of carbon capture in the iron and steel industry ranges between 50–100 USD/tonne CO_2. The cost of transport and storage is highly dependent of scale and transport distance with first estimates between 30–80 USD/tonne CO_2

(5). While the CO_2 price could double by 2030, the commercial viability of such enterprise will nonetheless rely on governmental support, establishing long-term value and safety of CO_2 storage, clear laws and regulations at both the Norwegian and European levels.

References

1. Eramet Norway (2021) Sustainability report 2020
2. IEA (2020) Iron and Steel Technology Roadmap, Paris, France. https://www.iea.org/reports/iron-and-steel-technology-roadmap
3. Norsk Industri (2016) The Norwegian Process Industries' Roadmap Combining Growth And Zero Emissions By 2050, English summary, Oslo, Norway
4. Global CCS Institute (2021) Global Status of CCS 2021. Technical Report, Melbourne, Australia
5. Kearns D et al (2021) Technology readiness and costs of CCS. Technical Report, Melbourne, Australia

Effect of Moisture and High Temperature to Separation Properties of Mixed Matrix Membranes

Dragutin Nedeljkovic

Abstract Recently, huge emissions of carbon dioxide have emerged as a major problem in different fields of engineering. Mixed matrix membranes are materials with huge potential for application in the field of carbon dioxide removal from the flue gases. Preliminary experiments have shown that dense composite membrane with polyethyleneoxide (PEO) as a matrix and zeolite powder as a dispersed phase with appropriate additive that serves as a homogenizer can be used. This type of membranes has shown good permeability of carbon dioxide and relatively low permeability for other gases (hydrogen, oxygen, nitrogen). The aim of this work is to test potential degradation of permeation properties and mechanical consistency of the membrane under repeated cycles of heating and cooling in presence of moisture. The experiments were performed at five different temperatures below melting or degradation point of polymers with three different partial pressures of water in combination with various gases. It was found that permeability of each gas will reach constant value after certain number of measurements.

Keywords Polymers · Composites · Carbon dioxide separation · Membrane

Introduction

With the increased demand for the energy, and the reliance of the industry to fossil and non-renewable fuels, mankind has recently been faced with the increased amount of flue gases formation. Waste gas (mainly CO_2) is emitted as products of both industrial and communal sources are the main source of the "greenhouse effect" and the global warming [1, 2]. Levels of waste gases are at level that threatens the survival and life of various living species on the Earth [3]. Typical producers of the carbon dioxide pollution include (but are not limited to): heating plants, process industry, power plants as well as transportation. In order to limit the production and emission

D. Nedeljkovic (✉)
College of Engineering and Technology, American University of the Middle East, Dasman, Kuwait
e-mail: Dragutin.Nedeljkovic@aum.edu.kw

A. Lazou et al. (eds.), *REWAS 2022: Developing Tomorrow's Technical Cycles (Volume I)*, The Minerals, Metals & Materials Series,
https://doi.org/10.1007/978-3-030-92563-5_61

of carbon dioxide, the United Nations initiated Paris Agreement (signed by virtually all UN members) in 2016. Despite all efforts, carbon dioxide level in the Earth's atmosphere reached level of 420 ppm in 2021, with expected growth in the following years [4]. In ideal case, human energy requirements would be fulfilled by renewable energy sources. However, those sources are not currently feasible alternative to the fossil fuels due to their availability and reliability. Therefore, main efforts should be addressed to reduction of carbon dioxide emission [5, 6]. Current process for separation of carbon dioxide from flue gases are based on cryogenic procedures (phase change based on a difference in boiling point of various gases with disadvantage of huge energy consumption); chemical adsorption. (This approach uses alkali solution for chemical reaction with carbon dioxide. Disadvantage of this process is high demand for complex equipment and environmentally hazardous chemicals) or physical adsorption (In this process, CO_2 is physically adsorbed on the surface of the adsorbent. Adsorbent must be regenerated prior to next cycle which significantly increase the recovery time, demand for energy and price.) [7, 8]. In a broader sense, carbon dioxide separation might be classified in one of the following categories: separation from nitrogen (flue and waste gas treatment); separation from hydrogen (syngas synthesis and, to certain extent, flue gas treatment) and separation from methane (natural gas production) [9, 10]. As this work is focused on the flue gas treatment, materials suitable for separation of CO_2 from nitrogen, oxygen, and hydrogen will be analyzed.

Separation process could also be performed by dense membranes that base their separation properties on solution/diffusion mechanism. This type of membrane is permeable for carbon dioxide, but not permeable for other gases commonly present in combustion products (nitrogen, oxygen, unburnt hydrocarbons, hydrogen). As the difference in particle sizes of different components occurs at the molecular level, conventional, sieve-based (porous) membrane is not suitable for this application [11–13]. Possible approach may be membrane with separation mechanism based on different permeability of different gases at the same temperature and pressure. The principle of this (non-porous or dense) type of membrane is that one of the components of the gas mixture (preferably CO_2) is adsorbed on the surface of membrane, dissolved in bulk of the membrane material and diffuses through the membrane driven by pressure gradient [14–16]. When it reaches permeate side, diffused molecule is detached from the surface. Dependence of the diffusion on temperature has been studied, and it is mostly determined by thermodynamics parameters of molecules of the solute and polymer chains of the membrane. Therefore, the main efforts in membrane selectivity are directed toward enhancing the solubility of carbon dioxide in comparison with other gases. In previous experiments it was observed that repeating units of poly(ethylene oxide)—PEO in polymer chains improves solubility of CO_2, while solubility of O_2 and N_2 remains relatively low [17]. Due to the different structure and size of carbon dioxide and hydrogen molecules (hydrogen is significantly smaller and non-polar), it is expected that diffusion coefficient of H_2 will be higher than diffusion coefficient of CO_2 if the temperature and pressure are the same (which is the case in mixture of those gases). Because of that, main efforts in improvement of membrane selectivity should be directed toward increasing of CO_2

solubility, while keeping the solubility of hydrogen as low as possible. Both oxygen and nitrogen have molecules smaller than CO_2, but the difference is not as high as in the case of hydrogen. Therefore, the difference in diffusivity will be smaller, but solubility of both nitrogen and oxygen should be kept as low as possible [18–21].

Good solubility properties of carbon dioxide in in PEO-based polymer can be theoretically confirmed by Hansen solubility rule that takes into account interaction between polymer chains and carbon dioxide on molecular level [22] This qualifies PEO and its co-polymers as good candidates for this application. In previous experiments with equivalent systems [23–27], a polymer available under commercial name PEBAX, supplied by Arkema have shown promising results. It is classified as thermoplastic elastomer with nylon-6 or nylon-12 as polyamide block. This block is mechanical carrier of the membrane providing the mechanical stiffness, while PEO block gives rubbery structure and provides diffusion. [28]. Chemical and physical properties of PEBAX can be fine-tuned by alternating the fraction of two blocks in co-polymer [29]. Based on the previous works, PEBAX 1657 was used for this experiment [23, 30]. However, as shown in previous experiments [22, 31, 32] permeation properties of pure co-polymer should be improved. Pure polymer shows relatively high permeability for both carbon dioxide and other gases, which results in low selectivity. Therefore, selectivity of carbon dioxide can be increased by adding different additives.

Possible additive that might improve solubility of carbon dioxide are zeolite powders. From the structural point of view, zeolites are frameworks of "caged-like" structures that can accommodate relatively huge molecules of carbon dioxide. Main specifications of each zeolite types are maximum sphere diameter (diameter of the biggest sphere that can be accommodated inside the framework) and maximum diffusion diameter (diameter of the biggest sphere that can diffuse through the pore opening). Chemically speaking, zeolites are mineral inorganic compounds of alumo-silicates with varying ratio of aluminum and silicone that may (or may not) contain different metals as well. In general, presence of zeolite powder increases the selectivity of membrane for carbon dioxide versus other gases [31].

In order to provide best permeation and separation results, zeolite particles should be evenly dispersed in the membrane bulk, free of any pin-holes, without formation of clusters and without voids on the contact surface between zeolite particles and polymer chains. This task is one of the main challenges in preparation of this type of membranes as non-polar, hydrophobic chains should be in a good contact with highly charged hydrophilic inorganic particles. To obtain membrane with smooth surface and homogenous distribution of particles, zeolite must be dispersed in the liquid which is good solvent for selected polymer. Possible approach to this challenge is to add additional homogenizing agent that would provide good contact of zeolites with polymer chains, at the same time preventing the formation of zeolite aggregates. In general, suitable additive should be miscible and compatible both with zeolite particles and polymer that is used for matrix. Possible approaches are either to use polymer that contains polar groups [33] or to use relatively small molecule that would pose as the "anchor" between polymer chains and zeolite particles. Some preliminary work has already been done in this field in dry conditions [23, 30]. The task of this paper is to test potential changes in properties of membrane at elevated temperatures, in the presence of the moisture and with repeated number of measurement cycles. The aim was to test if the permeation properties would be changed under repeated cycles of increased and decreased pressure.

Materials and Methods

In this paper, the influence of the repeated measuring cycles to the performance of the carbon dioxide separation membrane was tested. Matrix of the membrane was made of PEBAX 1657. This polymer is commercially available and was supplied by Arkema and used as received. The structure of PEBAX 1657 is presented on Fig. 1.

Zeolite powders used for this experiments were supplied by NanoScape and used as received. Characterization was done by supplier and all presented data regarding zeolite powder are from supplier, unless stated otherwise. Zeolite properties that influence diffusion and solution of gas in the membrane are defined by the direction of pores and by the maximal dimension of sphere that can diffuse through the opening. Specific surface is important on the macrolevel as it determines the surface that is potentially available for contact between zeolite and gas as well as surface that must be taken into consideration when membrane homogenization is analyzed. Powder used in this work has specific surface of approximately 800 m^2/g. For this measurement, IWS zeolite was chosen based on good results in previous experiments [23]. It contains three-dimensional relatively big pore openings. Maximal diameter of the sphere that can diffuse through this frame is 87 pm, and maximal size of the sphere that can be accommodated in the frame is 67 pm. In general, three-dimensional pores have shown the best properties as molecules can diffuse through them in any direction, regardless on the orientation of particle. Beside that, IWS have shown good results in analogous dry measurements, and it contains relatively high fraction of silica that should further increase solubility of carbon dioxide.

As a homogenizing agent, n-tetradecane trimethyl ammonium bromide (n-C14-TMABr – NTAB) was taken. It contains long, non-polar, hydrophobic chain that is miscible with polymer bulk. On the other hand, polar tetramethyl ammonium group is compatible with hydrophilic, electrically charged zeolite particle. By interacting with both polymer and zeolite, NTAB should provide good contact and prevent formation of voids between dispersed phase and matrix. NTAB was supplied by Sigma Aldrich and used as received.

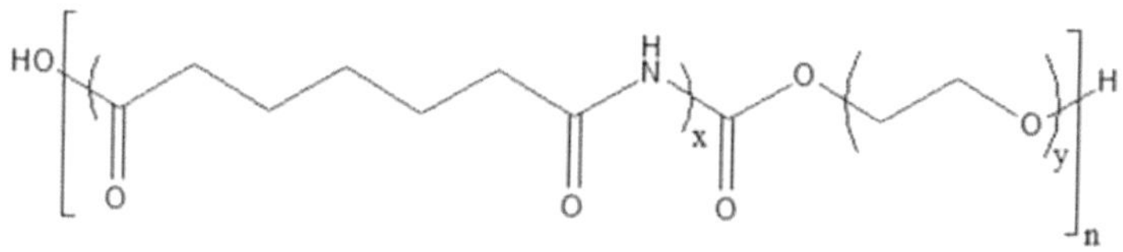

Fig. 1 Structure of the PEBAX 1657 used as a matrix for group of the membranes

Membranes used for measurements were prepared by the following procedure: Polymer (PEBAX 1657) was dissolved in mixture of water (70 wt%) and methanol (30 wt%) at elevated temperature (80 °C) under reflux. Zeolite powder was dissolved in the same mixture and NTAB was added. Homogenization was performed in ultrasound mixer (90 W, 40 kHz) and solutions of polymer and zeolite with additive were mixed and stirred overnight under same conditions (reflux at 80 °C). Mass fraction of IWS powder was 20% and fraction of additive 8% (both of fractions are mass fractions versus overall mass of the membrane). After overnight stirring viscous solution was casted on the plain Teflon surface bordered by Teflon ring, covered by non-woven textile and dried in working lab hood overnight. The aim of the Teflon was to prevent stitching of the membrane to the surface and to avoid potential damage of the membrane when it is removed from drier. If the drying was attempted at underpressure, evaporation of the solvent would be too quick, and bubbles might be formed in the bulk of the membrane decreasing its selectivity. Before the first round of testing, the membrane was placed on high vacuum line in order to remove any traces of residual solvent. For each set of the measurements, membrane was kept for 24 h in chamber with water at the appropriate temperature at atmospheric pressure, so liquid-water equilibrium could be achieved and the effects of the water to the membrane properties could be tested.

Permeability properties were determined applying the time lag method applying the solution-diffusion model which takes into account both solubility and diffusivity. The parameters were determined by the following equations [34–36]:

$$\alpha_{A/B} = \frac{P_A}{P_B} = \frac{D_A S_A}{D_B S_B}$$

$$D = \frac{l^2}{6\theta}$$

$$P = D \cdot S = \frac{V_p l\left(p_{p_2} - p_{p_1}\right)}{ART\Delta t\left(p_f - \frac{\left(p_{p_2} + p_{p_1}\right)}{2}\right)}$$

In those equations, S is solubility, D is diffusivity, P is permeability, $\alpha_{A/B/}$ is selectivity of the component A versus component B (defined as the ratio of permeability for gas A versus permeability for gas B), V_p is the permeate volume, l is the thickness of the membrane, R is the universal gas constant, Δt is for the time required for permeate pressure to increase from value p_{p1} to valuep_{p2}, p_f is feed pressure, θ is time lag.

Measurements were performed at 25, 50, 75, 100, and 125 °C. The membrane in measuring chamber was supported by the steel mash. Gas mixture was applied on the feed side, while the permeate side was kept at vacuum providing the pressure difference as a driving force for diffusion process. Gas whose permeability was measured was bubbled through the water which was heated to the measuring temperature. After reaching the constant pressure at the permeate side, vacuum was applied shortly to

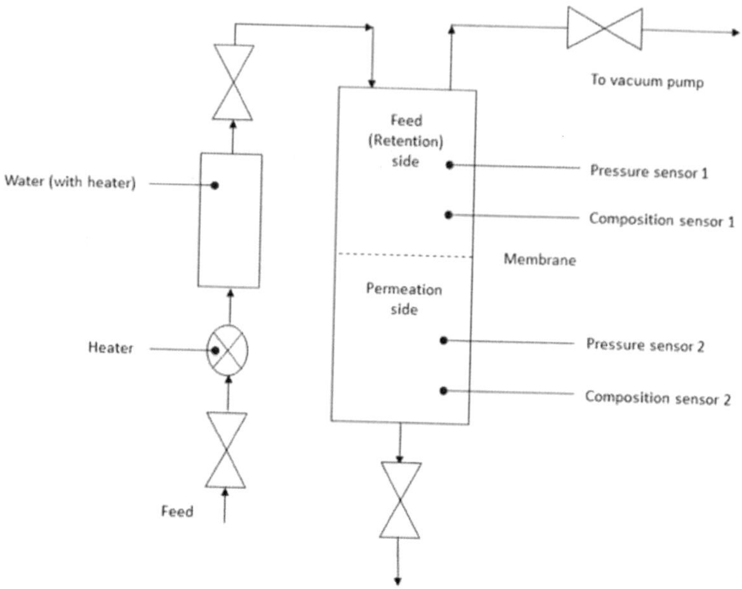

Fig. 2 Scheme of the apparatus for the permeability measurements

both of the sides before the next cycle. Permeability of every gas was measured in ten cycles at every temperature. Because of short time of vacuum application, diffused water that was accommodated in the bulk of the membrane was prevented from escaping, so potential influence of repeated vacuum-pressure cycles to the permeation process could be tested. The apparatus used for permeability measurements is presented on Fig. 2.

Beside gases commonly present in flue gases, helium measurement was performed as well. The aim of the helium measurements was to detect potential presence of the pin-hole or voids as its small, non-polar, perfectly rounded molecule has high diffusion coefficient. Gases were measured in a sequence that prevents formation of potentially flammable or explosive mixture (helium, hydrogen, nitrogen, oxygen, carbon dioxide). Selectivity of each gas was recalculated versus carbon dioxide.

Results and Discussion

All synthesized membranes were homogenous without observable particle aggregates, voids, or zones of different zeolite concentration. Membranes were smooth, transparent, or slightly opaque which indicated good contact between polymer and zeolite particle, as well as uniform distribution of zeolite powder in membrane bulk (white color of the membrane indicate void between polymer and zeolite particle which causes light refraction yielding non-transparent membrane). Thickness of membranes varied between 178 and 235 μm. All permeabilities are calculated and presented in Barrer (common permeation unit in the membrane community). Relation between Barrer and analogous SI unit is:

$$1 Barrer = 3.35 \times 10^{-16} \frac{m^3}{m^2 \cdot Pa \cdot s \cdot m}$$

(Permeation of 1 Barrer means that under pressure difference of 1 mmHg 1 cm^3 of oxygen at STP diffuses in 1 s through 1 cm thick membrane with the surface area of 1 cm^2, multiplied by 10^{-10}). Values for the vapor pressure of water at temperatures of measurement are presented in Table 1 [37].

As measurements were performed at the atmospheric pressure (101.3 kPa), steam at 125 °C was at the state of superheated steam. Results of the measurements for all gases at different temperatures are presented on Fig. 3.

As it can be seen on Fig. 3 permeability of each gas increases with the increased number of repeated measurements. The same trend is observed for every gas. Permeability of every gas will increase in first few cycles before reaching approximately

Table 1 Vapor pressure of water at the temperatures at which permeability was measured

Temperature (°C)	25	50	75	100	125
Vapor pressure (kPa)	3.17	12.35	38.6	102.3	232.6

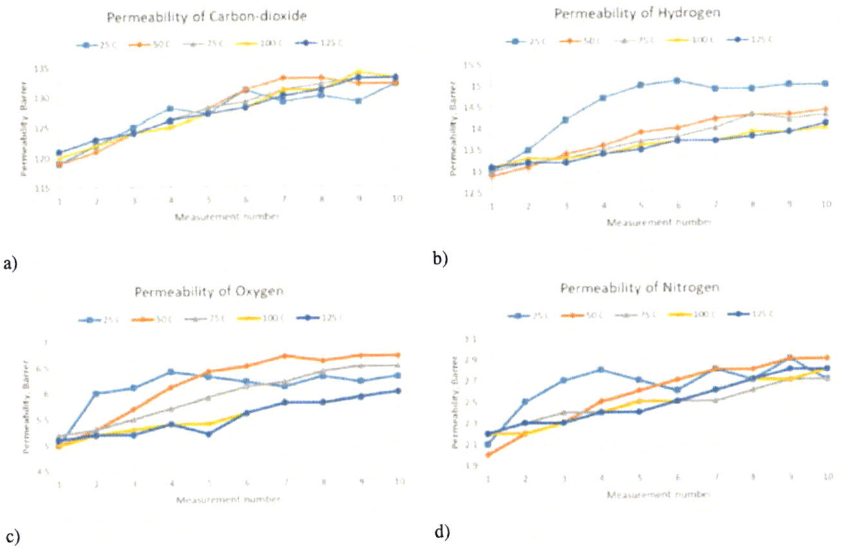

Fig. 3 Repeated permeability measurements of gases at different temperatures: **a** carbon dioxide; **b** hydrogen; **c** oxygen; **d** nitrogen

constant value. Figure 3a indicates that carbon dioxide reaches constant value after approximately 6–7 measurement cycles. Same trend is observed for hydrogen, but only at the lowest measuring temperature (room conditions). Analyzing the permeability of hydrogen at higher (above 50 °C) temperatures, it can be observed that permeability increases with every measurement. However, contrary to the room temperature conditions, constant value for permeability was not reached after ten measuring cycles. Following the trends of measurements for each of the temperatures, it is expected that constant value which is comparable to the permeability at 25 °C (15 Barrer) will be reached after approximately 20 measuring cycles. Analyzing data for permeability of oxygen (Fig. 3c) it can be observed that permeability slightly decreases with the increase of the temperature (and, thus, amount of water vapor). Constant permeability is apparently reached only at the temperature of 50 °C. Trend of increase in permeability with increased number of measurement cycles is observed, although not as prominent as in case of carbon dioxide and hydrogen. Assuming that maximum permeability has been reached at 50 °C and that similar trends will be followed, it is reasonable to conclude that constant permeability would be reached after 12 (at 75 °C) to 25 (at 125 °C) measurement cycle. Permeability measurement of nitrogen (Fig. 4d) has shown result similar to oxygen. Maximum permeability has been reached at 50 °C, and permeability increased with repeated measurements. However, no flat part of the curve (which would imply that maximum permeability has been reached) is observed at any temperature.

Comparing results of different gases at the same temperature, it can be observed that repeated number of measurement cycles causes increase in permeability for all measured gases. This effect is more obvious in the cases of oxygen and nitrogen

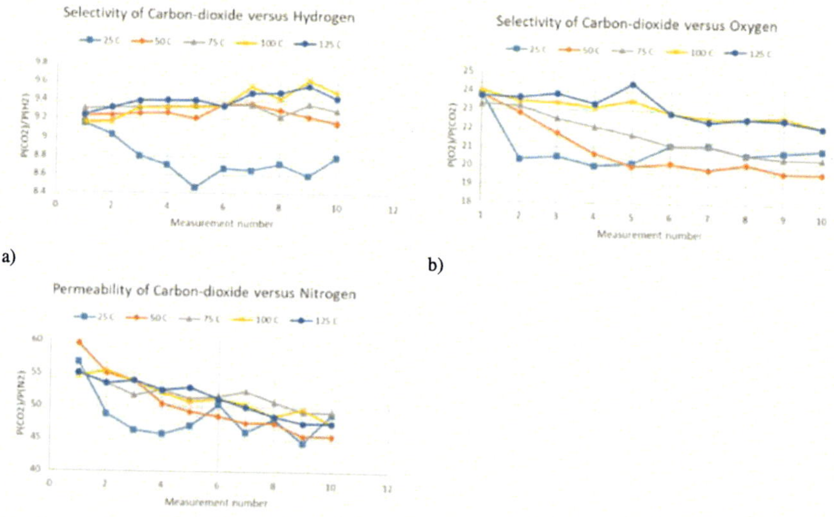

a)

b)

c)

Fig. 4 Selectivity versus carbon dioxide **a** hydrogen; **b** oxygen; **c** nitrogen than in the case of hydrogen. Potential explanation for this behavior is that after first measurement, membrane is already saturated with water vapor and measured gas, so the time lag is shorter and constant pressure on the permeate side is reached in a shorter time.

Selectivity data for hydrogen, oxygen, and nitrogen versus carbon dioxide at different temperatures is presented on Fig. 4.

Analyzing the selectivity of carbon dioxide versus hydrogen (Fig. 4a), it may be observed that the values for the room temperature are significantly lower in comparison with the elevated temperature. Possible explanation may be that the constant permeation value of hydrogen has not been reached after ten cycles of measurement for temperatures of 50 °C and above (see Fig. 3b). Based on the expected trends of permeability measurements for both carbon dioxide and hydrogen, it is reasonable to conclude that further repetition of measurements would lead to further decrease in selectivity. It can also be observed that after initial stagnation or slight increase, selectivity lines show decreasing trend, with increase in selectivity with increase of the temperature. Selectivity of carbon dioxide versus oxygen also decreases with the increased number of repetition cycles. Similar trend is observed like in the case of the permeability of oxygen. Constant value (selectivity of approximately 20) is obtained for lower temperatures (in particular for 50 °C, while for higher temperature decreasing trend is observed. Based on the curves, it is reasonable to conclude that similar value would be reached if sufficient number of measurements required to reach constant value for permeability of oxygen (as discussed in previous paragraph) were performed. Selectivity of carbon dioxide versus nitrogen (Fig. 4c) also decrease with increased number of cycles. It is also visible that selectivity is bit higher at elevated temperatures.

Comparing the results from Fig. 3 with results presented on Fig. 4, it can be observed that results are very similar, often within the margins of the measurement error. In most of the cases, permeability increases with repeated number of cycles, and after certain number of cycles, the value becomes constant. Carbon dioxide reaches constant permeability after fewest number of cycles, followed by hydrogen and oxygen, while nitrogen needs the highest number of cycles to reach constant permeability (it is very likely that required number of cycles is higher than 10 for nitrogen). Selectivity of nitrogen and oxygen decreases and selectivity of hydrogen remains constant.

Comparing results from different temperatures, it may be observed that increase of the temperature is followed by increase of the number of cycles required to reach constant permeability. At each temperature carbon dioxide was the first gas to reach constant permeability, and temperature had least effect to the number of required cycles (at each temperature constant value was reached after 6–8 cycles). Both hydrogen and oxygen have shown slow increase in required number of cycles with increased temperature, with hydrogen requiring lower number of cycles and both gases reaching constant values after 9–10 cycles at 100 °C. The temperature mostly affected nitrogen and comparing the numbers, it might be concluded that more than 10 cycles are required at 100 °C in order to reach constant value. At all temperatures selectivity of oxygen and nitrogen versus carbon dioxide decreases with increased number of measurement cycles. As a consequence of the different rates of changes in permeability of gases, selectivity of nitrogen decreases at a higher rate in comparison with oxygen. Selectivity of hydrogen is not significantly affected by repeated measurements, and remains constant, especially for temperatures above room temperature.

Another important outcome of the performed experiments is that exposure of membrane to sequential increase and decrease of pressure did not affect its performance. No degradation in structure of properties was observed at elevated temperatures and increased amount of water vapor. This gives good starting point for further investigation and potential application of this type of membranes. Based on the room temperature measurements, it can be concluded that selectivity will remain constant once the constant permeability for both carbon dioxide and other gas (hydrogen, oxygen, or nitrogen) is reached.

Conclusion

In this work, permeation properties of dense, composite membrane based on PEBAX polymer and IWS zeolite powder with NTAB as an additive were investigated. The aim of the work was to test if the permeation properties of the membrane would decrease if the membrane is exposed to repeated number of vacuum and pressure cycles at the elevated temperature in the presence of water vapor. The measurements were performed at five different temperatures that were below the decomposition point of the polymer used for the membrane. Four measurements were done at saturation pressure, and one with superheated steam. Each measurement consisted on ten cycles of vacuum and pressurizing. Obtained results are showing that with repeated measurements, permeability of each gas is increasing but at the different rates and after certain number of cycles, constant value for permeability is obtained. At each temperature, carbon dioxide required fewest number of cycles to reach constant permeability, followed by (in that order) hydrogen, oxygen, and nitrogen. At each temperature, selectivity of hydrogen versus carbon dioxide was approximately constant and the selectivities for oxygen and nitrogen decreased (selectivity for nitrogen decreased at the higher rate). As a main outcome of this work, no visible changes in appearance of the membrane were observed and no significant decay in properties was measured, so this may be potentially promising system for further investigation. Experiments performed in this paper show that systems that have previously been tested and measured in a single run measurement can sustain repeated cycles of measurement at the elevated temperatures, without losing its permeation and selectivity properties. As next direction for the research, similar measurements will be conducted with the mixture of gases that contains more than one gas in combination with water vapor, together with further increase in number of repetition cycles.

References

1. Desideri U, Corbelli R (1998) CO_2 capture in small size cogeneration plants: technical and economical considerations 39:857–867
2. Rao AB, Rubin ES (2002) A technical, economic, and environmental assessment of amine-based CO_2 capture technology for power plant greenhouse gas control. Environ Sci Technol 36:4467–4475
3. Mitsch WJ et al (2013) Wetlands, carbon, and climate change. Landsc Ecol 28:583–597
4. Walsh B et al (2017) Pathways for balancing CO_2 emissions and sinks. Nat Commun 8:14856
5. Chang P-H, Lee T-J, Chang Y-P, Chen S-Y (2013) CO_2 sorbents with scaffold-like CaAl layered double hydroxides as precursors for CO_2 capture at high temperatures. Chemsuschem 6:1076–1083
6. Kuppan CS, Chavali M (2019) CO_2 sequestration: processes and methodologies. In: Martínez LMT, Kharissova OV, Kharisov BI (eds) Handbook of ecomaterials. Springer, pp 619–668
7. Lin Y-J, Pan T-H, Wong S-H, Jang S-S (2011) Plantwide control of CO_2 capture by absorption and stripping using monoethanolamine solution. American Control Conference on O'Farrell Street, San Francisco, CA, USA June 29–July 01, 2011
8. Duarte GS, Schürer B, Voss C, Bathen D (2017) Adsorptive separation of CO_2 from flue gas by temperature swing adsorption processes. Chem Bio Eng Rev 4:277–288
9. Villalobos LF, Hilke R, Akhtar FH, Peinemann K-V (2018) Fabrication of polybenzimidazole/palladium nanoparticles hollow fiber membranes for hydrogen purification. Adv Energy Mater 8:1701567
10. Li J-R et al (2011) Carbon dioxide capture-related gas adsorption and separation in metal-organic frameworks. Coord Chem Rev 255:1791–1823
11. Vericella JJ et al (2015) Encapsulated liquid sorbents for carbon dioxide capture. Nat Commun 6:6124
12. Lee HJ, Kang SW (2020) CO_2 separation with polymer/aniline composite membranes. Polymers 12:1363
13. Živković LA, Pohar A, Likozar B, Nikačević NM (2016) Kinetics and reactor modeling for CaO sorption-enhanced high-temperature water–gas shift (SE–WGS) reaction for hydrogen production. Appl Energy 178:844–855
14. Živković L, Pohar A, Likozar B, Nikačević N (2019) Reactor conceptual design by optimization for hydrogen production through intensified sorption- and membrane-enhanced water-gas shift reaction. Chem Eng Sci 211:115174
15. Ješić D, Lašič Jurković D, Pohar A, Suhadolnik L, Likozar B (2021) Engineering photocatalytic and photoelectrocatalytic CO_2 reduction reactions: mechanisms, intrinsic kinetics, mass transfer resistances, reactors and multi-scale modelling simulations. Chem Eng J 407:126799
16. Nunes SP, Peinemann KV (2001) Membrane technology in the chemical industry. Wiley-VCH
17. Koros WJ, Fleming GK, Jordan SM, Kim TH, Hoehn HH (1988) Polymeric membrane materials for solution-diffusion based permeation separations. Prog Polym Sci 13:339–401
18. Chen JC, Feng X, Penlidis A (2005) Gas permeation through poly(ether-b-amide) (PEBAX 2533) block copolymer membranes. Sep Sci Technol 39:149–164
19. Bondar VI, Freeman BD, Pinnau I (1999) Gas sorption and characterization of poly(ether-b-amide) segmented block copolymers. J Polym Sci Part B: Polym Phys 37:2463–2475
20. Bondar VI, Freeman BD, Pinnau I (2000) Gas transport properties of poly(ether-b-amide) segmented block copolymers. J Polym Sci Part B: Polym Phys 38:2051–2062
21. Kupka V et al (2020) Well-blended PCL/PEO electrospun nanofibers with functional properties enhanced by plasma processing. Polymers 12:1403
22. Akhtar FH et al (2019) Scalable synthesis of amphiphilic copolymers for CO_2—and water-selective membranes: effect of copolymer composition and chain length. Macromolecules 52:6213–6226
23. Nedeljkovic D (2021) Homogenization of the dense composite membranes for carbon dioxide separation. In: Energy technology 2021: carbon dioxide management and other technologies; Springer International Publishing, 51–61

24. Lin H, Freeman BD (2004) Gas solubility, diffusivity and permeability in poly(ethylene oxide). J Membr Sci 239:105–117
25. Lin H, Freeman B (2005) Materials selection guidelines for membranes that remove CO_2 from gas mixtures. J Mol Struct 739:57–74
26. Jankowski A et al (2021) Polyimide-based membrane materials for CO_2 separation: a comparison of segmented and aromatic (co)polyimides. Membranes 11:274
27. Esposito E et al (2020) Glassy PEEK-WC vs. Rubbery Pebax®1657 Polymers: effect on the gas transport in CuNi-MOF based mixed matrix membranes. Appl Sci 10:1310
28. Baker RW (2012) Membrane technology and applications. John Wiley & Sons, pp 67–96
29. Baker RW (2002) Future directions of membrane gas separation technology. Ind Eng Chem Res. https://doi.org/10.1021/ie0108088
30. Nedeljkovic D (2021) The effect of the temperature and moisture to the permeation properties of PEO-based membranes for carbon-dioxide separation. Polymers 13:2053
31. Asghari M, Mosadegh M, Riasat Harami H (2018) Supported PEBA-zeolite 13X nano-composite membranes for gas separation: preparation, characterization and molecular dynamics simulation. Chem Eng Sci 187:67–78
32. Hernández-Martínez H et al (2021) Mixed matrix membranes based on fluoropolymers with *m*- and *p*-terphenyl fragments for gas separation applications. ACS Omega 6:4921–4931
33. Chen XY, Nik OG, Rodrigue D, Kaliaguine S (2012) Mixed matrix membranes of aminosilanes grafted FAU/EMT zeolite and cross-linked polyimide for CO_2/CH_4 separation. Polymer 53:3269–3280
34. Yoshino M, Ito K, Kita H, Okamoto K-I (2000) Effects of hard-segment polymers on CO_2/N_2 gas-separation properties of poly(ethylene oxide)-segmented copolymers. J Polym Sci Part B: Polym Phys 38:1707–1715
35. Buttersack C, Rudolph H, Mahrholz J, Buchholz K (1996) High specific interaction of polymers with the pores of hydrophobic zeolites. Langmuir 12:3101–3106
36. Car A, Stropnik C, Yave W, Peinemann K-V (2008) Tailor-made polymeric membranes based on segmented block copolymers for CO_2 separation. Adv Func Mater 18:2815–2823
37. Lide DR et al (2004) CRC handbook of chemistry and physics, vol 2661. CRC Press

Field Demonstration of the Reversa™ Mineral Carbonation Process Using Coal and Natural Gas Flue Gas Streams

Dale Prentice, Iman Mehdipour, Gabriel Falzone, Stephen Raab, Dante Simonetti, and Gaurav Sant

Abstract Concrete, a mixture composed of a cementation agent, mineral aggregates, and water has the potential to serve as a gigaton-scale sink for carbon dioxide (CO_2). This could make concrete the world's largest CO_2 utilization opportunity. CarbonBuilt's Reversa™ process, developed at UCLA's Institute for Carbon Management exploits simple acid-base chemistry to mineralize CO_2-dilute flue gas emissions into mineral carbonate-based cementation agents at ambient pressure, at flue gas temperatures, and without a need for carbon capture. The approach leverages innovations in the use of portlandite ($Ca(OH)_2$: calcium hydroxide, or slaked lime) which carbonates readily, and produces limestone ($CaCO_3$: calcium carbonate)—a potent cementation agent—upon its carbonation. Within the scope of a project sponsored by the US Department of Energy's Office of Fossil Energy, the Reversa technology was upscaled and demonstrated using a modularized pilot-plant at the Integrated Test Center (Gillette, WY) and National Carbon Capture Center (Wilsonville, AL) using coal- (~12 vol. % CO_2) and natural gas (~4 vol. % CO_2) flue gas streams.

D. Prentice (✉) · G. Sant
Department of Civil and Environmental Engineering, Laboratory for the Chemistry of Construction Materials (LC University of California), Los Angeles, CA 90095, USA
e-mail: daleprentice@ucla.edu

D. Prentice · D. Simonetti · G. Sant
Institute for Carbon Management (ICM), University of California, Los Angeles, CA 90095, USA

I. Mehdipour · S. Raab
CarbonBuilt Inc., 570 Westwood Plaza, Los Angeles, CA 90095, USA

G. Falzone
RCAM Technologies, Los Angeles, CA, USA

D. Simonetti
Department of Chemical and Biomolecular Engineering, University of California, Los Angeles, CA 90095, USA

G. Sant
Department of Materials Science and Engineering, University of California, Los Angeles, CA 90095, USA

California Nanosystems Institute (CNSI), University of California, Los Angeles, CA 90095, USA

A. Lazou et al. (eds.), *REWAS 2022: Developing Tomorrow's Technical Cycles (Volume I)*, The Minerals, Metals & Materials Series,
https://doi.org/10.1007/978-3-030-92563-5_62

The field demonstration led to the production of over 15,000 concrete masonry units (CMUs, also known as concrete blocks) and achieved: (1) a CO_2 utilization efficiency in excess of 75%, and (2) greater than 250 kg of CO_2 utilization per 13 tonnes of concrete (i.e., one production run). Importantly, based on rigorous third-party validation, the CMUs produced were confirmed to be compliant with all relevant industry specifications (ASTM C90). The success of this demonstration suggests that the pioneering Reversa technology is ready for commercialization.

Keywords CO_2 mineralization · Cement · CO_2 utilization

Introduction

Anthropogenic sources of carbon dioxide are generated from a number of sources, but key among these are ordinary Portland cement (OPC) production and combustion of fossil fuels [1]. Cement production is the largest global CO_2 source from the mineral decomposition of carbonates [1]. This is due to the clinkering process whereby limestone (mainly consisting of $CaCO_3$) is decomposed into CaO and CO_2, and combined with silica rich clays at high temperatures to form clinkers (i.e., the four key minerals that comprise cement) [2]. The high temperature range of 1400–1550 °C required for this process accounts for up to 60% of the generated CO_2 from cement production [3]. Combination of the limestone decomposition and thermal requirements of the clinkering process causes cement production to contribute 8–9% of annual global CO_2 emissions [1, 2, 4–6]. Combustion of fossil fuels (coal, oil, and gas) was shown to contribute a much larger portion of global CO_2 emissions. As of 2018, combustion of fossils accounted for 65% of global CO_2 emissions in 2018, where 41% was derived from stationary sources for electricity and heat generation and the other 24% was related to transport [7]. To reduce these contributions, key steps forward in CO_2 utilization technologies are required. Therefore, a CO_2 mineralization technology (CO_2Concrete) to reduce the OPC content in concrete, while utilizing flue gas emissions from fossil fuel combustion has been developed to address both areas simultaneously.

This CO_2Concrete technology utilizes low-carbon cementation agents produced by in situ CO_2 mineralization ("mineral carbonation reactions") to offer a promising alternative to OPC [8–12]. CO_2 mineralization relies upon the reaction of dissolved CO_2 with inorganic alkaline reactants to precipitate mineral carbonates (e.g., $CaCO_3$), which bind proximate particles and achieve cementation [9, 12–14]. Herein, a concrete *green body*, that is composed of a mixture of binder, water, and mineral aggregates, is exposed to CO_2 borne in industrial flue gas streams. This manner of CO_2 mineralization allows the production of construction components that feature equivalent engineering attributes as their OPC-based counterparts while featuring a much smaller embodied carbon intensity (eCI).

The reliability of the CO_2Concrete technology has proved effective for the production of concrete masonry units (CMUs) at bench scale, where the units exceeded the required 13.8 MPa compressive strength requirements [14, 15]. The study detailed herein will demonstrate the applicability of this technology at an industrial scale. In total, 12 production runs were completed at the Integrated Test Center (ITC) in Gillette, Wyoming using coal fired flue gas as the CO_2 source. Over the course of the production runs the CO_2 utilization as a function of time, 24-h CO_2 uptake, electricity usage, 28-d net area compressive strength and the eCI were recorded for each run. Collection of this data will be used to determine the success of the demonstration goals: (1) achieving in excess of 75% CO_2 utilization efficiency, (2) utilizing greater than 250 kg of CO_2 per production batch/run, and (3) ensuring compliance of CO_2Concrete blocks with industry standard specifications (ASTM C90 [15]).

Materials and Methods

The field demonstration at ITC required the CMUs to be produced off-site at the TCC materials concrete plant, where the CMUs were transported to ITC, then loaded into a custom-built carbonation chamber. A mixture of inorganic reactants (e.g., the *binder*), inert aggregates (concrete sand, #8 aggregate and washed fines), and water was used to produce "dry-cast" formulations suitable for the fabrication of concrete masonry units (CMUs; "concrete blocks"). The binder used consisted of commercially available portlandite ($Ca(OH)_2$) powder (Standard Hydrated Lime, Mississippi Lime Co.), ASTM C150-compliant ordinary portland cement (Type III OPC),[16], and an ASTM C618-compliant class C fly ash [17].

TCC materials produced structural hollow concrete masonry units (CMU) [15]. The overall dimensions of the blocks were 200 mm × 200 mm × 400 mm ($w \times h \times L$) with face-shell and web thicknesses of 32 mm and 25 mm, respectively. After forming, the fresh concrete blocks were pre-cured to gain sufficient strength (compressive strength $\sigma_c = 6 \pm 1$ MPa) to enable transport, handling and loading into the carbonation reactor.

The curing chamber at the ITC, consisted of a modified open-sided 40' shipping container with steel racks (40 in. wide × 98 in. deep × 76 in. tall), each capable of holding 108 blocks (around 1.73 metric tons of CO_2Concrete), comprising a total of 864 blocks (13.82 metric tons) of CO_2Concrete per production batch.

Four process cycles were used to complete the carbonation process: (1) drying cycle was used to remove water from the concrete, (2) carbonation cycle where conditioned flue gas was introduced to the curing chamber, (3) humidification cycle used to re-introduce water to the concrete to continue cement hydration, and (4) purge cycle to remove the flue gas from the chamber. Air was used during cycles (1), (3), and (4) at various temperatures and relative humidity. During the carbonation cycle, the inlet coal flue gas ($[CO_2] = 13 \pm 1$ v/v%) is combined with recycled flue gas from the curing chamber. This gas is conditioned using the chiller, heater, and humidification chamber. The chiller is used to reduce the relative humidity of the combined flue gas to 0–20% before being reheated to desired temperature.

CO_2 utilization as a function of time was calculated based on the change of CO_2 mass fraction of the outlet flue gas as compared to the inlet flue gas over time. Over the course of a 24 h production run, this value can be averaged to provide the 24 h CO_2 uptake value. Electricity usage of the chiller and process skid (heater, fans, and control HMI) was recorded to detail the energy usage of the process.

The performance of CO_2Concrete masonry units produced during system operations was assessed against the relevant industry standard for conventional concrete masonry units (ASTM C90 [15]) by TCC Materials as well as a certified testing lab (BASF Corporation, Construction Chemicals Division). The net area compressive strength of the CMUs was measured as per ASTM C140 at 28 days of age.

Embodied carbon intensity was calculated via the sum of the equivalent CO_2 for producing the raw materials, transport, energy used in the block making process, energy for the carbonation process and finally a reduction based on the CO_2 uptake.

Results

12 operational runs were completed fulfilling an operational duration of nearly 1350 h including a 35 day continuous operation period at nearly 90% up-time. Around 155 tonnes of concrete blocks were produced while achieving around 3 tonnes of CO_2 uptake. The CO_2 Concrete system's performance fulfilled our design specifications including (i) CO_2 utilization efficiency >75% (mass basis), and (ii) CO_2 utilization >0.25 tonnes per batch (see Fig. 1a). Importantly, these performance metrics were met using a "direct CO_2 utilization" process; i.e., without any need for flue-gas pretreatment (e.g., clean-up, enrichment). Thereafter, the CO_2 borne in the flue gas was utilized at sub-boiling temperature, at near-ambient pressure. The process conditions were optimized to (a) minimize the process's carbon (energy use) footprint, and (b) to achieve production price parity with traditional cement-based concrete blocks. Such optimizations were accomplished without sacrificing CO_2 uptake performance (see Fig. 1b), and optimizing humidification step (Fig. 1c).

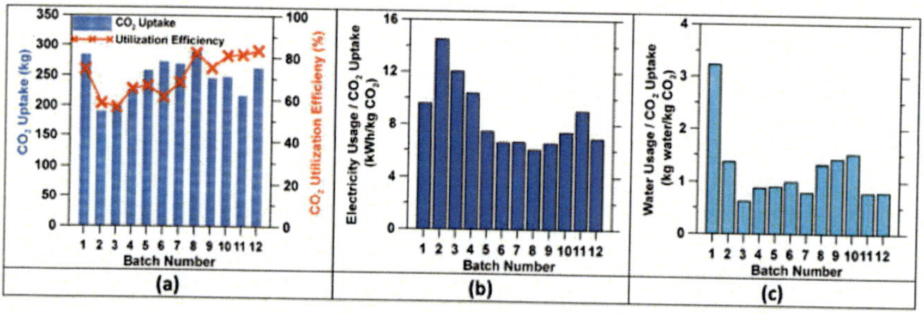

Fig. 1 Trends in batch-wise **a** CO_2 uptake and CO_2 utilization efficiency over the 24-h carbonation duration, **b** total energy usage per unit of CO_2 uptake, and **c** total water usage per unit of CO_2 uptake for pilot scale demonstration of the CO2Concrete system at Wyoming ITC host site

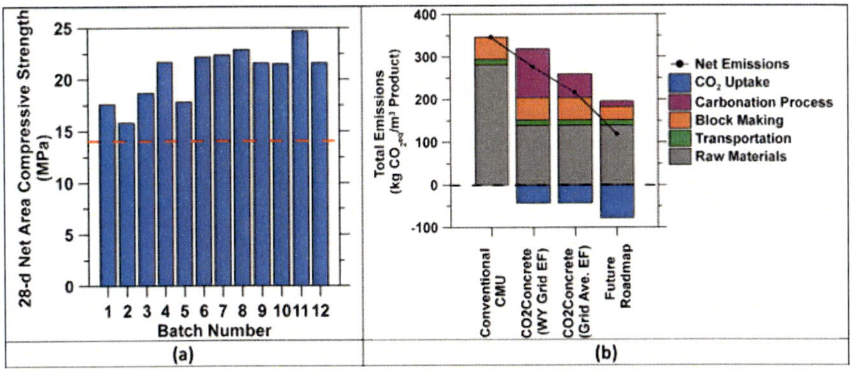

Fig. 2 **a** Average 28-day net area compressive strengths of CO_2 Concrete masonry units over 12 demonstration runs at Wyoming ITC. **b** Global warming potential (GWP, kg CO_2e/m^3 concrete) for CO_2 Concrete (estimated) and conventional OPC-concrete masonry, based on an industry-averaged environmental product declaration (EPD)

The compressive strength was measured for each batch (see Fig. 2a), with the average compressive strength determined to be 20.7 ± 2.5 MPa. This compressive strength of CO_2Concrete blocks easily fulfilled the compressive strength specified by ASTM C90 (i.e., 13.8 MPa). The water absorption values range from 142 to 168 kg/m^3, which is substantively lower than the ASTM C90 limits of 240 kg/m^3 for normal weight blocks [15]. Furthermore, compared to traditional concrete products and production practices, CO_2 Concrete™ offers a greatly reduced embodied carbon intensity (eCI, kg CO_2e/m^3), as shown in Fig. 2b. This reduction in carbon intensity is achieved while the compressive strength is around 40% greater than the requirements.

Conclusion

The system's performance fulfilled all design specifications: (1) achieving in excess of 75% CO_2 utilization efficiency, (2) utilizing greater than 250 kg of CO_2 per production batch/run, and (3) ensuring compliance of CO_2 Concrete blocks with industry standard specifications (ASTM C90 [15]). Furthermore, the CO_2 Concrete process conditions were optimized to reduce the energy input as compared to conventional cement-based concrete block. CO_2Concrete™ addresses the largest opportunity for CO_2 utilization (i.e., construction materials) by exploiting chemical reactions that do not require extrinsic energy inputs (i.e., thermodynamically downhill) even while using CO_2-dilute feedstocks (flue gases).

References

1. Andrew R (2018) Global CO_2 emissions from cement production. Earth Syst Sci Data Discuss 10:195–217. https://doi.org/10.5281/ZENODO.831455
2. Taylor HFW (1997) Cement chemistry, 2nd edn, Thomas Telford Publishing. https://doi.org/10.1680/cc.25929
3. IEA, Energy Technology Perspectives 2016. Towards Sustainable Urban Energy Systems
4. Mehta PK, Monteiro PJ (2014) Concrete: microstructure, properties, and materials, 3rd edn, New York: McGraw-Hill Education. https://www.accessengineeringlibrary.com/content/book/9780071797870
5. Gartner EM, Macphee DE (2011) A physico-chemical basis for novel cementitious binders. Cem Concr Res 41:736–749. https://doi.org/10.1016/j.cemconres.2011.03.006
6. Miller SA, Horvath A, Monteiro PJM (2016) Readily implementable techniques can cut annual CO_2 emissions from the production of concrete by over 20%. Environ Res Lett 11. https://doi.org/10.1088/1748-9326/11/7/074029
7. IEA, CO_2 emissions from fuel combustion: overview 2020
8. Wei Z, Wang B, Falzone G, La Plante EC, Okoronkwo MU, She Z, Oey T, Balonis M, Neithalath N, Pilon L (2018) Clinkering-free cementation by fly ash carbonation. J CO_2 Utiliz 23:117–127
9. National Academies of Sciences, Engineering, and Medicine, Gaseous Carbon Waste Streams Utilization: Status and Research Needs, The National Academies Press. Washington, DC (2018). https://doi.org/10.17226/25232
10. Mehdipour I, Falzone G, La Plante EC, Simonetti D, Neithalath N, Sant G (2019) How microstructure and pore moisture affect strength gain in Portlandite-enriched composites that mineralize CO_2. ACS Sustain Chem Eng 7:13053–13061
11. Jenewein C, Ruiz-Agudo C, Wasman S, Gower L, Cölfen H (2019) Development of a novel $CaCO_3$ PILP based cementation method for quartz sand. Cryst Eng Commun 21:2273–2280
12. Pan S-Y, Lai B, Ren Y (2019) Mechanistic insight into mineral carbonation and utilization in cement-based materials at solid–liquid interfaces. RSC Adv 9:31052–31061
13. Vance K, Falzone G, Pignatelli I, Bauchy M, Balonis M, Sant G (2015) Direct carbonation of $Ca(OH)_2$ using liquid and supercritical CO_2: implications for carbon-neutral cementation. Ind Eng Chem Res 54:8908–8918
14. Mehdipour I, Falzone G, Prentice D, Neithalath N, Simonetti D, Sant G (2021) The role of gas flow distributions on CO_2 mineralization within monolithic cemented composites: coupled CFD-factorial design approach, React. Chem Eng 6:494–504. https://doi.org/10.1039/D0RE00433B
15. ASTM C90—16a (2016) Standard specification for loadbearing concrete masonry units. ASTM International, West Conshohocken, PA
16. ASTM C150/C150M—18 (2018) Standard specification for portland cement. ASTM International, West Conshohocken, PA
17. ASTM C618—15 (2015) Standard specification for coal fly ash and raw or calcined natural pozzolan for use in concrete. ASTM International, West Conshohocken, PA

HIsarna: A Technology to Meet Both the Climate and Circularity Challenges for the Iron and Steel Industry

Johan van Boggelen, Hans Hage, Koen Meijer, and Christiaan Zeilstra

Abstract Over the past decade the European Union has set the ambition to become a carbon neutral and fully circular economy. To achieve this, it will be necessary to develop new technologies and processes that address both issues simultaneously. In this paper the HIsarna ironmaking technology is used as an example where the reduction of CO_2 emissions, the use of secondary raw materials as well as the valorisation of by-product streams is considered in the early stages of the process development. In the light of climate change, sustainability, resource efficiency and circularity these aspects should be part of the considerations for any new process being developed, even if their economic viability still needs to be established.

Keywords HIsarna · Smelting reduction · Ironmaking · Sustainability · CO_2 reduction · Circular economy

A Need to Change

In order to tackle climate change there is a global requirement to move to net zero CO_2 emissions across society. With the iron and steel industry responsible for 7% of the global total of CO_2 emissions from the energy system, reductions across the industry will have a major impact on global emissions [1, 2].

J. van Boggelen (✉)
Tata Steel IJmuiden BV, HIsarna, 1951 JZ Velsen-Noord, The Netherlands
e-mail: johan.van-boggelen@tatasteeleurope.com

H. Hage · K. Meijer · C. Zeilstra
Tata Steel Nederland Technology BV, Ironmaking, Steelmaking and Continuous Casting, 1951 JZ Velsen-Noord, The Netherlands
e-mail: hans.hage@tatasteeleurope.com

K. Meijer
e-mail: koen.meijer@tatasteeleurope.com

C. Zeilstra
e-mail: christiaan.zeilstra@tatasteeleurope.com

A. Lazou et al. (eds.), *REWAS 2022: Developing Tomorrow's Technical Cycles (Volume I)*, The Minerals, Metals & Materials Series,
https://doi.org/10.1007/978-3-030-92563-5_63

To stimulate action in the private sector, governments have set policies to encourage change in this direction. In Europe there is the Emissions Trading Scheme (ETS) [3] for CO_2 emission rights, which attaches a cost to CO_2 emissions. In the European Green Deal [4] clear objectives have been set to achieve net zero CO_2 emissions. It also contains objectives aimed at maximising resource efficiency and decoupling economic growth from resource use.

To achieve a net zero CO_2 emission steel industry, various new technologies are being developed and further new developments are to be expected in the future. With the introduction of new technologies it is likely than new revert and waste streams are created, for which there is not a clear solution yet. This would potentially compromise resource efficiency and circularity ambitions. Recovery, recycling, and reuse of materials will only take place at scale if the economic benefits outweigh the total cost of handling, processing, and administration. However, it is important to consider the options for reuse and recovery of new revert and waste streams at an early stage in the development of these new technologies. This is necessary in order to prevent the creation of a new problem for the next generation in our efforts to solve the current climate crisis.

HIsarna Technology

HIsarna is a new ironmaking process under development at Tata Steel in IJmuiden. The process leads to substantial intensification compared to the blast furnace (BF) route. No ore agglomeration or coke making is required as raw materials can de directly injected into the process in granular form (Fig. 1).

This results in a significant CO_2 emission reduction per tonne of steel [5]. In 2010 the HIsarna pilot plant was built. The main objective was to validate the process and demonstrate the CO_2 reduction capability. The plant has been in operation since 2011, initially in short campaigns of several months. Each campaign consisted of individual trial runs which lasted up to three days. Since 2017 the pilot plant has a full five shift rota, making it possible to operate the plant for longer runs at a time. In 2018 the plant was also integrated into the production route at the IJmuiden site. To date 11 800 tonnes of hot metal have been produced, 11 000 tonnes of which have been supplied to the BOF plant and used to produce steel. The longest continuous run is 19.5 days with an availability of 93% and an average hot metal production rate of 4.3 tonnes per hour.

Maximising Reduction of CO_2 Emissions

As already mentioned, the fact that iron ore agglomeration and coal coking are not required for HIsarna will result in a substantial CO_2 emission reduction. To achieve the objective of a net zero CO_2 emission process, further measures will be required.

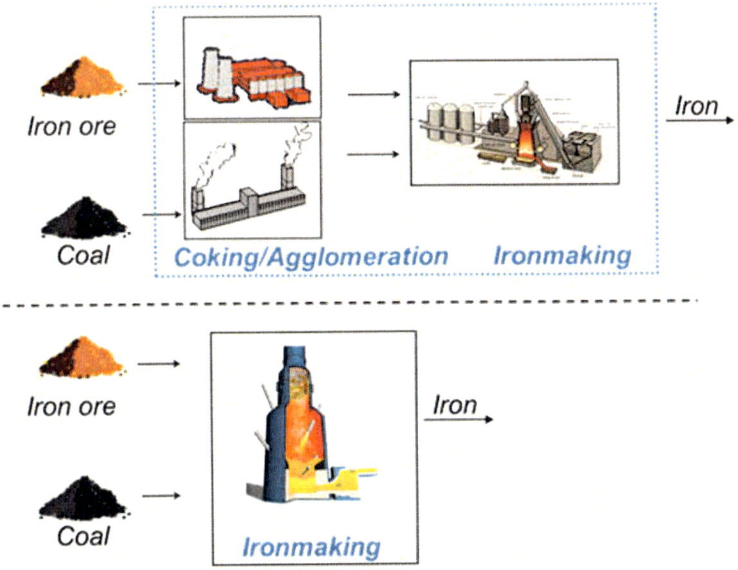

Fig. 1 HIsarna ironmaking route compared to the blast furnace process route

As the process gas from the HIsarna process is nearly pure CO_2, CO_2 capture would become relatively straightforward. Under normal operating conditions in the pilot plant CO_2 concentrations of more than 70% are measured at the reactor exit, with the other 30% made up largely of nitrogen. Most of this nitrogen comes from the carrier gasses used for ore and coal injection. It was already demonstrated that these carrier gasses can be replaced with compressed CO_2 [6]. This would result in an extremely concentrated CO_2 stream which can be easily captured for use or storage.

Even without carbon capture there is significant scope to achieve a net zero CO_2 emissions process. CO_2 emission reductions of 50% without carbon capture have already been demonstrated [6]. This was achieved by introducing up to 50% scrap into the reactor, in combination with more than 40% replacement of injection coal with sustainable biomass. The biomass used was in the form of charcoal. From a technical perspective it should be possible to achieve 100% replacement of coal with sustainable charcoal or biochar. The reactivity of charcoal is proven to be equal or better than coal [7]. An additional benefit of charcoal is the relatively low sulphur content. When combined with CO_2 capture, the use of biomass may even result in a nett negative CO_2 footprint.

A further option is to partly reduce the carbon input into the process by using natural gas, sustainable bio-gas, or hydrogen. Process analysis shows that a certain amount of solid carbon is needed to reduce the iron oxides in the slag and to carburise the hot metal. A large part of the injected carbon (coal) is used for heating purposes in the freeboard of the smelting reduction vessel. This provides scope to reduce the amount of solid carbon injected into the process and replace it with the injection of natural gas, sustainable bio-gas, or hydrogen. As a first step, tests are scheduled to partially replace the injected coal with natural gas.

HIsarna in a Zero Waste Society

By-product and waste streams from the iron and steel industry are widely used in a range of different applications already. Only a relatively small amount ends up as waste [8, 9]. However, often these applications are relatively low value.

At an early stage it was recognised that the HIsarna process opens new opportunities to use alternative and secondary raw materials. Preliminary tests were done with Zn containing streams and low grade ores [10].

Many steel products are made from galvanised steel, providing durable protection against corrosion. Recycling of galvanised steel scrap in the integrated steelmaking route (BOF-route) is generally low. The Zn would build up in process dust at concentrations too low for economic recovery. In many cases this process dust is recycled via the sinter plant and BF and the Zn can have a major impact on the BF process. This means that the Zn content of input materials has tight restrictions.

When Zn coated scrap is recycled via the EAF steelmaking route, the Zn again ends up in the dust. Here it is possible to achieve Zn concentrations that are suitable for recovery, but these zinc bearing dusts first have to be treated in a Waelz kiln to allow efficient zinc recovery. This step can be avoided in HIsarna. HIsarna can accept zinc bearing materials in the shape of galvanised scrap or as dust from BF (blast furnace), BOF (basic oxygen furnace), or EAF (electric arc furnace) plants. These materials can be injected as dry dust, briquettes, or granules.

HIsarna can allow for high Zn loads in the input material streams as it operates at a high temperature throughout the reactor. This ensures the Zn escapes the main process in gaseous form and turns into solid oxide when the process gas cools down. The Zn concentrate ultimately ends up in the bag filter of the gas cleaning plant. The objective of the ReclaMet project is to achieve a Zn enrichment in the process dust which makes direct use as a secondary raw material for Zn production possible (Fig. 2) [11].

The valorisation of slag is also considered, both through the recycling of slag from other processes in HIsarna as well as the application of HIsarna slag, which could become a product with similar applications to granulated blast furnace slag.

The HIsarna process requires flux additions in the form of CaO and MgO, to control the slag chemistry. As part of the LoCO2Fe project [6], trials were done to replace most of the lime with limestone and dolomite which is pre-mixed with the iron ore. These trials formed the basis for the decision to complete a large scale trial where limestone and dolomite were successfully replaced with recycled BOF slag [12]. This reduces the requirement for virgin minerals as fluxes in HIsarna to almost zero. Only a small amount of lime is required for injection, to trim the slag composition in the process. In addition a significant amount of the FeO in BOF slag

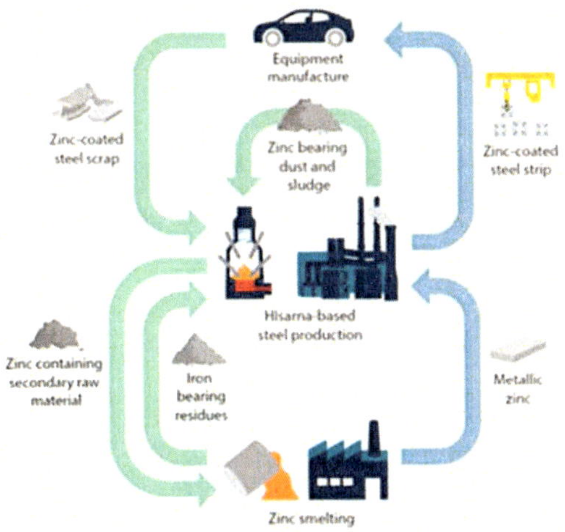

Fig. 2 ReclaMet, using HIsarna to close the Zn and Fe cycle

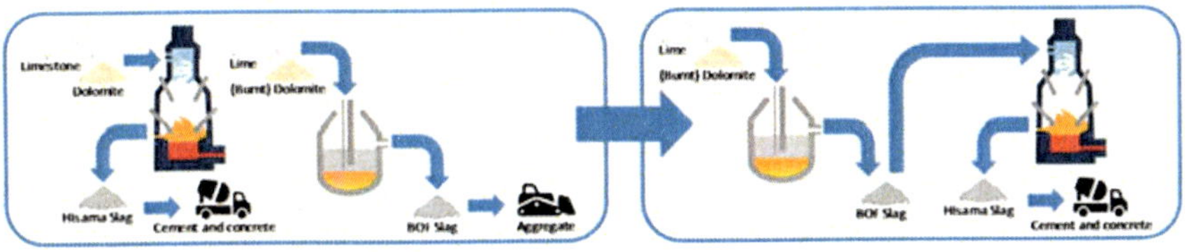

Fig. 3 Increasing resource efficiency and by-product value through HIsarna

is recovered in the HIsarna process, improving the overall yield of the HIsarna-BOF steelmaking route.

Using the BOF slag in HIsarna also increases the value of this slag. It transfers from a low value aggregate to a high value application in cement production (Fig. 3). The properties of HIsarna slag in cement and concrete are expected to be similar to granulated blast furnace slag. Slag granulation trials have been done at the HIsarna pilot plant. The granulated slag has subsequently been used to produce concrete test pieces. Testing is currently still ongoing.

Conclusions

To overcome the challenges related to climate change and circularity, new processes and technologies will be required. For any new process being developed, circularity requirements should be considered at an early stage, even if their economic viability still needs to be established.

Besides a significant reduction in carbon footprint, HIsarna offers clear possibilities to valorise by-products and waste streams to a greater extent than is currently the case.

Acknowledgements These HIsarna developments were supported by Tata Steel, ArcelorMittal, thyssenkrupp Stahl Europe, voestalpine Stahl and Paul Wurth. Previous work was supported by a larger group of companies, the ULCOS partners. It has received financial support from Horizon2020 (contract 654013), RFCS (contract RFSR-CT-2011-00002), the Dutch Government (contract DEI1160014), and the EIT Raw Materials (contract 17209).

References

1. IEA (2020) Iron and steel technology roadmap. IEA, Paris. https://www.iea.org/reports/iron-and-steel-technology-roadmap. Accessed 8 Oct 2021
2. IEA (2021) Net zero by 2050. IEA, Paris. https://www.iea.org/reports/net-zero-by-2050. Accessed 8 Oct 2021
3. European Commission. https://ec.europa.eu/clima/policies/ets_en. Accessed 8 Oct 2021
4. European Commission. https://ec.europa.eu/info/strategy/priorities-2019-2024/european-green-deal_en. Accessed 8 Oct 2021
5. Birat JP, Lorrain JP, de Lassat Y (2009) The 'CO_2 Tool': CO_2 emissions & energy consumption of existing & breakthrough steelmaking route. La Revue de Métallurgie 106(9):325–336
6. Commission E (2019) LoCO2Fe: development of a Low CO_2 iron and steelmaking integrated process route for a sustainable European steel industry. Ref Ares 2019:6822470
7. Khasraw D, Spooner S, Hage H, Meijer K, Li Z (2021) Evaluation of devolatilization behavior of different carbonaceous materials under rapid heating for the novel HIsarna ironmaking process. Fuel 292. https://doi.org/10.1016/j.fuel.2021.120329
8. Branca TA et al (2020) Reuse and recycling of by-products in the steel sector: recent achievements paving the way to circular economy and industrial symbiosis in Europe. Metals 10(3):345. https://doi.org/10.3390/met10030345
9. Birat JP (2020) Sustainable materials science—environmental metallurgy, vol 1. EDP Sciences, Les Ulis, France
10. Meijer HKA et al (2015) HIsarna Experimental Campaigns B and C. https://op.europa.eu/en/publication-detail/-/publication/2a1c611b-97f6-11e5-983e-01aa75ed71a1. Accessed 8 Oct 2021
11. Hage JLT, Zeilstra C, van Boggelen J, Meijer HKA, van der Stel J, Peeters T (2021) HIsarna: benefits of a new developed smelting-reduction option for ironmaking. In: Proceedings of of the iron & steel technology conference, 29 June–1 July 2021, Nashville, USA
12. Zeilstra C et al (2019) The use of LD-slag as a fluxing agent in the HIsarna process. In: Proceedings of the 6th International Slag Valorisation Symposium, 1–5 April 2019, Mechelen, Belgium, pp 325–328

Iron-Ore Reduction Using Green Hydrogen: A Study for Recycling Wastes in Egyptian Steel Industry

Abdelrahman A. Abouseada and Tarek M. Hatem

Abstract A new prospective in the Egyptian steelmaking industry using pure hydrogen to reduce waste iron ores in a two-stage fluidized bed reactor through direct reduction process is presented, modeled, and analyzed. The main advantage of applying this route to the steel industry is the enormous reduction in CO_2 emissions compared to today's dominant routes that rely on the blast furnace—basic oxygen furnace (BF/BOF). Moreover, the hydrogen direct reduction (H-DR) process could be directly applied for the reuse of the waste imported lump iron ores that are been crashed during the transportation process to a small fines particles; making them unusable in blast furnace processes unless sintering process is applied. A complete study to verify the applicability of this idea in Egypt has been investigated and comparing it to current fluidized bed reactor using syngas as the reducing agent.

Keywords Egyptian steelmaking industry · Direct reduction process · CO_2 emissions · Blast furnace—basic oxygen furnace · Waste imported lump iron ores

A. A. Abouseada · T. M. Hatem (✉)
Centre for Simulation Innovation and Advanced Manufacturing, The British University in Egypt, El-Sherouk City, Cairo 11837, Egypt
e-mail: tarek.hatem@bue.edu.eg

A. A. Abouseada
e-mail: abdelrahman.aboseada@bue.edu.eg

A. A. Abouseada
Chemical & Petrochemical Department, Faculty of Engineering, The British University in Egypt, El-Sherouk City, Cairo 11837, Egypt

T. M. Hatem
Microstructure Physics and Alloy Design Department, Max-Planck-Institut Fur Eisenforschung, 40237 Düsseldorf, Germany

A. Lazou et al. (eds.), *REWAS 2022: Developing Tomorrow's Technical Cycles (Volume I)*, The Minerals, Metals & Materials Series,
https://doi.org/10.1007/978-3-030-92563-5_64

Introduction

The world needs to find a way to make lots of steel while reducing the industry's carbon footprint. The Paris Agreement implies that all industrial sectors must reach zero emissions by 2060–2080, while the European Union seeks to achieve a 80–95% reduction of greenhouse gases by 2050 compared to 1990 [1]. And the UK government's independent Climate Change Committee has recommended that steelmaking should reach near zero carbon emissions by 2035. Therefore, Egyptian steelmaking industry should take some serious steps in achieving this target. There are five integrated steel companies in Egypt; the Egyptian iron and steel company Hadisolb in El-Tabbin, EZDK Company in Alexandria, Bishay steel company in El-Sadat city, EZ steel company in Suez, and Suez Company for steel production (El-Garhi MF) in El-Ain El-Sokhna. Among these five integrated companies the Egyptian iron and steel company, i.e., Hadisolb is the only one who adopts the blast furnace route. It is also the only company in Egypt, which has a sintering plant and uses local iron ores to produce pig iron in which Coke reduces the ore to iron in a blast furnace by reacting with oxygen to make carbon monoxide. One molecule of iron ore reacts with three molecules of carbon monoxide, leaving two iron atoms and three molecules of carbon dioxide. The other four integrated companies adopt either MIDREX technology or Danieli's HYL-ENERGERON technology that initiate with the natural gas DR route in which a mixture of carbon monoxide and hydrogen is used as the reductant of highly rich imported iron oxide pellets to produce sponge iron [2]. This route, compared to the blast furnace route, is 50% in CO_2 emissions. Yet a complete hydrogen-based route can reach far lower levels from 89 to 99% reduction in CO_2 emissions [3].

Discussion

As long as the four companies are producing sponge iron according to their designed capacities and so far as natural gas is available, about one million tons of highly rich iron oxide wastes almost free from Cl and MnO will be annually available as well. The following Table 1 illustrates the chemical composition of the highly rich iron oxide by-products and their annual production in EZDK Company in year 2015 [2].

These iron oxide wastes cannot be reused in their own plants as their particle sizes are not suitable any more for natural gas DR route of either MIDREX which treat

Table 1 Chemical composition of the highly rich iron oxide by-products and their annual production in EZDK Company

Material	Annual production 2015	Chemical composition					
		Fe %	CaO %	SiO_2%	Al_2O_3%	MnO%	Cl %
Oxide fines	200.000 tons	67.8	0.8	1.1	0.36	0.10	Nil

iron ore pellets of size 9–16 mm and can accept only 5% from the feed to be from the size 5–9 mm or HYL ENERGEON that treat Iron ore pellets of size greater than 5 mm [4]. Thus any iron ore with particle size lower than 5 mm is considered to be a waste. These wastes are been exported abroad; moreover, for instance EZDK gives these wastes to Hadisolb and getting instead the equivalent weight of pig iron. As Hadisolb has a sintering plant which convert these fines into sinters that afterward undergoes blast furnace route iron making. However, according to the new prospective of reducing the CO_2 emissions in steel industry, this process is considered a delaying step toward achieving the environmental target. Therefore, a promising alternative is been recommended in which these iron fine wastes could be considered a direct feedstock without prior agglomeration or sintering for a hydrogen-based direct reduction steelmaking process using Circored® technology, in which iron ore fines are directly reduced in a two-stage fluidized bed process that operates at low reduction temperatures using pure hydrogen as the reducing gas [5]. This route if applied will be the first in Egypt that is responsible for direct reduction of iron fines as the current direct reduction plants can only deal with either lump iron ores or pellets ores. Moreover, it will be the first in Egypt to use a complete hydrogen-based direct reduction process. Not as the current direct reduction plants that relies on reformed natural gas as the reductant agent [6].

The New Proposed Steelmaking Route

The first process of the route is the production of hydrogen by water electrolysis using CO_2-lean electricity. The challenge is to achieve massive production of H_2 in acceptable economic conditions [7]. The second process is the direct reduction of iron fines in a first reduction stage which is a circulating fluidized bed (CFB), then following a second reduction stage which is fluidized bed (FB) reactor operated with hydrogen only. The third process is the melting of the carbon-free direct reduced iron in an electric arc furnace to produce steel. The following Fig. 1 illustrates the hydrogen-based route for direct reduction of lump iron ores or pellets using the reduction shaft reactor of either MIDREX or HYL/Energiron (the left side of the figure); however, the hydrogen-based route for direct reduction of lump iron fines using fluidized bed reactors of Circored (the right side of the figure).

Conclusion

Steelmaking is one of the sectors contributing the most to the global greenhouse gas emissions; it is imperative to find a low-carbon process to keep global warming below 1.5 °C. There is a critical need to find new, low-carbon solutions for iron ore reduction. Hydrogen-based direct reduction is one of several promising options. A new prospective in the Egyptian steelmaking industry using pure hydrogen to

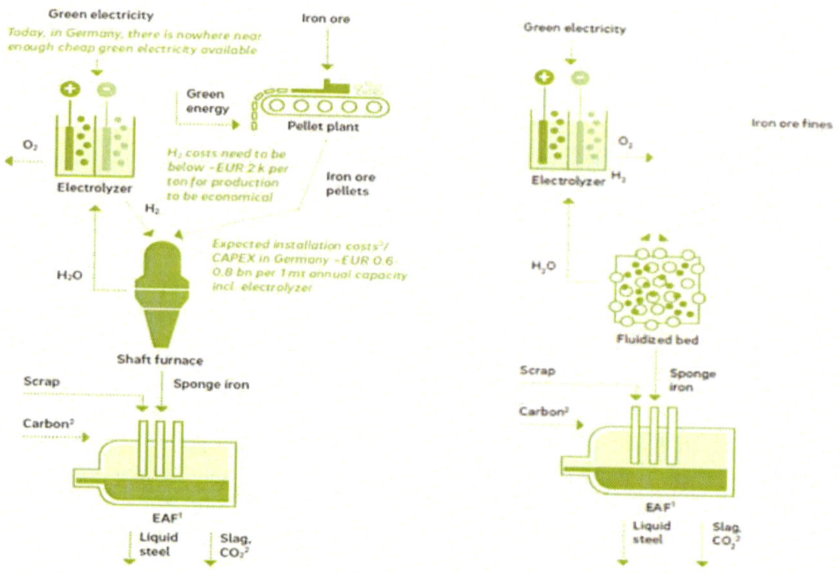

Fig. 1 Using hydrogen in the direct reduction—electric arc furnace route (DR-EAF)

reduce waste iron ore fines in a two-stage fluidized bed reactor through direct reduction process is presented, modeled, and analyzed [8]. The main advantage of applying this route to the steel industry is the enormous reduction in CO_2 emissions compared to today's dominant routes that rely on either the blast furnace—basic oxygen furnace (BF/BOF) or the reformed natural gas direct reduction route of MIDREX or HYL/Energiron technology. However, the hydrogen direct reduction (H-DR)—electric arc furnace (EAF) process using Circored technology could be directly the suitable solution for the reuse of the waste imported iron ore fines in Egypt that are in the current time either exported or exist as a feedstock for a blast furnace route after a sintering process is applied. A complete design to the process is to be proposed to verify the entitlement of applying this route in Egypt as it is a promising alternative for recycling these waste iron ore fines and producing liquid steel with very blameless environmental standings.

References

1. Vogl V, Åhman M, Nilsson LJ (2018) Assessment of hydrogen direct reduction for fossil-free steelmaking. J Clean Prod 203:736–745. https://doi.org/10.1016/j.jclepro.2018.08.279
2. Rassoul EMA, Galal AM, AbdelRassoul R (2019) The re-use of highly rich iron oxide wastes from integrated Dr steel companies in Egypt. https://www.researchgate.net/publication/335775022_THE_RE-USE_OF_HIGHLY_RICH_IRON_OXIDE_WASTES_FROM_INTEGRATED_DR_STEEL_COMPANIES_IN_EGYPT. Accessed Sep 06 2021

3. Patisson F, Mirgaux O (2020) Hydrogen ironmaking: how it works. Metals (Basel) 10(7):1–15. https://doi.org/10.3390/met10070922
4. Parisi DR, Laborde MA (2004) Modeling of counter current moving bed gas-solid reactor used in direct reduction of iron ore. Chem Eng J 104(1–3):35–43. https://doi.org/10.1016/j.cej.2004.08.001
5. Circored hydrogen-based reduction—Metso Outotec. https://www.mogroup.com/portfolio/circored-hydrogen-based-reduction/. Accessed 13 July 2021
6. MIDREX (2013) World direct reduction statistics. Midrex Technologies, Inc., p 14
7. Ranzani Da Costa A, Wagner D, Patisson F (2013) Modelling a new, low CO_2 emissions, hydrogen steelmaking process. J Clean Prod 46:27–35. https://doi.org/10.1016/j.jclepro.2012.07.045
8. Kinaci ME, Lichtenegger T, Schneiderbauer S (2020) A CFD-DEM model for the simulation of direct reduction of iron-ore in fluidized beds. Chem Eng Sci 227. https://doi.org/10.1016/j.ces.2020.115858

Toward Green Ferroalloys: Replacement of Fossil Reductants in the Pre-reduction Process of Chromite by Bio-Based Alternatives

Check for updates

M. Sommerfeld and B. Friedrich

Abstract The production of ferrochrome via carbothermic reduction in submerged arc furnaces is an energy-intensive process relying on the usage of coal and coke as reducing agents. The pre-reduction of chromite in a rotary kiln is currently carried out to decrease the specific electric energy consumption in the smelting furnace. However, as fossil reductants are still needed for reduction, CO_2 is emitted. The usage of bio-based carbon with a faster carbon cycle compared to fossil reductants could be an option to decrease the specific CO_2 footprint of Ferrochrome production. In this paper, the pre-reduction of chromite was investigated using various bio-based reducing agents and lignite coke as a fossil reference. Isothermal reduction trials were conducted at 1000, 1150, and 1300 °C and different holding times. While at lower temperatures the pre-reduction was insufficient, the bio-based reducing agents yield a degree of reaction between 61.0% and 65.4% at 1300 °C reaction times of 360 min. The highest degree of reaction is reached using coconut charcoal, followed by corn, olive, and bamboo charcoal. Coke results in the lowest degree of reaction with 51.9%. While the bio-based reducing agents performed similar after long reaction times, significant deviations were observed for shorter reaction times. X-ray diffraction was carried out to investigate the obtained product, which showed that the pre-reduction was mostly due to the formation of carbides, while the intensity of metals in the sample was rather low.

Keywords Pyrometallurgy · Sustainability · Ferroalloys · Ferrochrome · Pre-reduction

---

M. Sommerfeld (✉) · B. Friedrich
IME Process Metallurgy and Metal Recycling, Institute of RWTH Aachen University, Intzestraße 3, 52056 Aachen, Germany
e-mail: msommerfeld@ime-aachen.de

A. Lazou et al. (eds.), *REWAS 2022: Developing Tomorrow's Technical Cycles (Volume I)*, The Minerals, Metals & Materials Series,
https://doi.org/10.1007/978-3-030-92563-5_65

Introduction

The usage of fossil-reducing agents for the production of ferroalloys leads to the direct emission of CO_2. The production of ferrochromium solely was responsible for the emission of at least 15.5 million tons of CO_2 in 2016 due to the usage of reducing agents [1]. One option to reduce the emission of CO_2 could be the usage of bio-based carbon, which can be considered CO_2-neutral, if the same amount of biomass is recultivated, that is consumed [2]. Especially the iron and steel industry carried out a considerable amount of research regarding the usage of bio-based carbon [3–11], while less research was carried out considering the substitution of fossil carbon by bio-based carbon in the ferroalloy industry [1]. For the production of high carbon ferrochrome, several processes are currently industrially carried out: the smelting of chromite in open, semiclosed, or closed alternating current (AC) submerged arc furnaces (SAF), the smelting in open arc direct current (DC) furnaces and the pre-reduction of chromite in a rotary kiln followed by smelting in closed AC SAFs. The pre-reduction AC SAF process has two advantages: together with the DC furnace it has the highest chromium recovery, and has the lowest specific electric energy consumption (SEC) per ton of FeCr [12]. Therefore, the pre-reduction process is considered a viable option especially for countries with unsteady electricity supply or high prices for electric energy. Especially in recent years, a significant amount of research was carried out investigating the pre-reduction process of chromite [13–21]. Kleynhans et al. [14] determined, that the SEC decreases depending on the pre-reduction as shown in Eq. 1 [14].

$$\mathrm{SEC/(kWh/t)} = 3403.7\,\mathrm{kWh/t} - 21.5\,\mathrm{kWh/t} \cdot \mathrm{pre{-}reduction/\%} \tag{1}$$

Experimental

In this section, the used raw materials and methods are described.

Materials

In this study, a Turkish metallurgical-grade chromite concentrate (following named chromite) is used as a raw material. The chemical composition of the chromite ore determined by x-ray fluorescence spectroscopy (XRF) is shown in Fig. 1. A detailed characterization of the ore is presented elsewhere [22].

In addition, it was determined by Mössbauer spectroscopy, that 74% of iron is ferrous iron and 26% is ferric iron. These findings allowed the calculation of the amount of oxygen bound to iron and chromium, which must be removed in the

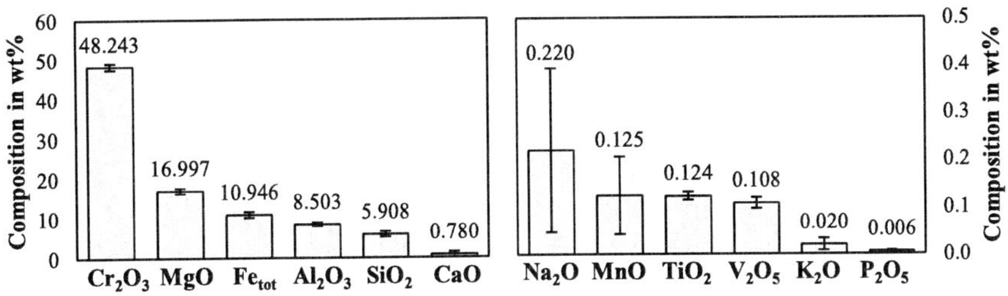

Fig. 1 Bulk chemical analysis of the used chromite sample

pre-reduction of chromite. The oxygen content that has to be removed is therefore 18.779 wt% in the chromite. Assuming that carbon will react completely with chromium- and iron oxide to carbon monoxide, 14.805 g of carbon per 100 g chromite are the stoichiometric carbon addition to remove oxygen bound to chromium and iron completely.

As reducing agents, charcoal made from coconut shells (following named coconut), corn cobs (following named corn), olive pomace (following named olive), bamboo (following named bamboo), and lignite coke made from Rhenish lignite (following named coke) are used. To calculate the required addition of reducing agents, it is necessary to determine the carbon content in each reducing agent. Table 1 shows the proximate-, ultimate-, P- and Cl-analysis of the reducing agents investigated in this study. For the calculation of the necessary addition of carbon, the carbon content as determined by the ultimate analysis is used.

According to the analysis shown in Table 1, the bio-based reducing agents have a significantly lower sulfur content compared to fossil coke, which is beneficial, since sulfur is an impurity in ferroalloys. However, the chlorine content of the bio-based reducing agents is higher compared to the chlorine content in coke, especially corn charcoal and olive charcoal have a high chlorine content. In an industrial operation,

Table 1 Proximate-, ultimate- and Cl-analysis of bio-based charcoal and coke used

All in wt%	Proximate analysis				Ultimate analysis					
Reducing agent	Fixed carbon	Total moisture	Ash	Volatile matter	C	H	N	S	O	Cl
Coconut charcoal	72.1	8.7	7.2	12.1	73.84	1.24	0.39	0.06	8.60	0.07
Corn charcoal	81.3	4.6	4.8	9.3	81.62	2.30	0.63	0.04	5.99	0.51
Olive charcoal	63.7	7.2	12.6	16.4	67.54	1.60	1.16	0.05	9.79	0.43
Bamboo charcoal	80.4	11.6	3.9	4.1	79.73	1.21	0.40	0.08	3.12	0.12
Lignite coke	87.5	0.5	9.0	3.0	89.0	0.4	0.4	0.5	0.7	0.03

Table 2 Ash analysis and basicity of bio-based charcoal and coke used

Reducing agent	Ash composition in wt%								Basicity
	Al_2O_3	CaO	Fe_2O_3	K_2O	MgO	Na_2O	P_2O_5	SiO_2	
Coconut charcoal	4.1	23.0	4.1	7.7	3.6	2.9	1.6	46.8	0.8
Corn charcoal	0.6	1.8	1.5	46.1	3.4	< 0.01	6.4	15.8	3.2
Olive charcoal	2.2	16.8	2.0	34.2	3.4	1.6	7.1	8.0	5.7
Bamboo charcoal	2.8	2.8	3.1	35.6	3.9	< 0.01	6.3	30.6	1.4
Lignite coke	3.1	35.4	10.9	0.8	15.8	6.7	0.2	2.2	13.1

this could lead to severe corrosion of equipment in contact with off-gas. Besides sulfur, further impurities could be introduced in the process from the ash. Therefore, the ash was further analyzed. Samples of the reducing agents weighing 100 g are burned in an oxidizing atmosphere at 815 °C until a constant mass is reached. Afterward, the ash is analyzed by XRF. Table 2 shows the determined chemical composition of the ash reported as oxides and the basicity of the ash samples as calculated by Eq. 2.

$$B = \frac{wt\%_{cao} \cdot wt\%_{Fe_2O_3} \cdot wt\%_{MgO} \cdot wt\%_{Na_2O} \cdot wt\%_{K_2O}}{wt\%_{SiO_2} \cdot wt\%_{Al_2O_3}} \quad (2)$$

Noticeable is the high CaO and MgO content in lignite coke, which results in the highest basicity of the lignite coke ash. In corn, olive, and bamboo ash, the prevalent compound is K_2O. Even though the analysis is reported as oxides, based on the high chlorine content in the samples shown in Table 1, potassium might be present at least partially as potassium chloride, which was also indicated by x-ray diffraction of the ash not presented in this article. The dominant component in coconut charcoal is SiO_2, yielding ash with the lowest basicity. During the smelting process after pre-reduction, the major oxides Al_2O_3, CaO, MgO, and SiO_2 from the ash will accumulate in the slag phase, while potassium and sodium will partially be volatilized during smelting as well. Especially since the potassium content in the biomass is quite high, this could lead to disadvantages during processing, like increased refractory corrosion. Iron oxide will be reduced either during pre-reduction or smelting as well as phosphorous, which is mostly transferred into the metal and partially into the slag. As phosphorous is an impurity in ferroalloys, this is another challenge to overcome, since all investigated ash samples contained more phosphorous than lignite coke.

Chromite was used in the trials as received and not pre-treated, while the reducing agents were ground in a vibratory disk mill for 20 s each. Table 3 shows the particle size of the raw materials as determined by dynamic imaging analysis.

While coconut, olive, and bamboo have a similar particle size, the same grinding time yielded finer corn charcoal and a coarser coke.

Table 3 Percentile particle sizes of raw materials used in this study

Raw material	Chromite	Corn charcoal	Coconut charcoal	Olive charcoal	Bamboo charcoal	Lignite coke
$x_{10,3}$ in µm	86.9	4.6	5.0	4.9	5.1	8.7
$x_{50,3}$ in µm	192.2	16.7	21.7	23.3	25.9	42.8
$x_{90,3}$ in µm	371.8	69.6	78.8	78.8	82.0	87.6

Methods

Mass loss trials were carried out using a Nabertherm HT 16/18 high-temperature furnace equipped with molybdenum disilicide heating elements. The initial sample mass per trial was 35.00 g, mixtures of ore with stoichiometric, sub-stoichiometric, and over-stoichiometric additions of reductants were investigated. The mass loss after every trial was measured, to determine the degree of reaction. Ore and reductants were also heated separately, to determine the individual mass loss without reduction reactions. The samples were placed in alumina crucibles (Ø = 50mm, h = 75mm) with a purity of 99.7%. To prevent oxidation of the samples, the furnace chamber was continuously flushed with 5 Nl/min argon in every part of the trial. The volume of the furnace chamber was 16 l. A maximum of six crucibles were placed inside the furnace per trial, to ensure that the temperature of the samples was uniform. Variable heating parameters in this investigation were the maximum temperature of each trial and the holding time at the maximum temperature. The heating time from room temperature to the maximum temperature was 3.5 h and the cooling time was not controlled. However, in the first hour after the trial, the furnace chamber cooled down by roughly 400 °C and in the subsequent hour by 200 °C.

To evaluate the degree of reaction, the mass loss based on the reduction was calculated using Eq. 3 subtracting the weight loss of the ore sample and reductant individually from the mass loss of the ore-reductant mixture. To consider the varying amount of ore and reductants in the mixtures, the mass loss of the pure compounds was calculated using Eq. 4. Finally, the degree of reaction was calculated per 100 g of chromite using Eq. 5. The maximum mass loss Δm_{Max} assumes, that oxygen bound to chromium and iron is completely removed as carbon monoxide. Based on the composition presented in Fig. 1, the maximum mass loss Δm_{max} per 100 g ore would be 32.863 g including oxygen removed from the ore and the removal of added carbon as carbon monoxide.

$$\Delta m_{Reduction} = \Delta m_{Mixture} - \Delta m_{Pure\,Compounds} \quad (3)$$

$$\Delta m_{Pure\,Compounds} = \frac{\Delta m_{Chromite}}{m_{Mixture}/m_{Chromite}} + \frac{\Delta m_{Reductant}}{m_{Mixture}/m_{Reductant}} \quad (4)$$

$$\text{Degree of Reaction} = 100\% \cdot \frac{\Delta m_{\text{Reduction}} \cdot 100\text{g} / m_{\text{Chromite}}}{\Delta m_{\text{Max}}} \tag{5}$$

Before the experimental work a thermochemical simulation of the pre-reduction process was carried out using the commercial modeling software FactSage™ 8.0 [23]. For the simulation the chromite composition presented in Fig. 1 including the ferric and ferrous ratio determined by Mössbauer spectroscopy was used, while carbon was used as the reducing agent. The databases FToxid, FactPS, and SGTE were used for the simulation. In contrast to Eq. 5, the removed oxygen per 100 g chromite was calculated by Eq. 6. The total mass of oxygen according to the simulation was used. The maximum amount of oxygen is based on the amount of oxygen bound to chromium and iron and is 18.779 g.

$$\text{Removed Oxygen} = 100\% \frac{\text{Mass of Oxygen in Atmosphere}}{m_{\text{Maximum Oxygen Removed}}} \tag{6}$$

Samples after the direct reduction of chromite and the initial ore were analyzed by x-ray diffraction with a “STADI MP” powder diffractometer made by “STOE&Cie GmbH” using the Kα1-radiation of molybdenum (wavelength = 0.70930Å). For the evaluation, HighScore Plus 4.9.0.27512 and the PDF-4 Axiom 2021 database were used.

Results and Discussion

In this chapter, the results of the thermochemical simulation, the experimental trials, the phase analysis of the reduced samples and the potential specific electrical energy saving are presented.

Thermochemical Simulation

Figure 2 shows the removed oxygen as calculated by Eq. 6 based on the thermochemical simulation in dependence of the temperature and the carbon addition.

According to Fig. 2, at temperatures below 1100 °C, a plateau is reached after the removal of 15–20% oxygen, which is mostly due to the reduction of iron oxides. At 1150 °C and 1200 °C, more oxygen is removed as chromium oxides start to react and a $(Cr, Fe)_7C_3$ phase is formed. However, a significant amount of chromium is still bound in a spinel phase. Above 1250 °C, the amount of the spinel phase and the chromium concentration is significantly reduced, in addition to the $(Cr, Fe)_7C_3$ phase, a $(Cr, Fe)_3C_2$ phase is formed in equilibrium with graphite. To examine the behavior of iron and chromium and iron more closely, Fig. 3 shows how iron and chromium are bonded at 1300 °C depending on the amount of carbon added.

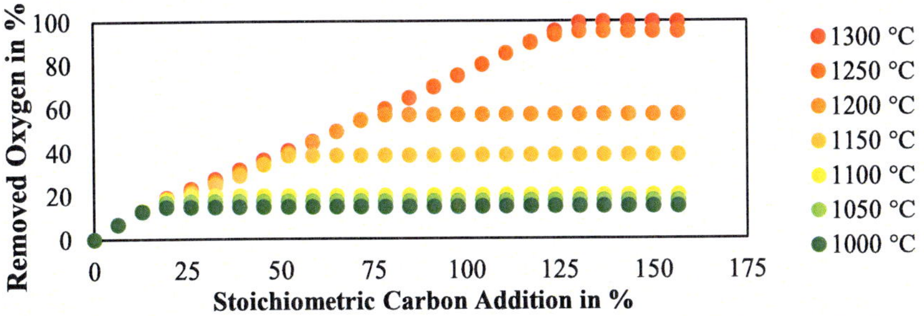

Fig. 2 Removed oxygen, dependent on the temperature and carbon addition

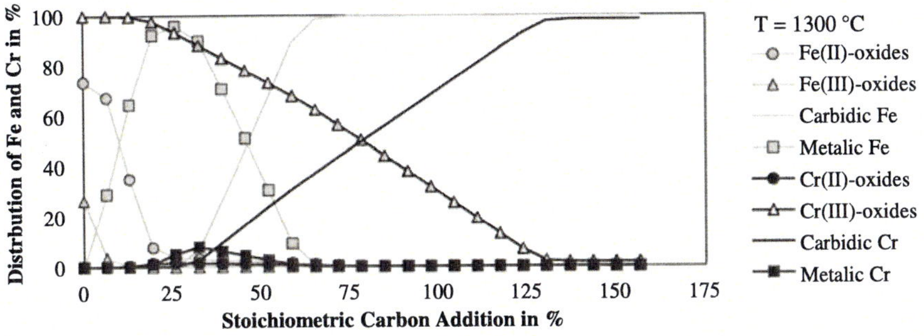

Fig. 3 Bonding of iron and chromium dependent on the carbon addition at 1300 °C

According to Fig. 3, adding 6.5% of the stoichiometric carbon requirement already leads to the formation of 29% iron in a metallic state, while trivalent iron oxides are reduced nearly completely. Adding 26% of carbon nearly reduces divalent iron completely and 5% of chromium is reduced into a metallic solution phase as well. Adding more carbon results in the reduction of chromium and $(Cr, Fe)_7C_3$ starts to form. In addition, metallic iron reacts to carbides as well. When the carbon addition exceeds 78.3%, no metallic phases are present anymore and iron and chromium are either bound as $(Cr, Fe)_7C_3$ or oxides. Above carbon additions of 130.5%, in addition to the $(Cr, Fe)_7C_3$ phase, a $(Cr, Fe)_3C_2$ phase starts to form and chromium and iron are not reduced anymore. The amount of iron not reduced in equilibrium with graphite is 0.02% and 1.71% of chromium is not reduced.

Degree of Reaction

The degree of reaction based on Eq. 5 was determined for various reducing agents, reaction temperatures and holding times for the solid-state reduction of chromite.

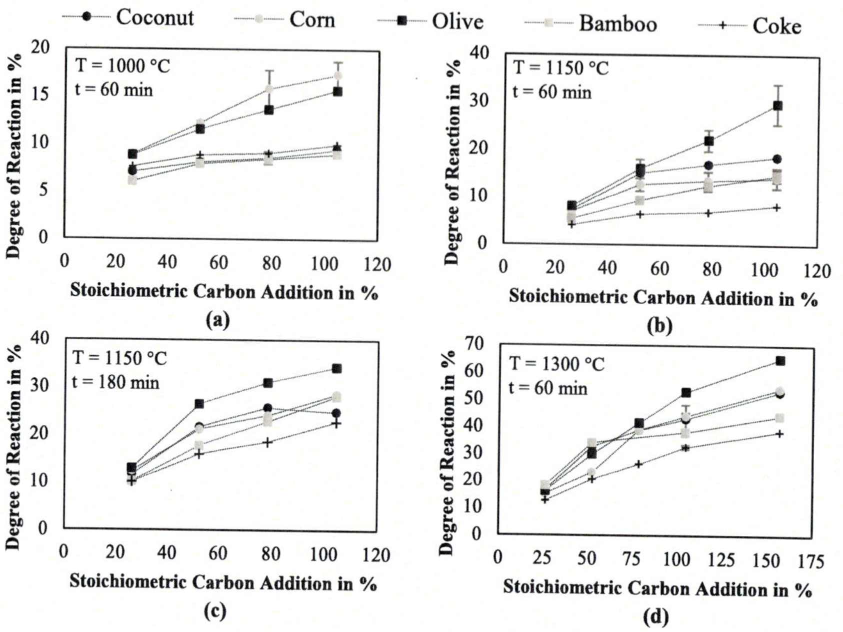

Fig. 4 Degree of reaction dependent on the carbon addition for various reducing agents and temperatures. **a** T = 1000 °C, t = 60 min, **b** T = 1150 °C, t = 60 min, **c** T = 1150 °C, t = 180 min, **d** T = 1300 °C, t = 60 min

Figure 4 shows the degree of reaction for various parameter combinations of preliminary trials.

In most cases, an increasing stoichiometric carbon addition also results in an increased degree of reaction. At 1000 °C using 104.43% carbon in the form of corn charcoal results in a degree of reaction of 17.4%, which is the highest degree of reaction and slightly higher than the theoretical removal of oxygen shown in Fig. 2. Olive charcoal results in a slightly lower degree of reaction, while the other three reducing agents yield a degree of reaction below 10%. Increasing the temperature to 1150 °C using the same holding time of 60 min results in a significant increase in the degree of reaction. At those parameters, olive charcoal yielded the highest degree of reaction. While the other bio-based reducing agents yielded a relatively similar degree of reaction, coke yielded a significantly lower degree of reaction. In Fig. 4c, the holding time is tripled compared to Fig. 4b, which yields a significantly increased degree of reaction for all reducing agents, except olive charcoal, where the degree of reaction increased only slightly. Coke still has the lowest degree of reaction, but is now closer to bamboo charcoal, coconut charcoal, and corn charcoal. In Fig. 4d, the holding time is 60 min again and the temperature is increased to 1300 °C, which increases the degree of reaction using olive charcoal in the best case to 65%, however, a stoichiometric carbon addition of 156.64% was used in this case. Using only 104.43% carbon yields a degree of reaction of 52.9% using olive charcoal, while the lowest degree of reaction was observed using coke with 32.6%.

As the degree of reaction at low temperatures was insufficient, more trials were carried out at 1300 °C with a constant carbon addition of 104.43% for different holding times. Higher temperatures were not investigated, to avoid partial melting of the mixtures, which could lead to damring formation in the pre-reduction process in rotary kilns [16].

During the first 90 min, olive charcoal yields the highest degree of reaction and coke the lowest. Corn and coconut charcoal behave similarly and slightly better than bamboo charcoal. A further increase in the reaction time especially yields an enhanced reaction for bamboo and coke, whereas the reaction of the other reducing agents increases less strongly. After 360 min, the bio-based reducing agents yield a degree of reaction between 61.0% and 65.4%. The highest degree of reaction is reached using coconut charcoal, followed by corn, olive, and bamboo charcoal. Coke results in the lowest degree of reaction with 51.9%. The lower degree of reaction of coke is in line with comparisons carried out by Kleynhans et al. [15], who investigated several fossil-reducing agents and determined, that the pre-reduction using coke was between 33 and 44%, while anthracite yielded a higher pre-reduction between 56 and 66% [15]. This was explained by the higher volatile content and therefore higher hydrogen content compared to the thermally treated coke [15]. The hydrogen content of the bio-based reducing agents varies between 1.21% and 2.30% and is 0.4% in the fossil coke, based on Table 1. The hydrogen-, and volatile content could be an explanation, why the bio-based reducing agents performed better compared to coke, but those properties cannot be used to explain the reducing ability of the bio-based reducing agents compared to each other. For example, coconut and bamboo have a similar hydrogen content, but they significantly differ in their degree of reaction.

Phase Analysis

To evaluate the solid phases generated in the trials, samples treated for 60 min with a stoichiometric carbon addition of 104.43% at different temperatures were analyzed by X-ray diffraction. In addition, the ore as received was analyzed by X-ray diffraction as well. Figure 6 shows the diffraction patterns of samples reduced with coconut charcoal and coke. Those reducing agents were chosen, as they yielded the highest and lowest degree of reaction as shown in Fig. 5.

Phases selected for the phase analysis were chromium carbide with the chemical formula Cr_7C_3 and the PDF-number 04-007-1045, an α-iron phase containing chromium with the chemical formula $Cr_{0.03}Fe_{0.97}$ and the PDF-number 04-004-2488, forsterite with the chemical formula Mg_2SiO_4 and the PDF-number 00-004-0768 and a magnesiochromite spinel with the chemical formula $Mg_{0.56}Ti_{0.01}Cr_{1.33}Fe_{0.51}Al_{0.59}O_4$ and the PDF-number 04-024-3779. In all reduced samples and in the raw ore, chromite and forsterite were identified. The intensity of chromite was the highest in all samples, however, spinels without chromite have peaks

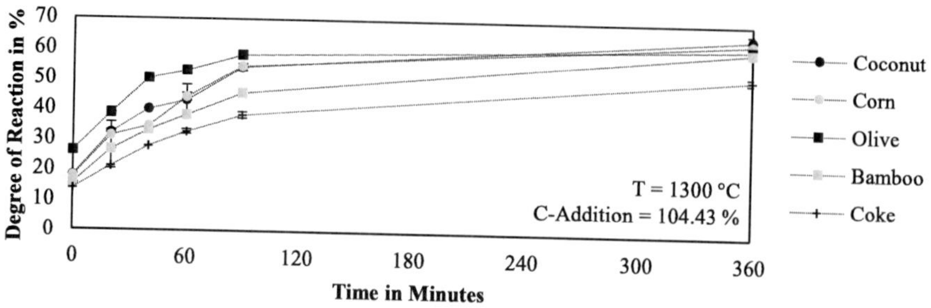

Fig. 5 Degree of reaction dependent on the holding time at 1300 °C for various reducing agents

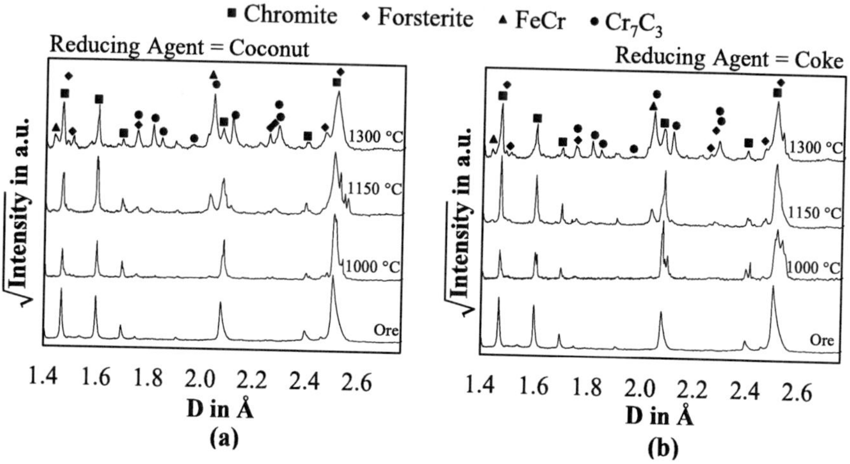

Fig. 6 X-ray diffraction pattern of raw ore and ore reduced at various temperatures with **a** coconut charcoal and **b** coke

at similar d-spacings, therefore it could be possible that a spinel without chromium is responsible for the measured intensities in the reduced samples. The observed forsterite peaks are weak in the raw ore and in the samples reduced at 1000 °C. In the samples reduced at higher temperatures, they become stronger, this is in line with the thermochemical simulation, that predicts an increasing amount of an olivine phase rich in Mg_2SiO_4. Peaks of the selected FeCr phase are already slightly visible at samples treated at 1000 °C, which was also predicted by the thermochemical simulation. Increasing the temperature leads to a slight increase in the intensity of the FeCr peaks. The majority of new peaks compared to the raw ore in the reduced samples at 1150 and 1300 °C can be explained by a chromium carbide phase, which has a relatively high intensity. Therefore, it can be assumed that the majority of pre-reduction does not result in the formation of a metallic phase, but rather a carbide phase instead.

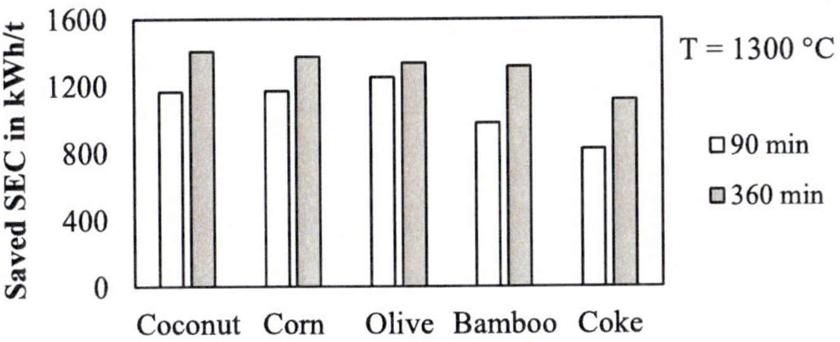

Fig. 7 Saved specific electrical energy consumption due to the pre-reduction at 1300 °C

Influence of the Pre-Reduction on the Specific Electrical Energy Consumption

Assuming, that the degree of reaction as calculated by Eq. 5 can be used to calculate the specific electrical energy consumption (SEC), the saved SEC in the subsequent smelting process can be calculated by Eq. 1. This is shown in Fig. 7 for the five reducing agents used at 1300 °C for the pre-reduction and for reaction times of 90 and 360 min.

It is clearly visible that any pre-reduction process results in a decrease of SEC. The highest decrease in SEC with 1373.8 kWh/t is possible with coconut charcoal at 360 min holding time. The lowest decrease of SEC with 822.0 kWh/t can be reached using coke as a reducing agent at 60 min holding time. The mean SEC savings in percent were 40% for the bio-based reducing agents and 32.8% for coke after 360 min.

Conclusion

The pre-reduction of chromite followed by the smelting in a submerged arc furnace is currently the industrial production route with the lowest specific electrical energy consumption. However, fossil carbon carriers are used as a reducing agent in this process, and this leads to direct CO_2 emissions. To decrease the specific CO_2 emissions for the production of ferrochrome, the suitability of bio-based reducing agents as an alternative to coke was investigated. In this work, the pre-reduction of chromite by several reducing agents was carried out for varying temperatures, reaction times, and carbon additions. While at 1000 and 1150 °C the degree of reaction was rather low, a reaction temperature of 1300 °C yielded a maximum degree of reaction up to 65.4% using coconut shell charcoal as a reducing agent, followed by corn cob charcoal with 63.9%, olive pomace charcoal with 62.1%, bamboo charcoal with 61.0%, and lignite coke with 51.9%. Especially after longer reaction times, the degree of reaction of the bio-based reducing agents was relatively similar, while for shorter durations significant differences in the degree of reaction were observable. The X-ray diffraction analysis of the pre-reduced samples further revealed, that the reaction occurring during the pre-reduction is mostly attributed to the formation of carbides with the stoichiometric formula Cr_7C_3, while the content of metals produced was significantly smaller.

It is estimated, that the pre-reduction of chromite can decrease the specific electrical energy consumption by 40% compared to the direct smelting of chromite. However, there are some issues to overcome before bio-based carbon can be a substitute in an industrial scale. As some samples of bio-based carbon contained significantly higher chlorine and phosphorus contents compared to coke. In the case of chlorine, this could be a challenge for equipment in contact with off-gas, which will likely contain elevated chlorine contents. In the case of phosphorous, this could be an issue regarding the quality of the produced ferrochrome, since phosphorous is an impurity in ferroalloys. Therefore, potential bio-based charcoals have to be carefully selected, to avoid problems due to the chemical composition. The sample with the lowest phosphorous and chlorine content is coconut shell charcoal, which also yielded the highest degree of reaction. One advantage of all bio-based carbon samples is a significantly lower sulfur content compared to fossil coke, which might yield lower sulfur contents in the final alloy and reduced SO_2-emission to the atmosphere.

References

1. Sommerfeld M, Friedrich B (2021) Replacing fossil carbon in the production of ferroalloys with a focus on bio-based carbon: a review. Minerals 11(11):1286. https://doi.org/10.3390/min11111286
2. Fick G, Mirgaux O, Neau P, Patisson F (2014) Using biomass for pig iron production: a technical, environmental and economical assessment. Waste Biomass Valorization 5(1):43–55. https://doi.org/10.1007/s12649-013-9223-1
3. Suopajärvi H, Umeki K, Mousa E, Hedayati A, Romar H, Kemppainen A, Wang C, Phounglamcheik A, Tuomikoski S, Norberg N, Andefors A, Öhman M, Lassi U, Fabritius T (2018) Use of biomass in integrated steelmaking—status quo, future needs and comparison to other low-CO_2 steel production technologies. Appl Energy 213:384–407. https://doi.org/10.1016/j.apenergy.2018.01.060
4. Mousa E, Wang C, Riesbeck J, Larsson M (2016) Biomass applications in iron and steel industry: an overview of challenges and opportunities. Renew Sustain Energy Rev 65:1247–1266. https://doi.org/10.1016/j.rser.2016.07.061
5. Quader MA, Ahmed S, Ghazilla RAR, Ahmed S, Dahari M (2015) A comprehensive review on energy efficient CO_2 breakthrough technologies for sustainable green iron and steel manufacturing. Renew Sustain Energy Rev 50:594–614. https://doi.org/10.1016/J.RSER.2015.05.026
6. Suopajärvi H, Kemppainen A, Haapakangas J, Fabritius T (2017) Extensive review of the opportunities to use biomass-based fuels in iron and steelmaking processes. J Clean Prod 148:709–734. https://doi.org/10.1016/j.jclepro.2017.02.029
7. Rosenfeld DC, Böhm H, Lindorfer J, Lehner M (2020) Scenario analysis of implementing a power-to-gas and biomass gasification system in an integrated steel plant: a techno-economic and environmental study. Renew Energy 147:1511–1524. https://doi.org/10.1016/J.RENENE.2019.09.053

8. Mandova H, Leduc S, Wang C, Wetterlund E, Patrizio P, Gale W, Kraxner F (2018) Possibilities for CO_2 emission reduction using biomass in European integrated steel plants. Biomass Bioenerg 115:231–243. https://doi.org/10.1016/j.biombioe.2018.04.021
9. Suopajärvi H, Pongráczb E, Fabritiusa T (2013) The potential of using biomass-based reducing agents in the blast furnace: a review of thermochemical conversion technologies and assessments related to sustainability. Renew Sustain Energy Rev 25:511–528. https://doi.org/10.1016/j.rser.2013.05.005
10. Gupta RC (2003) Woodchar as a sustainable reductant for ironmaking in the 21st century. Min Proc Ext Met Rev 24(3–4):203–231. https://doi.org/10.1080/714856822
11. Wei R, Zhang L, Cang D, Li J, Li X, Xu CC (2017) Current status and potential of biomass utilization in ferrous metallurgical industry. Renew Sustain Energy Rev 68:511–524. https://doi.org/10.1016/j.rser.2016.10.013
12. Basson J, Daavittila J (2013) High carbon ferrochrome technology. In: Gasik M (ed) Handbook of ferroalloys. Elsevier/Butterworth-Heinemann. Amsterdam, The Netherlands, pp 317–363
13. Kleynhans ELJ, Beukes JP, van Zyl PG, Fick JIJ (2017) Techno-economic feasibility of a pre-oxidation process to enhance prereduction of chromite. J S Afr Inst Min Metall 117(5):457–468. https://doi.org/10.17159/2411-9717/2017/v117n5a8
14. Kleynhans ELJ, Neizel BW, Beukes JP, van Zyl PG (2016) Utilisation of pre-oxidised ore in the pelletised chromite pre-reduction process. Miner Eng 92:114–124. https://doi.org/10.1016/j.mineng.2016.03.005
15. Kleynhans ELJ, Beukes JP, van Zyl PG, Bunt JR, Nkosi NSB, Venter M (2017) The effect of carbonaceous reductant selection on chromite pre-reduction. Metall Mater Trans B 48(2):827–840. https://doi.org/10.1007/s11663-016-0878-4
16. van Staden Y, Beukes JP, van Zyl PG, Ringdalen E, Tangstad M, Kleynhans ELJ, Bunt JR (2018) Damring formation during rotary kiln chromite pre-reduction: effects of pulverized carbonaceous fuel selection and partial pellet melting. Metall Mater Trans B 49(6):3488–3503. https://doi.org/10.1007/s11663-018-1376-7
17. Mohale GTM, Beukes JP, Kleynhans ELJ, van Zyl PG, Bunt JR, Tiedt LR, Venter AD, Jordaan A (2017) SEM image processing as an alternative method to determine chromite pre-reduction. J S Afr Inst Min Metall 117(11):1045–1052. https://doi.org/10.17159/2411-9717/2017/v117n11a9
18. Neizel BW, Beukes JP, van Zyl PG, Dawson NF (2013) Why is $CaCO_3$ not used as an additive in the pelletised chromite pre-reduction process? Miner Eng 45:115–120. https://doi.org/10.1016/j.mineng.2013.02.015
19. Paktunc D, Thibault Y, Sokhanvaran S, Yu D (2018) Influences of alkali fluxes on direct reduction of chromite for ferrochrome production. J S Afr Inst Min Metall 118(12). https://doi.org/10.17159/2411-9717/2018/v118n12a9
20. Sokhanvaran S, Paktunc D, Barnes A (2018) NaOH-assisted direct reduction of Ring of Fire chromite ores, and the associated implications for processing. J S Afr Inst Min Metall 118(6). https://doi.org/10.17159/2411-9717/2018/v118n6a4
21. Yu D, Paktunc D (2018) Kinetics and mechanisms of the carbothermic reduction of chromite in the presence of nickel. J Therm Anal Calorim 132(1):143–154. https://doi.org/10.1007/s10973-017-6936-6
22. Sommerfeld M, Friedrich B (2021) Proposition of a thermogravimetric method to measure the ferrous iron content in metallurgical-grade chromite. Minerals 12(2):109. https://doi.org/10.3390/min12020109
23. Bale CW, Bélisle E, Chartrand P, Decterov SA, Eriksson G, Gheribi AE, Hack K, Jung I-H, Kang Y-B, Melançon J, Pelton AD, Petersen S, Robelin C, Sangster J, Spencer P, van Ende M-A (2016) FactSage thermochemical software and databases, 2010–2016. Calphad 54:35–53. https://doi.org/10.1016/j.calphad.2016.05.002

Pyrometallurgy-Based Research Conducted at Mintek Towards Decarbonising the Metals Industry

Joalet Dalene Steenkamp, Pieter Johannes Andries Bezuidenhout, Itumeleng Thobadi, Lunia Malaka, Susanna Aletta Carolina Hockaday, Glen Michael Denton, Buhle Sinaye Xakalashe, Quinn Gareth Reynolds, Elias Matinde, Thokozile Penelope Kekana, Sonwabo Bambazala, and Aditya Kale

Abstract One of the focus areas for research conducted by the Pyrometallurgy Division at Mintek in South Africa is decarbonisation of the metals industry. In a country with a large pyrometallurgical industry, based primarily on carbonaceous primary reduction processes and electricity generated primarily by coal-fired power stations, the challenge can be viewed from a number of perspectives. The paper provides a summary of the projects currently underway, where the Technology Readiness Levels (TRL) ranges from TRL 1 to TRL8.

Keywords Decarbonising · Metals industry · Pyrometallurgy

Introduction

Mintek is a South African-based science council established by the Mineral Technology Act No. 30 of 1989 [1]. The objectives of Mintek are to promote mineral technology and foster the establishment and expansion of industries in the field of minerals and products derived therefrom, through research, development, and technology transfer. The history of Mintek goes back to 1934 with the Pyrometallurgy Division being one of the first to be established. The Pyrometallurgy Division at Mintek is well known for the research, development, and technology transfer of applications of direct current (DC) electric arc furnace (EAF) technology for metals production based on primary and secondary resources [2]. South Africa has a large

J. D. Steenkamp (✉) · P. J. A. Bezuidenhout · I. Thobadi · L. Malaka · G. M. Denton · B. S. Xakalashe · Q. G. Reynolds · E. Matinde · T. P. Kekana · S. Bambazala · A. Kale
200 Malibongwe Road, Randburg 2125, South Africa
e-mail: joalets@mintek.co.za

J. D. Steenkamp · E. Matinde
University of the Witwatersrand, 1 Jan Smuts Ave, Johannesburg 2000, South Africa

S. A. C. Hockaday · Q. G. Reynolds
University of Stellenbosch, Private Bag X1, Matieland, Stellenbosch 7602, South Africa

A. Lazou et al. (eds.), *REWAS 2022: Developing Tomorrow's Technical Cycles (Volume I)*, The Minerals, Metals & Materials Series,
https://doi.org/10.1007/978-3-030-92563-5_66

pyrometallurgical industry due to the abundant availability of raw materials and historically low prices of electricity [3]. Many of these are based on the carbonaceous primary reduction processes and electricity is generated primarily by coal-fired power stations. The challenge of decarbonising the Pyrometallurgy-based metals industry in South Africa can therefore be approached from either decarbonising reductants or decarbonising the energy mix, or both. The paper provides a summary of the projects currently underway at the Pyrometallurgy Division at Mintek to address this challenge, where the Technology Readiness Levels (TRL) ranges from TRL 1 to TRL8 [4].

Projects Focusing on Decarbonisation by Changing the Energy Mix

Energy Storage

To address the largest shortfall of renewable energy sources, namely the intermittent nature thereof, energy storage is required. With thermal energy storage being one of the most effective methods of storing energy and the majority of energy consumed being heat-based [5], this energy storage technology has the potential to ensure energy security build upon renewable energy sources. It is predicted by the International Energy Agency (IEA) that 11% of global electricity will be generated by CSP by the year 2050 [6], which speaks to the need for advancements in thermal storage technologies.

Mintek originally started to investigate molten salts for metallurgical applications, where these salts can be used for producing purer and carbon-free niche metals and alloys. Molten salt technologies can be used for the electrolytic production of niche metals such as Cr, Mn, and some rare-earth metals/alloys. During the initial investigation, the potential impact on Concentrated Solar Power (CSP) thermal storage technologies was also realised. Molten salt energy storage systems are currently the most applied CSP thermal storage technology on commercial power plants [7]. Molten salts for thermal storage have some inherent shortfalls: narrow range of temperatures in which liquid phases remain stable which complicate transport of the medium, as well as limiting operating temperatures and thermal properties such as high viscosities and low thermal conductivity [6]. To aid in the development of molten salts with improved energy storage capacity and better thermal stability, Mintek initiated a project focussing on and quantifying the thermal properties of thermal salts. These methods look at determining thermal properties such as melting point, heat capacity, thermal stability, and thermal conductivity. Preliminary results from the established laboratory methods are promising, which enabled the continuation of the project. Currently, theoretical selection of salt mixtures, using the FactSage thermochemical software package, is underway as well as molecular thermodynamic modelling/estimations of the respective thermal properties.

This technology is at TRL 1 level and ongoing.

The limitations of current molten salt technologies, together with the 800 °C target temperature for thermal storage systems set by the US Department of Energy [8], resulted in Mintek investigating alternative thermal storage research avenues. With molten salts typically having a safe operating temperature below 600 °C, the focus can be directed towards either a higher temperature sensible heat storage medium or latent heat storage mediums that possess much higher energy densities [7]. Mintek has proposed a project in which pyrometallurgical waste slag is to be investigated for a potential high-temperature sensible heat storage medium. Not only is slag a waste product, which will result in low material cost, it is also already processed enabling easy access and reduced material processing requirements. Aside from the physical appearance and sourcing advantages, the chemistry of these slags is well characterised and understood within the pyrometallurgy industry, making it an ideal candidate. Phase change materials (PCM) have also gained a lot of attention in recent times as potential energy storage mediums, owing to the higher energy densities associated with latent heat storage systems [5]. A special interest has been shown in PCMs that would allow operating temperatures in the range of 400–750 °C, as it could be directly matched with conventional steam turbine electricity generation systems [5]. As latent heat systems rely on the containment of molten materials at high temperatures, the pyrometallurgical experience and knowledge within Mintek is ideally positioned to aid in the development of these special thermal storage systems. A project has been proposed in which Mintek, in collaboration with some of the world's leading solar research facilities, is aimed at developing latent heat storage mediums that utilise molten metals and slag systems. These systems would not only allow for higher operating temperatures than the current state of the art CSP thermal storage systems but due to the higher energy densities, much smaller energy storage footprints are also predicted.

This technology is at TRL 1 with work to commence on approval of funding.

Solar Thermal Treatment of Manganese Ores

Given the advances made in renewable energy technologies [9–13], the direct solar thermal treatment of manganese ores is under investigation. Solar thermal technologies have been commercialised in water heating and steam production applications, including electricity production. More recently, studies have also investigated the use of solar thermal energy in the calcination of limestone and alumina [14, 15] and other minerals processing applications are under investigation [16–18]. Although the central tower technologies (including thermal storage) have been commercialised for power production, the direct treatment of solids has only recently found its first commercial applications [19]. Particle receiver technology [20], beam-down technology, and thermal storage technologies are key to the future of solar thermal applications in minerals processing.

Should solar thermal energy be integrated in the ferromanganese production chain and specifically in treatment of manganese ores before smelting? The most suitable solar technology for treating manganese ores at high temperature is the heliostat field and tower technology due to the high solar concentration ratios that can be achieved and the scalability of the technology. Work in this area has detailed the first direct solar treatment of manganese ores [21], the improvements realised with addition of air convection loops [22], the changes in the mineralogy of these samples, and the implications of these first experimental results. The solar thermal treatment of manganese ores was found to achieve thermal decomposition, calcination, and agglomeration to different degrees—serving as proof of concept that solar energy can replace fossil fuel combustion to achieve these aims.

This technology is now at TRL 3 and ongoing.

Integration of Preheating Step with AC or DC EAF Smelting Furnaces for the Processing of Manganese or Titanoferrous Ores

In contrast to the work discussed in Sect. "Solar Thermal Treatment of Manganese Ores", the PreMa project considers the indirect preheating of manganese ores before smelting using solar energy. A techno-economic study [23] has shown that the concept can realise reduced energy costs as well as reduced carbon dioxide emissions when compared to electric preheating and that this would be of most value to locations with high solar irradiance and high electricity costs. An experimental study on the heating of manganese ore with hot air is applicable to this aspect of solar thermal heating. The effect of integrating the preheating step, in the form of a rotary kiln, with SAF operations for the production of High Carbon Ferromanganese (HCFeMn) will further be demonstrated on pilot-scale at Mintek utilising the 300 kVA alternating current (AC) EAF. The PreMa project is reported on in more detail at the event by the project manager, Dr Eli Ringdalen, and will therefore not be discussed in more detail.

Once executed the technology should reach a TRL 7 level.

Mintek has over 45 years' experience in the research and development of titaniferous magnetite resources, available in abundance in South Africa, dating back to 1969. Mintek conducted in the order of 30 laboratory-scale projects, several pilot-scale prereduction or smelting trials (1–10 ton), and five smelting campaigns at the 100 ton demonstration scale resulting in Mintek developing and patenting the DC smelting of titaniferous magnetite. While various process options have been technically proven, their economics are markedly less attractive without a preheating or prereduction step to optimise electrical energy consumption. Although ore pre-treatment itself is not a technical challenge, the direct coupling of pre-treated titaniferous ore to a DC EAF has not been demonstrated. There is still a risk perception which provides a barrier to commercialisation. The TiMag project, funded by the

South African government, will demonstrate the direct coupling of the two stages. In a previous publication [24], Mintek reported on the basic engineering design of the pilot-scale facility, integrating a rotary kiln with a DC EAF, to evaluate the effect of preheating on electricity consumption in the EAF. The pilot facility is currently in the final stages of construction.

Once executed the technology should reach a TRL 8 level.

DC EAF Gasification of Discard Coal

Even though the developed world seems to be moving away from coal as the main energy source, it remains important in the developing world and South Africa is no exception. The process for mining and beneficiation of coal in South Africa produces just under 300 million tonnes of coal, and discards about 60 million tonnes per annum. The discard coal materials have marginal value and a calorific value below 16–19 MJ/kg, but still contain a significant amount of carbon that could be profitably used as a feedstock for the production of syngas by gasification with steam. Conventional gasification of coal is well known. The process energy is provided by the partial combustion of the feed coal with oxygen; thus this process relies on the coal calorific value. Plasma coal gasification is carried out using an external source of energy and can thus be carried out irrespective of the coal calorific value. The current research and development in plasma coal gasification is focused on plasma torches as a source of energy. Plasma torches have limited power and lifetime which has implications in scalability and availability of the gasification plant. For sustainable gasification of the vast dumps of discard coal in South Africa, economies of scale would be advantageous.

The DC arc plasma gasifier concept is currently under development at Mintek for testing the feasibility of producing syngas from waste coal fines and dry steam. The technology is intended for the valorisation of low-grade discard coals into market-quality syngas, which may then be used for large scale electrical power generation to address current grid supply challenges in South Africa. Other potential applications include feedstock for petro-chemical plants, fuel for transportation vehicles, domestic and industrial heating, and others. The proposed concept of a DC arc plasma gasifier was developed based on a plasma torch, metallurgical-based and entrained flow coal gasifiers. The principles of design and operation of the proposed gasifier are based on the concept of the Westinghouse Plasma Corporation (WPC) gasifier where the torches are substituted by graphite electrodes from which electric arc plasmas are generated [25]. The DC section of the gasifier will be designed considering Mintek's experience in the sealing of the furnace as carried out in the Mintek Thermal Magnesium Process. The DC EAF can be scaled up to sizes in excess of 100 MW. Furthermore, the use of graphite electrodes allow continuous operation of DC EAF. DC EAF are flexible with regard to size of feed material and type of feed materials.

The operability of the DC arc gasifier concept has been successfully tested at Mintek, through a combination of experimental work and modelling of the arc fundamentals (main critical elements). This work included (1) arc behaviour, (2) short burst gasification, as well as (3) modelling of the temperature distribution within the reactor. The 3.2 MVA pilot DC arc facility in Bay 1 at Mintek was used for the evaluation of the behaviour and stability of the arc in the presence of dry steam. The results obtained to date were sufficient to support and complete a South African provisional patent application (South African Patent Application No: 2016/01209). Small pilot-scale testwork as a proof of concept for coal gasification is planned.

This technology is now at TRL 2 and ongoing.

Projects Focusing on Decarbonisation Through Alternative Reductants

Methane Reduction

Silicomanganese (SiMn) is only produced in AC submerged arc furnaces (SAFs) and is electrical energy intensive requiring 3 500 and 4 500 kWh/ton alloy [26]. The production process produces 1.4 ton CO_2/tonne alloy [27] and requires raw materials to be larger than 6 mm to allow for gas permeability of the burden [26]. The SiManNat project investigates the potential for pre-reducing an agglomerate of manganese ore and quartz in natural gas, to form activated carbon and H_2, which will result in the solid state formation of manganese and silicon carbides at temperatures significantly lower than those associated with conventional smelting. The reduced agglomerate can subsequently be smelted at significantly lower electrical energy requirements, either in a conventional AC SAF or in a shallow burden DC EAF. The process will utilise manganese and quartz fines (−6 mm), and result in reduced CO_2 emissions and electrical energy required.

The project is currently at TRL 1 and ongoing.

Hydrogen Reduction

The use of hydrogen as an alternative reductant to carbon is foreseen as a possible way towards greener processes. Establishing hydrogen as a viable reductant for minerals beneficiation will increase the demand for hydrogen and advance the hydrogen economy. Hydrogen reduction of ores, particularly iron oxide, takes place at lower temperatures in comparison with carbothermic reduction. This could contribute to reduced energy requirements for the production of the metals of interest, compounding the benefits of employing this technology especially for an energy constrained country such as South Africa.

The production of steel and iron via a hydrogen reduction route has already been proposed and may become reality in the near future [28]. Hydrogen reduction of other metals has not been investigated to such detail and presents an opportunity for Mintek to be a global leader in this field. Mintek is investigating the development of hydrogen reduction in DC EAFs for metal production. In the context of South Africa, the targeted metals are those that already employ the DC EAF technology for their production and those that are of significant importance to the local economy. The anticipated results are reduced CO_2 emissions and reduced electrical energy consumption.

The project is currently at TRL 1 and ongoing.

Coke Produced from Discard Semi-Soft Coking Coal

The process of coke making in industrial coke ovens begins at the stage of mining hard coking coals and thereby the overall CO_2 emissions are a combination of the coal mining and coke-making (pyrolysis of hard coking coals) processes [29–31]. According to Miller [32], the underground coal mining process results in carbon emissions in the range 10–25 m^3 per ton of coal mined (i.e. 0.17–0.43 metric ton CO_2 equivalent per ton of coal mined), while the surface mining of coal results in the range of 0.3–2 m^3/ton of coal mined (i.e. 0.005–0.035 metric ton CO_2 equivalent per ton of coal mined) [32]. Hard coking coal, the precursor for high grade metallurgical coke, is mined using both underground and surface mining processes and hence has a combined contribution towards carbon emissions. In addition, during the coal mining stage, coal that contains a high ash content, that is not suitable for beneficiation, is stockpiled in open landfills and becomes a health hazard due to possible spontaneous combustion and further release of greenhouse gases such as CO_2 and methane. The stockpiled discard coal finds little economical use, however, based on studies reporting on the preparation of carbon from inorganic waste [33] and the co-pyrolysis of plastic waste with coking coal [29], discard semi-soft coking coal can be exploited for possible use as raw material feedstock in the coke-making process.

One of the studies at Mintek is looking into the on the beneficiation of discard semi-soft coking coal through upgrading its fixed carbon content and to possibly attain metallurgical coke properties suitable for application in ferroalloy production processes. Discard semi-soft coking coal contains ash contents ranging from 30 to 60% and fixed carbon content between 20 and 65%. The composition of the ash typically consists of metal oxides such as quartz (SiO_2), alumina (Al_2O_3), iron oxide (FeO), magnesium oxide (MgO), and calcium oxide (CaO). The high ash and mineral content in discard semi-soft coking coal renders it not suitable for use in the production of high quality coke, and in addition, semi-soft coking coal does not achieve the hardness strength, elasticity, and carbon content suitable for its use in the ferroalloy production process [34–38]. As a result of the poor coking properties, semi-soft coking coals, showing low ash content ($\leq$9%), are typically used as alternatives for thermal coal in energy production processes. In order to attain metallurgical coke properties, the choice of carbon material needs to show a microstructure of carbon that consists of a combination of ordered and disordered regions of carbon layers such that it is not completely graphitised nor amorphous. This intrinsic structural arrangement imparts special properties such as high mechanical/tensile strength at high temperature, high electrical resistivity, low CRI, and high thermal conductivity, to name a few [39].

Some oxides such as FeO and MgO, typically found in large quantities in coal ash, can serve as substrates or catalysts for growing carbon from hydrocarbon precursors at high temperatures. Furthermore, hydrocarbon materials such as waste plastic and biomass can be converted into 'hard carbon', a carbon material consisting of graphitic and non-graphitic layers of carbon, under inert/oxygen-free conditions at high temperatures. The combination of potential catalysts for carbon conversion in coal ash and high-temperature synthesis of 'hard carbon' under inert conditions shows promise to improve the carbon content of low-quality coal discards and impart coke-like properties through the synthesis of 'hard carbon' [40, 41].

In addition, studies [42–44] show that the catalytic pyrolysis (i.e. thermal decomposition in an inert environment) of plastics such as polyethylene terephthalate (PET) favours the formation of a higher yield of pyrolytic gas products (i.e. CO, CH_4, and H_2) due to the catalytic cracking mechanism of light polycyclic hydrocarbons and benzene derivatives present in the reaction mixture. Park et al. [42] further show that the pyrolytic gas yield is proportional to the catalyst loading and, therefore, there is potential in the co-pyrolysis of high ash discard semi-soft coking coals with waste plastics/biomass to produce increased yields of fuel gases in the coke gas oven (CO, CH_4, and H_2). The proposed materials may be achievable through the systematic investigation of the high-temperature co-pyrolysis of coal/plastic and coal/biomass composites using a gas-tight tube furnace. The successful preparation of metallurgical coke from discard semi-soft coking coal and waste plastics may revive the manufacturing of metallurgical coke in South Africa, which has dried up due to the lack of hard coking coal, and also potentially minimise the carbon emissions as it eliminates the mining process, only making use of the extensive stockpiles of around 1500 Mt of discard coal currently present in South African landfills.

The project is currently at TRL 2 and ongoing.

Aluminothermic Reduction for Silicon Production

Silicon (Si) is the second most common element in the Earth's crust, although it is hard to find in nature as a pure element. In natural form it is almost exclusively combined with oxygen as silicon dioxides or quartz. The current commercial carbothermic reduction process for the production of metallurgical grade silicon (MG-Si) uses carbon for reducing quartz at high temperature, achieved by electric arcs in an AC SAF. MG-Si is further processed to produce pure silicon. This process is energy intensive and has a high carbon footprint. Carbon is used as reductant and the carbon electrodes used, are consumed in the process of generating an electric arc in the SAF. The carbothermic reduction process generates around 4.68 ton CO_2 per ton of Si metal [45]. The total production volume of silicon worldwide in 2020 amounted to an estimated eight million metric tons [46].

The Norges Teknisk Naturvitenskapelige Universitet (NTNU) has developed a patented novel industrial process known as '*The SisAl process*' to produce Si by aluminothermic reduction, enabling a shift from the carbothermic process to a far more environmentally and economically alternative. SisAl is the acronym for **Si**licon production with low environmental impact using secondary **Al**uminium and silicon raw materials.

In the SisAl process quartz in slag is reduced though aluminothermic reduction that utilises secondary raw materials such as aluminium (Al) scrap and dross, as replacements for carbon reductants used in the present commercial process. Compared to the current silicon production, the SisAl process will be able to cut energy consumption by two-thirds, eliminate direct emissions of CO_2 and other harmful pollutants, and operate at a considerably lower cost. After successful demonstration of the SisAl process on laboratory scale the '***SisAl pilot***' project was initiated to demonstrate the novel industrial process at TRL 6–7 for production of Si in three different qualities, metallurgical grade (MG–Si, >98%), high purity silicon (HP Si, >99.9%), solar grade (SoG-Si >6 N) and Al-Si alloys along with Metallurgical Grade Alumina (MGA) and High Purity Alumina (HPA). The SisAl Pilot project is coordinated by NTNU. The consortium comprises 22 partners from 9 countries and Mintek is one of the consortium partners. Mintek's task is to demonstrate the SisAl process at Mintek's pilot plant on a 3.2 MW electric arc furnace aimed at MG–Si production. The campaign at Mintek will lay the basis for commercial process design.

The SisAl process can be described in following steps:

1. Combining silicon dioxide (preferably as less costly fines or sand rather than lump) and calcium oxide in a vessel at a temperature sufficient to form a molten calcium silicate slag with a $CaO:SiO_2$ mass ratio of approximately 1 (slag-making at 1600–1700 °C);
2. Introducing a source of aluminium metal (Al scrap, dross, etc.) to the calcium silicate slag to reduce the calcium silicate slag to Si metal and form a calcium aluminate slag; subsequently separating the Si metal from the calcium aluminate slag.

From the overall chemical reaction above, there should be 4/3 mol of Al to one mol of Si present. In practice, it is an advantage to keep the Al addition non-stoichiometric to achieve a slag with low SiO_2 concentration. This reduction reaction is advantageous as it yields liquid silicon at much lower temperatures around 1500–1600 °C compared to the silicon production in a SAF, where MG-Si is formed at around 2000 °C. As this reaction is exothermic, limited energy needs to be supplied as the heat of reaction melts the Al scrap or dross added into the reduction-refining step. In fact, the heat within the reduction furnaces can be controlled by the speed of addition of the Al to the process. Depending on the raw material purity and mix, different Si compositions can be produced (MG–Si, Al–Si). To achieve HP–Si, a second metal-slag reaction reactor is required in the process set-up, with also stricter requirements on raw material quality.

The main advantages of The SisAl process are the following:

- The SisAl process has a low environmental footprint by using secondary Al (with a low lifecycle CO_2 footprint) rather than carbon for reducing SiO_2 to Si, with a 3 times lower energy consumption and higher yield (no losses through SiO emission) than the SAF process. The SisAl process generates minimal NO_x, dust, PAH, and SO_2.
- CO_2 emission in SisAl process is less than 0.3 t/t Si as compared to 4.68 t/t Si in the existing commercial carbothermic process
- The SisAl process valorises dross material from the aluminium process and end-of-life, low value scrap into primary alumina feed for the Al electrolysis process
- The SisAl process is very versatile both in raw materials and products:
  - It can use quartz mining fines and fine (pure) quartz, which is not possible in the SAF Si process.
  - Al dross/scrap ratio can be varied to make different qualities or replaced with high purity Al to produce pure Si
- By recycling CaO from the intermediate slag by-product, the only raw material consumed in the SisAl process are quartz and aluminium.

The pilot plant campaign at Mintek will be conducted on a 3.2 MW electric arc furnace for a period of 2 weeks. The pilot campaign will demonstrate the SisAl process for production of Mg–Si. Locally sourced raw material will be used for the pilot campaign. The operational data consisting of heat and mass balance, energy consumption, and gas emission generated during the pilot campaign will be used for the process and equipment design for commercialisation of the process. Environmental impact, investment cost analysis, and economic analysis will be performed based on data collected from the pilot plant campaign at Mintek.

Once executed the project will reach a TRL 8 level.

Conclusions

In order to decarbonise the Metals Industry in South Africa, not only should the use of alternative reductants be considered but even more so a change in the energy mix. For example, in manganese ferroalloy production the amount of CO_2 produced by the process typically is 1.3 and 1.4 ton per ton of HCFeMn and SiMn alloy, respectively [27]. When produced in South Africa an additional 2.8 and 3.7 ton CO_2 per ton of HCFeMn and SiMn alloy, respectively, is added, increasing the amount of CO_2 associated three-to fourfold [47]. The challenge that lies ahead is enormous and Mintek is playing its part by conducting research and development work addressing it from various perspectives.

Acknowledgements The PreMa project is funded by the European Union's Horizon 2020 Research and Innovation Programme under Grant Agreement No 820561 and industry partners: Transalloys, Eramet, Ferroglobe, OFZ, and Outotec.

The SisAl Pilot project is funded by the European Union's Horizon 2020 Research and Innovation Programme under Grant Agreement No 869268.

The paper is published with permission from Mintek.

References

1. Mintek (1989) Mineral Technology act. https://www.mintek.co.za/wp-content/uploads/2015/07/Mineral-Technology-act.pdf. Statutes of the Republic of South Africa—Mining and Minerals. Accessed 10 Sept 2021
2. Jones RT (2014) DC arc furnaces—past, present, and future. Paper presented at the 143rd TMS annual meeting. San Diego, California
3. Jones RT (2019) Pyrometallurgy in South Africa. www.pyrometallurgy.co.za/PyroSA/index.htm. Accessed 10 Sept 2021
4. Heder M (2017) From NASA to EU: the evolution of the TRL scale in Public Sector Innovation. Innovation 22(2):1–23
5. Reed S, Sugo H, Kisi E (2018) High temperature thermal storage materials with high energy density and conductivity. Sol Energy 163:307–314
6. Caraballo A, Galán-Casado A, Caballero A, Serena S (2021) Molten salts for sensible thermal energy storage: a review and an energy performance analysis. Energies 14(4):1197
7. Heller L (2013) Literature review on heat transfer fluids and thermal energy storage systems in CSP plants. https://sterg.sun.ac.za/wp-content/uploads/2011/08/HTF_TESmed_Review_2013_05_311.pdf. Accessed 10 Sept 2021
8. Project Profile: High-efficiency thermal storage system for solar plants. https://www.energy.gov/eere/solar/project-profile-high-efficiency-thermal-storage-system-solar-plants. Accessed 10 Sept 2021
9. Philibert C (2017) Renewable energy for industry. https://www.nordicenergy.org/wp-content/uploads/2017/11/Renewable-Energy-for-Industry-Cedric-Philibert.pdf. Accessed 10 Sept 2021
10. Ho CK (2017) Advances in central receivers for concentrating solar applications. Sol Energy 152:38–56
11. Kodama T, Bellan S, Gokon N, Cho HS (2017) Particle reactors for solar thermochemical processes. Sol Energy 156:113–132
12. Koepf E, Alxneit I, Wieckert C, Meier A (2017) A review of high temperature solar driven reactor technology: 25 years of experience in research and development at the Paul Scherrer Institute. Sol Energy 188:620–651
13. Ekman BM, Brooks G, Rhamdhani MA (2014) A review: solar thermal reactors for materials production. Paper presented at the 143rd TMS annual meeting. San Diego, California
14. Flamant G, Benoit H, Jenke M, Santos AF, Tescari S, Moumin G, Rodriguez A, Azapagic A, Stamford L, Baeyens J, Boes Y, Pron F, Prouteau M, Dumont P, Abdenouri N, Mazouz H (2017) Solar processing of reactive particles up to 900 °C, the SOLPART project. Paper presented at SolarPACES 2017: international conference on concentrating solar power and chemical energy systems. Santiago, Chile
15. Davis D, Müller F, Saw WL, Steinfeld A, Nathan GJ (2017) Solar-driven alumina calcination for CO_2 mitigation and improved product quality. Green Chem 19(13):2992–3005
16. Eglinton T, Hinkley J, Beath A, Dell'Amico M (2013) Potential applications of concentrated solar thermal technologies in the australian minerals processing and extractive metallurgical industry. JOM 65(12):1710–1720

17. Hockaday L (2019) Solar thermal applications in minerals processing in South Africa. Paper presented at the 6th South African solar energy conference. East London, South Africa
18. Moreno-Leiva S, Valencia F, Haas J, Chudinzow D, Eltrop L (2017) Solar energy alternatives for copper production. Paper presented at SolarPACES 2017: international conference on concentrating solar power and chemical energy systems. Santiago, Chile
19. Kraemer S (2021) HelioHeat Commercializes DLR's 1000 °C solar receiver CentRec®. https://helioscsp.com/helioheat-commercializes-dlrs-1000c-solar-receiver-centrec/. Accessed 10 Sept 2021
20. Ho CK, Christian JM, Yellowhair J, Siegel N, Jeter S, GolobM, Abdel-Khalik SI, Nguyen C, Al-Ansary, H (2016) On-sun testing of an advanced falling particle receiver system. Paper presented at SolarPACES 2015: international conference on concentrating solar power and chemical energy systems. Cape Town, South Africa
21. Hockaday L, Reynolds QG, Dinter F, Harms T (2017) Solar thermal treatment of manganese ores. Paper presented at SolarPACES 2017: international conference on concentrating solar power and chemical energy systems. Santiago, Chile
22. Hockaday L, Dinter F, Reynolds QG, McGregor C (2018) A comparison of direct concentrating solar thermal treatment of manganese ores to fossil fuel based thermal treatments. In: SolarPACES 2019: international conference on concentrating solar power and chemical energy systems. Casablanca, Chile
23. Hockaday L, McKechnie T, Von Puttkamer MN, Lubkoll M (2020) The impact of solar resource characteristics on solar thermal Pre-heating of manganese ores. Paper presented at the 149th TMS annual meeting. San Diego, California
24. Steenkamp JD, Denton GM, Pieters T (2020) Upgrading pilot-scale facility at MINTEK to evaluate the effect of preheating on smelter operations. In: Paper presented at the 149th TMS annual meeting. San Diego, California
25. Kekana T, Akdogan G, Reynolds QR (2021) The feasibility of producing syngas from waste coal fines and steam using a DC arc gasifier. In: Paper presented at the XXX international mineral processing congress. Cape Town, South Africa
26. Tangstad M (2013) Chapter 7: manganese ferroalloy technology. In: Gasik M (ed) Handbook of ferroalloys—theory and technology. Butterworth-Heinemann, Oxford, United Kingdom, pp 221–266
27. Lindstad T, Olsen SE, Tranell G, Færden T, Lubetsky J (2007) Greenhouse gas emissions from ferroalloy production. Paper presented at INFACON XI New Delhi, India
28. Seftejani M, Schenk J (2018) Thermodynamic of liquid iron ore reduction by hydrogen thermal plasma. Metals 8:1–17
29. Nomura S (2018) Recent developments in cokemaking technologies in Japan. Fuel Process Technol 159:1–8
30. Díez MA, Alvarez R, Barriocanal C, (2002) Coal for metallurgical coke production: predictions of coke quality and future requirements for cokemaking. Int J Coal Geol 50(1):389–412
31. Alvarez R, Dıez M (2000) Chemistry of production of metallurgical coke. In: Marsh H, Rodrıguez-Reinoso F (eds) Sciences of carbon materials. Secretariado de Publicaciones, Alicante, Spain, pp 595–635
32. Miller BG (2016) Clean coal engineering technology. Butterworth-Heinemann, Oxford, United Kingdom
33. Rambau KM, Musyoka NM, Manyala N, Ren J, Langmi HW, Mathe MK (2018) Preparation of carbon nanofibers/tubes using waste tyres pyrolysis oil and coal fly ash derived catalyst. J Environ Sci Health, Part A 53(12):1115–1122
34. Mxinwa S, Deenanath ED, Robertson SW, Ndlovu S, Basson P (2020) The application of coal discards for acid mine drainage neutralization. J South Afr Inst Min Metall 120(9):531–540
35. Amaral Filho JR, Firpo BA, Broadhurst JL, Harrison STL (2020) On the feasibility of South African coal waste for production of 'FabSoil', a Technosol. Miner Eng 146:106059
36. Musyoka NM, Wdowin M, Rambau KM, Franus W, Panek R, Madej J, Czarna-Juszkiewicz (2020) Synthesis of activated carbon from high-carbon coal fly ash and its hydrogen storage application. Renew Energy 155:1264–1271

37. Jeffrey LS (2005) Characterization of the coal resources of South Africa. J South Afr Inst Min Metall 105(2):95–102
38. Lloyd P (2000) The potential of coal wastes in South Africa. J South Afr Inst Min Metall 100(1):69–72
39. Tangstad M, Beukes JP, Steenkamp JD, Ringdalen E (2019) Coal-based reducing agents in ferroalloys and silicon production. In: Suárez-Ruiz I, Diez MA, Rubiera F (eds) New trends in coal conversion—Combustion, gasification, emissions, and coking. Woodhead Publishing, Elsevier, Duxford, United Kingdom, pp 405–438
40. Lei C-M, Yuan W-L, Huang H-C, Ho S-W, Su C-J (2011) Synthesis and conductivity measurement of carbon spheres by catalytic CVD using non-magnetic metal complexes. Synthetic Met 161:1590–1595
41. Sharma S, Kalita G, Hirano R, Shinde SM, Papon R, Ohtani H, Tanemura M (2014) Synthesis of graphene crystals from solid waste plastic by chemical vapor deposition. Carbon 72:66–73
42. Park C, Kim S, Kwon Y, Jeong C, Cho Y, Lee C-G, Jung S, Choi K-Y, Lee J (2020) Pyrolysis of polyethylene terephthalate over carbon-supported Pd catalyst. Catalysts 10:496
43. Dhahak A, Grimmer C, Neumann A, Rüger C, Sklorz M, Streibel T, Zimmermann R, Mauviel G, Burkle-Vitzthum V (2020) Real time monitoring of slow pyrolysis of polyethylene terephthalate (PET) by different mass spectrometric techniques. Waste Manage 106:226–239
44. Nabgan W, Nabgan B, Tuan Abdullah TA, Jalil AA, Ul-Hamid A, Ikram M, Nordin AH, and Coelho A (2021) Production of hydrogen and valuable fuels from polyethylene terephthalate waste dissolved in phenol reforming and cracking reactions via Ni-Co/CeO_2 nano-catalyst. J Anal Appl Pyrolysis 154:105018
45. Monsen B, Lindstad T, Tuset JK (1998) CO_2 emissions from the production of Ferrosilicon and Silicon metal in Norway. In: Paper presented at the 56th Electric furnace conference. New Orleans, Louisiana, pp 371–378
46. Garside M (2021) Silicon—statistics & facts. https://www.statista.com/topics/1959/silicon/. Accessed 10 Sept 2021
47. Steenkamp JD, Chetty D, Singh A, Hockaday SAC, Denton GM (2020) From ore body to high temperature processing of complex ores: manganese—a South African perspective. JOM 72(10):3422–3435

Biomass as a CO_2-Neutral Carbon Substitute for Reduction Processes in Metallurgy

Christian Dornig and Jürgen Antrekowitsch

Abstract The increasing consumption of carbon-based products in recent decades and their impact on the environment and climate change is becoming more and more significant. To curb this problem, the use of pyrolysed biomass as a reducing agent substitute for carbon-based reduction processes can be part of the solution for a CO_2-neutral metallurgy. Previous research has shown that biochar can be a replacement for conventionally used carbon sources. However, the economic interest in this technology has gradually decreased, due to the falling prices for the CO_2-certificates about eight years ago. In the meantime, the prices for the certificates have risen strongly again and the topic is coming back into focus. In this publication, the use of biomass was investigated in an experimental setup and the selection, properties, and possible applications are discussed. The results show that biomass can be an adequate product for CO_2-neutral process control.

Keywords Biomass · CO_2-neutral · Reduction process

Introduction

Before the prices for CO_2-certificates dropped about ten years ago, there was a steady increase in the price. Just as in the last few years, especially in 2020 and 2021, there was a sharp rise again (Fig. 1). The growing impact of greenhouse gases on the climate and the resulting restrictions also require appropriate adaptation in industry.

Currently a lot of effort is being put into the use of hydrogen as a reductant for a CO_2-free metallurgical industry. However, in many cases it is not straightforward to realize, because the existing processes and related equipment are inoperative when applying a direct replacement of the used reducing agent by hydrogen. As another CO_2-neutral option, biomass can serve, e.g. as alternative for fossil-based reducing agent without major adaptation of the process.

C. Dornig (✉) · J. Antrekowitsch
Montanuniversität Leoben, Chair of Nonferrous Metallurgy, Franz Josef-Straße 18, A-8700 Leoben, Austria
e-mail: christian.dornig@unileoben.ac.at

A. Lazou et al. (eds.), *REWAS 2022: Developing Tomorrow's Technical Cycles (Volume I)*, The Minerals, Metals & Materials Series,
https://doi.org/10.1007/978-3-030-92563-5_67

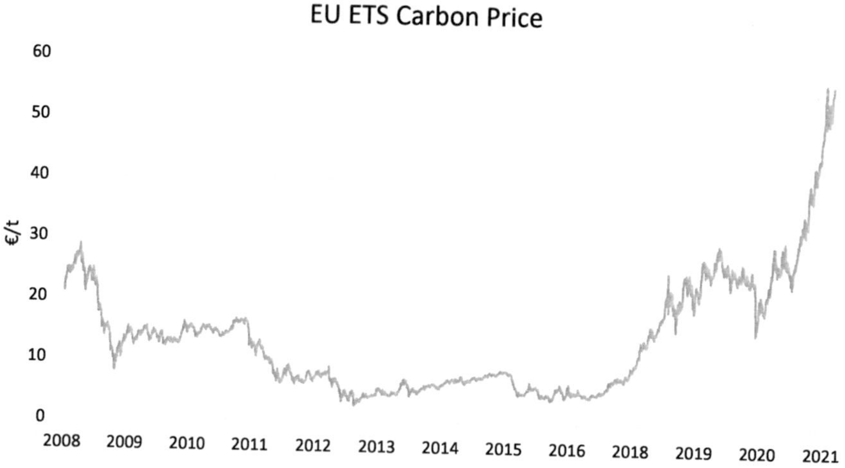

Fig. 1 Development of CO_2-Credit prices based on data from the International Carbon Action Partnership [1]

As mentioned before, the price for the CO_2-certificates started to increase a few years ago and so the efforts to research alternative reducing agents for metallurgical processes became more interesting.

The Chair of Nonferrous Metallurgy at Montanuniversität Leoben has now started to investigate the topic of biomass as a reducing agent substitute in more detail again. In the past, investigations have already been carried out, for example the application in a rotary kiln. This time, the research also focuses on the field of molten processes.

Many different types of biomasses have already been studied, while current investigations are focusing on waste from the forestry industry.

This paper recaps the results of the trials performed in the past and gives a short outlook on the future activities.

Definition of Biomass

Biomass describes all substances of organic origin or carbonaceous matter. This definition includes all plants and animals living in nature and their residues as well as the mortal remains, not yet fossil, phyto- and zoomass. Furthermore, the term also comprises materials that have been created by a substantial use of the biomass. This includes for example paper, pulp, slaughterhouse waste, vegetable oils, alcohols, and household wastes. According to the definition, fossil energy carriers do not belong to biomass. Turf belongs either to biomass or fossil energy sources, depending in the description [2].

Furthermore, biomass can be divided into primary and secondary products. Primary products result from the direct utilization of solar energy and thus contain the entire plant mass. This scope includes agricultural and forestry products (trees, grasses) and by-products or plant residues (straw, forest and industrial wood residues). The second area describes the secondary products, which use solar energy only indirectly. This biomass results from the decomposition or conversion of organic substances, for example in animals. In summary, this section includes the entire zoomass, the excrements, and the sewage sludge [2, 3].

This very comprehensive description of the definition demonstrate that biomass comes from various sources. Therefore, the products resulting from individual conversion steps are also very different [3].

Investigated Biomasses

In the past, various biomasses have been studied regarding their technical utility in metallurgical processes at the Chair of Nonferrous Metallurgy, Montanuniversität Leoben. The listed materials provide an overview of the previously investigated biomasses:

- Olive (stone, cuttings)
- Fruit cuttings
- Vine (cuttings/marc)
- Beer marc
- Oak tree cuttings
- Different types of eucalyptus and pine
- Wood residues
- Bark
- Wood chips
- Waste wood

Pyrolysis of Biomass

The biomass cannot be used directly due to its composition. To obtain a usable product, it is necessary to perform a thermal treatment, called pyrolysis.

In general, during the pyrolysis process, biomass is heated to a defined temperature in the absence of air or oxygen and kept at this temperature for a certain time. This treatment converts the biomass into charcoal, gas, and tar (Fig. 2). Via the temperature, the residence time and the heating rate, it is possible to influence the end products. In terms of temperature, pyrolysis can be divided into low (<500 °C), medium (500–800 °C), and high temperature pyrolysis (>800 °C) [4, 5].

Table 1 summarizes the different forms of pyrolysis existing. In the previous studies, the focus was on the resulting coal, so carbonization was preferred. This allows the production of a CO_2-neutral charcoal as a substitute for reducing agents, called biochar.

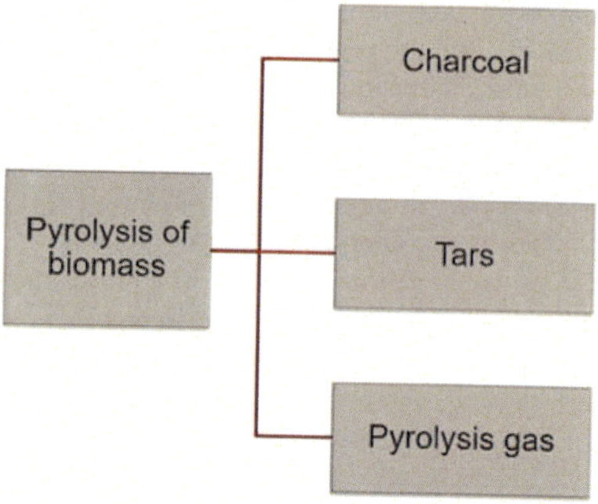

Fig. 2 Main products of pyrolysis

Table 1 Product yields of different biomass conversion techniques [7]

Process	Conditions	Liquid (%)	Char (%)	Gas (%)
Fast pyrolysis	Moderate temperature, short residence time, particularly vapour	75	12	13
Carbonization	Low temperature, very long residence time	30	35	35
Gasification	High temperature, long residence time	5	10	85

The resulting biochar is defined as biomass whose relative carbon content has been increased by a thermal treatment step [6].

Used Pyrolysis Unit

Due to the fact that the various types of biomasses are available with different moisture contents, a drying step has to be performed before the process of pyrolysis. After this treatment, the material is charged into the twin-screw reactor for conversion. With the semi-continuous aggregate, used at the Montanuniversität Leoben (Fig. 3), it is possible to charge up to 25 kg, depending on the specific weight. An external electrical heating system can be used to achieve a temperature up to 950 °C over several zones. The residence time is adjustable between a few minutes to a couple of hours by the twin screw rotation speed.

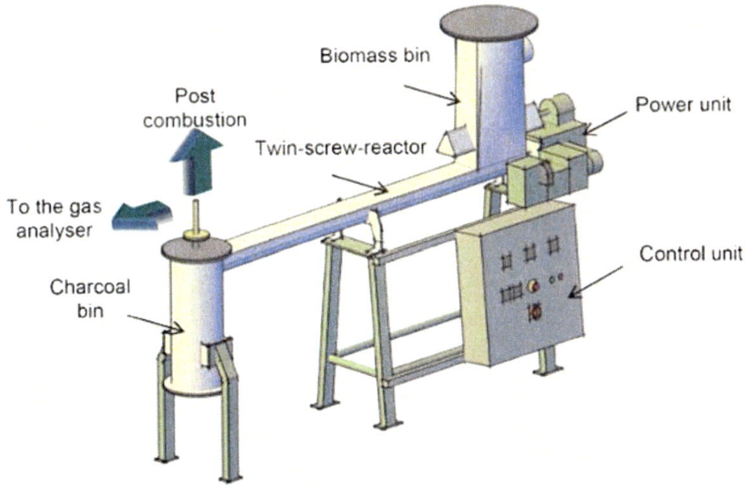

Fig. 3 Semi-continuous "twin screw reactor" [8]

Characterization of Biomass

For an evaluation of the produced biochar with regard to its usability in metallurgical aggregates, other characteristics are important in addition to the chemical composition. The most important properties for the various applications are:

- Coke reactivity
- Carbon content (C_{Fix})
- Volatiles
- Ash content
- Moisture content
- Sulphur content
- Specific mass
- Pressure resistance
- Specific surface

The difficulty is that the various metallurgical processes often require contradictory properties. For example, some processes require high reactivity, so that the carbon can react immediately (molten processes) and higher contents of volatiles may not be disturbing. In others, low reactivity is desired, because the carbon should not react directly after charging (Waelz process). Therefore, it has to be considered that the assessment of the parameters is different for every process and it has to be evaluated individually for each case which biomass could be suitable.

Investigations and Results on Pyrolysis

Not only the biomass itself or its composition determines the usability, also pre-treatments or adjustments during the pyrolysis process, as stated in the following section, have an impact.

Grain Size

For some biomasses it is necessary to perform a shredding before the pyrolysis. Thus, experiments were done which demonstrate the influence of the particle size on the pyrolysis result. As the values in Table 2 show, the charcoal yield does not change significantly between the different fractions. Furthermore, the content of volatile matter diminishes and the carbon content increases slightly. The remaining parameters show no significant change due to the particle size.

Table 2 Data from pyrolysis of fruit tree cuttings with different grain size [9]

Grain size [cm]	Charcoal yield [%]	C [%]	H [%]	N [%]	S [%]	Volatiles [%]	C_{Fix} [%]	H_2O [%]
<0.8	22.0	84.10	0.99	0.53	0.03	3.00	86.70	1.20
4–5	22.8	88.50	0.92	0.73	0.03	2.22	90.98	1.00
~10	21.3	91.20	0.99	0.50	0.03	1.85	94.05	1.40

Temperature and Residence Time

Figure 4 shows that on the one hand the use of higher temperatures leads to a reduction in the amount of volatile components and moisture content, and on the other hand, there is a greater proportion of ash and a substantially lower charcoal yield. In contrast, the longer residence time leads to an increase in charcoal yield.

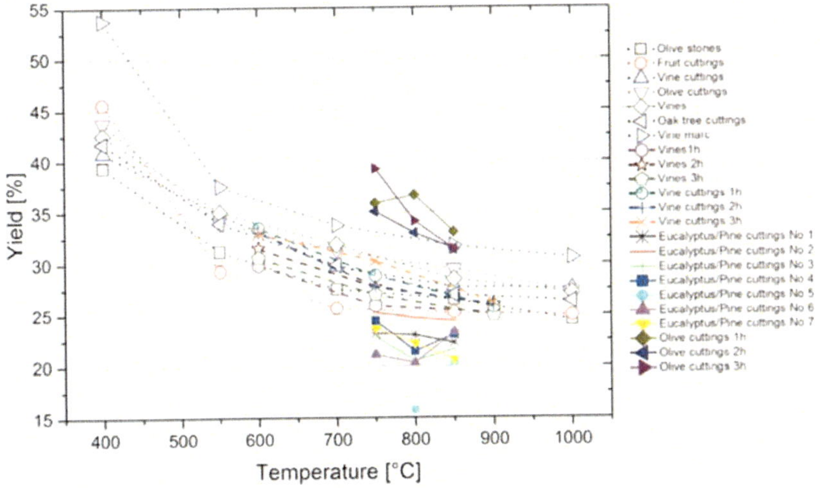

Fig. 4 Charcoal yield as a function of the temperature of various biomass and influence of the residence time according to [8–10]

It should be mentioned that the carbon content increases with a higher pyrolysis temperature because volatile components are removed faster and there is the chance for volatilization of components that generally need higher temperatures. Depending on the used biomass, different temperatures are necessary to reach defined fixed carbon contents [11].

Reactivity Measurements

The reactivity of a carbon substitute is a very important parameter for metallurgy and can determine whether it is possible to utilize charcoal in a certain metallurgical facility or not. The grain size, the pores, and the moisture content have an influence on the reactivity. Large grains and few pores reduce the reactivity of the charcoal, whereas small grains and many pores, especially open pores, increase the reactivity value due to the larger surface area. In the case of moisture content, higher values lead to activation of the coal, which increases the reactivity too. During heating, closed pores burst, because of evaporating water. All analysed biomasses show a decrease in reactivity with increasing temperature (Fig. 5). Depending on the intended application, it should be noted that the charcoal yield is lower with higher temperatures [9].

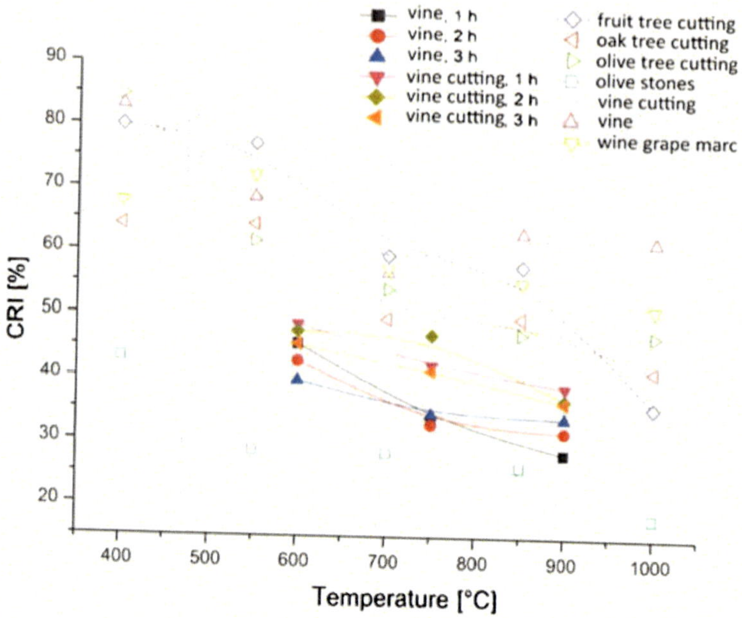

Fig. 5 (Coke Reactivity Index) as a function of pyrolysis temperature [9]

Biochar in Waelz Process

The Waelz process is a recycling process in which carbon is added as a reducing agent to regain zinc. At the trials carried out on laboratory scale, the carbon was substituted with biomass in order to determine the reduction capability in comparison with conventionally used petroleum coke. Steel mill dust from an electric arc furnace, lime (for adjustment of basicity and slag modification) and the carbon carrier were used to form pellets. The pellets were charged into a rotary kiln and treated at 1150 °C under an air atmosphere to simulate the reactions occurring in a Waelz kiln [9].

Table 3 Slag analyses with different charcoals at 1150 °C under air atmosphere [9]

Reducing agent	Zn (%)	Fe_{tot} (%)	Fe_{met} (%)	S (%)	C (%)	Zinc yield (%)
Petroleum coke	1.65	41,05	0.35	0.90	0.12	94.21
Charcoal out of vine cuttings (600 °C, 2 h)	0.22	42.55	0.31	0.53	0.07	99.23
Charcoal out of vine cuttings (750 °C, 2 h)	3.25	42.00	2.20	0.75	0.12	88.60
Charcoal out of vine cuttings (900 °C, 2 h)	1.55	40.61	0.30	0.32	0.06	94.56

The data obtained of the Waelz slag analyses from these trials (Table 3) show that the application of biomass as a carbon substitute is quite comparable to the use of petroleum coke. All blends deliver good results in terms of zinc yield as well as comparable values of carbon, zinc, and iron. With biochar, the sulphur content is lower, because biomass generally has a low sulphur content. The amount of biochar was calculated based on the carbon required for the reduction of the zinc and iron oxide and varied between 175.13 and 211.06 kg/t steel mill dust. In consideration of a stoichiometry factor of 1.15 and the resulting zinc yield, the requirement of biochar for the Waelz process based on laboratory scale experiments ranges from 642.64 to 735.47 kg/t zinc in steel mill dust [9].

However, one crucial point is the behaviour of the charcoal during the heating up within the first zone of a Waelz kiln. If the coal is too reactive, the emitted volatiles combust in the off-gas stream and increase the final off-gas temperature. Also a certain consumption of the charcoal before reaching the reduction zone would end up in a higher specific amount of required reducing agent. The ash portion introduced by the biochar passes into the Waelz slag. Therefore, an adjustment of the basicity has to be taken into consideration.

Regarding the carbon and nitrogen contents and their formation into greenhouse gases, it should be mentioned that the use of biochar shows hardly any differences compared to petroleum coke in the Waelz process. Besides the same required amount of carbon for the reduction, the nitrogen content of biochar (0.35–1.9%) is approximately the same as that of petroleum coke (0.2–1.5%) [9, 12, 13].

Usage of Pyrolysis Gas

As already mentioned at the beginning, not only coal is produced during pyrolysis. Also a certain amount of gas is generated depending on the type of pyrolysis. Due to the volatile components of the biomass, it is expected that the pyrolysis gas can also be beneficial. For this reason, the gas from the treatment of olive stones was analysed. As can be seen in Fig. 6, the composition of the gas changes due to cracking, reforming, decomposition, and the Boudouard reactions depending on the temperature [9, 11].

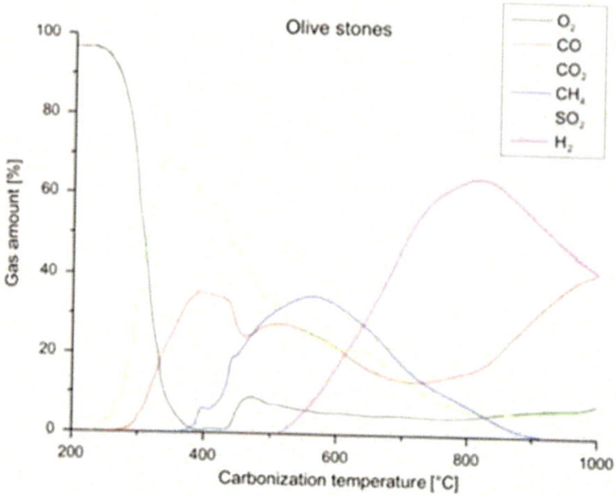

Fig. 6 Change in pyrolysis gas composition as a function of temperature [11]

The composition of the gas suggests, that it can be useful as a reducing gas in metallurgical processes. For this reason, investigations were performed with a synthetic gas (40% CO, 30% H_2, 15% CO_2, 15% CH_4). At this trials Waelz slag was reduced in a retort for different periods of time (1–3 h) at temperatures up to 1150 °C. Results showed that longer treatment times and especially higher temperatures lead to a better zinc yield and a better iron metallization grade [9].

Based on the synthesis gas trials, investigations were performed with pyrolysis gas from the twin-screw reactor. Due to the low iron metallization, carbon was added with a stoichiometric factor of 0.5. This combination resulted in good values for zinc reduction and iron metallization too [9].

Current Activities

At present, laboratory-scale investigations are being carried out concerning the behaviour of pyrolysed biomass for metallurgical processes based on gas–solid reactions.

For this purpose, a scheme was developed that includes a combination of a compulsory mixer and a tube furnace (Fig. 7). The input materials get homogeneously blended and micro-pellets are produced by means of the compulsory mixer. These pellets are filled in a ceramic shuttle and treated in the furnace under various conditions in several stages. The aim is to investigate the behaviour of the pyrolysed biomass during heating and reduction reactions in possible metallurgical aggregates. Experiments are already in progress and first results will be available in about six months.

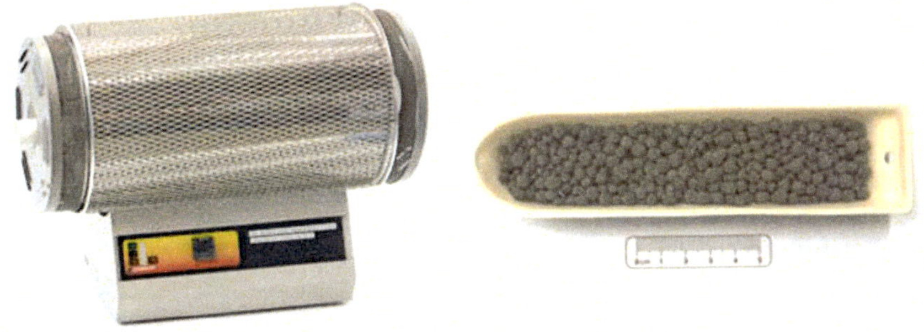

Fig. 7 Tube furnace at the Chair of Nonferrous Metallurgy, Montanuniversität Leoben (left) and ceramic shuttle with micro-pellets (right)

Summary and Outlook

After the prices of the CO_2-certificates have risen strongly over the last few years, and still rising, the interest in CO_2-neutral technologies is also increasing again. The topic is coming more and more into focus with regard to stricter regulations and the growing awareness of the industry too.

The previous experiments on pyrolysis of biomass and the use of biochar as a reducing agent substitute can represent a possibility for a replacement product and have shown that certain biomasses have a particular potential due to their properties. Special consideration should be given to the different characteristics of the biomass, such as reactivity, volatile components, ash content, and others.

Furthermore, the experiments with pyrolysis gas also show good results. The H_2 and CO contained in the pyrolysis gas is well suited for reduction processes. The use of the pyrolysis gas is also an important economic factor to make the use of biomass as a raw material more attractive.

Future research will focus more on the use of the products from the pyrolysis of biomass and their application in metallurgical aggregates. There is also the possibility to substitute only a part of the currently used carbon-based reducing agents with biochar. This can partially minimize the CO_2-emissions generated by fossil carbon carriers.

Acknowledgements COMMBY is a FFG Austria COMET project. Thank you to the funders Österreichische Forschungsförderungsgesellschaft (FFG), Steirische Wirtschaftsförderungsgesellschaft (SFG), Wirtschaft Burgenland GmbH, Land Steiermark an Land Burgenland.

References

1. International Carbon Action Partnership: Allowance Price Explorer (EU ETS Carbon Price). Internet: https://icapcarbonaction.com/en/ets-prices. Accessed 5 Sept 2021

2. Kaltschmitt M, Hartmann H (2000) Energie aus Biomasse. Springer, Berlin, Heidelberg, New York
3. Jørgensen SE, Fath BD (eds) (2008) Encyclopedia of ecology. Elsevier, Amsterdam
4. Wesselak V et al (2017) Handbuch regenerative Energietechnik. Springer, Berlin Heidelberg
5. Basu P (2010) Biomass gasification and pyrolysis. Academic Press, London
6. Scholz R, Beckmann M, Schulenburg F (2001) Abfallbehandlung in thermischen Verfahren. In: Bahadir M, Collins H-J, Hock B (eds) Teubner-Reihe Umwelt Abfall 2001. Teubner, Stuttgart
7. Quicker P, Weber K (2016) Biokohle. Springer Fachmedien, Wiesbaden
8. Bridgewater AV (2004) Biomass fast pyrolysis. Thermal Sci 8(2):21–50
9. Roessler G (2015) Implementation of alternative reducing agents in metallurgical operations. PhD thesis, Montanuniversität Leoben
10. Griessacher T (2012) Einsatzmöglichkeiten von Biomasse als Reduktionsmittel in metallurgischen Recyclingprozessen. PhD thesis, Montanuniversität Leoben
11. Roessler G (2011) Holzkohleherstellung sowie deren Charakterisierung hinsichtlich Ersatzreduktionsmittel in metallurgischen Prozessen. Master thesis, Montanuniversität Leoben
12. Griessacher T, Antrekowitsch J, Steinlechner S (2012) Charcoal from agricultural residues as alternative reducing agent in metal recycling. Biomass Bioenerg 39:139–146
13. Nanda S et al (2013) Characterization of North American lignocellulosic biomass and biochars in terms of their candidacy for alternate renewable fuels. Bioenerg Res 6:663–677
14. Predel H (2006) Petroleum coke. In: Ullmann's encyclopedia of industrial chemistry. Wiley-VCH Verlag GmbH & Co. KGaA, Weinheim, Germany

Techno-Economic Pre-feasibility Study of a Hydrogen Plasma-Based Ferromanganese Plant

Halvor Dalaker, Nils Eldrup, Roar Jensen, and Rannveig Kvande

Abstract Plasma species such as monoatomic hydrogen are much more reactive than diatomic hydrogen gas, $H_{2(g)}$. This means that plasma technology can be a way of introducing hydrogen as a reducing agent for producing metals and alloys, like manganese, where oxide reduction to the metallic state by hydrogen gas is not possible. While still an immature technology, it is nevertheless important to address questions regarding economic relevance. In this work, best estimates are used to describe a hypothetical plasma-based process and plant for manganese production. This description then makes the foundation for a techno-economic analysis comparing this hypothetical process with existing standards. It is found that at present, it seems plausible that hydrogen plasma technology can compete economically, and thus be a viable way to decarbonise manganese production. The most sensitive parameters appear to be hydrogen price, cost of CO_2-emissions, and to what extent excess energy from the plasma unit can be captured and used for pre-heating of the ore.

Keywords Economy · Plasma · Hydrogen · Sustainability · Manganese

Introduction

The metallurgical sector releases a lot of CO_2 globally, caused not only by energy consumption, but also directly resulting from the core processes of turning ores into metal by use of carbon as a reducing agent. Hydrogen is a candidate replacement reducing agent, as it can reduce some metal ores (e.g. iron ore):

$$MeO_x + XH_{2(g)} \rightarrow Me + XH_2O_{(g)}$$

H. Dalaker (✉) · R. Jensen · R. Kvande
SINTEF, Trondheim, Norway
e-mail: Halvor.Dalaker@sintef.no

N. Eldrup
SINTEF, Porsgrunn, Norway

A. Lazou et al. (eds.), *REWAS 2022: Developing Tomorrow's Technical Cycles (Volume I)*, The Minerals, Metals & Materials Series,
https://doi.org/10.1007/978-3-030-92563-5_68

Other metals like manganese (Mn) cannot easily be reduced by $H_2(g)$ since the reaction described above is not thermodynamically favourable. Now, this situation can be significantly altered by introducing more reactive hydrogen species, such as monoatomic and ionised hydrogen. These species can be produced through plasma generation in an electric arc and enable, e.g. the reaction:

$$MnO + 2H_{(g)} \rightarrow Mn + H_2O_{(g)}$$

SINTEF has initiated the project HyPla [1] to investigate the use of hydrogen plasma as a reducing agent in metal production.

A prerequisite for the introduction of hydrogen plasma in metal production is a good financial case. In this work we present a pre-feasibility study looking into the financial case for a possible industrial scale ferromanganese plant based on hydrogen plasma reduction. Due to the immaturity of the technology, the work is necessarily based on assumptions and estimates. The first goal of the current work is to establish whether a cost-effective plasma plant is within the realms of possibility. The second goal is to identify critical cost drivers, which can be useful guides in where to concentrate future activities. At each point, the best estimate based on our scientific understanding has been the basis for the assumptions made. However, sometimes trade-offs have been made in favour of a simpler solution when the gain in accuracy and impact on the overall outcome has been judged small. More detailed analyses will build from the current work once more experimental data becomes available.

A techno-economic analysis needs to be based on a description of a process, which means that such a process also needs to be designed. In the first sections of this paper, the layout of the process design is presented, together with the assumptions made and calculations relating to energy and mass balances. In the second part of this paper, the economic consequences are presented.

The Hypothetical Plasma-Based Ferromanganese Plant Operation, and Assumptions

The first constraint laid down in the current work was to base the modelled plant on a scale that was comparable to an existing running plant, and thus the plasma plant is modelled to have the annual alloy output of that of Eramet Norway's plant at Sauda, with a capacity of 265,000 tonnes per year [2].

Second, it was decided to split the manganese production into two separate processes: a hydrogen-based pre-reduction unit and a plasma-unit. This was done for mainly two reasons: 1-The off-gas from the plasma unit is expected to contain significant amounts of unutilised hydrogen, which is well suited for pre-reduction of manganese ores, and 2-it was thought best to use the generated plasma solely for the task where plasma is the only option, and not "waste" it on reducing iron oxides or higher order manganese oxides.

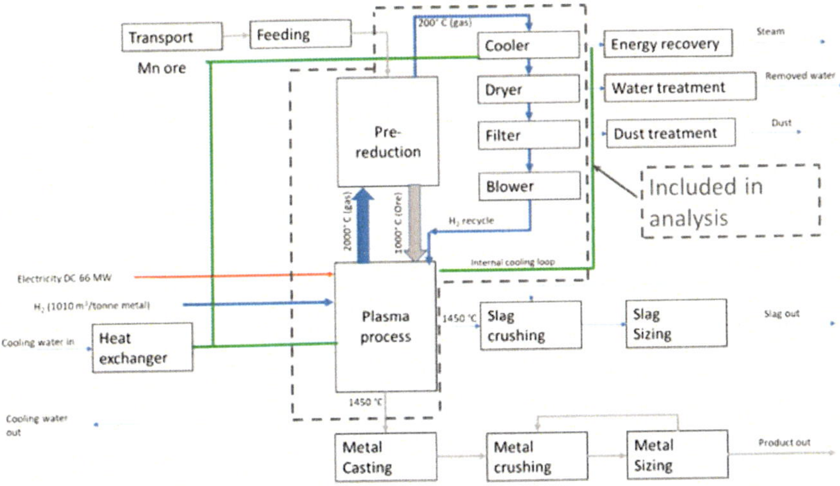

Fig. 1 Flowsheet of proposed plasma-based FeMn-plant. The components outside the dashed lines are assumed to be identical to today's process, and are not included in the techno-economic analysis.

Third, for simplicity as many parts of the process as possible was assumed to be identical to today's process, so that they could be left out of the techno-economic analysis. Broadly, this meant assuming that the handling of incoming raw material and outgoing products and by-products would be the same as today, such as the transport and charging of ore; the casting, crushing, and sizing of metal and slag; and the waste treatment of dust, etc. The rough flow-sheet of the plant is then as sketched in Fig. 1, where the scope of the investigation is also included.

In broad strokes, the operation of the plant is as follows: The ore is charged into the pre-reduction unit. As it descends towards the plasma unit, it meets hydrogen gas, which causes all iron oxides in the ore to be reduced to metallic iron, and all MnO_2 to be reduced to MnO. The other oxides in the ore remain unchanged. The pre-reduced ore then enters the plasma unit, where much of the MnO is reduced to Mn and slag and metal is then tapped in separate streams. The process and surrounding assumptions are expanded in the following.

Raw Material and Products

COMILOG ore is among the inputs at Eramet Sauda, and as it is rich in MnO_2 it is well suited to take advantage of the pre-reduction potential of unspent hydrogen. In this work it has been assumed that the input is 100% COMILOG ore. The composition of the ore is given in Table 1.

The temperature of the product alloy and slag at tapping are each assumed to be 1450 °C, which is similar to today's operation. Based on this assumption, a

Table 1 Chemical composition of COMILOG ore [3]

Wt% in wet ore				Wt% in dry ore			
H_2O	8.7	Al_2O_3	0.2	H_2O	0	Al_2O_3	0.3
XH_2O	4.9	MgO	0.2	XH_2O	5.4	MgO	0.2
MnO	2.9	CaO	0.6	MnO	3.2	CaO	0.2
MnO_2	69.4	BaO	0.1	MnO_2	76	BaO	0.7
Fe_2O_3	3.6	K_2O	0.1	Fe_2O_3	3.9	K_2O	0.11
FeO	3.7	P	0.2	FeO	4	P	0.1
SiO_2	0.3			SiO_2	5.5		
Total Mn in dry ore [Wt%]			50.5	Mn:Fe-ratio			18.5

flux of 50 kg CaO per ton of ore is used. Based on equilibrium calculations with FactSage [4], this flux gives a situation at 1450 °C in which the liquid metal is in equilibrium with a liquid, tappable slag. The fluxing has *not* been optimised for any other purpose, like Mn-yield. The resulting metal and slag compositions, as well as manganese distribution between the different streams, are shown in Table 2 below. For simplicity, a 100% up-time is assumed, which means that, with the material properties presented here, 67,575 kg of ore is processed per hour.

Table 2 Amounts and compositions of products, and distribution of manganese between them, as calculated using FactSage

Product	Amount (kg) per tonne of Wet ore	Dry ore	Composition (wt.%)										
Metal	448	487	Mn				Fe						
			90.8				9.2						
Slag	158	172	MnO	MnO_2	Fe_2O_3	FeO	SiO_2	Al_2O_3	MgO	CaO	BaO	K_2O	P
			30	0	0	0	32	1.8	1.2	30	4.1	0.6	0
Dust	30	33	MnO										
			100										
Manganese distribution (%)													
Metal			87										
Slag			8										
Dust			5										

The Pre-reduction Unit

On its way through the pre-reduction unit, the charge is slowly heated, and also undergoes several chemical reactions. For simplicity of calculating the energy balance, it is assumed that each reaction takes place at a single temperature, and each reaction's contribution to the energy balance is calculated using the enthalpy of reaction at that temperature. Between each temperature at which a chemical reaction takes place, the heat capacity of the charge is calculated as the weighted average of the heat capacities of all its components. Within each temperature interval, the heat capacity is assumed to be independent of temperature, and the value at the midway-point of the temperature interval is used as an approximation for the heat capacity across the entire temperature range.

The first reaction that occurs is the evaporation of water. The ore contains both moisture and crystal water which would be expected to evaporate at 100 °C and approximately 300 °C, respectively. Here the situation is simplified by a single evaporation step in which all water evaporates as H_2O at 200 °C, assuming a heat of evaporation of 36.5 kJ/mol. 200 °C is also the temperature used for the off-gas temperature at the top of the pre-reduction unit, which is a typical temperature in current operation. It is assumed that there is no thermal interaction between the steam generated from evaporation and the incoming ore.

Next come the reactions involving metal oxides. Only reactions with H_2 are included here, i.e. it is assumed no thermal decomposition of Fe_2O_3 or MnO_2. The chemical reactions are collected in Table 3. The charge is assumed to enter the pre-reduction unit at 25 °C, leave it at 1000 °C, and enter the plasma unit also at 1000 °C (no heat loss in transfer between units).

Manganese is fairly volatile, and in current operations, around 5% of incoming Mn in the ore is lost as dust [5]. As a first estimate, the dust loss for the plasma process was also set to 5%. This loss would represent Mn-fumes leaving from the top of the system, having first risen from the plasma unit through the pre-reduction unit. Since some of the rising fumes would likely condense in the pre-reduction unit, the fuming from the plasma unit to the pre-reduction unit was set at 25% of the incoming manganese, with 80% of this (so 20% of total Mn) condensing and re-entering the plasma-unit. It is further assumed that the manganese fumes from the plasma unit enter the pre-reduction unit as liquid droplets, 50% in the form of Mn(l) and 50% MnO(l), and that the only change these fumes undergo in the pre-reduction unit is solidification, with no oxidation taken into account. The solidification of both

Table 3 Oxide reduction taking place in the pre-reduction unit

Reaction	ΔH kJ/mol	T (°C)
$MnO_2 + \frac{1}{2} H_2 \rightarrow \frac{1}{2} Mn_2O_3 + \frac{1}{2} H_2O$	−83.3	500
$Fe_2O_3 + H_2 \rightarrow 2\,FeO + H_2O$	44.14	500
$Mn_2O_3 + H_2 \rightarrow 2\,MnO + H_2O$	−64.86	900
$FeO + H_2 \rightarrow Fe + H_2O$	14.84	900

species is assumed to take place at 1244 °C (melting point of manganese), and the heat of fusion is included in the energy balance, with heats of reaction of 12.1 kJ/mol (Mn) and 54.4 kJ/mol (MnO), respectively.

The H_2/H_2O gas mixture from the plasma unit is assumed to consist of 60% H_2 and 40% H_2O. It is further assumed to leave the plasma unit at 2000 °C, and with a 10% heat loss, to enter the pre-reduction unit at a temperature of 1773 °C. As this gas flows upwards counter current to the incoming ore, heat will be transferred from the gas to the ore, until the gas leaves the system at 200 °C. The efficiency of the heat transfer is not known and will depend on a lot of unknown factors like size of charge material, etc. In this first estimate, the heat transfer efficiency is set at 50%. During the ascent of the gas through the pre-reduction unit, it reacts with iron and manganese oxides, consuming hydrogen and producing steam. This effect has been ignored in the estimate of heat transfer from the gas, and the heat capacity of the gas has been assumed constant throughout its ascent.

As the gas leaves the top of the pre-reduction unit, it goes through cooling and filtering steps to remove dust and condense out water. The remaining hydrogen is then re-used in the plasma unit. It is assumed that 90% of the hydrogen leaving the top of the pre-reduction unit is successfully recycled in this manner.

For all energy calculations, a loss factor of 10% has been used. This means that for every positive contribution (from exothermic reactions, etc.), 90% of the theoretical value has been subtracted from the total energy requirement. For negative contributions (for heating and endothermic reactions), 110% of the theoretical value has been included as part of the energy requirement.

All the energy contributions are compared in Fig. 2 in terms of power. It is seen that the largest energy demand comes from heating of ore and evaporation of water, and also that the assumed 10% energy loss is a substantial contribution. On the

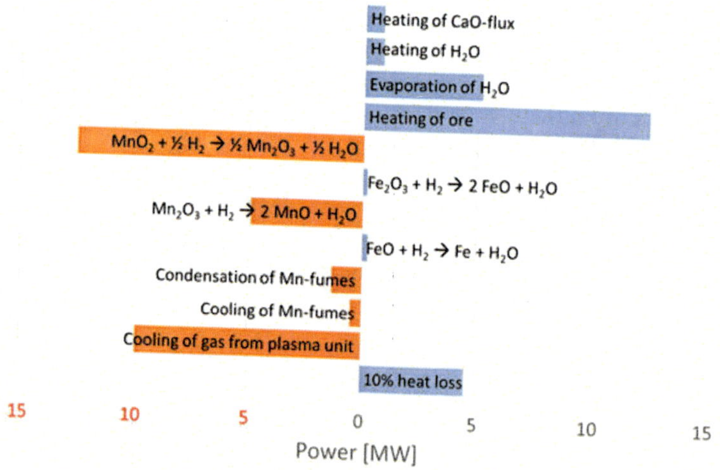

Fig. 2 Visualisation of the different contributions to the energy balance of the pre-reduction unit

other side of the equation, the pre-reduction of manganese oxides generates large amounts of energy, and large amounts of energy are also carried over through the gas-phase from the plasma unit. With these assumptions, the energy balance for the pre-reduction ends up with a 4 MW power surplus. The most sensitive parameter here is probably the heat transfer efficiency between the gas and the ore, and the associated temperature of the off-gas. However, even with heat transfer efficiency of only 27%, the pre-reduction unit has a power surplus.

Plasma Unit

Within the plasma unit, the main reactions occurring are the excitation and relaxation of plasma, and the reaction between plasma and MnO. The enthalpy contribution of the plasma-reduction consists of two steps: first the excitation of H_2 to 2H, followed by the reaction of MnO with H. However, energetically speaking, the overall reaction is

$$MnO + H_2 \rightarrow Mn + H_2O,$$

and this reaction has been used, with a total enthalpy of reaction of 99.2 kJ/mol.

The excitation/relaxation of unutilised plasma is an inefficiency, but the heat released from relaxation will contribute to the heating of raw materials and thus not go to waste, so this interaction has been ignored. The metallic iron and the slag-forming oxides are assumed to pass through the plasma unit without any other reactions taking place besides heating from 1000 °C to the tapping temperature of 1450 °C and the melting of the individual components.

The heat efficiency of the plasma unit has been estimated at 80%. Based on the heat in the mass-flows in and out of the plasma unit, the heat of phase transformation and the heat of reaction, the total energy demand of the plasma unit corresponds to a power input of 66 MW. The different contributions are visualised in Fig. 3.

Once again it is seen that a lot of heat is carried away in the gas-phase, and this matches the large contribution from the gas from the plasma unit in the pre-reduction energy balance, as seen in Fig. 2.

Techno-Economic Analysis

Scope

Based on the considerations of the previous section, a concept sketch was made of a 66 MW plasma-based manganese furnace. From this, the material costs could

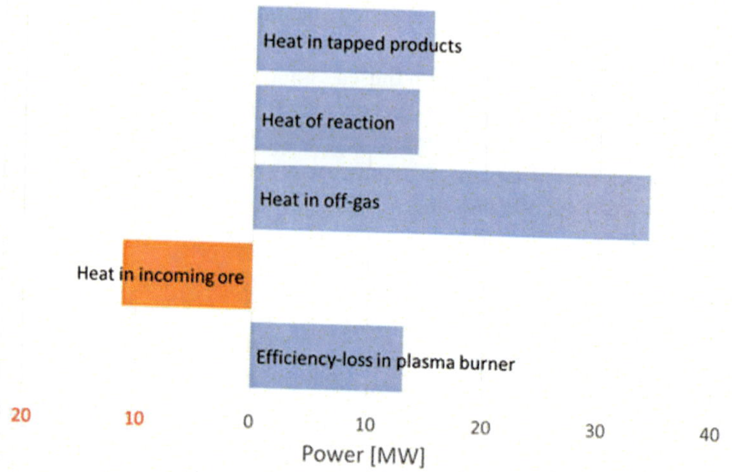

Fig. 3 Visualisation of the different contributions to the energy balance of the plasma unit

be estimated, taking into account steel for cooling jacket, framing and supports; chamotte and alumina bricks for insulation, copper for electrical conductors and DC-graphite electrodes for plasma generation.

The construction cost estimates included process engineering, engineering of mechanical components, plumbing, electrical work, instrumentation, foundation, insulation, procurement, construction management, project management, steel and concrete work. Commissioning and testing costs are also included. These cost estimates made the foundation for the CAPEX analyses.

The techno-economic analysis was performed by the use of Aspen In-plant Cost Estimator software [6] and in-house data. The discount rate is 10%, the repayment period is 10 years, and time of construction is assumed to be 1 year.

The process parameters and the concept sketch allowed for an estimate of cooling water consumption of 20.4t/hr. Taking the total cooling water use of Eramet Norway [7] and scaling it by the share of total production taking place at Sauda, the estimated cooling water consumption in the existing process is almost the same, at 19.8t/hr. It is therefore assumed that the costs associated with cooling water will be the same as in today's processes, and can be left out of the scope.

The CO_2-emissions for the current process were also based on taking total Eramet Norway numbers and scaling by the share of production taking place at Sauda. This gave an estimate of 1.157 tonnes CO_2-eq/tonne alloy. The plasma process is assumed to have zero process-related CO_2-emissions. Any CO_2-emissions related to production of electricity are ignored. In scenarios with a large non-renewable source of the energy mix, this could be taken into account by including CO_2 costs in the electricity costs.

The pre-reduction unit was estimated to have a large power surplus, depending on the heat transfer efficiency between gas and ore. It is thus assumed that no power is required to heat the pre-reduction unit, but neither is it assumed that any of the surplus heat is made use of (could perhaps be used to pre-heat ore prior to charging). Note that while there is significant leeway in terms of how efficient the heat transfer can be before the energy balance becomes negative (only 27% of the heat in the gas needs to be absorbed), with a lower heat transfer efficiency the off-gas will be hotter, the cooling requirements will be different, and it may no longer be correct to assume that this can be left out of the scope.

There are some differences in amounts and composition of the products between today's process and the plasma plant. As Table 2 shows, the projected manganese content of the product alloy is expected to be quite high. The fact that the process and thus the product are both free of carbon are also expected to increase the product value. However, the total tonnage of output is also somewhat lower, so the total revenue from product sales is assumed to be equal to the existing case. It has already been explained that the handling of incoming raw material and outgoing products and by-products are assumed identical to today. The remaining significant differences in operational costs between the two cases are then:

- Hydrogen as raw material for plasma plant (2.6 tonnes/hr).
- Metallurgical coke (8.7 tonnes/hr), anthracite (3.5 tonnes/hr), and electrode paste (0.56 tonnes/hr) as raw materials for existing plant.
- CO_2-emissions (37.7 tonnes/hr) for existing plant.

It is hard to find real data for the price of the carbon materials, since these are based on confidential contracts. In lieu of this information, prices have been taken from openly available sources, such as the U.S. Energy Information Administration [8].

Findings

The investment case can be analysed from three angles: 1-"OPEX only", where only the operational costs are considered; 2-"Full CAPEX", where both the OPEX and the full CAPEX for a new plasma plant are considered; 3-"OPEX plus *additional* CAPEX", where all OPEX is considered, but the only CAPEX included is that which is associated with additional costs of building a plasma plant compared to the capital cost of building a conventional plant. A negative outcome in the type-1 analysis means that the investment proposal can never be profitable. A positive outcome of a type-2 analysis on the other hand means that it would be profitable to scrap the existing plant and build a new plasma-based one. With a positive result in case 3, it would not necessarily be profitable to scrap an existing plant, but it *would* be profitable to choose the plasma-option over the conventional option in the case that a new plant would nevertheless be built, for example if a plant is at the end of its life or in an expansion of capacity.

Table 4 Breakdown of contributions to critical CO_2-price [NOK/tonne CO_2-eq] for positive investment decisions. The different scenarios are described in the text

	No sale of oxygen			Sale of oxygen at 1 NOK/kg		
	1-OPEX Only	2-Full CAPEX	3-OPEX and additional CAPEX	1-OPEX Only	2-Full CAPEX	3-OPEX and additional CAPEX
Hydrogen cost	1371	1371	1371	1371	1371	1371
Savings (coke etc.)	920	920	920	920	920	920
CAPEX	0	344	60	0	344	60
Sale of oxygen				548	548	548
Minimum cost of CO_2 (NOK/tonne)	451	795	511	−97	247	−37

The investment case for the proposed plant will depend on the price of hydrogen and the price of CO_2 emissions: For each hydrogen price, there will be a critical price of CO_2 emissions at which the investment becomes profitable in each scenario. The hydrogen volumes required in the current case would require a 140 MW electrolyser. Optimistic but realistic hydrogen price estimates from such an installation are in the 20–30 NOK/kg range (10NOK≈1€). This estimate is based on industry knowhow, future cost projections as well as the targets within the EU-H2020 call for an innovation action to build and demonstrate a 100 MW electrolyser [9]. In the first iteration, a price of 20 NOK/kg is used, and the results are presented in Table 4.

In the first three columns of Table 4, the critical CO_2 costs are presented for the three scenarios described above. The critical CO_2-price is in the range 450–800 NOK (45–80€). If the hydrogen is produced from an electrolyser, there will be 1 mol of oxygen gas produced per 2 mol of hydrogen. At the volumes required in an installation such as this, the volume of oxygen would be substantial. If this by-product could be sold for any value, it would have a large impact on the investment cases. As an example, the last three columns of Table 4 show how the sale of oxygen at 1NOK/kg impacts the calculation, resulting in negative values for the critical CO_2 price. In other words, such a scenario would be very profitable even with no price on CO_2. However, it will be challenging to find a buyer for such volumes of oxygen, especially if many industries decarbonise through hydrogen and supply of oxygen as by-product from H_2-electrolysers becomes greater than demand.

Table 5 demonstrates the sensitivity of the analysis to the hydrogen price. Here, the critical CO_2-price for all the three scenarios, with and without sale of oxygen, are listed for three different hydrogen prices. It is seen that an increase in hydrogen price of 50% from 20 to 30NOK increases the critical CO_2 price by much more than 50%.

The critical CO_2-prices from Tables 4 and 5 cover a wide range (0-1480NOK ≈0–150€). At the time of writing, the price in the EU is €56.86/tonne CO_2 [10], which would already make for a positive investment decision in several of the scenarios. Looking ahead to the time when the as-yet unproven plasma technology

Table 5 Critical CO_2-price [NOK/tonne CO_2-eq] for positive investment decisions at varying hydrogen costs. The different scenarios are described in the text

Hydrogen price [NOK/tonne]	No sale of oxygen			Sale of oxygen at 1 NOK/kg		
	1-OPEX Only	2-Full CAPEX	3-OPEX and additional CAPEX	1-OPEX Only	2-Full CAPEX	3-OPEX and additional CAPEX
20	451	795	511	−97	246	−36
25	794	1138	854	245	589	306
30	1137	1480	1197	589	932	649

would be mature enough for industrial deployment, estimates vary but prices are in general expected to increase significantly by 2030 (€90/tonne [11], €85/tonne [12], €71/tonne [13]) to 2040 ($140/ton[14]).

Based on all of this, at present there is no reason to believe that a working plasma-based manganese plant cannot become economically competitive.

Conclusions

A first techno-economic analysis has shown that a profitable plasma-based manganese plant is not outside the realms of possibility. Unsurprisingly, the investment decision will be highly sensitive to the hydrogen price and the cost of CO_2-emissions.

If hydrogen is produced on site in an electrolyser, then the sale of the co-produced oxygen can potentially have a large positive effect on the investment case, provided a buyer can be found and that prices of oxygen do not plummet as more and more H_2-electrolysers come online around the world.

From the technical side, the gas leaving the plasma unit is expected to carry large amounts of heat. To what extent this heat can be captured by the ore in the pre-reduction unit will have a significant impact on the economics.

Lastly it should be remembered that on the path to a more sustainable world that it may not always be a map drawn purely on economic considerations that will lead us to our goal.

Acknowledgements The authors wish to express their thanks to Professor Merete Tangstad at NTNU, and to our colleague Trine A. Larssen at SINTEF, for sharing their expertise on the manganese process and for fruitful input and discussions.

We are also indebted to Tina Andersen for her input and information regarding hydrogen.

References

1. SINTEF (2021) HyPla-Hydrogen plasma for CO2-free metal production. https://www.sintef.no/en/projects/2020/hypla-hydrogen-plasma-for-co2-free-metal-production/. Accessed 5 Sept 2021
2. Eramet Norway (2012) Environmental report, Miljørapport (in Norwegian). http://eramet.no/wp-content/uploads/miljorapport_norsk20121.pdf. Accessed 5 Sept 2021
3. Tangstad M, Calvert P, Brun H, Lindseth AG (2004) Use of Comilog ore in ferromanganese production. Paper presented at the Tenth International Ferroalloys Congress, Cape Town, South Africa
4. Bale CW, Bélisle E, Chartrand P, Decterov SA, Eriksson G, Gheribi AE, Hack K, Jung IH, Kang YB, Melançon J, Pelton AD, Petersen S, Robelin C, Sangster J, Van Ende M-A (2016) FactSage thermochemical software and databases, 2010–2016. Calphad 54:35–53. www.factsage.com
5. Olsen SE, Tangstad M, Lindstad T (2007) Production of manganese ferroalloys. Fagbokforlaget. ISBN- 978–8251921916
6. Aspen Technology Inc (2021) Aspen In-plant cost estimator. https://www.aspentech.com/en/products/pages/aspen-in-plant-cost-estimator. Accessed 5 Sept 2021
7. Eramet Norway (2018) Eramet Norway sustainability report 2018. https://2kaamo43d6qnc0etemarvpj6t-wpengine.netdna-ssl.com/wp-content/uploads/Eramet-Sustainability-Report-2018-ENG-low.pdf. Accessed 5 Sept 2021
8. U.S. Energy Information Administration (EIA) (2021) Coal prices and outlook. https://www.eia.gov/energyexplained/coal/prices-and-outlook.php. Accessed 5 Sept 2021
9. European Commission (2020) Horizon 2020-call, develop and demonstrate a 100 MW electrolyser upscaling the link between renewables and commercial/industrial applications, Topic ID: H2020-LC-GD-2020, https://ec.europa.eu/info/funding-tenders/opportunities/portal/screen/opportunities/topic-details/lc-gd-2-2-2020. Accessed 5 Sept 2021
10. Ember Climate (2021) Daily carbon prices. https://ember-climate.org/data/carbon-price-viewer/. Accessed 28 Aug 2021
11. Euractiv (2021) Analyst: EU carbon price on track to reach €90 by 2030. https://www.euractiv.com/section/emissions-trading-scheme/interview/analyst-eu-carbon-price-on-track-to-reach-e90-by-2030/. Accessed 5 Sept 2021
12. Bloomberg (2021) Europe CO_2 Prices may rise more than 50% by 2030, EU Draft Shows. https://www.bloomberg.com/news/articles/2021-06-29/europe-co2-prices-may-rise-more-than-50-by-2030-eu-draft-shows. Accessed 5 Sept 2021
13. CDC Climate Research (2014) Tendances Carbone. https://www.i4ce.org/wp-core/wp-content/uploads/2015/09/TC96_ENG.pdf. Accessed 5 Sept 2021
14. International Energy Agency (2020) World energy model, macro drivers. CO_2-prices. https://www.iea.org/reports/world-energy-model/macro-drivers#abstract. Accessed 5 Sept 2021

Solid Oxide Membrane (SOM)-Based Technology for Carbon-Free Efficient Production of Solar-Grade Silicon

Haoxuan Yan, Michelle Sugimoto, Adam Powell, and Uday Pal

Abstract State-of-the-art solar-grade silicon production is energy intensive and has a negative impact on the environment. Due to the robust and rapid growth of the Si-based photovoltaic (PV) industry, it is necessary to develop a greener technology for silicon production. Solid oxide membrane (SOM) electrolysis is a proven versatile green technology that can be developed to economically produce many important metal or metal compounds from their oxides. This work will discuss application of SOM electrolysis to produce solar-grade silicon from silica in a single-step resulting in net-zero-carbon emission. The high-temperature SOM electrolysis cell employs stable molten oxide-fluoride bath with silicon wafer cathode and stabilized zirconia membrane-based novel anodes. The cell design and process parameters are selected to enable silicon deposition on the Si wafer cathode. However, initially an electrochemical oxidation reaction occurred between silicon and oxygen that involved the cathode/flux/gas interfaces. An approach to successfully prevent this side reaction has been demonstrated. Electrochemical characterization of the SOM process is presented, and post-experimental characterization demonstrates Si deposits in the form of silicon carbide due to the use of graphite crucible and graphite current collector.

H. Yan · U. Pal (✉)
Division of Materials Science and Engineering, Boston University, 15 St Mary's St, Boston, MA, USA
e-mail: upal@bu.edu

H. Yan
e-mail: hyan910@bu.edu

M. Sugimoto · U. Pal
Department of Mechanical Engineering, Boston University, 110 Cummington Mall, Boston, MA, USA
e-mail: msugi@bu.edu

A. Powell
Department of Mechanical Engineering, Worcester Polytechnic Institute, 100 Institute Rd, Worcester, MA 01609, USA
e-mail: acpowell@wpi.edu

A. Lazou et al. (eds.), *REWAS 2022: Developing Tomorrow's Technical Cycles (Volume I)*, The Minerals, Metals & Materials Series,
https://doi.org/10.1007/978-3-030-92563-5_69

Keywords SOM electrolysis · Silicon production · Green technology

Introduction

Currently, industrial processes that produce silicon rely on multi-step batch units which are energy intensive, capital intensive, and environmentally unsound due to the heavy emission of pollutants and greenhouse gases [1, 2]. An innovative technology, solid oxide membrane (SOM) electrolysis, has been developed and successfully applied in extraction of metals or alloys from their oxides. Previous research that incorporated this technology has proven its feasibility producing a variety of metals including tantalum, titanium, ytterbium, magnesium, and aluminum [3–8]. In contrast to conventional silicon production methods, such as the Siemens process and fluidized bed reactors, the SOM technology has three main advantages: it is a single-step batch operation that is less energy and capital intensive, it does not release any direct greenhouse gases or halogenated emissions, and it yields pure O_2 as a valuable by-product [9]. It does so by efficiently electrolyzing silica in a well-engineered flux, the silicon is deposited at the cathode and the oxygen ions migrate through a ceramic solid oxygen ion conducting stabilized zirconia membrane to the anode where O_2 gas is evolved.

Three key innovations are needed for the successful operation of the SOM electrolytic cell: (i) the YSZ membrane needs to separate the anode from the flux. This prevents the oxygen gas produced at the anode from contacting the reduced metal at the cathode; (ii) a flux needs to be engineered that has high ionic conductivity, low electronic conductivity, low volatility, and high chemical stability in contact with the YSZ membrane; and (iii) a cathode current collector needs to be developed that is thermally and chemically stable in the SOM environment, has low resistance while contacting the silicon wafer cathode. This allows the cathode to be removed from the flux after the experiment. The first two innovations have been successfully implemented while developing a suitable cathode current collector still remains a challenge.

In this work, graphite is chosen as the cathode current collector because it has adequate electrical conductivity at high temperature, high chemical stability in contact with the flux, good machinability that enables secure connection and contact with the Si wafer cathode. The design details are described in the experimental section. For the ease of operation and to demonstrate process feasibility, graphite is also employed as the anode current collector with liquid silver serving as the anode. In this case the oxygen produced at the anode reacts with the carbon to form carbon monoxide (CO). Initially, an electrochemical oxidation reaction involving the silicon wafer, flux, graphite, and impurity oxygen in the gas phase consumed the silicon wafer cathode. The oxidation mechanism was identified and later mitigated by immersing the silicon wafer deeper into the flux. The SOM electrolysis deposited silicon which reacted with the carbon to form silicon carbide as predicted by the

Si-C-O phase stability diagram. Future work regarding stabilizing the pure Si phase free of carbon in the system is proposed.

Experimental Work

A proof-of-concept laboratory scale Si-SOM electrolysis was performed using the setup shown in Fig. 1. The setup consists of a graphite crucible that was heated to 1100 °C in forming gas (95% Argon–5%H2) to ensure an inert atmosphere. Inside the crucible, 450 g of powdered flux (45wt% MgF_2–55wt% CaF_2 with 5wt% SiO_2, 9wt% CaO, and 4wt% YF_3, suggested and verified by previous studies) was used to form the molten salt electrolyte [10–15]. All fluorides and oxides were pre-baked before melting, and salt flux without SiO_2 was pre-melted to prevent the formation of SiF_4(g). A one end-closed 8 mol% Yttria-stabilized zirconia (YSZ) membrane tube separated the flux from 5 g of liquid silver enclosed inside the YSZ tube. An alumina tube was used to attach YSZ membrane with gas-sealing ceramic paste (Aremco 552) for extra length. A graphite anode current collector was placed in the YSZ membrane tube and submerged in the liquid silver at the target temperature. A graphite rod is used as the current collector. A slit is machined at the end of the graphite rod and the silicon wafer is inserted and glued using graphite paste (Resbond 931). Two tungsten reference electrodes were inserted in the molten flux to monitor the electrochemical behavior of the flux during the electrolysis. The reference electrodes were electrically insulated by alumina tubes. A stainless-steel bubbling tube was submerged into the flux during the electrolysis to ensure that the molten flux is homogeneous and to reduce concentration polarization near the Si wafer cathode. The entire cell was operated in a vertical tube furnace with the cell placed in the center of the hot zone.

Si Wafer Stability Analysis—It was noted earlier that the Si wafer cathode experienced some oxidation or thinning when the Si wafer was removed from the flux after some initial SOM electrolysis experiments. Following this observation, a series of silicon wafer stability tests were performed. The graphite current collector with the silicon wafer cathode were heated to 1100 C in the reducing environment (95% Argon–5%H2) and immersed into the flux at different depths for 6 h. The flow rate

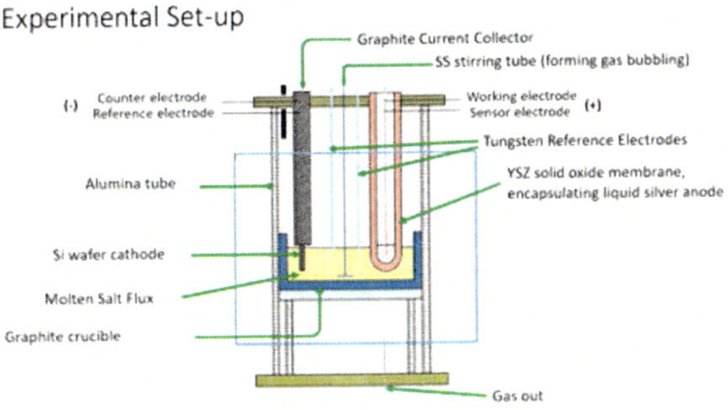

Fig. 1 Schematics of the SOM cell

of the forming gas was fixed at 50 ccm. In one of the tests the cathode current collector with the silicon wafer was placed above the flux for the same length of time (6 h) to evaluate the effect of trace amounts of oxygen in the system on the oxidation behavior of the silicon wafer. After these experiments, the silicon wafer was removed, sectioned, and mounted in epoxy to examine the microstructure under Zeiss Supra 55 scanning electron microscope equipped with EDAX EDS analyzer.

Electrochemical Measurements—Several electrochemical measurements were performed during the electrolysis. To monitor the behavior of the flux, the two tungsten rods were inserted into the flux periodically and electrochemical impedance spectroscopy (EIS) measurements were made. The current is also measured with small applied Potentiostatic hold below the dissociation potentials of oxides in the flux. These measurements were carried out with a Princeton Applied Research 263A Potentiostat and a Solartron 1250 frequency response analyzer. From these measurements the ionic and electronic transference numbers in the flux were determined. Also, prior to performing the SOM electrolysis, a potentiodynamic scan was performed across the anode and the cathode to determine the dissociation potential of silicon oxide and possibly other impurity oxides in the flux. The overall cell resistance was estimated from the linear portion of this scan. After these measurements were made, SOM electrolysis was performed above the dissociation potential of silica (0.95 V) for 8 h. Current versus time was recorded with a Princeton Applied Research 263A Potentiostat, and the gas bubbling from the anode inside the YSZ tube was also recorded using a digital mass flowmeter. After 8 h, when the electrolysis was complete, the anode was removed while the cathode remained inside the flux. Finally, the two tungsten reference electrodes were once again inserted into the flux to determine if there were any changes in the electrochemical behavior of the flux because of the electrolysis process. All electrochemical measurements were made at 1100 C under a forming gas (95% Argon–5%H2) flow rate of 50 ccm.

Post SOM electrolysis Characterization—After the experiment, the furnace is allowed to cool and the cathode with the solidified flux is removed and sectioned. The sectioned piece is mounted in epoxy and examined under the Zeiss Supra 55 scanning electron microscope (SEM) with the EDAX EDS system. A portion of the YSZ tube that contains the silver anode was also cross-sectioned and examined with the same instrument. The objective of the post characterization work is to determine the nature of the silicon deposition and ensure that the YSZ membrane was stable in contact with the flux.

Results and Discussion

Si Wafer Stability Analysis—As-received silicon wafer was sectioned and mounted without heating in the reducing environment to evaluate its thickness. The SEM/EDS examination showed the thickness to be 507 μm with no observable layer of silicon oxide (Fig. 2a). To determine the effect of the experimental gas environment in the reactor, the silicon wafer was heated to 1100 C in forming gas. It was found that

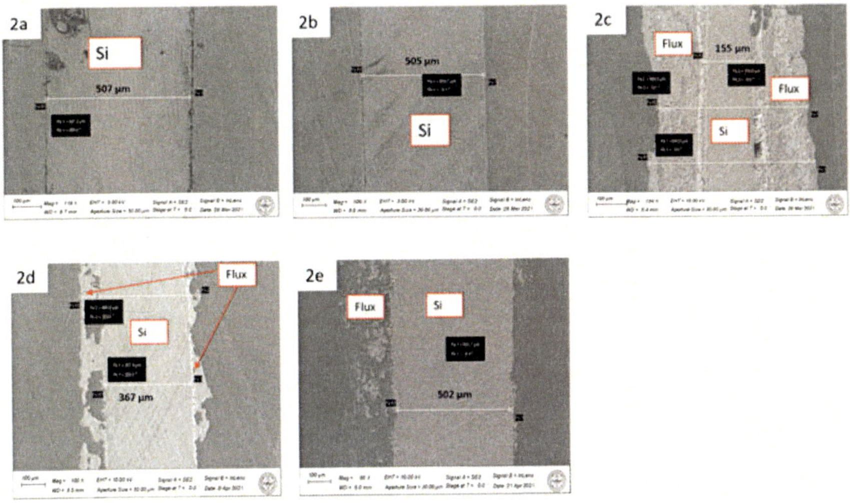

Fig. 2 SEM images of silicon wafer **a** as received, **b** above the flux at 1100 C for 6 h, **c** partially immersed in flux at 1100 C for 6 h, **d** completely immersed at 1.5 cm deep in the flux at 1100 C for 6 h, **e** completely immersed at 3 cm deep in the flux at 1100 C for 6 h

the thickness of the silicon wafer decreased by 4 µm and a very thin layer of silicon oxide formed on the surface (Fig. 2b). Next the silicon wafer was partially immersed in the flux. A significant amount of thinning was observed in the portion of the silicon wafer that was immersed in the flux. This portion of the wafer was only 155 µm thick (Fig. 2c). In subsequent experiments the silicon wafers were completely immersed in the flux to a depth of 1.5 cm and 3 cm, respectively. The silicon wafer that was 1.5 cm deep had a final thickness of 367 µm while the one that was 3 cm deep had a thickness of 503 µm. This indicates that the thinning of the wafer decreases with immersion depth in the flux.

Silicon wafer thinning mechanism—Fig. 2b suggests that the oxygen impurity in the forming gas environment reacts to form a very thin protective oxide scale that is only 3 µm thick on a silicon wafer that is originally 507 µm thick. However, when the silicon wafer is immersed in the flux the thinning continues and it is dependent on the immersion depth of the wafer. This mechanism is shown in Fig. 3. At the flux/gas interface the impurity oxygen in the gas phase is reduced in the presence of an electronic conductor (graphite or silicon) to oxygen ions. The silicon wafer is oxidized at the wafer/flux interface to silicon ions. Thus, the silicon wafer oxidation reaction is aided by the ionic flux while the rate of the reaction is controlled by the electron and the oxygen ion transfer between the oxidation and the reduction reaction sites. Hence it was seen that as the resistance to electron-oxygen ion transfer is increased by increasing the immersion depth. The rate of the silicon thinning also decreased.

To prevent silicon wafer thinning during SOM electrolysis, the immersion depth of the silicon wafer was always maintained at 4.5 cm from the flux/gas interface.

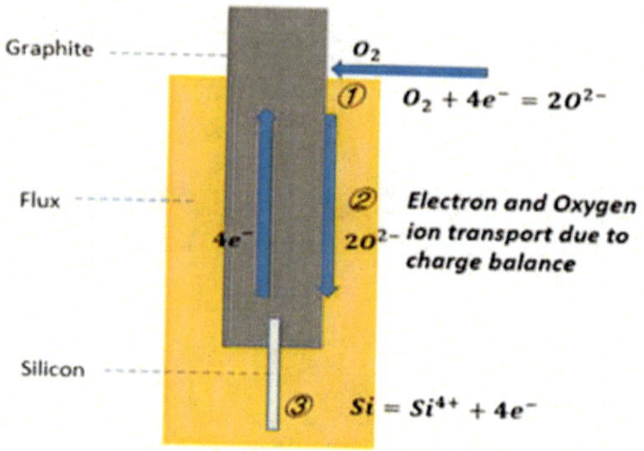

Fig. 3 Reaction mechanism of silicon wafer thinning

Electrochemical Measurements—The impedance measured between the two reference electrodes Fig. 4a provides the total salt resistance while the current measured between the electrodes at very low applied potentials (below the dissociation potentials of the oxides) provides the electronic resistance. From these two measured resistance values, it is possible to determine the ionic and electronic transference numbers [16]. Measurements made before and after electrolysis show that the flux remains primarily ionic throughout the experiment (see Table 1).

The potentiodynamic scan between the anode and the cathode confirms a dissociation potential of 0.51 V (Fig. 4a) corresponding to the following overall cell reaction: $SiO_2 + 2C = Si + 2CO(g)$ at 1100 C. From the linear portion of the current–potential scan the overall cell resistance was estimated to be around 0.5 Ω. In comparison with previous SOM cells, it appears that this cell also has low overall resistance, low electronic and high ionic conductivities, and is able to effectively dissociate SiO_2 [6, 7, 13].

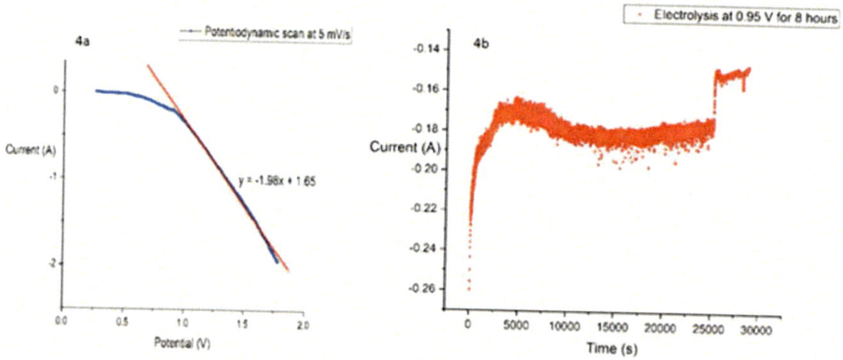

Fig. 4 Electrochemical measurements **a** Potentiodynamic scan at 5 mV/s across cathode and anode **b** Current–time curve during the electrolysis under an applied potential of 0.95 V for 8 h

Table 1 Electronic and ionic transferences number of the salt (flux) before and after the electrolysis

	Total resistance (Ω)	Electronic resistance (Ω)	Electronic transference number, t_e	Ionic transference number, t_i
Before electrolysis	0.11	42.144	2.6e-3	0.99740
After electrolysis	0.17	294.612	5.7e-4	0.99943

Figure 4b shows the electrolysis current with 0.95 V applied potential between the electrodes. During 8 h of electrolysis the current–time plot was stable. The measured changes in current and the recorded anodic gas evolution corresponded well with each other. The anodic gas bubbling was measured to be 4.1 ccm during the first 25,000 s and 3.7 ccm for the remainder of the time, in line with the small observed drop in the current drop from 0.17 to 0.14 A at the 25,000 s mark. The reason for this small drop in the current is not clear. Taking the integral of the current–time curve, 0.36 g of silicon should have been produced under Faradaic conditions which corresponds to a 3 mm thick planar deposit.

Post SOM electrolysis Characterization—The SEM images of sectioned silicon wafer are shown in Fig. 5a–c. Figure 5a shows the EDS mapping of the silicon/flux interface. It was observed that 20 μm thick silicon carbide deposited on the Si wafer cathode (Fig. 5b) instead of 3 mm as predicted based on Faradaic conversion of the

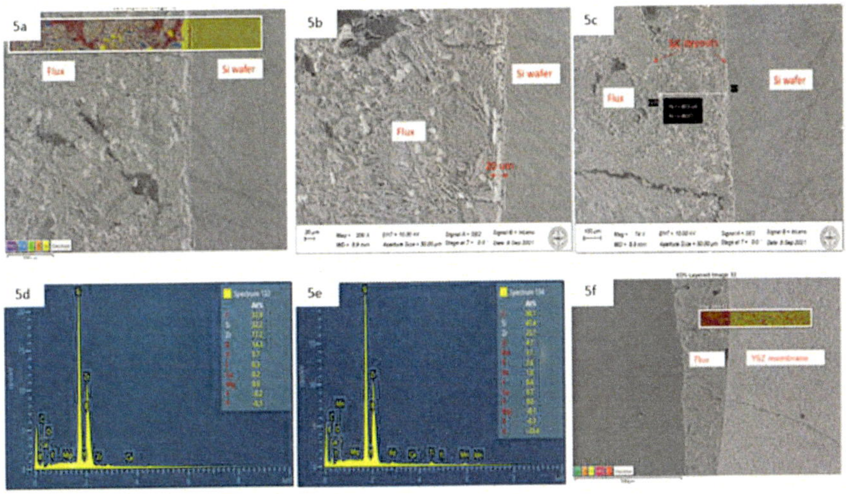

Fig. 5 SEM/EDS analysis of sectioned Si wafer and YSZ membrane **a** EDS mapping of the Si/Flux interface, **b** 20 μm thick SiC deposit on the Si wafer cathode, **c** dispersed SiC deposits in the flux adjacent to the Si wafer cathode, **d** EDX spectrum of the planar deposit close to Si wafer, **e** EDX spectrum of deposits dispersed in the flux adjacent to the Si wafer, **f** sectioned YSZ membrane in contact with flux along with EDS mapping post-electrolysis

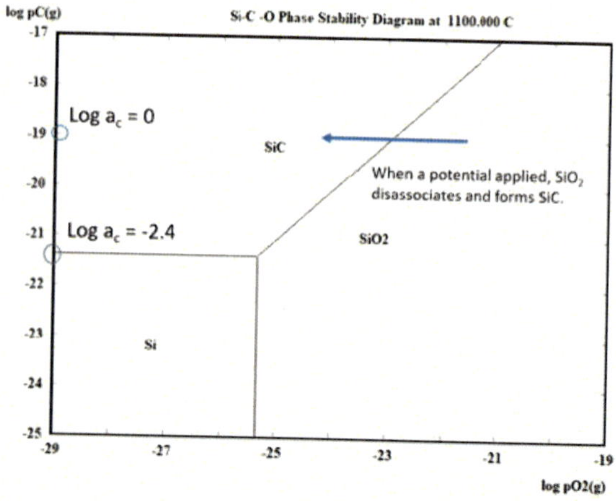

Fig. 6 Si-C-O phase stability diagram at 1100 C

electrolysis current. Some silicon carbide was also observed to be dispersed 450 μm in the flux adjacent to the silicon wafer (Fig. 5c). Further EDS point analysis on these two types of deposits indicates that they are mainly silicon carbide (Fig. 5d, e), with trace amounts of boron and phosphorous. Energy peaks of zirconium and phosphorous have strong overlap. The sources of boron and phosphorous are unclear. Furthermore, the section of the YSZ membrane in contact with the flux shows no diffusion of zirconium or yttrium into the flux (Fig. 5f). Thus, the YSZ membrane was stable in contact with the flux during the electrolysis. Therefore, it is unclear whether the observed Zr peak is real.

As previously discussed, silica dissociation occurred when the potential was greater than 0.51 V. It is likely that silicon was initially produced during the SOM electrolysis, but it reacted with carbon to form silicon carbide (The standard Nernst potential for reaction $SiO_2 + 3C = SiC + 3CO(g)$ at 1100 C is 0.32 V). This can also be explained from the phase stability diagram of Si-C-O (Fig. 6). In the presence of graphite (carbon at unit activity) when the applied potential is increased the oxygen activity decreases below 10^{-23} atm. and silica is reduced to SiC since Si is thermodynamically not stable. In order to reduce silica to Si the activity of carbon needs to be lower than $10^{-2.4}$.

Conclusions and Future Work

A state-of-the-art Solid Oxide Membrane Electrolytic Cell has been designed and assembled. An electrochemical reaction involving silicon wafer, flux, graphite, and impurity oxygen in the gas phase was shown to be responsible for the thinning of the wafer. By immersing the silicon wafer and the graphite current collector deep into the flux, this side reaction was significantly reduced. Subsequently SOM electrolysis of silica with graphite as the current collector was performed. The process was electrochemically characterized, and the deposit was microstructurally and compositionally characterized. The SOM cell has an overall low resistance, and the salt electrolyte remained primarily ionic during electrolysis. It was found that silicon was successfully produced, but it formed a carbide phase due to the presence of graphite in the system (crucible and the current collector). YSZ membrane was found to be stable in contact with the flux during electrolysis.

Future work will focus in four priority areas. The first will be to remove all carbon components from the system. That includes the crucible and the current collectors. The carbon crucible will be replaced by a stainless-steel crucible. The carbon cathode current collector will be replaced by refractory tungsten metal and the carbon anode current collector will be replaced by an inert oxygen current collector developed earlier in our group [17, 18]. With these modifications we hope to be able to deposit pure silicon on the silicon wafer cathode and produce pure oxygen at the anode. Next, we will perform fluid-dynamics simulation to identify ways to lower concentration or mass transfer polarization to sustain the electrolytic current and increase the thickness of the silicon deposit. To lower surface roughness of the silicon deposit, we also plan to perform periodic current reversal with auxiliary electrodes in the salt flux, like that practiced in the copper industry. Finally, for longer-term operation we plan to replace liquid silver anodes, which volatilize, with stable lanthanum nickelates.

Acknowledgements This work was supported, in part, by the US Department of Energy's Office of Energy Efficiency and Renewable Energy (EERE) under the Solar Energy Technologies Office Award No. DE-EE0008988 and by the US National Science Foundation under Award Nos. CMMI-1937829 and 1937818.

References

1. Bye G, Ceccaroli B (2014) Solar grade silicon: technology status and industrial trends. Sol Energy Mater Sol Cells 130:634–646
2. Goodrich A et al (2013) A wafer-based monocrystalline silicon photovoltaics road map: utilizing known technology improvement opportunities for further reductions in manufacturing costs. Sol Energy Mater Sol Cells 114:110–135
3. Krishnan A, Lu XG, Pal UB (2005) Solid Oxide Membrane (SOM) technology for environmentally sound production of tantalum metal and alloys from their oxide sources. Scand J Metall 34(5):293–301
4. Suput M, DeLucas R, Pati S, Ye G, Pal U, Powell AC (2008) Solid oxide membrane technology for environmentally sound production of titanium. Trans Section C, Mineral Process Extract Metall 117(2):118–122
5. Pal UB (2008) A lower carbon foot print process for electrolytic production of metals from oxides dissolved in molten salts. JOM 60(2):43–47
6. Guan X, Pal UB, Powell AC (2014) Energy-efficient and environmentally friendly solid oxide membrane electrolysis process for magnesium oxide reduction: experiment and modeling. Metallurg Mater Trans E, Mater Energy Syst 1(2):132–144
7. Su S, Pal U, Guan X (2017) Solid oxide membrane electrolysis process for aluminum production: experiment and modeling. J Electrochem Soc 164(4):F248–F255

8. Jiang Y, Xu J, Guan X, Pal UB, Basu SN (2013) Production of silicon by solid oxide membrane-based electrolysis process. Mater Res Soc Symp Proceed 1493:231–235
9. Villalon T Jr (2018) Zero-direct emission silicon production via solid oxide membrane electrolysis. Doctoral dissertation, Boston University. https://hdl.handle.net/2144/30729
10. Xu J, Lo B, Jiang Y, Pal U, Basu S (2014) Stability of yttria stabilized zirconia in molten oxy-fluorite flux for the production of silicon with the solid oxide membrane process. J Eur Ceram Soc 34(15):3887–3896
11. Lovering DG, Gale RJ (eds) (1987) Molten salt techniques, vol 3. Plenum Press, New York, NY
12. Prasad S (2000) Studies on the Hall-Heroult aluminum electrowinning process. J Braz Chem Soc 11(3):245–251
13. Su S (2016) Zero-direct-carbon-emission aluminum production by solid oxide membrane-based electrolysis process. Doctoral dissertation, Boston University. https://hdl.handle.net/2144/17080
14. Krishnan A (2006) Solid oxide membrane process for the direct reduction of magnesium from magnesium oxide. Doctoral dissertation, Boston University (2006)
15. Gratz E, Pati S, Milshtein J, Powell AC, Pal UB (2012) Control of Yttrium diffusion out of yttria stabilized zirconia during SOM electrolysis for magnesium production. Magnes Technol:499–503, Sept
16. Virkar AV (2005) Theoretical analysis of the role of interfaces in transport through oxygen ion and electron conducting membranes. J Power Sources 147(1–2):8–31
17. Guan X, Pal UB, Gopalan S, Powell AC (2013) LSM (La0.8Sr0.2MnO3)-Inconel inert anode current collector for solid oxide Membrane (SOM) Electrolysis. J Electrochem Soc 160(11): F1179–F1186, Sep
18. Guan X, Pal UB, Jiang Y, Su S (2016) Clean metals production by solid oxide membrane electrolysis process. J Sustain Metall:152–166, Feb

Part VII
REWAS 2022: Poster Session

Advanced Process of Waste Glass Bottle for the Production of Recycled Glass Aggregate and Cullet

Hoon Lee, Hansol Lee, and Kwanho Kim

Abstract In Korea, ~20% of waste glass bottles are buried in landfills every year. Many studies have been conducted to utilize this waste in foamed glass and cement production. However, more research is needed to find efficient methods to convert the waste to valuable materials, such as recycled fine aggregates. In this study, we designed an integrated crushing-grinding-separation process involving additional crushing equipment to handle large amounts of waste glass bottles and determined whether the products met recycled fine aggregate quality standards. Waste glass bottles were crushed to a size less than 10 mm using a pilot-scale shredder and roll, hammer, and vertical shaft impact (VSI) crushers. We found that the VSI crusher is the most suitable equipment, in terms of crushing performance and ease of operation, to convert the glass bottles to recycled fine aggregates. The crusher generated products exhibiting aspect ratios less than those of natural sand in most particle sizes. The proposed process is a promising way to convert glass wastes to recycled aggregates. We expect that incorporating optical-sorting-based techniques into the process will further increase the recycling rate of waste glass bottles, even from a mix of bottles with different colors.

Keywords Waste glass bottle · Recycled fine aggregate · Crushing · VSI crusher

H. Lee (✉) · K. Kim
Korea Institute of Geoscience and Mineral Resources, 124, Gwahak-ro, Yuseong-gu, Daejeon 34132, Republic of Korea
e-mail: hoonlee@kigam.re.kr

K. Kim
e-mail: khkim@kigam.re.kr

H. Lee
University of Science and Technology, 217, Gajeong-ro, Yuseong-gu, Daejeon 34113, Republic of Korea

A. Lazou et al. (eds.), *REWAS 2022: Developing Tomorrow's Technical Cycles (Volume I)*, The Minerals, Metals & Materials Series,
https://doi.org/10.1007/978-3-030-92563-5_70

Introduction

In 2019, ~610,000 tons of waste glass bottles were produced in Korea, and ~79.6% of them (480,000 tons) were recycled [1]. The recycling rate of waste glass bottles in Korea is above 75%, a relatively higher rate than that for other wastes. However, Korea's recycling rate is ~10% lower than that of top recycling countries like Germany and Japan. Since only the manufacturers with an annual shipment volume of 10 tons and above, sales revenue of 1 billion and above, and imports of 300 million Korean won and above in the preceding year are required to recycle, it is difficult to estimate the exact amount of waste glass bottles [2]; thus, it is estimated that the amount of waste glass bottles produced is more than expected. Furthermore, when the glass bottles are easily broken or different-colored bottles have been mixed, it is challenging to use them as raw materials for glass bottles with high recycling value; therefore, approximately more than 200,000 tons of waste glass bottles are buried in a landfill every year. Therefore, in this study, an integrated crushing-grinding-separation process has been designed to handle large amounts of waste glass bottles and to determine whether the products satisfied the recycled fine aggregate quality standards.

Waste Glass Bottle Recycling Process

Past Research on Waste Glass Bottle Recycling

Various studies have been conducted on recycling waste glass bottles that usually end up in a landfill, including studies on the use of waste glass in foamed glass production, cement production, concrete aggregate alternatives, and mortars [3–21]. These studies showed that the alkali-silica reaction (ASR), which occurs when concrete materials are mixed with waste glass, could be controlled by adjusting the particle size and by using additives. The glass enhances the physical properties of concrete via the pozzolanic reaction, which is a cement-hydration reaction. Thus, various methods can be applied to process and utilize waste glass bottles as a construction material.

Waste Glass Bottle Recycling Process in Korea

In Korea, waste glass bottles are collected by either local governments or private companies. After some impurities are initially removed during washing and hand-picking of the collected waste glass bottles, the bottles are delivered to intermediate companies that perform crushing and grinding. These companies apply various separation protocols to eliminate the impurities that were not removed during the previous

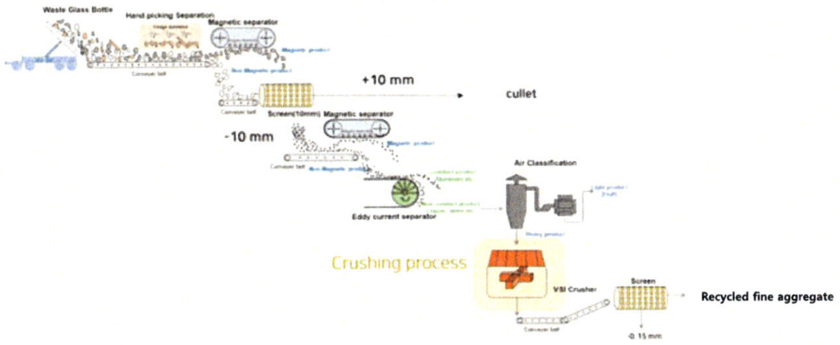

Fig. 1 Waste glass bottle recycling process

stage and to reproduce the waste glass bottles in the desired particle size by the crushing process. The subsequent processes are dictated by the final application. In this study, a new process was designed by adding a crushing equipment to reuse waste glass bottles as recycled aggregates. Based on this new design, the existing recycling process line of the recycling company currently treating the largest amount of waste glass bottles in Korea has been reconstructed, as shown in Fig. 1. First, ferrous metal pieces are removed from the separated waste glass bottles by a primary magnetic separator. Next, the bottles are separated by a trommel screen (screen size of 10 mm) to be used either as glass cullet for making glass bottles if the glass pieces are larger than 10 mm or as the recycled aggregates if they are smaller than 10 mm. Then, the ferrous and non-ferrous metals are removed from the separated glass pieces by magnetic and eddy-current separators, respectively, and light impurities like paper pieces are eliminated by an air classifier. Finally, the crushing equipment processes the glass to an average particle size of ~10 mm for the glass to be used for recycled fine aggregates that meet the quality standards of particle size distribution.

Crushing/Grinding Process for Waste Glass Bottles to Be Used as Recycled Aggregate and Determination of Quality Standard Adequacy

Crushing/Grinding Process and Particle Size Distribution

To reuse waste glass bottles as the glass for designing recycled fine aggregates, numerous quality standards must be satisfied, the most representative ones of which are the quality standards for impurities and particle size distribution. According to the Korea Standard (KS) F2576 test, the content of organic impurities, such as vinyl, plastic, wood, and paper, must be below 1.0% in the recycled aggregate sample. The quality standard for impurities has been met since most of the impurities were removed by the hand-picking, eddy-current, magnetic, and air separation processes, as shown in Fig. 1. Table 1 shows the particle size distribution quality standard of the glass for recycled fine aggregates (KS F2527 test) [22]. The impurity-free waste glass bottle did not meet the quality standard for particle size distribution (KS F2527 test), requiring an additional crushing process. In this study, waste glass bottles were crushed to a size less than 10 mm using a pilot-scale shredder and roll, hammer, and vertical shaft impact (VSI) crushers. The particle size distribution of the crushed products is shown in Fig. 2. Comparing the results of the crushing test with the standards revealed that the particle size distribution it did not meet the minimum particle size distribution standard due to insufficient crushing by the roll crusher and shredder. The shredder was considered unsuitable for materials with high hardness like glass because the equipment is designed to break down ductile and composite materials. The rolls of the roll crusher are spaced apart, which is also not suitable for crushing the particles of plate-type waste glass bottles. The hammer crusher, whose primary energy is the impact force, caused over-grinding due to the high kinetic energy of the rotating hammer. Since the hammer was worn away easily, it was considered unsuitable in terms of maintenance and repair. The VSI crusher, on the other hand, crushed the waste glass bottles sufficiently because it does not hit the samples directly but uses rotating impellers to disperse the sample over the wall, inducing crushing effects via the collision energy generated between particle–particle and particle–wall.

Table 1 Quality standard for particle size of recycled fine aggregate

Particle size, mm		10	5	2.5	1.2	0.6	0.6	0.15
Mass percentage of sieved particle, %	Max	100	100	100	90	65	35	15
	Min	100	90	80	50	25	10	2

Fig. 2 Particle size distribution of crushed waste glass bottle

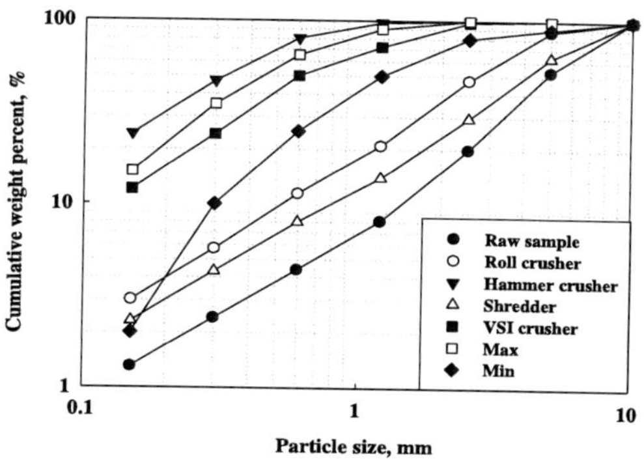

Aspect Ratio

According to literature, when waste glass bottles of different colors are crushed and used as recycled fine aggregates, the angularity and roundness are important properties to consider in the easy handling of a mixture of aggregates and binding materials, and can be determined by the aspect ratio (the ratio of the maximum measured length of a particle to the minimum measured length). In this study, the aspect ratio of waste glass bottles was measured before and after crushing using Image J, a common image processing program. As shown in Fig. 3, Image J rendered the sample into a digital image, enhanced the contrast of the image to remove the gray shades, and then converted them into a binary mode to clearly distinguish the particle boundary.

Figure 4 shows the aspect ratio for each particle size of the feeds and crushed products of the VSI crusher. Approximately 20% of natural sand particles have an aspect ratio of 1.5 and higher. In the feeds, ~18% of the relatively large particles of size 5.0–10.0 mm exhibited an aspect ratio of 1.5 and higher, similar to natural sand.

Fig. 3 Sample image analysis using Image J

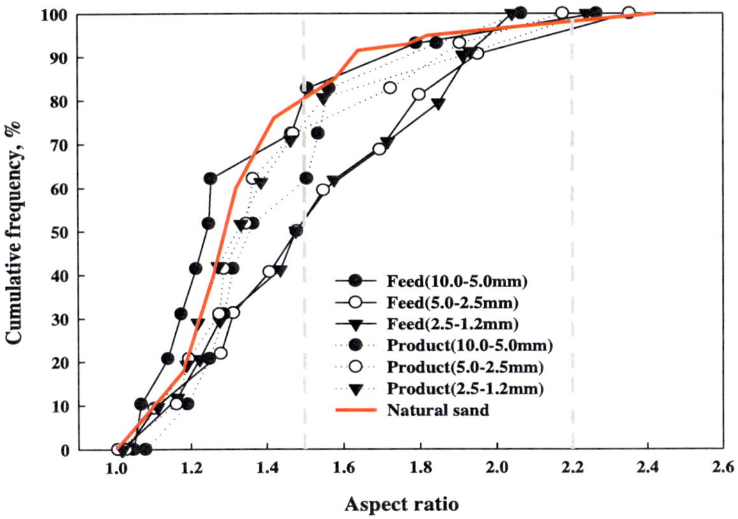

Fig. 4 Cumulative aspect ratio distribution

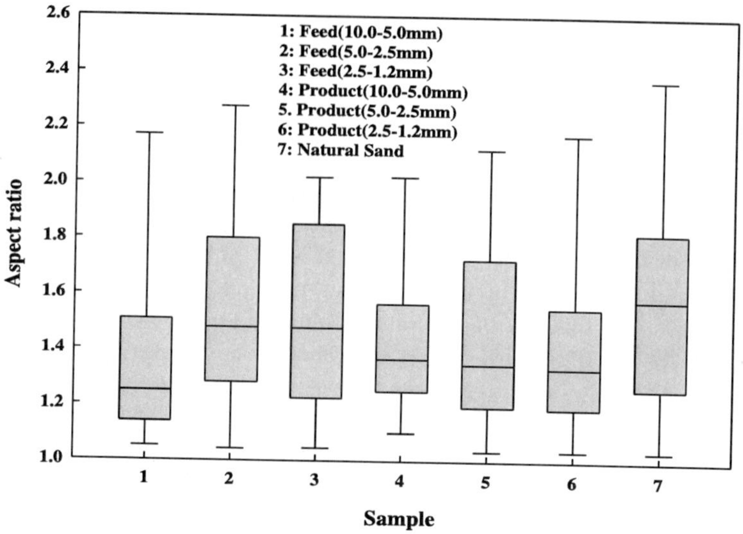

Fig. 5 Box plot of aspect ratios

However, ~50% of the particles of sizes 2.5–5.0 mm and 1.2–2.5 mm showed an aspect ratio of 1.5 and higher. In the products, the percentage of particles of size 5.0–10.0 mm with an aspect ratio of 1.5 and higher increased by ~22% compared to the particles of the feeds, but it decreased by ~20% for the particles of sizes 2.5–5.0 mm and 1.2–2.5 mm. The box plot in Fig. 5 shows that the crushed products exhibited aspect ratios less than those of natural sand in most particle sizes.

Future Research

The best way to add value to recycled waste glass bottles is to use them as cullet, a raw material for glass production. Cullet can only be used when it is composed of single-colored materials; however, since waste glass bottle pieces currently collected are of different colors, they are not reusable as the cullet. Therefore, sorting waste glass bottle pieces based on color using optical sorters and the process to separate and recycle waste glass bottles in terms of size and application should be studied in the future.

Conclusion

A process for recycling waste glass bottles as glass for recycled fine aggregates was proposed herein. The feasibility of forming waste glass bottles into recycled fine aggregates was investigated by identifying the optimal crushing equipment and measuring the products' aspect ratio. The VSI crusher is considered the most suitable equipment in terms of crushing performance and ease of operation, and it generated products that exhibited aspect ratios less than those of natural sand in most particle sizes. A large amount of waste glass bottles that are buried in a landfill every year can be recycled and used as a recycled fine aggregate using the proposed integrated process. If an optical-sorting-based treatment process for waste glass pieces of different colors is further added to the proposed process, it could significantly contribute to the improvement of the waste glass bottle recycling rate.

Acknowledgements This research was supported by the Korea Environmental Industry & Technology Institute (KEITI) as "The domestic waste recycling technology development project" (No. 2019002750001) and the Korea Evaluation Institute of Industrial Technology funded by the Ministry of Trade, Industry and Energy in Korea (Project No.: 20010817, 21-9806).

References

1. Ministry of Environment (2020) Korea environment corporation—Recycling performance by recycling designated businesses. https://www.data.go.kr/data/3043420/fileData.do. Accessed 26 Nov 2020
2. Korea Environment Corporation (2014) Guidelines for the fulfillment of recycling obligations. https://www.keco.or.kr/kr/business/resource/contentsid/1563/index.do. Accessed 1 July 2021
3. Bai J, Yang X, Xu S, Jing W, Yang J (2014) Preparation of foam glass from waste glass and fly ash. Mater Lett 136:52–54
4. König J, Petersen RR, Yue Y (2015) Fabrication of highly insulating foam glass made from CRT panel glass. Ceram Int 41(8):9793–9800
5. Chen M, Zhang FS, Zhu J (2009) Lead recovery and the feasibility of foam glass production from funnel glass of dismantled cathode ray tube through pyrovacuum process. J Hazard Mater 161(2):1109–1113
6. Hesky D, Aneziris CG, Groß U, Horn A (2015) Water and waterglass mixtures for foam glass production. Ceram Int 41(10):12604–12613
7. Chen G, Lee H, Young KL, Yue PL, Wong A, Tao T, Choi KK (2002) Glass recycling in cement production—An innovative approach. Waste Manag 22(7):747–753
8. Islam GS, Rahman M, Kazi N (2017) Waste glass powder as partial replacement of cement for sustainable concrete practice. Int J Sustain Built Environ 6(1):37–44
9. Polley C, Cramer SM, Cruz RV (1998) Potential for using waste glass in Portland cement concrete. J Mater Civ Eng 10(4):210–219
10. Vijayakumar G, Vishaliny H, Govindarajulu D (2013) Studies on glass powder as partial replacement of cement in concrete production. Int J Emerg Technol Adv Eng 3(2):153–157
11. Jin W, Meyer C, Baxter S (2000) "Glascrete"-concrete with glass aggregate. ACI Mater J 97(2):208–213
12. Idir R, Cyr M, Tagnit-Hamou A (2010) Use of fine glass as ASR inhibitor in glass aggregate mortars. Constr Build Mater 24(7):1309–1312

13. Gautam S, Srivastava V, Agarwal V (2012) Use of glass wastes as fine aggregate in Concrete. J Acad Indus Res 1(6):320–322
14. Corinaldesi V, Gnappi G, Moriconi G, Montenero A (2005) Reuse of ground waste glass as aggregate for mortars. Waste Manage 25(2):197–201
15. Kou S, Poon CS (2009) Properties of self-compacting concrete prepared with recycled glass aggregate. Cem Concr Compos 31(2):107–113
16. Shao Y, Lefort T, Moras S, Rodriguez D (2000) Studies on concrete containing ground waste glass. Cem Concr Res 30(1):91–100
17. Park SB, Lee BC (2004) Studies on expansion properties in mortar containing waste glass and fibers. Cem Concr Res 34(7):1145–1152
18. Lu JX, Duan ZH, Poon CS (2017) Combined use of waste glass powder and cullet in architectural mortar. Cem Concr Compos 82:34–44
19. Du H, Tan KH (2013) Use of waste glass as sand in mortar: Part II–Alkali–silica reaction and mitigation methods. Cem Concr Compos 35(1):118–126
20. Tan KH, Du H (2013) Use of waste glass as sand in mortar: Part I-Fresh, mechanical and durability properties. Cem Concr Compos 35(1):109–117
21. Lu JX, Poon CS (2018) Use of waste glass in alkali activated cement mortar. Constr Build Mater 160:399–407
22. Korea standard (2018) KSF2527 notice; Concrete aggregate. https://e-ks.kr/streamdocs/view/sd;streamdocsId=72059202121014111. Accessed 27 Dec 2018

Copper Recovery Assessment from Tubular Led Lamp Strips

Jonathan Tenório Vinhal, Gabriel Bastos Pacheco, Rafael Piumatti de Oliveira, and Denise Crocce Romano Espinosa

Abstract The technology of LED lamps has revolutionized the lighting market, stimulating its consumption. Consequently, a generation increasing of this type of waste is observed, LED lamps are composed of carcass, PCBs, strips and LEDs. The lamps have metals with economic potential for recycling. Al and Cu are the main metals in strips and can be recycled through hydrometallurgical techniques. Thus, this work aims to evaluate the possible recovery of Cu from tubular led lamps strips. The first stage of the study is the strips characterization through chemical analysis by ICP-EOS, and SEM. The next stage is the leaching thermodynamic simulation using FactSage 8.0 software. The strips characterization showed contents of 88% of Al and 8.5% of Cu. The thermodynamic simulation shows possible selective leaching of Cu against Al, under the following conditions: HNO_3, pH 3–3.8, Eh above 0.55, and 25 °C; and NH_3, pH 7–10, Eh −0.15 to 0.6, and 25 °C.

Keywords Waste from Electrical and Electronic Equipment (WEEE) · Led lamps · Cu recovery · Characterization · Thermodynamic simulation

Introduction

LED lamps have dominated the lighting market due to their greater energy efficiency compared to fluorescent and incandescent lamps. The growing consumption brings with it a problem related to their disposal at the end of their lifespan. Due to the recent development of LED lamp technology, there is still no specific and standardized process for their recycling. Furthermore, few studies have proposed recycling routes for this (WEEE), making necessary the seek for new sustainable recycling routes [1].

For the structuring of recycling processes, it is essential to know the parts that make up LED lamps. That can be made through a characterization study and assessment of information in the literature. Copper is one of the metals with great potential

J. T. Vinhal (✉) · G. B. Pacheco · R. P. de Oliveira · D. C. R. Espinosa
Department of Chemical Engineering, Polytechnic School of University of Sao Paulo, São Paulo, SP, Brazil
e-mail: jonathanvinhal@usp.br

A. Lazou et al. (eds.), *REWAS 2022: Developing Tomorrow's Technical Cycles (Volume I)*, The Minerals, Metals & Materials Series,
https://doi.org/10.1007/978-3-030-92563-5_71

for recovery in electrical and electronic waste [2]. In LED lamps, more specifically LED devices, it is also possible to identify rare earth elements, such as Y, Eu, and Ce, and critical metals such as In and Ga. The recovery of metals from secondary sources becomes attractive due to the difficulty in obtaining them [3, 4]. Not only do LEDs contain recoverable elements in the lamps, but also the electronics, and the supporting strips for the LEDs. The strips can be made of aluminum or polymeric material. The construction of the aluminum strips is made with a solder mask layer that covers the copper circuit, over a layer of dielectric material, supported by an aluminum strip. [5]

In addition to recovering metals from secondary sources, recycling LED lamps also has social benefits. As described in a study carried out in Australia [6], the material recovery process from electronic waste has a dismantling stage, requiring specialized labor. The recycling process of these residues directly reflects on the generation of employment and income [6]. Similar to the processing of TV LEDs [3], LED tube lamps require manual dismantling step to concentrate the fractions containing the elements of interest. In that way, the waste collection and pre-processing idealized in Australia is seen as useful for recycling LED lamps.

After pre-processing, one of the ways to perform urban mining is through hydrometallurgy. Leaching techniques are applied in extractive metallurgy using solutions for extracting metals from solid materials. The process consists of taking a metal from its solid form to an aqueous form. The literature has reported leaching technic as an alternative to recover copper from slags [7], mining waste [8] and WEEE, i.e., printed circuited board [9]. Normally, this process is used to ores, but it can also be applied to WEEE, as in the case of this work. Selective leaching is possible on solids with two or more metals contained, that means, the desired metals are solubilized while the others are kept in their solid form. In many cases this process is not perfect and needs further purification. The metal leaching technique can occur in both a basic and an acidic medium, depending on the material to be leached [10]. The literature has reported leaching technic to recover cooper from slags,

An important tool for hydrometallurgy is the Eh-pH diagram, also known as the Pourbaix diagram, or predominance diagrams. The axis of the diagram corresponding to the pH represents a measure of the H^+ ions present in the solution, while the other axis represents the redox potential. In each diagram region it will appear one or more species. This means that by thermodynamics, such species are predominant in that medium under the conditions covered by the region. The lines that divide each region are the balance lines. The applications of these diagrams are diverse, such as corrosion, passivation, water treatment, hydrometallurgy, and metal recovery [11].

HydraMedusa and HSC software are used to assemble predominance diagrams. The thermodynamic data provided by them facilitate and guide processes since it is possible to identify the phases formed under specific conditions. As mentioned by Botelho Junior et al. [12], the FactSage Software widely used in pyrometallurgical processes and little explored in hydrometallurgical studies has an equilibrium module for reaction in aqueous medium capable of supplying the formed products, as well as their quantities, and consumption of leaching agents. It is noteworthy that the software is also an important tool to assess the behavior of metals in different leaching medium, through the EpH module.

Therefore, given the background, this work aims to evaluate the possible recovery of Cu from tubular led lamps strips through two steps, strip characterization and thermodynamic simulation of leaching process.

Materials and Methods

To carry out this work, tubular LED lamps from CEDIR and from the maintenance of the Polytechnic School of the University of São Paulo were used. The lamps were dismantled and splitted into three parts: connectors; lamp tube, and strips. The target of this work was the strips.

A visual analysis intends that the strips are composed of layers, thus it was necessary for better understanding analysis by scanning electron microscopy associated with microanalysis by energy dispersive X-ray spectrometry (SEM–EDS, Phenom, proX model). A sample of the strip was embedded in resin to observe its profile and obtain a better-quality SEM image.

Chemical analysis of the strips was also performed. 250 mg of sample was weighed for microwave digestion (CEM, Mars model 6). Base on the elements identified in the SEM and according to Microwave Acid Digestion: Method Note Compendium [13], 9 mL of HNO_3 and 1 mL of HF were used. In addition, 0.45 g of H_3BO_3 were added after digestion to neutralize the free fluorides in the solution. Finally, the solution was volumed to 25 mL and analyzed by optical emission spectrometry with induced-coupled plasma (ICP-OES, Varian, model 725ES).

After strip characterization, thermodynamic simulations were carried out in different scenarios using FactSage software version 8.0 to identify the best conditions for recovering copper from the strips. The Pourbaix diagrams of aluminum and copper in water, as well as in a solution of HCl, H_2SO_4, and HNO_3 and NH_3 were assembled using the EpH module. Simulations of Cu and Al leaching were also carried out using the equilibrium module. The leachings studied were in acidic medium with HCl, H_2SO_4, and HNO_3 and in basic medium using NH_3. Figure 1 illustrates the interface of the EpH module used to build the Pourbaix diagrams.

In order to generate the Pourbaix diagrams, the EpH module on the FactSage 8.0 software was used. The database used was FactPS. It is possible in the elements tab to select all the elements involved in the reaction (metals or non-metals) as well as the number of metals. Oxygen and hydrogen were always present in non-metallic elements, whereas metals could have Al and/or Cu. For solutions containing some acid or NH_3, a new non-metallic element was added. When there were two metals in solution, an R in the range between 0 and 0.5 was selected on the metal mole fraction tab due to the characteristics of the strip, being R is equal to Cu/(Cu + Al). In the Species tab, the aqueous and solids options were selected and the molarity equal to 1 was standardized, indicating that the simulation would be performed with aqueous and solid phases. In this same section, the selection of the list of possible formed

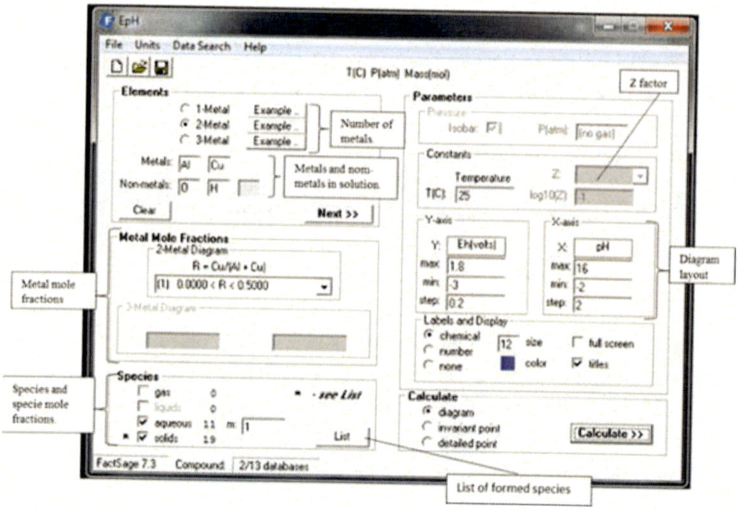

Fig. 1 FactSage 8.0 software EpH module interface

species was made, allowing, for instance, to differentiate the Pourbaix diagram in nitric acid medium from the diagram in ammoniacal medium. On the constants tab it was possible to determine the temperature and also select the type and the value of the "Z" factor. For all produced diagrams, the ranges on axis were kept in a way they would have the same dimensions, (Eh) ranging from −3 to 1.8 V in 0.2 step, and (pH) ranging from −2 to 16 at intervals of 2.

In diagrams building for acidic and basic medium, the non-metallic element present in the acid or base was selected. For the HCl leaching, chlorine was added, sulfur was added for H_2SO_4, and for the HNO_3 and NH_3 leaching, nitrogen was used. It should be noted that in these cases which is added a third non-metallic element to the system, a new option is enabled on the constants tab. This option is called "Z", it determines the amount of a species (ionic or not) in the solution. That means, from its definition it is possible to show the substance that is in the mixture. It is also worth noting that the values of Z are entered as logarithms of Z. For the case of HNO_3 the option of m(NO_3^-) was selected, that is, molarity of NO_3^-, for H_2SO_4, m(SO_4^{2-}) was used, for HCl m(Cl^-), and for NH_3 m(NH_3).

As it is a thermodynamic software, the simulation is carried out with all the existing compounds in the software database involving the added species. Therefore, some species that are unlikely to be formed were removed from the list of possible products in order to obtain a result closer to reality.

The diagrams were generated in different molarities, aqueous medium, and temperatures in order to generate a larger database for the analysis of the optimal process conditions. In general, aluminum hydrides (AlH_3), AlOOH and hydrated aluminas ($Al_2O_3(H_2O)$) were removed. For copper, the CuH_2O_2 species was removed. As already mentioned, the nitric and ammoniacal medium were differentiated by the adjustment and selection of species formed in the diagrams. For both

Table 1 Parameters evaluated in the building of Pourbaix diagrams

Metals	Acids/bases	Metal concentration	Acids/bases concentration	Temperature (°C)
Al	–	1 M	–	25, 40 e 55
Cu	–	1 M	–	25, 40 e 55
Al e Cu	–	1 M	–	25, 40 e 55
Al e Cu	HNO_3	1 M	0.5 e 1 M	25, 40 e 55
Al e Cu	H_2SO_4	1 M	0.5 e 1 M	25, 40 e 55
Al e Cu	HCl	1 M	0.5 e 1 M	25, 40 e 55
Al e Cu	NH_3	1 M	0.5 e 1 M	25, 40 e 55

cases, compounds involving only nitrogen and metals (AlN and CuN_3) were removed, as well as the compound $Cu_4(NO_3)_2(OH)_6$. Besides, for NH_3 copper complexes involving NH_3 were kept, while for HNO_3 all of them were removed, as $Cu(NH_3)^{2+}$.

Table 1 shows the parameters evaluated in the building of Pourbaix diagrams of copper and aluminum. They are acids and bases, aqueous medium concentrations, and temperature.

After the building of the predominance diagrams and their analysis, the optimal conditions for selective extraction of copper compared to aluminum by leaching were established. In addition, previous studies were consulted to guide and define the leaching parameters (solid/liquid ratio (25 and 50 g/L), temperature (25, 40 and 55 °C), sample mass (5 and 10 g) for a fixed leaching solution volume of 200 mL and concentration of the leaching agents (varying according to the selected leaching agent under optimal conditions) [14–21]. In that way, the equilibrium module on FactSage 8.0 was used to simulate the thermodynamic equilibrium of the residue leaching reactions under the optimal conditions established by the Pourbaix diagrams.

Given the initial composition of reagents and a list of possible products formed, the equilibrium module calculates, based on thermodynamic databases, the final composition of the species after reaching thermodynamic equilibrium. The reagents can be added in grams, moles, or pounds. It is also possible to select the initial temperature conditions of individually compounds. In simulation, it was opted not set the initial temperature conditions for compounds.

The list of possible formed products was the same used in the EpH module. However, more metallic compounds were used besides Al and Cu. From chemical analysis by ICP-OES, it is possible to determine the proportion of metals in the strip, thus adding Fe, Ti, Si, Ca, and Pb. On the Products tab, the selection of formed compounds is made from its phase, thus it is necessary to choose the possible products formed in the gaseous, aqueous, liquid, and solid phases.

Results and Discussion

Characterization

Table 2 shows the mass average values of the LED lamp components. Observing the results obtained after weighing the lamps, it is noted that the corresponding mass of each part of the lamp varies, depending on the brand, in some cases the glass used is thinner than others, while in other cases the strips are produced in polymeric material and not aluminum. Considering only the case of lamps that have strips which are made of aluminum, it is noted that the lamps have, on average, a total mass of 194.26 g, tubes 144.67 g, connectors 24.43 g, and strips 25.17 g. Tubes represent 74.47% of the total mass of the lamp, connectors 12.58% and strips 12.95%.

Through the SEM–EDS of the strip embedded in resin, Fig. 2b, it was possible to verify the structure of the strip. It was also seen that the dielectric layer has a thickness of around 100 μm, while the solder mask layer varies between 20 μm (in the thickest part) and 6 μm (in the thinnest part). Copper layer has a thickness of something around 22 μm. Figure 2b (point 3) indicates a notable presence of carbon and oxygen, so this fact suggests that the solder mask is formed from some type of organic compound. As mentioned by Oliveux et al. [22] and Sperança et al. [23], the soldering mask is an organic resin formed mainly by polyester, epoxy, or polypropylene. The whitish color of this layer is due to the presence of titanium oxide, as indicated by the EDS, titanium and oxygen are present in this portion of the material, so titanium oxide was probably used to produce the solder mask.

In addition to the solder mask layer, the expected distribution of aluminum and copper present in the strip can be seen. The aluminum layer works as bases structure and the copper layer as the electric circuit on the strip.

The dielectric layer image made it possible to indicate that it is composed of fiberglass as stated by Oliveux et al. [22]. The remarkable presence of silicon and oxygen in the lighter regions, Fig. 2b (point 5), and the presence of carbon in the darker regions, Fig. 2b (point 4), can be seen. The presence of Si and O, in addition to the cylindrical morphology of these portions, are indications of silica, that is, glass in the form of fibers.

Table 2 Lamp mass distribution

Sample	Lamp tube	Total mass (g)	Tube mass (g)	Strip mass (g)	Connectors (g)
1	Glass	193.1	158.0	3.2[a]	31.9
2	Glass	215.0	165.1	25.7	24.2
3	Glass	210.9	163.9	24.8	22.2
4	Glass	156.9	105.0	25.0	26.9

[a]Sample 1 has a polymer strip, that explains its lower mass compared to the others

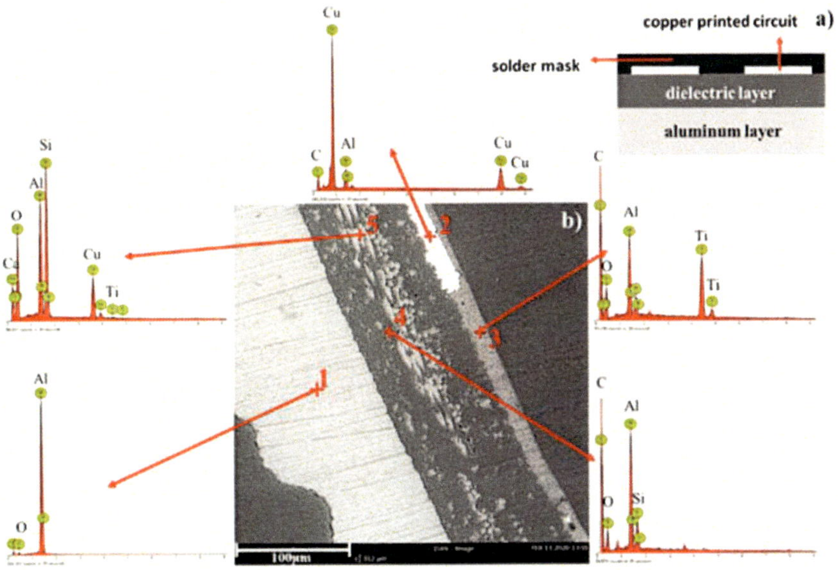

Fig. 2 **a** Schematic aluminum strip **b** Image obtained by SEM–EDS analysis of strip embedded in resin

It is noteworthy that during the process of polishing the sample of the embedded strips, the metallic parts suffered mechanical deformation and can be observed as contaminants in the microanalysis of adjacent layers. An example of this contamination is visible in the EDS in the region of the copper layer, Fig. 2b (point 2) where the presence of aluminum and carbon is found in the spectrum.

Based on the SEM–EDS analysis, it is understood that for a better leaching process, Al and Cu should be exposed to attack by the acidic or basic solution, which is not the case of the strip. Above the Cu circuits there is the overlay of the solder mask, so to expose Al and Cu to an acidic or basic attack it would be necessary to remove this layer. Such a process could be done through milling.

The chemical analysis corroborates the SEM–EDS, showing that most of the strip is composed of Al, around 88.1%. Copper corresponds to 8.5% of the strip. Therefore, taking the mass fraction of 12.95% of the lamps corresponding to the strip, at least 1.1% of the lamp mass is copper from the strip, and 11.4% aluminum. Aluminum comes from the strip matrix, while copper is part of the electrical junction circuits of the LEDs in the lamp. In addition to these two metals, there is also Pb, which can be linked to the solders present in the residue, Si probably coming from the dielectric material, that is, fiberglass, and Ti, a constituent of the white paints in the solder mask. Table 3 shows the chemical analysis of strip by ICP-OES.

An estimative of metals recovery from waste LED lamps is delicate due to the lack of data related to waste LED generated. The studies found in literature estimate the waste LED generated using other data as trade, sale, and lifespan. The few data of LED lamps waste collection, and different types and model of lamps make that work

Table 3 Chemical analysis of strip by ICP-OES

	Al	Ca	Cu	Fe	Pb	Si	Ti
%	88.1	0.8	8.5	0.5	0.1	1.4	0.5

even hard. Kumar et al. [1] have investigated the generation of LED lamps waste in Canada and the economic benefits through its recycling, showing an estimated contained value of Cu and Al for waste LED lamp equal to 1.6million USD, and 5.1million USD in 2021. That represents 174 tonnes of Cu and 1,774 tonnes of Al (values found based on metal price in 2021). Taking as reference the waste LED generated in 2021 Kumar et al. [1], the strip mass from Table 2, and the metal% from Table 3, around 43,5 tonnes of Cu and 451.2 tonnes of Al are potential to be recovered from the waste strips of LED lamps in Canada.

Thermodynamic Simulation

After strip characterization, the predominance diagrams were built. From Pourbaix diagrams of a system containing the two metals (Al and Cu) in water, it can be seen that both have wide ranges in which they can be leached by the formation of compounds or the metal in ionic form. Furthermore, it was possible to observe oxide formation bands, such as alumina. It is important to emphasize that, according to the diagrams, it is possible in a system involving Al and Cu by adding another compound, both in a basic and acidic medium. That facilitated the proposition of different compounds for their leaching, as can be seen in the diagrams involving HNO_3, HCl, H_2SO_4, and NH_3. It should also be noted that the higher the temperature, the more the equilibrium lines move toward acidic pHs.

Having in mind the study goal of recovery cooper by selective leaching, the sceneries analyses were made. The first system to be studied was HNO_3 (Fig. 3a). It is possible to observe that for a temperature of 25 °C, Eh above −1.75, and pH less than 3 the aluminum is solubilized in the form of Al3 +, for an Eh above 0.28 and a pH less than 3 copper also comes into solution, but these cases are undesirable according to the objective of the work. Despite this, there is a small range in which only Cu is solubilized in the form of Cu^{2+} and aluminum undergoes oxidation. Initially, this new condition is not appropriate for the purpose of the study due to aluminum oxidation, but aluminum can undergo passivation [24], that means, the passivation would protect the material from completely oxidizing. The higher the concentration of metals, the more the equilibrium lines shift toward acidic pHs, as in diagrams containing only Al and Cu. Based on the 0.5 M diagram at 25 °C it is possible to determine that this region is comprised between pH 3 and 4.1 and Eh above 0.45.

The second system to be analyzed was H_2SO_4, (Fig. 3b). From the Pourbaix diagram, it can be seen that copper is not leached throughout the pH and Eh range, and in some portions only aluminum is leached. According to Silvas et al. [25], Cu

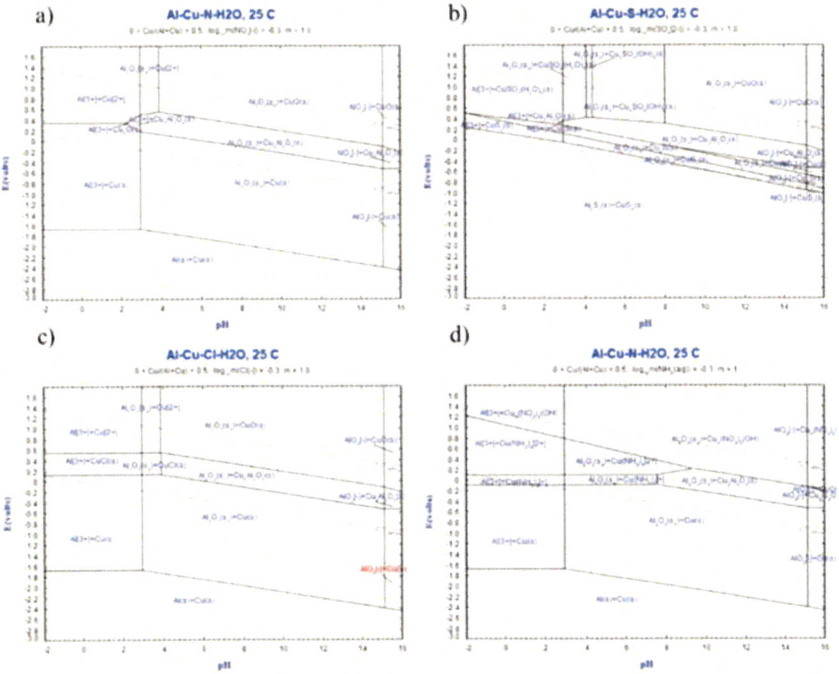

Fig. 3 Al and Cu Pourbaix diagram in 0.5 M of **a** HCl, **b** H_2SO_4, **c** HNO_3, and **d** NH_3 solution at 25 °C

can be leached in H_2SO_4 medium together with H_2O_2, but to carry out such a process it is necessary first leach the Al using only H_2SO_4, and then Cu with the adding of H_2O_2. Therefore, to a system with Al and Cu a leaching involving H_2SO_4 does not apply to the purpose of the work since the Al is in solution.

The third system to be analyzed is the acidic leaching using HCl, (Fig. 3c). As in the HNO_3 diagram, the equilibrium lines advance to acidic pHs with increasing temperature, which is an advantage, as it is possible to work at lower pHs and, therefore, use more HCl, thus increasing the speed of the reaction. As with leaching containing HNO_3 there are several bands where aluminum and copper are leached, however there is a small band where only copper is leached and aluminum is not. Based on the diagram for a temperature of 25 °C, this range is comprised between pH 3 and 4 and Eh above 0.55. Analyzing this region, it can be predicted that it will present the same problem in relation to the Eh range, as it encompasses values that are difficult to be reached. As can be seen in [23], systems containing Cl- ions are aggressive to aluminum, preventing its passivation, so this system would not enable selective copper leaching either.

The last system analyzed is the leaching system in a basic medium using NH_3. As can be seen in the studies of Oluokun and Otunniyi [19] and Sun et al. [21] the NH_3 solution can be made directly using NH_3 or using an ammonium salt. Looking at the diagrams, it can be seen that the selective copper leaching region in the center of the diagram grows with increasing NH_3 concentration. It is also noted that with increasing temperature the equilibrium lines shift to acidic pHs. Considering the objective of the work, there are two bands which copper is leached and aluminum remains in metallic form on strip. Considering the diagram of 0.5 M NH_3 at 25 °C, it can be seen that this range is between pH 3.25 and 10 and Eh between −0.15 V and 0.8 V, but because it is a leaching in a basic medium, if only the range after pH 7, then the new range is between pH 7 and 10 and Eh -0.15 V and 0.6 V. In this system, copper in solution would not be in the form of a pure ion, but would be in the form of $Cu(NH_3)_4^{2+}$. In addition, the hypothesis of aluminum passivation is also adopted. In the diagrams with acidic solutions, it can be seen that by raising the temperature the copper recovery process is favored, which is not true for the ammonia system. As the temperature increases, the equilibrium lines shift to acidic pHs, reducing the basic pH range, which is possible for carrying out the process. Furthermore, NH_3 is a volatile compound, so the increase in temperature facilitates its exit from the aqueous system, being extremely undesirable.

From the analysis made on built diagrams, the aqueous medium of nitric acid and ammonia were selected as favorable for the leaching process. Optimal regions were identified in the speciation diagrams, and applied to the leaching simulation using the equilibrium modulus. The regions found were between pH 3 and 3.8, Eh above 0.55 for HNO_3 at a temperature of 25 °C. For a system involving NH_3 the optimal region was between pH 7 and 10, Eh between −0.15 and 0.6 at a temperature of 25 °C. The first equilibrium modulus results to be analyzed is the HNO_3 system. In this scenario, it is possible to confirm what is observed in the Pourbaix diagram for HNO_3 at 25 °C, that is, at a pH close to 3 the Cu is leached in the form of Cu^{2+}. Despite a recovery of practically 100% of the copper, it can be seen that part of the Al (24.9%) also went into solution. This result provided by the software refers to thermodynamic equilibrium, disregarding any kinetic phenomena. At the beginning the acid molarity was 0.7 M, which corresponds to a pH below 3, and therefore, represents a region where aluminum is leached. However, it was considered for the leaching process that Al has a slower kinetics than Cu. Such consideration can be justified by the formed layer of Al_2O_3 on the metal surface, avoiding the aluminum inside to be leached (passivation). Therefore, Cu would first react with the acid raising the pH.

The simulations of ammoniacal system did not present the expected results from the analysis of the Pourbaix diagrams. When adding the reagents it was noticed that the expected copper complex ($Cu(NH_3)_4^{2+}$) was not formed, and the initial Cu reacted forming the CuO. The low fugacity of the compound given by the software's database could explain the non-formation of the complex. Although the simulation demonstrates that leaching by the formation of ammoniacal complexes would not be possible, it is known that in the literature the complexes would be formed [19, 21]. According to Oluokun and Otunniyi [19], H_2O_2 releases O_2 and this oxygen reacts with metallic Cu, forming CuO, which is presented in the simulation result. However, the reaction mechanism does not end with the formation of CuO, as proposed by the simulation. Oluokun and Otunniyi [19] suggest that CuO reacts with NH_3 and water, resulting in $Cu(NH_3)_4^{2+}$ and OH^-.

The analyzes made above are based on the fact that both Al and Cu are exposed to attack from the acidic or basic solution, however, as mentioned before, it would be necessary to remove the solder mask from the strip to achieve this condition.

Conclusion

From the above information, it is concluded that the selective leaching of copper against aluminum from the support strips of LED devices present in tubular lamps is possible due to the results of thermodynamic simulation. The favorable aqueous medium to this leaching is those of nitric acid and the ammoniacal medium. Therefore, the recovery of these metals in a recycling process is likely.

Lamp mass distribution, strips mass together with chemical analysis revealed that at least 1.1% of the lamp mass is copper from the strip, and 11.4% aluminum.

The construction structure of the strip observed in the MEV-EDS analysis points to a first layer formed by its matrix (Al), then a layer of dielectric material (composite with fiberglass), and finally an electric circuit, covered by a solder mask composed of a polymeric material. Removing the soldering mask and exposing the Cu to the leaching agent would be favorable for the hydrometallurgical process. Therefore, a prior grinding process would help with leaching.

The predominance diagrams obtained in the FactSage 8.0 software showed that in aqueous medium of HNO_3 and NH_3 there are optimal regions (for HNO_3 pH from 3 to 3.8 and Eh above 0.55 at 25 °C; for NH_3 pH of 7 to 10, Eh between −0.15 and 0.6 at 25 °C) which allows selective leaching of copper against aluminum. In these regions, copper is leached as ionic or ionic complex cooper and aluminum is oxidized to alumina. Despite the indication of aluminum oxidation, it is expected that in these medium it will undergo passivation, not compromising all the metallic aluminum in the residue. It was also seen that in sulfuric acid the selective leaching of copper is not possible and although HCl allows this selective leaching, the chloride ions would react with aluminum not allowing its passivation.

The results from the equilibrium module of the FactSage 8.0 software show that a leaching of 5 g of strip in an aqueous medium of 200 mL of 0.7 M HNO_3 leaches around 100% of copper from the strip and 24.9% of the aluminum. However, if there was a passivation of Al protecting the rest of the metal, only the Cu would be leached. For leaching in a basic medium using NH_3 it has not been proven to be effective for the purpose of the work. According to the results presented by the FactSage software the Cu would be in the form of CuO. Cu and NH_3 complexes are not calculated due to their low fugacity. Thus, it is not possible to confirm the selective leaching of Cu from an NH_3 solution through the software.

Acknowledgements We would like to thank University of Sao Paulo for support of this project, National Council for the Improvement of Higher Education (CAPES) and Counsel of Technological and Scientific Development (CNPq) for the financial support through doctorate grants, and Sao Paulo Research Foundation and Capes (FAPESP—grant: 2019/09669-7) for financial support.

References

1. Kumar A, Kuppusamy VK, Holuszko M, Song S, Loschiavo A (2019) LED lamps waste in Canada: generation and characterization. Resour Conserv Recycl 146:329–336. https://doi.org/10.1016/j.resconrec.2019.04.006 Jul
2. Xavier LH, Lins FAF (2018) Mineração urbana de resíduos eletroeletrônicos: Uma nova fronteira a explorar no Brasil. Bras Miner 1(379):22–26
3. Ruiz-Mercado GJ, Gonzalez MA, Smith RL, Meyer DE (2017) A conceptual chemical process for the recycling of Ce, Eu, and Y from LED flat panel displays. Resour Conserv Recycl 126:42–49
4. Nagy S, Bokányi L, Gombköto I, Magyar T (2017) Recycling of Gallium from end-of-life light emitting diodes. Arch Metall Mater 62(2):1161–1166
5. H. Information. LUXEON 2835 Line (2018)
6. Dias P, Bernardes AM, Huda N (2018) Waste electrical and electronic equipment (WEEE) management: an analysis on the australian e-waste recycling scheme. J Clean Prod 197:750–764
7. Khalid MK, Hamuyuni J, Agarwal V, Pihlasalo J, Haapalainen M, Lundström M (2019) Sulfuric acid leaching for capturing value from copper rich converter slag. J Clean Prod 215:1005–1013. https://doi.org/10.1016/J.JCLEPRO.2019.01.083
8. Terrones-Saeta JM, Suárez-Macías J, del Río FJL, Corpas-Iglesias FA (2020) Study of copper leaching from mining waste in acidic media, at ambient temperature and atmospheric pressure. J Miner 10:873. https://doi.org/10.3390/MIN10100873
9. Silvas FPC, Jiménez Correa MM, de Moraes MPKVT, Espinosa DCR, Tenório JAS (2015) Printed circuit board recycling: physical processing and copper extraction by selective leaching. J Waste Manag 46:503–510. https://doi.org/10.1016/J.WASMAN.2015.08.030
10. Gupta CK (2003) Chemical metallurgy principles and pratice, India, pp 343, 516–518
11. Huang HH (2016) The Eh-pH diagram and its advances. Metals 6:23. https://doi.org/10.3390/met6010023
12. Botelho Junior AB, Espinosa DCR, Tenório JAS (2020) The use of computational thermodynamic for yttrium recovery from rare earth elements-bearing residue. J Rare Earths. https://doi.org/10.1016/j.jre.2020.02.019
13. CEM (2018) Microwave acid digestion: method note compendium, pp 320
14. Kumar M, Lee JC, Kim MS, Jeong J, Yoo K (2014) Leaching of metals from waste printed circuit boards (WPCBs) using sulfuric and nitric acids. Environ Eng Manag J 13(10):2601–2607
15. Wang Z, Guo S, Ye C (2016) Leaching of copper from metal powders mechanically separated from waste printed circuit boards in chloride medium using hydrogen peroxide as oxidant. Procedia Environ Sci 31:917–924
16. Dutta D, Panda R, Kumari A, Goel S, Kumar M (2018) Sustainable recycling process for metals recovery from used printed circuit boards (PCBs). Sustain Mater Technol 17:e00066
17. Lee H, Mishra B (2018) Selective recovery and separation of copper and iron from fine materials of electronic waste processing. Miner Eng 123:1–7, Dec
18. Sohrab Hossain M, Naim Ahmad Yahaya A, Suhaila Yacob L, Zulkhairi Abdul Rahim M, Nadiah Mohamad Yusof N, Thomas Bachmann R (2018) Selective recovery of Copper from waste mobile phone printed circuit boards using Sulphuric acid leaching. Mater Today Proc 5(10):21698–21702
19. Oluokun OO, Otunniyi IO (2020) Kinetic analysis of Cu and Zn dissolution from printed circuit board physical processing dust under oxidative ammonia leaching. Hydrometallurgy 193:105320, Feb
20. Aracena A, Pérez F, CarvajaL D (2018) Leaching of cuprite through NH4OH in basic systems. Trans Nonferr Met Soc China (English Ed) 28(12):2545–2552
21. Sun ZHI, Xiao Y, Sietsma J, Agterhuis H, Visser G, Yang Y (2015) Hydrometallurgy Selective copper recovery from complex mixtures of end-of-life electronic products with ammonia-based solution. Hydrometallurgy 152:91–99
22. Oliveux G, Dandy LO, Leeke GA (2015) Degradation of a model epoxy resin by solvolysis routes. Polym Degrad Stab 118:96–103 Apr

23. Sperança MA, Virgilio A, Pereira-Filho ER, Batista de Aquino FW (2018) Determination of elemental content in solder mask samples used in printed circuit boards using different spectroanalytical techniques. Appl Spectrosc 72(8):1205–1214
24. Bruno GU (2014) Lixivação de alumínio durante o processo de tribocorrosão em meio ácido, básico e neutro. Universidade Federal do Rio Grande do Sul, Trabalho de diplomação
25. Silvas FPC, Jiménez Correa MM, Caldas MPK, de Moraes VT, Espinosa DCR, Tenório JAS (2015) Printed circuit board recycling: physical processing and copper extraction by selective leaching. Waste Manag 46:503–510

Fe-Si Alloy Preparation and Alumina Extraction from Red Mud and Silica Fume via Vacuum Carbothermal Reduction

Peng Wei, Wenzhou Yu, Hao Chen, and Joseph Emmanuel Nyarko-Appiah

Abstract With the increasing awareness of environmental protection, the disposal of solid waste such as red mud (RM) and silica fume (SF) has attracted widespread attention. In this study, a novel method by recycling RM and SF simultaneously was proposed for preparing Fe-Si alloy and extracting alumina. The process is mainly composed of vacuum carbothermal reduction and magnetic separation. The results show that Fe-Si alloy was successfully synthesized during the vacuum carbothermal reduction and it was effectively separated during the magnetic separation. Additionally, after the magnetic separation, the non-magnetic portion mainly contains alumina and it can be a potential raw material to produce alumina by the Bayer process. This process could recycle RM and SF simultaneously, and almost no waste residue was generated. Therefore, it provided a clean and sustainable route for synergistic recycling of RM and SF.

Keywords Fe-Si alloy · Alumina extraction · Red mud · Silica fume · Vacuum carbothermal reduction

Introduction

Red mud (RM) is a solid waste discharged in the production of alumina from bauxite ores [1]. In general, about 0.8–1.5 tons of RM is generated from each tons of alumina production by Bayer process, and 1.5–2.5 tons of RM is discharged from the sintering or combined process. [2] By 2018, the amount of cumulative RM had reached 4 billion tons globally [3]. However, RM is a corrosive substance because of its high alkalinity (containing 1–10% of Na_2O) [4], causing the serious environment problems due to its accumulation. At present, most of the RM were mainly disposed by direct dam storage [5]. However, these methods not only spend huge maintaining cost and occupy a vast of construction land, but also possibly result in the soil pollution and the groundwater pollution [6, 7]. Therefore, it is imminent to deal with RM

P. Wei · W. Yu (✉) · H. Chen · J. E. Nyarko-Appiah
College of Materials Science and Engineering, Chongqing University, Chongqing 400044, China
e-mail: yuwenzhoucd@cqu.edu.cn

A. Lazou et al. (eds.), *REWAS 2022: Developing Tomorrow's Technical Cycles (Volume I)*, The Minerals, Metals & Materials Series,
https://doi.org/10.1007/978-3-030-92563-5_72

effectively and economically. Recently, valuable components recovered from RM have attracted widespread attention all over the world. Actually, RM is rich in iron, aluminum, silicon, titanium, and other valuable metals. Statistically, the content of Fe_2O_3 in RM is approximately 30–40% [8], which implies the RM can be a potential resource for iron and steel production. In the last decades, many reports have attempt to extract iron from RM. Jamieson [9] investigated the extraction of iron from RM by direct magnetic separation. However, this method is restrained to the complexity of the mineralogical composition in RM, which is difficult to obtain high-grade Fe concentrate. Wei [10] presented the iron extraction from RM in the coal-based reduction with the assistance of sodium salt, and their results show that the Fe concentrate and recovery rate could achieve 79.32% and 85.43%, respectively. However, most of these methods focused only on iron component, and the recycling of other valuable components such as the aluminum, silicon, and titanium was not considered adequately. Therefore, more attention should be paid to the comprehensive recovery of valuable components from RM.

Silica fume (SF) is a kind of by-product, which derived from the metallurgical grade silicon (MG-Si) and Fe-Si alloy production process, and its main component is amorphous silica [11]. Owing to the growing demand of MG-Si and Fe-Si alloy, in 2018, the amount of SF discharged in China exceeded 2 million tons [12]. Up to now, SF is successfully utilized in the fields of concrete, cement, and refractory materials globally [13, 14]. However, the SF generated in China possessed the characteristic of low-purity and large particle, which is unfavorable for its large-scale application. Additionally, the accumulation of SF either cause the waste of land resources, or threaten to human health [15]. Therefore, it is significant to dispose SF in an efficient and clean method in China.

Considering the high-grade silicon content of SF and the high-grade iron content in RM, it is possible to use SF and RM as the silicon-bearing materials to produce Fe-Si alloys. Herein, we presented a synergistic recycling of RM and SF for the Fe-Si alloy preparation and alumina extraction via the vacuum carbothermal reduction process. In this work, the influencing conditions and the co-reduction mechanism for RM and SF were comprehensively investigated. Additionally, the effect of SF addition and the growth of Fe-Si alloy were studied. The proposed process takes full advantages of the valuable components recycling by solid wastes, and without waste residue generated. Therefore, it offered an efficient and sustainable route for synergistic recycling of RM and SF.

Experimental

Materials

The RM used in this study was collected from the Shandong Alumina Company, China, and the SF employed was obtained from Sichuan silicon company, China.

Table 1 Chemical compositions of RM and SF (wt%)

Element	Fe_2O_3	SiO_2	Al_2O_3	TiO_2	CaO	Na_2O
RM	37.27	18.04	23.55	8.51	2.35	8.50
SF	0.169	95.52	0.163	–	0.207	–

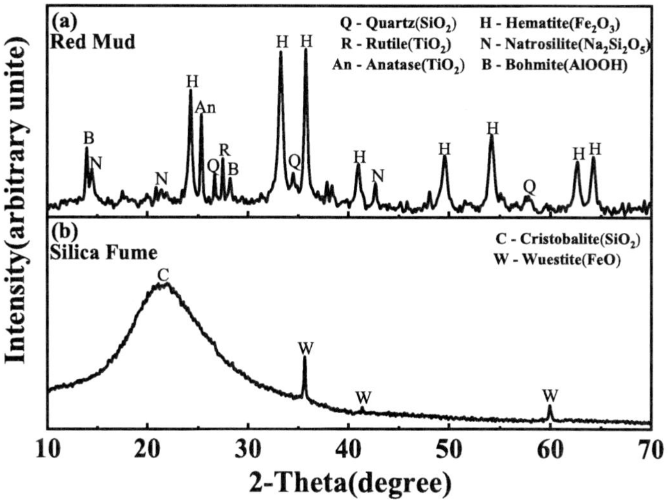

Fig. 1 X-ray diffraction pattern of RM and SF

According to the measurement of XRF (X-ray fluorescence) and XRD (X-ray diffraction), the chemical compositions and mineralogical components of RM and SF were shown in Table 1 and Fig. 1, respectively. It can be found from Table 1 that RM contains 37.27 wt% Fe_2O_3, 18.04 wt% SiO_2, 23.55 wt% Al_2O_3, 8.51 wt% TiO_2, 2.35 wt% CaO, and 8.5 wt% Na_2O, while SF was mainly consisted of 95.52 wt% SiO_2. Simultaneously, Fig. 1 shows that the major mineralogical phases of RM were hematite (Fe_2O_3), anatase (TiO_2), quartz (SiO_2), rutile (TiO_2), boehmite (γ-AlOOH), and natrosilite ($Na_2Si_2O_5$), while the SF were mainly consisted of cristobalite (SiO_2) and wuesite (FeO). Besides, the charcoal employed as the reductive agent in this study was bought from the market.

Methods

Firstly, the powdered RM, SF, and charcoal were mixed manually with 5:0:0.98, 4.5:0.5:1.12, 4:1:1.27, respectively. Thereafter, the three mixture samples were pressed into pellets (20 mm in diameter and 10 mm in height) in the pressure of 15 MPa. Next, the pellets were placed in drying oven to remove the moisture by drying at 393 K for 12 h. Subsequently, the pellets were poured into a corundum crucible, followed by carrying out the reduction process in the vacuum furnace. A pressure of 100 Pa (~0.001 atm) inside the furnace was set and the designated temperature (1523 $\pm$ 1 K) was gently reached with a heating rate of 10 K/min. After heating for 3 h, the reduced samples were cooled into the room temperature naturally inside the chamber under the vacuum conditions. After that, the reduced samples were ground in a ball mill to obtain fine powders (<74 μm). Magnetic separation was performed in a Magnetic Tube of XCGS-50 Davies to separate the reduced samples into two parts by using a magnetic field intensity of 0.1 T. The magnetic product was designated as Fe-Si alloy, while the aluminum and titanium components were concentrated in the non-magnetic product. The separation efficiency was calculated as:

$$\eta = \frac{m_1 \times (\beta + \alpha)}{m_0 \times (\gamma + \varepsilon)} \tag{1}$$

where η is the separation efficiency; m_1 is the total mass of the raw material; m_0 is the mass of the magnetic portion; β and α are the metallic iron and metallic silicon content in the magnetic part, respectively; γ and ε are the total iron and silicon content of the raw material, respectively.

Characterization

The mineralogical compositions of the raw material and the reduced samples were determined by X-ray diffraction (XRD, Rigaku, 2θ (10°–90°), Cu Kα as X-ray source), and the main chemical compositions of raw materials were measured by X-ray fluorescence spectrometry (XRF, Shimadzu XRF-1800). A scanning electron microscope (VEGA 3 LMH; TESCAN) was used to observe the microstructure of the reduced samples.

Thermodynamic Analysis

As depicted in Fig. 2, a thermodynamic analysis was calculated by FactSage 8.0 Reaction module (FToxid, FSstel, FactPS datebase) to theoretically clarify the vacuum carbothermal reduction behavior of SiO_2 and Fe_2O_3 in RM and SF. It can be found from Fig. 2a that the ΔG of reaction (1) decreases with the decline of CO partial pressure. When the partial pressure of CO decreased to 0.001 atm, the initial reduction temperature of SiO_2 can drop to 1042 K, which implies the vacuum condition can significantly promote the reduction of SiO_2. Additionally, by comparing ΔG of reactions (1), (2), (3), and (4), it claims that the formation sequence of Fe-Si alloys

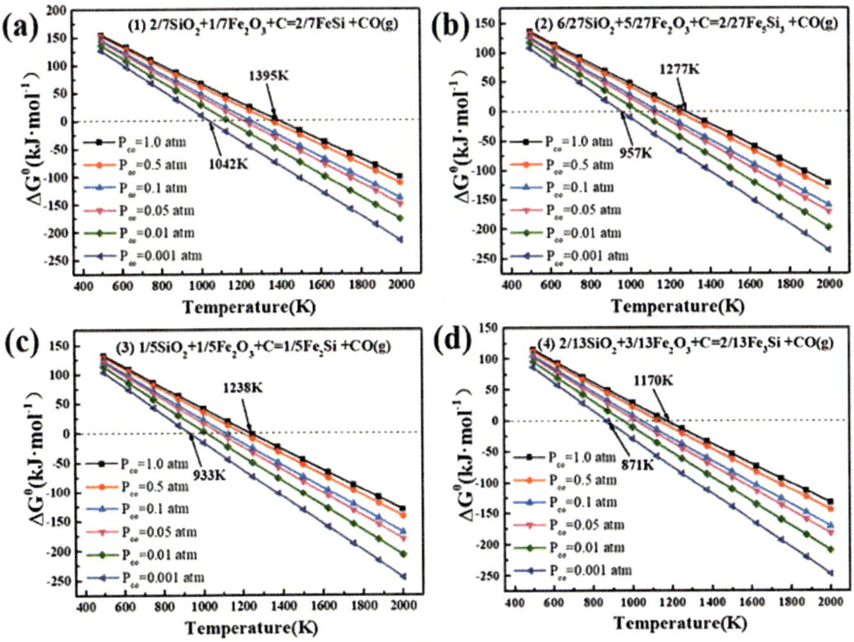

Fig. 2 Relationship between ΔG^{θ}-T of possible reactions at different pressures during the reduction process

is Fe_3Si > Fe_2Si > Fe_5Si_3 > FeSi, indicating that the more silicon-grade of Fe-Si alloy requires higher temperatures. These results proved that the Fe-Si alloys could be easily prepared under the vacuum condition.

Results and Discussion

Effects of the RM/SF Proportion on the Reduction

Based on the thermodynamic calculations, the total pressure of the vacuum carbothermal reduction process was predetermined at 0.001 atm (~100 Pa), and the reduction temperature and duration were designed at 1523 K and 3 h, respectively. Figure 3 displays the mineralogical evolutions of reduced samples with different SF proportion. It can be found that the main phases of the reduced samples were Fe-Si alloy (Fe_3Si), calcium aluminate ($CaO(Al_2O_3)_6$), alumina (Al_2O_3), and titanium carbide (TiC) in the absence of SF. As the proportion of SF increased to 10 wt%, the Fe_3Si phase transformed to FeSi phase and Fe_5Si_3 phase, indicating that the high-silicon content of Fe-Si alloys can be effectively synthesized. However, when the SF proportion was 20 wt%, the phase composition of Fe-Si alloys changed from Fe_5Si_3 and

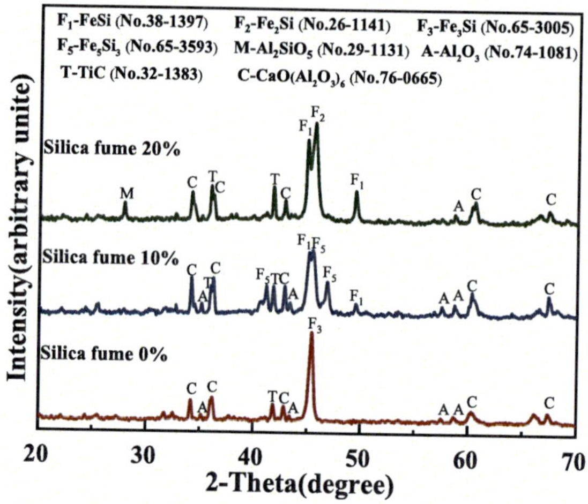

Fig. 3 X-ray patterns of reduced samples with different SF proportion at 1523 K for 3 h

FeSi phase to FeSi and Fe_2Si phase, implying that the silicon concentration of the Fe-Si alloy decreased with increasing the SF proportion. Simultaneously, it can be seen that the diffraction peaks of mullite (Al_2SiO_5) appeared, which suggested that the cristobalite (SiO_2) in SF cannot be reduced completely. This phenomenon should be attributed to the increase of cristobalite (SiO_2) content decreased the mass ratio of iron/silicon (Fe/Si) in the mixed samples. The study [16] researched the influences of Fe_2O_3 content in the reduction of quartz (SiO_2), and the results suggested that the reduction of quartz (SiO_2) could be facilitated with increasing the mass ratio of Fe/Si in mixture samples during the carbothermal reduction process. Similar in this work, the increased SiO_2 content in mixture decreased the mass ratio of Fe/Si, therefore, resulted the inefficient reduction of SiO_2.

Microstructure and Magnetic Separation Analysis

The microstructure of the reduced samples with different proportion of SF was presented in Fig. 4a, b. Obviously, the alloy particles showed a round or near-round structure in the presence of 10 wt% SF (Fig. 4a). However, when using 20 wt% SF proportion in the mixture (Fig. 4b), the features of the alloy particles become considerably small and dispersive. This phenomenon should be attributed to the occurrence of the high-melting-point material mullite (Al_2SiO_5) during the carbothermal reduction process, as shown in Fig. 3, which decreased the liquid phase and hindered the mass transfer. The separation efficiency was calculated by Eq. (1), as shown in Fig. 4c. It can be seen that when the SF proportion increased from 10 to 20 wt%, the separation efficiency was decreased from 80.52 to 63.27%, which indicates that the

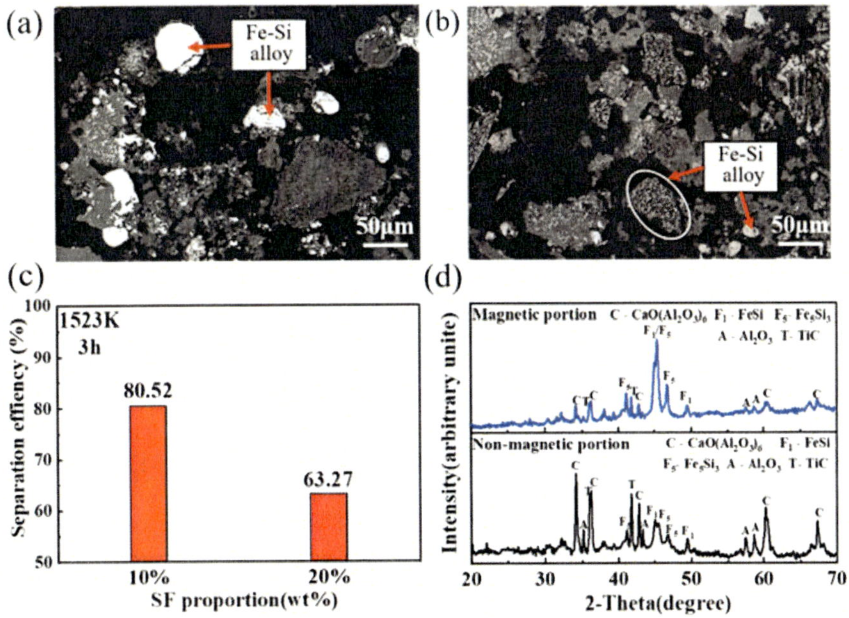

Fig. 4 **a–b** SEM image of reduced samples with different SF proportion: **a** 10 wt% **b** 20 wt%, **c** separation efficiency of samples in the presence of 10 wt% and 20 wt% SF, respectively, and **d** X-ray patterns of magnetic separation samples in the presence of 10 wt% SF proportion

excess proportion of SF (20 wt%) in the mixture was unfavorable for the magnetic separation of Fe-Si alloys and other components.

Figure 4d illustrated the mineral compositions of the magnetic product in the presence of 10 wt% SF proportion. The magnetic portion is mainly consisted of Fe-Si alloy (FeSi and Fe_5Si_3 phase) and a small amount of Al_2O_3, TiC and $CaO(Al_2O_3)_6$, which proves that the Fe-Si alloys can be efficiently separated by the magnetic separation process. Additionally, the separation efficiency could be further promoted by adding additive (e.g., sodium additive) under the reduction process or controlling the parameters (e.g., magnetic field intensity) during the magnetic separation process [17]. Meanwhile, the non-magnetic portion is mainly composed of Al_2O_3, TiC, and $CaO(Al_2O_3)_6$ phase, implying the aluminum and titanium components could be concentrated by magnetic separation. It should be noted that the Al_2O_3 and $CaO(Al_2O_3)_6$ can be dissolved in alkali liquor, and then employed to produce alumina by the Bayer process [18]. Thus, the non-magnetic product can be used as raw materials in the Bayer process for the alumina production. Furthermore, the titanium component was rich in the red mud, which could be extracted by nitric acid leaching process. Additionally, the sodium component in the mixture samples was reduced to Na gas [19] during the vacuum carbothermal reduction process, which can be recovered in the ash. Therefore, it was suggested that a sustainable and efficient route for synergistic recycling of RM and SF was obtained.

Conclusions

The synergistic recycling of RM and SF is proved feasible by lab-scale experiment to prepare Fe-Si alloy and extract valuable components, and the basic conclusions were summarized as follows:

(1) During the vacuum carbothermal reduction process, the cristobalite (SiO_2) in SF could easily be reduced and react with the hematite (Fe_2O_3) in RM to synthesize the Fe-Si alloys, in the 10 wt% SF proportion of the mixture. However, as the SF proportion increased to 20 wt%, the mullite (Al_2SiO_5) was generated due to the incomplete reduction of silicon oxide, resulting in the decrease of Si concentration in Fe-Si alloy. Simultaneously, the separation efficiency was decreased from 80.52 to 63.27%.

(2) The optimized reduction conditions were RM proportion of 90 wt%, SF proportion of 10 wt%, and reduction temperature and duration of 1523 K and 3 h, respectively. After magnetic separation, the Fe-Si alloys and other valuable components can be effectively separated. The magnetic portion is basically composed of Fe_5Si_3 and FeSi phase, while the non-magnetic portion is mainly consisted of Al_2O_3, TiC, and $CaO(Al_2O_3)_6$ phase, suggesting that an efficient enrichment of aluminum and titanium components form RM and SF was obtained.

Acknowledgements This research was supported by the National Natural Science Foundation of China (No. 52074058 and No. U2003215).

References

1. Sahu MK et al (2016) Chapter 19—Methods for utilization of red mud and its management. Environmental Materials and Waste 485–524
2. Wang SH et al (2021) Comprehensive utilization status of red mud in China: a critical review. J Clean Prod 289:125136
3. Rivera RM et al (2018) Extraction of rare earths from bauxite residue (red mud) by dry digestion followed by water leaching. Miner Eng 119:82–92
4. Lyu F et al (2021)Dealkalization processes of bauxite residue: a comprehensive review. J Hazardous Mater 403:123671
5. Evans K et al (2016) The history, challenges, and new developments in the management and use of Bauxite residue. J Sustain Metall 2:316–331
6. Liu ZB et al (2015) Metallurgical process for valuable elements recovery from red mud—A review. Chem Eng J 155:29–43
7. Khairul MA et al (2019) The composition, recycling and utilisation of Bayer red mud. Resour Conserv Recycl 141:483–498
8. Jayasankar K et al (2012) Production of pig iron from red mud waste fines using thermal plasma technology. Int J Miner Metall Mater 19:679–684
9. Jamieson E et al (2006) Magnetic separation of red sand to produce value. Miner Eng 19:1603–1605

10. Ding W et al (2021) Iron extraction from red mud using roasting with sodium salt. Mineral Process Extract Metall Rev 42:153–161
11. Wang YH et al (2018) Preparation of high silica microporous zeolite SSZ-13 using solid waste silica fume as silica source. Adv Powder Technol 29:1112–1118
12. Wang J et al (2020) Research progress in purification and application of industrial silica fume. Miner Rep 34:23081–23087
13. Mahmoud S et al (2017) A new approach for application of silica fume in concrete: wet granulation. Constr Build Mater 157:573–581
14. Khan MI et al (2000) Porosity and strength of PFA/SF/OPC ternary blended paste. Cem Concr Res 30:1225–1229
15. Pedro D et al (2017) Evaluation of high-performance concrete with recycled aggregates: use of densified silica fume as cement replacement. Constr Build Mater 147:803–814
16. Xue Y et al (2019) A novel process to extract alumina and prepare Fe-Si alloys from coal fly ash. Fuel Process Technol 185:151–157
17. Li GH et al (2012) Beneficiation of nickeliferous laterite by reduction roasting in the presence of sodium sulfate. Miner Eng 32:19–26
18. Azof FI et al (2019) Leaching characteristics and mechanism of the synthetic calcium-aluminate slags for alumina recovery. Hydrometallurgy 185:273–290
19. Lv XD et al (2019) Volatilisation behaviour of iron, silicon and magnesium during vacuum carbothermal reduction of ilmenite concentrate. Can Metall Q 58:419–426

Geopolymers Made of Construction and Demolition Waste: Current Trends and Perspectives

Angelica Cardoza and Henry A. Colorado

Abstract The review deals with geopolymers based on construction and demolition waste (CDW). A review of the research of the last 5 years (2015–2020) is presented in three different study areas: Energy—Business, Management, and Accounting—Environmental Sciences. This article initially emphasizes the environmental impact from the point of view of opportunities for this type of recycled materials. In a second part, it focuses on the experimental design of geopolymers, concentrating on the used materials and manufacturing methods. Finally, the third part focused on the mechanical properties obtained by different authors, carried out via a systematic review for providing a detailed summary of the different parameters found in the literature.

Keywords Geopolymer · Construction and demolition waste · Alkaline activation · Fly ash

Introduction

Ordinary Portland cement (OPC) is one of the most used materials in the construction industry due to its mechanical properties that make it an excellent material these type of applications. However, OPC has certain disadvantages, such as its high emission of greenhouse gases and their energy consumption [1]. It is estimated that approximately for every ton of OPC manufactured, between 0.85 and 1.0 tons of CO_2 are released [2]. As a green solution, OPC can be partially combined with inorganic polymers (geopolymer cements), discovered in 1958, which are materials formed from the alkaline activation of aluminosilicates [2, 3]. It has been shown in some research that the carbon footprint of these geopolymers is lower than that emitted by OPC [2, 4, 5]. But the high amount of alkaline silicate must be considered, since too high a quantity in geopolymers is not good with respect to the global environmental impact alert [5,

A. Cardoza · H. A. Colorado (✉)
CC Composites Laboratory, Universidad de Antioquia UdeA, Calle 70 No. 52-21, Medellín, Colombia
e-mail: henry.colorado@udea.edu.co

A. Lazou et al. (eds.), *REWAS 2022: Developing Tomorrow's Technical Cycles (Volume I)*, The Minerals, Metals & Materials Series,
https://doi.org/10.1007/978-3-030-92563-5_73

6]. The main challenges of this type of materials correspond to the properties they must have, such as mechanical behavior equal to or greater than OPC, fire resistance, non-emission of toxic fumes, resistance to acids and salts, low shrinkage, low thermal conductivity, and low pH [7].

Geopolymer cements do not require the processing at high temperature furnaces, therefore having less investment in plant and equipment [2]. The raw materials to produce geopolymers are aluminosilicates, which are geologically available resources around the world, making geopolymers quite attractive for acquisition. Currently, reductions of up to 80% in greenhouse gas emissions have been found with the use of geopolymers [2]. Joseph Davis states that these materials harden at room temperature and provide compressive strengths in the range of 20 MPa, in just 4 h, at a temperature of 20 °C, and compressive strengths after 28 days are in the range of 7–100 MPa [3, 8]. Among the main properties of these materials are resistance to sulfates, high resistance to early ages, resistance to freezing and thawing, low shrinkage, and resistance to corrosion [8, 9]. The main raw materials used in the preparation of geopolymers have been blast furnace slags rich in calcium, fly ash poor in calcium and clays, and as alkaline activators are alkali hydroxides, salts of weak acids, alkali silicates, aluminosilicates, alkali aluminosilicates, and salts of strong acids [10]. This research is focused in geopolymers based on demolition and construction waste (CDW) [11, 12]. The composition of CDW is very heterogeneous, mainly formed by mixtures of aggregates, concrete, stone, ceramic materials, gravel, sand, soil and, to a lesser extent, metals, glass, and plastics [13]. CDWs are one of the possible and promising raw materials for geopolymer cements. In Spain, it is estimated that a single person generates about 1,644 kg of waste per year, with around 75% of this waste deposited in landfills, only recycling from about 25 to 30% [14]. Green solutions for CDW is an important issue in Latin America, in which Colombia is just starting to implement environmental regulations, having serious problems in logistics [15], and in the application of circular economy strategies [16]. In Brazil, there are now intensive research in geopolymers involving many solid waste types [17].

Methodology

The methodology used in this article follows the specific guidelines for systematic reviews in supply chain management proposed by Durach et al. [18], and the sequence prescribed by Tranfield and Denyer [19]. For the systematic literature review, the research papers were selected excluding conference and proceedings papers. The search is carried out according to the steps summarized in Table 1. The time was delimited from January 2015 to July 2020 in the following areas: Energy—Business, Management and Accounting—Environmental Sciences.

The general results obtained for the search were 192 related publications, distributed in a period from 2010 to 2021, having a considerable peak in 2020, as can be seen in Fig. 1. Of these 192 documents, 137 are research articles, 18 are

Table 1 Search parameters

Terms	CDW, Alkali, Geopolymer, CDW				
Combined terms	Demolition and construction waste—Alkaline activation—Fly ash				
Search	Search string	Global	2015–2020	Articles	Areas
Search 1	(“cdw”) or (“CDW”) or ({demolition and construction waste}) or (“demolition and construction waste”) and (“geopolymer” or “alkali” or “alkaline activation”) or (“alkaline activation”)	192	180	132	46
Search 2	(“cdw”) or (“CDW”) or ({demolition and construction waste}) or (“demolition and construction waste”) and (“geopolymer” or “alkali” or “alkaline activation”) or (alkaline and activation) and not (“fly ash”)	43	27	16	5
Search 3	(“cdw”) or (“CDW”) or {demolition and construction waste} or (“demolition and construction waste”) and (“geopolymer” or “alkali” or “alkaline activation”) and not (“fly ash”)	18	16	7	4
Search 4	(“cdw”) or (“CDW”) or ({demolition and construction waste}) or (“demolition and construction waste”) and (“geopolymer” or “alkali” or “alkaline activation”) and (alkaline and activation) and not (“fly ash”)	5	4	1	0

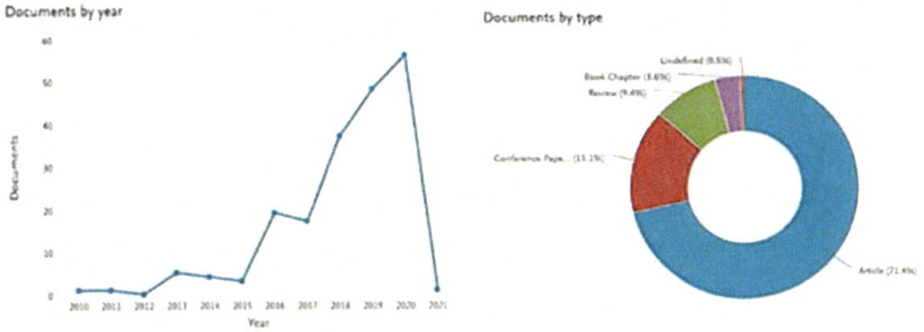

Fig. 1 Documents by year and type [20]

review articles, 7 book chapters, 29 correspond are conference articles, and 1 appears undefined.

The most prominent authors with this type of publications are Kostas Komnitsas with 6 papers, from the Technical University of Crete in Greece; and Rovnaníková Pavla with 6 papers as well, from the Czech Republic. The countries where it has been studied the most are China, Spain, and Portugal, see Fig. 2. According to the subject area, the documents where more research is presented are engineering, materials science, environmental sciences, and energy, see in Fig. 3. A two-step process was used to select studies for this review, with a total of 55 articles. In the first step, a total of 39 non-duplicate articles were identified, from which 14 articles were removed due to their irrelevant titles. Also, 13 were removed after verifying that the abstract was not within the scope of this review. The remaining 12 items are considered in the second step.

A full review of the 12 articles was carried out using the inclusion criteria. One article was excluded, since it only talks about the reuse of CDW as a partial replacement of OPC for the manufacture of mortar, but does not use the CDW for its activation and formation of geopolymer cement [21]. Another article is excluded as it speaks of CDW as a recycled aggregate of geopolymer concrete based on fly ash [22]. Another article was eliminated because the red brick CDW was used as a replacement in OPC mortars; in this study, the particle size of the CDW was reviewed as an influence on the OPC, and the authors managed to demonstrate that when the grinding time increased, particle size decreased, and surface area increased [23].

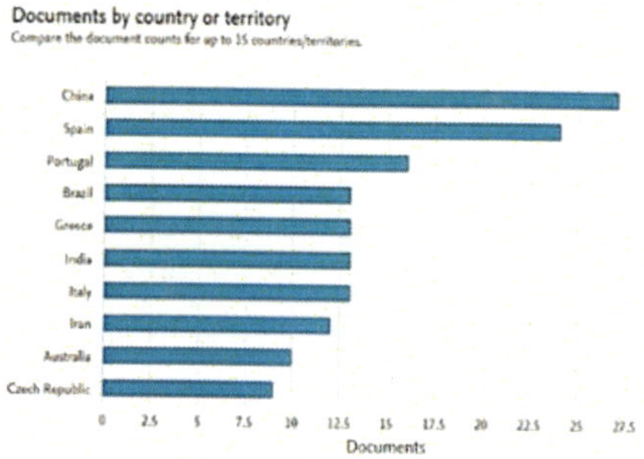

Fig. 2 Documents by country [20]

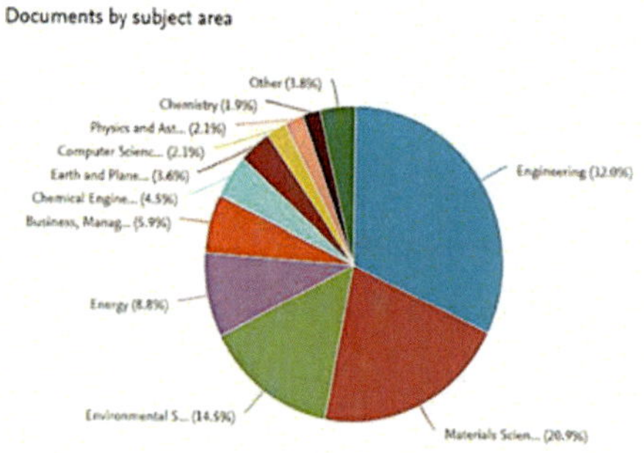

Fig. 3 Documents by area [24]

Fig. 4 Identification process

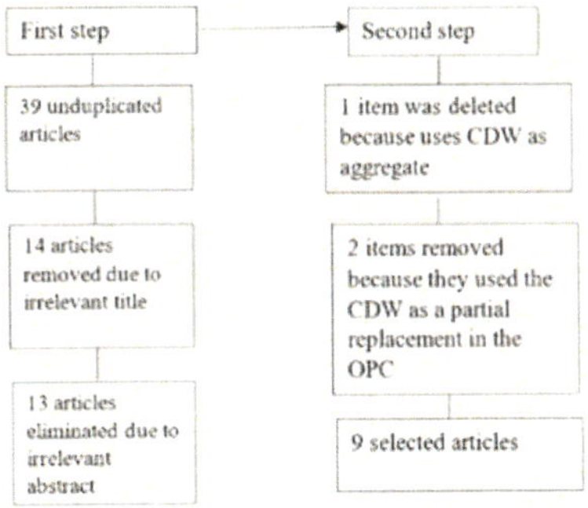

A total of 9 articles were identified to be included in this study. The identification process described is shown in Fig. 4.

Results and Discussion

This section is divided into four subsections. In the first, we present the selected studies, the actors, and general dimensions included in each of the studies that were identified. In the second subsection, it is presented the environmental impact in these studies. In the third section, it is presented the different formulations and techniques used, and finally it is analyzed the mechanical properties obtained in CDW-based geopolymers.

Selected Studies and General Results

The studies, the year of publication, the journal, and the main subject area are presented in Table 2. The studies selected using the methodology described above were published in the journals of Springer, Elsevier, Institution of Chemical Engineers, International Journal of Scientific and Technology Research.

Figure 5a shows that the area of greatest interest is environmental science. Figure 5b shows the trend in the publication of this type of articles, being higher in 2020.

For the articles in this search, we are based on reviewing the emphasis on the environmental impact of CDW, and the mechanical results obtained in geopolymers as a viable alternative, also in each of the materials used by the authors and in the methods or procedures for manufacture of geopolymers based on CDW. Likewise, the mechanical results obtained for these materials are mentioned.

Table 2 Selected studies

Authors	Year	Area	Publisher
Zawrah M.F., Gado R.A., Feltin N., Ducourtieux S., Devoille L	2016	Environmental	Institution of Chemical Engineers
Komnitsas K	2016	Environmental	Elsevier Ltd
Robayo-Salazar R.A., Mejía-Arcila J.M., Mejía de Gutiérrez R	2017	Management	Elsevier Ltd
Fořt J. et al	2018	Management—Energy	Elsevier Ltd
Bassani M., Tefa L., Coppola B., Palmero P	2019	Management—Energy—Environmental	Elsevier Ltd
Younis K.H., Salihi K.A., Ibrahim T.K	2020	Management	International Journal of Scientific and Technology Research
Tan J., Cai J., Li X., Pan J., Li J	2020	Management—Energy—Environmental	Elsevier Ltd
Rivera J., Castro F., Fernández-Jiménez A., Cristelo N	2020	Energy—Environmental science	Springer
Tang Z., Li W., Tam V.W.Y., Xue C	2020	Environmental	Elsevier B.V

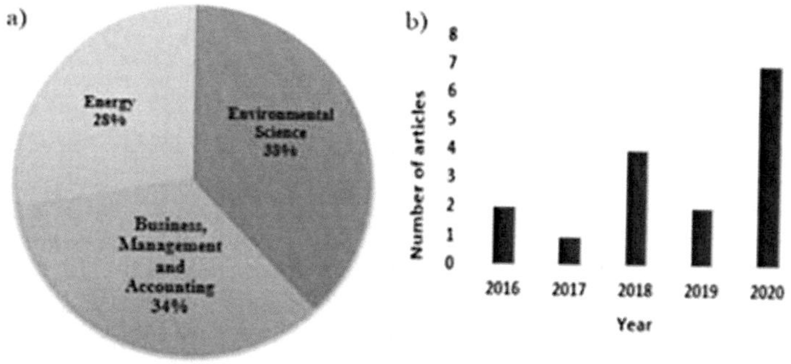

Fig. 5 Search trends. **a** Trend by area and **b** trend by year

Environment: Environmental Impact Analysis

To speak of geopolymers as a partial or total replacement for ordinary Portland cement, it is necessary to review the environmental impact of their production. Duxson et al. [7] affirm that the CO_2 emission from the production of geopolymers is lower than that generated by OPC: their geopolymer did not require a high temperature calcination step. McLellan et al. [25] indicated that the results of the emission of geopolymer gases depend on the source of the waste and its location, with a reduction between 44 and 64% in emissions compared to OPC. Habert et al. [6] affirm that it is true that the CO_2 emission is lower for the production of geopolymers but that there is an environmental impact in other aspects that contribute to global warming such as abiotic depletion, marine ecotoxicity, and acidification, which is due to the sodium silicate solution for activation. They recommend that the production of geopolymers must focus on the use of waste that cannot be recycled or reused in any other way, with an adequate molar ratio Si/Al to minimize the amount of sodium silicate, reducing its environmental effects. Oullet et al. [5] found that the main CO_2 emitter in geopolymers are the solids from the activating solution. They studied different compositions from the literature using the LCA methodology, concluding that the environmental impact of geopolymers is up to 75% in savings lower than that OPC.

In Spain, there are different regulations for the use and generation of CDW, such as Regulation 105/2008 that establishes the obligation to include waste management plans from the beginning of the projects, as a preliminary study to estimate the amount generated of waste [26]. Spain is one of the countries with the largest number of both public and private companies with this approach [27]. In Barranquilla (Colombia), Pacheco et al. [28] showed that there are many issues concerning waste management, treatment, and final disposal. Robayo et al. [29] achieved through reductions of 73% of the developed material in the global warming potential (kg * CO_2 * eq/MPa) compared to OPC, calculating the global warming possible (GWP) index of the geopolymer using the LCA methodology, and compared it with the GWP of the OPC. Robayo et al. [29] identified that the most polluting activators were sodium hydroxide and sodium silicate, but in turn with this alkaline activator they found a better mechanical performance for geopolymers, reaching 102 MPa. The reported carbon footprint for sodium hydroxide and sodium silicate is approximately 1.60 and 0.92 kg * CO_2 * eq * kg, while for other activators such as Na2CO3 and Na2SO4, they have a carbon footprint of 0, 22 and 0.098 kg * CO_2 * eq * kg, respectively, which shows that sodium hydroxide and silicate are 86.25% more polluting. For geopolymer-based mortars, the global warming potential (GPW) value is 265 kg * CO_2 * eq, while for ordinary Portland cement mortars, its value is 511 kg * CO_2 * eq, close to a 100% more. [29] Tan et al. [7, 25] performed a quantitative analysis of the manufactured geopolymer compared to carbon dioxide emissions. The authors [7, 25] analyzed previous studies by Yanget al. [31] and Song et al. [24] where CO_2 emissions divide them into three parts: materials, transportation, and production phases. Tan et al. [31] only took CO_2 emissions into account in production, as the other

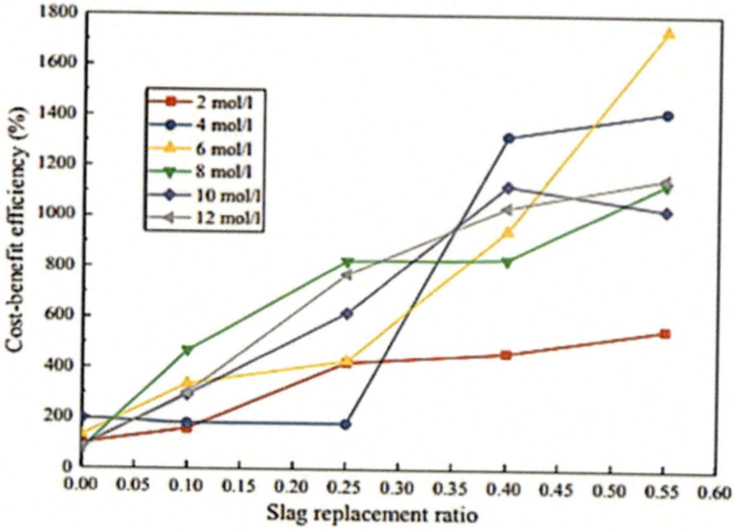

Fig. 6 Cost–benefit for different formulations [22]

two parts are much smaller compared to this one. To evaluate the eco-efficiency, the intensity of the CO_2 flow was considered as the relationship between the compressive strength of the concrete and the CO_2 emissions. The CO_2 emissions of the geopolymer developed by the authors consisted mainly in the manufacture of alkali solution, slag, and CDW powders; the latter was calculated by the electrical consumption of the crushing which was 0.182 kW h/kg, resulting in a CO_2 emission of 0.2457 (kg/kg). Tan et al. [31] also studied the cost of the geopolymer taking into account the materials used and energy consumption. Figure 6 shows the percentage of cost–benefit for the different formulations, where the authors managed to show the maximum percentage, replacement to 55% slag, and a concentration of 6 mol/L.

Fort et al. [32] showed that with the developed geopolymer, they achieve lower CO_2 emissions, lower energy consumption, and lower demand for materials, achieving up to 45% savings in energy and 72% savings in production of greenhouse gases compared to OPC pastes, see Fig. 7a.

Likewise, Fig. 7b shows CO_2 emissions of the different materials listed by the author, where the geopolymer emissions are lower compared to OPC. It can be concluded that the CDW-based geopolymer is a sustainable and environmentally friendly material, which manages to reduce the amount of greenhouse gases by being a substitute material for OPC, in addition to mitigating energy consumption.

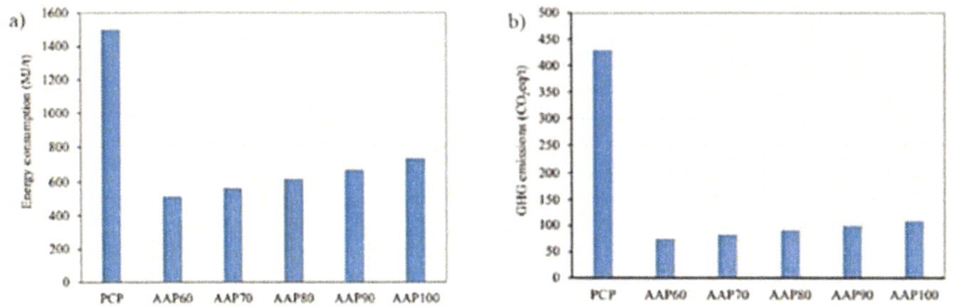

Fig. 7 **a** Energy consumed OPC and geopolymers. **b** CO_2 emissions for OPC and geopolymers [22]

Experimental Design: Materials and Methods

One of the main concerns regarding geopolymers is to find synthesis parameters that result in acceptable properties, according to the standards and regulations, that enable adequate uses. This section will detail materials, methods, and procedures used for the manufacturing of geopolymers based on CDW.

Raw Materials

The materials used by the different authors in this review are listed below. Fort et al. [32] used brick dust as a starting material. Komnitsas [33] uses marine sediments together with CDW (tiles, brick, and concrete) for alkali activation. Zawrah et al. [34] conducted a study on CDW composed of bricks with the addition of blast furnace slag for the production of geopolymers. Like Zawrah, Tan et al. [35] used CDW geopolymer composed primarily of 50% masonry and 42% concrete as recycled aggregate and calcium oxide-rich aggregate blast furnace slag to achieve a geopolymer matrix with a greater resistance to compression at an early age [31]. Robayo et al. [29] used raw material red brick and ordinary Portland cement; the CDW was ground in a hammer mill and then in a ball mill. Bassani et al. used CDW in fine particles; the authors used 5 different types of CDW that were recycled concrete, recycled ceramic (brick and ground tile), recycled asphalt, natural and excavated aggregates, and CDW fines (not classified) [36]. Table 3 summarizes the different materials, particle sizes, compositions, phases, and activators used.

Experimental Procedures

This section details the procedures or methods used by the authors to fabricate geopolymer materials. Komnitsas pulverized and homogenized both components to make the geopolymer. The activating solution was prepared by dissolving KOH

Table 3 Materials involved in the literature

Author	Material	Particle size	Chemical composition of CDW%	Alkaline activator	CDW crystalline phases
Zawrah et al. [34]	Red brick residue with addition of blast furnace slag	1100 nm	50.16 SiO_2, 15.95 Al_2O_3, 15.09 Fe_2O_3 4.39 CaO	Sodium silicate Sodium hydroxide	Quartz Albite Hematite
Komnitsas [33]	Marine sediments CDW (slab, brick, and concrete)	Concrete <190 μm Brick <140 μm Slab <140 μm	5.81 SiO_2 1.9 Al_2O_3 65.42 CaO 57.79 SiO_2 14.95 Al_2O_3 8.79 CaO 70.54 SiO_2 9.80 Al_2O_3 8.78 CaO	KOH Sodium silicate	Quartz Calcite albite Muscovite Kaolinite
Robayo-Salazar et al. [29]	Red brick residue with addition of Portland cement	24.25 μm	65.92 SiO_2 20.08 Al_2O_3 9.10 Fe_2O_3 0.73 CaO	NaOH, sodium carbonate Na2CO3, sodium sulfate Na2SO4, sodium silicate	Quartz, Hematite Muscovite Plagioclase-like structures
Fořt et al. [32]	CDW: brick	<0.125 mm	58.8 SiO_2, 19.6 Al_2O_3, 5.7 Fe_2O_3, 6.9 CaO	Sodium hydroxide Sodium silicate	Quartz, Hematite, Albite, Muscuvite
Tan et al. [31]	Mampostería 50% Hormigón 42% Ceramico 5% Vitreos 3%	100 mm	–48.2 SiO_2 19.8 Al_2O_3 8.6 CaO 6.1 Fe_2O_3, 5.7 MgO, 5.7 SO3	SiO_2/Na_2O	Quartz Calcite Gismondina Sporite

(continued)

Table 3 (continued)

Author	Material	Particle size	Chemical composition of CDW%	Alkaline activator	CDW crystalline phases
Bassani et al. [36]	– Recycled concrete – Crushed bricks and tiles – Recycled asphalt – Natural and excavated farms – CDW fines	63–125 mm <63 mm	32.9 SiO_2, 24.80 CaO 23.44 CaO 36.39 SiO_2, 38.7 CO2, 6.9 MgO 56.4 SiO_2, 18.61 Al_2O_3, 10.4 Fe_2O_3 44.3 SiO_2, 16.87 CO2 11.2 CaO 42.60 SiO_2,15.9 CO2 12.8 Al_2O_3	SiO_2/Na_2O	Quartz Calcite, Albite Fengita Cordierite Clinochlor Enstatite Iron silicate Magnesium Mica, Fengite, and Muscovite groups

in distilled water varying the morality from 2 to 12 mol and then adding the sodium silicate solution, which was left to rest 24 h before use. The solid–liquid ratio was varied between 0.25 and 0.35; the paste was left according to the formulations on a vibrating table to eliminate trapped air. It was demold after 2 h at room temperature, then taken to an oven at 80 °C for 24 h, and allowed to cure for 7 days [33]. Zawrah et al. [34] prepared the materials by grinding in a ball mill for 30 min, then passed to a vibrating screen to separate the size of the particles, and later grinding for 15 min in a planetary ball mill to homogenize and activate the residue. Sodium silicate and sodium hydroxide were used, with $SiO_2/Na_2O = 3.2$, and molar concentrations of NaOH 8 mol/L. The ratio between the two activators was 2.5, the solution was prepared 24 h before preparing the sample, and the liquid–powder ratio was set at 0.30 by weight. The samples were mixed by hand for 5 min, then poured into the mold and placed on a vibrating table for 1 min, demold after 24 h, and allowed to cure for up to 90 days at room temperature. Robayo et al. [29] used different alkaline activators to make the geopolymer among them, calcium carbonate, sodium sulfate, sodium hydroxide, and sodium silicate. For the optimal amount of activator, the authors partially added OPC in proportions from 5 to 30%. They used a liquid–solid ratio of 0.23. The pastes were mechanically mixed for 5 min and compacted on a vibrating table for 30 s. The samples were cured at room temperature for 24 h. The samples were placed in a curing chamber at a humidity of 80% until 7 and 28 days. Fort et al. [32] used brick dust generated by the crushing of thermal insulation bricks. This material did not require crushing and is ready to use, it was dried at 105 °C for 48 h, then sieved, and the molar ratio of the sodium silicate used by the author was 1.6 (SiO_2/Na_2O). The precursor was mixed and molded keeping it in a chamber at 20 °C and 50% humidity for 7 days. It is removed from the mold and allowed to cure for 28 days. Tan et al. [31] initially ground the CDW for 8 h in a ball mill to increase the fineness and surface of the material, and the molar ratio used by the authors of the alkaline activator (SiO_2/Na_2O) was from 1.6 to 1.9. The alkaline activator was stored for 24 h before preparing the samples, prepared at a speed of 140 rpm for 10 min, and then compacted on a shaking table for 15 s. They were left in a mold for 24 h at room temperature, and after being demold, they were left on a plastic membrane for 6 days before being analyzed. The authors did not use any thermal curing. Bassani et al. [37] ground the different types of CDW in a rotating drum to obtain fine powder and used fine material because the smaller the size, the faster the geopolymerization process develops and the higher the mechanical properties. The molar ratio of the sodium hydroxide was $SiO_2/Na_2O = 3.4$. The SiO_2/Na_2O had a concentration of 36%. The ratio between the two activators was 4: 1. The authors used the activator in three different ways (undiluted, diluted 50%, and diluted 75%) to study the influence on the reaction. 50wt% of the particles with size between 63 and 125 mm, and 50wt% of the particles with sizes smaller than 63 for the different samples were used. Solid–liquid ratios of 0.4 were used, while samples were cured at room temperature for 3, 7, and 28 days [36]. Each author carried out a different methodology to make the geopolymers, showing a very large variability in these processes. Dassekpo [38] mentions that more research is needed on the synthesis parameters for the manufacturing of geopolymers.

Mechanical Properties

The mechanical properties in this type of materials are very important due to their use. Table 4 lists the results of different research taken into consideration in this review.

Koministas [33] found that the best geopolymer formulation was the activation of tiles under a molarity of 8 mol, while the concrete showed a much lower activation potential due to its low contents of SiO_2 and Al_2O_3 necessary for activation. The high CaO content in concrete consumes KOH making it insufficient for dissolving aluminosilicates. For the mixture of tiles with marine sediments, the author finds a reduction in compressive strength explained by the decrease in the SiO_2/Al_2O_3 ratios. The sediments reached a lower resistance when they were activated. Zawrah et al. [17] managed to find that by adding the blast furnace slag to the CDW-based geopolymer, the porosity and water absorption were reduced due to the formation of new phases in the material matrix, and that the calcium oxide present in the slag was reduced to form hydrated calcium silicate products in the geopolymer matrix, good to improve the compressive strength and decrease the setting time. They also found an improvement in samples with longer curing times occur after the material has set [28, 33].

Robayo et al. [29] used several activators for the formation of activated alkaline cement, finding that the better mechanical behavior was achieved when sodium

Table 4 Mechanical properties of geopolymers

References	Formation conditions	Flexural strength	Compressive strength (MPa)	Curing time
[31]	Alkaline concentration 6 mol/l replacement 55% slag	8 MPa	70	7 days
[31]	Molar ratio 1.9 slag replacement 40%	–	58	7 days
[36]	CDW fines Undiluted activator	5 Mpa	12.8	28 days
[29]	CDW red brick plus 20% OPC NaOH + Na2SiO3	–	102.6	28 days
[34]	CDW red brick plus granulated blast furnace slag (GS60)	–	83 77	90 days 28 days
[33]	100% CDW tiles 90% CDW (tiles) 10% Souda sediments 90% CDW (tiles) 10% Patras sediments	–	42.7 40 41	7 days 7 days 7 days
[32]	AAP70	11.9 Mpa	41.9	28 days

hydroxide and sodium silicate were used, since the addition of sodium silicate increases by the presence of soluble silica species (Si–O-) [40]. Also, for activation with NaOH, resistances decreased with the addition of OPC, due to the loss of balance in the system as a function of pH and the high calcium content in the paste. Tan et al. [31] showed that due to the low calcium content of CDW, it is difficult to obtain high compressive strength at early ages without heat treatments, and by adding slag with a high calcium content, it fills the pores and voids, improving the resistance of the matrix. S. Puligilla et al. [41] show that the calcium content of the slag improves the geopolymerization process, due to the formation of the hydrated calcium and aluminum silicate gel (C-A-S–H). Authors made CDW geopolymer with 55% slag replacement due to the high compressive strength at an early age and the ambient temperature curing conditions.

Bassani et al. [36] demonstrated that CDW composed only of bricks and tiles and the unclassified CDW had higher fractions of highly reactive aluminosilicate and mica group phases, presenting higher mechanical strength. Thus, the final properties of geopolymers depend directly on many factors such as the type of activator, the manufacturing process, the starting material, and the particle size.

Conclusion

This review detailed resources for the elaboration of novel materials based on alkaline activation or as its coined name geopolymers, its main properties and disadvantages were named, as well as its use as an ecological alternative as a replacement for ordinary Portland cement. Several authors demonstrated with some phases presented in the CDW that this residue can be used for the formation of geopolymers, since they present reactions with the alkaline activator due to its siliceous-aluminous phases. With an addition of 30% of OPC to the brick-based geopolymer, the mechanical performances can be increased; in addition to the CDW, other residues, such as blast furnace slag, can be added and the compressive strength is notably favored. An increase in molarity of more than 4 mol in CDW-based geopolymers has no effect on mechanical performance. The better compressive strengths were obtained when high molar ratios of SiO_2/Al_2O_3 and $SiO_2/(Al_2O_3 + CaO)$ are present in the initial paste, whereas when SiO_2/Al_2O_3 is low, the resistances are low. It is concluded that the CDW-based geopolymer is a sustainable and environmentally friendly material, which manages to reduce the amount of greenhouse gases by being a substitute material for OPC, in addition to mitigating energy consumption for production.

References

1. He Z, Zhu X, Wang J, Mu M, Wang Y (2019) Comparison of CO2 emissions from OPC and recycled cement production. Constr Build Mater 211:965–973. https://doi.org/10.1016/j.con

buildmat.2019.03.289
2. Davidovits J (2015) False values on CO2 emission for geopolymer cement/concrete published in scientific papers. Geopolymer Inst Libr
3. Chen L, Wang Z, Wang Y, Feng J (2016) Preparation and properties of alkali activated metakaolin-based geopolymer. Materials (Basel) 9(9):1–12. https://doi.org/10.3390/ma9090767
4. Tailby J, MacKenzie KJD (2010) Structure and mechanical properties of aluminosilicate geopolymer composites with Portland cement and its constituent minerals. Cem Concr Res 40(5):787–794. https://doi.org/10.1016/j.cemconres.2009.12.003
5. Ouellet-Plamondon C, Habert G (2015) Life cycle assessment (LCA) of alkali-activated cements and concretes. In: Pacheco-Torgal F, Labrincha JA, Leonelli C, Palomo A Chindaprasirt P (eds) Handbook of alkali-activated cements, mortars and concretes, Woodhead Publishing. pp 663–686 https://doi.org/10.1016/C2013-0-16511-7
6. Habert G, D'Espinose De Lacaillerie JB, Roussel N (2011) An environmental evaluation of geopolymer based concrete production: reviewing current research trends. J Clean Prod 19(11):1229–1238. https://doi.org/10.1016/j.jclepro.2011.03.012
7. Duxson P, Fernández-Jiménez A, Provis JL, Lukey GC, Palomo A, Van Deventer JSJ (2007) Geopolymer technology: the current state of the art. J Mater Sci 42(9):2917–2933. https://doi.org/10.1007/s10853-006-0637-z
8. Davidovits J (1994) Properties of geopolymer cements. Alkaline Cem Concr KIEV Ukr 1–19. https://doi.org/10.1073/pnas.0811322106
9. Davidovits J (1991) Geopolymers—Inorganic polymeric new materials. J Therm Anal 37(8):1633–1656. https://doi.org/10.1007/BF01912193
10. Torres-Carrasco M, Puertas F (2017) Alkaline activation of different aluminosilicates as an alternative to Portland cement: alkali activated cements or geopolymers | La activación alcalina de diferentes aluminosilicatos como una alternativa al Cemento Portland: cementos activados alcaliname. Rev Ing Constr 32(2):5–12
11. Lampris C, Lupo R, Cheeseman CR (2009) Geopolymerisation of silt generated from construction and demolition waste washing plants. Waste Manag 29(1):368–373. https://doi.org/10.1016/j.wasman.2008.04.007
12. Reig L, Tashima MM, Borrachero MV, Monzó J, Cheeseman CR, Payá J (2013) Properties and microstructure of alkali-activated red clay brick waste. Constr Build Mater 43:98–106. https://doi.org/10.1016/j.conbuildmat.2013.01.031
13. Segura Sierpes Y, Borrachero Rosado MV, Monzó Balbuena JM, Payá Bernabeu J (2016) Preliminary studies on hydrated cement for its reuse in geopolymers. Dyna 83(196):229–238. https://doi.org/10.15446/dyna.v83n196.54189
14. Del Río M, Izquierdo P, Salto I, Santa Cruz J (2010) "La regulación jurídica de los residuos de construcción demolición (RCD) en España. El caso de la Comunidad de Madrid," *Inf. la Construcción*, vol. 62, no. 517, pp. 81–86. https://doi.org/10.3989/ic.08.059
15. Colorado D, Echeverry GI, Colorado H (2019) Logistics as an essential area for the development of the solid waste management in Colombia. Inf Técnico 83(2):131–154
16. Colorado HA, Echeverri-Lopera GI (2020) The solid waste in Colombia analyzed via gross domestic product: towards a sustainable economy. Rev Fac Ing 96:51–63. https://doi.org/10.17533/udea.redin.20191046
17. de Azevedo ARG, Marvila MT, Ali M, Khan MI, Masood F, Vieira CMF (2021) Effect of the addition and processing of glass polishing waste on the durability of geopolymeric mortars. Case Stud Constr Mater 15:e00662. https://doi.org/10.1016/j.cscm.2021.e00662
18. Durach CF, Kembro J, Wieland A (2017) A new paradigm for systematic literature reviews in supply chain management. J Supply Chain Manag 53(4):67–85. https://doi.org/10.1111/jscm.12145
19. Murata K, Wakabayashi K, Watanabe A (2014) Study on and instrument to assess knowledge supply chain systems using advanced kaizen activity in SMEs. Supply Chain Forum 15(2):20–32. https://doi.org/10.1080/16258312.2014.11517339

20. Scopus–Analyze search results (2020). [Online]. Available: https://aplicacionesbiblioteca.udea.edu.co:2125/term/analyzer.uri?sid=7abad18ee2539b0ee505458514f99b8f&origin=resultslist&src=s&s=%22cdw%22+or+%7BDEMOLITION+AND+CONSTRUCTION+WASTE%7D+OR+%28%22DEMOLITION+AND+CONSTRUCTION+WASTE%22%29+and+%22geopolymer%22+or. [Accessed: 07-Jun-2020].
21. Asensio E, Medina C, Frías M, Sánchez de Rojas MI (2020) Fired clay-based construction and demolition waste as pozzolanic addition in cements. Design of new eco-efficient cements. J Clean Prod 265. https://doi.org/10.1016/j.jclepro.2020.121610
22. Younis KH, Salihi KA, Ibrahim TK (2020) An overview of geo-polymer concrete including recycled aggregate. Int J Sci Technol Res 9(3):6239–6245
23. Zhao Y, Gao J, Liu C, Chen X, Xu Z (2020) The particle-size effect of waste clay brick powder on its pozzolanic activity and properties of blended cement. J Clean Prod 242. https://doi.org/10.1016/j.jclepro.2019.118521
24. López Ruiz LA, Roca Ramón X, Gassó Domingo S (2020) The circular economy in the construction and demolition waste sector—A review and an integrative model approach. J Clean Prod 248. https://doi.org/10.1016/j.jclepro.2019.119238
25. McLellan BC, Williams RP, Lay J, Van Riessen A, Corder GD (2011) Costs and carbon emissions for geopolymer pastes in comparison to ordinary portland cement. J Clean Prod 19(9–10):1080–1090. https://doi.org/10.1016/j.jclepro.2011.02.010
26. Morán del Pozo JM, Juan Valdés A, Aguado PJ, Guerra MI, Medina C (2011) Estado actual de la gestión de residuos de construcción y demolición: limitaciones," *Inf. la Construcción*, vol. 63, no. 521, pp. 89–95. https://doi.org/10.3989/ic.09.038
27. Pericot NG, Del Solar P (2013) Optimizing the recycling rate on construction waste: the approach of sustainability tools October 2014, p 13
28. Pacheco Bustos CA, Fuentes Pumarejo LG, Sánchez Cotte ÉH, Rondón Quintana HA (2017) Residuos de construcción y demolición (RCD), una perspectiva de aprovechamiento para la ciudad de barranquilla desde su modelo de gestión," *Ing. y Desarro.*, vol. 35, no. 2, pp. 533–555
29. Robayo-Salazar RA, Mejía-Arcila JM, Mejía de Gutiérrez R (2017) Eco-efficient alkali-activated cement based on red clay brick wastes suitable for the manufacturing of building materials. J Clean Prod 166:242–252. https://doi.org/10.1016/j.jclepro.2017.07.243
30. Song W, Yi J, Wu H, He X, Song Q, Yin J (2019) Effect of carbon fiber on mechanical properties and dimensional stability of concrete incorporated with granulated-blast furnace slag. J Clean Prod 238. https://doi.org/10.1016/j.jclepro.2019.117819
31. Tan J, Cai J, Li X, Pan J, Li J (2020) Development of eco-friendly geopolymers with ground mixed recycled aggregates and slag. J Clean Prod 256. https://doi.org/10.1016/j.jclepro.2020.120369
32. Fořt J et al (2018) Application of waste brick powder in alkali activated aluminosilicates: Functional and environmental aspects. J Clean Prod 194:714–725. https://doi.org/10.1016/j.jclepro.2018.05.181
33. Komnitsas K (2016) Co-valorization of marine sediments and construction and demolition wastes through alkali activation. J Environ Chem Eng 4(4):4661–4669. https://doi.org/10.1016/j.jece.2016.11.003
34. Zawrah MF, Gado RA, Feltin N, Ducourtieux S, Devoille L (2016) Recycling and utilization assessment of waste fired clay bricks (Grog) with granulated blast-furnace slag for geopolymer production. Process Saf Environ Prot 103:237–251. https://doi.org/10.1016/j.psep.2016.08.001
35. Nath P, Sarker PK (2014) Use of OPC to improve setting and early strength properties of low calcium fly ash geopolymer concrete cured at room temperature. Cem Concr Compos. https://doi.org/10.1016/j.cemconcomp.2014.08.008
36. Bassani M, Tefa L, Coppola B, Palmero P (2020) Alkali-activation of aggregate fi nes from construction and demolition waste: valorisation in view of road pavement subbase applications. J Clean Prod 234(2019):71–84. https://doi.org/10.1016/j.jclepro.2019.06.207
37. Assi LN, Eddie E, Ziehl P (2018) Effect of source and particle size distribution on the mechanical and microstructural properties of fly ash-based geopolymer concrete. Constr Build Mater 167:372–380. https://doi.org/10.1016/j.conbuildmat.2018.01.193

38. Dassekpo JBM, Zha X, Zhan J, Ning J (2017) The effects of the sequential addition of synthesis parameters on the performance of alkali activated fly ash mortar. Results Phys. 7(May):1506–1512. https://doi.org/10.1016/j.rinp.2017.04.019
39. Kourti I, Rani DA, Deegan D, Boccaccini AR, Cheeseman CR (2010) Production of geopolymers using glass produced from DC plasma treatment of air pollution control (APC) residues. J Hazard Mater 176(1–3):704–709. https://doi.org/10.1016/j.jhazmat.2009.11.089
40. Duxson P, Provis JL, Lukey GC, Mallicoat SW, Kriven WM, Van Deventer JSJ (2005) Understanding the relationship between geopolymer composition, microstructure and mechanical properties. Colloids Surf Physicochem Eng Asp 269(1–3):47–58. https://doi.org/10.1016/j.colsurfa.2005.06.060
41. Puligilla S, Mondal P (2013) Role of slag in microstructural development and hardening of fly ash-slag geopolymer. Cem Concr Res 43(1):70–80. https://doi.org/10.1016/j.cemconres.2012.10.004

KIGAM Technology for the Recovery of NdFeB Waste Magnet and Manufacturing Scraps

Kyeong Woo Chung, Ho-Sung Yoon, Chul-Joo Kim, Rina Kim, and Byunchul Lim

Abstract At KIGAM, the pilot plants have been developed and demonstrated two processes for recycling NdFeB waste magnets and manufacturing scraps. The former process uses thermal oxidation as a pretreatment before. The latter does caustic digestion and the thermal oxidation process. About 35% of the iron in feed was co-leached with REEs upon leaching of thermally oxidized products. Therefore, the separation process of REEs and non-REEs was carried out using the double-salt precipitation method to produce a high purity REEs solution applicable for solvent extraction. Instead, the leaching yield of iron upon leaching of caustic digestion and thermal oxidation products showed a value of 0.01%. Thus, the high purity REEs solution was obtained and directly applied to solvent extraction. Finally, REEs compounds of 99.9% or above could be obtained by solvent extraction using mixer settler and precipitation from raffinate and stripping solutions.

Keywords Recycling · Secondary recovery · NdFeB magnet

Introduction

Since the NdFeB magnet firstly appeared in early 1980, its utilization has become markedly grown in the high-tech industry [1]. Lately, its use is expected to vigorously expand in the electric vehicle and wind power industry to achieve carbon neutrality in 2050 [2]. The increased consumption will increase the future supply risk of REEs such and neodymium and dysprosium. Recycling is considered to be promising for securing supply stability of REEs [3]. In addition, the use of recycled materials also leads to the reduction of greenhouse gases' emission [4].

K. W. Chung (✉) · H.-S. Yoon · C.-J. Kim · R. Kim
Korea Institute of Geoscience and Mineral Resources (KIGAM), Daejeon 34132, Republic of Korea
e-mail: case7@kigam.re.kr

B. Lim
SungLim Rare Earth Metal Co. Ltd, Daegu 42714, Republic of Korea

A. Lazou et al. (eds.), *REWAS 2022: Developing Tomorrow's Technical Cycles (Volume I)*, The Minerals, Metals & Materials Series,
https://doi.org/10.1007/978-3-030-92563-5_74

At KIGAM, the recycling of NdFeB magnet/scrap based on hydrometallurgy has been studied since C-H Lee firstly reported the leaching behavior of thermally oxidized NdFeB powders in 1998 [5]. In the 2010s, process development has been intensively researched, and lately, the technology has been demonstrated in a pilot plant, which is ongoing. This paper presents the recycling technology developed at KIGAM for NdFeB waste magnets and scraps.

Process for Recycling of Waste Magnet

Figure 1 shows the recycling process for the NdFeB waste magnet consisting of peeling of plating layer, crush and grinding, thermal oxidation, sulfuric acid leaching, precipitation of rare earth double sulfate, conversion of double salt to REEs hydroxide, hydrochloric acid leaching of hydroxide, purification of REEs using solvent extraction, and synthesis of REEs compounds.

NdFeB magnet is prone to oxidize in the air even at room temperature, so usually coated with a metal of Ni or Zn. First, the plating layer was peeled off using cyanide solution. And they were crushed and ground to powders under 200um in size using a jaw crusher, roll mill, and vibration mill in sequence (Fig. 2). XRD data shows that powders were partially oxidized, giving rise to $NdFeO_3$ and γ-Fe_2O_3 during

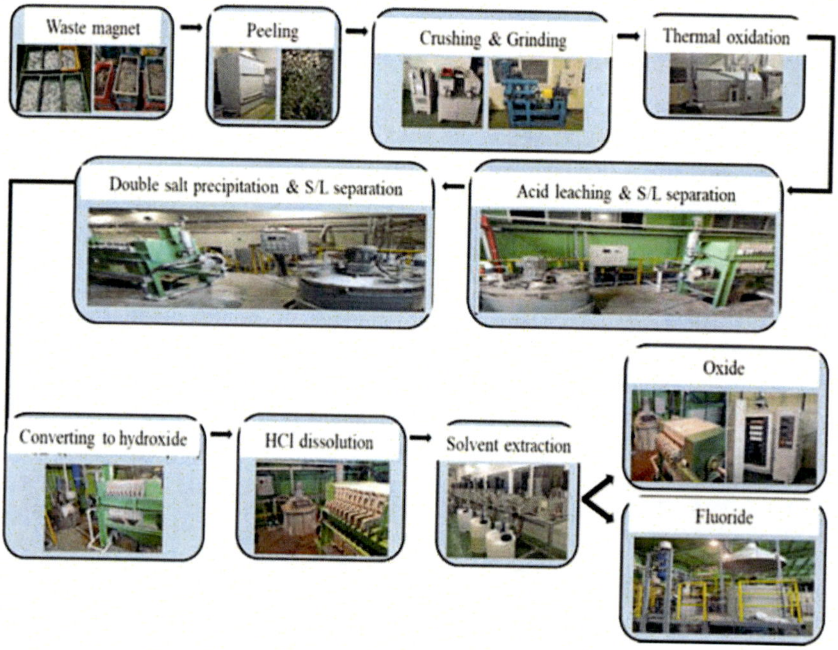

Fig. 1 Standard process for recycling process for NdFeB waste magnet

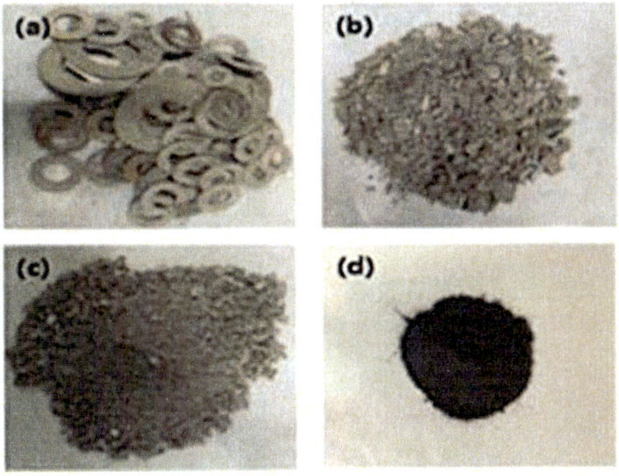

Fig. 2 Images of **a** peeled-off magnet and powders by **b** jaw crusher, **c** roll mill, and **d** vibration mill

crushing and grinding, as shown in Fig. 3. However, metallic iron was also observed. Therefore, thermal oxidation at 600 °C for 3 h using a rotary kiln was applied for ground powder which fully oxidized them. Metallic iron disappeared, and γ-Fe_2O_3 was mainly transformed to α-Fe_2O_3, which was expected to reduce iron leaching.

The resultant powders were leached in 2.5 M sulfuric acid solution at 60 °C for 3 h, giving rise to the leach liquor of ca. 28,000 mg/L of REEs and ca. 56,120 of Fe. The oxidation and leaching reactions are presented in Table 1.

After leaching, REEs were selectively recovered as a double salt with iron remaining in leach liquor by adding sodium sulfate into the leach liquor. The recovery yield of L-REEs showed 99.9%, irrespective of the input amount of sodium sulfate (Fig. 4). However, that of H-REEs was much affected by input amount. Seven equivalent of sodium sulfate was added in this study. Moreover, REEs double sulfate was

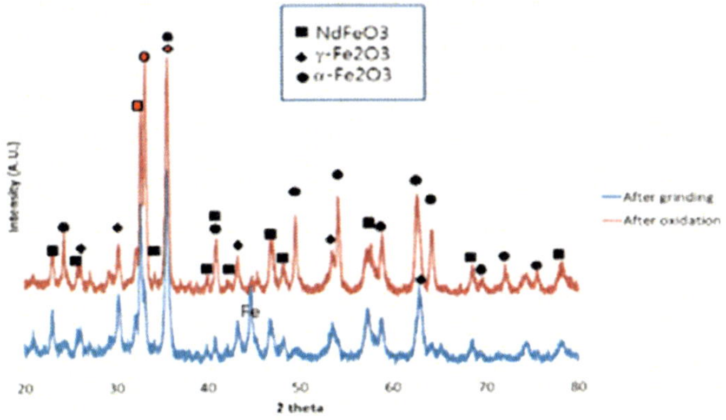

Fig. 3 XRD patterns of ground powders before and after thermal oxidation

Table 1 Chemical reactions involved with unit processes

Unit process	Chemical reaction
Thermal oxidation	$Nd + Fe + 3/2O_2 \rightarrow NdFeO_3$ $2Fe + 3/2O_2 \rightarrow Fe_2O_3$
Sulfuric acid leaching	$2NdFeO_3 + 6H_2SO_4 \rightarrow Nd_2(SO_4)_3 + Fe_2(SO_4)_3 + 6H_2O$
Double-salt precipitation	$Nd_2(SO_4)_3 + Na_2SO_4 \rightarrow 2Nd{\cdot}Na(SO_4)_2\downarrow$
Conversion to hydroxide	$2Nd{\cdot}Na(SO_4)_2 + 3NaOH \rightarrow Nd(OH)_3 + 2Na_2SO_4$
Hydrochloric dissolution	$Nd(OH)_3 + 3HCl \rightarrow NdCl_3 + 3H_2O$
Synthesis of REEs oxide	$2NdCl_3 + 3C_2O_4H_2 \rightarrow Nd_2(C_2O_4)_3\downarrow + 6HCl$ $Nd_2(C_2O_4)_3 + 3/2O_2 \rightarrow Nd_2O_3 + 6CO_2$

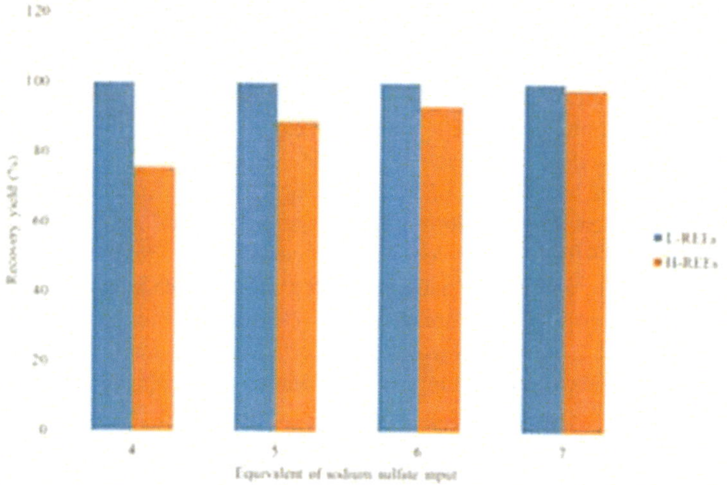

Fig. 4 Recovery yield of REEs with the input amount of sodium sulfate

re-dissolved in a hydrochloric acid solution of pH 3~3.5 after converting to REEs hydroxide in NaOH solution (see Table 1). It produced the high purity REEs chloride solution with an iron of below 100 mg/L.

The recovery yield of REEs in REEs chloride solution showed ca. 96% for L-REEs and ca. 91% for H-REEs, respectively. The lower value for H-REEs than L-REEs might be due to the lower precipitation yield of H-REEs double sulfate than L-REEs double sulfate (Table 2).

The REEs chloride solution was applied to the solvent extraction process for the purification of REEs. PC88A and kerosene were used as an extractant and a

Table 2 Recovery yield of rare earth elements in REEs chloride solution

	L-REEs		H-REEs		Total
	Pr	Nd	Tb	Dy	
Input amount (kg)	33.3	134.1	12.8	7.6	187.8
Recovered amount (kg)	32.2	129.5	11.7	7.0	180.4
Recovery yield (%)	96.8	96.6	91.5	91.2	96.1

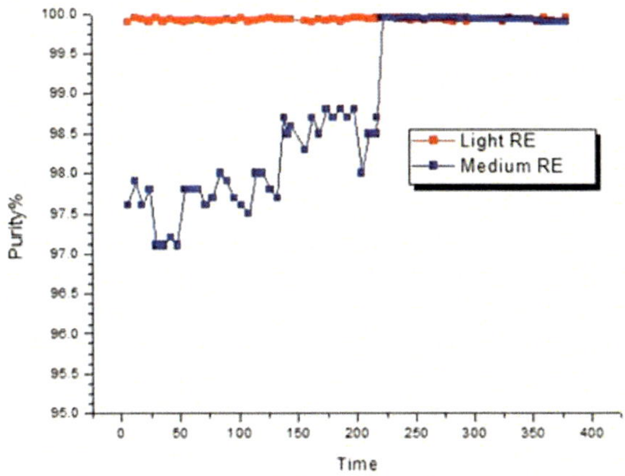

Fig. 5 Purities of light and heavy rare earth elements in the raffinate and stripping solution

diluent, respectively, where the concentration of PC88A was set at 0.5 M. Diluent and concentrated hydrochloric acids were used as scrubbing and stripping solutions (Fig. 5). The extraction, scrubbing, stripping, and washing stages were 4, 8, 3, and 2, respectively. The purity of L-REEs in raffinate initially quickly went over 99.9%. It means that H-REEs was easily well extracted under the condition used in this study. Instead, that of H-REEs showed ca. 98.5% after 200 h. It was attributed to the stripped L-REEs unscrubbed in organic solution at the initial stage before stabilizing the whole process. Thus, the stripping solution was exchanged with a fresh one after 220 h, so the purity of H-REEs immediately showed 99.9%.

Finally, REEs oxides were recovered through the precipitation and calcination of REEs oxalates (Fig. 6). REEs fluorides were synthesized by the addition of ammonium fluoride to the raffinate or stripping solution. The final products show the purity of 99.91% or above.

(a) (b)

Fig. 6 Images of didymium oxalate and oxide

Process for the NdFeB Scraps

Figure 7 shows the recycling process for the NdFeB scraps composed of caustic digestion, thermal oxidation, hydrochloric acid leaching, purification of REEs using solvent extraction, and synthesis of REEs compounds.

Unlike the ground powders of a waste magnet, manufacturing scrap exists unoxidized because coolant wraps it and prevents its oxidation by air. The caustic digestion of unoxidized powders in 50wt% NaOH at 100 °C for 1 h could decompose REEs and iron to their hydroxide and mixture of magnetite and metallic iron, and further oxidation at about 350 °C yielded the mixture of Nd_2O_3 and Fe_2O_3 [6]. The ternary oxide, $NdFeO_3$, was not formed; therefore, simultaneous dissolution of neodymium and iron from $NdFe_2O_3$ could be suppressed. Furthermore, Fe_2O_3 is highly resistant

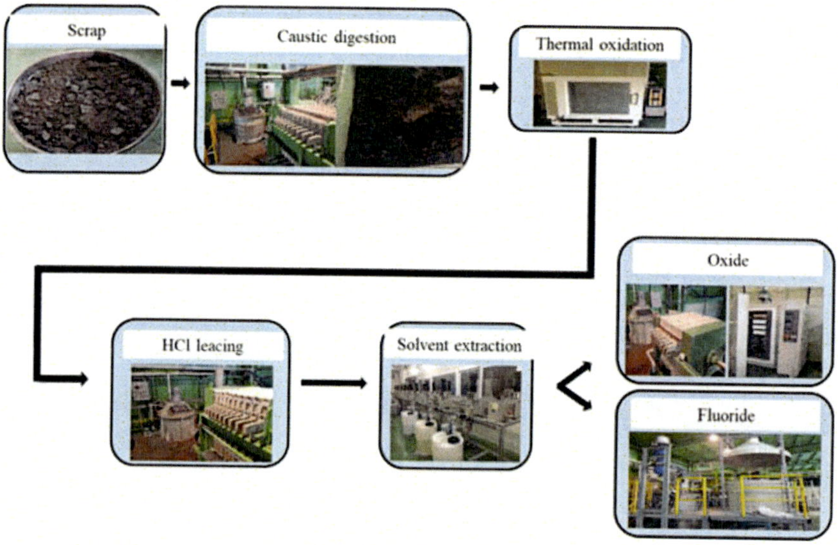

Fig. 7 Process for recycling process for NdFeB waste magnet

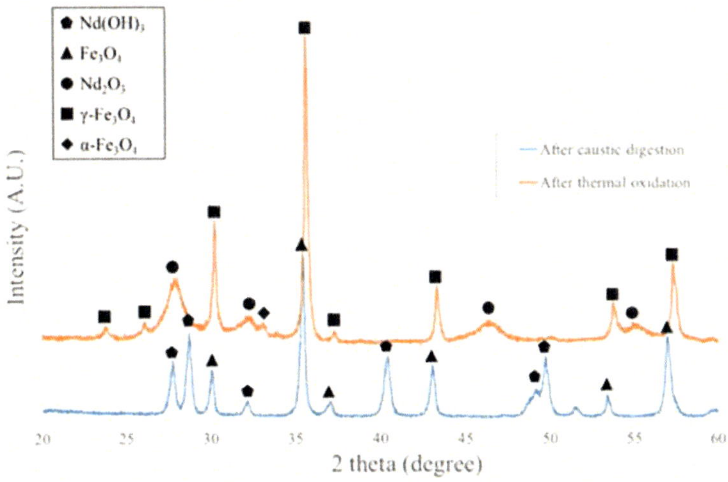

Fig. 8 XRD patterns of powders after caustic digestion and thermal oxidation

to the acid solution, unlike Nd_2O_3 which quickly dissolves in diluent acid solution. The leaching of caustic digestion and thermal oxidation powders in 0.45 M HCl at room temperature showed the highly selective leaching of REEs over iron, typically higher than 99% of neodymium leaching yield and lower than 0.1% of iron leaching yield. High purity of REEs chloride solution could be directly applied to the solvent extraction process, which simplified the recycling process of NdFeB scraps (Figs. 8 and 9).

Conclusions

At KIGAM, recycling processes from NdFeB waste magnet and scraps to REEs compound were researched and developed. The pilot plant scale test demonstrated that high recovery yields of REEs of over 90% and high purities of REEs compound of over 99.9% could be obtained using the developed process.

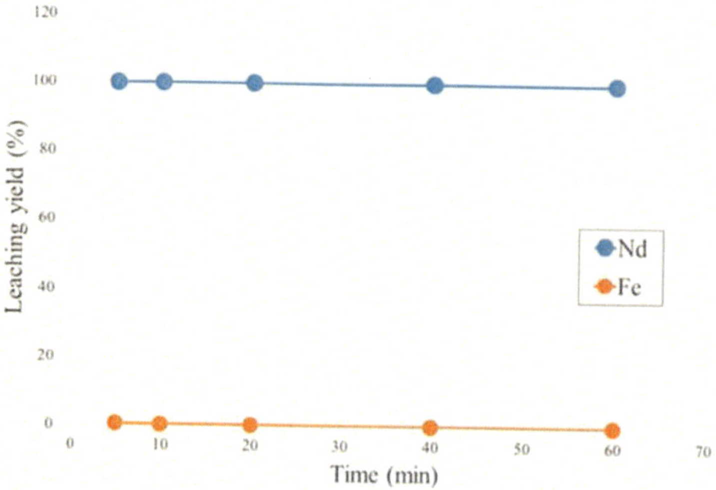

Fig. 9 Leaching yield of Nd and Fe from caustic digestion and thermal oxidation powders with time

Acknowledgements The research was supported by the Energy Resource Technology Developing Project of Korea Institute of Energy Technology Evaluation and Planning (KETEP) grant funded by Ministry of Trade, Industry, and Energy of Korea (MOTIE).

References

1. Yu LQ, Zhong XL, Zhang YP, Yan YG, Zhen YH, Zakotnik M (2011) Production and corrosion resistance of NdFeBZr magnets with an improved response to thermal variations during sintering. J Mang Magn Mater 323:1152–1155
2. Nakamura H (2018) The current and future status of rare earth permanent magnets. Scr Mater 154:273–276
3. Binnemans K, McGuiness P, Jones PT (2021) Rare-earth recycling needs market intervention. Nat Rev Mater 6:459–461
4. Jensen JP (2019) Evaluating the environmental impacts of recycling wind turbines. Wind Energy 22:316–326
5. Lee J-C, Kim W-B, Jeong J, Yoon IJ (1998) Extraction of Neodymium from Nd-Fe-B Magnet Scraps by Sulfuric acid. J Korean Inst Met Mater 36:967–971
6. Chung KW, Kim C-J, Yoon H-S (2015) Novel extraction process of rare earth elements from NdFeB powders via Alkaline treatment. Arch Metall Mater 60:1301–1305

Recovery of Copper, Iron, and Alumina from Metallurgical Waste by Use of Hydrogen

Casper van der Eijk and Halvor Dalaker

Abstract Substituting carbon with hydrogen is one of the few ways metal production can potentially become free of CO_2 emissions. Moreover, the metallurgical industry produces significant amounts of waste. The present work presents a circular concept that will be pursued in the HARARE project, based on increasing waste recovery by the use of hydrogen. The project will tackle two example cases: bauxite residue and copper smelter slags. The common theme is to use hydrogen to selectively reduce iron and copper, making it possible to extract these metals. Through a series of pyro and hydrometallurgical steps, as well as mechanical separation, it is also possible to recover secondary valuables like alumina, molybdenum, cobalt, nickel, zinc, and scandium. The final remaining residues can be valorised as building materials for a truly zero-waste concept. In this paper, the different process streams for the two example cases are laid out, including the valorisation of secondary material streams.

Keywords Hydrogen · Bauxite residue · Copper slag · Circular economy · Recycling

Introduction

The metallurgical industry is a large emitter of CO_2. In the EU27, it accounted for 162 million tonnes of direct CO_2 emissions in 2019 [1]. Of these, some 86 million tonnes are related to production of the energy used in the processes. While by no means trivial in practice, this energy can be replaced by emissions-free alternatives, and eliminating the energy-related portion of the CO_2 emissions is thus at least theoretically feasible with existing technologies. The remaining CO_2 emissions on the other hand—some 47% with EU numbers, and likely similar around the globe—are tied directly to the use of carbon in the industrial processes that turns ore into metal. Eliminating these emissions requires different processes from those of today, which is why metal production is listed as a "hard-to-abate sector" [2]. One of the

C. van der Eijk (✉) · H. Dalaker
SINTEF, Trondheim, Norway
e-mail: Casper.Eijk@sintef.no

A. Lazou et al. (eds.), *REWAS 2022: Developing Tomorrow's Technical Cycles (Volume I)*, The Minerals, Metals & Materials Series,
https://doi.org/10.1007/978-3-030-92563-5_75

few ways the process-related (i.e. non-energy) CO_2 emissions of metal production can be eliminated is to replace carbon with a different reducing agent, with hydrogen often being mentioned as the foremost candidate.

Another challenge of the metallurgical industry is the inefficiency that is found in its value chains. For example, 10–30% of the aluminium present in bauxite ore ends up discarded as bauxite residue (BR), and the discarded slags from copper smelting contain about 1% Cu—more than the concentration of Cu in ore from an average modern mine. These inefficiencies also indirectly increase the CO_2 emitted per ton of finished product, not to mention that they lead to waste problems, with millions of tonnes of metallurgical by-products piling up in landfills every year.

The HARARE project [3] (Hydrogen As the Reducing Agent in the REcovery of metals and minerals from metallurgical waste) will demonstrate sustainable pathways to produce metals using hydrogen as an enabler, for removing waste and valorising materials in carbon free processes. In this paper, we outline the proposed processes that will be investigated in the HARARE project.

Overall Concept

In both the slag from flash smelting of copper and the bauxite residue from alumina production in the Bayer process, the main component is iron oxide. At the core of the HARARE concept is the use of hydrogen gas to reduce the iron oxide to without process-related CO_2 emissions. Having removed the iron, the main components (copper and alumina, respectively) can be recovered and returned to their value chains. To ensure a zero-waste approach, potentially harmful elements must also be removed, and all other residues turned into useable resources like building materials.

The different approaches are outlined in Fig. 1.

The use of hydrogen for the recovery of metals from Cu slag and bauxite residue has already been described in literature. But for the case of copper, the emphasis was on raw materials rich in copper [4]. When bauxite residue is treated with hydrogen at temperatures above ca. 850 °C, hercynite ($FeAl_2O_4$) is formed [5]. The formation of this phase makes the separation of aluminium and iron difficult. So either the reduction temperature must be kept low, accepting that the iron oxides will not be

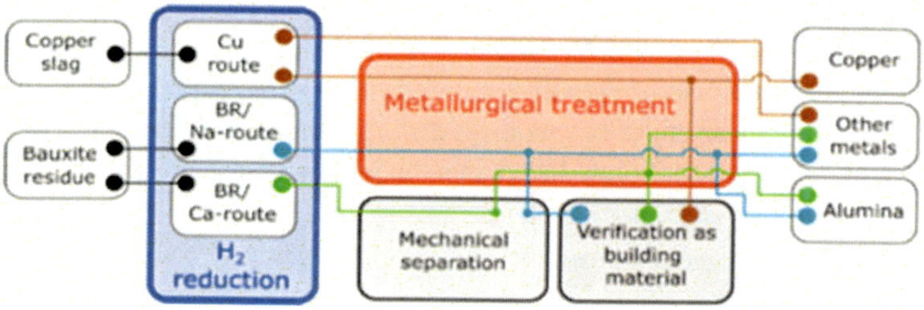

Fig. 1 Sketch of the overall idea

reduced to metallic iron but to magnetite (Na-route in HARARE) or Ca must be added leading to the formation of the Ca-Al-oxides which are more stable at high temperatures than hercynite (Ca-route in HARARE).

The Injection of Hydrogen to Refine Copper Slags

Slag from copper flash smelting contains not only slag-forming components, but also copper levels comparable to that in ores from modern mines (~1 wt%). In addition, it contains iron oxide (~50 wt%), alkalis and heavy metals (K, Na, Zn, Pb++, ~3–5 wt%), as well as smaller amounts of molybdenum, cobalt and nickel. The high iron content of the slag makes it challenging with traditional methods to extract the copper without also extracting a lot of iron and creating a low-Cu high-Fe alloy of little commercial relevance. Even without copper recovery, the alkalis and heavy metals make it difficult to valorise the slag as a construction material or as an iron source. The low-value options available for slag valorisation are as blast abrasion agent, cement compound, and river embankment. The slag is partly valorised through these means, but volumes are limited. Furthermore, legal limits for heavy metals will be lowered in the future, further restricting the opportunities available.

In HARARE, RWTH Aachen and Aurubis will develop selective reduction process based on hydrogen injection into the liquid slag. Using hydrogen as the reducing agent allows the reducing potential to be fine-tuned with the composition of the gas and the process temperature, controlling which metals will be reduced. The expected selectivity is up to 2,27 mCu/mFe in the metal phase. A Sankey diagram of the process is shown in Fig. 2. The process will have two steps:

- Step 1, lower reducing potential:

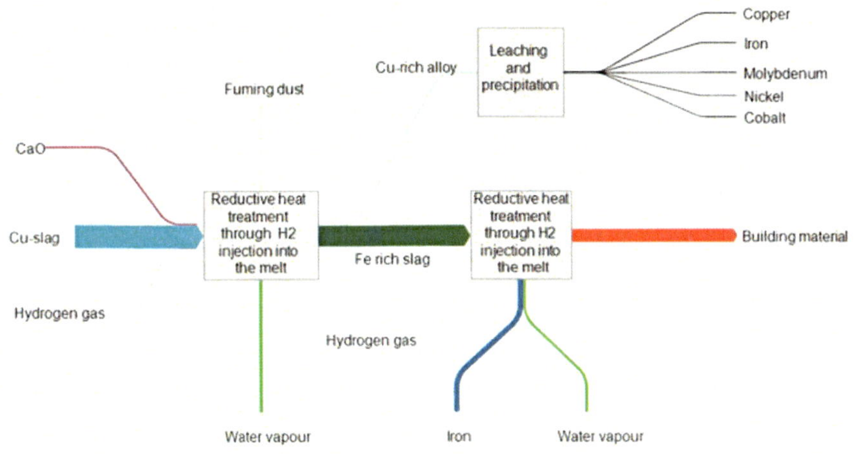

Fig. 2 Sankey diagram for the valorisation of copper slag

  - Copper will be reduced to liquid metal alloy along with smaller amounts of Fe, Mo, Co, and Ni.
  - Alkalis and heavy metals will be reduced to metallic fumes and collected as flue dust,
  - Iron will mostly remain in the slag in its oxidised state.
- Step 2, increased reducing potential:
  - Iron oxides are reduced to iron and removed as pig iron.

This process will separate the different valuable components of the slag and create the following products;

- High-value high-copper alloy.
- Pig iron
- Zn-rich flue dust as raw material for Zn production
- Slag free of iron, alkalis, and heavy metals, suitable as a construction material.

Recovery of Alumina from Bauxite Residue

In the production of aluminium, almost twice as much bauxite residue (BR) as aluminium metal is produced. Despite containing potentially valuable materials such as aluminium, silicon, and iron, as well as rare earth elements (REEs), BR is mostly landfilled. This is not only a waste of resources, but also presents a land-use problem, as BR takes up increasing amounts of land.

The iron oxide in the BR can be turned into iron by hydrogen reduction. In order to facilitate recovery of alumina and valorisation of other residues, fluxing of the BR prior to H_2 reduction is necessary. In the HARARE project, two routes for BR processing are pursued. The first route the "Ca-route", uses CaO as a flux, while the second route, the "Na-route", uses NaOH.

The Ca-route works at high temperature and thus fits best for BR qualities that are low in Ti because the Ti binds with Al in a perovskite structure at high temperatures. The Na-route on the other hand can tolerate BR qualities with high Ti, since it takes place at lower temperatures. Furthermore, the final residue of the Ca-route is easier to use as building material because with Ca-silicate content, it resembles normal cement. The two routes are described below.

Calcium Route for the Recovery of Alumina from BR

In the calcium route, NTNU and SINTEF will treat BR by hydrogen together with CaO to produce iron and a tailor-made slag of CaAl-oxides (mayenite) that is leachable at low temperatures. A Sankey diagram of the process is shown in Fig. 3. The process will take place below the melting temperature of iron, which means that the produced iron will be in the solid state and thus cannot be separated by a tapping

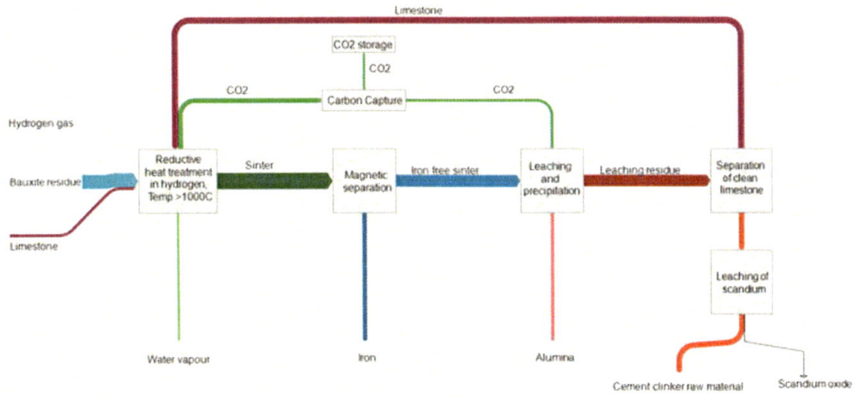

Fig. 3 Sankey diagram for the valorisation of BR through the Ca-route

process. The separation of the solid iron particles is one of the challenging parts of the process because the distribution the iron particles in the oxide is very fine. NTNU and ReSiTec will test different crushing, grinding, and separation techniques to liberate the metallic from the non-metallic phases, including magnetic separation and floatation.

After separation of the iron, the non-magnetic fraction will consist mainly of an easily leachable alumina-containing slag (mayenite). The alumina will be leached from the mayenite by NTUA, and Al-hydroxide will be precipitated from the solution and can be returned to the aluminium value chain.

The remaining residue at this stage is rich in limestone which will be recovered by flotation at NTNU to be recycled to the start of the process chain for a new roasting stage. During the roasting stage, CO_2 is generated from the limestone during the reductive treatment. The off-gas is a mixture of CO_2, H_2O, and unreacted H_2. From this gas mixture, the CO_2 can be easily concentrated.

Most of the CO_2 is reused in the process when the CO_2 is needed during precipitation, whilst the CaO is turned into $CaCO_3$ again. The CO_2 concentration for precipitation does not need to be 100%, so here is a possibility to use a diluted CO_2 stream from another industrial process while the concentrated CO_2 from the heat treatment is stored. The capture and storage of CO_2 is a possible add-on to the process which will not be included in the HARARE project work.

After the recovery of the limestone, the remaining residue consists of oxides suitable for use in building material, into which the Sc and other rare earth elements (REEs) in the BR will also have concentrated.

Sodium Route for the Recovery of Alumina from BR

BR can be recovered by roasting with carbon, smelting, and leaching [6]. Based on this work, KU Leuven will develop the process further into an electrically heated,

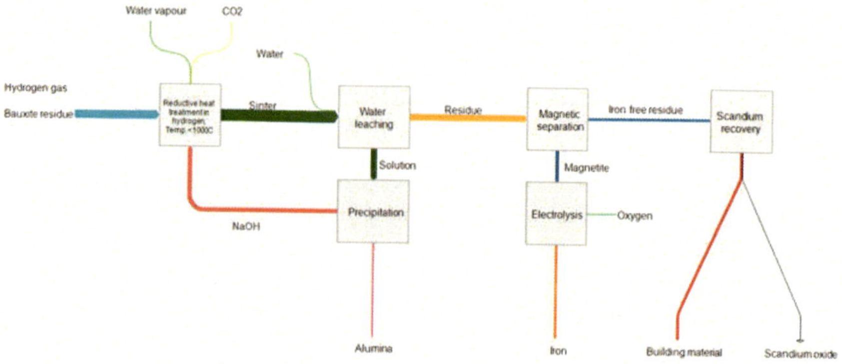

Fig. 4 Sankey diagram for the valorisation of BR through the Na-route

H_2-based reduction process, with no smelting. A Sankey diagram of the process is shown in Fig. 4.

Step (i) is alkali roasting of the BR with NaOH in H_2 atmosphere. This treatment will turn the aluminium oxide in the BR into water soluble $NaAlO_2$ and reduce the iron oxides. The temperature is lower in the Na-route than in the Ca-route, which means that the iron oxide is not reduced to metallic iron, but rather to magnetite, Fe_3O_4.

Step (ii) is water leaching. Sodium and aluminium will dissolve as Na^+ and Al^{3+} and can then be precipitated as hydroxides. $Al(OH)_3$ is returned to the aluminium value chain, while NaOH is recycled to the start of the Na-route and used with a fresh batch of BR.

The solid fraction in the leaching step consists of magnetite and a non-magnetic oxide-phase. These are separated by wet magnetic separation and can be used as raw material for iron and building materials respectively. Once again, the REEs of the BR are expected to be concentrated in the oxide residue (Fig. 4).

Secondary Material Streams and Zero-Waste Approach

The main focus of the HARARE processes is on the recovery of copper and alumina back into their respective value chains, but in a completely zero-waste approach, every material stream will be valorised. Some details on the approaches to valorise the secondary material streams are described below.

Separation of Copper-Alloy

The Cu-rich alloy produced from the copper slag also contains other metals like Fe, Mo, and Co. These can be separated through hydrometallurgical, electrometallurgical, or pyrometallurgical treatment. The best option will depend on the composition of the alloy, which is not only produced to maximize the Cu content in the alloy, but also to minimize the Cu lost to the slag. NTUA will perform the hydrometallurgical treatment and electrometallurgical treatment, which are both based on dissolution followed by selective precipitation, while NTNU will perform pyrometallurgical treatment based on vacuum refining.

Rare Earth Elements, Scandium

BR contains significant amounts of REEs, in particular scandium. In both the Ca-route and the Na-route, these valuable materials are expected to concentrate towards the non-ferrous oxide-phase after recovery of the alumina- and iron-rich phases.

From the Ca-route, NTUA will treat remaining oxide-phase after limestone recovery with mineral acids (HNO_3, HCl, or H_2SO_4) and ionic liquids to recover REE and Sc by selective leaching.

From the Na-route, the non-magnetic leachate is a mixture of an amorphous silica-rich phase and a perovskite ($CaTiO_3$) which contains most of the scandium and REE. By dissolving the silica-phase at high temperatures and pressures, KU Leuven will recover the valuable $CaTiO_3$ phase, an "artificial deposit" of REE.

Zn and Pb from Flue Dust

During the hydrogen treatment of the Cu slag, the volatile metals Zn and Pb will end up as oxides in the flue dust. The recovery of zinc and lead from such oxidic powders is part of the common, industrial zinc winning and recycling practices. There, zinc metal or oxide is leached with sulphuric acid, while lead oxide stays unsolved or metallic lead precipitates as lead sulphate. RWTH will demonstrate that the recovered flue dust from the HARARE process is compatible with these recovery techniques.

Electrolytic Refining of Iron from Magnetite

The magnetic phase at the end of the Na-route is rich in magnetite, Fe_3O_4, but cannot be used as a raw material in blast furnaces due to its high Na content. NTUA will use it as a raw material for electrolysis of iron, by using a process similar to that developed in the Horizon2020 project called SIDERWIN [7] in which NTUA and Mytilineos are partners.

Building Materials

Even with the substantial efforts of product- and by-product valorisation, significant fractions from the HARARE value chain will remain as non-metallic residues. In order to determine the best way to valorise these as building materials, KU Leuven will correlate the structure of the residues with their predisposition for dissolution through reactivity protocols, produce high-performance alkali-activated or blended cements, and deliver predictive models on the properties of the binders.

To demonstrate that there are no adverse effects on the surrounding environment of using these materials in construction, SINTEF will investigate the leaching of harmful substances from the valorised residues.

Outlook

The HARARE project offers innovative solutions for materials streams that generate over 160 Mt/year of residues that are either landfilled or used for low-end applications (2020 data). Furthermore, these streams contain valuable metallic and non-metallic phases that are not valorised in any way. The solutions proposed in Harare are based on hydrogen, will thus be free of process-related CO_2 emissions, and look very attractive from an environmental perspective. The ecological sustainability will be verified through LCA, lead by AdMiRIS. Since a positive environmental impact is not sufficient for real-world implementation, AdMiRIS will also evaluate techno-economic viability and value creation assessment, together with SINTEF.

To some members of the general public, hydrogen is associated with safety concerns and fear of explosions. In developing the concepts described herein, both in the laboratory and pilot scale, the expertise of Linde will ensure that such risks are managed in a responsible manner, and that the processes will be safe. Communications of results will be aimed at the general public by popularizing the information in national and local media as well as in social media including screen casts and videos. This will raise awareness among the general public about the applications, potential, and benefits to society of the developed technologies.

The project kicked off in June 2021 and will run for 48 months. We look forward to publishing results as they become available.

Acknowledgements This work has been performed as part of the HARARE project. This project has received funding from the European Union's Horizon2020 research and innovation programme under grant agreement no. 958307.

The authors also acknowledge the input and collaboration of our project partners: AdMiRIS, Aurubis, KU Leuven, Linde, Mytilineos, Norwegian University of Science and Technology (NTNU), The National Technical University of Athens (NTUA), ReSiTEC, RWTH Aachen.

References

1. https://www.eea.europa.eu/data-and-maps/data/data-viewers/greenhouse-gases-viewer
2. Mission Possible (2018) Reaching net-zero carbon emissions from harder-to-abate sectors, report, Energy Transitions Commission
3. HARARE project, Hydrogen as the reducing agent in the recovery of metals and minerals from metallurgical waste. https://h2020harare.eu/
4. Qu G, Wei Y, Li B, Wang H, Yang Y, McLean A (2020) Distribution of copper and iron components with hydrogen reduction of copper slag. J Alloy Compd (824). https://doi.org/10.1016/j.jallcom.2020.153910
5. Lazou A, van der Eijk C, Balomenos E, Kolbeinsen L, Safarian J (2020) On the direct reduction phenomena of bauxite ore using H_2 Gas in a fixed bed reactor. J Sustain Metall (6):227–238. https://doi.org/10.1007/s40831-020-00268-5
6. Borra CR, Blanpain B, Pontikes Y, Binnemans K, Van Gerven T (2017) Recovery of rare earths and major metals from bauxite residue (red mud) by alkali roasting, smelting, and leaching. J Sustain Metall 3(2):393–404 https://doi.org/10.1007/s40831-016-0103-3

7. SIDERWIN- Development of new methodologies for industrial CO_2-free steel production by electrowinning. https://www.siderwin-spire.eu/

Recovery of Lithium from Black Cathode Active Materials of Discarded Lithium-Ion Batteries

Pankaj Kumar Choubey, Rukshana Parween, Rekha Panda, Om Shankar Dinkar, and Manis Kumar Jha

Abstract Lithium (Li) is the lightest energy critical element used in manufacturing of active cathode material of lithium-ion batteries (LIBs). Thus, the consumption of lithium is constantly increasing in the LIBs. Meanwhile, LIBs become obsolete after reaching its end-of-life resulting in the generation of huge amount of spent LIBs. Present study reports the roasting and leaching process for selective recovery of Li from active cathode material. To optimize the process parameters viz. roasting temperature, time, and mass ratio studies were carried out varying the experimental conditions for the conversion of lithium oxide to lithium sulfate from the complex of lithium cobalt oxide. It was found that cathode material converted into lithium sulfate at 750 °C in two hours maintaining mass ratio of $LiCoO_2/Na_2SO_4$: 1/0.5. Subsequently, 99.1% Li was leached from roasted product at 75 °C in de-ionized water within two hours. Further, Li can be precipitated as lithium carbonate using sodium carbonate.

Keywords Spent LIBs · Active cathode material · Roasting · Leaching · Lithium

Introduction

Lithium-ion batteries (LIBs) are mainly used in the electronic products due to their higher energy storage capacity than other rechargeable batteries. Thus, the demand of LIBs is constantly increasing; as a result, massive amount of LIBs becomes obsolete after reaching their end-of-life. Approximately 25 billion units (500 thousand tons) of spent LIBs were generated in 2020 [1]. It has been expected that more than eleven million tons of spent LIBs will become obsolete by the end of 2030 [2]. These spent LIBs contain valuable metals and materials {lithium (Li), copper (Cu), manganese (Mn), cobalt (Co), nickel (Ni)} and materials (graphite) [3]. Therefore, it is required to explore a sustainable process for the extraction of metals from spent

P. K. Choubey · R. Parween · R. Panda · O. S. Dinkar · M. K. Jha (✉)
Metal Extraction and Recycling Division, CSIR-National Metallurgical Laboratory, Jamshedpur 831007, India
e-mail: mkjha@nmlindia.org

A. Lazou et al. (eds.), *REWAS 2022: Developing Tomorrow's Technical Cycles (Volume I)*, The Minerals, Metals & Materials Series,
https://doi.org/10.1007/978-3-030-92563-5_76

LIBs, which will not only minimize the dependency on primary resources but also protect the environment. Usually, LIBs were pre-treated by heat and mechanical or mechano-chemical techniques to concentrate the metallic values for subsequent recovery of metals. Sun and Qiu [4] separated the cathode material by decomposing the binder and electrolyte of spent LIBs using the vacuum pyrolysis method in a range of temperature 500 to 600 °C. The cathode material was further dissolved in 2 M sulfuric acid solution at 80 °C to leach Co and Li [5]. In subsequent study, ascorbic acid was mixed with sulfuric acid to enhance the dissolution efficiency of cobalt and lithium. It was observed that 95.7% Li and 93.8% Co leached with 2 M H_2SO_4 in presence of 0.11 M ascorbic acid at 80 °C [6].

Alternatively, studies were also conducted in organic acids like citric, malic, oxalic, succinic acid, etc. to dissolve the lithium and cobalt from spent LIBs. The 98% Li and 97% Co were dissolved in oxalic acid at 90 °C in 150 min, while cobalt was precipitated as cobalt oxalate [7]. Subsequently, 97.8% Ni, 97.6% Co, 97.3% Mn, and 98% of Li leached in1.0 mol/L DL-malic acid at 80 °C in 30 min with addition of 4 vol% H_2O_2 [8]. The biodegradable trichloroacetic acid (TCA) is further used as a leachant with addition of 4% H_2O_2 as the reducing agent to dissolve Co, Ni, Mn, and Li leaving Al remained in the residue. Chen et al. [9] found that 99% Li, 91% Ni, 92% Co, and 94% Mn were dissolved in 1.5 mol/L citric acid at 80 °C, while D glucose was added as a reducing agent. Further, metals (Cu, Ni, Co, Mn, and Li) were selectively extracted by liquid–liquid extraction to purify the leach liquor for subsequent extraction of metals. Wang et al. [10] used the di-(2-ethylhexyl) phosphoric acid (D2EHPA) for extraction of Mn at phase ratio (O/A): 1/1, and then cobalt was extracted with 2-ethylhexyl hydrogen-2-ethylhexyl phosphonate (PC-88A) at pH 4.25. Shuya et al. [11] extracted Ni, Co, and Mn with neodecanoic acid/ versatic 10 at pH: 6 leaving Li in the solution. Recently, benzoyltrifluoroacetone (HBTA), green kerosene, and trioctylphosphine oxide (TOPO) have been used to separate the nickel, cobalt, and lithium from leach liquor. It was found that 88% Co was extracted with a mixture of extractants (0.2 M TOPO and 0.2 M HBTA) at O/A ratio 1/1 having high separation factor values of cobalt extraction with respect to nickel Co/Ni:57 and lithium Co/Li: 173. The literature review showed that selective extraction of Li from spent LIBs faces challenges due to co-extraction of Co and other metals.

In a view of above aspect, present paper reports the roasting and leaching studies for selective extraction of Li from discarded Li-ion batteries. Studies were performed at various process parameters such roasting temperature, mass ratio of black powder and sodium sulfate to selectively convert the insoluble phase of Li into its soluble species for extraction of lithium. Further, leaching studies were performed at different conditions to optimize the parameters for selective leaching of Li from roasted sample.

Table 1 Composition of dried black powder of spent LIBs (wt. %)

Li	Co	Mn	Ni	Cu	Graphite	Others
3.37	16.17	5.9	1.34	2.7	47.7	22.7

Experimental

Materials

Waste LIBs of mobile phones were used for the experimental studies. At first, discarded LIBs were put in saturated solution of sodium chloride for 4 to 5 h to completely discharge it to avoid any explosion during pre-treatment. The discharge LIBs were processed by mechanical pretreatment (crushing) followed by flotation to separate the black powder containing lithium and cobalt. The black materials of waste LIBs were further dissolved in aqua regia to analyze the metals contents using inductively coupled plasma optical emission spectroscopy (ICP-OES). The typical analysis of black powder used for recovery of Li is presented in Table 1.

Methodology

Roasting Procedure

Roasting studies of black powder were conducted in a temperature controlled muffle furnace in a range of temperature varied from 500 to 800 °C using sodium sulfate as roasting agent. The black powder was mixed with different amount of sodium sulfate in presence of lubricant (low density oil) to homogenize the mixture of black powder and sodium sulfate. Thereafter, mixture of black powder and sodium sulfate was kept in the muffle furnace at fixed temperature to oxidize the lithium oxide to lithium sulfate from the complex of Li-Co oxide. The roasted black powder was further used for the leaching studies to recover the lithium.

Leaching Procedure

Leaching experiments were carried out in a Pyrex leaching reactor (Capacity: 1L) equipped with condenser facility to avoid the loss of solution during heating of solution. A temperature controlled hot plate having sensor facility was used to control the solution temperature throughout the leaching studies of lithium from roasted samples. During the leaching, solution was taken at different interval of time (5 to 60 min) to analyze the Li in the in the solution. The leached residue obtained after the leaching was further dried at 120 °C for 120 min to analyze the material balance.

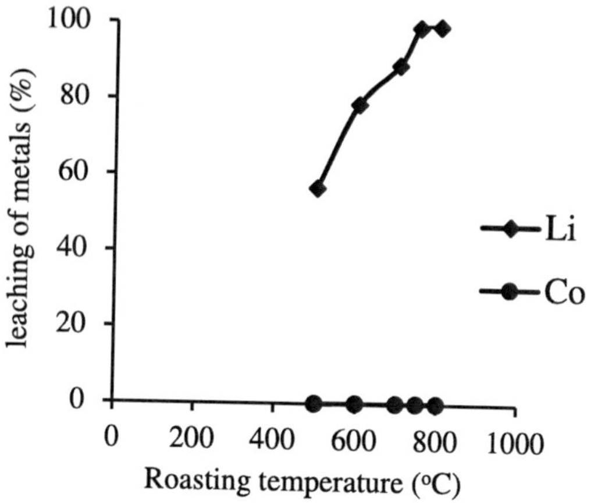

Fig. 1 Roasting of black powder at different temperature [mass ratio of black powder and sodium sulfate: 1/0.5, reaction time: 120 min]

Results and Discussion

Roasting Studies

For selective recovery of lithium from black powder of spent LIBs, roasting studies of black powder were carried out with sodium sulfate at different temperature varied from 500 to 800 °C to transform the lithium cobalt oxide into lithium sulfate. Result showed (Fig. 1) that percentage conversion of lithium cobalt oxide to water soluble lithium sulfate increases with increase in roasting temperature. Approximately, 57% black powder was converted to lithium sulfate at 500 °C in presence of sodium sulfate while complete conversion of black powder to lithium sulfate occurred at 750 °C in 2 h maintaining mass ratio of black powder to sodium sulfate 1/0.5.

Leaching of Lithium From Roasted Black Powder

In order to optimize the parameters for selective leaching of lithium from roasted black powder, leaching experiments were carried out at different parameters such as reaction time, temperature, S/L ratio, etc. as discussed below.

Effect of Reaction Time

In order to optimize the equilibrium time for dissolution of Li from roasted black powder, experiments were conducted at various times varying from 15 to 150 min at 75 °C while maintaining the pulp density 100 g/L. Result (Fig. 2) shows that leaching of Li increases with increase in reaction time, 19.2% Li was dissolved in 15 min,

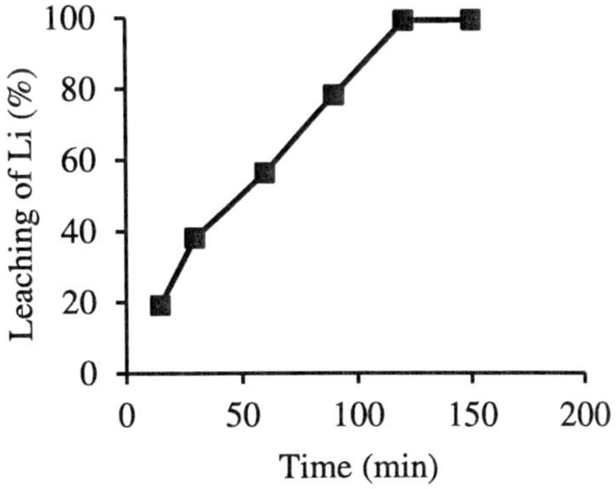

Fig. 2 Effect of reaction time on Li leaching from roasted black powder [leachant: distilled water, solid: 100 g roasted black powder, temperature: 75 °C]

while 99.1% Li dissolved in 120 min at 75 °C. Further, increase in reaction time has no effect on the dissolution efficiency of Li. It indicates that reaction attained the equilibrium in 120 min; therefore, 120 min reaction time has been chosen as optimal reaction time for dissolution of Li from roasted black powder.

Effect of Temperature

Studies were performed at different temperature varying from 30 to 90 °C to optimize the temperature for maximum leaching of Li from roasted black powder. It was found that leaching of Li increases with rise in solution temperature. ~15.2% Li was dissolved at 30 °C in 120 min at pulp density 100 g/L while complete dissolution of Li occurred at 75 °C under same experimental conditions. Leaching of Li was found to increase with rise in the solution temperature is due to increase in rate of reaction. Above 75 °C (Fig. 3), temperature has no major effect on the enhancement of lithium leaching efficiency. Therefore, 75 °C has been chosen as optimum temperature for

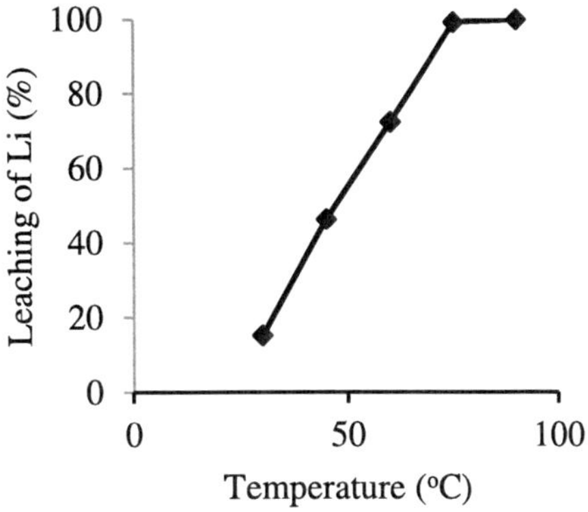

Fig. 3 Effect of temperature on leaching of Li from roasted black powder [leachant: distilled water, solid: 100 g roasted black powder, temperature: 25 to 90 °C, reaction time: 120 min]

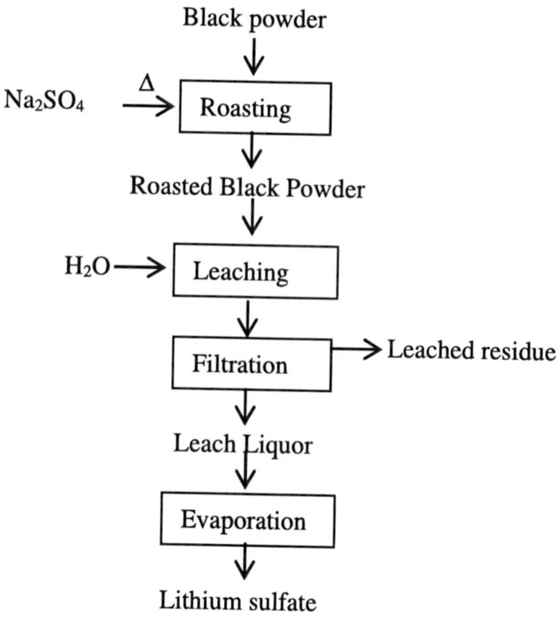

Fig. 4 Developed process flow sheet for selective recovery of Li from spent LIBs

maximum leaching of Li from roasted black powder. Finally, a process flow sheet has been drawn for selective recovery of lithium from spent LIBs as shown in Fig. 4.

Conclusions

On the basis of laboratory scale roasting and leaching studies, the following conclusions have been made for selective recovery of lithium from black powder of LIBs as given below.

1. It was found that water insoluble species of lithium cobalt oxide of black powder converted to water soluble species of lithium at 750 °C in 120 min using sodium sulfate as a roasting agent at mass ratio of black powder to sodium sulfate 1/0.5.
2. More than 99% lithium was selectively dissolved from the roasted black powder at 75 °C in 120 at pulp density 100 g/L while Co, Mn, Cu and Ni remained in the leached residue.
3. The obtained pure solution of Li can be used for recovery of lithium in form of lithium carbonate and lithium sulfate using precipitation or evaporation process.

Acknowledgements The authors are thankful to the Director, CSIR-National Metallurgical Laboratory, Jamshedpur for giving the permission to publish this paper.

References

1. Zhao Q, Hu L, li, W., Liu, C., Jiang, M., Shi, J., (2020) Recovery and Regeneration of Spent Lithium-Ion Batteries from New Energy Vehicles. Front Chem 8:807
2. https://www.orfonline.org/expert-speak/recycling-liion-batteries-opportunities-challenges-68409/.
3. Choubey PK, Dinkar OS, Panda R, Kumari, a., Jha, m.k., Pathak, D.D., (2021) Selective extraction and separation of Li, Co and Mn from leach liquor of discarded lithium ion batteries (LIBs). Waste Manag 21:452–457
4. Sun L, Qui Q, K. (2011) Vacuum pyrolysis and hydrometallurgical process for the recovery of valuable metals from spent lithium-ion batteries. J. Hazar. Mater. 194(30):378–384
5. Jha MK, Kumari A, Jha AK, Kumar V, Hait J, Pandey BD (2013) Recovery of lithium and cobalt from waste lithium ion batteries of mobile phone. Waste Manage 33:1890–1897
6. Peng C, Hamuyuni J, Wilson BP, Lundstrom M (2018) Selective reductive leaching of cobalt and lithium from industrially crushed waste Li-ion batteries in sulfuric acid system. Waste Manag 76:582–590
7. Zeng X, Li J, Shen B (2015) Novel approach to recover cobalt and lithium from spent lithium-ion battery using oxalic acid. J. Hazar. Mater. 15:295112–295118
8. Ning P, Meng Q, Dong P, Duan J, Xu M, Lin Y, Zhang Y (2020) Recycling of cathode material from spent lithium ion batteries using an ultrasound-assisted DL-malic acid leaching system. Waste Mange. 103(15):2–60
9. Chen, X., Chen, Y., Zhou, T., Liu, D., Hu, H., Fan, S., 2015. Hydrometallurgical recovery of metal values from sulfuric acid leaching liquor of spent lithium-ion batteries. Waste Manag. 38 (1), 349e356.
10. Feng, W., Rong, S., Xu, J., Kang, M., Recovery of cobalt from spent lithium ion batteries using sulphuric acid leaching followed by solid–liquid separation and solvent extraction., RSC Adv., 2016, 6, 85303.
11. Shuya, L., Yang, C., Xuefeng, C., Wei, S., Yaqing, W., Yue, Y., 2020. Separation of lithium and transition metals from leachate of spent lithium-ion batteries by solvent extraction method with Versatic 10. Sep. Purif. Technol. 250(1), 117258.

Recovery of Rare Earth Elements from Nd-Fe-B Magnet Through Selective Chlorination Using Zinc Chloride

Kyung-Hwan Lim, Chan Choi, Gyeonghye Moon, Tae-Hyuk Lee, and Jungshin Kang

Abstract The selective recovery of rare earth elements from Nd-Fe-B magnets through a novel selective chlorination process using zinc chloride was investigated. A Nd-Fe-B magnet powder and zinc chloride mixture in an alumina crucible was positioned in a gas-tight quartz tube. This quartz tube was placed in an electric furnace preheated to 1000 K for 1.5 h for the reactions. After the experiments, a mixture of metallic iron and neodymium chloride was produced owing to the selective chlorination of rare earth elements in the magnet powder. In addition, the chlorination efficiencies of neodymium, dysprosium, and praseodymium were 96.5%, 57.2%, and 97.6%, respectively, under certain conditions. Therefore, it was demonstrated that the novel selective chlorination using zinc chloride developed in this study is feasible for the efficient recycling of Nd-Fe-B magnets.

Keywords Nd-Fe-B magnet · Selective chlorination · Recovery of rare earth elements

Introduction

Neodymium (Nd)–iron (Fe)–boron (B) magnets with addition of small amount of dysprosium (Dy) have excellent properties such as high coercivity even at high temperatures [1]. Because of these properties, Nd magnets can be used in various applications such as hybrid and electric cars. The demand for green cars is expected

K-H. Lim · C. Choi · T-H. Lee · J. Kang (✉)
Korea Institute of Geoscience and Mineral Resources, 124 Gwahak-ro Yuseong-gu, Daejeon 34132, Korea
e-mail: jskang@kigam.re.kr

G. Moon
HANNAE For T, 22 Sandan 3-ro, 5-gil, Seongmun-myeon, Dangjin-si, Chungcheongnam-do, Korea

J. Kang
University of Science and Technology, 217 Gajeong-ro Yuseong-gu, Daejeon 34113, Korea

A. Lazou et al. (eds.), *REWAS 2022: Developing Tomorrow's Technical Cycles (Volume I)*, The Minerals, Metals & Materials Series,
https://doi.org/10.1007/978-3-030-92563-5_77

to increase in the future to meet the need of carbon dioxide (CO_2) gas emission reduction; therefore, the utilization of Nd magnets will increase.

Rare earth elements (REEs) are extracted from several mineral resources such as bastnaesite, monazite, and ion adsorption deposits. Among REEs, light rare earth elements (LREEs), such as Nd and praseodymium (Pr), are extracted from bastnaesite and monazite. These mineral resources are abundant in nature and also distributed around the world [2]. However, heavy rare earth elements (HREEs), such as Dy and terbium (Tb), can only be extracted economically from ion adsorption deposits in the southern parts of China [2]. This is the one reason that China monopolizes the world production of REEs.

During the extraction of LREEs from monazite, environmental threats and radiation risks are encountered, because monazite contains radioactive elements such as uranium (U) and thorium (Th) [2]. Therefore, proper treatments of these radioactive elements generated during the extraction are necessary to preserve the environment. In contrast, ion adsorption deposits contain very little amount of radioactive elements because these elements are washed out because of weathering [2]. However, there are still environmental concerns during the extraction of REEs because they are extracted via elution using ammonium sulfate ($(NH_4)_2SO_4$).

To produce REEs with preservation of environment, recycling of spent Nd magnets is required. However, only Nd scraps generated in magnet production plants are commercially used as feedstock for recycling because these scraps contain a large amount of REEs and processing costs are minimum [2, 3]. Although spent Nd magnets are not recycled at present, it is necessary to develop an efficient and environmentally sound process for the recovery of REEs from spent magnets when the increase of the use of Nd magnets is considered.

Numerous methods have been developed to recover REEs from spent magnets [4–9]. Among those, the selective recovery of REEs through the chlorination of spent magnets using metal chloride is promising [8, 9]. When the selective chlorination of spent magnets using ferrous chloride ($FeCl_2$) or magnesium chloride ($MgCl_2$) is utilized, only REEs are recovered as chloride without chlorination of Fe in magnets. Shirayama reported that Nd in a spent magnet was selectively chlorinated to neodymium chloride ($NdCl_3$) by reacting with $MgCl_2$ at 1273 K [8]. In addition, approximately 80% of Nd and Dy was recovered from the initial magnet after the reactions for 12 h [8]. Hua reported that the rate-determining step of the selective chlorination using $MgCl_2$ is the diffusion of $MgCl_2$ into an ash layer consisting of Fe metal remained [10].

In this study, a selective chlorination process using zinc chloride ($ZnCl_2$) was investigated to recover REEs from Nd magnets, as shown in Fig. 1 [11]. When $ZnCl_2$ is used as the chlorinating agent, reaction temperature can be decreased because of its low melting point. In addition, the reaction rate can be increased because low melting temperature of zinc (Zn) metal produced by the chlorination reaction is alloyed with the ash layer of Fe metal. Furthermore, rare earth chlorides such as $NdCl_3$ can be separated from the residue obtained after selective chlorination by utilizing different vapor pressures at high temperatures.

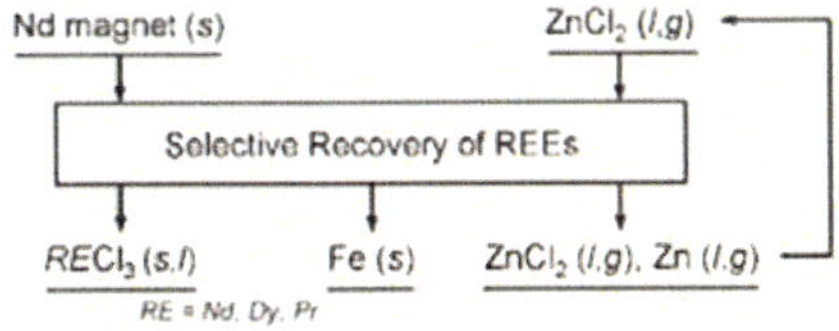

Fig. 1 Flowchart of the selective chlorination of Nd magnet using $ZnCl_2$ investigated in this study

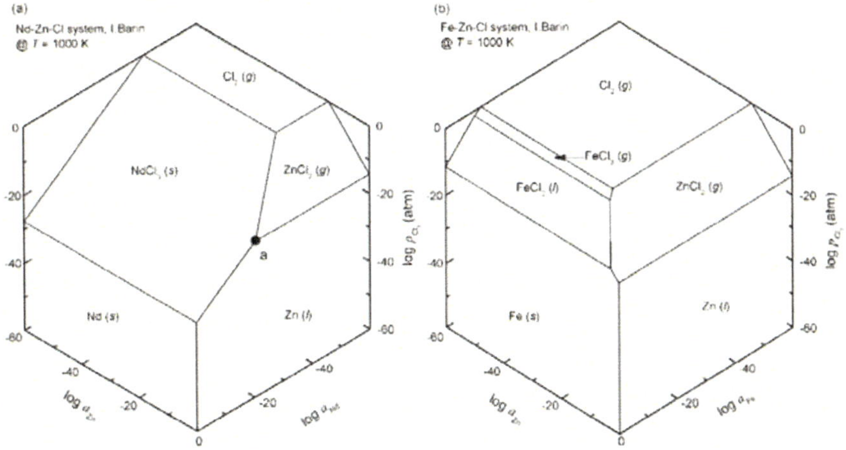

Fig. 2 Chemical potential diagrams of the **a** Nd–Zn–Cl and **b** Fe–Zn–Cl systems at 1000 K (the boiling point of $ZnCl_2$ is 999.5 K [12] or 1005 K [14] depending on the reference consulted, Reprinted by permission from Springer Nature: Springer, *Journal of Sustainable Metallurgy*, Selective Chlorination of Rare Earth Elements from a Nd–Fe–B Magnet Using Zinc Chloride [15]

Mechanisms of Selective Chlorination of Nd-Fe-B Magnet Using $ZnCl_2$

Figure 2a, b show the chemical potential diagrams of Nd–Zn–Cl and Fe–Zn–Cl systems, respectively, at 1000 K [12, 13]. $ZnCl_2$ (*l*,*g*) equilibrates with Zn (*l*) and $NdCl_3$ (*s*) at point a in Fig. 2a while Nd (*s*) does not coexist with $ZnCl_2$ (*l*,*g*) at 1000 K. Therefore, when a sufficient amount of $ZnCl_2$ is used as a chlorinating agent, Nd in the magnets will be chlorinated to $NdCl_3$ by reacting with $ZnCl_2$ at 1000 K, as shown in Eq. (1). In addition, Dy and Pr in the magnets also will be chlorinated to dysprosium chloride ($DyCl_3$) and praseodymium chloride ($PrCl_3$) by reacting with $ZnCl_2$ at 1000 K, as shown in Eqs. (2) and (3), respectively. On the other hand, the chemical equilibrium point for $FeCl_x$ ($x = 2,3$, *l*,*g*), $ZnCl_2$ (*l*,*g*), and Zn (*l*) does not exist, as shown in Fig. 2b. As a result, even though a sufficient amount of $ZnCl_2$ is used for the reactions, Fe in magnets will not be chlorinated

$$2\text{Nd}(s,\ \text{in Nd-Fe-B magnet}) + 3\text{ZnCl}_2(l, g) = 2\text{NdCl}_3(s) + 3\text{Zn}(l)$$
$$\Delta G_r^\circ = -785.0\ \text{kJ at 1000 K [12]} \tag{1}$$

$$2\text{Dy}(s,\ \text{in Nd-Fe-B magnet}) + 3\text{ZnCl}_2(l, g) = 2\text{DyCl}_3(l) + 3\text{Zn}(l)$$
$$\Delta G_r^\circ = -676.0\ \text{kJ at 1000K [12]} \tag{2}$$

$$2Pr\ (s,\ \text{in Nd-Fe-B magnet}) + 3\text{ZnCl}_2(l, g) = 2\text{PrCl}_3(s) + 3\text{Zn}(l)$$
$$\Delta G_r^\circ = -811.6\ \text{kJ at 1000 K [12]} \tag{3}$$

Experimental

Figure 3 shows a schematic of the experimental apparatus, and Table 1 lists the experimental of the selective chlorination process using $ZnCl_2$ at 1000 K for 1.5 h. A mixture of Nd-Fe-B magnet powder and anhydrous $ZnCl_2$ placed in an alumina (Al_2O_3) crucibles was positioned in a quartz holder inside a glove box. Afterward, the quartz holder was removed from the glove box and placed at the bottom of a gas-tight quartz reactor. The gas inside the reactor was replaced with Ar gas through evacuation and filling of Ar gas. The pressure inside the reactor was maintained at 1 atm during the experiments by flowing Ar gas. After the completion of atmosphere control, the reactor was placed in an electric furnace, which was preheated to 1000 K. When the reaction time was finished, the reactor was immediately removed from the furnace and cooled to room temperature. The samples obtained after the selective chlorination reaction were subjected to deionized (DI) water leaching at room temperature for 0.5 h, and the residues obtained after the leaching were dried at 343 K for 0.5 h.

Table 1 Experimental conditions of the selective chlorination process using $ZnCl_2$ at 1000 K for 1.5 h

Exp. no[a]	Weight of magnet, w_{magnet}/g	Particle size, d_{magnet}/μm	Reaction time, t_r/h
190812b-1	1.000	<75	1.5
190812b-2	1.003	75–150	1.5
190812b-3	1.001	150–300	1.5
190812b-4	1.000	300–600	1.5

a Experimental conditions;
Weight of zinc chloride, $w\ ZnCl_2 = 2.72$ g
Reaction temperature, $T = 1000$ K

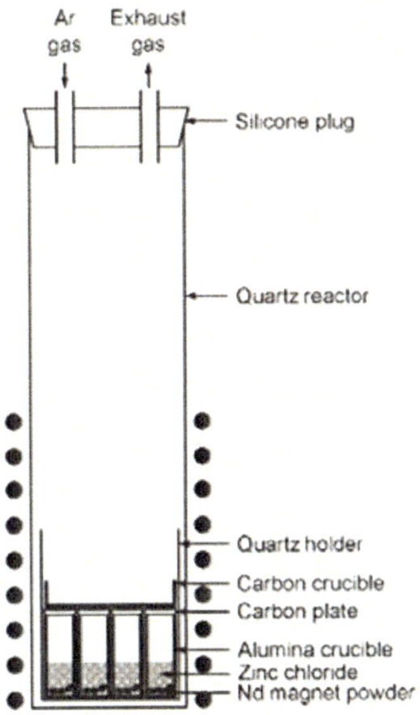

Fig. 3 Schematic of the experimental apparatus (Reprinted by permission from Springer Nature: Springer, *Journal of Sustainable Metallurgy*, Selective Chlorination of Rare Earth Elements from a Nd-Fe-B Magnet Using Zinc Chloride, Kyung-Hwan Lim et al. 2021)

Results and Discussion

Table 2 lists the composition of the Nd magnet used as feedstock. In addition, Fig. 4 shows the results of XRD analysis of the residues obtained after the selective chlorination reactions without water leaching. As expected from the thermodynamic analysis, $NdCl_3$ and Fe (Zn) were identified after the experiments. $DyCl_3$ and $PrCl_3$ were not identified because the feedstock contains low concentrations of these elements, as shown in Table 2.

Further, Fig. 5 shows the influence of the particle size of the feedstock on the chlorination efficiencies of Nd, Dy, and Pr after the experiments. When the particle size of the feedstock used for selective chlorination using $ZnCl_2$ increased from <75 μm to 300–600 μm, the chlorination efficiencies of Nd, Dy, and Pr increased from 93.8%, 36.3%, and 96.0% to 96.5%, 57.2%, and 97.6%, respectively. In addition, the total chlorination efficiency of Nd, Dy, and Pr was 92.1% when the particle size of the feedstock was 300–600 μm. As the particle size decreased, the chlorination efficiencies of REEs decreased because the concentration of oxygen in the feedstock increased from 0.188 mass% at 300–600 μm to 0.467 mass% at <75 μm. REEs in

Table 2 Composition of the Nd-Fe-B magnet used as feedstock

	Concentration of element *i*, C_i (mass%)[a]			
	Nd	Dy	Pr	Fe
Nd magnet feed	20.7	3.71	6.62	66.5

[a] Determined by ICP-MS and ICP-OES analysis

the magnet powder feedstock easily react with oxygen during processing because these elements are reactive. However, the oxides of REEs could not be chlorinated to rare earth chlorides by reacting with $ZnCl_2$ at 1000 K. For example, when the oxides of Nd and Dy react with $ZnCl_2$ at 1000 K, neodymium oxychloride (NdOCl) and dysprosium oxychloride (DyOCl) are produced, as shown in Eqs. (4) and (5), respectively [12, 15]. As a result, the chlorination efficiencies of REEs increased when the particle size of the feedstock was increased.

$$\begin{aligned} &Nd_2O_3(s) + ZnCl_2(l, g) = 2NdOCl(s) + ZnO(s) \\ &\Delta G_r^\circ = -81.3 \text{ kJ at } 1000\text{K}\,[12] \end{aligned} \tag{4}$$

$$\begin{aligned} &Dy_2O_3(s) + ZnCl_2(l, g) = 2DyOCl(s) + ZnO(s) \\ &\Delta G_r^\circ = -4.2 \text{ kJ at } 1000\text{K}\,[12, 15] \end{aligned} \tag{5}$$

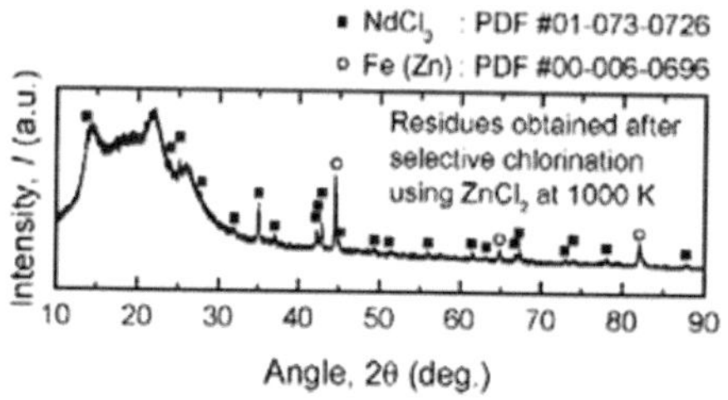

Fig. 4 Results of XRD analysis of the residues obtained after the selective chlorination reactions (Reprinted by permission from Springer Nature: Springer, *Journal of Sustainable Metallurgy*, Selective Chlorination of Rare Earth Elements from a Nd-Fe-B Magnet Using Zinc Chloride, (Kyung-Hwan Lim et al. 2021)

Fig. 5 Influence of particle size of the feedstock on the chlorination efficiencies of Nd, Dy, and Pr

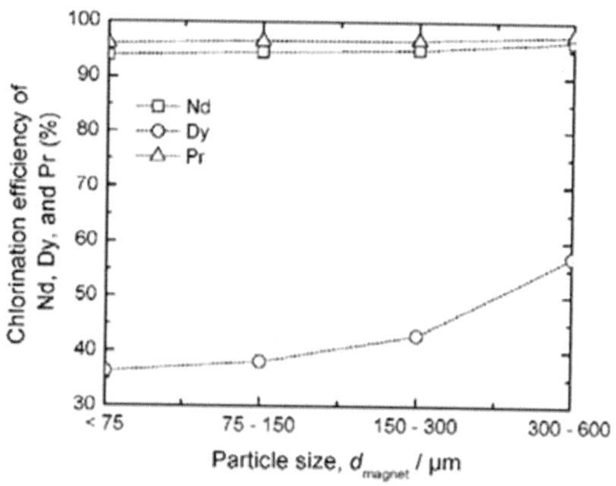

Conclusions

The selective chlorination process using $ZnCl_2$ to recover Nd, Dy, and Pr from Nd magnets was investigated in this study. When the Nd magnet reacted with $ZnCl_2$ at 1000 K, Nd in the feedstock chlorinated to $NdCl_3$; however, Fe in the feedstock did not. In addition, the chlorination efficiencies of Nd, Dy, and Pr reached 96.5%, 57.2%, and 97.6%, respectively, when the particle size of the feedstock used for the chlorination reaction at 1000 K for 1.5 h was 300–600 μm. As a result, 92.1% of REEs was recovered from the Nd magnets through the selective chlorination process.

Acknowledgements This research was supported by the Korea Evaluation Institute of Industrial Technology, the Korean Ministry of Industry (Project No. 20000970, 21-9805)

References

1. Haider SK, Lee J-Y, Kim D, Kang YS (2020) Eco-friendly facile three-step recycling method of $(Nd\text{-}RE)_2Fe_{14}B$ magnet sludge and enhancement of $(BH)_{max}$ by ball milling in ethanol. ACS Sustainable Chem Eng 8:8156–8163
2. Takeda O, Okabe TH (2014) Current status on resource and recycling technology for rare earths. Metall Mater Trans E 1:160–173
3. Akahori T, Miyamoto Y, Saeki T, Okamoto M, Okabe TH (2017) Optimum conditions for extracting rare earth metals from waste magnets by using molten magnesium. J Alloys Compd 703:337–343
4. Okabe TH, Takeda O, Fukuda K, Umetsu Y (2003) Direct extraction and recovery of neodymium metal from magnet scrap. Mater Trans 44:798–801
5. Takeda O, Okabe TH, Umetsu Y (2006) Recovery of neodymium from a mixture of magnet scrap and other scrap. J Alloys Compd 408–412:387–390
6. Kobayashi S, Kobayashi K, Nohira T, Hagiwara R, Oishi T, Konishi H (2011) Electrochemical formation of Nd-Ni alloys in molten $LiF\text{-}CaF_2\text{-}NdF_3$. J Electrochem Soc 158:E142–E146
7. Saito T, Sato H, Ozawa S, Yu J, Motegi T (2003) The extraction of Nd from waste Nd-Fe-B alloys by the glass slag method. J Alloys Compd 353:189–193
8. Shirayama S, Okabe TH (2018) Selective extraction and recovery of Nd and Dy from Nd-Fe-B magnet scrap by utilizing molten $MgCl_2$. Metall Mater Trans B 49:1067–1077
9. Uda T (2002) Recovery of rare earths from magnet sludge by $FeCl_2$. Mater Trans 43:55–62
10. Hua Z, Wang J, Wang L, Zhao Z, Li X, Xiao Y, Yang Y (2014) Selective extraction of rare earth elements from NdFeB scrap by molten chlorides. ACS Sustainable Chem Eng 2:2536–2543
11. Lim K-H, Choi C, Moon G, Lee T-H, Kang J (2021) Selective Chlorination of Rare Earth Elements from a Nd-Fe-B Magnet Using Zinc Chloride. J Sustain Metall 7:794–805
12. Barin I (1995) Thermochemical data of pure substances. VCH Verlagsgesellschaft mbH, Weinheim
13. Hatada N (2014) Chesta: Software for Creating Chemical Potential Diagram, version 3.2.6.9, http://www.aqua.mtl.kyoto-u.ac.jp/wordpress/chestaEng.html
14. Windholz M, Budavari S, Blumetti RF, Otterbein E.S (1983) The Merch Index – An encyclopedia of chemicals, drugs, and biologicals, 3rd ed. Rahway N.J. USA, pp. 9932
15. Jacob KT, Dixit A, Rajput A (2016) Stability field diagrams for Ln-O-Cl systems. Bull Mater Sci 39:603–611

Recycling of Automobile Discarded Ceramic Converters for Pt-Group Metals' Recovery Through Pressure CN-Leaching

Sadia Ilyas, Hyunjung Kim, and Rajiv Ranjan Srivastava

Abstract Recycling of automobile catalytic converters was studied using the cyanidation at an elevated temperature and pressure that dissolving Pt-group metals (PGMs) followed by their liquid–liquid separation with ionic-liquid Cyphos IL101. PGMs were efficiently dissolved by autoclaving of the spent catalysts sample preconditioned in 2.0 M NaOH solution at 90 °C for 60 min duration. The cyanide used in the study was metabolically produced by *Chromobacterium violaceum*. Under the optimized autoclave leaching process at temperature 150 °C, pO_2 200 psi, and time 120 min, approximately 90% PGMs could be dissolved. The residual cyanide can be subjected to the biodegradation within the bacterial life-cycle.

Keywords Automotive catalysts · Platinum group metals · Pressure leaching · Biocyanidation

Introduction

Autocatalytic converters are refractory in nature due to the large mass fraction (on average about 43 wt.%) of cordierite ceramic ($Mg_2Al_4Si_5O_{18}$) therein. They usually applied to provide the stable basic skeleton in honeycomb shape of the catalytic converters [1]. A thin porous layer of γ-Al_2O_3 washcoated onto cordierite provides

S. Ilyas (✉) · H. Kim
Department of Mineral Resources and Energy Engineering, Jeonbuk National University, Jeonju, Jeonbuk 54896, Republic of Korea
e-mail: sadiailyas1@yahoo.com

H. Kim
e-mail: kshjkim@jbnu.ac.kr

H. Kim
Department of Environment and Energy, Jeonbuk National University, Jeonju, Jeonbuk 54896, Republic of Korea

R. R. Srivastava
Center for Advanced Chemistry, Institute of Research and Development, Duy Tan University, Da Nang 550000, Vietnam

A. Lazou et al. (eds.), *REWAS 2022: Developing Tomorrow's Technical Cycles (Volume I)*, The Minerals, Metals & Materials Series,
https://doi.org/10.1007/978-3-030-92563-5_78

the effective surface area for dispersing the active metals for the catalytic conversion of exhausted emissions to their less harmful substance (i.e., CO_2, N_2, and H_2O) [2, 3]. Due to the exhibition of desired oxidation performance along with a better thermal stability, the platinum group metals (PGMs that includes Pt, Pd, and Rh) are applied since the inception of automotive catalysts for capturing the gaseous emissions [4, 5]. The indispensable role of PGMs in autocatalytic converters are leading to consume about 35% Pt, and over 80% Pd and Rh of their total market demands in 2019 [6]. Henceforth, PGMs' recycling from exhausted catalytic converters is imperative due to a fact that their amount in exhausted catalysts (about a total of 0.1 wt.%) is much higher than the primary ores [7, 8]. PGMs' recycling coupled with the concept of circular economy can lead to dissociate the economic development from the use of natural minerals. [8]

In the recent times, biotechnology has emerged as a good alternative with low carbon footprint in extractive metallurgy from ores [9, 10]; however, infant in the field of PGMs' recycling from refractory materials like autocatalytic converters as they are hard-to-leach noble metals. Although literature survey reveals that PGMs' property to form cyanide complexes has been utilized by some researchers with low efficiency of Pt and Pd about 80–85%, and Rh around 70–75% [11, 12]. To the best of our knowledge, no effective biocyanidation-leaching of PGMs is reported yet except some simulated chemical leaching with NaCN instead of using the actual biogenic cyanide in the system, which has been examined in this study under the elevated temperature and partial oxygen pressure (pO_2).

Experimental

The spent catalytic converter sample that containing Pt 756 ppm, Pd 367.5 ppm, and Rh 170.8 ppm was used in this study. A cyanide-producing bacterium, *C. violaceum* DSM30191^T maintained in YP medium at 30 °C was accumulation for growth and biogenic cyanide production. After 16 days of cyanide entrapped in NaOH solution, a total of 5.8 g/L NaCN was obtained in the solution of pH value 11.6, which was used as lixiviant for PGMs' pressure leaching in autoclave (500 mL Parr autoclave with (Ni–Cr)Alloy 600 reactor). 40 g sample in 200 mL biogenic cyanide solution of predetermined concentration was charged in the reactor under the fixed agitation speed of 200 rpm and duration of 1 h, while the temperature, pressure, and NaCN concentration was varied. To vary the NaCN concentration, duration of biocyanidation process was accordingly adjusted along with maintaining the pH value 11.6 as invariable. The oxygen pressure was maintained by externally supplying the industrial grade O_2 after reaching at the desired temperature and considering the *in*-situ pressure on that temperature. At the end of the experiment, the oxygen supply was stopped and the autoclave was water-cooled. Then, the solution was filtered using a 0.45 μm membrane filter and collecting the filtrate for PGMs' analysis using an inductively coupled plasma mass spectrometry (ICP-MS; MSS 01, SPECTRO

Analytical Instruments GmbH, Germany). All leaching tests were carried out in triplicates, and the average results are reported herein.

Results and Discussion

Effect of Temperature

Temperature has a direct relation with pressure for the enhancement of rate of reaction. Hence, pressure leaching of PGMs using biogenic cyanide lixiviant was first investigated at different temperatures elevated in the rage of 100–200 °C, while keeping the other parameters constant (NaCN concentration, 5.8 g/L; pO_2, 150 psi; initial pH, 11.6; and time, 60 min). Results shown in Fig. 1 depict a positive effect of temperature on PGMs' dissolution. As can be seen, at the elevated temperature of 100 °C and oxygen pressure of 150 psi, PGMs' dissolution in cyanide solution was not more than 85%. Dissolution efficiency reached to the maximum of 91% Pt, 94% Pd, and 87% Rh at 150 °C, thereafter, increasing temperature showed a decline in PGMs' dissolution (87% Pt, 81% Pd, and 84% Rh at 150 °C). Comparatively, Pt and Rh dissolution decreased slightly in contrast to a steep decline of Pd, which can be corroborated to non-stability of $Pd(CN)_4$ [2–12] and cyanide decomposition at higher temperatures [13].

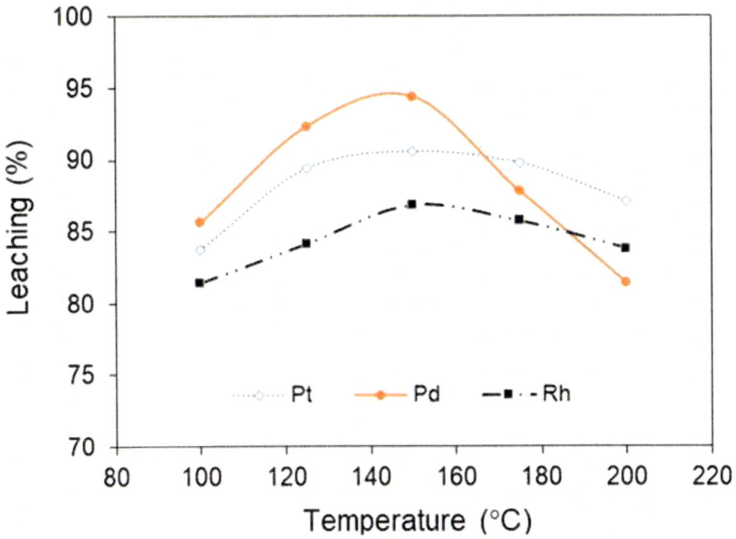

Fig. 1 Effect of temperature on pressure leaching behaviour of PGMs from NaOH-preconditioned autocatalytic converter at NaCN concentration, 5.8 g/L; pO_2, 150 psi; biocyanide solution pH, 11.6; and time, 60 min

Effect of Partial Oxygen Pressure

For autoclaving, in particular to dissolve difficult-to-leach metals from a refractory material, the role of pressure becomes very important to liberate the desired metals into solution. Hence, PGMs' dissolution in biogenic cyanide was examined at various oxygen pressures in the range of 50–250 psi at a fixed temperature of 150 °C. Results in Fig. 2 depict an increasing trend for PGMs' dissolution with elevating oxygen pressure to the closed system. The biocyanide leaching of PGMs was found to increase significantly (Pt, from 83 to 94%; Pd, from 86 to > 6%; and Rh, from 76 to ~ 90%) with change in pO_2 from 50 to 200 psi. This behaviour can be understood by breaking the surface chemical barrier to attack directly on PGMs particles by the biogenic cyanide solution that enhance the dissolution efficiency [8]. Moreover, the catalytic role of oxygen in cyanide can also be corroborated to the exhibited behaviour through the cyanidation reactions as given in Eqs. (1–3) [13]. A slight decrease in Pt and Rh efficiency observed at pO_2 of 250 psi presumably due to the surface oxide passivation of both. Since the maximum dissolution was achieved with a pO_2 of 200 psi, it was maintained throughout the experimental sets.

$$2Pt + 8NaCN + O_2 + 2H_2O \leftrightarrow 2Na_2[Pt(CN)_4] + 4NaOH \quad (1)$$

$$2Pd + 8NaCN + O_2 + 2H_2O \leftrightarrow 2Na_2[Pd(CN)_4] + 4NaOH \quad (2)$$

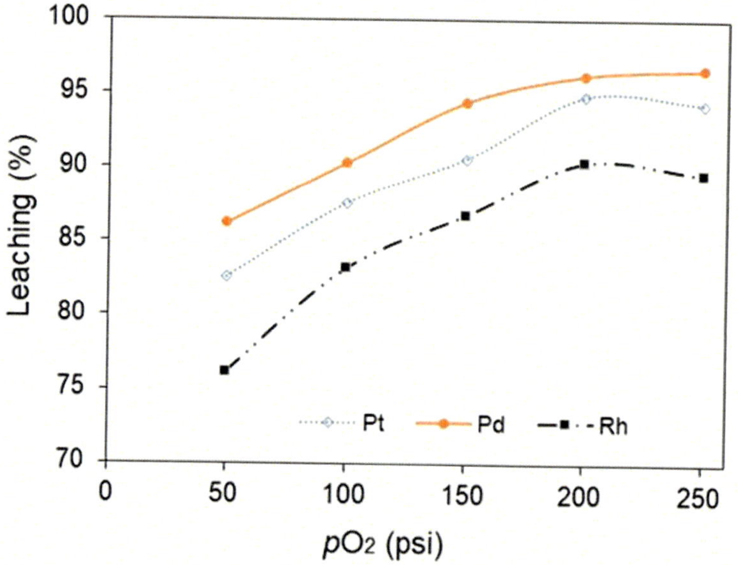

Fig. 2 Effect of oxygen pressure on autoclave leaching of PGMs from NaOH-preconditioned autocatalytic converter at temperature, 150 °C; NaCN concentration, 5.8 g/L; biocyanide solution pH, 11.6; and time, 60 min

$$4Rh + 24NaCN + 3O_2 + 6H_2O \leftrightarrow 4Na_3[Rh(CN)_6] + 12NaOH \quad (3)$$

Effect of NaCN Concentration

To form PGMs-cyanide complex, the optimization of cyanide concentration for driving the Eqs. (5–7) was carried out in the range of 0.63 g/L to 6.43 g/L NaCN at the constant temperature (150 °C) and pO_2 (200 psi). As can be seen from Fig. 3, PGMs' dissolution progressed with increasing concentration of cyanide as the lixiviant medium. The biocyanide leaching of PGMs was found to change significantly (Pt, from 73% to > 95%; Pd, from 82 to 96%; and Rh, from 68% to ~ 91%) with change in NaCN concentration from 0.63 g/L to 6.43 g/L. This change in behaviour of PGMs' dissolution indicates a change in the rate mechanism from the rate of diffusion at low cyanide concentration to the chemically controlled at a higher cyanide in solution [8, 14, 15]. The maximum leaching efficiency having been achieved at 5.82 g/L NaCN solution captured through the microbial-driven HCN synthesis was optimized, at which, the PGMs' concentration was found to be Pt, 157.5 mg/L; Pd, 77.6 mg/L; and Rh, 33.9 mg/L.

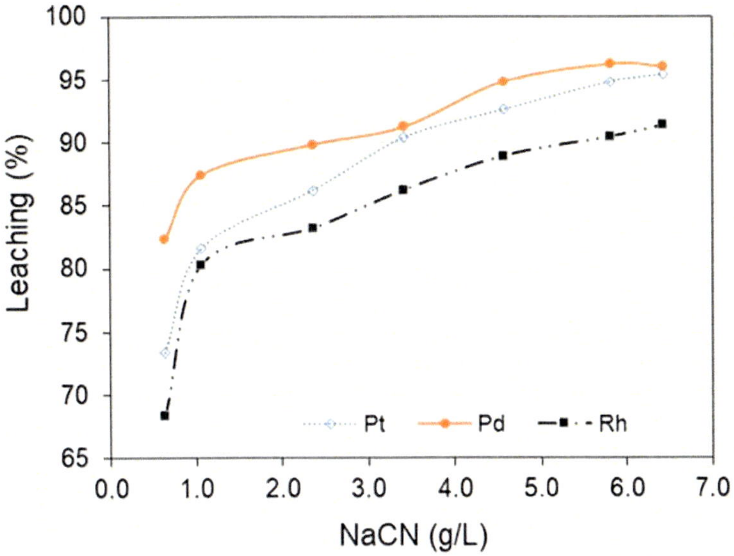

Fig. 3 Effect of NaCN concentration on pressure leaching of PGMs from NaOH-preconditioned autocatalytic converter at temperature, 150 °C; pO_2, 200 psi; biocyanide solution pH, 11.6; and time, 60 min

Conclusion

In the present study, a pressure autoclave leaching of PGMs from catalytic converters was examined that yielded 91% Pt, 94% Pd, and 87% Rh at 150 °C. Leaching efficiency showed progress at 250 psi pO_2 and 5.8 g/L NaCN concentration, yielding 94% Pt, > 96% Pd, and ~ 90% Rh. This behaviour can be understood by breaking the surface chemical barrier to attack directly on PGMs particles by the biogenic cyanide solution that enhance the dissolution efficiency. Moreover, the catalytic role of oxygen in cyanide system could also be established, whereas the change in PGMs' leaching behaviour with cyanide concentration indicated a change in the rate mechanism from the rate of diffusion at low cyanide concentration to the chemically controlled at a higher cyanide in the solution.

Acknowledgements This work was supported by the Brain Pool Program through the National Research Foundation of Korea (NRF) funded by the Ministry of Science and ICT (Grant No. 2019H1D3A2A02101993 and 2021H1D3A2A01100016) and Basic Science Research Program through the National Research Foundation of Korea (NRF) funded by the Ministry of Education (Project No. 2020R1I1A1A01074249).

Conflict of Interest The authors declare that they have no conflict of interest.

References

1. Ilyas S, Srivastava RR, Kim H, Cheema HA (2020) Sep Purif Technol 248, 117029
2. Singh MP, Raj S (2016) IJSER 7:927
3. Lox ES (2008) Automotive exhaust treatment.. In: Ertl G, Knözinger H, Schüth F, Weitkamp J (eds) Handbook of heterogeneous catalysis, p 2274. KGaA: Wiley VCH Verlag GmbH & Co., Weinheim
4. Russell A, Epling WS (2011) Catal Rev 53:337
5. Twigg MV (2006) Catal Today 117:407
6. Johnson Matthey PGM market report, February 2020. http://www.platinum.matthey.com/services/market-research/pgm-market-reports (accessed 26 April 2020).
7. Steinlechner S, Antrekowitsch J (2013) PGM recycling from catalysts in a closed hydrometallurgical loop with an optional cerium recovery. In: REWAS 2013: enabling materials resource sustainability (eds.) Kvithyld A, Meskers C, Kirchain R, Krumdick G, Mishra B, Reuter M, Wang C, Schlesinger M, Gaustad G, Lados D, Spangenberger J Cham: Springer, p 361
8. Ilyas S, Kim H, Srivastava RR (2021) Sustainable Urban Mining of Precious Metals. CRC Press, Boca Raton
9. Ilyas S, Chi R, Bhatti HN, Bhatti IA, Ghauri MA (2012) Bioprocess Biosyst Eng 35, 433
10. You J, Solongo SK, Gomez-Flores A, Choi S, Zhao H, Urik M, Ilyas S, Kim H (2020) Bioresour Technol 307, 123181
11. Atkinson GB (1992) US Patent 5160711
12. Desmond DP (1991) High-temperature cyanide leaching of platinum group metals from automobile catalysts — laboratory test. in Report of Investigations, vol. RI-9384, United States Bureau of Mines
13. Chen J, Huang K (2006) Hydrometallurgy 82:164

14. Habashi F (1969) Principles of Extractive Metallurgy, vol I. Gordon and Breach, Science Publishers Inc., New York
15. Munir H, Srivastava RR, Kim H, Ilyas S, Khosa MK, Yameen B (2020) J Chem Technol Biotechnol 95:2286

Selective Separation of Molybdenum from Leaching Solution of Spent Catalyst by Solvent Extraction with TBP

Kunpeng Shi, Yanfang Huang, Guihong Han, and Shengpeng Su

Abstract Separation and recovery of molybdenum and vanadium from leaching solution of spent catalyst are of great significance for resource utilization of spent catalyst. In this work, selective separation of molybdenum was comprehensively investigated by tributyl phosphate (TBP) with the emphasis on the hydrochloric acid concentration, TBP concentration, reaction time, and oil–water ratio. The results indicated that molybdenum was effectively extracted by 40% (v/v) TBP. More than 95.2% molybdenum was extracted, while the extraction percentage of vanadium was negligible under the optimal conditions (hydrochloric acid concentration of 2 mol/L, reaction time of 10 min, and phase ratio (O/A) of 1/2). Furthermore, the molybdenum in loaded organic phase can be stripped using 0.1 mol/L sodium hydroxide, which achieves the selective separation of molybdenum and vanadium.

Keywords Selective separation · Molybdenum · Vanadium · Solvent extraction

Introduction

Molybdenum and vanadium are rare strategic elements on Earth and coexist with nickel, tungsten, copper, and other elements in nature [1]. They are important rare elements [2]. Molybdenum and vanadium have high melting point, high wear resistance, high tensile strength and other properties, aerospace, catalyst, steel industry, and other high-tech fields are widely used as important metals [3]. Molybdenum and vanadium coexist with other worth materials in diverse resources [4]. Since their primary resources are depleting gradually, recovery of them from different secondary resources is under exploration by the metallurgical industries. These secondary resources include different solid wastes such as spent catalysts, waste batteries, waste electronic equipment, municipal solid wastes, discarded scraps, and slags [5–7]. For example, the deactivated spent oil hydrotreating catalyst is rich in many valuable metal elements such as vanadium, molybdenum and aluminium [7,

K. Shi · Y. Huang · G. Han (✉) · S. Su
School of Chemical Engineering, Zhengzhou University, 450001 Zhengzhou, P.R. China
e-mail: hanguihong@zzu.edu.cn

A. Lazou et al. (eds.), *REWAS 2022: Developing Tomorrow's Technical Cycles (Volume I)*, The Minerals, Metals & Materials Series,
https://doi.org/10.1007/978-3-030-92563-5_79

Table 1 A summary on the solvent extraction of Mo and V by different extractants

Extractant	Aqueous solution	Selectivity	Reference
D2EHPA	H_2SO_4, pH = 0.8, Mo = 4.45 g/L, V = 0.82 g/L	Mo over V	[17]
D2EHPA + TBP	HNO_3, pH = 0.4, Mo = 9 g/L, V = 0.1 g/L	Mo over V	[11]
Cyphos IL 101	H_2SO_4, pH = 0.5, Mo = 0.5 g/L, V = 0.5 g/L	Mo over V	[18]
Aliquat 336	NaOH, pH = 8.5, Mo = 20 g/L, V = 1 g/L	V over Mo	[8]
LIX-84	H_2SO_4, pH = 0.6, Mo = 0.6 g/L, V = 9 g/L	V over Mo	[5]
P507-N235	H_2SO_4, pH = 1.8, Mo = 0.72 g/L, V = 0.75 g/L	Both Mo and V	[19]
TBP	HCl, 2 mol/L, Mo = 1 g/L, V = 1 g/L	Mo over V	This Study

8]. Using the secondary resources as metallic feed stocks to the metallurgical process has benefits.

A selective extraction separation method for molybdenum and vanadium was established by using leaching solution of HDS spent catalyst. Because of analogous physicochemical properties, the extraction and separation process for molybdenum and vanadium are difficult [9]. Among the common methods for separation and purification of Mo(VI) and V(V), solvent extraction is thought to be better than other techniques, because of the initial concentration of the metal is higher than 1 g/L [10, 11]. In recent years, many researchers researched the process for extraction of molybdenum and vanadium from the leach solution [12–14]. The process details with consideration of different conditions from various extractants were summarized in Table 1. There has been little experimental report on the separation of Mo (VI) from V (V) by TBP up to now; however, there are plenty experiments using TBP as a modifier [15, 16]. In order to obtain a better extraction rate, some important parameters, such as the hydrochloric acid concentration, TBP concentration, contact time, and oil–water ratio for the extraction of Mo(VI) from V(V) by TBP were considered.

Experimental

Chemical Agents

The commercial extractant TBP (99.5% purity) was purchased from Yakuri Pure Chemical. Sulfonated kerosene was provided by Alfa Aesar as the diluent (98%

purity). The organic phase was prepared using TBP extractant with concentration between 20 and 45% (v/v) diluted in kerosene. The aqueous solutions were prepared by dissolving 1 g/L and 1 g/L of $Na_2MoO_4{\cdot}2H_2O$ (Aladdin, AR, 99.0%) and Na_3VO_4 (Macklin, AR) in distilled water, respectively to imitate leach solution [19, 20]. Hydrochloric acid was used to adjust the pH of the aqueous solution. Sodium hydroxide was utilized as a stripping solvent, which was supplied by Sigma-Aldrich.

Analysis Apparatus

The concentrations of Mo(VI) and V(V) ions in the aqueous phase were measured by ICP-OES instrument (ICAP PRO, Thermo Scientific). Also, concentrations of ions in the organic phase were obtained with mass balance. In addition, the mechanical shaker (TQZ-312, ShangHai Jinghong) was used to contact the two phases at the equilibrium condition.

Experimental Procedure

The experiments in one stages (extraction) were performed by mixing 40 mL of feed solution and organic phase (O/A = 1) using a mechanical shaker with 150 rpm shaking rate. After shaking, the combinations of two phases were transferred to a separation funnel at 25 ± 1 °C. Then, the metal concentrations in the aqueous phase were determined by ICP-OES analysis.

Results and Discussion

Mo(VI) and V(V) Speciation in Aqueous Solution

In the solvent extraction, the information on the distribution of the metal ions is important in choosing an appropriate extractant. The distribution of species formed depends on the concentration of metals and ligands and solution pH [21]. The anionic species of molybdenum and vanadium, such as $Mo_8O_{26}{}^{4-}$, $HMoO^{4-}$, $H_2Mo_6O_{21}{}^{4-}$, and $HV_{10}O_{28}{}^{5-}$ are exist in the pH range of 2–6, and $MoO_4{}^{2-}$ the main complex in the range of pH > 6; $HV_{10}O_{28}{}^{5-}$, $V_4O_{12}{}^{4-}$ exist at 2.5 < pH < 10, and the $HVO_4{}^{2-}$ is exist at pH > 10 [22]. The cationic species of molybdenum and vanadium, such as $MoO_2{}^{2+}$ $VO_2{}^{+}$, are present at pH less than 2 [23]. A small amount of $VOCl_2$ exists at 2 M HCl solution indicated that the molybdenyl cation ($MoO_2{}^{2+}$) might combine with chloride ion to form extractable neutral chloro complex (MoO_2Cl_2) in HCl solution [23–25].

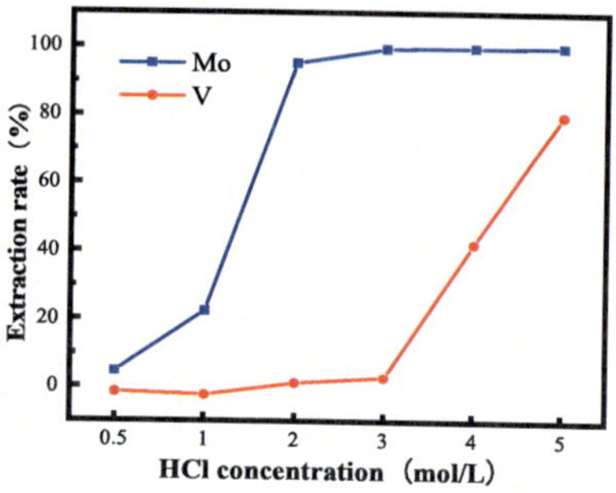

Fig. 1 Effect of different hydrochloric acid concentrations in molybdenum and vanadium extract

Effect of HCl Concentration

In the recovery of valuable metals from spent HDS catalyst by hydrometallurgical method, acid leaching is often employed to dissolve these metals from the catalyst [26–28]. Generally, the concentration of Mo(VI) and V(V) in the acid leaching solution of the secondary resources is around 1 g/L [19, 26]. Therefore, the concentration of Mo(VI) and V(V) in the synthetic solution was fixed at 1 g/L, the contact time is 10 min, and the concentration of HCl was varied from 0.5 to 5 M. The neutral species of Mo(VI) are predominant at these acid concentrations. Therefore, TBP is chosen as the extractant in present study. The concentration of TBP was fixed at 40% (v/v), and the volume ratio of aqueous and organic phase was unity. The extraction of Mo(VI) and V(V) is shown in Fig. 1. The extraction percentage of Mo(VI) increased from 4.75% to 95.32%, respectively with the increase in HCl concentration from 0.5 to 2 M and then was nearly constant with the further increase in HCl concentration. The extraction of V(V) was negligible in the HCl concentration range of 1–3 M and then increased from 2.94% to 79.87% with the further increase in HCl concentration to 5 M. In concentrated HCl solution, neutral species of Mo(VI), such as MoO_2Cl_2, can exist and their mole fraction increases with the increase in HCl concentration [16, 29]. Therefore, the extraction of Mo increased with the increase in HCl concentration. Although most of V(V) exist as VO_2^+ at pH less than 2 [21], the low extraction percentage of V(V) by TBP in the HCl concentration range from 1 to 3 M was attributed to the existence of V(V) as cationic species in this HCl solution.

Based on above results, Mo(VI) and V(V) could be separated by TBP at HCl concentration 2 M-3 M. Therefore, the extraction behavior of Mo(VI) and V(V) at different TBP concentration was investigated in further experiments.

Effect of TBP Concentration

Then, the extraction percentages of Mo(VI) and V(V) in simulated leaching solution were researched by changing TBP concentration at 2 M HCl concentration. In this experiment, the concentration of the two metals was fixed at 1 g/L, the contact time

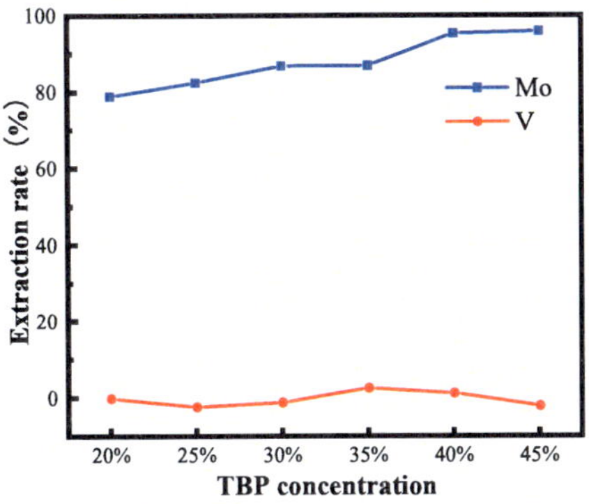

Fig. 2 Extraction of molybdenum and vanadium at different TBP concentrations under 2 mol/L HCl

is 10 min, and that of TBP was varied from 20 to 45% (v/v). The volume ratio of aqueous and organic phase was equal. The variation in the extraction of Mo(VI) and V(V) with different TBP concentration at 2 M HCl is shown in Fig. 2. The extraction percentage of Mo(VI) increased from 78.91% to 95.92% respectively, with an increase in TBP concentration from 20 to 45% (v/v), while the V(V) cannot be extracted by TBP. At low pH, most cationic Mo(VI) species such as MoO_2^{2+} is predominant. This species can form MoO_2Cl_2 with chloride ion, the extraction of MoO_2Cl_2 by TBP from aqueous acid solutions [29, 30]. Most of V(V) exist as VO_2^+ at pH less than 2, so the extraction percentage of V(V) was nearly negligible for any TBP concentration [31, 32].

The solvent extraction reaction of Mo(VI) by TBP from HCl solution can be represented as

$$MoOCl_{2(aq)} + 2TBP_{(org)} = 2TBP \cdot MoCl_{2(org)}$$

where subscript (aq) and (org) represent aqueous and organic phase.

The solvent extraction behavior of the two metals was investigated at 2 M HCl solutions. Figure 3 shows the variation in the extraction of Mo(VI) and V(V) with TBP. It can be ascribed to the coordination of the pair electron on $- P = O$ group in TBP molecules with the neutral species of Mo(VI) [33]. The extraction data from 2 M HCl solution indicate that it is possible to remain V(V) in the raffinate by solvent extraction with TBP; thus, the association of MoO_2Cl_2 with the pair electron on TBP would be easier than that of V [16].

Effect of Organic to Aqueous Phase Ratio

To find out a suitable organic to aqueous phase ratio, the extraction of Mo(VI) and V(V) was analyzed using different O/A ratio, extending from 1/1 to 1/5 with 40% (v/v) TBP maintaining acid concentration of leach liquor 2.0 mol/L. The aqueous

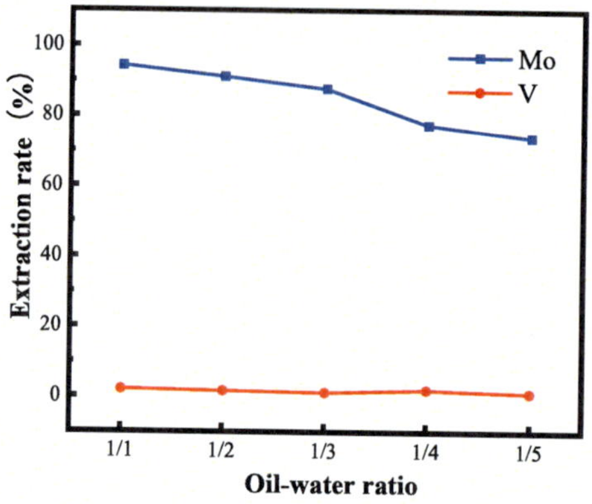

Fig. 3 Extraction changing curve with different oil–water ratio of Mo(VI) and V(V)

solution containing Mo–1 g/L and V–1 g/L was used for the experiment, and the contact time is 10 min. The results are shown in Fig. 3. It was observed that with increase of O/A phase ratio from 1 to 0.2, extraction of Mo decreased from 94.11% to 74.07%. When the oil–water ratio was 1/2, the molybdenum extraction rate reached 91.16%, while the extraction rate of vanadium is negligible; in order to reduce the extraction stages and simplify the operation process, the extraction ratio O/A = 1/2 was selected.

Effect of Contact Time

The influence of contact time on the extraction rate of molybdenum and vanadium was discussed. The concentration of Mo(VI) and V(V) was fixed at 1 g/L and the TBP-kerosene system with an organic phase concentration of 40%, O/A = 1:2, a series of extraction experiments at different reaction time from 5 to 40 min were conducted under the conditions of normal temperature and 2 M HCl concentration. The results are shown in Fig. 4. It can be seen from Fig. 5 that when the contact time is 10 min, the extraction of molybdenum reaches equilibrium, and the contact time is increased continuously, which has little effect on the extraction rate of molybdenum. The contact time has little effect on the extraction rate of vanadium; consider reducing energy consumption and improving efficiency, so the contact time for experiments is selected as 10 min.

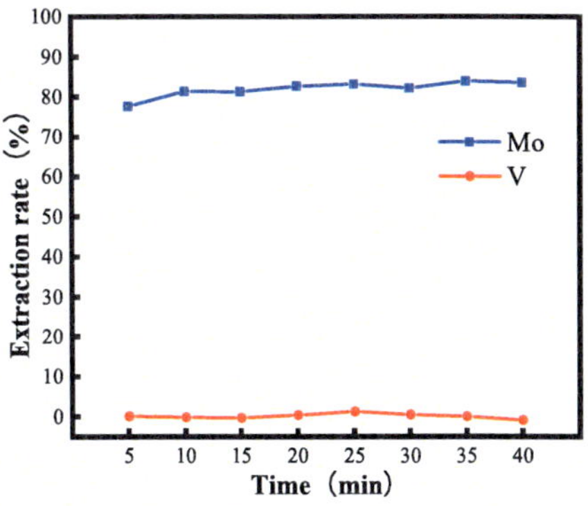

Fig. 4 Influence of contact time on Mo(VI) and V(V) extraction rate

Conclusion

The separation of Mo(VI) and V(V) from the hydrochloric acid solution containing 1 g/L of Mo(VI) and V(V) was investigated using TBP. In the HCl concentration range from 0.5 to 3 M, no V(V) was extracted by TBP, and thus, it was possible to leave V(V) in the raffinate. The extraction behavior of Mo(VI) is affected by HCl concentration. When HCl concentration was 2 M, the extraction percentage of Mo(VI) was much higher than that of V(V), indicating the possibility of separating the two metals by adjusting TBP concentration. It was able to separate Mo(VI) by stripping with dilute NaOH solution from the loaded TBP. Our research indicates that solvent extraction of the two metals from 2 M HCl solution by TBP is advantageous to separate the two metals.

Acknowledgements This work was financially supported by the Natural Science Foundation of China (U2004215, No. 51974280 and No. 51774252), the Educational Commission Fund of Henan Province of China (No. 20HASTIT012).

References

1. Wang M, Chen B, Sheng H (2018) Extraction of molybdenum and nickel from Ni-Mo ore by acid leaching combined with chlorate oxidation and phosphate complexation. J. Minerals Engineering. 124:63–67
2. Knyaz'Kina OV, Kuznetsova GG, Travkin V (2010) Extraction of molybdenum with bis(2,4,4-trimethylpentyl)phosphine acid (Cyanex-272). J. Russian Journal of Non-Ferrous Metals. 51:451–456
3. R.R Moskalyk a, A.M Alfantazi b, (2003) Processing of vanadium: a review. J. Minerals Engineering. 16:793–805
4. Nguyen TH, Lee MS (2014) Separation of Vanadium and Tungsten from Sodium Molybdate Solution by Solvent Extraction. J. Industrial and Engineering Chemistry Research. 53:8608–8614

5. Dp A, Ap A, Dpks A (2020) Recovery of Mo, V and Ni from spent catalyst using leaching and solvent extraction. J. Materials today: proceedings. 30:322–325
6. Liu J, Qiu Z, Yang J (2016) Recovery of Mo and Ni from spent acrylonitrile catalysts using an oxidation leaching–chemical precipitation technique. J. Hydrometallurgy. 164:64–70
7. Wang H, Feng Y, Li H (2020) Recovery of vanadium from acid leaching solutions of spent oil hydrotreating catalyst using solvent extraction with D2EHPA (P204), J. Hydrometallurgy. 195: 105404.
8. Zeng L, Chu YC (2010) Recovery of molybdenum and vanadium from synthetic sulphuric acid leach solutions of spent hydrodesulphurisation catalysts using solvent extraction. J. Hydrometallurgy. 101:141–147
9. Li Q, Li Z, Xiao L (2009) Completely removing vanadium from ammonium molybdate solution using chelating ion exchange resins. J. Hydrometallurgy. 98:287–290
10. Nguyen TH, Lee MS (2015) Separation of Molybdenum(VI) and Tungsten(VI) from Sulphate Solutions by Solvent Extraction with LIX 63 and PC 88A. J. Hydrometallurgy. 155:51–55
11. Shakib B, Torab-Mostaedi M, Outokesh M (2019) Extraction of Molybdenum (VI) and Vanadium (V) from Nitrate Solutions Using Coupling of Acid and Solvating Extractants. J. International Journal of Engineering. 32:1366–1371
12. Hosseinzadeh M, Alizadeh M, Ranjbar M (2014) Improvement of the solvent extraction of rhenium from molybdenite roasting dust leaching solution using counter-current extraction by a mixer-settler extractor. J. International Journal of Engineering. 27:651–658
13. Chen B, Xian P, Huang S (2017) Extraction of molybdenum from acidic leach solution of Ni–Mo ore by solvent extraction using tertiary amine N235. J. Canadian Metallurgical Quarterly. 56:426–431
14. Zhao Z, Min G, Mei Z (2015) Extraction of molybdenum and vanadium from the spent diesel exhaust catalyst by ammonia leaching method–ScienceDirect. J. Journal of Hazardous Materials. 286:402–409
15. Chen W S, Liang W C, Lee C H (2018) Study of Molybdenum Extraction from Alkali Roasted and Water Leaching of Ferro-Molybdenum Slag by Using TOA and TBP, J. IOP Conference Series Earth and Environmental Science. 182: 012015.
16. Truong HT, LEE M S, (2017) Separation of rhenium(VII), molybdenum(VI), and vanadium(V) from hydrochloric acid solution by solvent extraction with TBP. J. Geosystem Engineering. 20:224–230
17. Biswas RK, Wakihara M, Taniguchi M (1985) Recovery of vanadium and molybdenum from heavy oil desulphurization waste catalyst. J. Hydrometallurgy. 14:219–230
18. Zhu Z, Tulpatowicz K, Pranolo Y (2015) Solvent extraction of molybdenum and vanadium from sulphate solutions with Cyphos IL 101. J. Hydrometallurgy. 154:72–77
19. Li H, Feng Y, Wang H (2020) Separation of V (V) and Mo (VI) in roasting-water leaching solution of spent hydrodesulfurization catalyst by co-extraction using P507-N235 extractant, J. Separation and Purification Technology. 248: 117135.
20. Kim HI, Lee KW, Mishra D (2014) Separation and recovery of vanadium from leached solution of spent residuehydrodesulfurization (RHDS) catalyst using solvent extraction. J. Journal of Industrial and Engineering Chemistry. 20:4457–4462
21. Zeng L, Chu Y (2009) A literature review of the recovery of molybdenum and vanadium from spent hydrodesulphurisation catalysts Part II: Separation and purification. J. Hydrometallurgy. 98:10–20
22. Zhang J, l, Zhang LF, (2016) Thermodynamic Equilibrium of Mo(VI)-V(V)-H_2O System and Its Application for Deep Removal of Vanadium from Molybdate Solution. J. CHINESE JOURNAL OF RARE METALS. 40:701–707
23. Olazabal MA, Orive MM, Fernandez LA (2007) Selective Extraction of Vanadium (V) from Solutions Containing Molybdenum (VI) by Ammonium Salts Dissolved in Toluene. J. Solvent Extraction and Ion Exchange. 10:623–635
24. Nguyen TH, Lee MS (2016) A review on the separation of molybdenum, tungsten, and vanadium from leach liquors of diverse resources by solvent extraction. J. Geosystem Engineering. 19:247–259

25. Nekovar P, Schrotterova D (2000) Extraction of V(V), Mo(VI) and W(VI) polynuclear species by primene JMT. J. Chemical Engineering Journal. 79:299–233
26. Kim HI, Lee KW, Mishra D (2015) Separation of molybdenum and vanadium from oxalate leached solution of spent residue hydrodesulfurization (RHDS) catalyst by liquid–liquid extraction using amine extractant. J. Journal of Industrial and Engineering Chemistry. 21:1265–1269
27. Padhan E, Sarangi K (2014) Separation of molybdenum and cobalt from spent catalyst using Cyanex 272 and Cyanex 301. J. International Journal of Mineral Processing. 127:52–61
28. Le MN, Man SL (2020) Separation of Al(III), Mo(VI), Ni(II), and V(V) from model hydrochloric acid leach solutions of spent petroleum catalyst by solvent extraction. J. Journal of Chemical Technology and Biotechnology. 95:2886–2897
29. Sato T, Watanabe H, Suzuki H (1990) Liquid-liquid extraction of molybdenum(VI) from aqueous acid solutions by TBP and TOPO. J. Hydrometallurgy. 23:297–308
30. Basualto C, Marchese J, Valenzuela F (2003) Extraction of molybdenum by a supported liquid membrane method. J. Talanta. 59:999–1007
31. Nguyen TH, Lee MS (2015) Solvent extraction of vanadium(V) from sulfate solutions using LIX 63 and PC 88A. J. Journal of Industrial and Engineering Chemistry. 31:118–123
32. Wang L, Zhang Y, Tao L (2013) Comparison of ion exchange and solvent extraction in recovering vanadium from sulfuric acid leach solutions of stone coal. J. Hydrometallurgy. 131–132:1–7
33. Ojo JO, Ipinmoroti KO, Adeeyinwo CE (2008) Solvent extraction of molybdenum (VI) from diluted and concentrated hydrochloric acid. J. Global Journal of Pure and Applied Sciences. 14:289–294

Synthesis and Characterization of Cu-Ni Bimetallic Catalysts Support on GO, rGO, and NGO

Chengrui Wang, Yanhong Fang, Guangfen Liang, Huamei Duan, Dengfu Chen, and Mujun Long

Abstract CO_2 activation plays an important role in CO_2 utilization. Thus, the catalyst consisting of Cu-Ni supported on graphene oxide (GO), ammonia modified graphene (NGO), and reduced graphene oxide (rGO) were synthesized. Their properties were analysed by BET, XPS, TEM, and TG-DSC. The specific surface area of supports followed the order of ammonia modified graphene (NGO) > graphene oxide (GO) > reduced graphene oxide (rGO). Cu^0 existed in rGO and NGO supported catalysts. 29.5% of Cu^{2+} was reduced to Cu^+ or Cu^0 in rGO. In NGO, 30% of Cu^{2+} was reduced. Most of the Cu and Ni was dispersed uniformly on these two supports. In GO, some particles were sintered, which was composed of Cu and Ni with a size up to 100 nm. In rGO and NGO, the metal particle size was less than 50 nm. The CO_2 activation energy was determined by TG-ESC experiment, and the calculation was done by Ozawa method. The results showed that CuNi-rGO and CuNi-NGO could activate CO_2, and the activation energy (E) was 78.26 and 91.30 kJ·mol^{-1}, respectively. Compared with literature, these catalysts could reduce the activation energy (E) by 48%.

Keywords CO_2 activation · Catalysts · Commercial graphene · Activation energy

Introduction

An important substance to the continuation of life on earth, CO_2 has very stable chemical properties. The carbon balance of nature creates a comfortable environment for the reproduction of species [1], and the natural carbon cycle involves about 90 Gt of CO_2 [2]. However, because of human activities, CO_2 emissions has exceeded this amount [3]. This impacts the human living environment considerably [4–6].

C. Wang · Y. Fang · G. Liang · H. Duan (✉) · D. Chen · M. Long
College of Materials Science and Engineering, Chongqing University, Chongqing 400044, China
e-mail: duanhuamei@cqu.edu.cn

Chongqing Key Laboratory of Vanadium-Titanium Metallurgy and New Materials, Chongqing University, Chongqing 400044, China

A. Lazou et al. (eds.), *REWAS 2022: Developing Tomorrow's Technical Cycles (Volume I)*, The Minerals, Metals & Materials Series,
https://doi.org/10.1007/978-3-030-92563-5_80

Many countries and international organizations have adopted now policies to reduce CO_2 emissions. However, high investment and low return in the short term threatens the realization of goals. Therefore, it is necessary to develop new technologies to reduce the net CO_2 emissions and decrease with the CO_2 emissions. At present, these methods include physical storage [7] and chemical activation [8]. Compared with physical storage, chemical activation may not only eliminate the problem of excessive CO_2 emission, but the chemical raw materials may also produce economic value.

Among chemical activation methods, the Chemisorption method based on transition metals is one of the most commonly methods. Behner et al.[9] explored the adsorption and activation of CO_2 on Fe(110), Fe(100), and Fe(111), and CO_2 was adsorbed on gradient Fe(110) and Fe(111) planes at –196 and –133 °C. Then, CO_2 decomposed into O and CO. Erik et al. [10] estimated the adsorption heat of CO^{2-} on Pd(111), Cu(111), Ni(111), and Fe(111) surfaces by UBI-QEP method and discussed the activation pathways of CO_2 on different metal surfaces. The results showed that the stability of CO^{2-} on four transition metal surfaces was Fe > Ni > Cu > Pd, and CO^{2-} was easily produced on the surface of Fe and Ni. The activation of CO_2 by chemical catalysis is based on adsorption theory[11]. Guo Jianzhong et al.[12] found CO_2 and CH_4 could be co-adsorbed and activated on Ni/SiO_2 catalyst with different particle sizes. Ni of smaller particle sizes could promote the reaction between surface oxygen species and CO_2. Liu et al. [13] investigated the chemical adsorption state of CO_2 on Co, Fe, Ni surface, and explored the process of CO_2 into CO on the transition metal surface. The results showed that some electrons could transfer from transition metal to CO_2 on the transition metal surface, which causes the bond bend and stretch, to then form CO^{2-}. The chemical adsorption on catalyst was usually aided with high pressure and temperature [14–16]. CO_2 adsorption on metal oxide surface could then form different adsorption species. Zhao Yunpeng et al.[17] took CuO-ZnO-Al_2O_3 as catalyst to study the CO_2 hydrogenation to methanol. The temperature and reaction space velocity was 240 °C and 2400 h^{-1}, $V(H_2)/V(CO_2) = 2.6/1$, the CO_2 conversion rate was 9.0%, and methanol selectivity was 14.1%. To enhance the performance of the catalyst, the catalyst was usually supported. Compared with traditional supports, graphene and its derivatives had better properties [18–23], such as excellent electrical conductivity and a specific surface area of up to 2630 $m^2.g^{-1}$. So, in this paper graphene was chosen as the support of the catalyst.

In this paper, Cu, Ni, and Cu-Ni supported on graphene oxide (GO), reduced graphene oxide (rGO), and ammonia modified graphene (NGO) were synthesized to explore their properties. First, the catalyst was prepared by then impregnation method. Then, the specific surface area of three supports was measured by Brunner–Emmet–Teller (BET) method. X-ray photoelectron spectrometer (XPS) analysis was used to determine the phase composition. The distribution and size of the particles on the support could be characterized by Transmission Electron Microscope (TEM). Finally, the activation energies were calculated by thermogravimetric and differential scanning calorimetry analysis (TG-DSC).

Experimental

Reagents and Instruments

The chemicals used in this paper include graphite (≥99.95%, Linen); sulfuric acid (95.0–8.0% H_2SO_4, Chuandong); sodium nitrate (≥99.0% $NaNO_3$, Chuandong); potassium permanganate (≥99.5% $KMnO_4$, Boyi); hydrogen peroxide (≥30.0% H_2O_2, Chuandong); hydrochloric acid (36.0 –8.0% HCl, Chuandong); barium chloride (10% BaCl, Alfaesha); anhydrous ethanol (≥99.7% C_2H_6O, Chuandong); N-methyl pyrrolidone (≥99.0% C_5H_9NO, Alighting); ascorbic acid (≥99.7% $C_6H_8O_6$, Kelong); ammonia (25 –28% $NH_3{\cdot}H_2O$, Chuandong); copper nitrate (≥99.0% $Cu(NO_3)_2{\cdot}3H_2O$, Kelong); and nickel nitrate (≥98.0% $Ni(NO_3)_2{\cdot}6H_2O$, Kelong). All chemicals were used as purchased without further treatment.

Brunner–Emmet–Teller method (Micromeritics ASAP2020M + C, USA); XPS spectrogram analysis (ESCALAB 250Xi, Seymour Fraser); thermogravimetry–differential scanning calorimetry analysis (STA44PF3, Germany); transmission electron microscope analysis (Talos F200S, The United States of America); intelligent triple magnetic heating agitator (ZNcl-DLS 230*2303, Tenheng Instrument); CNC ultrasonic cleaner (KQ2200DB, Ultrasonic Instrument); desktop high-speed centrifuge (TG16-WS, Xiangyi); vacuum drying oven (DZF-6030, Jinghong); rotary evaporator (RE-52, Yarong); and tube furnace (1100–60, Haoyue) were used for characterization and preparation of the catalysts.

Experimental Method

All supports were purchased from the company of Aladdin–Holdings Group. The catalyst was prepared by impregnation method. First, the ratio of metal mass to support was 1:10 (the molar ratio of Cu and Ni is 2). Cu and Ni metal nitrate were dissolved in anhydrous ethanol, and the support was submerged in the solution. After that, the solution was put into the rotary evaporation apparatus under vacuum to remove the anhydrous ethanol. The resulting solid was put into a tube furnace. The calcination was carried out at 573 K for 2 h with a ramping rate of 5 K/min under 5% H_2 in N_2 atmosphere. The sample was collected after cooling to room temperature. These catalysts were denoted as CuNi-GO, CuNi-rGO, and CuNi-NGO.

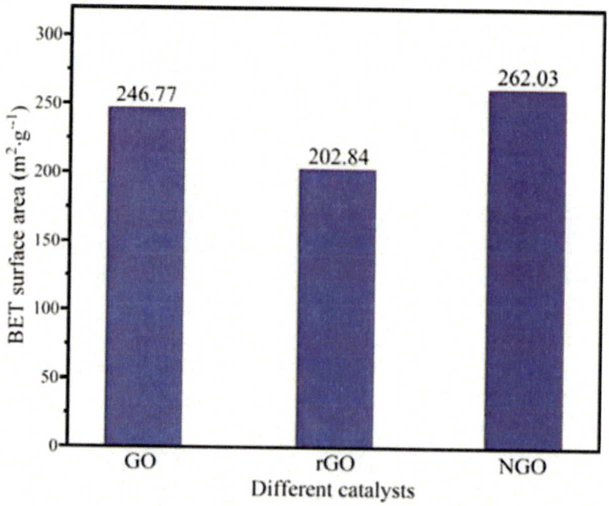

Fig. 1 BET analysis: **a** GO; **b** rGO; **c** NGO

Results and Discussion

BET Analysis

The specific surface area of support has a very important impact to the dispersion of active components and the reaction performance of catalyst. A large specific surface area is conducive to the dispersion of active components and the diffusion of gas components in the catalyst. In our work, the specific surface area of three supports was tested by BET. As shown in Fig. 1, the specific surface area of NGO was the largest (262.03 $m^2 \cdot g^{-1}$). The addition of alkali solution (NH_3H_2O) strengthened the electrostatic repulsion between the layers and may have increased the specific surface area of the support. The specific surface area of GO was 246.77 $m^2 \cdot g^{-1}$, and the rGO had a surface area of 202.84 $m^2 \cdot g^{-1}$. This might be due to the excessive addition of C_5H_9NO during the preparing of rGO, which made graphene agglomerate and reduced the specific surface area.

XPS Analysis

In order to understand the difference of active components on the commercial support catalysts, XPS analysis was used. All peak positions can be found in the XPS database [24]. Figure 2a–c were CuNi-GO, CuNi-rGO, and CuNi-NGO Cu2p orbital X-ray photoelectron spectroscopy. According to the Cu2p energy spectrum, the copper ion in CuNi-GO existed in the form of Cu^+ and Cu^{2+}. In rGO and NGO, all the Cu^{2+}

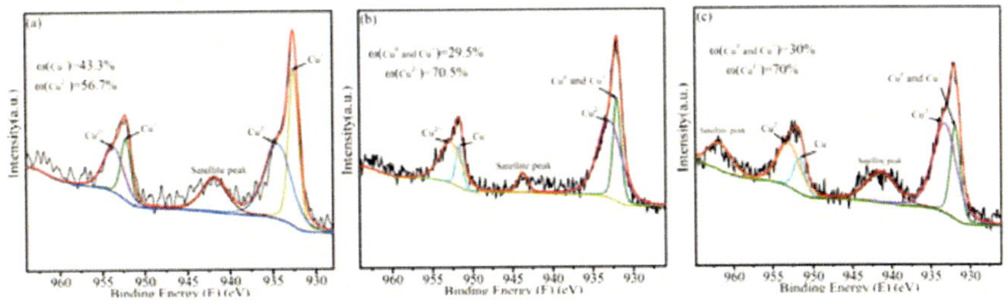

Fig. 2 XPS analysis: **a** Cu2p orbitals of CuNi-GO; **b** Cu2p orbitals of CuNi-rGO; and **c** Cu2p orbitals of CuNi-NGO

were reduced to Cu^{+} and Cu^{0}. This means the chemical reactions below, (1)–(3), occurred. There wasn't a Cu^{2+} diffraction peak in CuNi-rGO and CuNi-NGO, which is consistent with our previous report [25].

$$2Cu(NO_3)_2(s) \triangleq 2CuO(s) + 4NO_2(g) + O_2(g) \tag{1}$$

$$2CuO(s) + H_2(g) \triangleq Cu_2O(s) + 2H_2O(g) \tag{2}$$

$$CuO(s) + H_2(g) \triangleq Cu(s) + H_2O(g) \tag{3}$$

$$2Ni(NO_3)_2(s) \triangleq 2NiO(s) + 4NO_2(g) + O_2(g) \tag{4}$$

In Fig. 2a, according to the statistics of the main peak areas of Cu^{+} and Cu^{2+}, the contents of Cu^{+} and Cu^{2+} in CuNi-GO were 43.3% and 56.7%, respectively. In Fig. 2b, it could be seen that the contents of reduced copper (Cu^{0} and Cu^{+}) and unreduced copper (Cu^{2+}) in CuNi-rGO were 29.5% and 70.5%, respectively. In CuNi-NGO, from Fig. 2c, the reduced copper and unreduced copper were 30% and 70%, respectively. The active component (Cu^{0}) [26, 27] existed in CuNi-rGO and CuNi-NGO.

TEM Analysis

Figure 3a–i shows the TEM figure of CuNi-GO, CuNi-rGO, and CuNi-NGO of commercial support catalysts and EDS mapping of Cu and Ni. Figure 3a–c showed the TEM figure of CuNi-GO. From this figure, it can be seen that some particles were sintered and agglomerated into larger particles. Some particles in CuNi-GO were larger than 50 nm. According to the EDS mapping in Fig. 3b, c, large particles mainly contained Cu and Ni elements. Almost all the Cu present aggregated. Part of the Ni aggregated on GO, and the Ni particles were dispersed on GO evenly.

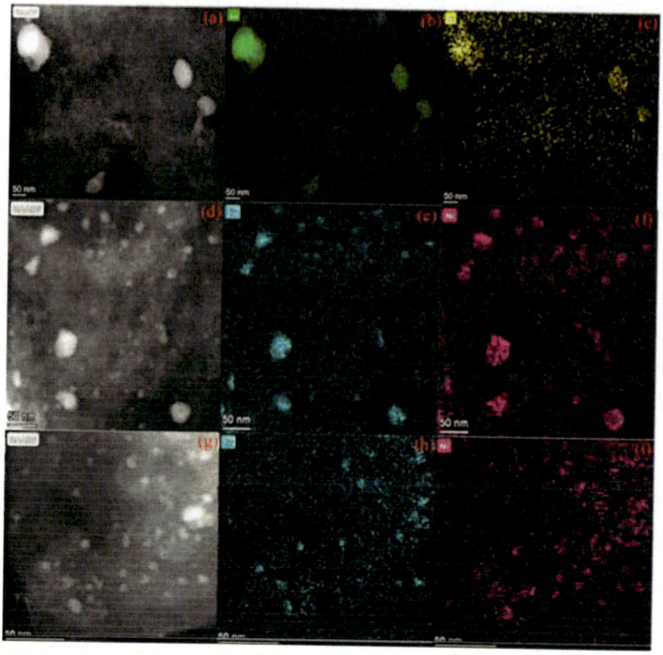

Fig. 3 TEM images of CuNi-GO (**a**), CuNi-rGO (**d**), and CuNi-NGO (**g**), the EDS mappings of CuNi-GO (**b–c**), CuNi-rGO (**e–f**), and CuNi-NGO (**h–i**)

According to XPS analysis, the main metal components were (Cu^{2+} and/or Cu^{+}) and Ni^{2+}.

In CuNi-rGO and CuNi-NGO, it can be seen that most of metal components are dispersed throughout the supports evenly. Some metal components in rGO and NGO were agglomerated, and the size of most particles is less than 50 nm. The EDS figure shows that the metal components were also composed of Cu and Ni elements. Referring to the XPS analysis, these components were composed of (Cu^{0} and/or Cu^{+}) and Ni^{2+}, while flake crystals are present across the three catalyst. The largest component size was CuNi-GO, followed by CuNi-rGO, and the smallest was CuNi-NGO.

CO_2 Activation Analysis

In order to explore the CO_2 activation performance of these catalysts, thermodynamic analysis of catalysts was conducted by TG-DSC and TG-DTG. In the DTG curve, the downward peaks indicate that a reaction has taken place there. Figure 4a–f were the TG-DSC and TG-DTG curves of the three commercial support catalysts.

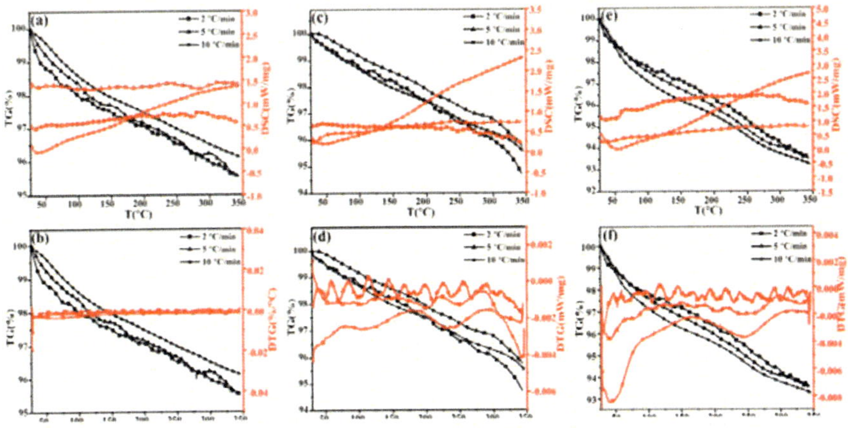

Fig. 4 TG-DSC and TG-DTG curve: **a** and **b** CuNi-GO; **c** and **d** TG-DTG curve of CuNi-rGO; and **e** and **f** TG-DTG curve of CuNi-NGO

Figure 4a, b show the TG-DSC and TG-DTG curves of CuNi-GO. From these figures, it can be seen that there is no exothermic peak, which means CuNi-GO had low CO_2 activation strength. For catalysts supported by rGO and NGO, CuNi-rGO and CuNi-NGO had endothermic peaks in the TG-DTG curves. Since the decomposition temperature of hydroxyl was 200–300 °C, the exothermic peak of CuNi-rGO corresponds to the absorption of CO_2. For CuNi-rGO, it was calcined at 300 °C so most the functional group should decompose. Therefore, in Fig. 4c, d, the exothermic peak is also the absorption of CO_2. Compared with CuNi-GO, Fig. 3d showed that the CuNi-rGO had better CO_2 activation ability. Figure 4e, f show that CuNi-NGO exhibit an exothermic peak around 230 °C, which is the exothermic peak generated by adsorption of CO_2. This is similar with CuNi-rGO, which also shows that the activation center of methanol from CO_2 hydrogenation is Cu^0. There is a small exothermic peak in the TG-DTG (Fig. 4f) curve of CuNi-NGO, which also means it exhibits CO_2 activation ability. The exothermic peak data is shown in Table 1.

The high activation energy (E) of the catalysts shows the degree of difficulty and reaction rate for CO_2 activation, and it may be calculated by Ozawa method [28]. The Ozawa equation was as follows.

$$\frac{d(\ln\beta)}{d(1/T_P)} = -\frac{1.052E}{R} \tag{5}$$

R (8.314 $J{\cdot}mol^{-1}{\cdot}K^{-1}$) is the ideal gas constant, β is the temperature ramp rate, and T_P is the temperature at the peak of DSC curve. With different heating rates, a set of DSC curves and T_P is generated, to draw a line directly by $\ln\beta$ and $1/T_P$. The linear slope is used to calculate E. The data of heating rate and peak temperature are shown in Table 1. Through the equations of (5), a diagonal line is obtained, which is shown in Fig. 5. Due to the weak activation property of CuNi-GO, we did not

Table 1 TG-DSC exothermic peak data of CuNi-rGO and CuNi-NGO

Sample	Heating rate β/°C·min^{-1}	DSC peak temperature T_p/°C
CuNi-rGO	2	216
	5	225
	10	232
CuNi-NGO	2	222
	5	230
	10	238

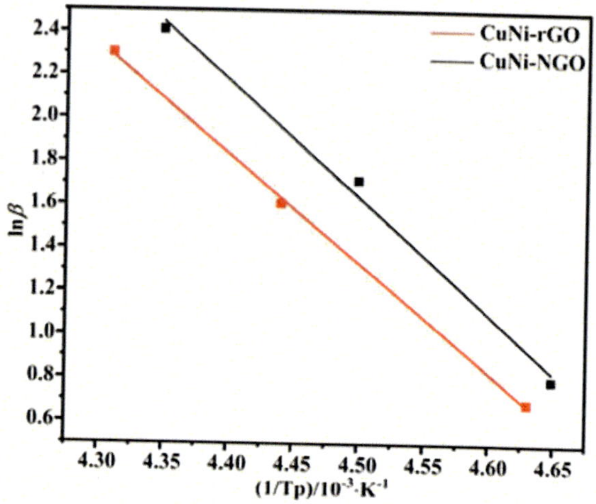

Fig. 5 Kinetic fitting curve of CuNi-rGO and CuNi-NGO

calculate their *E*.

Figure 5 shows the kinetic fitting curve of CuNi-rGO and CuNi-NGO. From this figure, the *E* of CuNi-rGO and CuNi-NGO are calculated to be 78.26 and 91.30 kJ·mol^{-1}, respectively. The *E* of CO_2 is 152.256 kJ·mol^{-1} [29]. So, the commercial support catalyst of CuNi-rGO and CuNi-NGO could decrease the *E* of CO_2.

Conclusion

This paper investigated the structure and properties of Cu-Ni bimetallic catalyst. Through BET analysis, we found the specific surface area of support followed the order of NGO > GO > rGO. The addition of alkaline solution may control the electrostatic repulsion between the layers to increase the specific surface area of the support. 29.5% of Cu^{2+} was reduced to Cu^0 or Cu^+ on CuNi-rGO, while on CuNi-NGO was 30%. Compared with CuNi-GO, the size of most particles in CuNi-rGO and

CuNi-NGO was smaller. Through TG-DSC analysis, we found that CuNi-rGO and CuNi-NGO could activate CO_2, and the E were 78.26 and 91.30 $kJ\cdot mol^{-1}$, respectively. These findings provide useful information for designing and optimizing Cu-Ni bimetallic catalysts supported on graphene derivatives to activate CO_2.

Acknowledgements The authors gratefully acknowledge the financial support provided by the National Natural Science Foundation of China, project No.51704048, 52074051, the Natural Science Foundation of Chongqing (No.cstc2020jcyj-msxmX0794), and the fundamental funds for the Central Universities (No.2020CDJ-LHSS-010). The authors would like to thank Analytical and Testing Center of Chongqing University for BET, XPS, TEM, and TG-DSC analysis.

References

1. Pierrehumbert RT (2004) High levels of atmospheric carbon dioxide necessary for the termination of global glaciation. Nature 429(6992):646–649
2. Woodard DL, Davis SJ, Randerson JT (2019) Economic carbon cycle feedbacks may offset additional warming from natural feedbacks. Proc Natl Acad Sci 116(3):759–764
3. Smol JP (2012) Climate Change: A planet in flux. Nature (London) 483(7387):S12–S15
4. Ortega-Huerta MA, Townsend Peterson A, Hannah L, Hughes L, Green RE, Beaumont LJ et al (2004) Extinction risk from climate change. Nature 427(6970):145–148
5. Lacis AA, Schmidt GA, Rind D, Ruedy RA (2010) Atmospheric CO_2: Principal Control Knob Governing Earth's Temperature. Science (American Association for the Advancement of Science). 330(6002):356–359
6. Pandolfi JM, Connolly SR, Marshall DJ, Cohen AL (2011) Projecting Coral Reef Futures Under Global Warming and Ocean Acidification. Science (American Association for the Advancement of Science). 333(6041):418–422
7. Mac Dowell N, Fennell PS, Shah N, Maitland GC (2017) The role of CO2 capture and utilization in mitigating climate change. Nat Clim Chang 7(4):243–249
8. Sun Q, Liu Z (2011) Mechanism and kinetics for methanol synthesis from CO2/H2 over Cu and Cu/oxide surfaces: Recent investigations by first-principles-based simulation. Front Chem China 6(3):164–172
9. Behner H, Spiess W, Wedler G, Borgmann D (1986) Interaction of carbon dioxide with Fe(110), stepped Fe(100) and Fe(111). Surface ence. 175(2):276–286
10. Vesselli E, Baraldi A, Comelli G, Lizzit S, Rosei R (2004) Ethanol Decomposition: C?C Cleavage Selectivity on Rh(111). Chemphyschem A European Journal of Chemical Physics & Physical Chemistry. 5(8):1133
11. Li J, Deng TY, YANG L, et al (2013) Research progress of adsorption activation and catalytic hydrogenation of CO2[J]. Chemical Industry and Engineering Progress. 32(2):341–342
12. Guo J, Hou Z, Gao J, Zheng X (2007) DRIFTS Study on Adsorption and Activation of CH4 and CO2 on Ni/SiO2 Catalyst with Various Ni Particle Sizes. Chin J Catal 28(1):22–26
13. Cong L, Cundari TR, Wilson AK (2012) CO2 Reduction on Transition Metal (Fe, Co, Ni, and Cu) Surfaces: In Comparison with Homogeneous Catalysis. Journal of Physical Chemistry C. 116(9):5681–8.
14. Wang S, Mao D, Guo X, Wu G, Lu G (2009) Dimethyl ether synthesis via CO 2 hydrogenation over CuO–TiO 2–ZrO 2/HZSM-5 bifunctional catalysts. Catal Commun 10(10):1367–1370
15. Zhang Q, Zuo YZ, Han MH, Wang JF, Jin Y, Wei F (2010) Long carbon nanotubes intercrossed Cu/Zn/Al/Zr catalyst for CO/CO2 hydrogenation to methanol/dimethyl ether. Catal Today 150(1–2):55–60
16. Krogman JP, Foxman BM, Thomas CM (2011) Activation of CO2 by a heterobimetallic Zr/Co complex. J Am Chem Soc 133(37):14582

17. ZHAO Y, JIA L, XIN G, et al (2009) Study on the synthesis of methanol by low pressure hydrogenation of carbon dioxide over CuO-ZnO-Al_2O_3 catalyst [J]. Natural Gas Chemical Industry. 6.
18. Spyrou K, Gournis D, Rudolf P (2013) Hydrogen Storage in Graphene-Based Materials: Efforts Towards Enhanced Hydrogen Absorption. Ecs Journal of Solid State Science & Technology. 2(10):M3160–M3169
19. Huang B, Li Z, Liu Z, Zhou G, Hao S, Wu J et al (2008) Adsorption of gas molecules on graphene nanoribbons and its implication for nano-scale molecule sensor. J Phys Chem C 112(35):13442–13446
20. Shokuhi Rad A, Zareyee D (2016) Adsorption properties of SO2 and O3 molecules on Pt-decorated graphene: A theoretical study. Vacuum 130:113–118
21. Goharibajestani Z, Yurum A, Yurum Y (2019) Effect of Transition Metal Oxide Nanoparticles on Gas Adsorption Properties of Graphene Nanocomposites. Applied Surface Science. 475(MAY 1):1070–6.
22. Yang LWNRSRT (2011) Unique hydrogen adsorption properties of graphene. AIChE J 57(10):2902–2908
23. Alisultanov ZZ, Meilanov RP, Nukhov AK (2013) On the role of lattice defects in the formation of adsorption properties of graphene. Tech Phys Lett 39(2):171–174
24. Wang Jianqi (1992) Introduction to XPS/XAES/UPS[M] //introduction to XPS/XAES/UPS. National Defense Industry Press.
25. Wang C, Fang Y, Liang G, Lv X, Duan H, Li Y, et al (2021) Mechanistic study of Cu-Ni bimetallic catalysts supported by graphene derivatives for hydrogenation of CO2 to methanol. Journal of CO2 utilization. 49:101542.
26. Li C, Yuan X, Fujimoto K (2014) Development of highly stable catalyst for methanol synthesis from carbon dioxide. Appl Catal A 469:306–311
27. Liu H, Chen T, Yang G, Wang G-Y (2017) Investigation of Active Center of Cu-Based Catalyst for Low Temperature Methanol Synthesis from Syngas in Liquid Phase: The Contribution of Cu+ and Cu0. ChemistrySelect (Weinheim). 2(26):8000–8007
28. Nobuyoshi (2013) ChemInform Abstract: Ozawa's Kinetic Method for Analyzing Thermoanalytical Curves - History and Theoretical Fundamentals. Chem Inform. 44(48).
29. Dietz L, Piccinin S, Maestri M (2016) Mechanistic Insights into CO2 Activation via Reverse Water-Gas Shift on Metal Surfaces. J Phys Chem C 119(9):4959–4966

The Research Status and Progress on the Utilization of Coal Fly Ash: A Review

Joseph Emmanuel Nyarko-Appiah, Wenzhou Yu, Peng Wei, and Hao Chen

Abstract Coal fly ash (CFA), solid waste from the thermal power plant, is gaining attention as an alumina content substitute, but its production has long been considered to be problematic in environmental management. Many researchers have proposed and developed utilization technologies that can help to solve these problems by utilizing large amounts of the CFA. This review aims at pointing out aspects for CFA utilization; one aspect of the utilization technology that has had a great impact on the material society is how it affects the environment, and the other is the recovery of rich minerals. The current research status, progress, and future scope of research on CFA will be reviewed in this paper based on comprehensive research data. More research is required to provide a better understanding of the reaction mechanism and economic advantages of CFA. This review paper provides existing opportunities to motivate researchers to be involved in this study.

Keywords Coal fly ash (CFA) · Environmental effect · Recycling and secondary recovery · Economic advantages

Introduction

With the intensified coal consumption, the coal fly ash (CFA) is an industrial by-product produced in the thermal power plant; it has an alumina (Al_2O_3) content that ranges between 20 and 50% and is comparable to mid or low-grade bauxite ores [1–3]. A major percentage of the total CFA production is used in construction, paving, filling, and even agricultural soil amendment [4, 5], and the remaining percentage has caused serious environmental and health problems due to its open-air stockpiling [6, 7]. The global average utilization rate of CFA is estimated to be 60%; the current utilization rates have been estimated at 70% for China, 56% for the United States, 85% for Korea, and 63% for India [8, 9]. Therefore, it is essential to expand the utilization of CFA to decrease its damage. In recent years, there have been general concerns

J. E. Nyarko-Appiah · W. Yu (✉) · P. Wei · H. Chen
College of Materials Science and Engineering, Chongqing University, Chongqing 400044, China
e-mail: yuwenzhoucd@cqu.edu.cn

A. Lazou et al. (eds.), *REWAS 2022: Developing Tomorrow's Technical Cycles (Volume I)*, The Minerals, Metals & Materials Series,
https://doi.org/10.1007/978-3-030-92563-5_81

about the impacts of CFA and the continual increase in the price of bauxite around the world because of its supply shortage. The increasing awareness of environmental protection and ecological balance has prompted the utilization of CFA [10, 11].

In consideration of the high alumina content, CFA can be utilized as a substitute for bauxite as a potential and economical source of alumina. Hence, the alumina recovery from CFA has attracted much attention [3]. The recovery of alumina from CFA is an extensive alternative to convert waste into a new source of aluminium [12, 13] and high value-added products (cement, concrete) that bring substantial economic and ecological benefits [14]. The traditional process for alumina production has been the Bayer process [15], which is still generally used in industries. However, due to the scarce bauxite resources, extraction of alumina from alumina-rich resources as well as the establishment of other novel processes to improve the aluminium production is being explored. The typical metallurgical processes for alumina extraction from CFA are the alkali-sintering process (limestone or soda-lime sintering) and the acid leaching process [16, 17]. All these processes have their advantages and limitations. According to research studies, this paper discusses the research status and progress on the utilization of CFA, summarizing the technical challenges and development of the metallurgical methods, hoping to provide an insightful CFA utilization.

Current Status and Utilization of CFA

Production of CFA

The natural resources, global economic and population growth require immerse energy consumption which is likely to be based on coal. Given that coal contributes almost one-third of global energy, coal power plants are the workhorse in the electricity industry [18, 19]. Despite the rapid development of hydroelectric power and other new energy technologies for industries, coal remains the dominant fuel for power production [20]. The importance of recycling CFA, a solid waste produced from power plants, is by the scale of global CFA production and the consequent waste generation. Figure 1 shows a chart summarising the global production and utilization of CFA since 2000. There are many good reasons to view CFA as a resource, rather than a waste material. However, although the CFA utilization rate is increasing, the accumulation of CFA continues to increase [21].

Utilization of CFA

In recent years, the utilization of CFA as raw materials as well as the recovery of usable materials from CFA in various sectors is the current global trend. According to research data, the CFA production rate outweighs consumption due to the increased

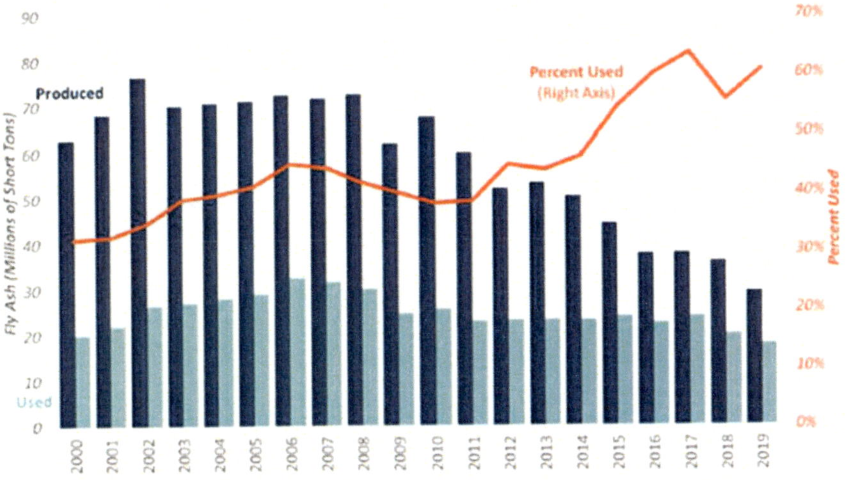

Fig. 1 CFA production and utilization [22]

amount of energy being generated by coal-fired power plants [23, 24]. The CFA open-air stockpiling contributes to the air, water, and soil pollution with heavy metals (such as mercury, cadmium, and arsenic) that causes human health problems and various environmental issues [10, 11]. These issues hinder the ecological cycles if not well disposed of, therefore, good waste management practice as an alternative to industrial resource, process, or application of CFA is needed for a healthy environment. However, the utilization of CFA in various countries depends on the CFA generation, physical and chemical properties, environmental protection policies, materials scarcity, technology, and the structure of industries [8, 25]. The utilization of CFA illustrated in Fig. 2 indicates that construction and building materials are the most successful uses of CFA in various countries because of its rich oxide material [26, 27]. On the other hand, CFA is a valuable secondary source for raw materials like alumina which can satisfy the increasing demand of the aluminium industry [28, 29]. Therefore, to ensure sustainable economic ecosystems, it is important to reduce the negative impact of CFA on the environment and produce high value-added products.

CFA is a fine powdery particle composed of both organic and inorganic materials [29]. Also, CFA is a complex material because of the large number of individual minerals and mineral groups recognized in it [30]. The inorganic materials also include toxic and hazardous elements such as As, Pb, Ni, and Cr that are harmful to the environment. CFA composition varies due to coal source, combustion methods, and the type of furnace used [29]. The major oxides present in CFA from different countries are summarised in Table 1. According to Table 1, the main elements of all CFA are SiO_2, Al_2O_3, and Fe_2O_3. Globally, CFA is grouped into two chemical types, specifically Class F and Class C, for their industrial utilization. The CFA containing more than 70 wt.% of the main elements is classified as Class F, while Class C contains 50–70 wt.% [31]. Class C typically has CaO higher than Class F; higher

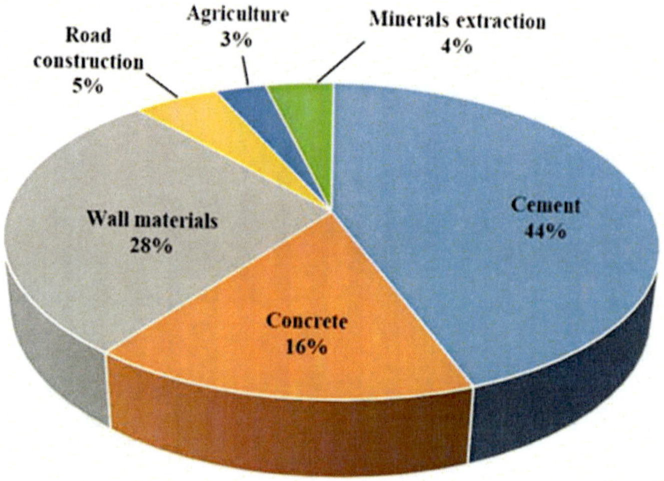

Fig. 2 Utilization of CFA [21]

Table 1 Chemical composition of CFA [31]

Country	Chemical composition (wt.%)								
	SiO_2	Al_2O_3	Fe_2O_3	CaO	MgO	SO_3	TiO_2	Na_2O	K_2O
China	35–57	18–55	2.3–19	1.1–7	0.7–4.8	1–2.9	0.2–0.7	0.6–1.3	0.8–0.9
USA	34–58	19–28	3.2–25	0.7–22	0.5–4.8	0.1–2.1	1–1.6	0.2–1.8	0.9–2.9
Japan	53–63	18–26	4.2–5.7	2–8.1	0.9–2.4	0.3–1.4	0.8–1.2	1.1–2.1	0.6–2.7
France	47–51	26–34	6.9–8.8	2.3–3.3	1.5–2.2	0.1–0.6	na	2.3–6.4	na
South Africa	46–67	21–27	2.4–4.7	6.4–9.8	1.9–2.7	na	1.2–1.6	0–1.3	0.5–1
India	50–59	14–32	2.7–16	0.6–9	0.1–2.3	na	0.3–2.7	0.2–1.2	0.2–4.7

CaO levels have cementitious properties used in building materials. From Table 1, it can be noticed that China's CFA has a relatively wider Al_2O_3 content range than other countries listed.

Alumina Recovery Methods

CFA is produced at high temperatures in a power plant and contains mullite (a major source of alumina), quartz, and some amorphous phases. Alumina recovery from CFA was first proposed in the nineteenth century [32] because of the high content found to exist in CFA, and since then, governments and scientists all over the world have proposed many recovery methods in this field. The utilization of the CFA depends greatly on the effectiveness and economic improvement of the technology to process or recycle them. There are many technical methods for extracting alumina as well as other minerals from CFA. The most popular methods of extraction are the alkali-sintering process (limestone or soda-lime sintering) and the acid leaching process, as well as their combinations.

Alkali-Sintering Process

An efficient and mature method to recover alumina from CFA is the alkali-sintering process. The CFA is sintered with limestone or soda-lime to form sodium aluminate and calcium silicate. The sinter is then leached with water or caustic soda solution to dissolve the soluble sodium aluminate [33]. The resulting slurry is separated, and the solution is decomposed as in the Bayer process or is treated with carbon dioxide to precipitate hydrated alumina. This extraction process is divided into lime sintering, lime-soda sintering, pre-desilication, and lime-soda sinter combination method.

Acid Leaching Process

The acid leaching process is a simplified extraction process with lower consumption of energy [4]. The oxides are leached with a sulfuric acid or sodium carbonate solvent at ambient temperature and atmospheric pressure [34]. It is also done by pressure acid-leaching [35] whereby pressure and temperature are increased to accelerate the operation. The overdosed sulfuric acid is separated from the solids with filtration, and the aluminum sulfate product is left on the surface of leached CFA. The aluminum sulfate crystals are obtained by concentrating the aluminum sulfate solution. The alumina powder that had been prepared is used in the Bayer process, which is the mature process to produce the metallurgical level of alumina in the industry [36].

Comparative Analyses on Some Alumina Recovery Technologies

Guo et al. [37] and Liu et al. [38] proposed an alumina extraction method from CFA via the activation-acid leaching method. Their results showed that Na_2CO_3 reacted with the Al and Si-containing minerals at the calcination conditions to form nepheline ($NaAlSiO_4$), zeolite [Na_6 $(AlSiO_4)_6$], and sodium silicate (Na_2SiO_3). However, the industrial utilization of the technology has been limited by the high consumption of Na_2CO_3, that is, 0.8–1.2 tons of Na_2CO_3 needed for activating 1 ton of CFA [37]. Although the acid leaching method has low alumina extraction efficiency [35, 39], it has lower energy consumption and avoids large waste residues. For efficient utilization of alumina and silicon resources in CFA, the acid-leached residues which contain a large amount of reactive amorphous SiO_2 could be further utilized to manufacture silica products. On the other hand, acid leaching has high impurity content, such as calcium oxide and iron oxide, which are difficult to remove from the leaching solution [40]. Many researchers are focused on improving this method; however, the demand for corrosion-resistant equipment during the acid leaching limits the industry application of this process [41].

The soda-lime sintering method has operational tolerance and a mature system conducted to activate the CFA [42, 43] but is limited by high sintering temperature and a large amount of waste residues generation [44]. The mullite ($Al_6Si_2O_{13}$) and quartz (SiO_2) in the CFA react with CaO or $CaCO_3$, forming calcium aluminate and calcium silicate, respectively. The calcium silicate formed is hardly dissolved which is helpful for the separation of alumina and silicon [3]. Moreover, the formation of calcium aluminates is unavoidable in this process. Alkali-sintering processes can achieve high alumina extraction efficiency, but the alkali solutions used in these processes may cause a high caustic ratio in the final solution [45]. On the other hand, the use of additives can lower the calcination temperature during heating and promote the extraction of Al_2O_3 [46]. Comparing lime-soda sintering with the lime sintering method, the lime sintering method produces less residue with a high extraction rate but the lime-soda sintering process is complex, and the cost of soda is also high [47, 48]. Therefore, sintering processes are considered to have significant potential for industrial use as they are similar to the currently used extraction technologies in the aluminium plants [28].

Several processes for extracting alumina from CFA have been proposed. Xue et al. [49] proposed a clean process for extracting alumina and silica simultaneously from CFA via vacuum carbothermal reduction. This process proved to be effective having high extraction efficiency by employing Fe_2O_3 as an additive to activate mullite and react with SiO_2 in CFA to produce Fe-Si alloy under vacuum conditions. The prepared Fe-Si alloy can be used in steel making as a raw material. Compared with traditional extraction processes, the energy consumption and waste residues were further reduced by employing vacuum carbothermal technology [50, 51]. Additionally, the vacuum carbothermal reduction process has a lower reaction temperature (1423–473 K) than that of the sintering process (1573–1673 K) [5] and is more environmentally friendly. However, there was some difficulty in separating these particles thoroughly due to the small particle size of the produced Fe-Si alloy, thereby reducing the quality of alumina.

Guo et al. [52] also proposed the pyrosulfate calcining method for recovering alumina from CFA; this method has got attention due to the recyclable raw materials and low energy consumption. The pyrosulfate calcining method breaks the stable structure of the mineral phase to enhance the alumina extraction rate. This process deals with the large number of waste residues which significantly reduce environmental pollution and extraction costs [52]. Comparing the pyrosulfate method to other extraction methods, the pyrosulfate method has a simple process, mild reaction conditions, recyclability, and high extraction efficiency but has a relatively high cost [48]. In the past few years, researchers mainly focused on the alumina extraction rate and the phase transformations by using alumina extracting processes. However, the Al_2O_3 and SiO_2 phase transformation, alumina extraction rate improvement, and extraction methods are still unclear.

Conclusion

The utilization of CFA is a big concern as governments and researchers are constantly improving on the harmless disposal and utilization of CFA in some other engineering applications. Undoubtedly, the use of CFA for engineering purposes will save costly materials and also result in a decrease in land degradation, air pollution, and water contamination. On the other hand, employing some novel processes for the recovery of valuable metals from CFA has potential economic and environmental problems due to their extraction cost and large waste residues generation. Therefore, further research studies are needed to improve the utilization of CFA in optimising valuable metal recovery technologies and value-added products. The recent metallurgical recycling technologies have their advantages and limitations that can be developed upon to improve the efficiency rate. Also, new recycling technologies can be developed, by using a combination of recovery processes, to turn the utilization of CFA into an economic reality.

Acknowledgements The authors acknowledge the support from the Natural Science Foundation of China (No. 52074058 and No. U20032151).

References

1. Xu DH, Li HQ, Bao WJ, Wang CY (2016) A new process of extracting alumina from high-alumina coal fly ash in NH4HSO4+H2SO4 mixed solution. Hydrometallurgy 165:336–344. https://doi.org/10.1016/j.hydromet.2015.12.010
2. Dai SF, Zhao L, Peng SP, Chou CL, Wang XB, Zhang Y, Li D, Sun YY (2010) Abundances and distribution of minerals and elements in high-alumina coal fly ash from the Jungar power plant, Inner Mongolia. China. International Journal of Coal Geology 81(4):320–332. https://doi.org/10.1016/j.coal.2009.03.005
3. Yao Z, Xia M, Sarker P, Chen T (2014) A review of the alumina recovery from coal fly ash, with a focus in China. Fuel 120:74–85. https://doi.org/10.1016/j.fuel.2013.12.003
4. Guo YX, Zhao ZS, Zhao Q, Cheng FQ (2017) Novel process of alumina extraction from coal fly ash by pre-desilicating-Na2CO3 activation-Acid leaching technique. Hydrometallurgy 169:418–425. https://doi.org/10.1016/j.hydromet.2017.02.021
5. Ding J, Ma SH, Shen S, Xie ZL, Zheng SL (2017) Zhang Y. Research and industrialization progress of recovering alumina from fly ash: A concise review. Waste Manage 60:375–387. https://doi.org/10.1016/j.wasman.2016.06.009
6. van der Merwe EM, Gray CL, Castleman BA, Mohamed S, Kruger RA, Doucet FJ (2017) Ammonium sulphate and/or ammonium bisulphate as extracting agents for the recovery of aluminium from ultrafine coal fly ash. Hydrometallurgy 171:185–190. https://doi.org/10.1016/j.hydromet.2017.05.015

7. Zacco A, Borgese L, Gianoncelli A, Struis RPWJ, Depero LE, Bontempi E (2014) Review of fly ash inertisation treatments and recycling. Environ Chem Lett 12:153–175. https://doi.org/10.1007/s10311-014-0454-6
8. Jin SX, Zhao ZH, Jiang SF, Sun J, Pan HB, Jiang L (2021) Comparison and summary of relevant standards for comprehensive utilization of fly ash at home and abroad. Paper presented at IOP Conference Series: Earth and Environmental Science 621:2021. https://doi.org/10.1088/1755-1315/621/1/012006
9. Harris D., Heidrich C., Feuerborn J. (2019) Global aspects on Coal Combustion Products. Paper presented at Coaltrans Conference, 01 September 2019. https://www.coaltrans.com/insights/article/global-aspects-on-coal-combustion-products. Accessed 29 August 2021
10. Gianoncelli A., Zacco A., Struis R., Borgese L., Depero L., Bontempi E. (2013) Fly ash pollutants, treatment and recycling. In Lichtfouse E., Schwarzbauer J., Robert D. (ed) Pollutant diseases, remediation and recycling. Environmental chemistry for a sustainable world, Springer, Cham, Volume 4, pp. 103–213
11. Han Y, Xie H, Liu W, Li H, Wang M, Chen X, Liao X, Yan N (2016) Assessment of pollution of potentially harmful elements in soils surrounding a municipal solid waste incinerator, China. Front Environ Sci Eng 10(6):7. https://doi.org/10.1007/s11783-016-0873-7
12. Guo CB, Zou JJ, Wei CD, Jiang YS (2013) Comparative study on extracting alumina from circulating fluidized-bed and pulverized-coal fly ashes through salt activation. Energy Fuels 27(12):7868–7875. https://doi.org/10.1021/ef401659e
13. Yan F, Jiang JG, Liu N, Gao YC, Meng Y, Li KM, Chen XJ (2018) Green synthesis of mesoporous γ-Al2O3 from coal fly ash with simultaneous on-site utilization of CO2. J Hazard Mater 359:535–543. https://doi.org/10.1016/j.jhazmat.2018.07.104
14. Sun LY, Luo K, Fan JR, Lu HL (2017) Experimental study of extracting alumina from coal fly ash using fluidized beds at high temperature. Fuel 199:22–27. https://doi.org/10.1016/j.fuel.2017.02.073
15. Farjana SH, Huda N, Mahmud PM (2019) Impacts of aluminum production: A cradle to gate investigation using life-cycle assessment. Sci Total Environ 663:958–970. https://doi.org/10.1016/j.scitotenv.2019.01.400
16. Mutlu BK, Cantoni B, Turolla A, Antonelli M, Hsu-Kim H, Wiesner MR (2018) Application of nanofiltration for rare earth elements recovery from coal fly ash leachate: Performance and cost evaluation. Chem Eng J 349:309–317. https://doi.org/10.1016/j.cej.2018.05.080
17. Shemi A, Mpana R, Ndlovu S, van Dyk LD, Sibanda V, Seepe L (2012) Alternative techniques for extracting alumina from coal fly ash. Miner Eng 34:30–37. https://doi.org/10.1016/j.mineng.2012.04.007
18. IEA (2020) In: Fuel and Technologies; Coal. https://www.iea.org/fuels-and-technologies/coal. Accessed 7 August 2020
19. Prafulla KS, Kangjoo K, Powell MA, Equeenuddin SM (2016) Recovery of metals and other beneficial products from coal fly ash: a sustainable approach for fly ash management. International Journal of Coal Science and Technology 3(3):267–283. https://doi.org/10.1007/s40789-016-0141-2
20. Marinina O, Nevskaya M, Jonek-Kowalska I, Wolniak R, Marinin M (2021) Recycling of coal fly ash as an example of an efficient circular economy: a stakeholder approach. Energies 14. https://doi.org/10.3390/en14123597
21. Harris D (2017) Ash as an internationally traded commodity. In: Robl T, Oberlink A, Jones R (eds) Coal combustion products (CCP's). Woodhead Publishing, pp 509–529
22. American Coal Ash Association, 2020. In: Fly ash use in concrete increases slightly as overall coal ash recycling rate declines. https://acaa-usa.org/wp-content/uploads/coal-combustion-products-use/Coal-Ash-Production-and-Use.pdf. Accessed 31 August 2021
23. Adegoke KA, Oyewole RO, Lasisi BM, Bello OS (2017) Abatement of organic pollutants using fly ash based adsorbents. Water Sci Technol 76(10):2580–2592. https://doi.org/10.2166/wst.2017.437
24. Ghazali N, Muthusamy K, Ahmad SW (2019) Utilization of Fly Ash in Construction. Paper presented at IOP Conference Series: Materials Science and Engineering 601:2019. https://doi.org/10.1088/1757-899X/601/1/012023

25. Koukouzas NK, Zeng R, Perdikatsis V, Xu W, Kakaras EK (2006) Mineralogy and geochemistry of Greek and Chinese coal fly ash. Fuel 85(16):2301–2309. https://doi.org/10.1016/j.fuel.2006.02.019
26. Dwivedi A, Jain MK (2014) Fly ash-waste management and overview: a review. Recent Research in Science and Technology 6(1):30–35
27. Tiwari MK, Bajpai S, Dewangan UK (2016) Fly ash utilization a brief review in Indian context. International Research Journal of Engineering and Technology 3(4):949–956
28. Liu K., Xue J., Zhu J. (2012) Extracting alumina from coal fly ash using acid sintering-leaching process. In: Suarez C. (ed) TMS Light Metals 2012. Springer, Cham, pp. 201–206. doi:https://doi.org/10.1007/978-3-319-48179-1_36
29. Han F.L., Wu L. (2019) Industrial Solid Waste Recycling in Western China, 1st ed., Springer, Singapore. doi:https://doi.org/10.1007/978-981-13-8086-0
30. Vassilev SV, Vassileva CG (2007) A new approach for the classification of coal fly ashes based on their origin, composition, properties, and behaviour. Fuel 86(10–11):1490–1512. https://doi.org/10.1016/j.fuel.2006.11.020
31. Bhatt A, Priyadarshini S, Mohanakrishnan AA, Abri A, Sattler M, Techapaphawit S (2019) Physical, chemical, and geotechnical properties of coal fly ash: A global review. Case Studies in Construction Materials 11. https://doi.org/10.1016/j.cscm.2019.e00263
32. Luo Y, Wu YH, Ma SH, Zheng SL, Zhang Y, Chu PK (2021) Utilization of coal fly ash in China: a mini-review on challenges and future directions. Environ Sci Pollut Res 28:18727–18740. https://doi.org/10.1007/s11356-020-08864-4
33. In: Definition of lime-soda sinter process. https://www.mindat.org/glossary/lime-soda_sinter_process. Accessed 04 January 2021
34. Zhu PW, Dai H, Han L, Xu XL, Cheng LM, Wang QH, Shi ZL (2015) Aluminum extraction from coal ash by a two-step acid leaching method. Journal of Zhejiang University-SCIENCE A 16:161–169. https://doi.org/10.1631/jzus.A1400195
35. Wu CY, Yu HF, Zhang HF (2012) Extraction of aluminum by pressure acid-leaching method from coal fly ash. Transactions of nonferrous metals society of China 22(9):2282–2288. https://doi.org/10.1016/S1003-6326(11)61461-1
36. Li LS, Wu YS, Liu YY, Zhai YC (2011) Extraction of alumina from coal fly ash with sulfuric acid leaching method. Chin J Process Eng 11(2):254–258
37. Guo YX, Li YY, Cheng F, Wang M, Wang XM (2013) Role of additives in improved thermal activation of coal fly ash for alumina extraction. Fuel Process Technol 110:114–121. https://doi.org/10.1016/j.fuproc.2012.12.003
38. Liu DD, Fang L, Guo YX, Yan KZ, Yao C, Cheng FQ (2018) Effects of calcium oxide and ferric oxide on the process of alumina extraction of coal fly ash activated by sodium carbonate. Hydrometallurgy 179:149–156. https://doi.org/10.1016/j.hydromet.2018.04.017
39. Valeev D, Kunilova I, Alpatov A, Mikhailova A, Goldberg M, Kondratiev A (2019) Complex utilization of ekibastuz brown coal fly ash: Iron & & Carbon separation and Aluminium extraction. J Clean Prod 218:192–201. https://doi.org/10.1016/j.jclepro.2019.01.342
40. Matjie RH, Bunt JR, van Heerden JHP (2005) Extraction of alumina from coal fly ash generated from a selected low rank bituminous South African coal. Miner Eng 18(3):299–310. https://doi.org/10.1016/j.mineng.2004.06.013
41. Nayak N, Panda CR (2010) Aluminium extraction and leaching characteristics of Talcher thermal power station fly ash with sulphuric acid. Fuel 89:53–58. https://doi.org/10.1016/j.fuel.2009.07.019
42. Awual MR, Miyazaki Y, Taguchi T, Shiwaku H, Yaita T (2016) Encapsulation of cesium from contaminated water with highly selective facial organic-inorganic mesoporous hybrid adsorbent. Chem Eng J 291:128–137. https://doi.org/10.1016/j.cej.2016.01.109
43. Shahat A, Hassan HM, Azzazy HM, El-Sharkawy E, Abdou HM, Awual MR (2018) Novel hierarchical composite adsorbent for selective lead (II) ions capturing from wastewater samples. Chem Eng J 332:377–386. https://doi.org/10.1016/j.cej.2017.09.040
44. Samavati A, Fattahi M, Khorasheh F (2013) Modeling of Pt-Sn/gamma-Al2O3 deactivation in propane dehydrogenation with oxygenated additives. Korean J Chem Eng 30:55–61. https://doi.org/10.1007/s11814-012-0095-z

45. Guo CB, Zou JJ, Ma SH, Yang JL, Wang KH (2019) Alumina extraction from coal fly ash via low-temperature potassium bisulfate calcination. Minerals 9(10):585. https://doi.org/10.3390/min9100585
46. Jiang ZQ, Ma HW, Yang J, Ma X, Yuan JY (2015) Thermal decomposition mechanism of desilication coal fly ash by low-lime sinter method for alumina extraction. Ferroelectrics 486(1):143–155. https://doi.org/10.1080/00150193.2015.1100878
47. Ding J, Ma SH, Xie ZL, Wang XH, Zheng SL, Zhang Y (2019) Formation mechanism of an undesirable by product in the mild hydro-chemical process for the extraction of alumina from fly ash and its mitigation. Hydrometallurgy 186:292–300. https://doi.org/10.1016/j.hydromet.2019.04.012
48. Guo CB, Zhao L, Yang JL, Wang KH, Zou JJ (2020) A novel perspective process for alumina extraction from coal fly ash via potassium pyrosulfate calcination activation method. J Clean Prod 271. https://doi.org/10.1016/j.jclepro.2020.122703
49. Xue Y, Yu W, Mei J, Jiang W, Lv X (2019) A clean process for alumina extraction and ferrosilicon alloy preparation from coal fly ash via vacuum thermal reduction. J Clean Prod 240. https://doi.org/10.1016/j.jclepro.2019.118262
50. Feng YB, Yang B, Dai YN (2012) Carbothermal reduction-chlorination-disproportionation of alumina in vacuum. Transactions of Nonferrous Metals Society of China 22(1):215–221. https://doi.org/10.1016/S1003-6326(11)61163-1
51. Kruesi M, Galvez M, Halman M, Steinfeld A (2011) Solar aluminum production by vacuum carbothermal reduction of alumina-Thermodynamic and experimental analyses. Metall and Mater Trans B 42:254–260. https://doi.org/10.1007/s11663-010-9461-6
52. Guo YX, Li J, Yan KZ, Cao LQ, Cheng FQ (2019) A prospective process for alumina extraction via the co-treatment of coal fly ash and bauxite red mud: Investigation of the process. Hydrometallurgy 186:98–104. https://doi.org/10.1016/j.hydromet.2019.04.011

Author Index

A. Lazou et al. (eds.), *REWAS 2022: Developing Tomorrow's Technical Cycles (Volume I)*, The Minerals, Metals & Materials Series,
https://doi.org/10.1007/978-3-030-92563-5

Subject Index

A. Lazou et al. (eds.), *REWAS 2022: Developing Tomorrow's Technical Cycles (Volume I)*, The Minerals, Metals & Materials Series,
https://doi.org/10.1007/978-3-030-92563-5